HYPERFINE INTERACTIONS (C)

Hyperfine Interactions (C)

Volume 5

HYPERFINE INTERACTIONS (C)

Proceedings of the International Conference on the Applications of the Mössbauer Effect, (ICAME 2001) September 2–7, 2001, Oxford, U.K.

Edited by

Michael F. Thomas

Department of Physics,
University of Liverpool,
Liverpool, U.K.

John M. Williams

Department of Physics,
University of Sheffield,
Scheffield, U.K.

and

Terence C. Gibb

Department of Chemistry,
University of Leeds,
Leeds, U.K.

KLUWER ACADEMIC PUBLISHERS
DORDRECHT / BOSTON / LONDON

A C.I.P. Catalogue record for this book is available from the Library of Congress.

ISBN 1-4020-1087-7

Published by Kluwer Academic Publishers,
P.O. Box 17, 3300 AA Dordrecht, The Netherlands.

Sold and distributed in North, Central and South America
by Kluwer Academic Publishers,
101 Philip Drive, Norwell, MA 02061, U.S.A.

In all other countries, sold and distributed
by Kluwer Academic Publishers,
P.O. Box 322, 3300 AH Dordrecht, The Netherlands.

Printed on acid-free paper

Printed in the Netherlands.

Preface

The 26th International Conference on the Applications of the Mossbauer Effect (ICAME2001) was held from 2nd to 7th September 2001 in Oxford.
The Conference was attended by 290 participants who contributed the 215 manuscripts published in Hyperfine interactions and in Hyperfine Interactions (C).

An extensive program of invited presentations included talks on the vibrational density of states in Myoglobin, magnetic multilayers, ball-milling products, synchrotron studies, high pressure phenomena, spin transitions in polymeric compounds, extra-terrestrial planetary surfaces and the current status of the gamma-ray laser.
Contributed presentations were grouped into Advances in techniques, Amorphous, nano-crystalline and small particle systems, Applications in magnetism and lattice dynamics, Biological and medical applications, Chemical applications: structure and bonding, Earth science, Mineralogy and archaeology, Industrial applications, catalysis and corrosion, Materials science and metallurgy and Surfaces, interfaces, thin films and multilayers.

In his closing remarks, Guido Langouche said that while Mossbauer Spectroscopy is more and more just one of the techniques available in materials science, it remains an extremely powerful tool for magnetism studies. There were presentations in which, among the results of a number of complementary techniques it was the Mossbauer data that was the key to understanding the problem. The new growth area is Synchrotron studies, possibly open to only a minority of experimenters, which promised exciting results in new resonances, holography and studies of phonon densities of states.

The organizers wish to acknowledge the following institutions and organizations that have provided financial support to ICAME2001
The University of Liverpool
The Open University
The University of Wolverhampton
The Royal Society of Chemistry: Mossbauer Spectroscopy Discussion Group
Ultra-Physics Limited

The organizers wish to thank the staff of the Institute of Physics for all their work in the administering and running of this Conference.

The Proceedings Editors wish to thank the many participants who cheerfully and expertly refereed the many manuscripts.
We would like to extend particular thanks to the Editorial Co-coordinator, Julie Wardman, for her efficiency and humour in keeping track of all contributions.

Proceedings Editors:
Michael Thomas
Terry Gibb
John Williams
ICAME2001

CONTENTS

Excitation of the nuclear resonance in ^{61}Ni at 67.4 keV by Synchrotron Radiation

H.-C. Wille, M. Gerken, E. Gerdau, Yu.V. Shvyd'ko and H.D. Rüter
Institut für Experimentalphysik, Universität Hamburg, D-22761 Hamburg

H. Franz
Hamburger Synchrotronstrahlungslabor HASYLAB, Notkestr. 85, D-22603 Hamburg, Germany

Abstract. We report on the first observation of the excitation of the nuclear resonance in ^{61}Ni by synchrotron radiation. Ni is an element very important in the field of magnetism, chemistry, biology and material science. The energy and the lifetime of the nuclear transition were determined to be 67.419(8) keV and 7.6(4) ns respectively.

Nuclear resonant scattering is a tool for structural and dynamical studies as it provides information on hyperfine fields and phonon spectra. During 16 years, following the first successful experiment with ^{57}Fe in 1984 [1], resonant scattering from several other isotopes was observed.

^{61}Ni has a natural abundance of 1.14 %. The considered nuclear transition of M1+E2 multipolarity takes places between two states of spin and parity $5/2^-$ and $3/2^-$ respectively. 12 transitions between the magnetic hyperfine levels are possible. The measured transition energy of 67.419 keV and lifetime of about 7.6 ns correspond to a natural energetic width of about $\Gamma = 87\,neV$. The conversion coefficient is 0.13. The first verification of the transition at the 67.4 keV in ^{61}Ni was done in a classical Mössbauer experiment and is reported by Obenshain and Wegener in 1961 [2]. The lifetime of the two parent isotops ^{61}Co and ^{61}Cu for excited ^{61}Ni are 99 minutes and 200 minutes respectivly, what makes classical investigations rather difficult. Therefore the possibility of using synchrotron radiation for the excitation process of the nuclei is very desirable. There are several technologically important alloys based on Ni (superalloys). Mössbauer spectroscopy may give a broad amount of information similar to the well known case of ^{57}Fe . Local magnetic fields, electric field gradients and element specific vibrational properties are worthwhile to be studied. Due to the short wavelength of this transition and the accordingly low Lamb-Mössbauer-factor f the observation of quasielastic scattering will be extremely difficult. With high resolution monochromators phonon energy spectra should be accessible which is a very attractive application in our opinion. Most important for biophysical experiments is the fact, that Ni is the reactive center for several metallo-proteins or enzymes. Also in this context information about hyperfine fields giving information on the local environment and vibrational properties may be achieved.

Synchrotron sources of the third generation provide high countrates in the region above 50 keV. Efficient detection systems for x-ray energies above 50 keV and nuclear transition lifetimes smaller than 10 ns are still a vacancy in detector physics. In our case a standard avalanche silicon diode could be used detecting the low energy conversion x-rays at 7.5 keV as described below.

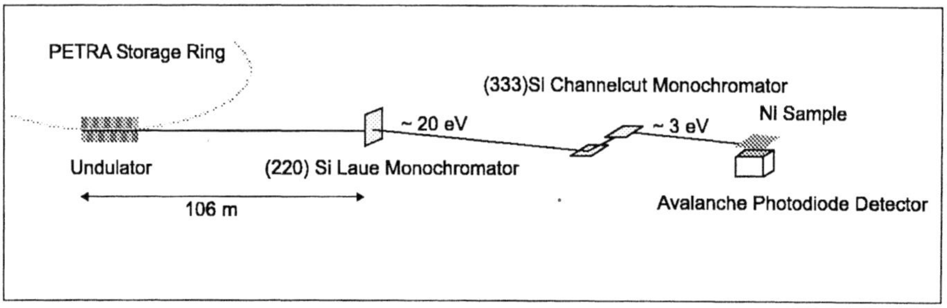

Figure 1: Experimental setup for inelastic nuclear resonant scattering in ^{61}Ni

The measurements were performed at the high energy part of the undulator beamline at the 12 GeV e^-/e^+ PETRA II storage ring at DESY, Hamburg, Germany, which is dedicated to experiments with x-ray energies above 65 keV. The undulator supplies a radiation spectrum of high intensity up to 200 keV. Figure 1 shows the experimental setup.

PETRA was operated with 40 mA maximum current in a 40 bunch mode, where the circulating positron bunches are separated by 192 ns.

A silicon single crystal in Laue geometry at the (220) reflection provides x rays in an energy band of about 20 eV width, which is monochromatised further down to about 3 eV by a (3 3 3) silicon channelcut crystal. The channelcut monochromator was calibrated in energy by measuring the Tantal k-edge with the same setup. A sample of metallic Ni powder enriched in ^{61}Ni to 85 % was used. The ^{61}Ni sample was mounted in about 5 mm distance from the silicon avalanche photodiode detector (APD) and inclined about 5° with respect to the beam.

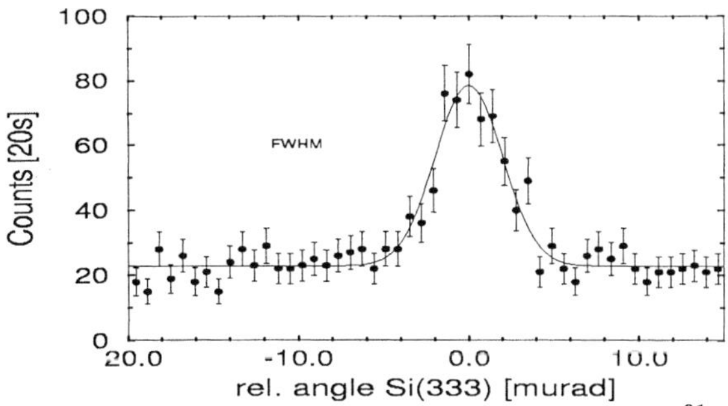

Figure 2: Energy scan with resonance peak from ^{61}Ni .

Detector signals from the electronically scattered incoming synchrotron radiation bunches are electronically blocked. Only delayed quanta beginning from 18 ns after the exciting pulses are counted. All measurements were performed at room temperature where the Lamb-Mössbauer-factor is about 0.01. This means that the recoilless fraction of the scattering is near to zero. Due to this the emission takes places nearly homogeniously into 4π. The energy of the incoming quanta is tuned by rocking the (3 3 3) channel cut crystal.

Figure 2 shows the countrate of delayed quanta vs. the incident angle of the beam on the (3 3 3) crystal. There the resonance effect can clearly be seen. The peak countrate was about 4 Hz, while the background counterate was about 1 Hz.

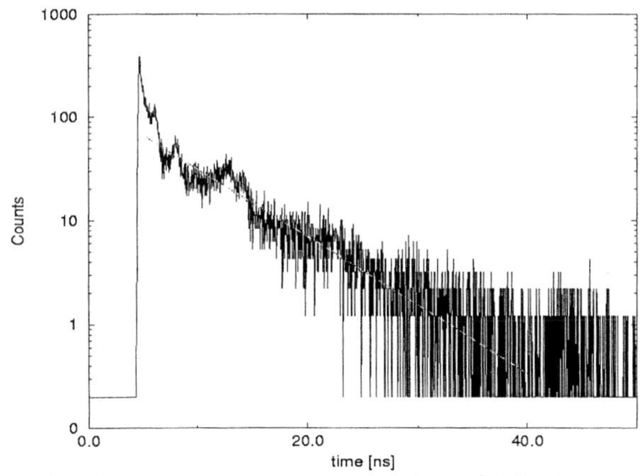

Figure 3: Nuclear decay in time space from ^{61}Ni after excitation.

With the well determined value of the edge energy of Tantal the energy of the transition was determined to be 67.419(8) keV. Furthermore spectra of the nuclear decay in time space where measured using a standard setup for time resolved nuclear resonant spectroscopy. One of the decay spectra is shown in figure 3. As can be seen there, some spurious bunches up to 15 ns after the main bunch overlay the resonance effect in that region. A fit in the time window from t = 20 ns to t = 100 ns using a function of the form $A \times \exp \frac{-t}{\tau} + B$ resulted in a value for the lifetime of $\tau =$ 7.6(4) ns. Our measurements have shown, that nuclear resonant scattering of synchrotron radiation at the 67.419 keV transition of ^{61}Ni is possible. We hope that the way to many exciting applications is open and that they will follow soon.

References

[1] E. Gerdau, R. Rüffer, H. Winkler, W. Tolksdorf, C.P. Klages, and J.P. Hannon. Nuclear Bragg Diffraction of Synchrotron Radiation in Yttrium Iron Garnet. *Phys. Rev. Lett.*, 54:835, 1985.

[2] F.E. Obenshain and H.H.F. Wegener. *Phys. Rev.*, 121:1344, 1961.

Address for Offprints: Kluwer Prepress Department
P.O. Box 990
3300 AZ Dordrecht
The Netherlands

Multibeam laser selection of nuclides for gamma-laser, Moessbauer spectroscopy and other practice

S.V.Karyagin

Semenov Institute of Chemical Physics of RAS, Gol'danskii Department of Substance Structure

119991, Kosygina street 4, Moscow, Russia; e-mail: <karyagin@chph.ras.ru>

New, «multibeam», principal for the acceleration of a selection of nuclei, atoms, molecules is represented. It could do a high progress in different spheres from γ-laser to nuclear energetics.

The idea of γ-laser (GL) on base of selective laser ionisation of atoms with excited nuclei was suggested at 1973 by Letokhov [1] encouraged by seeming perspectives for the Laser Isotope (Isomer) Separation (LIS) methods. But now the very deep drawbacks for all known LIS schemes, e.g., Atomic Vapour LIS (AVLIS), are revealed [2]. They turned up in truth to be very slow for many industrial and scientific aims, GL included. The reasons for such setback of LIS are the recharging of goal ions in an atomic vapour and the other adverse effects [3].

Present paper is devoted to a new, so called «*multibeam*» (MB) general principle of selection (separation), which could abolish above and other hindrances in many known selection methods, LIS included. MB-principle was first introduced at 1995 by Karyagin [3] in application to new scheme of GL [4] named later as Soft Prompt Transplantation of Excited Nuclei (SPTEN) [5]. The MB-principle was extrapolated from GL to other spheres, e.g., the MB-selection of stable and long-lived isotopes, isomers, radionuclides, rare atoms, excited atoms or molecules, chromatography of giant albumen molecules at one distinct conformation, etc. [4-6].

As on MB-principle an atomic (or molecular) beam of big cross size $\sim 10^1$ - 10^2 cm is to be splintered into $\sim 10^6$ - 10^9 parallel microbeams by use of *multibeam emitter* (MBE) at the input of the selecting device [4 - 6]. Properly MBE must be constructed as like as a big microchannel plate with high developed, high oriented, deeply engraved microrelief (fig.1). MBE could be joint together with the special ion-optical and laser systems. A big cross size and a developed surface of MBE lead to essential increase for amount of atoms at the input of selection. A structure of an atomic flow, as a bunch of the parallel microbeams, leads to a new class of wide formatted but well short focused ion-optics [4-7]. In a result it is possible to accelerate and then to retard the selected ions and to get the good parameters: maximal voltage ~1 *MeV*; delivery time ~ 1 *μs*; an output end energy ~ 0.5 keV; the running cross section of the ionic flow could be compressed from an initial size ~10^3 - 10^4 cm^2 (visual MBE surface) to the size of the end target (length ~ 1 cm, width ~ 10^{-4} *cm*) at the delivery ~ 10% with account of aberrations; delivery path ~ 10^2 cm.

Note that the delivery path of any device of AVLIS-type needs be ~ 10^5 cm in order to get the same selectivity at the same input surface ~ 10^4 cm^2. But even at such big length the efficiency of AVLIS will be by ~ 3 orders less than in case of the MB-methods. So the MB-LIS is much more powerful and compact than AVLIS.

Hence an atomic mixture density at selective ionisation [1,4] could be by some orders less than in AVLIS. So a charge exchange and other adverse effects, e.g. plasma's polarisation effects [3] could be suppressed by some orders [4-6].

There are two classes of MBE: *Passive MBE* has only one function, the splintering of a single beam into a bunch of parallel microbeams. *Active MBE* (AMBE) has two functions: the creation of goal nuclei and the splintering of a beam. The AMBE must be irradiated from a strong source [4], e.g., *aperiodical pulse reactor* [9,10]. In AMBE due to Szcilard-Chalmers effect a relative concentration of goal nuclei in a beam is very high at the beginning of selection.[4] E.g., in MB-LIS it is by 5-9 orders more than in AVLIS. [4] Moreover, the MB-LIS holds: a more effective transversal geometry of a selective laser ionisation with a high selectivity due to negligible Doppler dispersion due to a high orientation of MB [4-7]; relative concentration of goal nuclei ~ 0.9 could be achieved at the end target [4]; it is possible to work with short-lived goal radio-nuclides due to auto-cleaning of MB from non-goal nuclei [5]; the selection cycle for a renovation (mainly a cleaning of vacuum) is shorter by some orders than in AVLIS [5-7].

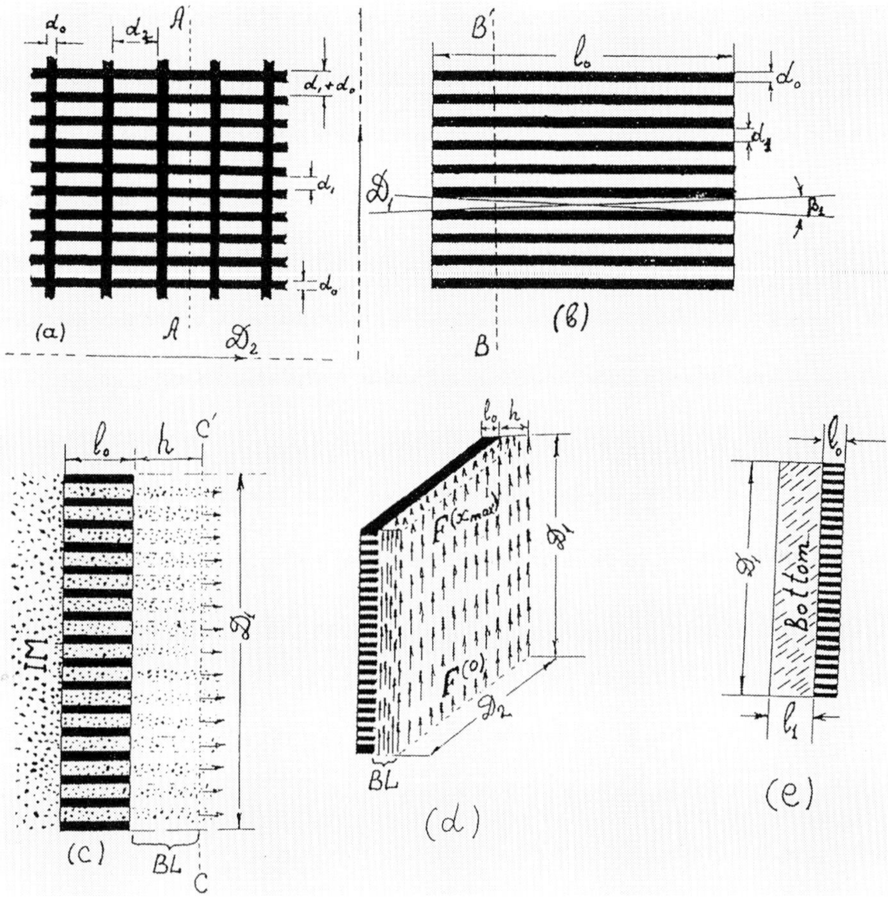

Fig. 1. The construction of MBE [4-7]: (*a*), (*b*), (*c*) are the layers of the fragment of the MBE at three planes; (*c*) is relevant to PMBE, the right is an initial flow, the left is a multibeam; (*d*) is a laser radiation along MBE in MB-LIS; (e) is a layer of AMBE. The sizes of MBE are $d_1 \sim d_0 \sim 10^{-3}$ cm $<< a_2 \sim 0.1$ cm $<< l_0 \sim h \sim 0.1 - 1$ cm $<< D_1, D_2 \sim 10^1 - 10^2$ cm.

The simplest MBE-types for non-γ-laser sphere (see above) could be realised long before first direct GL-experiments and could be used, e.g., in the *Moessbauer Effect Applications* as the *high super-intensive pulse sources of Moessbauer γ-radiation from excited nuclei* at life $\tau_+ > 10^{-7}$ *s*. The advantages of that MB-sources are linked with their super-high intensity even without any γ-lasing ($\sim 10^{13}/\tau_+$ decays, e.g., 10^8 Curie at the time-life $\tau_+ \sim 10^{-5}$ *s*) and *wide amount of different nuclei for such MB-spectroscopy*. They are very important in a research of short-lived processes, in case of the super-small amounts of substance, in criminalistics, etc..

Semenov Institute of Chemical Physics holds the experimental basis, know-how for the mintage of MBE and takes lead for the formation of the International Co-operation for the development and creation of MBE for the various goals, γ-laser included, on base of modern technologies.

References

[1] V.S. Letokhov, JETP 1973(64)1555; Sov.Phys.-JETP 1973(37)787.

[2] Editors' Preface in: Proc. IV All-Russian Int. Sc. Conf. Phys.-Chem. Processes at Selection, 1999, Zvenigorod; V.Yu. Baranov, Yu.A. Kolesnikov, eds. CNIIATOMINFORM 2000, p. 1.

[3] Chen F.F., Phys. Fluids, 25(1982)2385.

[4] S.V. Karyagin, Las. Phys. 5(1995)343.

[5] S.V.Karyagin, in: Proc. IV All-Russian Int. Sc. Conf. Phys.-Chem. Processes at Selection, 1999, Zvenigorod; V.Yu. Baranov, Yu.A. Kolesnikov, eds. CNIIATOMINFORM, 2000, p. 187.

[6] S.V.Karyagin, Proc. SPIE, 2001, Vol. 4352, p. 234.

[7] S.V.Karyagin, Izv. Ran., ser. fiz., Russian, Vol. 65, №11, in print.

[8] A.A.Sysoev, I.V.Shchekina and S.V.Karyagin, Hyp. Int. 107(1997)481.

[9] Yu.B. Khariton, A.M. Voinov, V.F. Kiseliov, V.F.Kolesov, et al., "Aperiodic pulse research reactors". Problems of Modern experimental and theoretical physics, ed. by A.P. Alexandrov, Leningrad, Nauka: 1984 (in Russian).

[10] V.F.Kolesov, «Aperiodical Pulse Reactors», Sarov, RFNC-VNIIEF: 1999 (in Russian)..

Baseline investigation in transmission Mössbauer spectroscopy using resonance detectors

Y. Maltsev[1], H. Mehner[1], M. Menzel[1], N. Maslova[1], B. Rogozev[2], E. Kruglov[3]

[1]*Federal Institute for Materials Research and Testing (BAM)*
D-12489, Berlin-Adlershof, Richard-Willstätter Straße 11, Germany
[2]*Radium Institute, 194021, St. Petersburg, 2-nd Murinsky ave. 28, Russia*
[3]*A. F. Ioffe Physico-Technical Institute*
194021, St. Petersburg, Polytekhnicheskaya street 26, Russia

It was recognized, that the baseline in transmission Mössbauer spectroscopy using resonance detectors reveals a specific distortion. This distortion was investigated from both the experimental and theoretical point of view. It was shown, that this specific distortion caused by an energy shift of the gamma-quanta that were Rayleigh- scattered from the moving sample.

Keywords: Mössbauer spectroscopy, resonance detector, Rayleigh scattering of Mössbauer radiation

1. Introduction

Resonance detectors (RD) are known since 1963, when Mitrofanov [1] suggested introducing a substance containing a Mössbauer isotope into the sensitive area of a detector. The use of the RD in transmission Mössbauer spectroscopy improves the resolution of spectral lines up to 1.4 times and increases the intensity of the resonance effect several times. Finally, RD reduces the time of the measurements.

In Mössbauer spectroscopic experiments to investigate samples with low content (<0.01%) of a resonance isotope, it was noticed that the baseline of the Mössbauer spectrum measured by RD in transmission geometry reveals a specific distortion (figure 1). The observed distortion is not caused by a modulation of a solid angle, because the distance between the source and the RD is fixed.

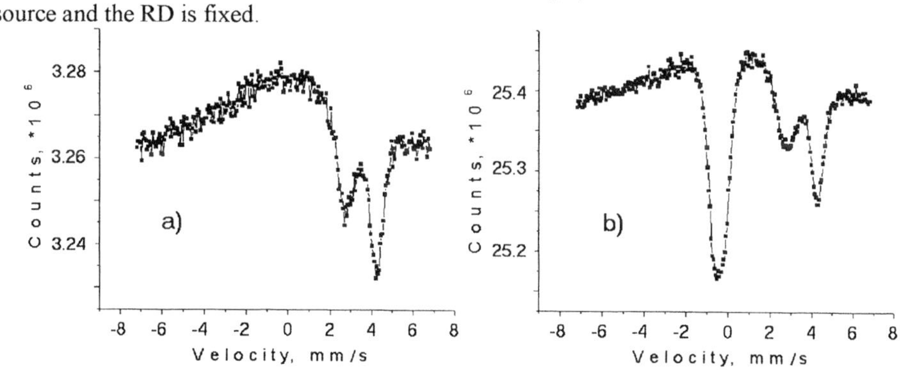

Figure 1: Mössbauer spectra of fluoride-phosphate glass containing 0.5 % tin
a) - melted under reductive conditions in a graphite crucible
b) - melted under oxidative conditions in a platinum crucible

The observed distortion was not strong. In our cases it was less than 0.7% (figure 1a) and less than 0.3% (figure 1b) of the baseline. Nevertheless, for samples with low content of a Mössbauer isotope this distortion leads to a significant error in the phase analysis, because the ratio of the line intensities was disturbed.

2. Experiments

To investigate the distortions, the samples were made of aluminium (0.9 mm and 1.5 mm thick), copper (0.2 mm thick) and Plexiglas (3.0 mm and 6.0 mm thick). A 3-mCi Sn^{119m} source in $CaSnO_3$ matrix and a resonance scintillation detector (RSD) with a $BaSnO_3$ converter were used all produced by Ritverc Company. The diameters of the active part of the source, the sensitive area of the RSD, and the samples were 8 mm, 20 mm, and 22 mm, respectively. The distance between the source and the RSD was 14 mm, and the sample was in the middle position.

We investigated the gamma-quanta count rate dependencies on the velocities of the samples in parallel and perpendicular direction to the gamma-ray beam. The profiles of experimental dependencies were practically the same for different thickness of the samples, but were different for copper, aluminum, and Plexiglas.

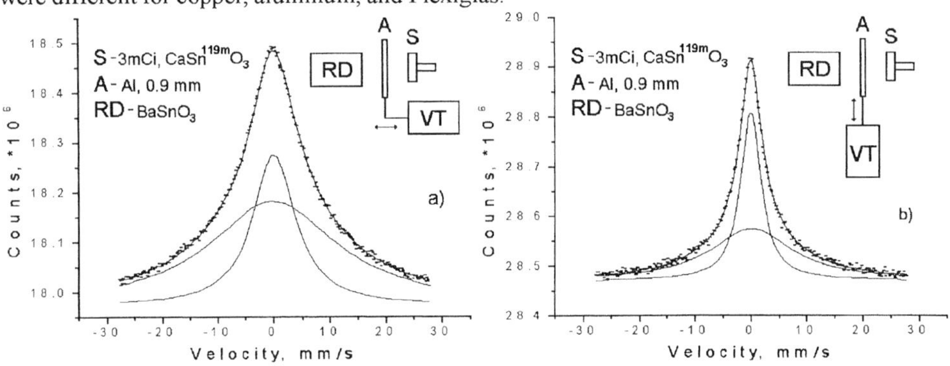

Figure 2. The dependence of the RD count rate on the velocity of an aluminium sample for parallel (a) and perpendicular (b) motion in respect to the axis of gamma-ray beam

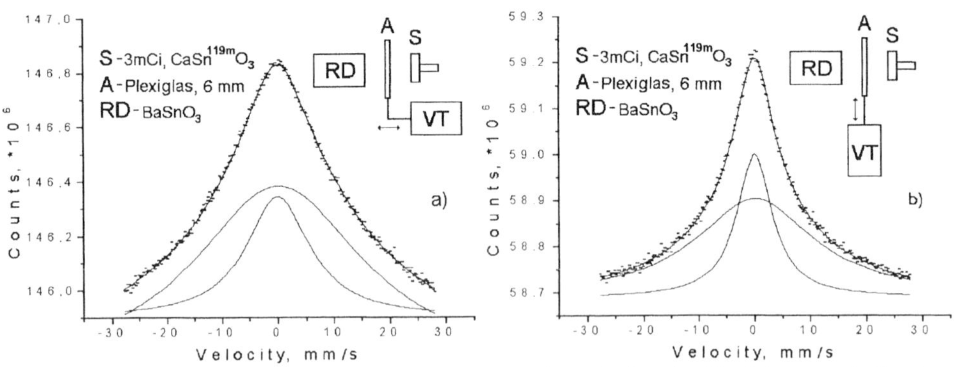

Figure 3. The dependence of the RD count rate on the velocity of a Plexiglas sample for parallel (a) and perpendicular (b) motion in respect to the axis of gamma-ray beam

In figures 2 and 3 the count rate dependencies on the velocities of the sample in parallel (a) and perpendicular (b) direction to the gamma-ray beam were shown for samples made of aluminium and Plexiglas. One can see the shape of the distribution depends on the material of the sample. It was found out that these experimental dependencies can be well approximated by two Lorentz-profile distributions, and the solid lines on figures 2 and 3 show the result of least square fitting using two Lorentz-profiles.

3. Discussion

When a photon with the energy E_1 is Rayleigh-scattered from a moving heavy particle it changes the energy [2]:

$$E_2 = E_1 \frac{1-\frac{\vec{v}}{c}\vec{\Omega}_1}{1-\frac{\vec{v}}{c}\vec{\Omega}_2};$$

E_1, $\vec{\Omega}_1$ - the energy and the direction of the photon before scattering
E_2, $\vec{\Omega}_2$ - the energy and the direction of the photon after scattering
$\vec{v}$ - the velocity of the particle before scattering
c - velocity of light

From this formula it is easy to find the Doppler energy shifts $E_2 - E_1 = E_1 \cdot v/c \cdot (\cos\varphi - 1)$, and $E_2 - E_1 = E_1 \cdot v/c \cdot \sin\varphi$ for the cases when the sample velocity $\vec{v}$ is parallel to the direction of the incident gamma-ray beam $\vec{\Omega}_1$ (case "a"), and $\vec{v}$ is perpendicular to $\vec{\Omega}_1$ (case "b"), respectively (φ – the angle between $\vec{\Omega}_2$ and $\vec{\Omega}_1$; $v/c \ll 1$). For example, the Sn^{119m} isotope (E_1=23.8 keV; $\Gamma_{nat}=2.47\cdot10^{-8}$ eV), for v=15 mm/s and $\varphi=25^0$, gives $E_2 - E_1 = -4.5\cdot\Gamma_{nat}$ (case "a"), and $E_2 - E_1 = 20\cdot\Gamma_{nat}$ (case "b"). These energy shifts make the scattered gamma-quanta practically not registerable by RD.

The expression for the count rate $n_{RD}(\vec{v})$ in the RD in the case of axial symmetry can be written as:

$$n_{RD}(\vec{v}) = \int_{-\infty}^{+\infty}\int_{\Omega_2}\int_{\Omega_1} I(\vec{v}, E, \vec{\Omega}_1, \vec{\Omega}_2)\cdot\varepsilon(E)\,d\vec{\Omega}_1 d\vec{\Omega}_2 dE\,,$$

where:

$$\varepsilon(E) = \frac{C_1}{E^2 + \left(\frac{\Gamma_{RD}}{2}\right)^2}$$

- registration efficiency of the RD;

C_1 – constant;

Γ_{RD} - the width of the RD absorption line;

$I(\vec{v}, E, \vec{\Omega}_1, \vec{\Omega}_2)$ – intensity of the photons with energy E, emitted by a source in the direction $\vec{\Omega}_1$, and scattered in the direction $\vec{\Omega}_2$ from a sample moving with velocity $\vec{v}$.

The term $I(\vec{v}, E, \vec{\Omega}_1, \vec{\Omega}_2)$ includes the Rayleigh scattering form-factors for sample atoms [4], which are responsible for differences in the count rate distributions for different materials (see figures 2 and 3).

The integral for $n_{RD}(v)$ cannot be solved analytically. In the present work we calculated it by Monte-Carlo simulations for real sizes of source, detector, and samples from different materials. The cross sections for gamma-ray scattering were taken from [3]. The form-factors were taken from [4].

The result of the Monte-Carlo simulation for an aluminium sample is shown in the figure 4. From this figure it can be concluded, that the suggested model is in a good agreement with the experimental data.

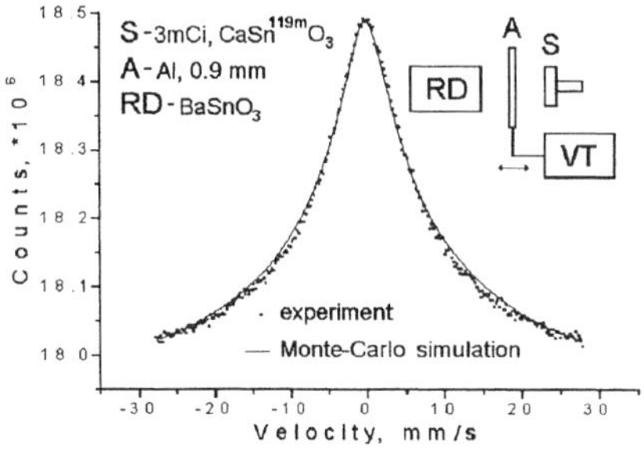

Figure 4.
The comparison of the experimental and Monte-Carlo simulated dependencies of the count rate on the velocity for an aluminium sample

In the velocity range of ±30 mm/s the experimental distributions can be well approximated by two Lorentz- shape profiles with maximum at 0 mm/s and the widths of $\Gamma_1 > 10 \cdot \Gamma_{nat.}$ and $\Gamma_2 > 50 \cdot \Gamma_{nat.}$ (solid curves on the figures 2 and 3). The Lorentz- shape distributions are available in all fitting programs, so, the baseline distortion can be easily corrected.

4. Conclusions

1. The distortion of the baseline of Mössbauer spectra measured in transmission geometry using RD is caused by the sensitivity of the RD to the energy shift of gamma-quanta Rayleigh scattered from a moving sample. Neglecting this effect may cause wrong values of the spectra lines intensities.
2. To minimise the influence of this effect on the Mössbauer spectrum one should increase the distance between the source and the RD.
3. Both, the experimental dependencies of the count rate on the velocity of nonresonant samples and Monte-Carlo simulations, show the strong influence of a Rayleigh scattering form factor on the shape of the experimental distribution.
4. The distortion of the baseline of the Mössbauer spectra measured in a narrow transmission geometry using the RD can be described and corrected precisely by a superposition of two Lorentz-profile distributions that should be involved in the spectrum fitting procedure.
5. The sensitivity of RD to the energy shift of Rayleigh scattered gamma-quanta broadens the field of applications of Rayleigh Scattering of Mössbauer Radiation (RSMR) e.g. for investigation of form factors, local motions in samples, etc.

References:

[1] K.P. Mitrofanov, N.V. Illarionava, V.S. Shpinel, Pribori i Tech. Eksper. N3 (1963), p. 49
[2] R. Sunaev, E. Churazov, Astronomy Letters, V.22, N.10 (1996).
[3]. E. Storm, H. Israel, in: Photon Cross Sections from 0.001 to 100 MeV for Elements 1 Through 100, Los Alamos Scientific Laboratory, New Mexico, 1967.
[4] H.P. Hanson, F.Herman, J.D. Lea, and S. Snillman, Acta Crystallographica, V.17, part 8 (1964), p.1040.

Perforated diamond anvils for research at very high pressures

R. D. Taylor[a,*], M. P. Pasternak[a,b], A. Dadashev[b] and G. Kh. Rozenberg[b]

[a] *MST-10, MS-K764, Los Alamos National Laboratory, Los Alamos NM 87545, USA.*

[b] *School of Physics and Astronomy, Tel Aviv University, 69978 Tel Aviv, ISRAEL.*

Diamond anvil cells allow static pressures on small samples up to a few hundred GPa. The anvils also serve as an optical window to the sample. Although the anvils are rather transparent for most types of radiation, the absorption of γ- or soft x-rays in a few mm of diamond can prevent or seriously degrade the quality of the Mössbauer spectrum. Using smaller (thinner) diamonds helps, but then the ultimate pressure suffers. Several schemes are proposed and tested to reduce the pathway (absorption) by drilling away most of the anvil in the axial path, yet retaining its strength and stability.

Key words: High pressures, Diamond anvil cells, Mössbauer spectroscopy

1. INTRODUCTION

Most of the recent progress in high pressure science may be attributable to the development and use of the diamond anvil cell (DAC) [1, 2]. With a table to culet tip area ratio of ~1000, pressures up to over 200 GPa can be generated by an opposing diamond anvil pair held in a simple structure that also supplies the modest closure force. Figure 1 is a generic DAC showing the backing plates, anvils, and a gasket made of a high-Z hard metal into which a sample cavity is formed. A hydrostatic-pressure medium and ruby chips, for *in situ* determination of P, are added to the sample [1]. The select ~ 1/3 carat, brilliant-cut natural diamonds modified with 16-sided faceting and a beveled anvil tip that are used to reach these high pressures, are about 2.6-mm thick. Attenuation of a pair of such anvils can seriously degrade Mössbauer experiments, especially for energies below ~20 keV. Experience has showed that smaller, and therefore thinner diamonds break before reaching 20-30 GPa. An alternative is to remove much of the diamond in the path through drilling or perforation.

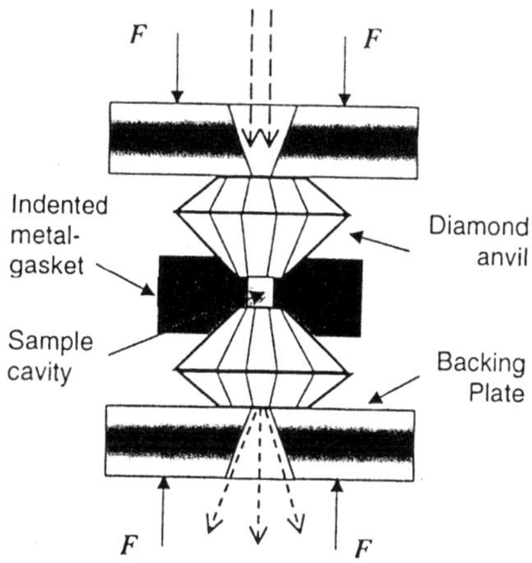

Figure 1. Essential components of a diamond anvil cell. The closure and force mechanism vary for different DACs models.

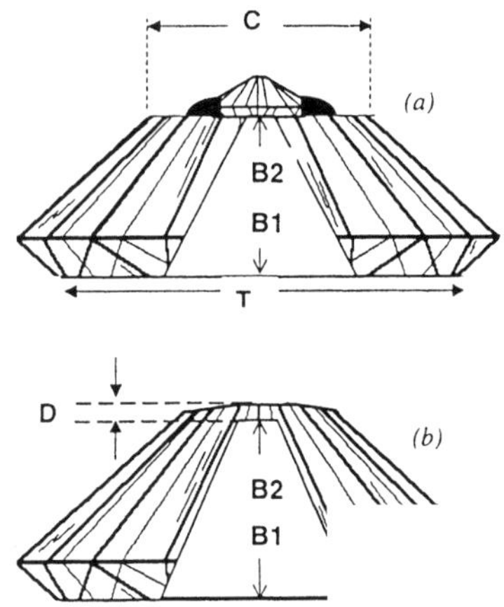

Figure 2. (a) Perforat backing plate with mir attached. (b) Partially perforated diamond anvil.

This approach has been used by Bassett *et al.*[3] for soft x rays in DAC's up to 0.8 GPa. We also show that by using fully penetrated diamonds as a backing plate support for miniature (1/100 carat) diamond anvils, one can achieve pressures of 100 GPa, and furthermore, this combination is sturdier than the equivalent, undrilled, conventional anvil.

Many MS (mostly ^{57}Fe, ^{119}Sn, ^{129}I, and ^{151}Eu) and XRD experiments [4] have been carried out despite the attenuation of a pair of 0.3 carat anvils; however we shall show that for ^{57}Fe MS dramatic improvements are possible using perforated diamonds. The transmission of a pair of regular 0.3 carat (5 mm) for 14.4 keV is ~22%. With a combination of perforated diamonds and a pair of miniature anvils of 1.6mm this becomes 62%. A more detailed report of this work and its applications to other spectroscopies has recently been published [5].

2. EXPERIMENTAL

Experiments were conducted to test the performance of the fully perforated anvils, designated as a *diamond backing plate* (DBP), combined with the *miniature anvils* (MA), and the *partially perforated anvil* (PPA), up to 100 GPa. Four 0.3-, three 0.012-, and one 0.1-carat diamond anvils of the modified-brilliant-cut design were obtained from *D'ANVILS* for the present work. All anvils were selected from gem quality stock with minimal internal flaws. Three of the 0.3 carat anvils, 1.6-mm high (H), 3.7-mm diam. table (T), and 1.8-mm diam. culets (C), were chosen for laser drilling. The conical holes, polished throughout, were 1-, 1.5-, and 2-mm diam. at the table (B1) and 0.3-mm diam. at the culet face (B2) (see figure 2a). The fourth 0.3 carat anvil, with H = 2.2 mm, T = 3.7 mm, and C = 1.8 mm beveled to give a 0.2-mm flat was *partially* drilled, leaving a residual wall of 0.5-mm thickness under the culet with 2.0-mm B1 diam. and 0.3-mm B2 diam. (see figure 2b). The specifications of the three miniature anvils were: H = 0.8 mm, T = 1.1 mm, and C = 0.2 or 0.3 mm.

Various combinations of regular anvil, PPA, and DBP+MA were tested using a miniature piston-cylinder DAC that has a flat and a hemispheric tungsten carbide backing plate allowing final cell alignment. The MA was centered and epoxied to the DBP. Rhenium gaskets, preindented to 25 μm and drilled to form a sample space half the culet diam., were filled with ruby chips and liquid Ar for pressure tests and with Fe foil for MS studies.

3. RESULTS

Except for one MA, all components survived the 100 GPa tests. Of primary interest here, we show the enhanced MS of ^{57}Fe in iron in a DAC with a PPA and a culet matching DBP+MA anvil. The latter is required in order to view the sample and its ruby; as the hole in the PPA is not of optical quality. These MS results (figure 3a) are compared with that using regular 0.3 carat anvils (figure 3b). The large improvement reflects improved count rate and improved "good/bad" counts in the 14.4 keV window.

It was also found that wider effective apertures were possible with the DBP+MA arrangement than with the regular anvils. DAC's with regular 0.3-carat anvils mounted on WC backing plates with an aperture greater than 1.2 mm failed to reach 40 GPa. In contrast 100 GPa pressures were reached with 1.0-2.0-mm conical apertures. Wide exit apertures are generally important, especially with MS and XRD.

Following this technological development high pressure MS with low-energy γ-rays such as ^{181}Ta (6.3 keV) and ^{83}Kr (9.4 keV) will be possible. This also opens exciting venues for *nuclear inelastic scattering* using the 6.3 keV Fe K-x-rays in dedicated synchrotron beams.

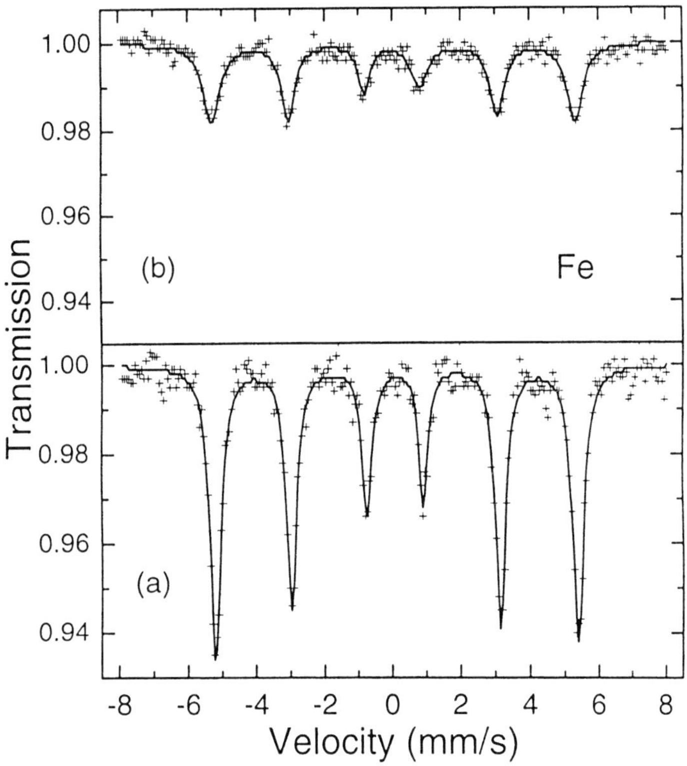

Figure 3 (a) Room temperature MS of 25μm natural iron foil using perforated diamonds, a PPF and a DBP+MA anvil arrangement, in a miniature piston/cylinder DAC. The MS in (b) is for the same arrangement except regular 0.3 carat anvils were used. Pressures < 1 GPa.

Acknowledgement: This research was supported in part by the BSF grant 1998003 and by Israeli Science Foundation grant 36/00.

References:

* email: rdt@lanl.gov

[1] A. Jayaraman, Rev. Mod. Phys. 55 (1983) 65.
[2] H. K Mao and R. Hemley, Rev. Mineralogy 37 (1998) 4.
[3] W. A. Bassett, A. J. Anderson, R. A. Mayanovic, and I. M. Chou, Chemical Geology, 167 (2000) 3, and Z. Kristallog. 215 (2000) 711.
[4] M. P. Pasternak and R. D. Taylor in *Mössbauer Spectroscopy Applied to Magnetism and Materials Science, Vol. 2*, G. Long and F. Grandjean, eds. (Plenum Press, New York 1996) Chapter 8, p167ff.
[5] A. Dadashev, M. P. Pasternak, G. Kh. Rozenberg and R. D. Taylor, Rev. Sci. Instr. 72 (2001) 2633.

Nuclear Excitation of ^{127}Iodine using Synchrotron Radiation

Y. Yoda [a], S. Kishimoto [b], X.W. Zhang [b], M. Seto [c] and S. Kikuta [a]

[a] *Japan Synchrotron Radiation Research Institute, 1-1-1 Kouto, Mikazuki-cho, Sayo-gun, Hyogo 679-5198, Japan*

[b] *Photon Factory, High Energy Accelerator Research Organization, 1-1 Oho, Tsukuba-shi, Ibaraki 305-0801, Japan*

[c] *Research Reactor Institute, Kyoto University, Kumatori-cho, Sennan-gun, Osaka 590-0494, Japan*

The nuclear excitation of 57.6 keV level in ^{127}I using synchrotron radiation was first observed. The obtained lifetimes agree well with the previously reported value. ^{127}I will be more promising than ^{129}I for the study of the local vibrational states using nuclear resonant inelastic scattering once the high-resolution monochromator for this energy range is developed.

The iodine is widely used medically and industrially and one of the key elements in materials and biological sciences. There are two Mössbauer isotopes of ^{127}I with 57.61 keV transition and ^{129}I with 27.78 keV transition and their Mössbauer spectroscopy has so far offered the useful information [1, 2] especially in the chemical bonding. Nuclear resonant scattering using synchrotron radiation (SR) has specific features. A lot of novel applications such as nuclear inelastic scattering [3] have been created [4]. Recently the excitation of nuclei with relatively higher transition energies such as ^{40}K [5], ^{121}Sb [6] and ^{61}Ni [7] has been reported using synchrotron radiation generated by the large storage ring to evaluate the possibility of application use and ^{197}Au to calibrate the cross section [8]. The iodine is also the target nuclei whose nuclear resonant scattering study using SR is expected. We report the first excitation of ^{127}I using SR and compare ^{127}I and ^{129}I from the point of view of SR use.

The experiments were performed at BL09XU, SPring-8 [9]. The storage ring was operated in a 203 bunch mode with 23.6 nsec intervals and its current was 100 mA - 70 mA. As shown in Fig. 1, Si 333 reflections of the beamline double-crystal monochromator were adopted to monochromatize the X-rays generated from the in-vacuum undulator. An Al absorber with 15 mm thickness was placed to eliminate the first harmonic X-rays through the Si 111 reflections. It is checked by the current dependence of the Si-PIN photodiode detector on the Al thickness that the intensity of 19.2 keV X-ray is less than 10^{-3} of that of 57.6 keV X-rays. The beam size was restricted to 2.0 (h) x 2.5 (v) mm^2 by a slit. The measured incident flux was about 7×10^{10} cps. The sample of NaI crystal is a tablet with a size of 10 x 10 x 0.5 mm^3. The natural abundance of ^{127}I is 100%. Two arrangements were adopted in turn to measure the resonant excitation of ^{127}I. The incoherent scattering geometry was taken in both arrangements. In the first arrangement two elements avalanche photo-diode (APD) with a sensitive area of 1 x 0.5 mm^2 and a depletion layer of 50 μm was placed in the air. In the second arrangement NaI and APD with a sensitive area of 3 mm in diameter and a depletion layer of 30 μm were placed in a vacuum chamber to detect the conversion electrons from ^{127}I as well as X-rays [8].

Fig. 2 shows the delayed intensity from NaI in the second arrangement scanning the angle of Si 333 monochromator, which changes the incident X-ray energy. A clear peak of about 5 cps was observed at the expected monochromator angle. Its width (FWHM) is 1.0x 10^{-3} degree from the Gaussian fit, which shows the incident X-ray energy width of 9.6 eV. The obtained delayed intensity in the first arrangement is about 1 cps. Fig. 3 shows the time spectra measured (a) in the first arrangement and (b) in the second arrangement. Both spectra show exponential decays at least in the time range after 5 nsec, which reflect the incoherent nature of the nuclear resonant scattering in these arrangements. The half-life obtained from the linear fits of the 6 – 12 nsec data are 1.98+-0.08 nsec in Fig.3 (a) and 1.91+-0.05 nsec in Fig. 3 (b) which agree well with the previous data of 1.95 +- 0.01 [10].

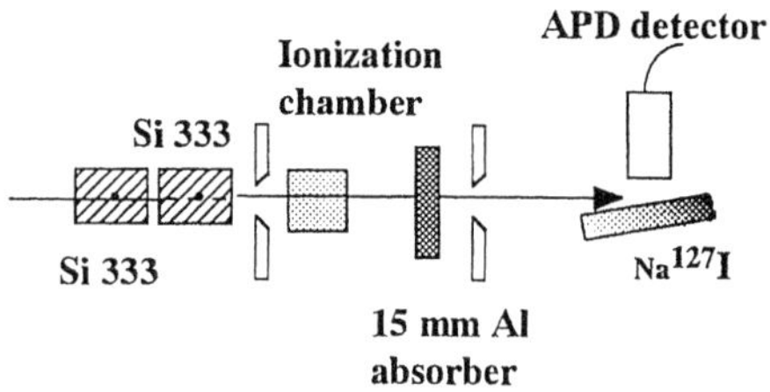

Figure 1. Schematic view of the experimental setup for the nuclear excitation of 57.6 keV level in ^{127}I using SR. The first harmonic X-rays of the monochromator are reduced with the Aluminum absorber. The natural abundance of ^{127}I is 100 %.

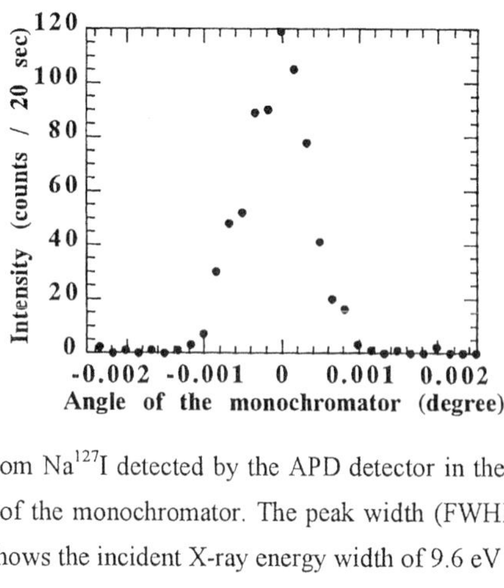

Figure 2. Delayed intensity from Na^{127}I detected by the APD detector in the vacuum chamber as a function of the relative angle of the monochromator. The peak width (FWHM) is 1.0x 10^{-3} degree from the Gaussian fit, which shows the incident X-ray energy width of 9.6 eV

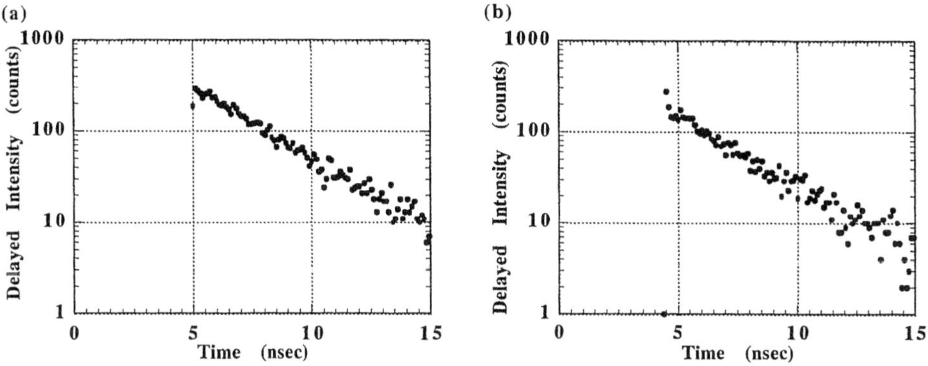

Figure 3. Time spectra from $Na^{127}I$ detected by the APD detector (a) in the air (b) in the vacuum chamber. The half-life obtained from the linear fits of the 6 – 12 nsec data are 1.98+-0.08 nsec in (a) and 1.91+-0.05 nsec in (b).

It is well known that ^{129}I has been more widely used than ^{127}I in spite of the crucial inconvenience of the source and absorber, because ^{129}I has narrower natural linewidth and offers more reliable data than ^{127}I. This situation will not change even at SR when it is used for coherent nuclear resonant scattering to study the hyperfine interactions. But it will possibly change when it is used for nuclear resonant inelastic scattering to study the vibrational states.

We will compare ^{127}I and ^{129}I in the following seven points assuming it is used for nuclear inelastic scattering. The property of the two Mössbauer isotopes is listed in Table 1. 1) Linewidth: Intensity of incoherent nuclear resonant scattering increases linearly with the linewidth if an enough fast detector is used. 2) Cross section: The lower cross section of ^{127}I due to its higher energy can be covered by the thicker sample. 3) Recoilless fraction: The smaller recoilless fraction of ^{127}I at the same temperature makes the inelastic signal increase. 4) Background noise: ^{129}I causes β-decay with a half-life of 1.57 x 10^7 year accompanying the emission of 39.6 keV γ-ray and Xe fluorescent X-rays (< 35 keV). It means that the simple substance of ^{129}I with a cubic shape of only 100 μm^3 emits more than 68 counts/sec X-rays in the whole solid angle. And in the incoherent setting, the detector is usually placed to make the solid angle as large as possible, although the actually detected background noise is reduced by the detector efficiency and time-gating. 5) SR source: Collimated X-rays of 57.6 keV from the undulator can be obtainable at SPring-8 and PETRA so far. 6) High-resolution monochromator: high-resolution monochromator for ^{129}I is not so difficult to make. On the other hand that for X-rays above 40 keV has not been developed yet. 7) Detector: The energies of the detected X-rays in case of ^{127}I which is mostly composed of fluorescent X-rays (<33 keV) is almost same as that in case of ^{129}I. So the detector efficiency is almost same in both isotopes.

There is possible problem of inherent background noise when using ^{129}I. So the focusing of incident X-rays and a small amount of the sample will be essential to overcome the background noise when ^{129}I is used. On the other hand, the development of the high-resolution monochromator for 57.6 keV is a key for using ^{127}I. It is not easy, but not impossible.

In summary, we have observed the excitation of ^{127}I using SR. ^{127}I is a good candidate for the study of the local vibrational states using nuclear resonant inelastic scattering.

Table 1. Property of the two Mössbauer isotopes of iodine relevant to the nuclear resonant inelastic scattering to study the local vibrational states of iodine.

Isotope	Energy (keV)	Half life (nsec)	Natural abundance (%)	Internal conversion coefficient	Cross section ($x10^{-20}$ cm^2)
^{127}I	57.61	1.95	100	3.75	20.7
^{129}I	27.78	16.8	0, radioactive	5.0	39.7

Acknowledgment

The authors thank Dr. Kitao for the useful discussions. They acknowledge the Beamline and Accelerator Group of SPring-8 for the operation and the effort of obtaining pure bunches. This work was performed with the approval of Japan Synchrotron Radiation Research Institute (Proposal No. 2000A0215-ND-np)

References

[1] R.V. Parish, Mössbauer spectroscopy Applied to Inorganic Chemistry, ed. by G. J. Long (Plenum Press, N. Y., 1987) and references therein.

[2] N. N. Greenwood and T. C. Gibb, Mössbauer spectroscopy (Chapman & Hall, London, 1971) and references therein.

[3]M. Seto, Y. Yoda, S. Kikuta, X. W. Zhang and M. Ando, Phys. Rev. Lett., 74 (1995) 3828.

[4] E. Gerdau and H. de Waard (Eds.), Hyp. Interact. 123/124 (1999) .

[5] M. Seto, S. Kitao, Y. Kobayashi, R. Haruki, T. Mitsui, Y. Yoda, X. W. Zhang, and Yu. Maeda, Phys. Rev. Lett. 84 (2000) 566.

[6] S. Kitao, T. Mitsui, M. Seto, J. Phys. Soc. Jpn 69 (2000) 683.

[7] This conference.

[8] S. Kishimoto, Y. Yoda, M. Seto, Y. Kobayashi, S. Kitao, R. Haruki, T. Kawauchi, K. Fukutani, and T. Okano, Phys. Rev. Lett. 85 (2000) 1831.

[9] Y. Yoda, M. Yabashi, K. Izumi, X. W. Zhang, S. Kishimoto, S. Kitao, M. Seto, T. Mitsui, T. Harami, Y. Imai and S. Kikuta, Nucl. Instrum. Methods A. in press.

[10] Table of Isotopes, eds. R. B. Firestone, V. S. Shirley and S.Y. Frank Chu (Wiley, New York, 1996) 8th ed.

Multiple Mössbauer Spectra Data Acquisition with an In-situ Cyclic Voltammetry System

T L Greaves[1], J D Cashion[1], A L Benci[1], N K Jaggi[1,3] C Hogan[2] and A M Bond[2]

[1] School of Physics and Materials Engineering, Monash University, Vic 3800, Australia
[2] School of Chemistry, Monash University, Vic 3800, Australia
[3] Physics Department, Illinois Wesleyan University, Bloomington IL 61702-2900, US

A Mössbauer spectroscopy system has been developed which is capable of taking multiple Mössbauer spectra which are related in time. Its present design is for operation in conjunction with an in-situ electrochemical cell, such that up to 256 Mössbauer spectra can be taken in a dedicated PC. These spectra are uniformly distributed around a cyclic voltammetry loop which is continually retraced until sufficient statistics are accumulated. This system was used to acquire four spectra of a Prussian-blue-modified graphite paste electrode, while the electrode was involved in an electrochemical process. The spectra show differences due to the reduction and oxidation of the iron in Prussian blue.

Keywords: Cyclic voltammetry, Spectroelectrochemistry, Prussian blue.

1. Introduction

Conventionally Mössbauer Spectroscopy involves the acquisition of a single spectrum completely before another is started. This imposes some limitations on the technique of Mössbauer Spectroscopy which are not inherent in the Mössbauer Effect, but only in its applications. Multiple Mössbauer Spectroscopy involves the acquisition of more than one spectrum during the same time period. This can be very beneficial in some circumstances, enabling the retention of information that would otherwise be lost, such as changes in the iron species across a sample, or throughout a time varying sample.

The previously developed Imaging Multiple Mössbauer Spectrometer (Smith et al., 1992), is capable of taking up to 256 spectra in a 1-dimensional array across a thin sample. The benefit of the imaging system is that it retains phase information of the iron compounds across the sample.

The present Time Dependent Multiple Mössbauer Spectrometer (TDMMS) was designed based upon the Imaging system to take multiple spectra over time throughout a cyclic process. From the series of spectra the iron species present at different stages of the cyclic process can be determined. Currently the system is designed to operate in conjunction with a specially designed in-situ electrochemical cell, in which a cyclic voltammetry process occurs while spectra are taken of the working electrode.

The Prussian-blue-modified graphite electrode was chosen for testing the equipment due to it being well characterised electrochemically (Dostal et al., 1995), and by Mössbauer spectroscopy (Reguera et al., 1992). The electrochemistry of soluble Prussian blue, $KFe^{III}[Fe^{II}(CN)_6]$, is complicated due to the two different iron sites present, which both can undergo oxidation and reduction. The reduction product is Everitt's salt, $K_2Fe^{II}[Fe^{II}(CN)_6]$, and the oxidation product Prussian yellow, $Fe^{III}Fe^{III}(CN)_6]$ (Dostal et al., 1995). In this investigation the TDMMS has been used to acquire four spectra of a Prussian-blue-modified graphite paste electrode while it is undergoing a cyclic voltammetry process.

2. Experimental

The TDMMS consists of a cyclic voltammeter coupled to a Mössbauer spectrometer, as shown in Figure 1. The cyclic voltammeter is driven by a triangular voltage applied to the in-situ electrochemical cell. This voltage is also converted into a sawtooth form, which ensures there is a unique relationship between this potential and any point in the cyclic voltammetry sweep. The arrival of a pulse corresponding to a 14.4 keV gamma triggers a sampling of the sawtooth potential by an analogue to digital converter. This digital signal is processed to determine into which spectrum number the count should be stored. The velocity information determines the channel number of the spectrum in the usual manner.

Currently the spectrometer is capable of taking up to 256 spectra at room temperature. It is designed flexibly so that the input can be changed to any other cyclic process, such as a magnetic hysteresis loop.

The in-situ electrochemical cell used for these experiments was specifically designed to enable Mössbauer spectra to be taken of the working electrode while the electrochemical process is taking place. The design of the cell has been described elsewhere (Greaves et al.).

Insoluble Prussian blue powder was prepared according to the procedure in (Dostal et al., (1995) for Prussian blue I, and was electrochemically converted into soluble Prussian blue. Graphite powder was combined with the Prussian blue in a 5:1 ratio by weight, and 280mg of the mixed powders were stirred with approximately 200μL Teflon wax, then quickly transferred to the perspex support. Once dry, the surface was polished with fine sandpaper and rinsed with distilled water. A Ag/AgCl reference electrode and platinum mesh counter electrode were used with 4M KNO_3 as the electrolyte. The cyclic voltammeter swept through the voltage range between –0.25 to 1.60 V, at a scanning rate of 5mV/s.

3. Results

A cyclic voltammogram of the electrode is shown in Fig. 2. Traversing the trace clockwise from –0.25V we can see the features corresponding to the oxidation of the high spin iron site from Fe^{2+} to Fe^{3+} at 0.6V, and the reduction pair to this at 0.17V on the lower branch. The expected peaks for the oxidation and reduction of the ferrocyanide site were not seen, which was probably due to the positive potential not going sufficiently far. The data acquisition too place over 6 days, and the stability of the cyclic voltammetry trace was monitored on a chart recorder, with the voltages of the peaks remaining constant.

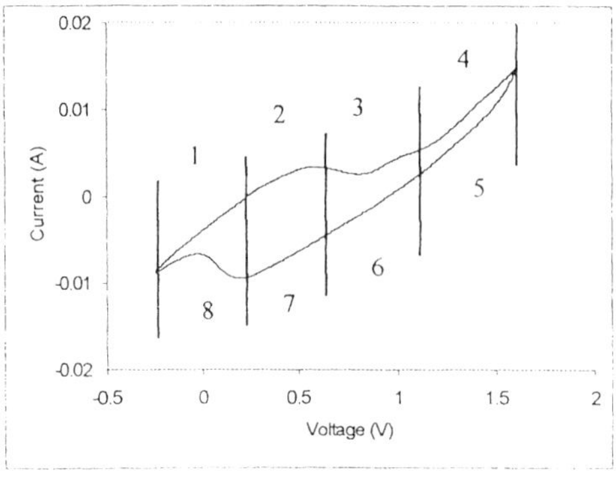

Fig. 2: Cyclic voltammogram of the working electrode between –0.25 to 1.6V.

Eight Mössbauer spectra were acquired at room temperature using a constant acceleration drive. These spectra were then combined to give four non-equivalent spectra in the potential ranges (a) –0.25 to 0.21V, (b) 0.21 to 0.68V, (c) 0.68 to1.14V, (d) 1.14 to 1.6V. These spectra are shown in Fig. 3, and the mean fitting parameters in Fig. 4.

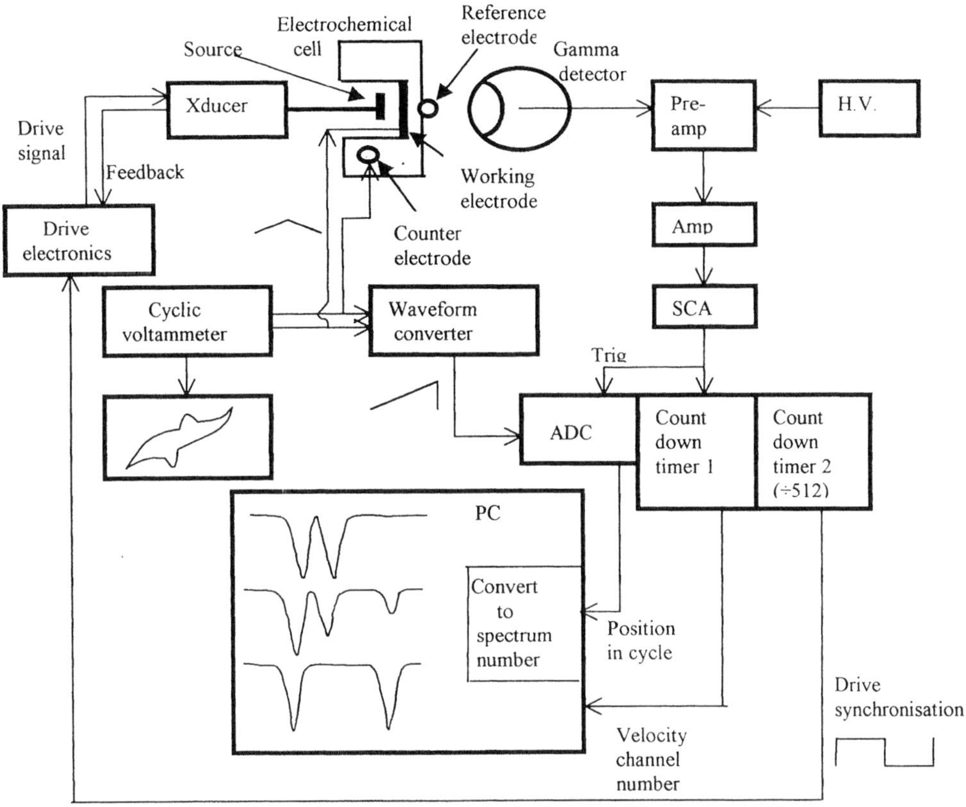

Fig. 1: Schematic diagram of the equipment.

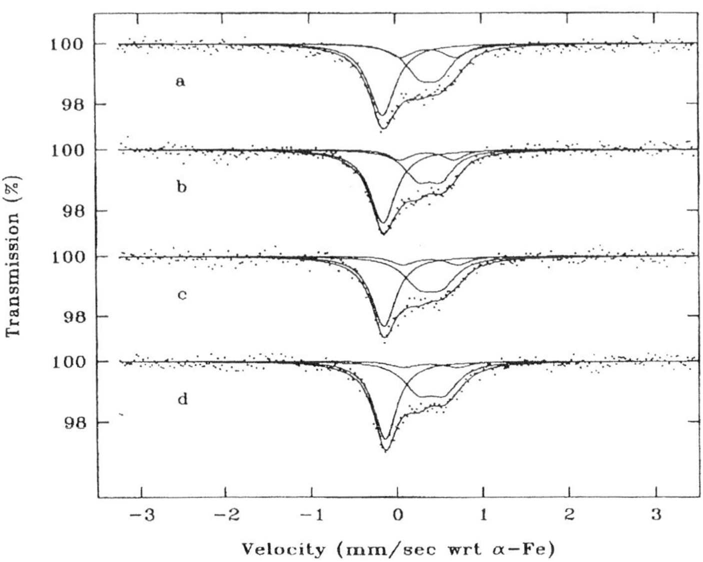

Fig. 3: Mössbauer spectra of the cyclic system, for the four different potential ranges.

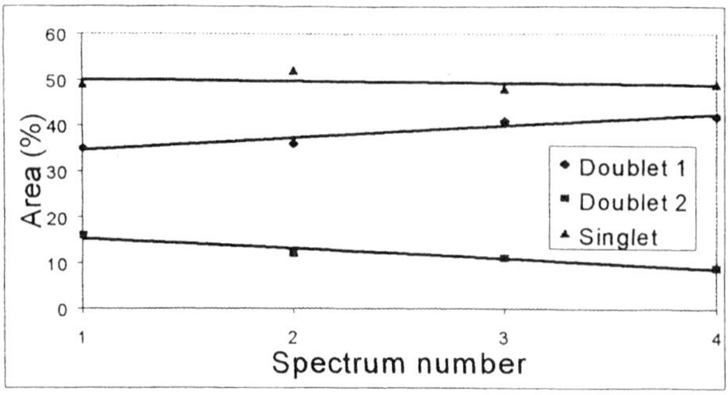

	IS (mm/s)	QS (mm/s)	Γ (mm/s)
Doublet 1	0.39	0.27	0.40
Doublet 2	0.38	0.64	0.35
Singlet	-0.15		0.35

Figure 4: Mean fitting parameters for the spectra. The singlet and doublet 1 correspond to the standard parameters for soluble Prussian Blue.

4. Discussion

The four Mössbauer spectra obtained showed a general trend which was consistent with the partial reduction and re-oxidation of soluble Prussian blue during the cyclic process. The parameters for the low spin Fe^{II} site and the high spin Fe^{III} site present in all four spectra agree well with the parameters for soluble Prussian blue as given in (Reguera et al., 1992). The doublet with the larger quadrupole splitting decreased with increasing applied potential. This doublet corresponds to the reduced high spin iron site. The results are consistent with the cyclic voltammetry data. No changes were detectable in the spectra for the oxidation of the low spin ferrocyanide site.

No previous Mössbauer spectra have been taken of the electrochemically produced oxidation and reduction products of Prussian blue. This is probably due to their instability with no applied potential. The instability of the products is not an issue with the TDMMS, since the process is continually cycling. This new system extends the types of possible spectra which can be taken using Mössbauer spectroscopy.

5. Conclusion

Multiple Mössbauer spectra were obtained throughout a periodic cycling of an electrochemical system consisting of a Prussian-blue-modified graphite paste electrode. The Mössbauer spectra taken of the Prussian blue showed partial reduction of the high spin iron site at 0.5V, but no change due to the oxidation on the low spin iron site.

Acknowledgements

We are pleased to acknowledge helpful discussions with J. Ozaki and the financial support of the Australian Research Council and the Monash Research Fund.

References:

P.R. Smith, J. D. Cashion, L. J. Brown. Hyp. Int. 71 (1992) 1503.

A. Dostal, B. Meyer, F. Scholz, U. Schröder, A. M. Bond, F. Marken, S. J. Shaw. J. Phys. Chem. 99 (1995) 2096-2103.

E. Reguera, J. Fernández-Bertrán, A. Dago, D. Díaz. Hyp. Int. 73 (1992) 295-308.

T. L. Greaves, J. D. Cashion and J. Ozaki. To be published.

Sapphire X-ray Resonator.
First experience and results

Yu.V. Shvyd'ko, E. Gerdau, M. Lerche, M. Lucht and H.-C. Wille
Institut für Experimentalphysik, Universität Hamburg, D-22761 Hamburg

E.E. Alp
Advanced Photon Source, Argonne National Laboratory, Argonne, IL 60439

P. Becker
Physikalisch-Technische Bundesanstalt, Postfach 3345, D-38023 Braunschweig

Abstract. μeV energy resolution in the hard X-ray range could be achieved by employing X-ray resonators. An X-ray resonator with independently adjustable crystal mirrors has been designed and its time response was analyzed. Time response of the resonator with a decay time of 0.9 ns is observed, which corresponds to a resonance width of $\simeq 0.73$ μeV.

Fabry-Pérot resonators are indispensable to modern visible-light optics. Such instruments have been used over the 20th century to perform high-precision wavelength measurements.

High resolution instruments of this kind would also be useful for electromagnetic radiation in the x-ray spectral range. A spectral resolution in the μeV range could be achieved. They could be applied in metrology to bridge optical and x-ray (or γ-ray) wavelengths. They could be used as interference filters of very high spectral resolution for studying, e.g., dynamics of solids, liquids, polimers, or macroscopic biological molecules.

Up to now x-ray resonators remain an exciting and challenging unsolved problem. The current studies are aimed at creating a test device and studying the physics of x-ray resonators.

The main components of a Fabry-Pérot resonator are two mirrors of high reflectivity arranged parallel and separated by a gap $d_{\rm g}$. The system becomes transparent despite the high reflectivity of each mirror when the resonance condition for a standing wave formation in the gap is fulfilled.

$$E = \frac{hc}{2d_{\rm g}\sin\theta}\, n \qquad n = 1, 2, 3, ... \tag{1}$$

The transmission resonances can be observed by varying the gap $d_{\rm g}$, the x-ray energy E, the angle of incidence θ, or by observing the time response of the resonator. On the energy scale the resonances are spaced by a free spectral range $E_{\rm f} = hc/2d_{\rm g}\sin\theta$. The width of the resonances Γ is determined by the mirror reflectivity R. The higher the reflectivity the sharper the resonances - $\Gamma = E_{\rm f}/F$. Here $F = \pi\sqrt{R}/(1-R)$ is the so-called finesse of the resonator. The time response has a characteristic duration $\tau = \hbar/\Gamma$. Due to the large number of equally spaced resonances the time response is modulated by oscillations (quantum beats) of period $t_f = h/E_f \equiv 2d_g \sin\theta/c$. We have studied this time response to characterize the resonators.

Mirrors of high reflectivity are a pivotal element of x-ray resonators. Bragg diffraction from crystals at normal incidence is used. Normal incidence ($\theta = \pi/2$) is crucial for achieving a sufficiently large angular acceptance of the resonances. Crystals with cubic symmetry, like Si, suffer from low reflectivity at normal incidence. The latter is due to multiple beam Bragg diffraction, which originates in crystals of cubic symmetry whenever the Bragg backscattering conditions for the incident radiation is fulfilled. Hexagonal sapphire (Al_2O_3) crystals are free from this drawback [1]. However, on average they unfortunately have lower quality than Si crystals. High quality sapphire mirrors have been selected by using x-ray topography [2] and reflectivity measurements.

Sapphire crystals with Bragg reflection (0 0 0 30) were used as normal incidence mirrors for 14.315 keV x-rays. A resonator with independently adjustable mirrors was designed. The technical design of the x-ray resonator was aimed at allowing (i) parallel alignment of the mirrors with nanoradian accuracy; (ii) temperature stability and homogeneity within 10 mK. A gap of $d_g = 50$ mm was chosen. With such a gap the free spectral range becomes $E_f = 12.4$ μeV and the quantum beats period is $t_f = 330$ ps.

Experiments were performed at the wiggler beamline BW4 at HASYLAB. Two x-ray resonators with pairs of mirrors of different thickness were tested. The right panels in Fig. 1 show the energy dependencies of the reflectivity of the mirrors (dark grey) and the spectral distribution of the incident x-rays (light grey). The resonator mirrors had thicknesses of 54 μm - Fig. 1a and 39 μm - Fig. 1b, respectively. We observe the reflectivity $R = 0.68$ in Fig. 1b and $R = 0.82$ in Fig. 1a. Thus, in agreement with the dynamical theory, the thicker the crystals the higher is the reflectivity.

Incident radiation pulses of a 150 ps duration (crosses, bottom left) excited the resonators. The left panels in Fig. 1 show time responses of the resonators measured with a detector of 100 ps time resolution. The decay is exponential with superimposed beating. The thicker the mirrors and accordingly the higher their reflectivity the longer the decay time is observed to be. One should note that an opposite behaviour has been observed in the attempt to build an x-ray resonator consisting of silicon mirrors [3]. More than 30 successively reflected beams are observed in Fig. 1a. All together they contribute to the multiple beam interference which results in narrow resonances. The number 30 is limited only by the intensity of the x-ray source used. It should be

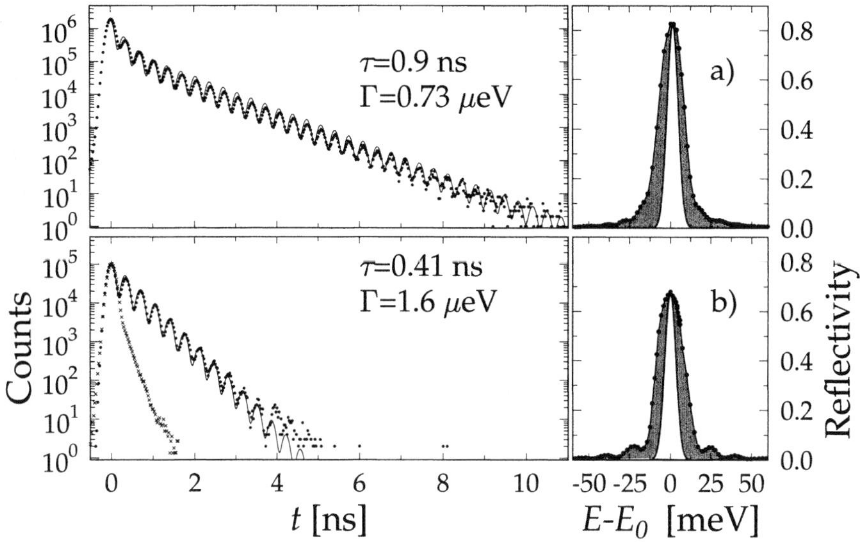

Figure 1. Time responses of x-ray resonators (left panels, circles) and the energy dependences of reflectivity of their mirrors (right panels, dark grey). Resonators with sapphire mirrors of different thicknesses were tested - 54μm (a) and 39μm (b). The time dependence of the incident photons is shown as crosses (bottom, left), while their spectral distribution is shown in light grey (right).

noted, that to obtain the exponential decay with the longest decay time, as depicted in Fig. 1, the mirrors had to be adjusted parallel with sub-μrad accuracy and the temperature of the mirrors was kept equal to within 10 mK.

The longest response time observed $\tau \simeq 0.9$ ns corresponds to an energy width Γ of the transmission resonances of $\simeq 0.73$ μeV. The long response time or alternatively the large number of observed multiple reflected beams are the necessary conditions for the formation of sharp interference resonances. Direct observation of the resonance behaviour using ^{57}Fe Mössbauer-radiation is in progress.

References

[1] Shvyd'ko, Yu.V. and Gerdau, E.: 1999, *Hyperfine Interact.* **123/124**, 741-776

[2] Chen, W.M., McNally,P.J., Shvyd'ko, Yu.V., Tuomi, T., Lerche, M., Danilewsky, A.N., Kanatharana, J., Lowney, D., O'Hare, M., Knuutila, L., Riikonen, J. and Rantamäki, R.: 2001 , *Phys. Stat. Sol. (a)* **186**, No.3, 365-371

[3] Liss, K.-D., Hock, R., Gomm, M., Waibel, B., Magerl, A., Krisch, M., and Tucoulou, R.: 2000 *Nature* **404**, 371.

Address for Offprints: Kluwer Prepress Department
P.O. Box 990
3300 AZ Dordrecht
The Netherlands

Glass dynamics and scaling behaviour under pressure using quasielastic nuclear forward scattering

T. Asthalter[1], I. Sergueev[2,3], H. Franz[4], W. Petry[2], K. Messel[2,4], R. Verbeni[3]

1: Institut f. Physikalische Chemie, Universität Stuttgart, Pfaffenwaldring 55, D-70569 Stuttgart
2: Institut f. Experimentalphysik E13, Physik Department, Techn. Universität München, James-Franck-Str. 1, D-87547 Garching
3: European Synchrotron Radiation Facility, B.P. 220, F-38043 Grenoble
4: Hasylab/DESY, Notkestr. 85, D-22607 Hamburg

Abstract

We have investigated the influence of pressure on the α-relaxation in dibutylphthalate/ferrocene using quasielastic nuclear forward scattering of synchrotron radiation, in a temperature range between 150 and 250 K and for pressures up to 4 kbar. As expected, the dynamics is considerably slowed down when increasing pressure. Checking the pressure-temperature superposition principle for the relaxation rates proves that the scaling on equal density does not hold for the dynamics on atomic length scales.

1 Introduction

For practical reasons, combined studies of glass dynamics where both temperature and pressure are varied are rather scarce in literature. Such studies are nevertheless invaluable if one wishes to test the validity of various theories, such as mode-coupling theory [1,2] or the still largely popular free-volume model [3] for structural α-relaxation in the extended Vogel-Fulcher-Tammann parametrization, where the pressure dependence is introduced, e.g., by the Tammann-Tait equation of state [4]. Specific-heat spectroscopy of glassy ortho-terphenyl (OTP) at two different temperatures [5], for example, provided a strong evidence for the failure of free-volume models, whereas another study of dibutyl phthalate by dielectric spectroscopy at one fixed temperature [6] is suggested to support their validity. The time window of specific-heat spectroscopy is limited to large relaxation times above $\sim 10^{-5}$ s; quasielastic nuclear forward scattering (QNFS) of synchrotron radiation extends the accessible range down to $\sim 10^{-8}$ s. Following recent atmospheric-pressure investigations of dibutylphthalate doped with 5 atomic % ferrocene (enriched to 95.3 % in ^{57}Fe) [7] and experiments on the same system in confined geometries [8], we have therefore applied the QNFS technique to this system at various pressures and temperatures.

2 Experiment

The QNFS experiment was carried out at the beamline ID22N at the European Synchrotron Radiation Facility [9]. A standard Si(111) high-heat load monochromator with

LN_2 cooling and a 3.5 meV four-bounce high-resolution monochromator [10] served to reduce the detector load. A seven-hole Beryllium compound refractive lens [11] in the beamline front-end collimated the beam, thus optimizing the flux through the HRM [12]. A two-crystal bender [13, 14] having a flat first and a sagitally bent second Si (1 1 1) crystal was used to focus the beam horizontally down to a spot size of less than 0.5 mm. The pressure apparatus [15] consisted of a closed-cycle cryostat with a specially designed cell having entrance and exit diamond windows of 1 mm thickness, respectively, a spindle and a Nova-Swiss pressure gauge. A sketch of the setup is shown in Fig. 1.

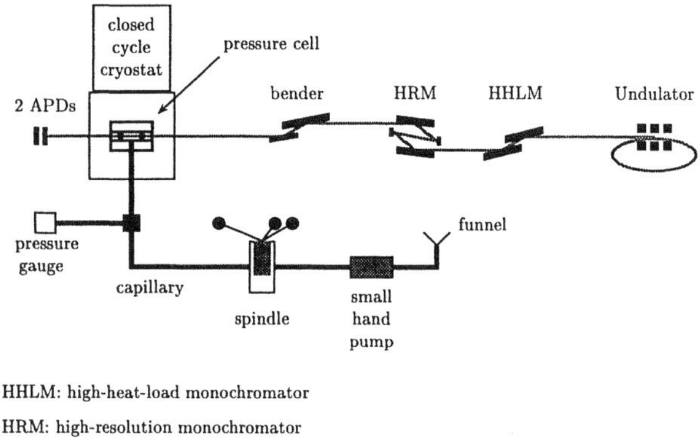

Figure 1: Experimental setup for the QNFS pressure experiment

Forward count rates were in the range of 15 Hz (160 K/2 kbar; 180 K/4 kbar) down to 0.5 Hz (220 K/2 kbar; 250 K/4 kbar). The spectra, which showed a simple quantum beat structure owing to the quadrupole splitting of ferrocene (cf. [7]), were treated as described in [8], assuming a Kohlrausch-Williams-Watts relaxation function with a stretching parameter of $\beta = 0.5$. In order to separate the contributions from the Lamb-Mößbauer factor and the relaxational dynamics, delayed rocking curves were recorded for each T, P point. After integration over time the QNFS spectrum can be factorized into two terms: one is the intensity within the resonant bandwidth that has impinged upon the sample during the measurement, the other one is a "response function" $R(f_{\mathrm{LM}}, \tau_D)$ that only depends on the properties of the sample and not on the experimental conditions which may change during the integration time. By recording the *delayed rocking curve* of the high-resolution monochromator, we obtain a quantity that is proportional to this response function. With the help of low-temperature spectra where $\tau_D = \infty$ this enables us to separate f_{LM} and τ_D also for the high-T measurements [8].

3 Results and discussion

The resulting Lamb-Mößbauer factors and diffusion rates are shown in Fig. 3 (left and right side, respectively). We observe the expected stiffening of the "lattice" as well as a shift of the onset of relaxation to higher temperatures.

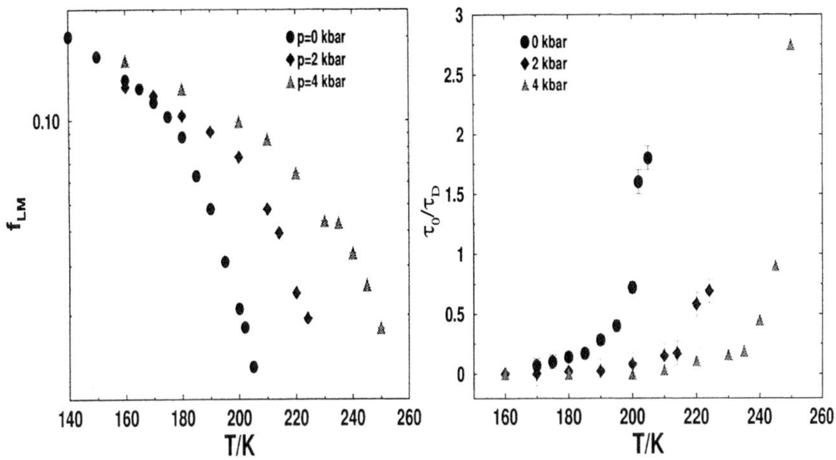

Figure 2: Lamb-Mößbauer factors (left) and diffusion rates (right) of ferrocene/dibutylphthalate for three different pressures

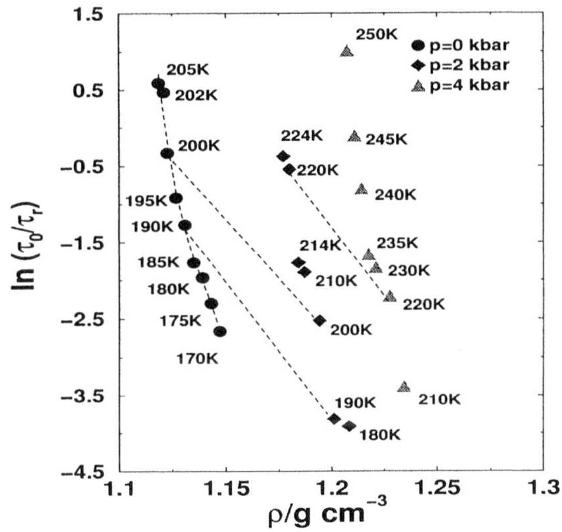

Figure 3: Diffusion rates of ferrocene/dibutylphthalate scaled versus density. Circles: atmospheric pressure, triangles: 2 kbar, diamonds: 4 kbar. The atmospheric-pressure data and some isotherms are provided with guides to the eye (dotted lines)

Using known V, P data and thermal expansion coefficients and applying the Tammann-Tait equation [4] to calculate the density, we obtain the plot of the diffusion rates versus density shown in Fig. 3. We can conclude – in analogy to the study of OTP [5] (see Fig. 5 in this paper) – that the diffusion times do not show the same scaling behaviour, depending on whether density is changed by applying pressure or by heating. Thus the validity of the free-volume model, which explains the dynamics in viscous liquids near to the glass transition in terms of an excess free volume only, seems to be disproved also in our case.

References

[1] U. Bengtzelius, W. Götze, A. Sjölander, J. Phys. C 17 (1984) 5915

[2] W. Götze, L. Sjögren, Transp. Theory Stat. Phys. 24 (1995) 801

[3] M. H. Cohen, G. S. Grest, Phys. Rev. B 20 (1979) 1077

[4] G. Tammann, Z. Phys. Chem. 17 (1895) 620

[5] H. Leyser, A. Schulte, W. Doster, W. Petry, Phys. Rev. E 51 (1995) 5899

[6] M. Paluch, J.Zioło, P. Habdas, S. R. Rzoska, W. Suchański, High Temp. High Press. 29 (1997) 155

[7] A. Meyer, H. Franz, J. Wuttke, W. Petry, N. Wiele, R. Rüffer, C. Hübsch, Z. Phys. B 103 (1997) 479

[8] T. Asthalter, I. Sergueev, H. Franz, R. Rüffer, W. Petry, K. Messel, P. Härter, A. Huwe, Eur. Phys. J. B 22 (2001) 301

[9] T. Asthalter, R. Rüffer, W. Petry, Jahresbericht E13, TUM 1998

[10] design A. I. Chumakov, outer reflection: Si (3 3 3), inner reflection: Si (9 7 5)

[11] A. Snigirev, V. Kohn, I. Snigireva, B. Lengeler, Nature 384 (1996) 49

[12] A. I. Chumakov, R. Rüffer, O. Leupold, A. Barla, H. Thiess, T. Asthalter, B. P. Doyle, A. Snigirev, A. Q. R. Baron, Appl. Phys. Lett. 77 (2000) 31

[13] D. W. Berreman, J. Stamatoff, S. J. Kennedy, Appl. Opt. 16 (1977) 2081

[14] B. Jenninger, Internal report about the OFIR bender, ESRF 1994

[15] A. Mermet, A. Cunsolo, E. Duval, M. Krisch, C. Masciovecchio, S. Perghem, G. Ruocco, F. Sette, R. Verbeni, G. Viliani, Phys. Rev. Lett. 80 (1998) 4205

MECHANICALLY ASSISTED NITRIDATION IN THE Zr-Fe SYSTEM

M.A. Bab, L.C. Damonte and L. Mendoza-Zélis*
Depto. De Física, Universidad Nacional de La Plata, CC67, 1900, La Plata, Argentina
**Comisión de Investigaciones Científicas de la Provincia de Buenos Aires*

Abstract. Crystalline and amorphous Zr-Fe alloys have been milled under N_2 in an oscillatory mill. The structural changes produced by the mechanically assisted gas-solid reaction were investigated by means of Mössbauer spectroscopy and X-ray diffraction. Regardless of the initial composition and structure, the final product after prolonged milling was a mixture of ZrN and almost pure Fe, evidencing the tendency of Fe to segregate and that of Zr to combine with N. Several intermediate states were observed including $Zr_{1-x}Fe_x$ N.

1. Introduction

Mechanical milling (MM) has achieved widespread relevance as a tool for producing nanocystalline structures and phase compositions no attainable by conventional processing techniques. It might found interesting applications in the production of metal-ceramic nanocomposites, combining the metal properties with the hardness, thermal conductivity and stability of transition metal nitrides[1].

Several metallic nitrides may be prepared by mechanically assisted gas-solid reactions induced by milling pure metals in a nitrogen atmosphere [2][3]. During MM, the powder particles are fractured and grains are refined producing fresh clean surfaces where N_2 may be chemisorbed. MM also assists nitrogen diffusion and structural transformations until a homogeneous nitride phase is formed[4]. While the Fe nitridation by this method is rather difficult[5], Zr and other refractory metals, easily form the respective nitrides. Structural factors seem to play an important role in this concern. On the other hand, Zr forms with Fe several compounds and amorphous alloys in a wide compositional range.

Based on the above considerations we undertook experimental studies on the nitridation by MM of a series of samples in the Zr-Fe system. Mössbauer spectroscopy (MS), X-ray diffraction (XRD) and N absorption kinetics measurements were used to investigate how structure and alloying affects the behaviour of Zr.

2. Experimental

The MM was performed in an oscillatory mill using a cylindrical steel vial O-ring sealed and connected to a N_2 reservoir with one steel ball (ϕ = 1.2cm). Samples of 500mg of the following materials were milled: Zr (99.9%), Fe (99.9%), $ZrFe_2$ (cF24), $Zr_{64}Al_7Cu_{17}Ni_{10}Fe_2$ amorphous alloy and a sample of nominal composition $Zr_{63}Fe_{37}$ constituted by Zr_2Fe (tI12) with 10wt% $ZrFe_2$. The gas pressure during MM was measured by a strain transducer, and the instant number of absorbed N atoms per Zr (n) was evaluated from the system volume ($160cm^3$), the pressure change and the Zr mass.

MS experiments were carried out at room temperature with a conventional spectrometer in transmission geometry, employing a ^{57}CoRh source. XRD patterns were obtained with a Philips PW1710 apparatus (CuK_α) and fitted by the Rietveld method using pseudo-Voigt profiles.

3. Results and Discussion

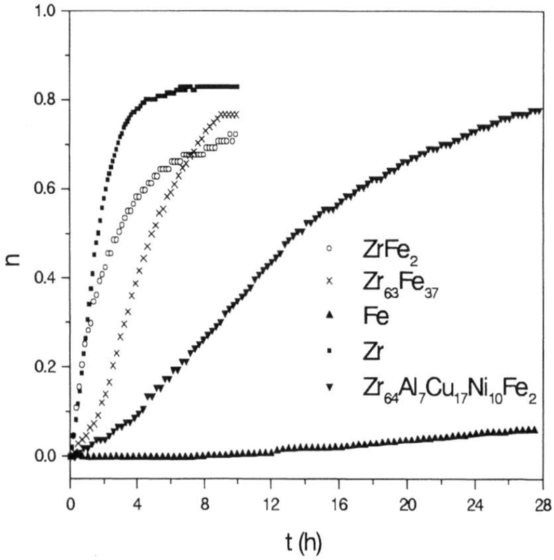

Fig. 1: n vs. t_m for the samples studied

To study the N absorption process, we follow the variation of n with milling time (t_m) for the various samples investigated (fig. 1). We notice the following features: i) Zr, like Hf [4] presents a sigmoidal absorption curve in which the acceleration stage is related to the rate of N diffusion through grain boundaries and its insertion at octahedral sites in free surfaces; ii) the amorphous sample presents a much slower absorption rate than the crystalline ones probably as a consequence of its compact atomic packing and the absence of grain boundaries; iii) the absorption rate is higher for $ZrFe_2$ than for $Zr_{63}Fe_{37}$ in agreement with the extreme fragility of the former, which results in a drastic reduction in grain size with MM and the associated increase in free surface for absorption; iv) all these curves spread to a saturation value about n = 0.8. The figure includes the result for Fe which appears as a sluggish absorber.

The structural changes that accompany this N incorporation are evident in the MS and XRD experiments.

$Zr_{64}Al_7Cu_{17}Ni_{10}Fe_2$: the spectrum before MM (fig. 2) shows the quadrupole splitting distribution characteristic of these amorphous alloys. For t_m = 10h a distributed doublet with the reverse asymmetry appears together with a minority sextet α-Fe like. The XRD pattern shows only the ZrN reflections and a broad line at the position of the main reflections of Fe and Ni. After t_m= 28h a complex magnetic spectrum develops that may be fitted with two hyperfine field distributions centered around 30T and 20T, assigned to Fe-rich and M(Al,Ni,Zr)-rich FeM precipitates.

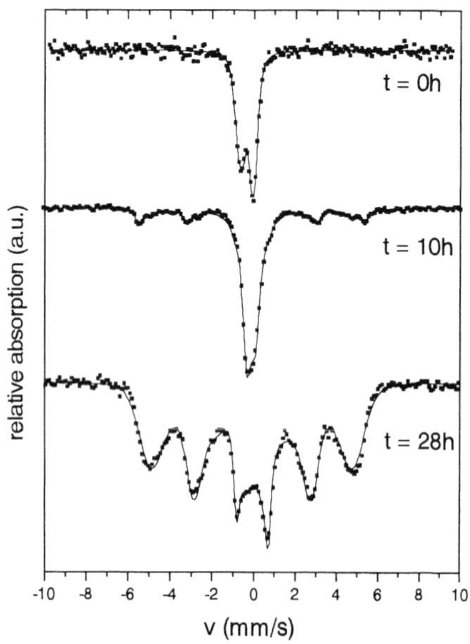

Fig. 2: $Zr_{64}Al_7Cu_{17}Ni_{10}Fe_2$ MS spectra for various t_m

$Zr_{63}Fe_{37}$: the spectrum before MM (fig. 3) consists of a doublet attributed to the main phase (Zr_2Fe) and two sextets characteristic of $ZrFe_2$ (see below)[5]. Within the experimental error, the relative area of the paramagnetic component remains constant (80\%) under MM, while a magnetic interaction, α-Fe like,

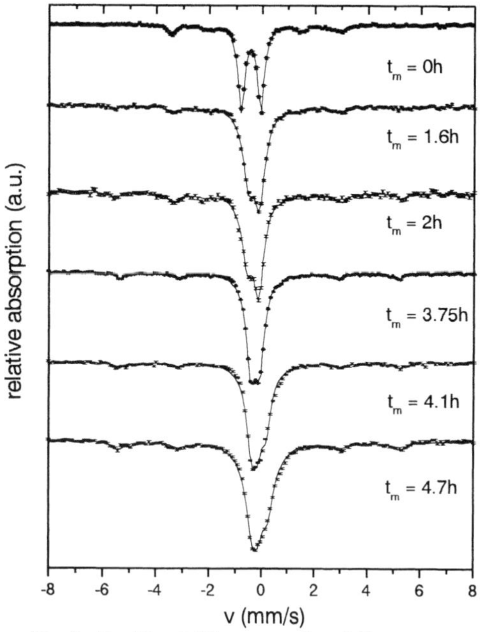

Fig.3: $Zr_{63}Fe_{37}$ MS spectra for different t_m

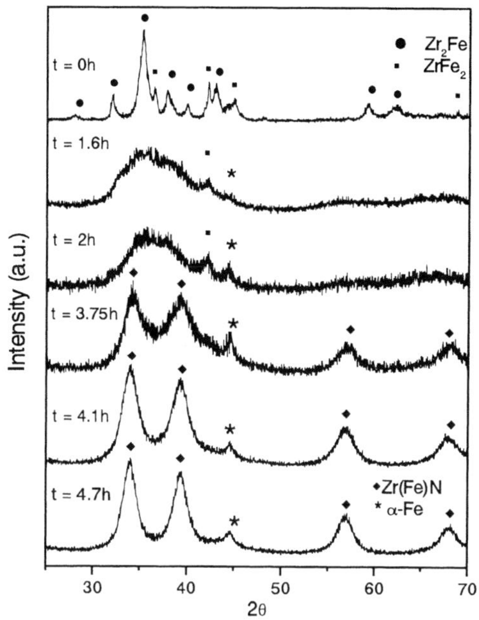

Fig.4: $Zr_{63}Fe_{37}$ diffractograms for different t_m

gradually replaces that of $ZrFe_2$, which disappears near 3.75h. For the paramagnetic component, the fitted parameters after short t_m are consistent with an amorphous phase with mean composition ($Zr_{67}Fe_{33}$)[7] [8] and at t_m = 3.75h a change occurs indicative of an important short range order variation. For higher t_m the spectrum asymmetry is reversed and the isomer shift increases with N content as a result of the hybridization between Fe 3d and N 2p states. From XRD (fig 4) we note that, initially an amorphous phase is formed with n = 0.12 and that for 3.75 h (n = 0.37) δZrN (cF8) appears. Afterwards, only the δZrN reflections are present together with a small peak at 44.6°C consistent with the observed α-Fe sextet. These facts indicate a structural transformation to Zr(Fe)N with Fe randomly replacing Zr. This conclusion is supported by the agreement between the volume fraction of δ phase estimated from XRD (92%) and that obtained from MS for Zr(Fe)N, (assuming the same f factor for this phase and for Zr_2Fe). The grain size of δZr(Fe)N crystallites was estimated in 3nm. After prolonged milling (10 h) the Fe peak

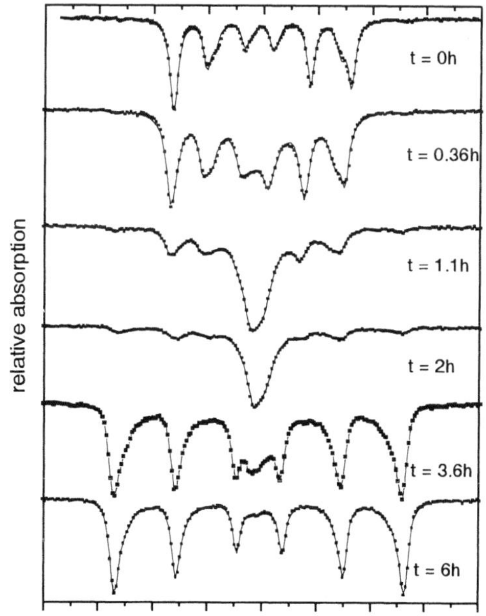

Fig.5: $ZrFe_2$ MS spectra for different t_m

in the XRD pattern (not shown) increases indicating the Fe segregation while MS shows the corresponding increase in the magnetic component.

$ZrFe_2$: the spectrum before MM (fig. 5) may be fitted with two sextets (19.9T and 18.3T) originated in the two possible angles between the main electric field gradient component and the magnetization. Under MM two components develop, a doublet that may be attributed to superparamagnetic $ZrFe_2$ or to an amorphous phase and a sextet α-Fe like. For t_m = 2h, the sextets from $ZrFe_2$ have nearly disappeared. For longer t_m, the spectra are complex showing: i) a distribution of hyperfine fields close to α-Fe values; ii) a low field distribution probably originated in grain boundaries and iii) an unresolved doublet. From these results and the corresponding XRD patterns we suggest that the incipient amorphization process was inhibited by the Zr+N reaction and the accompanying Fe segregation. ZrN and α-Fe grain sizes were estimated in 3 nm.

4. Conclusions

All the studied Zr-Fe alloys and compounds were unstable under MM in N_2. Zr rich alloys formed δZrN following a reaction kinetics governed by the rate of mechanical energy supply [4]. In the amorphous the absence of grain boundaries delays the absorption evidencing the role of microstructure in the whole process. We highlight the formation of δZr(Fe)N, among the observed intermediate phases. To our knowledge, this $Zr_{1-x}Fe_xN$ phase in which Fe randomly replaces Zr in the known nitride, has not been reported previously.

Acknowledgements

We are recognized to J. Eckert (IFW Dresden, Germany) who kindly supplied the amorphous sample.

References

[1] C. R. Zhou, K. Lu and Xu, Mat. Sci. Forum, **343-346** (2000) 643.
[2] A. Calka and J.S. Willians, Mat. Sci. Forum, **88-90** (1992) 787
[3] L. Mendoza-Zélis, M.A. Bab, L.C. Damonte and F. H. Sánchez, Mat. Sci. Forum, **312-314** (1998) 179.
[4] M.A. Bab, L. Mendoza-Zélis and L.C. Damonte, Acta Materialia, in press
[5] J. Rawers, D. Cook and T. Kin, Phil. Mag. A, **78** (1998) 965.
[6] F. Aubertin, U. Gonser, S. Campbell and H. Wagner, Z. Metallkde, **76** (1985) 237.
[7] C. Michaelsen and E. Helistern, J. Appl. Phys. **62** (1985) 1.
[8] K.M. Unruh and C. L. Chien, Phys. Rev. B, **30** (1984) 4968.

STRUCTURAL STUDY OF Fe-Cr-Co ALLOYS OBTAINED BY MECHANICAL ALLOYING.

F. Z. Bentayeb[1], S. Alleg[1], B. Bouzabata[1], N. Ayari[1] and J. M. Greneche[2].

[1]Département de Physique, Université de Annaba, B.P.12, 23000 Annaba, Algérie.
[2]Laboratoire de Physique de l'Etat Condensé – UMR 6087, Université du Maine, 72085 Le Mans, Cedex 9 France.

Abstract

Nanostructured Fe-Cr-Co alloys were prepared by mechanical alloying (MA) of elemental powders in the following proportions: $Fe_{57}Cr_{31}Co_{12}$, $(Fe,Cr)_{88}Co_{12}$ and $(Fe,Co)_{69}Cr_{31}$. ^{57}Fe Mössbauer and X-ray diffraction (XRD) are used to model the atomic scale alloying mechanism which occurs as a function of milling conditions.

Key words : Mössbauer spectrometry, mechanical alloying, Fe-Cr-Co alloys.

Introduction

Nanostructured materials offer a range of remarkable properties and corresponding potential applications beyond those available for crystalline or amorphous materials. MA induced by high energy milling is used to produce a wide range of nanomaterials: supersaturated solid solutions (SSS), metastable nanocrystalline and amorphous phases. The introduction of a large number of crystal defects upon mechanical milling of a mixture of powders provides the conditions required for the solid state reaction, leading to alloying effect. When metals, solutions or mixtures of phases having simple crystal structures such as fcc, bcc or hcp, are intensively milled a large amount of internal energy may be stored and this can lead to a destabilisation of the lattice, yielding the amorphous phase or other metastable crystalline structures. It is thus understandable that these nanostructures are far from thermodynamic equilibrium and the accurate characterisation of the milled powders is very important in determining the processes which occur during MA. The presence of γ-Fe precipitates in nanostructured ball milled Fe powders has been suggested in [1]. The existence of this metastable structure has been interpreted as a result of epitaxial relationship between nanocrystals and grain boundary regions. The formation of fcc-Fe has also been observed in milled $Fe_{50}Cu_{50}$ by Jiang et al. [2]: Fe atoms coherently precipitate in the fcc-Cu matrix giving rise to fcc Fe clusters. Macri et al. [3] also suggested that the fcc Fe phase is consumed and the Fe atoms progressively incorporated into fcc-Cu matrix. Pietro et al. [4] reported that the 12% of remaining Fe in $Fe_{30}Co_{20}Cu_{50}$ alloy milled for 80h exhibit two different structures: bcc and body tetragonal centred. The allotropic transformation of Co induced by ball milling has been studied by Fenineche et al. [5]: they showed that the hcp to fcc transformation is determined by the mill intensity. As expected from the phase diagram of Co, this structural transition can occur at high pressure and temperature. The aim of this work is to study the effect of milling conditions on structural changes which take place during MA of the nanostructured $Fe_{57}Cr_{31}Co_{12}$ alloy prepared in 3 different ways. A combination of XRD and Mössbauer spectrometry is used to follow the atomic interdiffusion phenomenon.

Experimental Section

3 series of the $Fe_{57}Cr_{31}Co_{12}$ alloy have been milled for different times ranged from 1 to 12h: (i) using elemental powders of Fe, Cr and Co, (ii) starting from mixture of $Fe_{69}Cr_{31}$ alloy and pure Co powder and (iii) starting from the mixture of $Fe_{88}Co_{12}$ and pure Cr powder. The

$Fe_{57}Cr_{31}Co_{12}$, $(Fe,Cr)_{88}Co_{12}$ and $(Fe,Co)_{69}Cr_{31}$ series are labelled as samples I, II and III, respectively. The powder mixtures were ball-milled in a planetary ball mill type Fritsch Pulverisette 7, under argon atmosphere using stainless steel vial and balls with a powder to-ball ratio R= 1/35 and a rotation speed of about 1000 rpm. The XRD patterns have been performed with Cu-Kα radiation. ^{57}Fe Mössbauer spectra were obtained at 300K in a standard transmission geometry using a ^{57}Co source in Rh matrix. The Mössbauer spectra were reproduced by means of a discrete distribution of hyperfine magnetic fields, according to a fitting procedure initially proposed by Varret et al. [6].

Results and discussions

Some XRD patterns of $Fe_{57}Cr_{31}Co_{12}$ are shown in Fig. 1. Before milling, the diffractogram displays the Bragg peaks corresponding to bcc Fe, bcc Cr, fcc and hcp Co. After 1h of milling, the peaks of fcc Co disappear and the diffraction peaks of both bcc Fe and bcc Cr are still resolved. The Bragg peaks become broadened while their intensities gradually decrease with milling time, indicating a refinement of the crystallite size and an increase of microstrains. The disappearing of Co peaks after 6h of milling can be explained by the interdiffusion of Cr and Co atoms within the Fe matrix leading to the formation of Fe-Cr-Co SSS with different structural types (bcc, fcc and hcp), tetragonal CrCo and Cr oxide. A careful analysis using Rietveld refinement [7] provides evidence that the SSS reactions in this ternary system are rather complex. In fact, upon milling, the interdiffusion of Fe, Cr and Co results in a bcc Fe-Cr-Co SSS with a lattice parameter a = 2,869Å. In addition to this, some Fe crystalline grains may adopt the hcp and fcc structure by plastic deformation occurred by milling process and the diffusion of Cr and Co into this structures results in formation of Fe-Cr-Co SSS with hcp and fcc structure, respectively. On the other hand, the diffusion of Fe and Cr atoms into fcc-Co matrix can also lead to the formation of fcc Fe-Cr-Co. Further, the intermixing of Cr and Co atoms in each other results in the formation of a known tetragonal CrCo phase. It is noted that very small particles of Cr_2O_3 are detected.

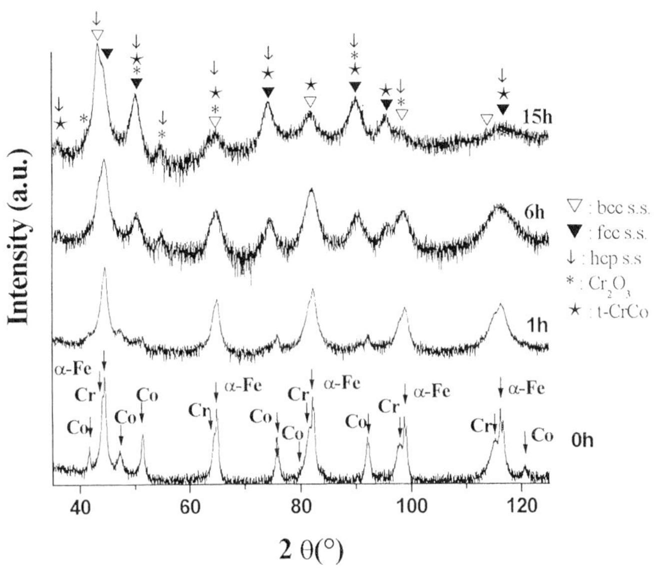

Figure 1: X-ray diffraction patterns of sample I for various milling times.

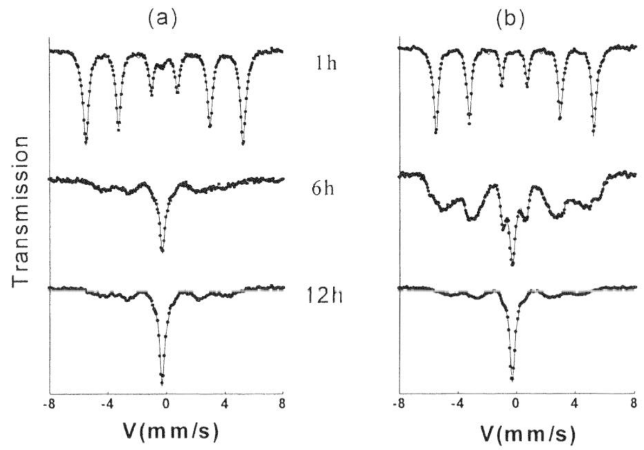

Figure 2 : Mössbauer spectra of samples I (a) and III (b) as a function of milling time.

Mössbauer spectra of sample I are shown in figure 2-a. After 1h of milling, the Mössbauer lines are broadened while a single line appears in the centre of the spectrum. As the milling time increases, one observes a fast broadening of magnetic sextet which is assigned to disordered bcc Fe-Cr-Co SSS, and an increase of the paramagnetic component which reaches a relative fraction of about 50% after 12h of milling. This component is attributed to fcc/hcp Fe-Cr-Co phases and/or to Cr-rich α' phase, since that all these phases are paramagnetic. Similar results are obtained for sample II. However, a different behaviour is observed for sample III at first and intermediate stages of alloying. Indeed, after 6h of milling (fig. 2-b), the spectrum exhibits 15% of paramagnetic component and a large sextet which is due to a random distribution of Fe, Cr and Co atoms. For longer milling times, the Mössbauer spectra of the 3 samples show the same structural features: about 50% of magnetic component relative to bcc-FeCrCo SSS and 50% of paramagnetic component relative to fcc, hcp Fe-Cr-Co SSS and/or the Cr-rich phase.

From these results, it can be concluded that the milling process induces, after 12h of milling the same FeCrCo solid solution, whatever the initial conditions. It seems that the differences at first and intermediate stage of alloying are due, principally, to the presence of Co. In fact, contrary to Fe-Cr system where Cr and Fe have the same crystallographic structure, Co crystallises into hcp structure under ambient conditions. Upon milling, both the plastic deformations, fracture and cold welding effect allow the transformation of the hcp Co into a cubic phase and a possible mutual diffusion of Co, Cr and Fe in fcc and/or bcc structures.

Acknowledgments:

This work was supported by the Ministère de l'Enseignement Supérieur et de la Recherche Scientifique (Algérie) and the French-Algerian Cooperation program CMEP under Project no. 99MDU449.

References

[1] L. Del Bianco, A. Hernando, E. Bonetti, E. Navarro, Phys. Rev. B 56 (1997) 8894.
[2] J. Z. Jiang, U. Gonser, C. Gente, R. Bormann, Appl. Phys. Lett. 63 (1993) 1056; 2768
[3] P. P. Macri, P. Rose, R. Frattini, S. Enzo, G. Principi, W. X. Hu, N. Cowlam, J. Appl. Phys. 76 (7) (1994) 4061.

[4] C. Pietro, A. de Bernabé, N. Gay-Sanz. M. Vàzquez, Seong-Cho Yu, J. Non-Crystalline Solids 246 (1999) 169.
[5] N. E. Fenineche, O. El Kedim and E. Gaffet, JMNM. 7 (2000) 41.
[6] F. Varret and J. Teillet, unpublished Mosfit program.
[7] L. Lutterotti and P. Scardi, J. Appl. Crystallogr. 23 (1990) 246.

A MÖSSBAUER STUDY OF NANOCRYSTALLISATION UNDER CONTINUOUS HEATING CONDITIONS

Pere Bruna[1], D. Crespo[1], M.T. Clavaguera-Mora[2],

N. Clavaguera[3], E. Pineda[4], T. Pradell[4]

[1]Departament de Física Aplicada, Universitat Politècnica de Catalunya.
Campus Nord UPC, Mòdul B4, 08034-Barcelona, SPAIN

[2]Grup de Física de Materials I, Departament de Física, Facultat de Ciències,
Universitat Autònoma de Barcelona, 08193-Bellaterra, SPAIN

[3]Grup de Física de l'Estat Sòlid, Departament d'ECM, Facultat de Física,
Universitat de Barcelona, Diagonal 647, 08028-Barcelona, SPAIN

[4]E.U.E.T.A.B, C.E.I.B. Universitat Politècnica de Catalunya.
Urgell 187, 08036-Barcelona, SPAIN

Abstract The kinetics of nanocrystallisation of a Finemet metallic glass of composition $Fe_{73.5}Si_{17.5}B_5Cu_1Nb_3$ under isothermal and continuous heating conditions has already been widely studied. However, the kinetics of the nanocrystalline precipitation in Finemet alloys under a constant heating rate has never been studied by Mössbauer spectroscopy. Mössbauer analysis has the capability of giving information on either the short range order of the amorphous phase and of the nanocrystalline precipitates. The present study examines the differences and similarities in the kinetics of the nanocrystallisation and on the final nanostructured obtained for both isothermally heated and continuously heated materials.

1.Introduction

Nanostructured Fe based alloys obtained by a controlled crystallisation of metallic glasses (Fe-Zr-B and Fe-Si-Cu-Nb-B based alloys) show extremely good soft magnetic properties even better than their metallic glass precursors. The nanostructure is obtained including in the initial composition of the metallic glass a small content of atoms of big size like Zr, Nb, etc. which reduce the growth rate of the precipitates and stabilize the untransformed glassy matrix. This results in a two step crystallisation process; during the first, α-Fe(Si) nanocrystals (bcc) are formed and, during the second, the rest of the amorphous matrix transforms giving mainly Fe based borides and small quantities of other phases. A high nucleation rate and a low growth rate are desirable to obtain a small grain sized particulate system. For the particular case of Finemet alloys, copper clusterization formation seems to be the responsible for the high nucleation rate of the primary phase at relatively low temperatures.

The kinetics of nanocrystallisation of a Finemet metallic glass of composition $Fe_{73.5}Si_{17.5}B_5Cu_1Nb_3$ under isothermal conditions was already studied[1,2]. The hyperfine parameters related to the amorphous phase change due to the copper segregation previous to the nanocrystalline precipitation and during the nanoprecipitation due to the compositional changes of the amorphous matrix. The Fe content and order of the nanocrystalline precipitates changed along the transformation.

The thermodynamical parameters such as nucleation and growth rates, transformed fraction

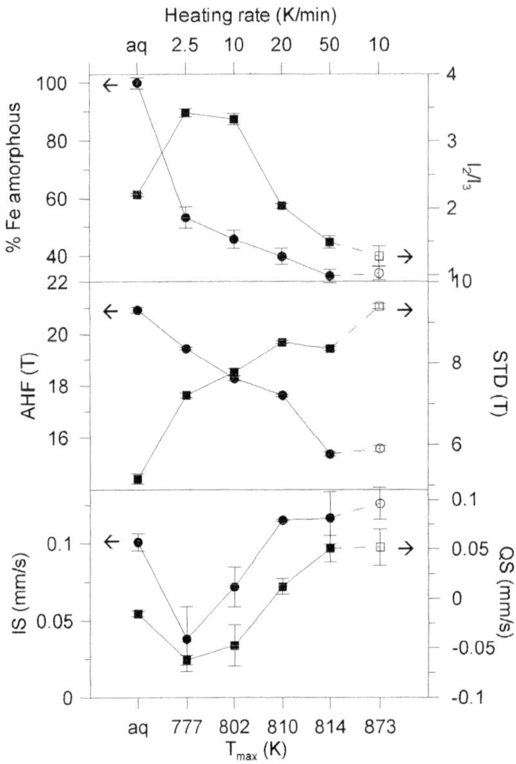

Figure 1 *Hyperfine parameters of the magnetic hyperfine distribution related to the untransformed amorphous phase for the different thermal treatments: at% of Fe and I_2/I_3 ratio (top), Average Magnetic Hyperfine field (AHF) and Standard Deviation (STD) (center) and Average Isomer Shift (IS) and Quadrupole Splitting (QS) (bottom).*

at the end of the primary precipitation and composition of the nanocrystallites are temperature dependent. Therefore, the kinetics of the transformation during continuous heating thermal path may also be evaluated if some of this temperature dependent parameters are known. In the present study, the similarities and differences in the nanocrystalline precipitation for isothermal and continuous heating regimes are compared.

2.Experimental

The Finemet alloy was heated at several heating rates, 2.5K/min, 10K/min, 20K/min and 50K/min, and cooled down at 20K/min. Maximum temperatures reached were for 777K for 2.5K/min, 802K and 875K for 10K/min, 810K for 20K/min, 814K for 50K/min. Thermal treatments were realized on a DTA and the samples remained 2 min at the highest temperature due to thermal inertia. In all cases the primary nanocrystallization was fulfilled, and the influence of the thermal paths in the final nanocrystalline materials was studied.

2.Results and discussion

Room temperature TMS data were fitted by using an histogram magnetic hyperfine-field distribution[3] including a linear correlation between the isomer shift and the magnetic hyperfine field[4]. The crystalline emerging phase has been fitted by five subspectra corresponding to the different Fe containing sites related to the $Fe_{3+x}Si_{1-x}$ DO_3 structure, the I_2/I_3 has been also fitted but settled equal for all the subspectra. The fitted magnetic hyperfine fields are 32.4(1)T, 31.2(1)T, 28.8(1)T, 24.8(1)T and 19.8(1)T and the isomer shifts fitted are 0.03(1)mm/s, 0.06(1)mm/s, 0.14(1)mm/s, 0.18(1)mm/s and 0.24(1)mm/s for the 8NN, 7NN, 6NN, 5NN and 4NN respectively.

The hyperfine parameters fitted are given in figures 1 and 2 as a function of the heating rate and maximum temperature reached, and compared to the original amorphous material. Figure 1 shows the hyperfine parameters fitted for the amorphous phase and Figure 2 the relative Fe content on the different crystalline sites. Although the maximum temperature reached give a complete crystallisation in all the cases, comparison between the results obtained for the two samples heated at 10 K/min up to 802 K and 873 K respectively show very different values of the hyperfine parameters of both the amorphous untransformed phase and of the nanocrystalline emerging phase. Therefore, the final phase structure is determined by the maximum temperature reached, the heating rate playing a secondary role.

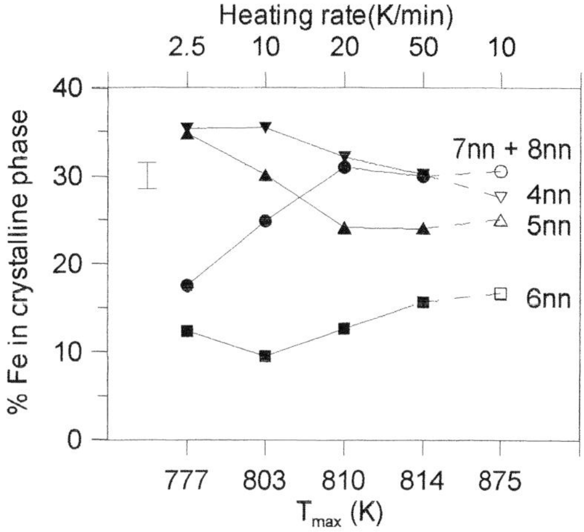

Figure 2 *Relative % of Fe in the different sites of the crystalline phase.*

Figure 1 (top) shows a decrease in the percentage of Fe in the amorphous phase as the maximum temperature increases, indicating an increase in the maximum transformed fraction at the end of the primary crystallization. The ratio I_2/I_3, which increases in the samples annealed up to 777K and 802K, abruptly decreases between 802K and 810K. This shows a two-stage process; low temperature heating increases the orientation of magnetic domains parallel to the surface, while high temperature annealing destroys this preferential orientation and gives rise to a nearly-randomly oriented structure of the magnetic domains. The nanocrystalline phase has a similar value of the I_2/I_3 ratio, indicating that both, amorphous and crystalline phases are coupled. This temperature dependent behaviour was also observed under isothermal annealing conditions [1,2]. Figure 1 (center) shows that the average hyperfine field (AHF) decreases monotonically while the width of the hyperfine magnetic field distribution (STD) increases with the heating rate (maximum heating temperature), showing the increase in non-Fe first neighbours for the Fe atoms remaining amorphous phase. This change in the SRO is also shown by the changes in the quadrupolar splitting (QS) and the isomer shift (IS), shown at the bottom of the Figure. These changes are similar to those observed under isothermal annealing a 763K when increasing the annealing time[1,2], showing the link existing between the maximum annealing temperature for continuously heated materials and the annealing time in isothermally annealed materials.

Figure 2 shows the Fe-Fe coordination number in the crystallized phase. A considerable change can be clearly seen between the samples heated at 802K and 810K, the latter being more close to the percentages of nearest neighbours corresponding to a DO_3, Si defective, $Fe_{3+x}Si_{1-x}$ structure [1]. Note, however, that the Figure shows relative percentages, which combined with the results of Figure 1 (top) show that the total amount of 7+8NN coordination increases while the amount of 4-6NN coordination remains nearly constant at the different temperatures. A similar behaviour was found after isothermal annealing[2]. It is likely that the random distribution of vacancies is obtained by interdiffusion inside of the crystallites, which is highly dependent on the atom mobilities, which in turn depend on temperature. This results show that the time needed for this interdiffusion process at low temperatures is extremely long and is even longer than the time needed for full crystallization.

In summary, it was observed that the key parameter defining the nanostructure developed in the primary crystallization is the maximum annealing temperature. The maximum transformed phase at the end of the primary crystallization increases with the maximum annealing temperature. The preferred magnetic orientation of both the domains existing in the untransformed amorphous phase and the crystallites are coupled and they are also temperature dependent, nearly parallel to the surface for low annealing temperatures but losing this ordering for higher annealing temperatures. However, the interdiffusion process responsible for the crystalline structure of the crystalline phase depends mainly on the mobility of the Si and Fe atoms, and thus it gives rise to similar structures for short annealing times at high annealing temperatures and for large annealing times at low temperatures. The characteristic time of this interdiffusion process can be much larger than the characteristic time of the primary crystallization, and thus can be one of the reasons of the large tails observed in calorimetric measurements.

References

[1] T. Pradell, N. Clavaguera, J. Zhu, and M.T. Clavaguera-Mora, J. Phys.: Condens. Matter, 7 (1995) 4129.
[2] T. Pradell, J. Zhu, N. Clavaguera and M.T. Clavaguera-Mora, J. Appl. Phys. 83 (1998) 5171.
[3] R.A. Brand, Normos Programs, Lab. Ang. Physik. Universitaet Duisbøurg.
[4] R.E. Vandenberghe, D. Gryffroy, and E. De Grave, Nucl. Instrum. Methods Phys. Res. B, 26 (1987) 603.

Acknowledgements This work was financed by CICYT, grant MAT98-0672. PB was supported by a Beca de formación de investigadores from Ministerio de Ciencia y Tecnologia, SPAIN.

Investigation of Eu^{3+} Site Occupancy and Eu-O Covalency in Nanocrystalline Y_2O_3 by Mössbauer Spectroscopy

G. Concas, F. Congiu, G. Spano, M. Bettinelli[a] and A. Speghini[a]
Dipartimento di Fisica, Università di Cagliari and Istituto Nazionale per la Fisica della Materia, S.P. Monserrato-Sestu km 0.700, I-09042 Monserrato (CA), Italy
[a] Dipartimento Scientifico e Tecnologico, Università di Verona, Ca' Vignal, Strada le Grazie 15, I-37134 Verona, Italy

Abstract

A sample of nanocrystalline $Y_{1.8}Eu_{0.2}O_3$ was examined by ^{151}Eu Mössbauer spectroscopy. The degree of covalency of the Eu-O bond has been studied. The spectrum of the cubic $Y_{1.8}Eu_{0.2}O_3$ sample has been resolved into 2 contributions due to europium in the C_{3i} and C_2 sites, for the first time in ^{151}Eu Mössbauer spectroscopy. The degree of covalency and the electric field gradient of the 2 sites has been compared. The occupancy, by the lanthanide ion, of the more and less symmetric sites in the cubic structure of $Y_{1.8}Eu_{0.2}O_3$ has been investigated and discussed.

Lanthanide-doped yttrium oxides are important materials, which find numerous applications in disparate fields. In particular, cubic Y_2O_3 doped with Eu^{3+} is a well-known luminescent material [1], whilst cubic $Y_{2-x}Ln_xO_3$ systems form an interesting class of diluted magnetic semiconductors, and yttria and rare earth oxide based catalysts find applications in the selective oxidation of ethane to ethene. In this work we report on the estimation of the covalent component of the Eu-O bond in the $Y_{1.8}Eu_{0.2}O_3$ nanocrystalline sample and the evaluation of the occupancy by the Eu^{3+} ions of the 2 available cation sites in the cubic structure of nanocrystalline $Y_{1.8}Eu_{0.2}O_3$.

Nanocrystalline Y_2O_3 powders doped with 10% Eu_2O_3 ($Y_{1.8}Eu_{0.2}O_3$) were prepared as described elsewhere [2]. Mössbauer absorption spectra were obtained using a standard transmission geometry, with a $^{151}SmF_3$ source of activity 3.7 GBq. A calibration was performed using a source of ^{57}Co in rhodium and a metallic iron foil (6 μm thick) as the absorber. The full width at half maximum (FWHM) of the crystalline absorption peak, determined with our source, was measured using $Cs_2NaEuCl_6$. The isomer shift values have been referred to EuF_3. Measurements on the crystalline samples were carried out at room temperature on powders with an absorber thickness of 3.8 mg/cm^2 of Eu. The absorber was contained in a Plexiglas holder. The absorption spectra were analysed following the procedure described elsewhere [3]. The quality of the fits was tested using the usual chi-squared (χ^2) test and a weighted form of the Durbin-Watson d statistics, that was recently introduced in the analysis of Mössbauer spectra [3].

The crystal structure of the samples under investigation was checked by powder X-ray diffraction [4, 5]. These measurements confirmed that the sample is single phase and that $Y_{1.8}Eu_{0.2}O_3$ has been obtained as the cubic C [6] polymorph. Moreover, X-ray diffraction showed that the sample was nanocrystalline with sizes in the range 20÷100 nm. In addition, the luminescence spectra of nanocrystalline $Y_{1.8}Eu_{0.2}O_3$ [7] is typical of Eu^{3+} doped C-phase Y_2O_3 [8].

The width of the crystalline, i.e. in an ordered environment, absorption peak for trivalent europium, determined with our source, has been measured using $Cs_2NaEuCl_6$, which contains Eu^{3+} in a octahedral site (cubic symmetry) [9]. In a site with cubic symmetry, the gradient of the electric field is zero and therefore there is a single absorption line [10]. The FWHM has been measured with an absorber thickness of 3.8 mg/cm^2 of Eu; this value correspond to an effective thickness t = 1, when calculated using the recoilless fraction of the source, f = 0.6 [10]. The measured FWHM is 1.76 ± 0.01 mm/s with t = 1.

The line width measured for other samples, with the same source, must be compared with the width measured for $Cs_2NaEuCl_6$ with the same effective thickness, i.e. approximately with the same superficial density of Eu. Widths measured with distinct sources of $^{151}SmF_3$ cannot be compared, because of the different features of the commercial sources [10].

The Mössbauer absorption spectrum of the $Y_{1.8}Eu_{0.2}O_3$ nanocrystalline sample is shown in Fig. 1. The parameters obtained from the fit with a single line are given in Table 1. The value of the isomer shift obtained from this fit represents an approximated average of the shift values of the two cation sites of the Y_2O_3 cubic structure. The FWHM of the peak (3.02 mm/s) is much larger than the crystalline width (1.76 mm/s); this width results from the unresolved quadrupole splitting and from the presence of two unresolved contributions due to the different sites. The value of the reduced chi-squared (4.7) shows that the fit is unsatisfactory.

The spectrum of the $Y_{1.8}Eu_{0.2}O_3$ sample has been also fitted using two contributions for the two sites. The 1st contribution is associated with the site with point symmetry C_{3i}; the asymmetry parameter η has been fixed to zero (8 allowed transitions) because this site has a threefold symmetry axis. The 2nd contribution is associated with the C_2 site; the η parameter is free (12 allowed transitions). The FWHM of each component absorption lines, corresponding to each allowed transition, has been fixed to the crystalline line width for effective thickness $t = 1$ (1.76 mm/s).

Fig. 1 shows (solid curve) the data calculated with the fit with 2 contributions for $Y_{1.8}Eu_{0.2}O_3$; the inner curves show the contribution of each site. Table 1 gives the best fit Mössbauer parameters and the control parameters; the reduced chi-squared is 1.0. The values of d and Q_d point out that the consecutive residuals are not correlated. Therefore this last is a good fit of the experimental spectrum. It results that the relative area of the contribution due to the C_{3i} site is about equal, within the experimental error, to that due to the C_2 site (49% and 51% respectively).

The spectra of the nanocrystalline samples here investigated have values of the isomer shift corresponding to an oxidation number +3; the values near to 1 mm/s point towards a strong covalency of the Eu-O bond for both nanocrystalline hosts [10]. In the $Y_{1.8}Eu_{0.2}O_3$ sample, the C_{3i} site has a larger isomer shift than the C_2 site; the former site appears to be characterised by a stronger covalent character. The axial component of the electric field gradient is much smaller in the more symmetric site, because the value of the quadrupole interaction parameter is about half.

The europium dopant ion is distributed about equally between the C_{3i} and the C_2 site; therefore in nanocrystalline Y_2O_3 it occupies preferentially the C_{3i} sites, which are only 1/4 of the available sites.

References

[1] G. Blasse and B. C. Grabmaier, "Luminescent Materials", Springer-Verlag, Berlin, 1994.

[2] G. Concas, C. Muntoni, G. Spano, M. Bettinelli and A. Speghini, Z. Naturforsch. **56a**, 267 (2001).

[3] G. Concas, F. Congiu, C. Muntoni, M. Bettinelli and A. Speghini, Phys. Rev. **B 53**, 6197 (1996).

[4] G. Fagherazzi, S. Polizzi, M. Bettinelli and A. Speghini, J. Mater. Res. **15**, 586 (2000).

[5] S. Polizzi, G. Fagherazzi, M. Battagliarin, M. Bettinelli and A. Speghini, J. Mater. Res., in press.

[6] H. O'Connor and T. M. Valentine, Acta Crystallogr. **25,** 2140 (1969).

[7] M. Bettinelli et al., unpublished results.

[8] T. Ye, Z. Guiwen, Z. Weiping, and X. Shangda, Mater. Res. Bull. **32**, 501 (1997).

[9] L. R. Morss, M. Siegal, L. Stenger and N. Edelstein, Inorg. Chem. **9,** 1771 (1970).

[10] F. Grandjean and G. J. Long, in: "Mössbauer Spectroscopy Applied to Inorganic Chemistry", Vol. 3, ed. G. J. Long and F. Grandjean, Plenum Press, New York, 1989, p. 513.

Table 1. δ is the isomer shift with respect to EuF_3, Γ is the FWHM, eQV_{zz} is the quadrupole interaction parameter, η is the asymmetry parameter and Area is the relative area. The d and Q_d Durbin-Watson parameter and the chi-squared parameter χ^2 are also reported. Statistical errors are given in parenthesis as errors on the last digit.

	Fit 1	Fit 2	
		Site C_{3i}	Site C_2
δ (mm/s)	1.05(1)	1.23(3)	0.97(1)
Γ (mm/s)	3.02(1)	1.76	1.76
eQV_{zz} (mm/s)		-5.5(5)	-10.7(3)
η			1.0(1)
Area (%)		49(1)	51(1)
d	0.4	1.8	
Q_d	1.6	1.7	
χ^2	4.7	1.0	

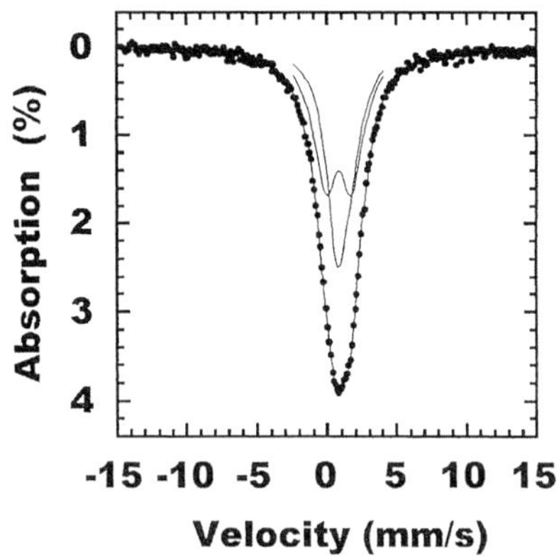

Fig. 1. Absorption spectrum of the $Y_{1.8}Eu_{0.2}O_3$ sample. Experimental data are reported as dots, the solid curves show the fit with 2 contributions and the contribution of each site.

Mössbauer Studies of the tetravalent doped Magnetite

A M Gismelseed[1], A A Yousif[1], A D Al-Rawas[1], M E Elzain[1], I Ayub[2], H. M. Widatallah[3], F. J. Berry[2]

[1]Physics Department, College of Science Sultan Qaboos University Box 36 code 123 Oman.
[2]Department of Chemistry, The Open University, Walton Hall MK7 6AA, UK
[3]Physics Department, University of Khartoum, Box 321, Khartoum, Sudan

Hydro-thermally prepared Sn^{4+} and Ti^{4+} doped -magnetite samples are characterized through scanning electron microscopy and X-ray diffraction (XRD). The samples are studied using ^{57}Fe Mössbauer spectroscopy in the temperature range from liquid nitrogen to room temperature. The XRD pattern shows that the samples are in pure spinel structure. The room temperature Mössbauer spectrum of Sn^{4+} -Fe_3O_4 exhibits a broadened magnetic sextet superimposed on a paramagnetic doublet while the spectrum of Ti^{4+} -Fe_3O_4 is very broad with a larger paramagnetic component. However, at liquid nitrogen temperature the two spectra are similar. The influence of the doping of each cation on the iron sites is discussed.

Introduction

The Magnetite (Fe_3O_4) crystallizes in the spinel structure, in which the oxygen ions form a closed-packed cubic lattice and the Fe ions are located in two different kinds of interstices between the oxygen ions, namely the tetrahedral sites (A) and the octahedral sites [B]. The network of both the tetrahedral and octahedral cations are antiferromagnetically coupled and show inverse spinel structure with formula $(Fe^{3+})[Fe^{3+}Fe^{2+}]O_4$ for pure magnetite. The compound was studied very intensively by numerous techniques aimed at understanding the crystallographic transition known as Verwey transition (occurs at temperature T_v=122K for stoichiometric magnetite) since the subtleties of this transition reveal when the system goes off stoichiometry, or is slightly doped [1,2]. Mössbauer measurement and interatomic potential calculations carried on Sn^{4+}/T^{4+} doped magnetite show that the substitution of such dopants favors the octahedral site and is accompanied by reduction of Fe^{3+} to Fe^{2+} for charge balancing [3,4]. These findings are consistent with the results of the X-ray photoelectron spectroscopic data sited in [5]. However, despite these extensive studies certain aspects on these properties are still not clearly understood.

Since Sn^{4+} and Ti^{4+} are magnetically inactive dopants and both have the preferences of entering the octahedral site, the hyperfine field and the lattice properties of Sn- and Ti-doped magnetite samples will be studied using ^{57}Fe Mössbauer spectroscopy technique.

Experimental

The studied samples were prepared and treated hydrothermally as described in [6]. X-ray diffraction (XRD) data were recorded with a Siemens D5000 diffractometer in reflection mode using CuKα radiation. The Mössbauer spectra were obtained on (25mg) powdered samples in the temperature range 78 – 300 K using a constant acceleration Mössbauer spectrometer with a 50 mCi ^{57}Co in Rh source. The spectrometer was calibrated with α-Fe foil spectrum at room temperature. Low temperature measurements were performed using a liquid nitrogen flow cryostat.

Results and Discussion:

X-ray diffraction analysis showed that the two compounds are monophasic spinel type with cubic structure at room temperature. The lattice parameter for the tin-dopant sample of a=8.4524 Å is larger than that for the Ti-dopant of 8.3264 Å and of the pure magnetite 8.3777 Å. This difference reflects the larger size of Sn^{4+} (0.69 Å) in sixfold oxygen coordination as compared to the size of Ti^{4+} (0.61 Å) and of Fe^{3+} (0.65 Å).

The Mössbauer spectrum of stoichiometeric magnetite Fe_3O_4 usually results in two well-known distinct sextets with typical hyperfine interaction parameters for Fe^{3+} and $Fe^{2.5+}$ and with an area ratio of about 1.9 assigned to sites (A) and [B] [7]. However, due to the two possible orientations of the EFG with respect to the hyperfine field, two-sextets assigned to the [B] site are adequate [8].

Selected Mössbauer spectra of the compound $Fe_{2.91}Sn_{0.09}O_4$ measured in the temperature range 78 – 300K are displayed in Figure 1a. It is clear that at intermediate temperature, the lines of the tetrahedral site are sharp compared to the B site profiles. The fitting parameters obtained for this site are identical to the parameters of the (A) site in pure magnetite, implying that the tetrahedral site is essentially not altered by the Sn^{4+} ions. On the other hand, it can be seen that the incorporation of tin in magnetite dramatically affects the [B] sublattice, changing the shape of the [B] site profiles. This demonstrates that Sn^{4+} ions have a clear preference for the octahedral site. Tracing the kinks on the side of the first line over the whole range of the measured temperatures, the [B] site can perfectly be fitted with optimum chi square, by assuming three sextets noted as B_0, B_1, and B_2. The sextets B_o and B_1 are assigned to Fe^{2+}/Fe^{3+} hopping on octahedral site since their parameters are close to the two sextets described for pure magnetite [7]. The third sextet B_2 is characterized by a higher isomer shift (0.82 mm/sec at 78K) and low magnetic hyperfine field, and can be attributed to the interaction of Fe^{2+} and Fe^{3+} on octahedral sites with tin ions. The resulting Mössbauer fitting parameters are collected in Table I.

Table I: Mössbauer parameters of Sn- doped Fe_3O_4 at given temperatures

Sample	300K			78K				
IS (mm/s)	Δ (mm/s)	H_{hf} (T)	A (%)	IS (mm/s)	QS (m/s)	H_{hf} (T)	A (%)	Site
0.36	-0.10	49.1	32	0.51	-0.11	52.4	35	[A]
0.25	0.17	47.6	24	0.35	0.12	50.4	32	B_0 [B]
0.54	-0.04	45.4	20	0.57	-0.16	48.3	19	B_1 [B]
0.69	-0.18	42.7	13	0.82	-0.30	45.1	11	B_2 [B]
0.33	0.59		11	0.33	0.45		3	[B]

Figure 1b shows the Mössbauer spectra of $Fe_{2.8}Ti_{0.2}O_4$ at different temperatures. The spectra were fitted in a similar manner as that for Sn- dopant sample but with considerable line broadening. Their hyperfine interaction parameters are collected in Table II. The line shape of the magnetic sextet at room temperature is very broad and difficult to fit it with the same number of sites, therefore a model of superposition of the paramagnetic doublet and hyperfine field distribution on site A and B independently.

The small doublet in the Mössbauer spectrum, which account for not more than 3.5% at 78K was found to increase with increasing temperature. The increasing rate is faster for Ti-dopant. The occurrence of a paramagnetic component superimposed on a broadened magnetic sextet in magnetite with diamagnetic substitution indicates various competing short-range order effects with long-range order. Superparamagnetism has been cited as the cause for the phenomenon in many systems such as Li-Ti ferrites [9]. The effect is observable in Ti –doped magnetite because of their particles size as detected by TEM techniques (TEM results: 8 nm for Ti and 55 nm for Sn).

Table II: Mössbauer parameters of Ti- doped Fe_3O_4 at given temperatures.

Sample	160K			78K				
IS (mm/s)	Δ (mm/s)	H_{hf} (T)	A (%)	IS (mm/s)	Δ (mm/s)	H_{hf} (T)	A (%)	Site
0.39	-0.02	50.4	34	0.47	-0.07	52.0	34	[A]
0.28	0.03	47.6	23	0.38	0.25	50.2	25	B_0 [B]
0.34	-0.08	43.0	17	0.45	-0.08	48.5	26	B_1 [B]
0.59	-0.50	41.8	12	0.65	-0.38	44.8	11	B_2 [B]
0.26	0.46		14	0.26	0.46		4	[B]

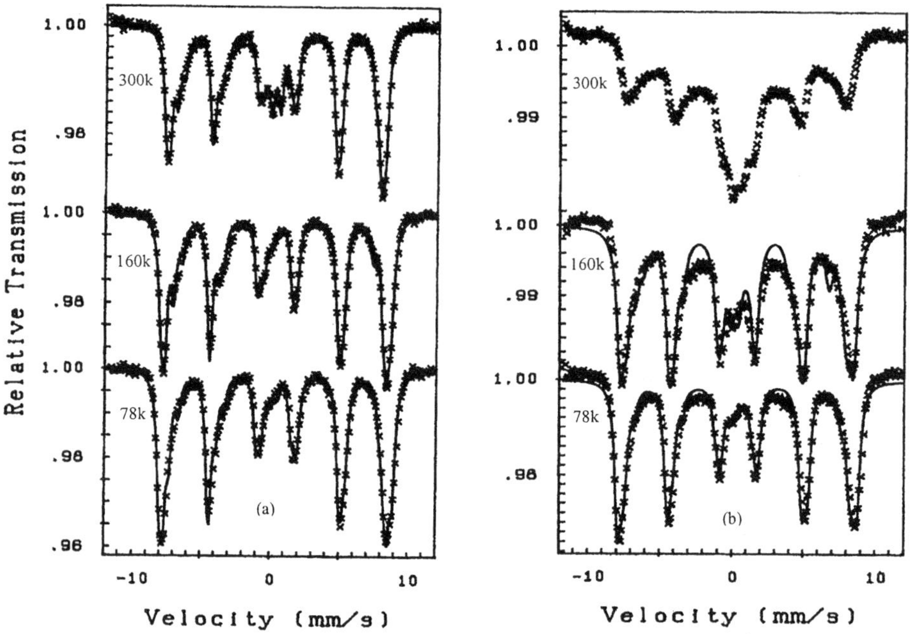

Figure 1: Mössbauer spectra measured at 300, 160 and 78K for (a) Sn- (b) Ti- doped magnetite

References:

[1] L. Häggström, H Annersten, T Ericsson, R Wäppling W Karner and S Bjarman, Hyp. Int. 5 (1978) 2101.

[2] R M Persoons, E De Grave and R E Vandenberghe, H, Hyp. Int. 54 (1990) 655 – 660.

[3] F J Berry, and Ö Helgason, Hyp. Int. 126 (2000) 269-275.

[4] E Moore, A Bohorquez, F Berry, Ö Helgason and J F Marco J Phys Chem solids 62 (2001) 1277.

[5] F J Berry, C Greave, Ö Helgason and J McManus, J Mater. Chem. 9, (1999) 223-226

[6] F J Berry, R Bilsborrow, Ö Helgason, J Marco and S Skinner, Polyhedron 17, (1998) 149

[7] L Diamandescu, D Mihaila- Tarabasanu, V Teodoreescu, N Popescu- Pogrion, Materials Letters, 37 (1998) 340 – 348.

[8] E De Grave, R Persoons, R Vandenbergh and P de Bakker, Phys Rev B, 47 (1993) 5881.

[9] A A Yousif, M E Elzain, S A Mazen, H H Sutherland M H Abdalla and S F Masour, J Phys: condens Matter 6 (1994) 5717.

Mössbauer Spectra of Iron Fine Particles by Computer Simulation

Kazuko Fukumura and Takayuki Kobayashi

Department of Physics, Shiga University of Medical Science, Otsu, Shiga 520-2192, Japan.

Growth of iron fine particles in a sapphire crystal was simulated with a computer and the expected Mössbauer spectra were calculated at various temperatures considering the influence of superparamagnetic relaxation and collective magnetic excitation. These spectra were compared with the experimentally observed spectra. The agreement between them was fairly well at lower temperatures.

Keywords : fine particle, CEMS, computer simulation, iron, implantation

1. Introduction

Several years ago, Conversion electron Mössbauer spectroscopy (CEMS) investigation of the specimen of ^{57}Fe-implanted sapphire crystal (α-Al_2O_3) at temperatures between 14 K and 300 K (room temperature) was performed, where the implantation dose was between $4x10^{16}$ and $1x10^{17}$ Fe/cm^2 [1]. It was confirmed that these conversion electron Mössbauer (CEM) spectra have lines attributed to both Fe^{2+} and metallic iron (α-iron). At a dose of $1x10^{17}$ Fe/cm^2, the spectrum obtained at 14 K is dominated by a six-line component due to α-iron particles. In the spectra removed the Fe^{2+}-component shown in Fig. 1, the line-intensity and hyperfine field of the magnetic component decrease with increasing temperature. The growth of iron fine particles in a sapphire crystal was also simulated with a computer in our previous study [1] and the expected Mössbauer spectra were calculated by considering the influence of the nearest and next-nearest neighbors on the magnetic hyperfine field of the iron atoms at very low temperature according to the method of Rancourt and Ping [2]. The simulated spectra agreed well with the experimentally observed spectra

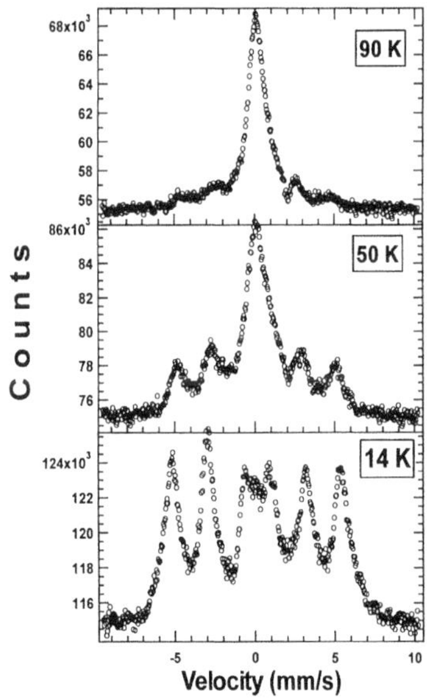

Fig.1 CEM spectra removed Fe^{2+}-component at the dose of $1x10^{17}$ Fe/cm^2.

at 14 K, but the spectra at higher temperatures were not simulated. In the present study, Mössbauer spectra at various temperatures are intended to calculate considering the influence of superparamagnetic relaxation and collective magnetic excitation.

2. Calculation Method

2.1 Formation of Fine Particles

The simulation procedures are described in details in our previous work [1]. When iron atoms implanted in a sapphire lattice form small particles of α-iron, the lattice around the particles is probably deformed tremendously, and many defects and vacancies are produced in the vicinity of the particles. An iron atom coming close to a particle is most likely trapped at the defects or vacancies to make the particle grow more. In the calculations, it was assumed that an iron atom generated in the matrix will come to rest at a some site in a small domain denoted by a cube with $(2L+1)^3$ sites in the matrix, where L is an integer. Alpha-iron has the bcc-type structure and its primitive lattice is orthorhombic. The computer simulation was performed with periodic boundary conditions in a matrix with 70x70x70 sites. Since the lattice constant of metallic iron is 0.286 nm, the number of atoms generated in the matrix with 70x70x70 sites are approximately 116000 when the implantation dose is $1\text{x}10^{17}$ Fe/cm^2.

Calculating the numbers of the nearest and next-nearest neighbors, hyperfine field at an iron site was calculated.

2.2 Calculation of Mössbauer Spectra

The observed Mössbauer spectra given in Fig. 1 show a gradual collapse of the magnetic hyperfine splitting due to superparamagnetic relaxation. Relaxation time, τ, is expressed as

$$\tau = \tau_0 \exp(KV / k_B T),$$

where K is the anisotropy energy constant, V the volume of a fine particle, k_B Boltzmann's factor, T absolute temperature and τ_0 is of the order of 10^{-11} s [3]. The occurrence rate of superparamagnetic relaxation, p, is assumed that it may increase rapidly with the temperature, such as

$$p = p_0 \exp(-KV / k_B T),$$

where p_0 is an arbitrary parameter and p is limited at most to unity.

The small fluctuations of the magnetization vector around an easy direction, i.e., collective magnetic excitations, may also have an influence on the Mössbauer spectra [4]. The mean value of hyperfine field is given by

$$H_{hf}(V,T) = H_{hf}(V = \infty, T) < \cos\theta >_T .$$

The occurrence rate of collective magnetic excitation is estimated with

$$q = 1 - p.$$

3. Results and discussions

At very low temperature, the observed spectra were reproduced well when L is a random integer between 0 and 10 [1]. When L is fixed, for example, when L=2 or 4, the distribution of the particle-size is too narrow, while the distribution is broad for random integer of L, which may reflect a great variety of surroundings of implanted atoms just before coming to rest in the sapphire lattice.

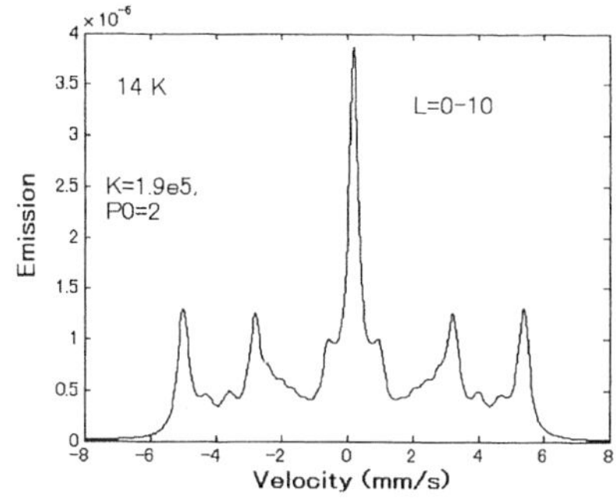

Fig.2 Calculated CEM spectra when the implantation dose is 1x10[17] Fe/cm^2, L=0-10 and K= 1.9x10^5 J/m^3.

The calculated Mössbauer spectrum is shown in Fig. 2 at 14 K in the case of L=0-10. Even at 14 K the intensity of the six-line magnetic component is too small compared with the observed spectrum shown in Fig. 1. This indicates that there exist only few particles with big size in the matrix. In Fig. 3, the size distribution of particles is shown in the cases of L=0-10 and L=3-10. The horizontal and vertical axes are the cube root of the number of atoms in a single particle, d, and the number of particles, respectively. In the case of L=3-10, a broad peak is noticed around d=13 and, at lower d (d<8), the number of particles increases monotonously with decreasing d. When L=0-10, the peak shifts to lower d and the monotonously increasing distribution of the number of particles at furthermore lower d exceeds that in the case of L=3-10, which means there are much smaller particles when L=0-10 than when L=3-10. It is, therefore, reasonable to introduce a characteristic number defined by the mean value of d in the broad peak, which is a measure of particle size in the matrix. The anisotropy energy constant K of a fine particle is related to the particle size, and their relation was formulated empirically by Bødker *et al.* [5]. The value of K for the characteristic number in the present cases was estimated with this empirical formula. In Fig. 4, the Mössbauer spectra calculated at temperatures from 14 K to 90 K are shown in the case of L=3-10. The spectrum at 14K in Fig. 4 is sufficiently similar to the experimentally observed spectrum in Fig. 1, but, at 90 K, the magnetic split lines are too much conspicuous, which indicates there are too many larger particles in the matrix. Similar calculations

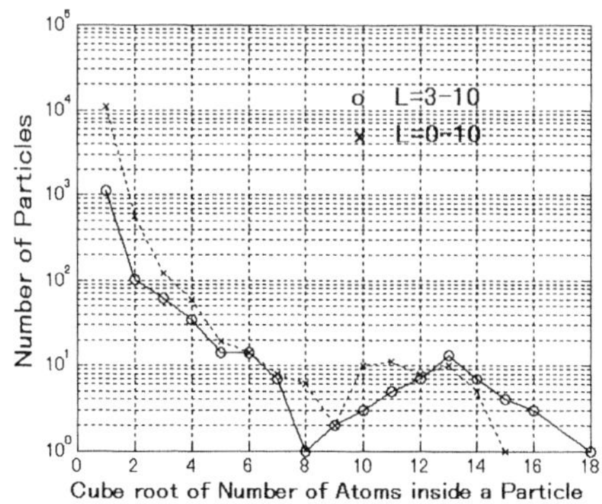

Fig.3 Distribution of the size of particles in the cases of L=0-10 and L=3-10.

were also performed for the doses of 7x10[16] and 4x10[16] Fe/cm[2], and their spectra agree fairly well with the experimentally observed spectra at lower temperatures.

The calculation of Mössbauer spectra described here is based on the very simple assumptions, and the reproduced spectra agree fairly well with the experimentally observed spectra at lower temperatures, but the agreement between them at higher temperatures is insufficient. Further investigation is required to reproduce the experimentally observed spectra and to understand the formation process of fine particles.

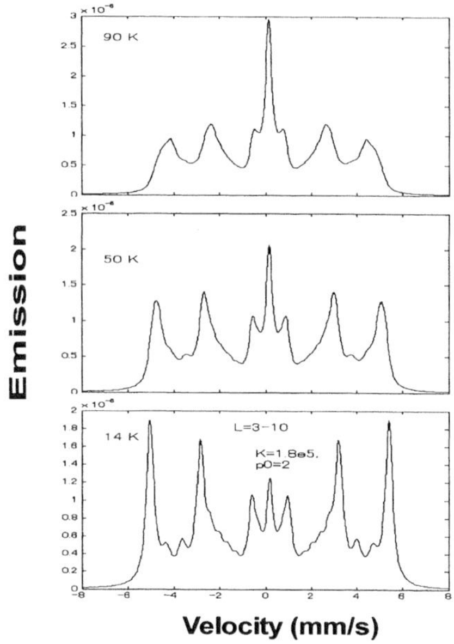

Fig.4 Calculated CEM spectra when the implantation dose is 1x10[17] Fe/cm[2], L=3-10 and K=1.8x10[5] J/m[3].

References

[1] T. Kobayashi, A. Nakanishi, K. Fukumura and G. Langouche, J. Appl. Phys. 83(1998)4631.

[2] D.G. Rancourt and J. Y. Ping, Nucl. Instr. and Meth. B58(1991)85.

[3] H. H. Wickman and G. K. Wertheim, in: Chemical Applications of Mössbauer Spectroscopy, ed. V.I. Gol'danskii and R.H. Herber, Academic Press, 1968, p. 582.

[4] S. Mørup, J. A. Dumesic and H. Topsøe, in: Application of Mössbauer spectroscopy, vol. II, ed. R. L. Cohen, Academic Press, 1980.

[5] F. Bødker, S. Mørup and S. Linderoth, Phys. Rev. Lett. 72(1994)282.

Structure and magnetic properties of the Fe(Co,Ni)ZrB systems prepared by the rapid quenching and mechanical alloying methods

A. Grabias[a], M. Kopcewicz[a], D. Oleszak[b], J. Latuch[b] and T. Kulik[b]

[a] *Institute of Electronic Materials Technology, Wólczynska 133, 01-919 Warszawa, Poland*

[b] *Department of Materials Science and Engineering, Warsaw University of Technology, Woloska 141, 02-507 Warszawa, Poland*

Transmission Mössbauer spectroscopy is applied to study the structure and magnetic properties of the Fe(Co,Ni)ZrB alloys prepared by the rapid quenching and ball milling techniques. There are significant differences in the structure of these alloys that depend on the preparation method. The powders prepared from the amorphous ribbons remain amorphous after milling, while powders obtained from the crystalline elements form the bcc iron-based solid solutions.

1. Introduction

Amorphous and nanocrystalline iron-based alloys are important magnetic materials due to their excellent soft magnetic properties [1]. It has been reported that certain multicomponent iron-based systems can be obtained as bulk amorphous alloys by a copper mold casting method [2,3]. These alloys are characterized by a wide supercooled liquid region allowing the reduction of the cooling rate required for the formation of a metastable amorphous phase.

In the present study we prepared two sets of the Fe(Co,Ni)ZrB systems in the form of ribbons and powders. The amorphous ribbons were obtained by the rapid quenching method. The powder samples were prepared by the high-energy ball milling of the as-quenched amorphous ribbons and of the crystalline elemental powders. Transmission Mössbauer spectroscopy is applied to determine the hyperfine field distributions of the amorphous alloys as well as to investigate phase transformations in the iron-containing powders as a function of milling time and alloy composition.

2. Experimental details

Amorphous $Fe_{70-x-y}Co_xNi_yZr_{10}B_{20}$ and $Fe_{80-x-y}Co_xNi_yZr_7B_{13}$ ($x = 0$ or 10, and $y = 0$, 7 or 10) alloys were prepared by the melt spinning technique in the form of ribbons about 30 μm thick and 4 mm wide. The amorphous state of the melt-spun ribbons was confirmed by the X-ray diffraction and transmission Mössbauer spectroscopy measurements. In order to obtain a bulk amorphous alloy the ribbons were exposed to the high-energy ball milling. The milling was performed under an argon atmosphere in the Spex vibrational mill using steel vials and balls. On the other hand, the powder samples with the same compositions as the ribbons were also prepared by high-energy ball milling of the powder mixtures of the pure crystalline elements.

All samples were characterized at room temperature by transmission Mössbauer spectroscopy using a 25 mCi ^{57}Co-in-Rh source. The spectra were analysed with the NORMOS-SITE and NORMOS-DIST programs [4]. The relative contents of the phases present in the samples were calculated as a ratio of the relevant subspectral area to the total spectral area, assuming a similar Debye-Waller factor for each phase. The isomer shifts are given with respect to the standard α-Fe foil.

3. Results

Figure 1a shows the Mössbauer spectrum of the as-quenched $Fe_{60}Co_{10}Ni_{10}Zr_7B_{13}$ ribbon. This spectrum is typical for an amorphous, ferromagnetic alloy. The hyperfine field distribution extracted from the spectrum reveals a dominating broadened peak accompanied by a smaller one at lower hyperfine fields (Fig. 1a'). After high-energy ball milling up to 10 hours (Figs. 1b-1d) the hyperfine field distributions reveal an additional peak at about 33 T which can be easily assigned to the formation of a certain amount of the bcc Fe (Figs. 1b'-1d'). The relative content of this phase is about 4% after 10 hours of milling.

Very similar features of the Mössbauer spectra are observed for all other Fe(Co,Ni)ZrB amorphous ribbons. There are, however, some differences in the average hyperfine fields that result mostly from the differences in the alloy composition. The lowest average hyperfine field calculated from the distributions is observed for the $Fe_{63}Ni_7Zr_{10}B_{20}$ alloy, $\langle H_{hf} \rangle = 15.51$ T, while the highest one – for $Fe_{60}Co_{10}Ni_{10}Zr_7B_{13}$, $\langle H_{hf} \rangle = 22.65$ T. Cobalt is known to increase significantly the hyperfine field of both the amorphous and nanocrystalline iron-containing phases [5]. The contribution of the peak related to the bcc Fe formed due to the milling increases slightly with increasing milling time. Its relative content does not exceed 5% after 10 hours of milling.

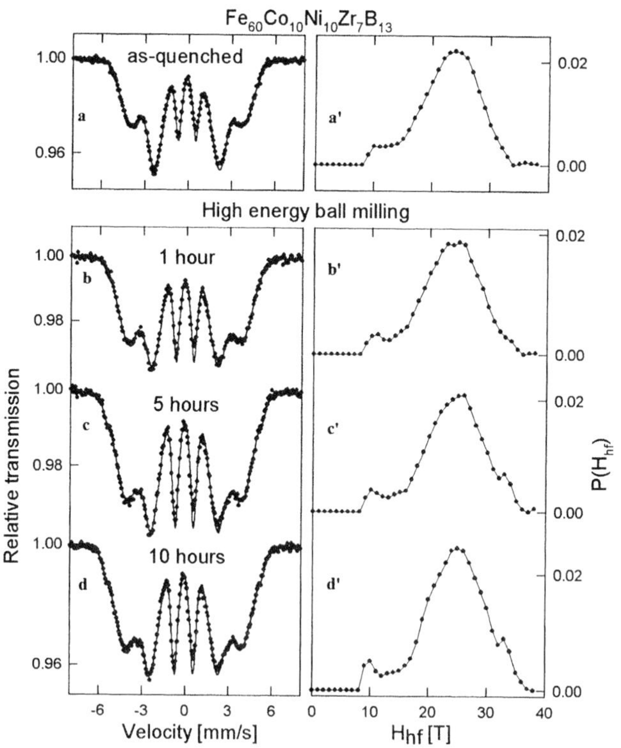

Figure 1. Mössbauer spectra and the corresponding hyperfine field distributions obtained for the $Fe_{60}Co_{10}Ni_{10}Zr_7B_{13}$ alloy in the as-quenched state and after high-energy ball milling.

The Mössbauer results obtained for the $Fe_{60}Co_{10}Ni_{10}Zr_7B_{13}$ system prepared by high-energy ball milling of the elemental powders are shown in figures 2a-2d. The spectrum recorded after 5 hours of milling contains two magnetic components: the dominating one with hyperfine field, H_{hf}, of 33.0 T and isomer shift, δ, of 0.00 mm/s corresponding to the bcc-Fe phase, and the other one with H_{hf}=35.1 T and δ=0.02 mm/s assigned to the bcc-FeCo phase [5], (Fig. 2a). The increase of the milling time causes further formation of the bcc-FeCo phase at the expense of the bcc-Fe component (Figs. 2b-2d). The lines of the FeCo sextet are broadened, indicating the formation of the bcc solid solution. The relative content of this phase after 90 hours of milling is about 60%. The spectra recorded after 10 hours of milling and longer contain also four less intense spectral components: one with H_{hf}=30.5 T and δ=0.03 mm/s, and three sextets related to the tetragonal Fe_3B phase [6], (H_{hf1}=28.4 T, δ_1=0.08 mm/s; H_{hf2}=25.6 T and δ_2=0.06 mm/s; H_{hf3}=21.2 T and δ_3=0.09 mm/s). The assignment of the sextet with H_{hf} of about 31 T, appearing in the spectra of the FeZrB-based systems is controversial. It was usually related to the Fe atoms located at the interface of the bcc-Fe nanograins [7]. On the other hand, the same reduction of the hyperfine field can result also from a small amount of Zr atoms dissoluted in the bcc structure of iron [8], what is very probable in the case of the inhomogeneous powder samples.

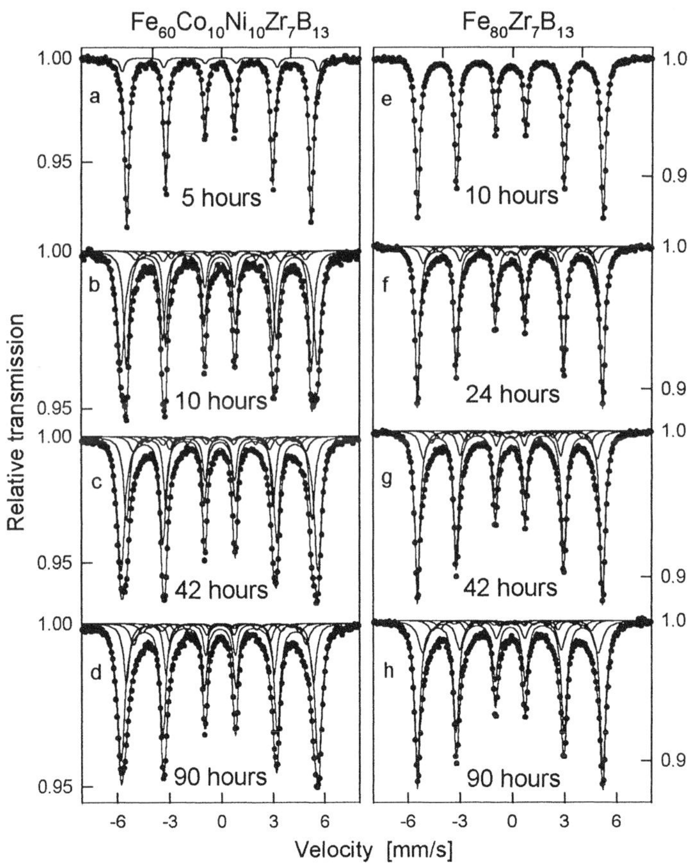

Figure 2. Mössbauer spectra of the $Fe_{60}Co_{10}Ni_{10}Zr_7B_{13}$ and $Fe_{80}Zr_7B_{13}$ alloys prepared by high-energy ball milling of the mixture of the crystalline, elemental powders.

The Mössbauer spectra obtained for the $Fe_{80}Zr_7B_{13}$ system prepared by high-energy ball milling of the mixture of metals and boron crystals are shown in figures 2e-2h as a function of milling time. The phases formed due to the milling are the same as in the case of the $Fe_{60}Co_{10}Ni_{10}Zr_7B_{13}$ sample, except for the FeCo phase. The phases start to form after milling times exceeding 10 hours (Figs. 2f-2h). There is an additional paramagnetic component observed in the spectra of the $Fe_{80}Zr_7B_{13}$ samples. The hyperfine parameters (the quadrupole splitting of 0.50 mm/s and δ=0.02 mm/s) suggest the formation of the amorphous, iron-rich FeZr phase [9]. However, its relative spectral contribution is not significant (up to 3% after 24 hours of milling, Fig. 2f).

4. Conclusions

The amorphous Fe(Co,Ni)ZrB systems were produced in the form of ribbon. The powders obtained after the high-energy ball milling of the amorphous ribbons remain amorphous. A small amount of the bcc iron is observed for longer milling times. The powders prepared by high-energy ball milling of the mixture of the pure elements are completely crystalline. The iron-based solid solutions are formed preferably.

Acknowledgement

This work was supported by the Polish Scientific Research Committee under the grant No. KBN 7 T08A 03917.

References

[1] A. Makino, A. Inoue and T. Mizushima, Mater. Trans. JIM 41 (2000) 1471.
[2] A. Inoue and J.S.Gook, Mater. Trans. JIM 36 (1995) 1282.
[3] A. Inoue, A. Makino and T. Mizushima, J. Magn. Magn. Mat. 215 (2000) 246.
[4] R.A. Brand, J. Lauer and D.M. Herlach, J. Phys. F: Met. Phys. 13 (1983) 675.
[5] R. Brüning, K. Samwer, C. Kuhrt and L. Schultz, J. Appl. Phys. 72 (1992) 2978.
[6] G. LeCaër and J.M. Dubois, Phys. Status Solidi A 64 (1981) 275.
[7] M. Miglierini and J.M. Greneche, J. Phys.: Condens. Matter 9 (1997) 2303.
[8] J. Balogh, L. Bujdoso, D. Kaptas, T. Kemeny, I. Vincze, S. Szabo and D.L. Beke, Phys. Rev. B 61 (2000) 4109.
[9] R. Pizarro, J.S. Garitaonandia, F. Plazaola, J.M. Barandiaran and J.M. Greneche, J. Phys.: Condens. Matter 12 (2000) 3101.

Mössbauer and electrical properties of sulfur doped basalt glass

M Y Hassaan, M M El-Desoky[1], N S Gomaa[2], N A Eissa[3]

Department of Physics, University of Al- Azhar, Cairo, 11884 Cairo, Egypt
[1]Department of Physics, University of Suez Canal, El- Arish, Egypt
[2]Department of Physics, University of Al-Azhar (Girls' Branch), Cairo, 11884, Egypt
[3]Department of Physics, University of Al-Azhar, Cairo, 11884, Egypt

Mössbauer and dc conductivity measurements were carried out on five samples. The first sample is the basalt rock and the second sample is the basalt glass without sulfur addition and the other three samples are basalt glass doped with different percentages of sulfur [5,10 and 15] mass% in addition. The basalt glasses were prepared by melting basalt rock starting from powder form in an electric furnace at [1550] °C in air, for half an hour, then quenched in air. The results show that the addition of sulfur to the basalt glass reduces the ionic states of iron oxide, which are present in basalt glasses, where the ferric ions completely disappeared at 10 mass% S content and the ferrous ions are the only iron site present. The relation between the electrical conductivity and activation energy was discussed and correlated with Mössbauer results.

1. Introduction

The electronic conductivity of semiconducting oxide glasses containing transition metal ions has been studied extensively [1]. Basalt glass is considered as a type of oxide glasses, which is composed of SiO_2 as a glass former. Because, basalt glass contains about 13% Fe_2O_3, as a transition metal oxide, the iron oxide in different redox states are dispersed in the glass matrix and the charge transfer can take place between Fe^{2+} and Fe^{3+} ions [2, 3]. Mössbauer spectrometer can recognize strictly these two states of iron ions inside different solids as well as basalt glass. Helgason et.al [4] used Mössbauer spectroscopy in conjunction with melting experiments at controlled oxygen pressures to determine redox reactions in basaltic melts at 1300°C. On the other hand, sulfur has a good effect in reducing glasses [5]. Hassaan established an easy and economic method for preparing a completely reduced basalt glass using sulfur [6]. Electrical conduction can take place by the transfer of electrons from the ion of low valence state to that of high valence state, therefore the ratio $M_{reduced}/M_{total}$, where M represents the transition – metal ions i.e. iron, becomes one of the factors affecting electrical conductivity. Other important factors are the concentrations of transition – metal oxide, glass former and glass modifier as well as melting conditions. The transport of electrons in these glasses is usually termed "small polaron hopping". A general formula for the electrical conductivity of these glasses was proposed by Mott [7, 8] in which the conductivity, σ, is given by

$$\sigma = \frac{\nu_0 e^2 c(1-c)}{kTR} \exp(-2\alpha R)\exp(-W/kT) \qquad (1)$$

where ν_0 is a phonon frequency, α is the rate of the wave function decay, c is the ratio of ion concentration in the low valency state to the total concentration of transition – metal ions, R is the average hopping distance, W is the activation energy, e is the electronic charge, k is Boltzmann's constant and T is the absolute temperature.

Equation (1) can be compared to be common Arrhenius equation

$$\sigma = \frac{\sigma_0}{T} \exp(-W/kT) \qquad (2)$$

where

$$\sigma_0 = \frac{\nu_0 e^2 c(1-c)}{kR} \exp(-2\alpha R) \qquad (3)$$

and

$$W \simeq W_H + \tfrac{1}{2} W_D \qquad (4)$$

W_H is the hopping energy and W_D is the disorder energy arising from the energy difference of neighbours between two hopping sites [7]. It was thought to be interesting to study the electronic conductivity (d.c.) of basalt glass as a function of reduction effect of sulfur, which affects markedly ferrous/total iron ratio in basalt glass [9]. Therefore, this work deals with studying the reduction effect of sulfur on the d.c. conductivity of basalt glass.

2. Experimental

Four basalt glasses were prepared by melting the powder of basalt rock with different sulfur (0, 5, 10 and 15 wt%) content in a platinum crucible at 1550 °C in air for 30 min. Each melt was quenched in air. X-ray diffraction measurements were performed using CuK_α radiation. The basalt glass samples were pulverized for measuring the Mössbauer spectra by the constant acceleration method. ^{57}Co (Rh) was used as a gamma ray source. To measure the d.c. conductivity, the samples were highly polished and the two surfaces of each sample were coated with silver paste then it was measured using a constant d.c. voltage with potential probe method. The sample temperature was measured using a calibrated copper-constantan thermocouple. The data were collected from room temperature up to 570 K.

3. Result and discussions

Chemical analysis of the used basalt rock is shown in table (1). The x-ray diffraction results demonstrated the glassy state formation and show that the short range order increases as the sulfur addition increases.

Table 1.Chemical analysis of basalt rock.

L. O. I.	SiO_2	Al_2O_3	Fe_2O_3	CaO	MgO	SO_3	K_2O	Na_2O	TiO_2	Cl	Total
5.40	46.32	13.60	10.70	12.97	5.75	0.06	1.03	1.08	2.50	0.00	99.41

3.1. Mössbauer results

Figure (1) shows the Mössbauer of the used basalt rock. The spectrum can be analysed into four peaks; two sites due to Fe^{3+} with octahedral (Oh) and tetrahedral (Td) symmetry, and the other two due to Fe^{2+} with Oh and Td symmetry. Parameters of these sites are summarized in table 2. Mössbauer spectra of basalt glasses with and without sulfur are shown in Fig. 2. Corresponding Mössbauer parameters are

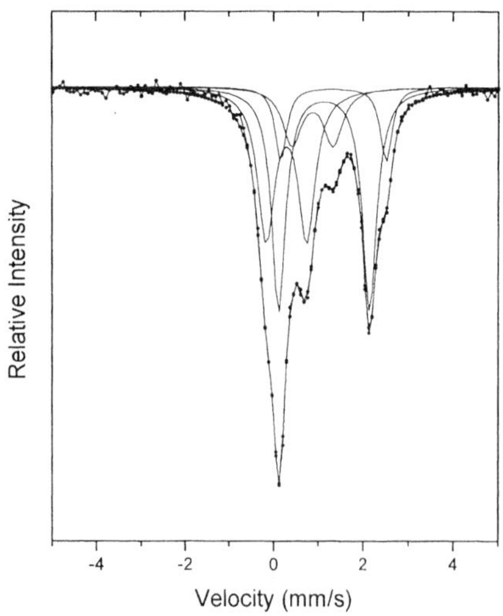

Fig.1 Mössbauer spectrum of basalt Rock.

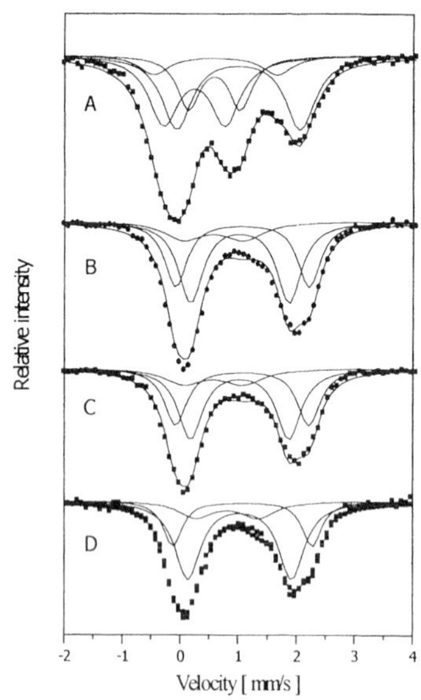

Fig.2 Mössbauer spectra of basalt glass A-pure B- with 5 wt%sulfur, C- with 10 wt% sulfur and D- with 15 wt% sulfur.

summarized in table 2. Pure basalt glass consists of four doublets; two of them represent Fe^{2+}(Oh) sites and the one represents Fe^{2+}(Td). The fourth doublet belongs to Fe^{3+}(Td). Sample 3 containing 5 mass% sulfur have four iron sites (see Fig. 2B), although there is a slight difference in the relative absorption area when compared with sulfur-free sample 2 (Fig. 2A). The fraction of Fe^{3+} in the sample 3 (Fig. 2B) was estimated to be only 5.8 %; *i.e.*, the fraction of Fe^{2+} was 94.2 %. In the sample 4 containing 10 mass% sulfur reduction of iron caused by the sulfur was completed and the doublet due to Fe^{3+} disappeared completely (see Fig. 2C). Figures 2c and 2d show that only the Fe^{2+} sites are present in the basalt glass sample when the concentration of sulfur is equal to or more than 10 mass%. Two of them represent Fe^{2+} with (Oh) symmetry, while the third one represents Fe^{2+}(Td) site.

Mössbauer spectrum of sample 5 (Fig. 2D), which contains 15 mass% sulfur, is essentially the same as that of sample 4 (Fig. 3A). It is noteworthy that the sulfur content shows a linear relationship with the Fe^{2+} fraction in the Mössbauer spectrum of basalt glasses. This result means that 7.5 mass% of sulfur is large enough to completely reduce the iron in basalt glass.

Table 2. Mössbauer parameters of basalt rock and basalt glass with and without sulfur.

Sample no.	Fe^{3+}(Td) mms^{-1}				Fe^{2+}(Td)* . mms^{-1}				Fe^{2+}(Oh) mms^{-1}				Fe^{2+}(Oh) mms^{-1}			
	I.S	Q.S	L.W	R.I %	I.S	Q.S	L.W	R.I%	I.S	Q.S	L.W	R.I%	I.S	Q.S	L.W	R.I%
Basalt rock 1	0.29	0.81	0.3	33.5	0.88	0.92	0.38	13.9	1.11	2.03	0.49	41.5	1.34	2.35	0.45	11.2
2	0.24	1.06	0.61	33.7	0.6	0.9	0.46	19.7	0.6	2.12	0.64	10.03	0.99	2.12	0.6	36.6
3	0.13	0.8	0.42	5.8	0.77	1.2	0.61	13.1	1.02	1.72	0.51	47.7	1.07	2.3	0.44	33.3
4	-	-	-	-	0.57	1.04	0.8	16.6	1.02	1.7	0.5	47.7	1.07	2.3	0.46	35.7
5	-	-	-	-	0.81	1.08	0.71	16.4	1.03	1.78	0.53	58.4	1.07	2.4	0.4	25.2

* This phase is Fe^{2+}(Oh) in case of basalt rock only.

3.2. D.c. conductivity results

The temperature dependence of the electrical conductivity for glass samples is shown in Fig. (3). In this figure, ln (σT) is plotted against 1/T the reciprocal of the temperature of the sample under measurement.

It is observed that the conductivity increases smoothly with increasing temperature, indicating temperature dependent activation energy characteristic of hopping conduction. However, above 400 K the conductivity changes almost linearly with inverse temperature so that the activation energy in the high temperature limit is almost independent of temperature.

The compositional dependence of the electrical conductivity at various temperatures for the basalt glass samples is shown in Fig. (4), where the conductivity increases with increasing S content.

This may be due to the increase of $Fe^{2+}/\Sigma Fe$ ratio in glass samples, which is determined from the Mössbauer results. On the other hand, as the present glasses have been synthesized in ambient conditions at high temperature, it is possible that the oxygen ions that are replaced by sulfur may have transformed into molecules of oxygen, contributing to charge neutrality, that finally escape to the atmosphere. However, the radius of the sulfur ion is greater than the radius of oxygen ion and the electronegatively of sulfur is smaller then that of oxygen, this leads to decrease of activation energy and increase of the conductivity with an increase of sulfur content [10].

In Eq. (1), there are two important factors which are attributable to the magnitude of the electrical conductivity. Pre-exponential factor and activation energy. With regard to the pre-exponential factor, the value of c(1-c) in eq. (3) calculated from Mössbauer results. Further, R is almost the same in all of the present glasses, which is calculated by using the relation [7]

$$R = (4/3\ \pi N)^{-1/3} \qquad (5)$$

Hence, it is concluded that in the pre-exponential factor, the tunneling term, exp (-2 αR), contributes most predominantly to the electrical conductivity. Then the following analysis is carried out in order to ascertain which is decisive for the electrical conductivity in the present case, α or another important factor w. The following equation is derived from eqs. (1) and (2):

$$\ln(\sigma T) = -2\,\alpha R + \ln\,[\nu_o\, e^2\, c(1-c)/kR] \qquad (6)$$

when the third term on the right hand of eq. (6) is not so sensitive to the variation of the glass composition, the linear relation between ln (σT) and w at constant temperature T should be obtained and the slope of the line is equal to $-1/kT$ in the case which αR is independent of the composition [11]. The plots of ln (σ T) vs. w for the present glasses is shown in Fig. (5). The solid lines are drawn by analyzing the experimental data using the method of least squares. This plot shows a straight line in nature, (at three concentration of S), which indicates that Eq. (2) is confirmed. From the slope of the plot the value of temperature (T_M) is estimated, which is very different to the fixed temperature (T_o = 473 K) chosen.

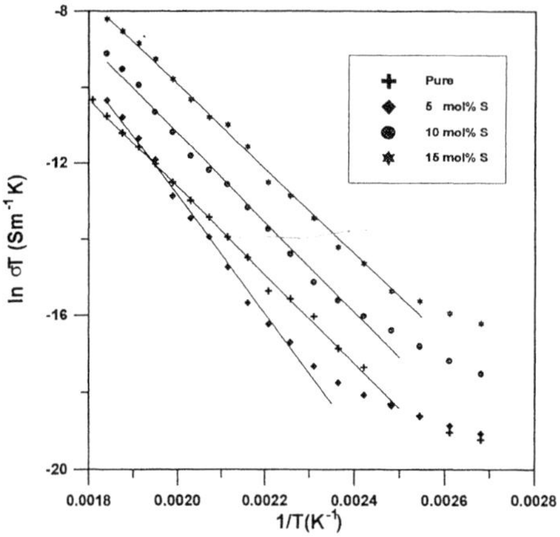

Fig.3 Plot of ln σT vs. 1/T for the glass samples.

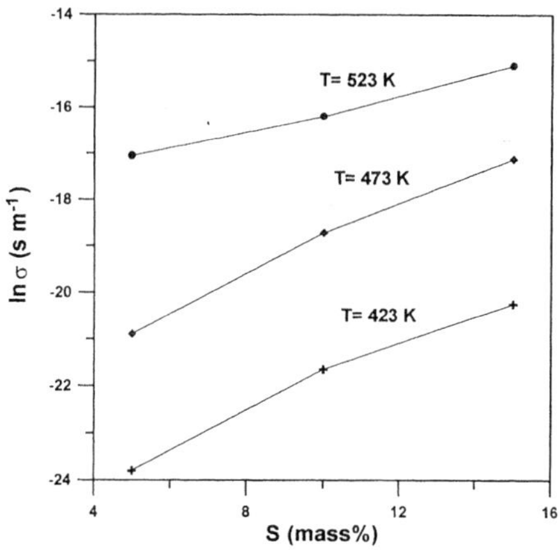

Fig.4 Plot of ln σ vs. S mass% for the glass samples.

The discrepancy between T_0 and T_M observed is ascribable to the compositional variation of the $Fe^{2+}/\Sigma Fe$ ratio which leads to the compositional dependence of the third term in Eq. (5). This indicates that the hopping conduction is non-adiabatic in nature [12]. On the other hand, the discrepancy between (T_o)and (T_M) indicates that αR is dependent of the composition for all glass samples. Therefore, it is now concluded that the electrical conductivity values of the basalt glasses is ascribable to the difference of the tunneling term or the pre-exponential and the difference of the activation energy. Application of the Mott theory has indicated that the polaron term, w_H, lies between 0.3 and0.35 eV. Since the total

Table 3.Density (d), average hopping distance (R), number of transition metal ions (N) and activation energy (w).

Wt % (S)	d (gm/cm^3)	R (A^o)	N x 10^{22}	W (eV)
0	2.738	1.40	8.6	0.9
5	2.735	1.40	8.6	1.32
10	2.735	1.40	8.6	1
15	2.695	1.41	8.46	0.97

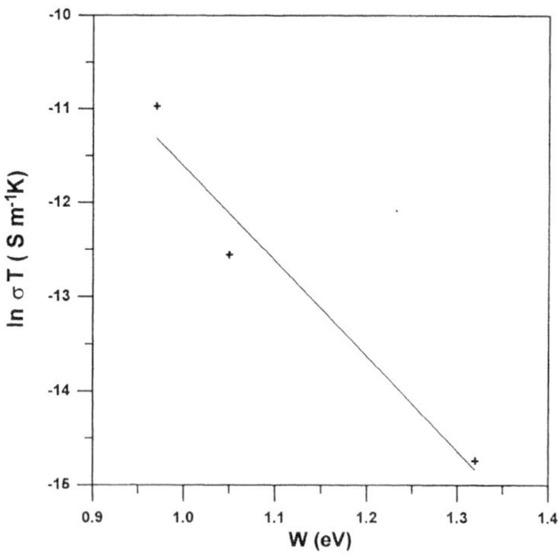

Fig. 5 Plot of ln σT vs. W for three glass samples.

activation energies have values of 0.9 to 1.32 eV. (see Table 2). The derived w_D value is higher than the estimated theoretical disorder term (< 0.1 eV) for the randomly distributed impurities in a broad – band semiconductor [11, 12] . This implies that an additional term , Δu, describing structure differences between iron ions should appear in the activation energy expression [8] such that

$$W = W_H + \tfrac{1}{2}W_D + \Delta U \qquad (7)$$

Finally, basalt glasses containing different concentrations of S show a good agreement with eq. (1) for non –adiabatic transport of small polarons. This means that the tunneling factor, exp (-2 αR),should be taken into consideration in eq. (1)[11].

References

[1] D. Virgo and B. O. Mysen, Phys. Chem. Mineral. 12 (1985) 65.
[2] A. K. Bandyopadhyay and J. O. Isard, J. Phys. D. Appl. Phys. 10 (1977) L99.
[3] A. K. Bandyopadhyay and J. O. Isard, "Amorphous and Liquid Semiconductors", edited by W. E. Spear (C.I.C.L., University of Edinburgh, 1977) 683 .
[4] O. Helgason, S. Steinthorsson, Hyperfine Interactions 70 (1992) 985-988.
[5] M. Y. Hassaan, M. M. El-Desoky, S. M. Salem and S. H. Salah, J. Radioanalytical and Nuclear Chemistry249 (2001) 595.
[6] M. Y. Hassaan, "An Easy and Economic Method for Preparing a completely Reduced Basalt Glass", Submitted for publication.
[7] N. F. Mott, J. Non-cryst. Solids.1 (1968) 1.
[8] I. G. Austin, N. F. Mott, Adv. Phys. 18 (1969) 41.
[9] B. V. R. Chowdari, P. Promoda, K.U. Mari, Solid State Ionic 113 (1998) 665.
B. V. R. Chowdari, K. F. Mok, J. M. Xie, R. Gopalakrishnan, J. Non-cryst. Solids 160 (1993) 73.
[10] V. K. Dhawan, A. Mansingh and J. Non-cryst. Solids 51 (1982) 87.
[11] G. Austin, J. Non-cryst. Solids 2 (1970) 474.
[12] Leon Murawsky, J. Mater. Sci. 17 (1982) 2155.

^{129}I-MÖSSBAUER STUDY OF IODINE-DOPED SINGLE-WALLED CARBON NANOTUBES

S. Kitao[1], M. Seto[1], Y. Kobayashi[1], R. Haruki[1], S. Masubuchi[2], S. Kazama[3], H. Kataura[4], Y. Maniwa[4], S. Suzuki[5] and Y. Achiba[5]

[1]Research Reactor Institute, Kyoto University, Kumatori, Osaka 590-0494, Japan,
[2]Tokyo Medical University, 6-1-1 Shinjuku, Shinjuku-ku, Tokyo 160-8402, Japan,
[3]Department of Physics, Chuo University, 1-13-27 Kasuga, Bunkyo-ku, Tokyo 112-8551, Japan,
[4]Department of Physics, Tokyo Metropolitan University, 1-1 Minami-Osawa, Hachioji, Tokyo 192-0397, Japan,
[5]Department of Chemistry, Tokyo Metropolitan University, 1-1 Minami-Osawa, Hachioji, Tokyo 192-0397, Japan

Abstract The ^{129}I-Mössbauer spectroscopy was applied to the iodine-doped single-walled carbon nanotubes (SWNTs) in the region of heavy and medium doping. Two kinds of SWNTs with different synthesis methods were used as samples. For both samples, the iodine in SWNTs in these doping levels was proved to work as electron acceptors in the forms of triiodides (I_3^-) and pentaiodides (I_5^-) and to exist as iodine molecules (I_2) in addition. Moreover, monoiodide ions (I^-) appear additionally. As the doping level decreases, the proportions of triiodide and monoiodide ions increase, whereas those of pentaiodide ions and iodine molecules decrease. This tendency is in common for both kinds of SWNTs. However, the charge distribution of iodides and the ratios of iodine forms are a little different between two kinds of SWNTs, probably because of difference in morphology and remaining catalysts.

1. Introduction

Carbon nanotubes are graphitic carbon tubules with a diameter from a few to a few tens of nanometers, which were discovered by Iijima *et al.* in 1991 [1]. Since the carbon nanotubes have unique physical properties and prospective applications, they have been investigated vigorously [2]. Among them, the single-walled carbon nanotube (SWNT) is a favorable specimen to simplify the physical phenomena. The intercalation with iodine is one of the most effective ways to promote the electronic properties. The doped iodine plays a role of either electron donor or acceptor depending on host materials. In conducting polymers such as polyacetylene and polythiophene, the iodine exists as the electron acceptor in the forms of polyiodides [3, 4], whereas the iodine in fullerenes acts as the electron donor [5]. Since carbon nanotubes have a bundle structure like polymers, iodine may exist as polyiodides. On the contrary, since they are constructed with carbon sheets as in fullerenes, there is a possibility of existence as electron donors. A Raman study on the iodine-intercalated SWNTs [6] reported that the intercalated iodine enters into the interstitial channels in SWNT rope lattice and that the iodine is charged as polyiodides (I_3^- and/or I_5^-). However, the Raman study is not always quantitative and it has silent species such as monoiodides. Therefore, the charge transfer between the iodine and the SWNT has not been clarified well yet.

In this report, we applied the ^{129}I-Mössbauer spectroscopy to the iodine-doped SWNTs for understanding the chemical structure of iodine and for evaluating the amount of charge transferred. We used two kinds of SWNTs with different synthesis methods to know the difference of iodine species when the diameters of nanotubes differ. To investigate the evolution of the iodine species as the change of doping levels, samples with two different doping levels were studied for both SWNTs.

2. Experimental

Two kinds of SWNTs were prepared: one is synthesized by the laser vaporization with Ni-Co catalysts at Rice University (hereafter, R-SWNT) [7], and the other one is synthesized by the arc-discharge method with Ni-Y catalysts at Tokyo Metropolitan University (T-SWNT). R-SWNT is mainly (8,8) tubes, while T-SWNT is a mixture of thicker tubes such as (8,8) and (10,10) tubes. Remaining catalysts are said to be below 10 weight-% in R-SWNT, while more catalysts remain in T-SWNT. After a heat treatment at 350°C for 1 hour to remove remaining carbon soot, the SWNTs were doped with $^{129}I_2$ in vapor phase. The molecular $^{129}I_2$ was prepared by decomposing $Pd^{129}I_2$, which had been obtained from precipitation by adding HCl and $PdCl_2$ into a $Na^{129}I$-aqueous solution. A $Zn^{129}Te$ Mössbauer source with a halflife of 69.6 minutes was prepared by 1-hour neutron irradiation of $Zn^{128}Te$ at the Kyoto University Reactor. To elongate the source lifetime, a $Mg_3{}^{129m}TeO_6$ source with a halflife of 33.6 days was also prepared by a 300-hour neutron irradiation of $Mg_3{}^{128}TeO_6$ and annealing at 700°C. The Mössbauer spectra were measured at 10 K. The doping level was evaluated by weight uptake.

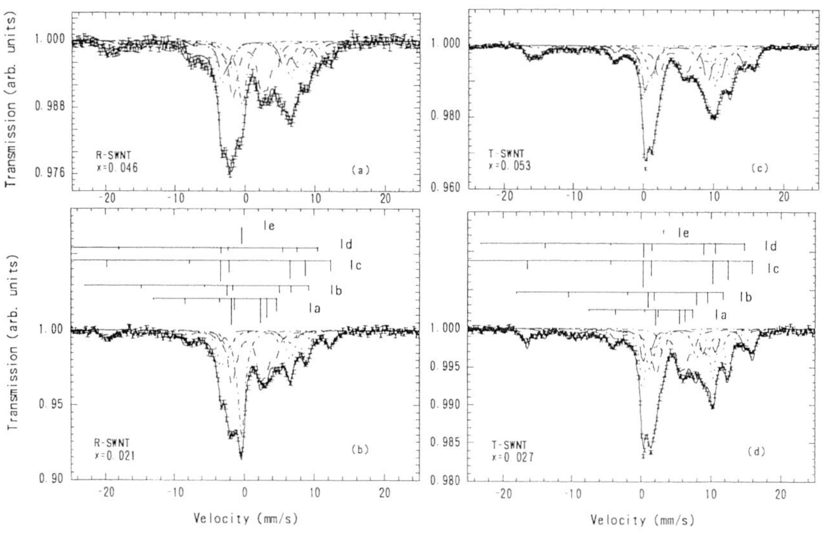

Fig. 1 *^{129}I-Mössbauer spectra of iodine-doped single-walled carbon nanotubes, R-SWNT with doping levels of CIx, (a) x=0.046 (b) x=0.021 relative to ZnTe source and T-SWNT with doping levels of (c) x=0.053 (d) x=0.027 relative to Mg_3TeO_6 source. Stick diagrams denote peak positions and intensity ratios of each iodine species.*

3. Results and Discussion

We measured ^{129}I-Mössbauer spectra of the iodine-doped R-SWNTs in the heavily and mediumly doped region by using $Zn^{129}Te$ source (Fig.1(a) and (b)). Iodine-doped T-SWNTs are measured using $Mg_3{}^{129m}TeO_6$ source in the same manner (Fig.1(c) and (d)). The spectra were analyzed with four quadrupole octets (Ia, Ib, Ic, Id) and a singlet (Ie). The obtained typical Mössbauer parameters for mediumly-doped R- and T-SWNTs are tabulated in Table I. The isomer shifts and the quadrupole coupling constants for heavy doping were similar to those corresponding for medium doping.

Table I. Typical Mössbauer parameters of iodine-doped SWNT in mediumly-doped region measured at 10 K.

Sample	Iodine species	δ [a] [mm/s]	e^2qQ [b] [MHz]	Linewidth [mm/s]	h_p	U_p	i [c]	Relative area [d]
R-SWNT	Ia	0.13	-1021	1.3	0.45	0.45	-0.55	1
x=0.021	Ib	1.06	-1879	1.2	1.07	0.82	-0.18	0.35
	Ic	1.29	-2500	1.3	1.22	1.09	0.09	0.81
	Id	0.83	-2233	1.2	0.91	0.97	-0.03	0.21
	Ie	-0.35	0	1.3	0.13	0	-1	0.65
T-SWNT	Ia	3.68	-874	1.3	0.49	0.38	-0.62	1
x=0.027	Ib	4.27	-1742	1.2	0.88	0.76	-0.24	0.87
	Ic	4.92	-2538	1.0	1.31	1.11	0.11	1.35
	Id	4.40	-2182	1.1	0.97	0.95	-0.05	0.76
	Ie	3.18	0	1.2	0.16	0	-1	0.19

a) The isomer shift relative to ZnTe in R-SWNT and to Mg_3TeO_6 in T-SWNT. b) The quadrupole coupling constant converted to the value of ^{127}I. c) The charge estimated from $i = U_p - 1$. d) The absorption area relative to Ia.

Based upon Townes and Dailey's treatment [8], the isomer shift (δ) and the quadrupole coupling constant (e^2qQ) are empirically expressed by the electron population of the iodine atom. When an iodine atom makes bonds with neighboring atoms, the bonding interaction leaves holes in the valence shell from full $5s^25p^6$ configuration of iodide ion (I^-). The observed δ in Mössbauer spectra is related with the numbers of holes in 5s-shell (h_s) and in 5p-shell (h_p) as follows:

$$\delta = -9.2\, h_s + 1.5\, h_p - 0.54 \qquad (1)$$

in mm/s unit relative to ZnTe. The δ of Mg_3TeO_6 source is shifted by 3.49 mm/s relative to ZnTe. The e^2qQ indicates the electric field gradient at the iodine nucleus. The electric field gradient arises from the imbalance of p-electron (U_p):

$$U_p = h_z - (h_x + h_y) / 2, \qquad (2)$$

where h_x, h_y and h_z are the holes in $5p_x$, $5p_y$ and $5p_z$ orbital; hence, $h_p = h_x + h_y + h_z$. If it is assumed that the electric field gradient arises only from the U_p, the e^2qQ converted to ^{127}I value in MHz unit is obtained by

$$e^2qQ = U_p\ (-2292.7). \qquad (3)$$

If π-bonding is assumed to be negligible, i.e. $h_x = h_y = 0$, the asymmetry parameter (η), which is given by

$$\eta = 3\,(h_x - h_y) / (2\, U_p), \qquad (4)$$

becomes zero. Hence $U_p = h_p$. Furthermore, if it is assumed that s-electron is not used in the bonding, i.e. $h_s = 0$, the charge on iodine atom (i) is estimated as

$$i = h_p - 1 = U_p - 1. \qquad (5)$$

We adopted the i value from U_p, because the error of fitting for e^2qQ is less than that for δ. The obtained values for each species are shown in Table I.

The spectra were successfully fit with these species assuming η to be near zero. This fact indicates that the assumption is adequate and the iodine is almost linearly aligned. Among these species, Ia, Ib, and Ic, whose charges are about -0.6, -0.2, and +0.1, respectively, are considered as constituents of triiodides (I_3^-) and/or pentaiodides (I_5^-) as in following configurations: Ia-Ic-Ia (I_3^-) and Ib-Ic-Ia-Ic-Ib (I_5^-). The other species, Id is considered to be iodine molecules (I_2); and Ie is monoiodide ions (I^-). The ratio of iodine forms and transferred charges are summarized in Table II.

Table II. Ratios of iodine forms and transferred charge.

Sample	Doping level x in CI_x	Ratios of iodine forms in atomic percent $I_3^- : I_5^- : I_2 : I^-$	Total transfered charge q in C^{+q} $[I_x]^{-q}$	Transferred charge per iodine atom q/x
R-SWNT	0.046	30 : 38 : 21 : 11	0.013	0.29
	0.021	42 : 30 : 7 : 21	0.0086	0.41
T-SWNT	0.053	8 : 63 : 29 : 0	0.0081	0.15
	0.027	19 : 59 : 17 : 4	0.0060	0.22

We revealed that the iodine in SWNTs in these doping level exists in the forms of I_3^-, I_5^- and I_2. The fact that iodine exists in the forms of polyiodides indicates that a charge transfer has occurred by iodine doping, where iodine acts as electron acceptors, as in conducting polymers. Moreover, I^- was proved to appear additionally. This fact had not been found in other spectroscopy so far. It is also revealed that, as the doping level decreases, the proportions of I_3^- and I^- increase, whereas those of I_5^- and I_2 decrease. Therefore, when doping level increases, total transferred charge increases while transferred charge per iodine atom decreases. This tendency is in common even if the synthesis method of SWNT is different. However, the ratios of iodine forms have a little difference between R- and T-SWNT. The iodine in T-SWNT is more likely to be as the form of I_5^-, while iodine in the form of I_3^- exists more likely as R-SWNT. Since the diameter of SWNT differs in two kinds of samples, it is considered that iodine configuration is a little different morphologically. Moreover, it may be due to remaining catalysts, since remaining catalysts in T-SWNT are said to be more than those in R-SWNT.

References

[1] S. Iijima, Nature 354 (1991) 56.

[2] For example, M.S. Dresselhaus, G. Dresselhaus and P.C. Eklund, *Science of Fullerenes and Carbon Nanotubes* (Academic Press, San Diego, 1996) Chap. 19.

[3] T. Matsuyama, H. Sakai, H. Yamaoka, Yu. Maeda and H. Shirakawa, J. Phys. Soc. Jpn. 52 (1983) 2238.

[4] S. Kitao, T. Matsuyama, M. Seto, Yu. Maeda, Y.F. Hsia, S. Masubuchi and S. Kazama, Hyperfine Interact. 93 (1994) 1439.

[5] M. Seto, Yu. Maeda, T. Matsuyama, H. Yamaoka and H. Sakai, Synth. Met. 55-57 (1993) 3167.

[6] L. Grigorian, K.A. Williams, S. Fang, G.U. Sumanasekera, A.L. Loper, E.C. Dickey, S.J. Pennycook and P.C. Eklund, Phys. Rev. Let. 80 (1998) 5560.

[7] A.G. Rinzler, J. Liu, H. Dai, P. Nikolaev, C.B. Huffman, F.J. Rodriguez-Macias, P.J. Boul, A.H. Lu, D. Heymann, D.T. Colbert, R.S. Lee, J.E. Fischer, A.M. Rao, P.C. Eklund, R.E. Smalley, Appl. Phys. A 67 (1998) 29.

[8] R.V. Parish, in *Mössbauer Spectroscopy Applied to Inorganic Chemistry*, Vol.2, ed. G. J. Long (Plenum Press, New York, 1987) Chap. 9.

Acknowledgements

This work was partly supported by Grant-in-Aid for Encouragement of Young Scientists no. 12740180 from the Ministry of Education, Culture, Sports, Science and Technology of Japan.

^{197}Au MÖSSBAUER STUDY OF Au NANO-PARTICLES

Y. Kobayashi [a], M. Seto [a], S. Kitao [a], R. Haruki [a] and T. Teranishi [b]
[a] *Research Reactor Institute, Kyoto University, Kumatori, Osaka 590-0494, Japan*
[b] *School of Material Science, Japan Advanced Institute of Science and Technology, Tatsunokuchi, Ishikawa 923-1292, Japan*

Abstract: The ^{197}Au Mössbauer measurement have been performed on Au nano-particles surrounded with the surfactant. We take the spectra of the nano-particles with the reference, and obtain the recoilless fraction from the area ratio of the subspectra of the nano-particles and the reference. From the temperature dependence of the recoilless fractions, the Debye temperatures are obtained as 110, 120 and 160 K on 2.4, 3.3 and 5.7 nm nano-particles, respectively. The reduction of Debye temperature is due to the inter-particle vibrations.

Introduction

The Mössbauer measurement is useful technique for the study of nano-particles [1]. The recoilless fraction is important parameter on Mössbauer spectroscopy. It contains important information of the motions of the lattices and particles. Some kind of Au small particle supported on metal oxide shows catalytic activity [2]. On catalytic activity, the lattice vibration plays an important role in absorption and dissociation of molecule. In this time, we measured the recoilless fraction of unsupported nano-particles for the basic data of the supported particles [3].

On the study of nano-particles, the uniformity of particle size is important. In these days, we can obtain the well size controlled nano-particles by chemical method [4]. The particles were made by precipitation in the solution including surfactant. The particles were overlaid with surfactant, so the particles are usually stable at room temperature. It is convenient for measurements.

Experimental Procedure

The Au nano-particles were prepared by the reduction of $H(AuCl_4)\cdot 4H_2O$ with $NaBH_4$ in the presence of dodecanethiol (DT) after transferring $AuCl_4^-$ from water to toluene phase with tetraoctylammonium bromide [4]. The sizes of particle were controlled by the temperature of solution and the strength of acid. The average diameters and standard deviations of the specimens are 2.4 nm (σ = 0.54 nm), 3.3 nm (σ = 0.20 nm) and 5.7 nm (σ = 0.69 nm). These diameters were obtained from the observation using an electron microscope. The specimens are clayey on 2.4 nm and 3.3nm, and powdery on 5.7 nm. This difference is due to the weight ratio of the particles and the surfactant.

For ^{197}Au Mössbauer measurement, the γ-ray source were ^{197}Pt in Pt metal (98% enriched with ^{196}Pt) prepared by the neutron irradiation by Kyoto University Reactor, and the absorbers were the

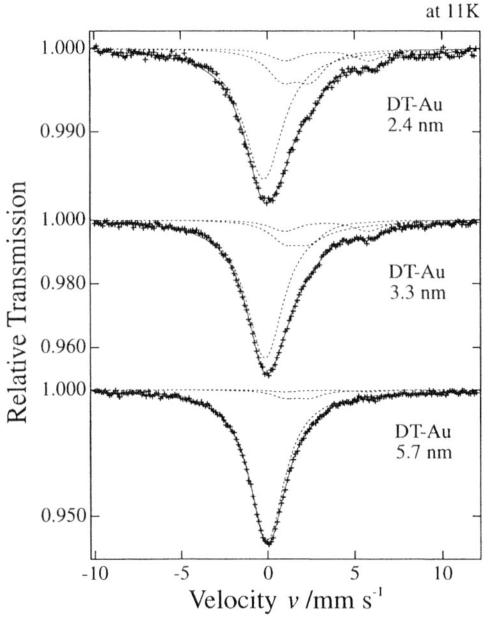

Fig. 1. ^{197}Au Mössbauer spectra of the Au nano-particles overlaid with dodecanethiol (DT).

nano-particle specimens. The source and the specimens were cooled by the helium refrigerator. Temperatures of the specimens were controlled by the heater attached on the specimen holder and cold head of the refrigerator. The zero velocity positions of the spectra are peak positions of pure bulk Au.

There are some difficulties for measurement of the recoilless fractions using short lifetime source, for example, a saturation of a detector, an approximations of background noise. Thus, we used the reference material to measure the recoilless fraction of the nano-particle. The reference was $Na(AuCl_4)\cdot 2H_2O$. From the point of stability, the Au foil is better for the reference. However, the nano-particle specimens include zero isomer shift subspectrum, and the spectrum from Au foil reference masks this subspectrum. Thus we use $Na(AuCl_4)$ for the reference. The Debye temperature of $Na(AuCl_4)$ reference is obtained 120K by the comparison with the spectrum of bulk Au foil.

Results and Discussion

Fig.1 shows the observed spectra of the specimens at 11 K. Each spectrum includes one singlet having nearly zero isomer shift and two doublet having positive isomer shifts. The doublets are due to the Au atoms on the surface of nano-particles. Considering the effect from the ligand of

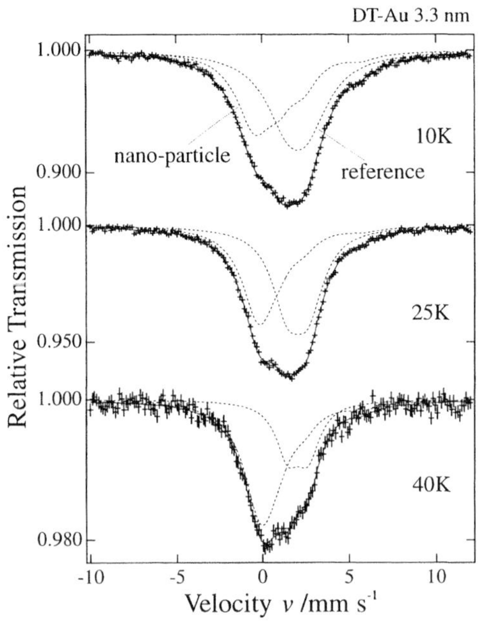

Fig. 2. Temperature dependence of the ^{197}Au Mössbauer spectra of the 3.3 nm Au nano-particles and reference ($Na(AuCl_4)$).

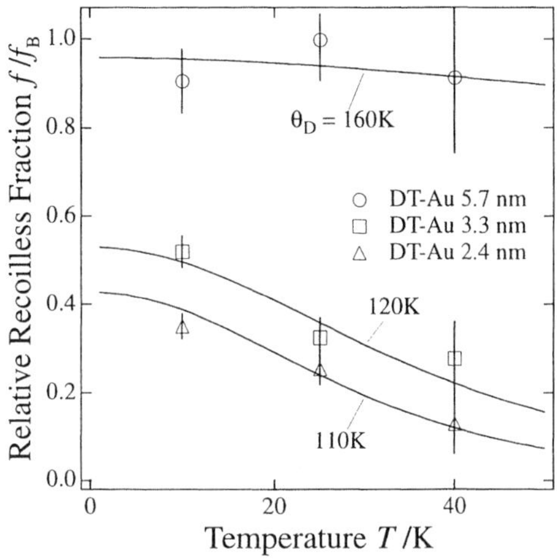

Fig. 3. Temperature dependence of the relative recoilless fractions of the nano-particles. f_B is the recoilless fraction of bulk Au.

the surfactant, the subspectrum having larger quadrupole splitting is owing to the Au atoms on surface coordinating with the ligand and another is owing to the Au atoms on the surface not coordinating them. The zero isomer shift singlet is owing to the Au atoms not on the surface; it is due to the inner shell of the nano-particles. The area ratio of each component shows good agreement with the Au atoms ratio obtained from the particle size.

Hori *et al.* reported magnetic moments on Pd and Au nano-particles [5]. However, the obvious magnetically split was not observed on this measurement. The hyperfine magnetic field is too small to observe a splitting.

Fig. 2 shows the temperature dependence of the spectra of 3.3 nm nano-particles and reference. Before this measurement, we observed the spectra of the nano-particle without the reference at same temperature. The area ratio of the reference decreases with increase of temperature. This result shows the Debye temperature of the nano-particle is higher than the reference. The area ratios between nano-particle and the reference are easily converted to the nano-particle and Au foil using the result of the reference and Au foil. Fig. 3 shows the relative recoilless fraction ratio of the nano-particles (f) and bulk Au (f_B). The 5.7 nm diameter particles show bulk like Debye temperature, and smaller particles show lower Debye temperature. Buffat *et al.* reports the decrease of the melting point on small particles [6]. Their result suggests the softening of the lattice of the particles. In our Mössbauer results, the area ratio of the subspectra of nano-particles changes almost equally; the Debye temperatures of whole nano-particles are decrease. The subspectra from the inner shells have nearly zero isomer shifts, so these components show bulk like electronic states. However, the recoilless fraction of this component is much lower than bulk. It is difficult to conclude that the softening of lattices does not change the electronic state. This decrease of Debye temperature is due to the inter-particle vibrations. Smit *et al.* discuss the effect of inter-particle and intra-particle vibration [7]. Unfortunately, our data are not enough temperature points to apply their model.

Conclusion

We obtained the Debye temperatures of the nano-particles. Applying the normal Debye model, the Debye temperatures are 110, 120 and 160K on 2.4, 3.3 and 5.7 nm nano-particles, respectively. The Debye temperatures decrease with the particle size. This decrease is mainly due to the inter-particle vibrations. The obvious magnetically split was not observed on this measurement.

References

[1] M. P. A. Viegers and J. M. Trooster, Phys. Rev. B, **15** (1977) 72.
[2] Haruta, N. Yamada, T. Kobayashi and S. Iijima, J. Catal., **115** (1989) 301.
[3] Kobayashi, S. Nasu, S. Tsubota and M. Haruta, Hyperfine Int, **126** (2000) 95.
[4] T. Teranishi, H. Hori, M. Miyake, J. Phys. Chem. B, **101** (1997) 5774.
[5] H. Hori, T. Teranishi, Y. Nakae, Y. Seino, M. Miyake, S. Yamada, Phys. Lett. A, **263** (1999) 406.
[6] Ph. Buffat and J-P. Borel, Phys. Rev. A, **13** (1976) 2287.
[7] H. H. A. Smit, R. C. Thiel and L. J. de Jongh, Z. Phys. D, **12** (1989) 193.

Mössbauer study of $Sm_2Fe_{17-x}Si_x$.

A. Nandra, C. Djega-Mariadassou, L. Bessais,
LCMTR, UPR209 CNRS, 2/8 rue Henri Dunant, B.P. 28, F-94320 Thiais, France

J. M. Grenèche,
Laboratoire de Physique de l'état condensé, UMR CNRS 6087, Université du Maine, F-72085 Le Mans, France

and E. Burzo
Facultatea de Fizica, Universitatea Babes-Bolyai, 3400 Cluj-Napoca, Romania

September 10, 2001

Abstract. $Sm_2Fe_{17-x}Si_x$ ($x \leq 2$) prepared by mechanical alloying and subsequent annealing. They are single phase with the $R\bar{3}m$ Th_2Zn_{17}-type structure. The best Rietveld agreement factor assesses for a preferential occupation of Si on the $18h$ site. These results are corroborated by the Mössbauer spectroscopy measurements obtained from 295K down to 77K. The Mössbauer spectra analysis is based on the correlation between the Wigner-Seitz volume and the isomer shift evolution taking into account the statistical distribution of Si in the $18h$ site.

1. Introduction

R_2T_{17} (R = rare-earth, T = transition metal) compounds still attract much attention since it was shown that interstitial elements such as C and N could improve their magnetic properties [1–3]. Noticeable improvements can be obtained by partial substitution of p elements for Fe. Si, Al, Ga atoms appear to play a key role in the stabilization of carbides with, simultaneously, an increase of Curie temperature T_C and make the carbides promising candidates for permanent magnets [4]. This paper is devoted to the special case of the Si substitution for which many basic ambiguities have appeared in recent years. It is found that c parameter is increasing [5] or decreasing with Si content [6]. Moreover, the nature of the site occupied by Si in the crystallographic $R\bar{3}m$ Sm_2Fe_{17} structure is not clearly established. These fundamental discrepancies must be cleared up to improve the understanding of the bonding and the magnetic properties. We have thus decided to investigate the structural properties of the $Sm_2Fe_{17-x}Si_x$ alloys from careful powder x-ray diffraction measurements with Rietveld refinements and ^{57}Fe Mössbauer spectrometry studies.

2. Experiment

The samples were prepared by high-energy ball milling (5 h) and subsequent annealing (30 min at 1423 K). The x-ray patterns were recorded by means of a

Bruker diffractometer. The unit cell parameters were measured with Si as the standard (a = 0.54308 nm) leading to a unit cell parameter accuracy of ± 1 x 10^{-4}nm.The data treatment was carried out by a Rietveld refinement from the FULLPROF computer code. The magnetic ordering temperatures T_C were determined using a differential sample magnetometer under an applied field of 1000 Oe with sample ca. 10 mg sealed in vacuum in small silica tubes in order to prevent oxidation of the sample during heating. The Mössbauer spectra were obtained from room temperature down to 77 K. The resulting spectra were least-square fit as discussed below. The estimated errors are ± 0.1 T for hyperfine fields (H_{HF}), ± 0.01 mm/s for isomer shifts (δ) and ± 0.01 mm/s for quadrupole interaction (2ϵ).

3. Results and discussion

The x-ray diffraction diagrams of the $Sm_2Fe_{17-x}Si_x$ series show that the sample are all of rhombohedral Th_2Zn_{17}-type structure (space group : $R\bar{3}m$). No iron-based additional phase is observed as expected from the small initial Sm excess. Both unit cell parameters c and a contract with x quasi-linearly. A simple steric effect of Fe substitution (r_{Fe} = 0.127 nm) for a smaller atom can nevertheless explain such behaviour here if one takes into account the Si covalent radius equal to 0.117nm. We have based the Rietveld analysis on several considerations as following: (i) The Si $6c$ site location has been systematically ignored due to the relative abundance of the Fe-dumbbell site appearing systematically at the positive velocity side of all Mössbauer spectra as it will be shown below. The corresponding sextet area of 12% and 13% for x equal to 0.5 and 2 is consistent with this assumption in agreement with a lack of silicium in this site. (ii) The $9d$ crystallographic site with the smallest Wigner-Seitz cell (WSC) volume in the Sm_2Fe_{17} phase, clearly far below the $18f$ and $18h$ sites, can be avoided by Si with a high probability since larger sites are available. It follows that Si can enter the medium WSC volumes, with close values, i.e. the f and h sites constituted by the nearest-neighbor to d sites. (iii) Both $18f$ and $18h$ sites have the same number of nearest neighbors. Consequently, we have chosen to refine the occupancy parameters of those sites, keeping in mind that their respective WSC volumes are close. The refinement shows that Si atoms mainly prefer $18h$ sites up to $x = 2$ in agreement with the theoretical results [7, 8] on $Nd_2Fe_{17-x}X_x$. The Curie temperature increases with Si concentration. The increase in Curie temperature of 180 K is achieved in $Sm_2Fe_{14}Si_3$ compared to the compounds without Si.

Fig. 1a shows the experimental spectra as a function of x at 77 K as an example. The same features are observed on spectra recorded at temperature range 77 K to 300 K. They are complex spectra, constituted of broadened sextets resulting from the convolution of numerous sub-spectra. No amount of extra α(Fe-

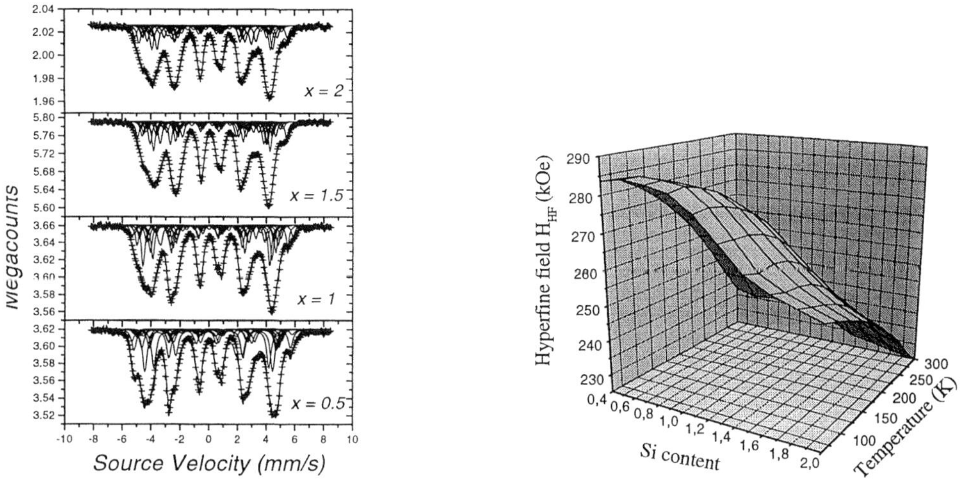

Figure 1. (a) The 77 K Mössbauer spectra of $Sm_2Fe_{16}Si$. (b) The dependence of the mean hyperfine field on temperature and Si concentration.

Si) is observed and a detailed analysis of the spectra is possible since a pertinent model is proposed.

The relative areas of the sextets were calculated according to the statistical distribution of Si in $18h$ site. Due to the basal orientation of the magnetization in Sm_2Fe_{17}, for the $9d$, $18h$ and $18f$ sites, the angle between the hyperfine field and the electric-field gradient will be non-zero. Each of these sites will split into two magnetically non-equivalent sites, with an abundance ratio 2:1.

In the fitting procedure, as the samples were random powders, the line intensities of the sextets were assumed to obey the 3:2:1 relationship in agreement with the lack of texture deduced from the x-ray diffractograms. One common line-width of 0.27 mm/s was used for all sextets. The Lamb-Mössbauer absorption factor was assumed to be the same for all sites. We neglected the site abundances lower than 1.5%. All along the fitting process, the site abundances were maintained as fixed parameters while δ, H_{HF} and 2ϵ were considered as free. In the first step of the procedure, the hyperfine field assignment to one crystallographic family was based upon the physical fact that the highest fields would belong to $6c$ sites and the lowest to $9d$, according to their respective theoretical abundances. In a second step, the mean δ value was attributed to each of the hyperfine fields taking into account the correlation between δ and WSC volume, and the fit was performed with all parameters free, except the abundances.

For a given $x \geq 1$, the sequence observed for the isomer shift is $9d < 18h < 18f < 6c$, while for $x < 1$ one can see that $\delta\{18f\} < \delta\{18h\}$, which is an agreement with WSC volume calculation. With increasing x, one notices that the less the number of $18h$ neighbors, the less the increase of δ, which is in agreement with our predictions. The quasi-constancy of $18h$ corroborates the location of Si in the $18h$ site. The mean hyperfine field decreases with x as shown in Fig. 1b. The hyperfine field is dominated by the Fermi contact term. This contains a positive term H_{4s} arising from a self-polarisation of $4s$-conduction-electrons, and a negative core-electrons ($1s$, $2s$, $3s$) polarisation term H_{ce} produced by $3d$ electrons via exchange interaction. H_{ce} decreases slightly with Si content due to a magnetic dilution effect, while H_{4s} increases with the Fe moment per atom. Qualitatively, it means that the resulting hyperfine field decreases.

4. CONCLUSIONS

The substitution of Fe for Si in $Sm_2Fe_{17-x}Si_x$ ($x \leq 2$) alloys leads to a contraction of both unit cell parameters a and c. The Rietveld analysis shows that Si atoms are located in $18h$ sites. The Mössbauer spectra from 77 K up to 300 K were analyzed on the basis of the statistical distribution of Si atoms. The Wigner-Seitz cell volumes have been calculated with the rhombohedral description of the unit cell in order to distinguish the Si atom position from the Fe. The site assignment of the hyperfine parameters was ruled by the correlation between the Wigner-Seitz cell volume and the isomer shift δ. As a result for $x \geq 1$, the δ sequence changes from $9d < 18f < 18h < 6c$ to $9d < 18h < 18f < 6c$. With increasing x, it appears that the isomer shift of $18h$ site remains quasi constant. The $18h$ site is the less affected by Si substitution. This fact corroborates that Si atoms are located in the later site in agreement with the Rietveld analysis. The magnetic hyperfine field obeys to the following classification $6c > 9d > 18f > 18h$ in agreement with the number of Fe neighbors in the corresponding site. The decrease with x of the weighted average hyperfine field can be explained by the competition between two Fermi contact terms of opposite signs: the negative core-polarization term reflects the Fe moment behavior and the positive term corresponds to the conduction electrons contribution enhanced by the Si substitution.

References

1. D. B. de Mooj, K. H. J. Buschow, and A. M. Tsvelik, J. Less-Common. Met. **142**, 349 (1988).
2. M. Katter, J. Wecker, L. Schultz, and R. Grossinger, J. Magn. Magn. Mater. **92**, L14 (1990).
3. J. M. D. Coey and J. Sun Hong, J. Magn. Magn. Mater. **87**, L251 (1990).
4. Z. W. Li and A. H. Morrish, J. Phys. Condens. Matter **6**, L283 (1994).
5. Z. W. Li, X. Z. Zhou and A. H. Morrish, Phys. Rev. B **51**, 2891 (1995).
6. C. Lin,Y. X. Sun, Z. X. Liu, H. W. Jiang and Z. H. Liu, IEEE Trans. Magn. **MAG-28**, 2844 (1992).
7. W. Y. Ching and Ming-Zhu Huang, J. Appl. Phys. **79**, 4602 (1996).
8. M. Z. Huang, W. Y. Ching and Z. Q. Gu, J. Appl. Phys. **81**, 5112 (1997).

Charge state of rare earth ions and glass transition in the glassy $Al_{85}Ni_{10}R_5$ alloys

Hideya Onodera*, Takaomi Itoi†, Kentaro Indoh, Takeshi Takamizawa, Aya Tobo and Akihisa Inoue
Institute for Materials Research, Tohoku University, Sendai 980-8577, Japan

September 3, 2001

Abstract. A glassy $Al_{85}Ni_{10}La_5$ alloy has a supercooled-liquid region of 16 K between a glass transition temperature at 526 K and a crystallization temperature at 542 K, while $Al_{85}Ni_{10}Eu_5$ has no supercooled-liquid region and crystallizes at 514 K. ^{151}Eu Mössbauer spectroscopic study reveals that the Eu ions in $Al_{85}Ni_{10}Eu_5$ are divalent, although the charge state of La ions is naturally considered to be trivalent in $Al_{85}Ni_{10}La_5$. It is supposed that the thermal stability of this glassy alloy system strongly depends on not only the ionic radius of rare earth element but also the electronic state.

Keywords: glassy alloy, Mössbauer effect, charge state, glass transition, supercooled liquid, spin glass

1. Introduction

Glassy alloys with a wide temperature region of supercooled-liquid state have attracted significant interest, since these alloys have a high glass-forming ability which was found in the several alloy systems. In these alloys, the wider the temperature range of the supercooled-liquid state, the more stable the amorphous state against crystallization. This ability makes it possible to produce large-scale bulk glassy alloys by casting at low cooling rates (Inoue et al., 1991; Inoue et al., 1997; Inoue and Zhang, 1995; Kui et al., 1984). Our interest have been aroused by the appearance of wide supercooled-liquid regions not only for practical application but also from the viewpoint of fundamental comprehension as to why such glassy alloys have the remarkable thermal stability against crystallization.

The glassy Al-Ni-La alloy system has been found to have wide supercooled-liquid regions up to 60 K around 55%La and 20%Ni (Inoue et al., 1990; Inoue et al., 1989). This alloy system has another concentration region around 5%La and 10%Ni at which the supercooled-liquid state exists in a region of 20 K (Inoue et al., 1990). In the present work,

* Corresponding author. e-mail address: onodera@imr.edu.

† Present address: Dep. Electr. & Electron. Engn. Kitami Institute of Technology, Kitami, 090-8507, Japan.

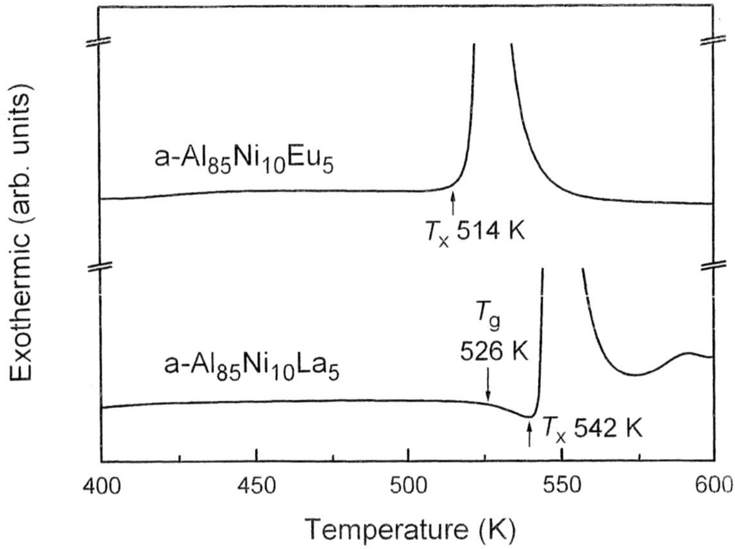

Figure 1. Differential scanning calorimetric curves of the glassy $Al_{85}Ni_{10}Eu_5$ and $Al_{85}Ni_{10}La_5$ alloys. The arrows show the crystallization temperature, T_x, and the glass transition temperature, T_g.

we succeeded to make the glassy alloys, $Al_{85}Ni_{10}La_5$ and $Al_{85}Ni_{10}Eu_5$. The glassy state of the two alloys were examined by differential scanning calorimetry (DSC), magnetic susceptibility experiment and ^{151}Eu Mössbauer spectroscopy.

2. Experimental

Alloy ingots were prepared by conventional arc-melting method using Al, Ni, La and Eu metals of 99.9% purity in an argon atmosphere. Glassy ribbons with cross sections of about 1.5×0.02 mm^2 were prepared from the melts by single-roller melt-spinning method in an argon atmosphere. The glassy states were examined by X-ray diffraction. The thermal stability associated with the glass transition and crystallization was examined by DSC at a heating rate of 0.33 K/s. The ^{151}Eu Mössbauer effect measurements were carried out at 4.2, 77.3 and 297 K with a sinusoidal velocity wave in transmission mode. A standard radioactive source $^{151}SmF_3$ of 3.7 GBq was used at room temperature. The DC and AC magnetic susceptibilities were measured using a SQUID magnetometer in the temperature range 2.0–350 K.

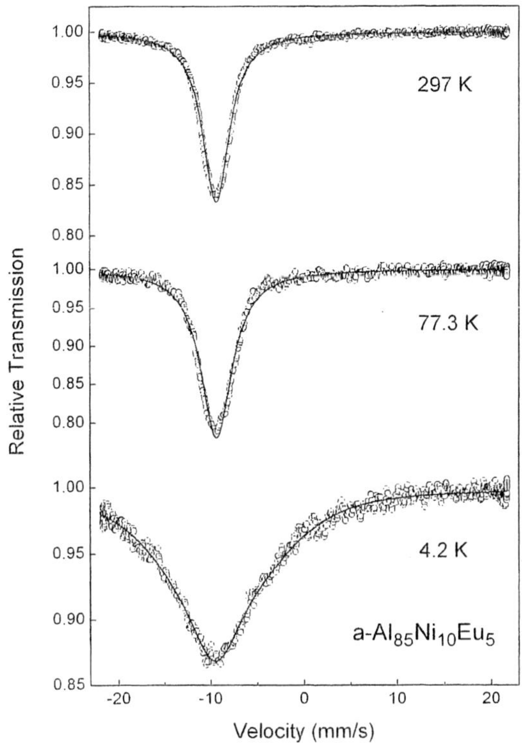

Figure 2. ^{151}Eu Mössbauer spectra of the glassy $Al_{85}Ni_{10}Eu_5$ alloy at 4.2, 77.3 and 297 K. The circles and solid lines shows the observed and fitted spectra, respectively.

3. Results and discussion

Figure 1 shows the DSC curves of the glassy $Al_{85}Ni_{10}R_5$ alloys (R=Eu and La). $Al_{85}Ni_{10}La_5$ undergoes a glass transition at T_g= 526 K and crystallizes at T_x= 542 K. The supercooled-liquid state exists in a region of T_x= 16 K which is narrower than the reported region of 20 K (Inoue et al., 1990). On the other hand, $Al_{85}Ni_{10}Eu_5$ starts to crystallize at T_x= 514 K with no glass transition.

The ^{151}Eu Mössbauer spectra at 4.2, 77.3 and 297 K are shown in Fig.2 where the circles and solid lines show the observed and calculated spectra, respectively. The spectra at 77.3 and 297 K are satisfactorily interpreted by a single Lorentzian absorption peak at −9.4 mm/s which is relative to EuF_3. The spectrum at 4.2 K is rather broader that those at 77.3 and 297 K. Although this broadening is attributable to magnetic hyperfine interactions, a single set of hyperfine parameters gives an unsatifactory analysis. Here, we tentatively analyze it by assuming two subspectra with the same isomer shift: the one is a singlet with a halfwidth of $\Gamma = 7.2$ mm/s, and the other is a Zeeman-splitted one of $\Gamma = 9.1$ mm/s and a magnetic hyperfine field of $H_{hf} = 13.4$ T. The isomer shift common to two subspectra is $\delta = -9.4$ mm/s

The glassy $Al_{85}Ni_{10}Eu_5$ alloy undergoes a spin-glass transition at T_{sg}= 3.3 K, where the DC magnetic susceptibilities below T_{sg} measured with field cooling (FC) differ from those with zero-field cooling (ZFC). The AC susceptibility experiments also give typical spin-glass behavior that the transition temperature T_{sg} depends on AC frequency. It is characteristic of spin glasses that the distributed magnetic hyperfine fields appears at temperatures above T_{sg} (Gurewitz and Atzmony, 1982; Ito et al., 1996). Therefore, it is supposed that the magnetic hyperfine fields in $Al_{85}Ni_{10}Eu_5$ distribute at 4.2 K with an averaged field of about 8 T.

The value of isomer shift definitely reveals that the Eu charge state is divalent in the glassy $Al_{85}Ni_{10}Eu_5$ alloy. Surely, the spin-glass transition originates in magnetic interactions between the Eu^{2+} ions with a magnetic moment of 7 μ_B, since the Pauli-paramagnetic behavior of $Al_{85}Ni_{10}La_5$ indicates that the Ni atoms carry no magnetic moment. The La atoms in $Al_{85}Ni_{10}La_5$ is considered to be trivalent, because the La ions in a metallic system are in general trivalent.

It is remarkable that $Al_{85}Ni_{10}La_5^{3+}$ has a supercooled-liquid region and $Al_{85}Ni_{10}Eu_5^{2+}$ has no supercooled-liquid region, that is, the thermal stability of this alloy system strongly depends on the charge state of a minor constituent. In glassy $Mg_{65}Cu_{25}R_{10}$ alloys, similarly, no supercooled-liquid region appears when R= Eu (Kim, 1992). However, $Mg_{65}Cu_{25}Yb_{10}$ in which the Yb ion is divalent has a supercooled-liquid region comparable with those of alloys with light rare earths, and its crystallization temperature is rather lower than those of other rare earths. Consequently, the stability of glassy state strongly depends on not only the ionic radius of the rare earth element but also the electronic state.

References

Inoue A., A. Kato, T. Zhang, S. G. Kim and T. Masumoto. *Mater. Trans. JIM*, 32, 609–616 1991.

Inoue A., N. Nishiyama and H. Kimura. *Mater. Trans. JIM*, 38, 179–183 1997.

Inoue A. and T. Zhang. *Mater. Trans. JIM*, 36, 1184–1187 1995.

Kui H. W., A. L. Greer and D. Turnbull. *Appl. Phys. Lett.*, 45, 615-616 1984.

Inoue A., T. Zhang and T. Masumoto. *Mater. Trans. JIM*, 31, 425–428 1990.

Inoue A., T. Zhang and T. Masumoto. *Mater. Trans. JIM*, 30, 965–972 1989.

Gurewitz E. and U. Atzmony. *Phys. Rev. B*, 26, 6093–6098 1982.

Ito A., S. Morimoto and H. A. Katori. *J. Phys. Soc. Jpn*, 65, 3331–3335 1996.

Kim S. G. *Doctor Thesis, Tohoku University*, 1992.

Magnetic Interactions between Nanoparticles of Different Materials.

Christopher W. Ostenfeld and Steen Mørup,
Department of Physics, Building 307, Technical University of Denmark,
DK-2800 Kgs. Lyngby, Denmark.

Samples consisting of nanoparticles of α-Fe_2O_3 and mixtures of α-Fe_2O_3 and NiO or CoO have been studied by Mössbauer spectroscopy in the temperature range 20-320 K. Nanoparticles of α-Fe_2O_3, which normally do not have a Morin transition above 5 K, exhibited a Morin transition when mixed with NiO nanoparticles. Pure hematite particles, coated with oleic acid, showed a typical superparamagnetic behaviour at higher temperatures, whereas a sample of uncoated particles was less affected by relaxation, indicating a strong influence of inter-particle interactions. Admixture of NiO nanoparticles resulted in a smaller influence of inter-particle interactions on the spectra of uncoated particles. Admixture of CoO nanoparticles did not induce a Morin transition in α-Fe_2O_3 particles, but unlike NiO particles, CoO nanoparticles significantly slowed down the superparamagnetic relaxation.

1. Introduction.

Exchange coupling between different magnetic materials is a subject which has attracted considerable attention because of the application of exchange coupled materials in, for example, spin valves. Most of the studies have been carried out on layered materials, in which the structure can be well controlled. Only a few studies of interacting nanoparticles of different magnetic materials have been published [1-3]. We have recently performed detailed studies of non-interacting and interacting hematite (α-Fe_2O_3) nanoparticles [4-6]. Below the Morin transition temperature ($\approx$ 263 K) bulk hematite is antiferromagnetic (AFM) with the sublattice magnetisation parallel to the <111>-direction [7]. Above this temperature, the sublattice magnetisation directions are in the (111)-plane with a small deviation from antiparallel orientation of the sublattices, resulting in a weakly ferromagnetic (WFM) state [7]. Several studies have shown that the Morin transition is absent in hematite nanoparticles such that they are in the WFM state at least down to 4.2 K. The present work is an extension of our previous work on interacting hematite nanoparticles [5], where in this work the influence of interaction with NiO and CoO nanoparticles on the magnetic properties of hematite nanoparticles is studied.

2. Experimental.

Nanoparticles of hematite with an average size of about 11 nm were synthesised by a modified gel-sol method developed by Sugimoto [8]. NiO nanoparticles were prepared by thermal decomposition of $Ni(OH)_2$ at 600 K for 3 h [9]. These NiO particles are disc-shaped with an average diameter of about 17 nm and a thickness of about 3 nm. CoO nanoparticles with an average particle size of about 12 nm were prepared by ball-milling a mixture of Co_3O_4 and metallic Co in the appropriate ratio

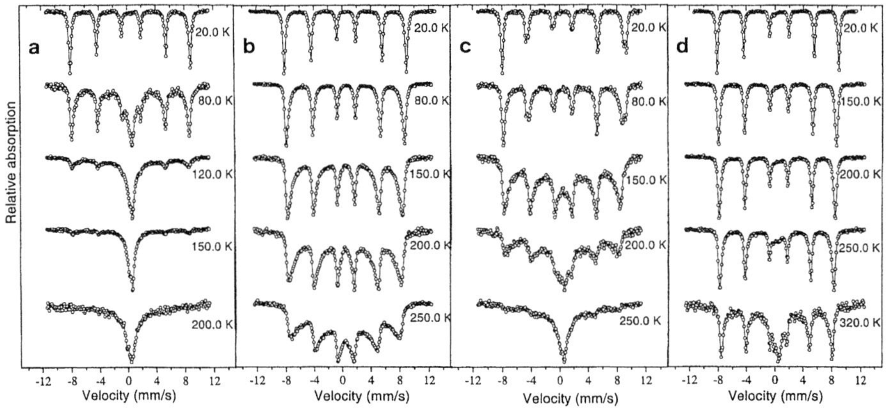

Figure 1. Mössbauer spectra of hematite particles at the indicated temperatures. (a): coated particles. (b): uncoated particles. (c): mixed with NiO nanoparticles. (d): mixed with CoO nanoparticles.

in an argon atmosphere. A sample of hematite particles was coated with oleic acid and suspended in heptane in order to minimise inter-particle interactions. Dry samples of pure hematite nanoparticles and mixtures of hematite and NiO or CoO particles (with weight ratio 1:1) were prepared by exposing aqueous suspensions of the particles to ultrasound followed by drying at room temperature. Mössbauer spectra were obtained in the temperature range 20-320 K.

3. Results and discussion.

Mössbauer spectra of pure hematite particles (coated and uncoated) and samples of mixtures of hematite nanoparticles and nanoparticles of NiO and CoO are shown in Fig. 1. At the lowest temperatures, the spectra of the pure hematite particles (a and b) consist of a sextet with parameters characteristic of the WFM state, i.e. there is no indication of a Morin transition in these samples, in accordance with earlier studies of hematite nanoparticles [4,6,7,10]. At higher temperatures, the spectra of the coated particles (a) show a typical superparamagnetic behaviour, i.e. a doublet due to rapidly relaxing particles appears and its relative area increases with increasing temperature. For $T \geq 150$ K, all the particles exhibit fast superparamagnetic relaxation. In the spectra of the sample of dried, uncoated hematite particles (b), the lines of the sextet become broadened with increasing temperature, but even at 250 K the sextet has only partially collapsed. The difference between the relaxation behaviour of the two samples can be understood as follows: In the sample of non-interacting nanoparticles, the relaxation is governed by an anisotropy energy barrier, and the relaxation time for magnetisation reversals is given by $\tau = \tau_0 \exp(KV/kT)$, where K is the magnetic anisotropy constant, V is the particle volume, k is Boltzmann's constant, and T is the temperature. With a typical value of τ_0 of the order of 10^{-11} s [6] and a typical particle size distribution, only a small fraction of the particles will have relaxation times of the order of a nanosecond, which give rise to line broadening. Therefore, the spectra consist of sextets and doublets with relatively sharp lines. If inter-particle interactions dominate in a sample, the interaction field due to interactions with neighbouring particles dominate the magnetic energy. In this case, the sublattice magnetisation direction may fluctuate rapidly in directions around that of the interaction field, and this leads a reduced hyperfine field and, because of the variations in interaction fields, to a broadening of the lines of the sextet [5,11], as seen in Fig. 1(b).

The spectra of the α-Fe_2O_3/NiO sample, Fig.1(c), show some new features. First, the low-temperature spectra consist of two sextets, indicating that some of the hematite has undergone a Morin transition (about 65 % of the spectral area at 20 K) such that this part of the hematite is in the AFM state. Studies of mixtures of nanoparticles of α-Fe_2O_3 and NiO with other particle sizes showed a similar effect. A close analysis of the spectra showed, however, that the quadrupole shift of the WFM component was close to zero, whereas it is about -0.1 mms^{-1} in bulk WFM hematite. Moreover, the lines of this component are broadened considerably (up to about 1 mms^{-1} for lines 1 and 6). Similar values for the quadrupole splitting have been reported in other studies of hematite particles [12]. Earlier studies have shown that the Morin transition at the surface and in the interior of hematite particles can take place at different temperatures [13,14]. Therefore, in a certain temperature range, the surface and the interior of the particles may be separated by domain walls with a magnetic structure, which gradually changes from WFM to AFM. In such a domain wall, the angle between the electric field gradient and the magnetic hyperfine field may vary continuously, and this may explain the values of the quadrupole shift as well as the line broadening. We also studied the sample at 80 K in applied magnetic fields up to 0.7 T, but neither the relative area of the WFM component nor the spectral parameters were affected by the external fields. At higher temperatures, e.g. at 250 K, the spectra of the α-Fe_2O_3/NiO sample are more collapsed than the spectra of the uncoated α-Fe_2O_3 sample. This indicates that the inter-particle interactions become weaker when the hematite particles are mixed with NiO particles.

The spectra of the α-Fe_2O_3/CoO sample, Fig.1(d), show other new features. In contrast to the α-Fe_2O_3/NiO sample, there is no indication of a Morin transition even at the lowest temperatures. At higher temperatures, the spectra of the α-Fe_2O_3/CoO sample differ from both the dried α-Fe_2O_3 sample and the α-Fe_2O_3/NiO sample. Instead of the gradual line broadening with increasing temperature, seen in the two other dried samples (Fig. 1 b and c), the α-Fe_2O_3/CoO sample shows a coexistence of a sextet and a doublet with relatively sharp lines above 200 K. This behaviour is more similar to the coated sample for which the magnetic relaxation is governed by the magnetic anisotropy energy barrier. However, the blocking temperature, T_B, (defined as the temperature at which 50% of the sextet has collapsed to a doublet) of the α-Fe_2O_3/CoO sample is above 320 K. This is much higher than that of the coated α-Fe_2O_3 sample ($T_B \approx 100$ K), and the relaxation is also more suppressed than in the dry α-Fe_2O_3 sample. The coexistence of a sextet and a doublet with relatively sharp lines in the spectra the α-Fe_2O_3/CoO sample suggests that fluctuations of the magnetisation directions in the α-Fe_2O_3 particles are governed by an anisotropy energy barrier rather than by an interaction field due to inter-particle interactions. A sample of α-Fe_2O_3 particles, which was exposed to Co ions during aqueous suspension, showed much faster relaxation. Therefore, the suppression of superparamagnetic relaxation in the α-Fe_2O_3/CoO sample seems not to be due to Co ions in the surface, which might increase the surface anisotropy. It might be expected that the suppression of relaxation due to magnetic inter-particle interactions should take place only below the Neél temperature of bulk CoO (291 K). However, the interaction of CoO with a magnetic iron oxide can increase the Neél temperature near the interface [15], and therefore magnetic interaction effects may be important also above 291 K. It is possible that a strong interaction of hematite particles with CoO particles, which have larger magnetic anisotropy, could result in a situation where the sublattice magnetisation directions of the hematite particles are bound to the sublattice magnetisation directions of the CoO particles, such that the relaxation of the hematite particles follows the relaxation of the CoO particles.

4. Conclusions.

The studies of samples of interacting α-Fe_2O_3, α-Fe_2O_3/NiO and α-Fe_2O_3/CoO nanoparticles show several interesting features. At low temperatures, interaction with NiO nanoparticles results in a Morin transition in the α-Fe_2O_3 nanoparticles. At higher temperatures, the presence of NiO particles results in faster relaxation of the α-Fe_2O_3 particles. Interaction with CoO nanoparticles does not introduce a Morin transition in the α-Fe_2O_3 particles, but slows down the superparamagnetic relaxation at higher temperatures in a way which suggests that CoO and α-Fe_2O_3 relax together in a way which is governed by the larger anisotropy of the CoO particles. Further studies are in progress in order to elucidate these phenomena.

Acknowledgement.

The work was supported by the Danish Technical Research Council.

References.

1. D.S. Geoghegan, P.G. McCormick and R. Street, Materials Science Forum, **179-181** (1995) 629.
2. J. Sort *et al.*, Appl. Phys. Lett., **75** (1999) 3177.
3. J. Sort *et al.* , J. Magn. Magn. Mater., **219** (2000) 53.
4. F. Bødker, M.F. Hansen, C. Bender Koch, K. Lefmann and S. Mørup, Phys. Rev. B, **61** (2000) 6826.
5. M.F. Hansen, C. Bender Koch and S. Mørup, Phys. Rev. B, **62** (2000) 1124.
6. F. Bødker and S. Mørup, Europhys. Lett. **52** (2000) 217.
7. A.H.Morrish, Canted Antiferromagnetism: Hematite (World Scientific, Singapore, 1994).
8. T. Sugimoto, Y. Wang, H. Itoh, and A. Muramatsu, Colloid Surfaces A: Physicochem. Eng. Aspects, **134** (1998) 265.
9. F. Bødker, M.F. Hansen, C. Bender Koch and S. Mørup, J. Magn. Magn. Mater., **221** (2000) 32.
10. W. Kündig, H. Bömmel, G. Constabaris, and R.H. Lindquist, Phys. Rev., **142** (1966) 327.
11. S. Mørup, M.B. Madsen, J. Franck, J. Villadsen, and C.J.W. Koch, J.Magn. Magn. Mater. **40** (1983) 163.
12. A.E. Verbeck, E. De Grave, and R.E. Vandenberghe, Hyperfine Interact., **28** (1986) 639.
13. O. Nicolov, T. Ruskov, T. Tomov, and A. Toshev, Hyperfine Interact., **39** (1988) 409.
14. E. De Grave, C. Dauwe, L.H. Bowen, and R.E. Vandenberghe, Hyperfine Interact. (C), **1** (1996) 286.
15. P.J. van der Zaag, *et al.* Phys. Rev. Lett. **84** (2000) 6102.

SYNTHESIS AND MAGNETIC PROPERTIES OF MAGNESIUM-DOPED $Li_{0.5}Fe_{2.5}O_4$ NANOCRYSTALLINE PARTICLES

H. M. Widatallah[1], F. J. Berry[2], A. M. Gismelseed[3], C. Johnson[2], M. Pekala[4] and A. A. Yousif[3]

[1]*Department of Physics, University of Khartoum, P.O. Box 321, Khartoum, Sudan.*
[2]*Department of Chemistry, Open University, Walton Hall, Milton Keynes, MK7 6AA, UK.*
[3]*Physics Department, Sultan Qaboos University, PO Box 36 Al Khod 123, Oman.*
[4]*Department of Chemistry, Warsaw University, Pl-02-089-Warsaw, Poland.*

The structure and magnetic properties of nanocrystalline Mg-doped-$Li_{0.5}Fe_{2.5}O_4$ particles synthesised by the ball milling of polycrystalline sample prepared at 600°C have been examined. The magnetisation of the nanoparticles was found to decrease with decreasing crystallite size. The Curie temperatures were found to be lower than those of the corresponding bulk mareials. These properties are discussed in terms of surface magnetism and superparamagnetism.

1-Introduction:

The synthesis of spinel-related ferrite nanoparticles has attracted growing interest due to their rather peculiar properties and technological potential [1-5]. Such nanoparticle materials are increasingly being processed by mechanical techniques [3-5]. We have recently synthesised Mg-doped-$Li_{0.5}Fe_{2.5}O_4$ at 600°C [6] and we report here on the structural and magnetic properties of nanoparticles synthesised by submitting this material to ball milling.

2- Experimental:

A mixture of hydrothermally prepared corundum-related Mg-substituted α-Fe_2O_3 [7] and Li_2CO_3 in a molar ratio of 7: 1 was calcined in air at 600°C (12h). This led to the formation of spinel-related Mg-doped-$Li_{0.5}Fe_{2.5}O_4$ with a cation ratio Li : Fe : Mg = 11 : 61 : 6. The Mg^{2+} ions were found to substitute for Fe^{3+} at tetrahedral A sites, thus changing the A- to B-site Fe^{3+} occupancy ratio from 1: 1.5 as in pure $Li_{0.5}Fe_{2.5}O_4$ to 1: 2 [6]. The material (hereafter S0) was dry-milled in air for different periods of time in a Retsch PM400 planetary ball mill with stainless steel vials (250 ml) and balls (20mm) at 200 rpm. The powder-to-ball weight ratio was 1: 20. Samples were labelled S1, S7, S15, S38 and S50 where the number indicates the number of hours of milling time. X-ray powder diffraction (XRD) patterns were recorded with a Siemens D5000 diffractometer using Cu$K\alpha$ radiation. Rietveld refinement of the X-ray powder diffraction patterns was performed using the program WinProf [8]. Magnetisation was measured with a Faraday balance in a magnetic field of 1.5T. ^{57}Fe Mössbauer spectra were recorded at 298K and 78.5K using a 25 mCi Co/Rh source. Isomer shifts are quoted relative to that of α-Fe at 298K.

3- Results and Discussion:

The XRD patterns of the Mg-doped $Li_{0.5}Fe_{2.5}O_4$ samples are shown in Figure 1The results show that the spinel phase is preserved in the milled samples and that milling does not induce any new phases. The decrease in the intensities of the reflection peaks and their broadening with increasing milling time reflects an increasing structural disorder and smaller particles. Rietveld refinement of the XRD patterns of all milled samples was performed using a model involving a crystalline spinel phase (core) with a bulk-like cation distribution [6] and a disordered spinel-related phase (surface). The respective volume ratio of both phases dramatically changed from 1.5: 1 for S1 (1h) to ~1: 4 for samples obtained at higher milling times signifying the important role the disordered surface may have in determining the properties of these samples. The average crystallite size <d> was determined from the 311-reflection peak using the Scherrer equation. The values obtained for <d> in each sample and the corresponding lattice constant *a* are given in Table 1. The increase of the lattice constant with decreasing crystallite size to 13nm (S38) may be related to the surface disordered structure and to strain within the particles [2].

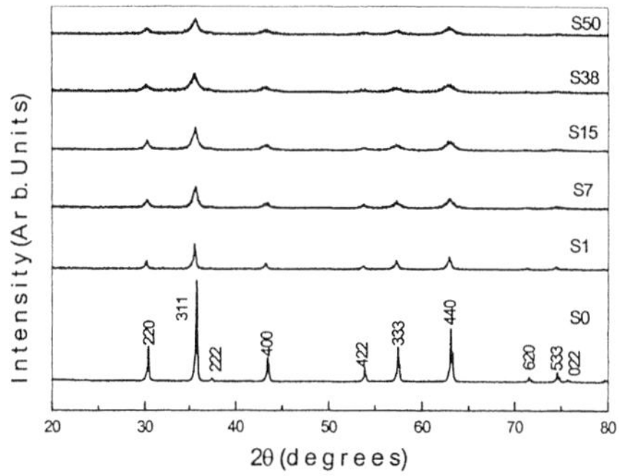

Figure 1: X-ray powder diffraction patterns of the pristine (S0)and milled Mg-doped-$Li_{0.5}Fe_{2.5}O_4$ samples (S1-S5).

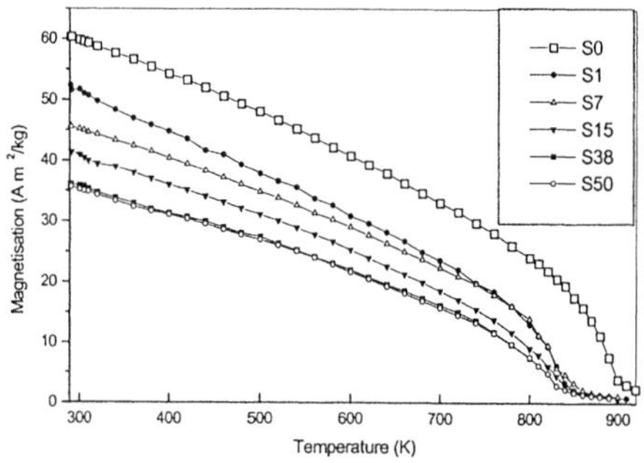

Figure 2: The temperature variation of the magnetisation of the pristine (S0)and milled Mg-doped-$Li_{0.5}Fe_{2.5}O_4$ samples (S1-S50)

Table 1: Average crystallite size <d> and lattice constant *a* of the Mg-doped-$Li_{0.5}Fe_{2.5}O_4$ samples

Sample	S0	S1	S7	S15	S38	S50
<d> (nm)	1135	38	22	17	13	11
a (Å)	8.3410	8.3416	8.3451	8.3490	8.3517	8.3469

Figure 2 shows the temperature variation of the magnetisation for the different Mg-doped $Li_{0.5}Fe_{2.5}O_4$ samples. At room temperature the magnetisation decreases with decreasing crystallite size from 60.3 Am^2/kg for S0 (1135nm) to 35.6 Am^2/kg for S50 (11nm). The magnetisation also decreases monotonically with increasing temperature. The Curie temperature T_c of the nanocrystalline particles was reduced to ~ 840K relative to 900K for the non-milled sample S0. To elucidate the origin of these properties, we have recorded the Mössbauer spectra of the Mg-$Li_{0.5}Fe_{2.5}O_4$ nanoparticles at 298K and 78.5K. The results are shown in Figure 3 (a) and (b), respectively. The corresponding hyperfine parameters are given in Table 2. The spectrum at 298K for each sample was fitted by the superposition of two partially unresolved magnetic sextets and a central doublet. The two sextets are assigned to Fe^{3+} at the A and B sites. The quadrupole splitting for both sextets is negligible and their isomer shifts are slightly higher than those obtained for the non-milled sample [6] which probably is due to more Fe atoms in or near the surface layer with coordination different from that of the core. Generally the magnetic hyperfine fields decrease with decreasing crystallite size which could be associated with the reduction in atomic neighbours of the Fe^{3+} surface ions. Such a surface effect should also result in asymmetric line shapes [9]. This asymmetry becomes less pronounced in the spectra as the crystallite size gets smaller suggesting that in addition to the aforesaid surface effect, the hyperfine field reduction is partially due to a size effect. This can be explained in terms of the collective magnetic excitations model [10].

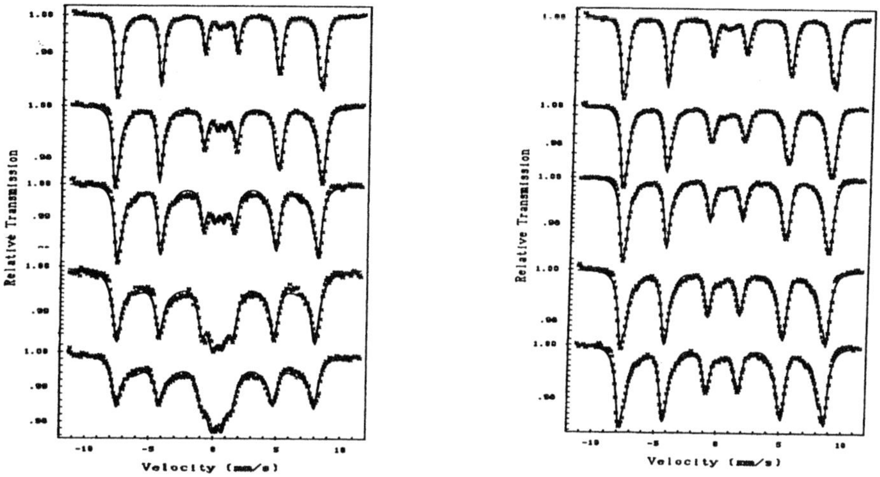

(a) (b)

Figure 3: The Mössbauer spectra of the nanocrystalline Mg-doped-$Li_{0.5}Fe_{2.5}O_4$ samples at 298K (a) and 78.5K (b). The top-to-bottom order of the samples is S1, S7, S15, S38 and S50.

Table 2: Mössbauer parameters of the nanocrystalline Mg-doped-$Li_{0.5}Fe_{2.5}O_4$ samples at 298K and 78.5K. (δ =Isomer shift; Q.S.= Quadrupole splitting; H= Hyperfine magnetic field).

Temp. (K)	Sample	Sextet A		Sextet B		Area	Central doublet		
		δ (mm/s)	H(T)	δ (mm/s)	H(T)	(A+B) %	δ (mm/s)	Q.S. (mm/s)	Area (%)
	S1	0.25	48.4	0.30	50.2	95.7	0.30	0.51	4
	S7	0.28	48.4	0.32	50.4	94.0	0.32	0.54	6
298	S15	0.29	46.0	0.33	48.9	89.1	0.32	0.67	11
	S38	0.28	40.5	0.36	48.2	84.4	0.32	0.77	16
	S50	0.28	40.7	0.35	47.9	80.7	0.32	0.80	19
	S1	0.34	51.2	0.40	53.4	96.7	0.33	0.48	3
	S7	0.37	50.2	0.44	52.5	97.4	0.27	0.59	3
78.5	S15	0.40	49.2	0.45	51.7	97.6	0.32	0.63	2
	S38	0.37	48.0	0.40	51.2	98.2	0.34	0.49	2
	S50	0.37	47.3	0.40	51.0	98.6	0.36	0.49	2

According to this model the magnetisation is thermally excited to oscillate about the direction of the minimum energy leading to a particle-size-dependent hyperfine field which is smaller than that of the bulk. The central doublet in the Mössbauer spectrum whose area increases from 4% for sample S1 (38nm) to 19% for sample S50 (11nm) results form superparamagnetic relaxation of particles having "effective" volumes with blocking temperatures T_B<298K [2,4,11].

In the Mössbauer spectra recorded at 78.5K (Figure 3(b)) the superparamagnetic relaxation is almost suppressed. The spectra consist of two magnetic sextets whose lines broaden as the crystallite size decreases. Some residual relaxation effects are still present as evidenced by the very weak unresolved doublet, which may be due to particles with superparamagnetic blocking temperatures lower than 78.5K. The differences between the isomer shift values for both sextets at 78.5K and 298K are due to the second order Doppler shift [12]. The broadening of the spectral lines with decreasing crystallite size may be related to a distribution of quadrupolar interactions arising from the disorder in the [3]. The ratio of the Fe^{3+} ions on the A site to those on the B site in the nanocrystalline particles obtained from the areas of the Mössbauer sextets at 78.5K varied between ~1: 1.45 to 1: 1.2 relative to 1: 2 for the non-milled sample [6]. This is indicative of milling-induced cationic movement between both sites in the disordered spinel-related surface. As the magnetic moments on each site are antiparallel with the dominant contribution to magnetisation coming from the B site, the substitution of Fe^{3+} ions on that site by Mg^{2+} ions, as reflected in the above ratios, explains the decrease in the magnetisation and Curie temperature.

Acknowledgement: HMW thanks SIFNS (Sudan), SIDA (Sweden) and the Abdus Salam ICTP for financial and research support. MP is supported by the Polish research grant No. 120-501/SPUB-77 (Poland).

References:

[1] A. Goldman, Modern Ferrite Technology, Van Nostrand Reinhold, New York, 1990.
[2]H. Yang, Z. Wang, M. Zhao, J. Wang, D. Han, H. Luo and L. Wang, Mater. Chem. Phys., 48 (1997) 60.
[3] M. Mahmoud, H. H. Hamdeh, J. C. Ho, M. J. O'Shea and J. C. Walker, J. Magn.Magn. Mater. , 220 (2000) 139.
[4]G. F. Goya, J. Mater. Sci. Lett., 16 (1997) 563.
[5]H. Guerault and J-M Greneche, J. Phys.: Condens. Matter, 12 (2000) 4791.
[6]H.M. Widatallah, C. Johnson, F. J. Berry and M. Pekala, Solid State Commun. (in press).
[7]F. J. Berry, C. Greaves, J. McManus, M. Mortimer, G. Oates, J. Solid State. Chem., 130 (1997) 272.
[8]A. Jouanneaux, "WinProf", Int. Union Crystallogr. Newsletter, 21 (1999) 13.
[9] K. Haneda and A. H. Morrish, Phys. Lett., 64A (1977)259.
[10]S. Morup and H. Topsoe, Appl. Phys., 11 (1976) 63.
[11] J. M. D. Coey and D. Khalafalla, Phys. Stat. Sol. (a) 11 (1972) 229.
[12] V. Sepelak, K. Tkacova, V. V. Boldyrev, S. Wibmann, K. D. Becker, Physica B, 234-236 (1997) 617.

Crystallization Process of α-Fe/$Nd_2Fe_{14}B$ Type Nanocomposite Magnet Alloys

M. Yamasaki[a], M. Hamano[b] and T. Kobayashi[c]

[a] *Toda Kogyo Corporation, Otake, Hiroshima 739-0652, Japan*
[b] *Mikuni-Makino Industrial Co., Ltd., Ashigarakami-gun, Kanagawa, 258-0019. Japan*
[c] *Department of Physics, Shiga University of Medical Science, Otsu, Shiga 520-2192, Japan*

The crystallization process of α-Fe/$Nd_2Fe_{14}B$ type nanocomposite magnet alloys has been studied with the x-ray diffraction method and Mössbauer spectroscopy. Melt-spun alloy ribbons with the compositions of $Nd_8Fe_{81}Co_4Nb_{0.5}B_{6.5}$ and $Nd_9Fe_{79}Co_4Nb_{1.5}B_{6.5}$ were annealed at 600, 650, 700 and 750 °C for the former, and at 700 °C for the latter, respectively. In the case of $Nd_8Fe_{81}Co_4Nb_{0.5}B_{6.5}$, the α-Fe and $Nd_2Fe_{14}B$ phases appear in the very early stage of the crystallization process, that is at 650 °C. On the other hand, the alloy of $Nd_9Fe_{79}Co_4Nb_{1.5}B_{6.5}$ annealed at 700 °C has fewer α-Fe and more Fe_3B phases compared with the case of $Nd_8Fe_{81}Co_4Nb_{0.5}B_{6.5}$. Discussions taking notice of the atomic ratio of Fe in the above alloys lead us to the conclusion that α-Fe and $Nd_2Fe_{14}B$ are crystallized at comparatively low annealing temperature when the Fe ratio is high in the alloys, while fewer α-Fe and more Fe_3B phases appear when the Fe ratio is low.

Keywords: nanocomposite magnet, Mössbauer spectroscopy, crystallization process, intermediate metastable phase

1. Introduction

Nanocomposite magnetic materials which are mixtures of nanometer-scale soft and hard phases have relatively high coercivity with high remanence due to the magnetic exchange coupling between these minute grains. This magnetic exchange coupling arises effectively through an intergranular phase between the soft and hard grains. Nanocomposite magnetic materials of α-Fe/$Nd_2Fe_{14}B$ type are usually prepared by heat-treating the amorphous alloys. In this case, it is known that the growth of soft and hard grains can be suppressed by adding excess Nd atoms and a small amount of Nb atoms, and consequently magnetic performance is enhanced. In order to develop a new material with improved magnetic performance, therefore, it is decisively important to elucidate the formation process of the soft and hard phases and the intergranular phase as well as the magnetic properties of these phases.

Many studies have been carried out for investigating the formation process of the soft and hard phases and the intergranular phase [1,2], and it has been observed that some intermediate metastable phases, such as t-Fe_3B and $Nd_2Fe_{23}B_3$, appear at the early stage of heat-treating of the alloys and they decompose into $Nd_2Fe_{14}B$ and α-Fe at higher temperatures. We have also obtained a similar result in the recent work of the alloys with the composition of $Nd_8Fe_{80}Co_4Nb_{1.5}B_{6.5}$ [3]. In the present study, we investigate the influence of the atomic composition of alloys on the formation process of the intermediate phases.

2. Experimental

The melt-spun alloys with the compositions of $Nd_8Fe_{81}Co_4Nb_{0.5}B_{6.5}$ (A1) and $Nd_9Fe_{79}Co_4Nb_{1.5}B_{6.5}$ (A2) were prepared by the single roller melt-spinning process in a similar way to our previous work [4]. The crystallization process of the melt-spun ribbons, which were as quenched and heat-treated at 600, 650, 700 and 750 °C for the composition A1 and at 700 °C for the composition A2, was followed by Mössbauer spectroscopy. Crystallization of the ribbons was confirmed by x-ray diffractometry (Fe-Kα).

3. Results and Discussions

3.1. X-ray Diffraction Analysis

The x-ray diffraction patterns of the alloys with the A1 composition shown in Fig. 1 indicate that α-Fe is formed already before heat-treating the ribbon. This is reflected, in the Mössbauer spectrum of the as-quenched ribbon, in the lines at around ±5.5 mm/s. Indistinct diffraction peaks due to $Nd_2Fe_{14}B$ and Fe_3B are noticed at 600 °C, and the peaks due to Fe_3B grow conspicuously at 650 °C. They decrease at higher temperatures, while the peaks due to α-Fe and $Nd_2Fe_{14}B$ increase.

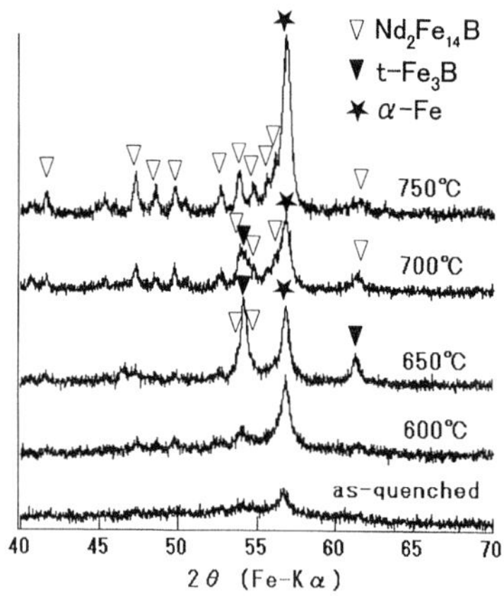

Fig.1 X-ray diffraction patterns of the alloys with the A1 composition.

The diffraction patterns of the alloys with the A2 composition change also with temperature in a similar way, but the relative contribution of α-Fe to the diffraction patterns is limited compared with the case of A1.

3.2. Mössbauer Analysis

The Mössbauer spectra of two specimens as-quenched and heat-treated at 600 °C with the A1 composition were analyzed by supposing a sextet of α-Fe, for the as-quenched alloy, and several sextets corresponding to α-Fe and $Nd_2Fe_{14}B$, for the heat-treated alloy, besides two components with hyperfine-field distribution corresponding to the amorphous phases in the alloys. No reasonable result was obtained at 600 °C from the analysis supposing the sextets due to Fe_3B.

Several constraints were put on the analysis of the Mössbauer spectra. The relative-intensity ratios of the sextets were assumed to be 3:2:1:1:2:3, that is, a random orientation of nanocrystalline grains was assumed. The widths of all sextets were assumed the same. The isomer-shift (δ) and quadrupole-splitting (Δ) of α-Fe were fixed to zero mm/s referring to α-Fe metal at room temperature. The intensity ratios of the components corresponding to the Fe sites $16k_2$, $16k_1$, $8j_2$, $8j_1$, 4c and 4e in the $Nd_2Fe_{14}B$ phase were fixed to 16:16:8:8:4:4 according to the ratio of Fe occupancy. There are three Fe sites in Fe_3B phase [5] and, in this study, the ratio of line-intensity of the Fe sites in the Fe_3B phase were assumed as 1:1:1 in proportion to the number of Fe sites. It was reported that the intensity ratio of Fe_3B in bulk is different from 1:1:1 [5], but the situation in the present study is quite different from that of the bulk case: the crystal growth is insufficient and/or the grains are fine. In the present study, we can, therefore, find no definite reason to shift the ratio from 1:1:1.

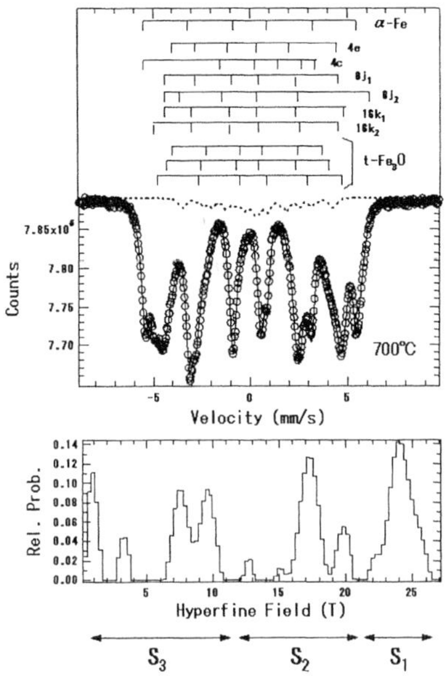

Fig. 2 Mössbauer spectrum of the alloy with the A1 composition heat-treated at 700 °C.

The patterns in Fig. 1 shows that the alloys are almost crystallized at 650-750 °C, but the Mössbauer spectra are not simulated well by supposing the sextets only due to α-Fe, $Nd_2Fe_{14}B$ and Fe_3B. They have other complicated components as shown in Fig. 2 for the case of the alloy heat-treated at 700 °C with the A1 composition. The line-positions of the sextets due to α-Fe, $Nd_2Fe_{14}B$ and Fe_3B are given in the bar diagrams above the spectrum. The other complicated components are represented by a broken line in the spectrum, the field distribution of which is shown below the spectrum. This distribution is far from that due to a typical amorphous phase, and may suggest the existence of several layers on the surface of grains of α-Fe, $Nd_2Fe_{14}B$ and Fe_3B, that is the intergranular phase, as studied in our previous work [4]. The spectra were analyzed again by supposing, besides the sextets due to α-Fe, $Nd_2Fe_{14}B$ and Fe_3B, further two sextets S1 and S2 representative of the components in the field ranges of 21-36 T and 11-21 T, respectively, and a component with hyperfine-field distribution S3 below 11 T.

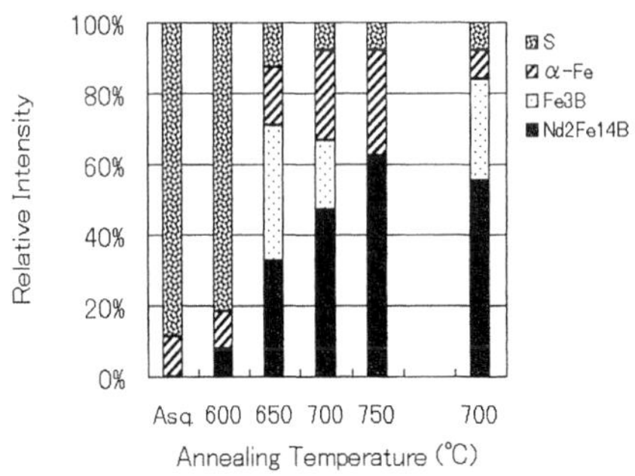

Fig. 3 Relative intensities of the components in the Mössbauer spectra for the A1 and A2 (the right end) compositions.

The relative intensities of α-Fe, $Nd_2Fe_{14}B$, Fe_3B and S(=S1+S2+S3) obtained for the compositions A1 and A2 are shown in Fig. 3 as a function of heat-treated temperature. The results shown in Fig. 3 are consistent with the x-ray diffraction patterns. Similar results obtained in our recent study for the alloys with the composition of $Nd_8Fe_{80}Co_4Nb_{1.5}B_{6.5}$ (A3) [4] are shown in Fig. 4. At 600 °C, some crystallized grains are formed in the alloy, but we cannot say in which phase of α-Fe, $Nd_2Fe_{14}B$ and Fe_3B the grains are probably because the grains grow insufficiently. Figures 3 and 4 show that the temperature at which α-Fe and $Nd_2Fe_{14}B$ are formed is lower for A1 compared with that for A3. In the case of A1, α-Fe is already formed before heat-treating and $Nd_2Fe_{14}B$ starts to crystallize at 600 °C, while they are not formed in the A3 alloy until the alloy is heat-treated at 650 °C. The intensity of $Nd_2Fe_{14}B$ is almost the same in the A1 and A3 alloys at 750 or 760 °C, but the A1 alloy has more α-Fe phase and, therefore, less S phase compared with the A3 alloy.

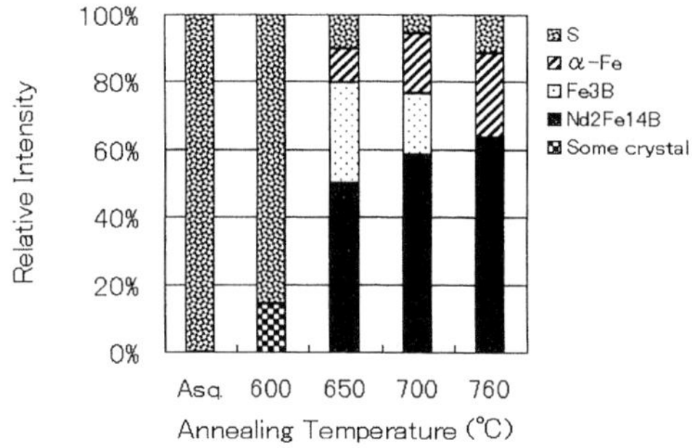

Fig. 4 Relative intensities of the components in the Mössbauer spectra for the A3 composition.

At 700 °C, the α-Fe phase is less and the $Nd_2Fe_{14}B$ and Fe_3B phases are more in the A2 alloy than in the A1 alloy. Comparing with the A3 alloy, the α-Fe phase is less and the Fe_3B phase is more in the A2 alloy, but the intensity of $Nd_2Fe_{14}B$ is almost similar in A2 and A3 alloys.

It is easy to understand these results as a whole when we focus on the ratio of iron atoms in the alloys. Comparing with the A3 alloy, the A1 alloy has less Nb atoms, i.e., more Fe atoms. In this case, the α-Fe phase crystallizes in the first stage of heat-treating and, subsequently, the $Nd_2Fe_{14}B$ phase does. On the other hand, there is less α-Fe and more Fe_3B phases in the A2 alloy having more Nd, i.e., less Fe, than the A3 alloy. Summarizing all the results obtained in this study, it is suggested that the relative amount of α-Fe and Fe_3B phases in a heat-treated alloy is closely related to the amount of Fe atoms more than the amount of Nd and Nb atoms in the alloy.

References

[1] A. Inoue, A. Takeuchi, A. Makino and T. Masumoto, Mater. Trans., JIM, **36-5**(1995)676.
[2] Yandong Gao, Shaoqing Zhang and Bocao Liu, J. Magn. Magn. Mater., **208**(2000)158.
[3] will be published somewhere.
[4] T. Kobayashi, M. Yamasaki and M. Hamano, J. Appl. Phys., **87**(2000)6579.
[5] T. Hinomura, S. Nasu, H. Kanekiyo and S. Hirosawa, J. Japan Inst. Metals, **61**(1997)184.

Mössbauer and small polaron hopping conduction studies of alkali borogermanate glasses containing iron

M M El-Desoky , J Lipka[1] and M Y Hassaan[2]

Physics Department, Faculty of Education, Suez Canal University, El-Arish,Egypt.
[1]Department of Nuclear Physics and technology, Faculty of Electrical Engineering and Information Technology, Ilkoviova 3, Bratislava, 81219, Slovak Republic.
[2] Physics Department, Faculty of Science, Al-Azhar University, Nasr City,Cairo, Egypt.

Mössbauer, DTA and dc conductivity of glasses in system $Na_2B_4O_7 - GeO_2 - Fe_2O_3$ prepared by press quenching was studied at temperatures between 323 and 573 K. `$T_g$-Δ rule` i.e. a linear relationship between the glass transition temperature (T_g) and Mössbauer quadrupole splitting (Δ) is applied to the glass samples. A slope of the straight line obtained from the plot of T_g vs Δ (i.e. 798°C/ mm $s^{-1})^{-1}$ for the glass samples implies that the Fe^{3+} atoms substitute tetrahedral GeO_4 atoms and play a role of network former. A linear decrease of Δ proves that the symmetry of distorted FeO_4 tetrahedra was linearly increased with Fe_2O_3 content. The dc conductivity was found to be decrease as the Fe_2O_3 content was increased while the activation energy increases with decreasing Fe_2O_3 content. From the conductivity temperature relation, it was found that small polaron hopping model was applicable at temperature above 350K ; the electrical conduction at T > 350K was due to nonadiabatic small polaron hopping of electrons between iron ions.

1.Introduction

Mössbauer spectroscopy is probably the most useful technique in structural studies of iron in a glass matrix. The Mössbauer effect provides essentially local information on the electric and magnetic field at a nuclear site Fe^{57} and the mean square velocity and displacement of the resonating nucleus. The isomer shift (δ) and quadrupole interaction (Δ) are determined mainly by the immediate neighbours, so it should be possible to discover something of the bonding coordination number and local structure around the Fe ions using this technique.

Oxide glasses containing transition metal ions with different valence states are categorized as amorphous semiconductors. The electronic conduction of the glass is describable in terms of the small polaron hopping conduction where an electron in the transition metal ion with lower valence state transfers to that with higher valence state by the aid of the optical phonon [1,2]. The mechanism of hopping conduction at high temperatures was discussed quantitatively by Austin and Mott [3], who proposed a theoretical relation between conductivity and temperature by considering the probability that an electron moves from one site to the adjacent site with the aid of the phonons.

The purposes of the present study were three fold. The first was to study the effect of Fe_2O_3 concentration on the Mössbauer parameters and the relationship between the quadrupole splitting (Δ) and the glass transition temperature (T_g). The second was to investigate the effect of Fe_2O_3 concentration on the electrical conductivity. The third was to clarify the mechanism of electrical conduction for $Na_2B_4O_7 - GeO_2 - Fe_2O_3$ glasses and to discuss the Fe_2O_3 concentration dependence of the conduction.

2- Experimental

The structureless x-ray diffraction (XRD) spectra obtained from a Philips PW 1710 diffractometer suggest that the samples are indeed amorphous in nature. Differential thermal analysis (DTA) studies of these samples using a Shimadzu 50 thermal analyzer give the glass transition temperature(T_g). The glass samples were pulverized for measuring the Mössbauer spectra by the constant acceleration method, and they were fitted with discrete doublets program. ^{57}Co (Rh) was used as a gamma ray source.

Glass samples of compositions (75) $Na_2B_4O_7$ – (25-x) GeO_2 – xFe_2O_3 were prepared from the reagent grade chemicals $Na_2B_4O_7.10H_2O$, GeO_2 and Fe_2O_3 for x = 0-25 mol%. The appropriate mixtures of these chemicals in 10 g batches were melted for 1 h in air in platinum crucibles in an electrical furnace at temperatures in range 1100-1150°C depending on compositions. The melts were then quenched by pressing between two brass plates.

Electrical conductivity measurements at 323 to 573K were conducted on samples that were polished and electrodes of silver paste were painted on both faces of the polished samples. The conductivity measurements were carried out by using a Chen Hwa 1061 Lcz meter in the frequency range 40 Hz to 200 kHz. Ohmic contact between the glass sample and the electrode was confirmed for all measurements. The conductivity was obtained using the complex impedance plot. The experimental error of the common logarithm of the conductivity value was estimated to be ± 0.05.

3.Results and Discussion

Translation of the fragments in the supercooled liquid becomes frozen at T_g and then a solid state "glass" is formed. Temperature dependence of physical properties changes remarkably at T_g. A linear relationship, termed '$T_g - \Delta$ rule' [4,5] exists between the quadrupole splitting (Δ) of Fe^{3+} which occupied substitutional sites of network former (NWF) in several oxide glasses. The '$T_g - \Delta$ rule' is expressed by

$$T_g = a\,\Delta + b \qquad (1)$$

The Δ value is in proportion to the distortion of Fe^{3+} site, since it has a symmetric electron configuration of $3d^5$ in the outer-most orbital and hence the contribution of these valence electrons on the electric field gradient can be ignored at the nuclear site. The 'T_g -Δ rule' implies that T_g of oxide glass is in proportion to the distortion of NWF oxygen polyhedra e.g., FeO_4 and FeO_6 [6].

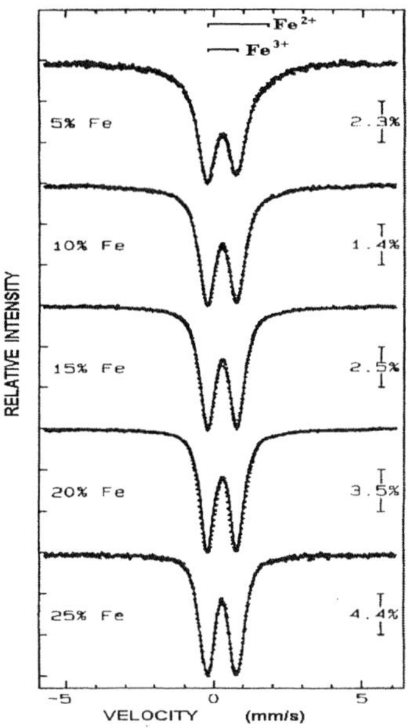

Fig.1 Room temperature Mössbauer spectra of the glass samples .

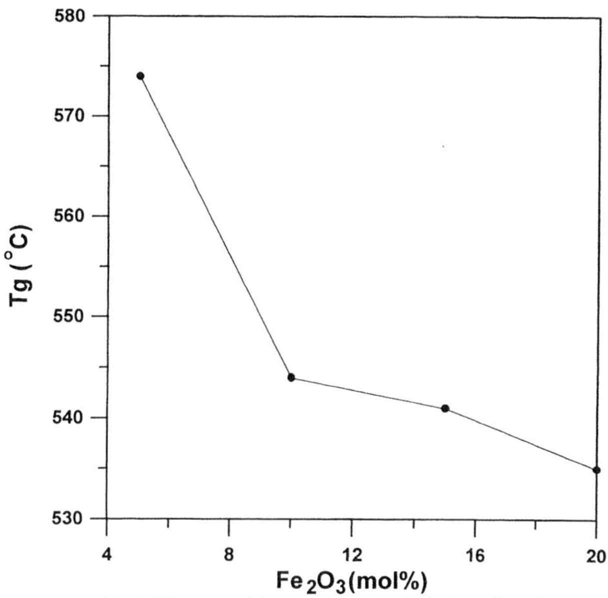

Fig.2 Glass transition temperature (T_g) plotted against Fe_2O_3 content.

Fig.1 shows the Mössbauer spectra of the glass samples containing increasing content of Fe_2O_3. The analysis of these spectra revealed that each spectrum has two doublets. One of them represents the Fe^{3+} in the tetrahedral state [$\delta = 0.262 - 0.279$ and $\Delta = 1.04 - 0.82$ mm s^{-1}]. The other doublet represents the Fe^{2+} site [$\delta = 0.814 - 0.95$ and $\Delta = 2.002 - 2.992$ mm s^{-1}]. The spectra showed also that as the Fe_2O_3 content was increased, the Fe^{2+} state decreased and the Fe^{3+} state increased gradually. In the sample containing 25 mol% of Fe_2O_3 the Fe^{2+} state disappeared and replaced by F^{3+} in the octahedral state [$\delta = 0.378$ and $\Delta = 1.306$ mm s^{-1}].

The glass transition temperatures T_g ,were determined using a heating rate 20 °C min^{-1}. Fig.2 shows T_g occurs values as a function of Fe_2O_3 lie between 574 and 535 °C. From the figure it is clear that a step by step decrease of T_g when the Fe_2O_3 content is increased gradually. This is may be due to an increased fraction of GeO_4 tetrahedra and to nonbridging oxygen (NBO) in the GeO_4 tetrahedra, as observed previously [7,8].

Change of the Mössbauer parameters for the glass samples are shown in Fig.3. The value of isomer shift (δ) for the glass samples increases almost linearly from 0.262 to 0.277 mm s^{-1} when Fe_2O_3 content increases from 5 to 20 mol%. The quadrupole splitting values (Δ) of the glass samples decrease gradually from 1.047 to 1.013 mm s^{-1}, reflecting a gradual decrease in the distortion of FeO_4 tetrahedra [6]. On the other hand, a linear decrease of Δ proves that the symmetry of distorted FeO_4 tetrahedra was linearly increased with Fe_2O_3 content. Symmetry of GeO_4 tetrahedra will also be increased with the Fe_2O_3 content, since Fe^{3+} preferentially occupies the sites of GeO_4 tetrahedra in alkali germante glasses [9] we can understand from Fig. 3 that the step by step substitution of Fe_2O_3 for GeO_2 results in an decrease in distortion of GeO_4 tetrahedra.

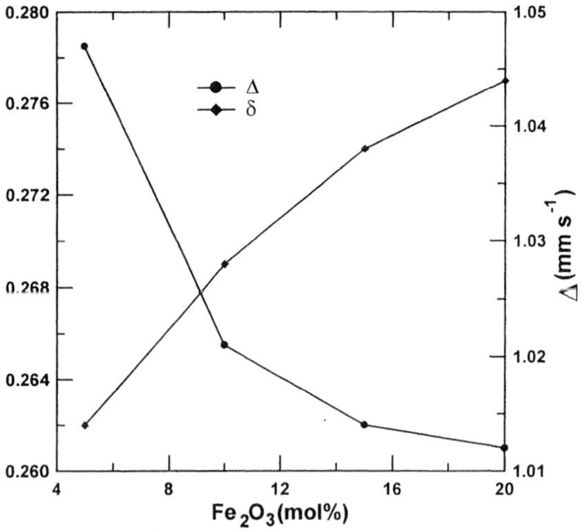

Fig.3. Plot of isomer shift (δ) and quadrupole splitting (Δ) against Fe_2O_3 content.

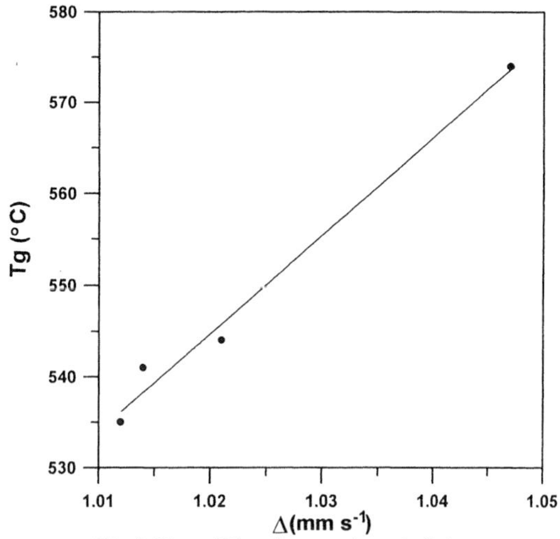

Fig.4. Plot of T_g values against Δof glass Samples.

The site occupation of Fe^{3+} in these glasses can be determined from the Tg – Δ rule. The linear relationship between T_g and Δ is simply expressed by Eq. (1). It was proved that the slope, a, and the intercept, b, were 798 and -548 °C/(mm s^{-1}) respectively, when Fe^{3+} occupies the NWF sites. The plot of T_g values against the Δ of Fe^{3+} is illustrated in Fig. 4. The value of 'a' obtained from this figure suggests that Fe^{3+} exists at substitutional sites of tetrahedral GeO_4 as NWF, having highly covalent Fe^{3+}- O bonds [5]. Values of line width, being in a range of 0.783 – 0.524 mm s^{-1}, indicate that distorted FeO_4 tetrahedra constitute the 'glassy' phase.

These experimental results manifest that Fe^{3+} atoms occupy tetrahedral NWF sites, irrespective of the Fe_2O_3 content. It is generally known that the network structure of oxide glass, e.g., the coordination number of NWF, fraction of NBO, dimension of the network, etc., is regulated by the alkali metal oxide content [10]. The linear change of T_g and Δ, illustrated in Fig. 4 does not reflect such a change of network, but does the local distortion of the network forming teterahedra i.e., FeO_4, BO_4, and GeO_4.

The dc conductivity (σ) of $Na_2B_4O_7$ – GeO_2 – Fe_2O_3 system were measured. The temperature dependence of dc conductivity is shown in Fig.5. It is found from this figure that dc conductivity is dependent on the amount of Fe_2O_3 and decrease with increasing Fe_2O_3 content. It is also observed from the figure that for all glass compositions the dc conductivity shows an activated behaviour above 350 K and below this temperature the conductivity exhibits a nonlinear characteristic, indicating a temperature dependent activation energy, which decreases with the decrease in temperature.

It is also observed that at all temperatures the conductivity of the glass composition containing GeO_2 and Fe_2O_3 is higher and that of the glass composition containing GeO_2 is lower than that of the glass composition containing B_2O_3 and Fe_2O_3. These observations clearly indicate that the glass forming oxides other than the B_2O_3 contribute significantly in the conduction processes.

The marked decrease in the dc conductivity will be associated with the increased distortion of the structural units, which caused a decreased probability of the electron (polaron) hopping from Fe^{2+} to Fe^{3+} . Increase of the energy level for the distorted Fe^{3+} will increase the activation energy for the polaron hopping.

According to Austin and Mott [3] for the small polaron hopping conduction the electrical conductivity and temperature for the glass containing TMO is related by equation

$$\sigma=[\nu_o R^2e^2C(1-C)/kT]\exp(-2\alpha R)\exp(-W/kT) \qquad (2)$$

where ν_o is the optical phonon frequency, α is the wave function decay constant, C is the ratio of the ion concentration in the low valency state to the total concentration of the transition metal ions, R the is hopping distance and W is the activation energy where

$$W = W_H + W_{D/2} \text{ for } T > \theta_D/2$$
$$= W_D \text{ for } T > \theta_D/4 \qquad (3)$$

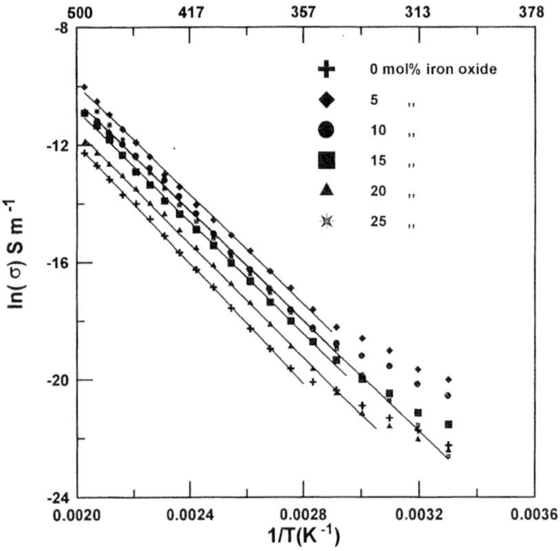

Fig.5. The dc conductivity plotted as a function of inverse temperature for the glass samples .

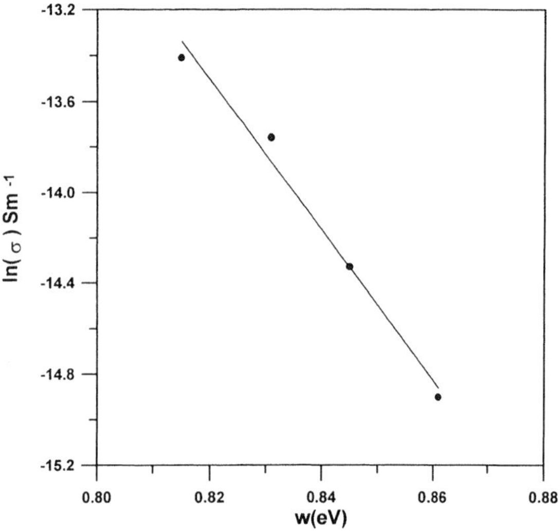

Fig.6. The dc conductivity plotted as a function of activation energy for the glass samples.

Where W_H is the polaron hopping energy and W_D is the disorder energy difference of the neighbours between two hopping sites and, θ_D is the Debye temperature given by $k\theta_D = h\nu_o$ .A typical value of W_D calculated from theory, is 0.1 eV.

When the overlap integral between sites exp $(-\alpha R)$ approaches 1, the hopping is the adiabatic and it is mainly controlled by the activation energy. The dc conductivity is given by

$$\sigma = [\nu_o R^2 e^2 C(1-C)/kT]\exp(-W/kT) \qquad (4)$$

To examine the nature of hopping conduction (adiabatic or non-adiabatic), the condition given by Sayer and Mansingh [11] is applied. The exploration of Eq.4 is done by plotting ln σ versus W (Fig. 5) at a fixed temperature (T=423 K) for various glass compositions (Fig.6). This plot shows a straight line nature, then it indicates that Eq. (4) is valid. The temperature T_e estimated from the slope of such a plot would be close to T if the conduction occurs in the adiabatic regime. However, if T_e is very different from T, the conduction is nonadiabatic. It is observed from Fig. 6 that the estimated temperature T_e = 342 K is very different from T = 423 K and thus conduction may be in the nonadiabatic regime.

Conclusions

A linear relationship was obtained between the glass transition temperature and quadrupole splitting (Δ). A slope of the straight line obtained from the plot of T_g vs Δ (i.e. 798°C/ mm $s^{-1})^{-1}$ for the glass samples implies that the Fe^{3+} atoms substitute tetrahedral GeO_4 atoms and play a role of network former. Electrical conduction of the glasses was confirmed to be nonadiabatic small polaron hopping.

References

[1] N.F. Mott, Conduction in Non-Crystalline Materials (Clarendon Press, Oxford, 1987); J. Non. Cryst. Solids 1(1968) 1.
[2] N.F. Mott and E.A. Davis, Electronic Processes in Non-Crystalline Materials, 2nd Ed (Clarendon Oxford 1979).
[3] I.G. Austin and N.F. Mott Adv. Phys. 18(1969) 41.
[4] T. Nishida, H. Ide and Y. Takashima, Bull. Chem. Soc. Jpn, 63 (1990) 548.
[5] T. Nishida, J. Non. Cryst. Solids. 177 (1994) 451.
[6] T. Nishida, J. Iwashita and S. Kubuki, J. Cer. Soc. Jpn 107 [5] (1999) 408.
[7] K. El-Egili and H. Doweidar, Phys. Chem. Glasses 39 (6) (1998) 332.
[8] H. Jain and S. Krishnaswami, Solid State Ionics 105 (1998) 129.
[9] H. Jain, X. Lu, J. Non Cryst. Solids 196 (1996) 290.
[10] N.J. Kreidi "Glass Science and Technology" Ed by D. R. Uhlmann and N.J. Kreidl, Vol. 1 "Glass. Forming Systems", Academic Press, New York (1983) pp. 105-299.
[11] M. Sayer and A. Monsingh, Phys. Rev. B6 (1972) 4626.

Mössbauer Studies of Nanosize $MnFe_2O_4$ Particles

N. S. Gajbhiye [a*], G. Balaji [a], B. Stahl [b] and M. Ghafari [b]

[a] *Department of Chemistry, Indian Institute of Technology, Kanpur - 208016, U.P. India.*

[b] *Materials Science Department, Darmstadt University of Technology, Petersenstr. 23, 64287 Darmstadt, Germany.*

Nanosize $MnFe_2O_4$ particles were studied by ^{57}Fe Mössbauer spectroscopy from room temperature to 10 K, to investigate the cation distribution and interaction effects. As synthesized samples showed a higher degree of inversion than the bulk. The relative of intensities of the sextets suggests that Fe^{3+} is distributed in two sites as 48.9% (A) and 51.1% (B) and the corresponding amount of Mn^{2+} cations occupy the B sites. With the decrease of measurement temperature from 300 to 10 K, the spectra showed the gradual appearance of sextet component and the disappearance of doublet. Nearly 10% reduction of hyperfine field was observed as compared to bulk. The Mössbauer spectroscopy results are consistent with SQUID magnetization measurements and the blocking temperatures for 9.0 nm $MnFe_2O_4$ particles were found at 250 K and near to room temperature respectively. In our system of strongly interacting nanoparticles, the magnetic ordering resembles a spin glass like state since the interactions are random in nature at the surface of the $MnFe_2O_4$ particles.

Key words: Cation distribution, Superparamagnetism, Ferrites, Fine Particles

1. Introduction

Magnetic nanoparticles have received much attention due to their interesting properties, which differ considerably from the corresponding bulk materials [1]. Apart from fundamental interest they are important for the technological applications - like ferro-fluids, magnetic recording media, catalysts, and magnetic refrigeration. The magnetic nanosize particles are superparamagnetic at some finite temperature. The magnetic properties of spinel ferrites are governed by cation distributions among the two lattices namely A and B sites [2]. These aspects are very important to study them from the basic point of view.

For the non-interacting nanoparticles with uniaxial anisotropy, the superparamagnetic relaxation time τ, is given by Neel's expression [3],

$$\tau = \tau_0 \exp(KV/kT) \tag{1}$$

where τ_0 is of the order of 10^{-10}-10^{-12} s, k is the Boltzmann constant, T is the temperature, K is the magnetic anisotropy constant and V is the volume. Superparamagnetic relaxation is influenced by the size of the particles and their distribution, and the magnetic interactions among the particles [4]. Superparamagnetic relaxation can be nicely studied by Mössbauer spectroscopy. It is reported the literature that the blocking temperature measured from the magnetic studies (i.e. the peak in ZFC curve) is shifted to higher temperatures with the increase of strength of interactions. On the contrary, in Mössbauer measurements, the blocking temperature was found to decrease when the interparticle distance is decreased and a spin-glass like state at low temperatures was proposed [5].

The motivation of this work is to study the cation distributions and interaction effects in nanocrystalline $MnFe_2O_4$ particles by Mössbauer spectroscopy. $MnFe_2O_4$ is an excellent system to study the cation distribution since it exhibits magnetic ordering at low temperatures.

2. Experimental

Nanosize $MnFe_2O_4$ particles were synthesized by the thermal decomposition of manganese iron citrate precursor in an inert atmosphere [6]. The precursor was decomposed at 623 K for different duration of heating time to obtain particles having sizes 7, 9, and 11 nm. X-ray diffraction (Rich Seifert model Isodebyeflex 2002) was used to confirm the formation of mono phase nanosize $MnFe_2O_4$ particles. The

*Presently at Institute of Nanotechnology, Forschungszentrum Karlsruhe, P.O.Box 3640, D-76021 Karlsruhe, Germany.

crystallite size was calculated from the x-ray line broadening of (311) plane reflection. Particle size and morphology were characterized by TEM (Philips model E-301). Magnetic measurements were performed using the SQUID magnetometer (Quantum Design). Mössbauer spectroscopic measurements were carried out by using ^{57}Co source in Rhodium matrix. The spectrum was then analyzed by using software PC-MOS supplied by CMTE Electronik, Germany. The values are presented relative to natural iron.

3. Results and Discussion

$MnFe_2O_4$ particles crystallize in cubic spinel structure with a = 0.83 nm (bulk value is 0.85 nm). Electron micrographs, (TEM) indicated that particles are spherical in shape and the agglomerates of 26.0 nm. Particle size calculated from BET method (48 nm) is more than that measured from the XRD and TEM - micrographs.

The value of Curie temperature (T_C) of 7, 9, 11 nm particles were found to be 673 K that is larger than the bulk value (T_C = 603 K). This can be attributed to the non-equilibrium cation distribution present in our as prepared samples, i. e. more Mn^{2+} ions are present in the B sublattice [7]. We have performed Mössbauer spectroscopic measurements on these nanoparticles in order to probe the particle size effect, cation distribution, and magnetic interaction.

3.1 Mössbauer Studies

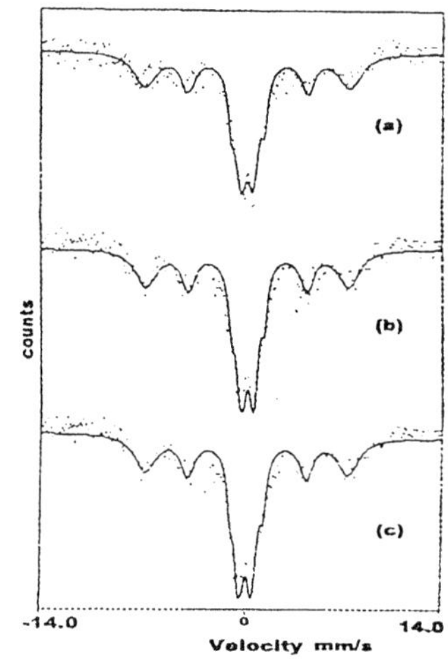

Fig. 1: Room temperature Mössbauer spectrum of $MnFe_2O_4$ particles having sizes (a) 7, (b) 9, and (c) 11 nm respectively.

Table 1: Mössbauer parameters of $MnFe_2O_4$ particles having particle sizes 7, 9, and 11 nm.

Sample Size (nm)	Isomer shift (mm/s)			quadrupole splitting (mm/s)			hyperfine field (T)	
	A	B	D*	A	B	D*	A	B
7	0.41	0.43	0.38	0.013	-0.001	0.82	43.4	46.9
9	0.41	0.43	0.38	0.009	-0.005	0.83	43.4	46.1
11	0.41	0.44	0.37	0.006	-0.005	0.83	44.0	46.4
Bulk	0.40	0.44	-	0.024	0.008	-	49.1	52.4

where D* indicates the doublet component of the spectra.

The isomer shift and quadrupole splitting are nearly the same for all the particles and are given in Table 1. The magnitude of hyperfine field were found to be nearly 10% less than the bulk value. This may be due to the fine particle nature [4,8]. The spectra of these samples are composed of both doublet and sextets and this coexistence indicates the presence of superparamagnetic fractions in the samples. With the increase of size, the relative amount of superparamagnetic fractions decreases and also the anisotropy energy barrier increases leading to more blocking of particle moments.

The A and B site occupancies are calculated from the intensity ratios of two sextets for 9 nm particles. The Mössbauer measurements were carried out at low temperatures for the determination of the cation distribution in nanostructured $MnFe_2O_4$ particles (Fig. 2). Mössbauer spectrum recorded at 10 K showed a sextet pattern corresponding to A and B sites. With the increase of temperature as 70, 150, 200, 300 K, the sextet component gradually collapses and a doublet appears. The sextets and doublets in these spectra were fitted and the Mössbauer parameters are given in Table 2. The lines become broad and asymmetric as the temperature is increased from 10 K to 300 K (Fig. 2).

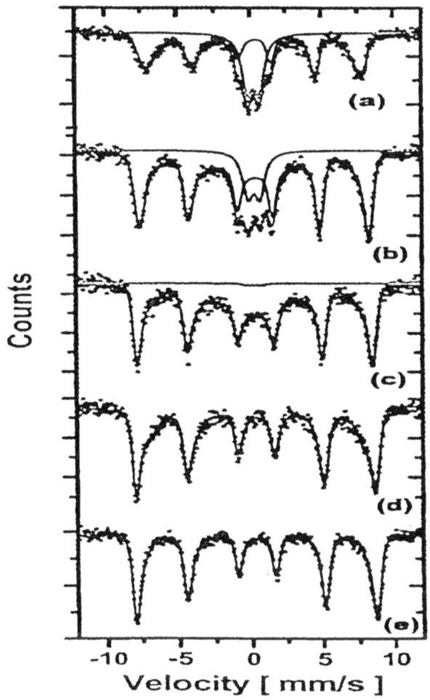

Fig. 2: Mössbauer spectrum of 9 nm $MnFe_2O_4$ particles at (a) 300 K (b) 200 K (c) 150 K (d) 70 K and (e) 10 K.

Table 2: Mossbauer parameters of 9.0 nm $MnFe_2O_4$ particles at different temperatures. D^* indicates doublet component.

Temperature (K)	Hyperfine field (T)		Intensity field (%)		
	A	B	A	B	D^*
300	46.1	42.4	15.2	20.9	63.9
200	49.8	46.2	29.0	30.2	40.4
150	50.5	47.4	39.1	40.8	20.1
80	51.1	48.2	48.9	51.1	-
10	51.5	48.7	48.9	51.1	-

The relative of intensities of the sextets (Fig. 2 (e)) suggests that Fe^{3+} is distributed in two sites as 48.9% (A) and 51.1% (B). This means, the corresponding amount of Mn^{2+} cations occupy the B sites. For this ferrite composition, in magnetic measurements, T_c was found to be 673 K which is more than the bulk value of 603 K. The occupancy of Iron sites in the bulk materials is 20% (A) and 80% (B). Therefore, we conclude from the Mössbauer spectroscopy and magnetic measurements that synthesized $MnFe_2O_4$ samples show a higher degree of inversion as compared to the bulk materials. Also, the observed 10% decrease of hyperfine field corroborates such cation distributions in A and B sites.

The blocking temperatures is sensitive to the time constant of the instrument and so different blocking temperature can be expected from Mössbauer and magnetic measurements [4]. The typical time constant of Mössbauer and magnetic instruments are of the order of 10^{-9} and 10 s respectively. For the 9 nm particles, the blocking temperature of 250 K was found from Mössbauer measurements; whereas in SQUID measurements (time constant was kept 1 s), the blocking temperature was found near to the room temperature (to be published separately). In our system of strongly interacting nanoparticles, the blocking temperatures determined from these instrumental techniques are nearly in the same temperature range. To explain this, magnetic ordering resemble a spin glass like state since the interactions are random in nature at the surface of the $MnFe_2O_4$ particles [4,9,10].

4. Conclusion

Nanosize $MnFe_2O_4$ particles synthesized by precursor technique at 623 K showed higher degree of inversion compared to the bulk sample. The Mössbauer spectroscopy results are consistent with magnetization measurements, which are studied separately. Blocking temperature for 9.0 nm $MnFe_2O_4$ particles were found near to room temperature. Strong interactions may lead to a spin glass like ordering at the surface of the nanoparticles.

Acknowledgement

We acknowledge financial assistance during investigation by CSIR, New Delhi. One of us (NSG) wishes to thank Professor H. Gleiter, Institute of Nanotechnology, Forschungszentrum, Karlsuhe for the encouragement.

References

[1] J.L. Dormann and D. Fiorani, eds., *Magnetic Properties of Fine Particles* (North - Holland, Amsterdam, 1992).
[2] V.A.M. Brabers, Phys. Rev. Lett. 68 (1992) 3112.
[3] L. Neel, Ann. Geophys. 5 (1949) 99.
[4] S. Morup, Europhys. Lett. 28 (1994) 671.
[5] S. Morup, F. Bodker, P.V. Hendricksen and S. Linderoth, Phys. Rev. B 52 (1995) 287.
[6] N.S. Gajbhiye and G. Balaji, Thermochimica Acta. (accepted).
[7] J.P. Chen, C.M. Sorenson, K.J. Klabunde, G.C. Hadjipanayis, E. Devlin, and A. Kostikas, Phys. Rev. B 54 (1996) 9288.
[8] F. Bodker, S. Morup, and S. Linderoth, Phys. Rev. Lett. 72 (1994) 282.
[9] R.W. Chantrell, M. El-Hilo and K. O'Grady, IEEE Trans. Mag-27 (1991) 3570.
[10] W. Luo, S.R. Nagel, T.F. Rosenbaum and R.E. Rosensweig, Phys. Rev. Lett. 67 (1991) 2721.
[11] B. Martinez, X. Obradors, Ll. Balcells, A. Rouanet, and C. Monty, Phys. Rev. Lett. 80 (1998) 181.

Advanced double coated metal particle tapes studied by Mossbauer spectroscopy and magnetic measurements

V.Kuncser[1,2], M.Vopsaroiu[3], B.Sahoo[1], P.R.Bissell[3] and W.Keune[1]
1. Laboratorium für Angewandte Physik, Gerhard-Mercator-Universität , 47048 Duisburg, Germany
2. National Institute for Materials Physics, Bucharest-Magurele, Romania
3. University of Central Lancashire, Magnetic Materials Research, Preston, PR1 2HE, UK
* Correspondence: Laboratorium für Angewandte Physik, Gerhard-Mercator-Universität , 47048 Duisburg, Germany

Samples from a new generation of metal particle tapes have been analysed by Mössbauer spectroscopy and magnetic measurements. The Mössbauer results show the presence of both metallic α-Fe(Co) fine particles and a very important fraction of iron oxide. An out-of-plane distribution of the Fe spins within a mean angle of around 26° has been evidenced by Mössbauer spectroscopy, in acceptable agreement with the magnetic data. The spin dispersion seems to be related especially to dynamical effects.

Keywords: metallic nanoparticles, magnetic tapes, phase and spin distribution, Mössbauer spectroscopy

1. Introduction

The metal particle (MP) tapes still remain a very attractive alternative for the high density recording media [1] due to a very basic recording principle and a very low cost. Elongated iron based metal particles provide large remanent magnetisation and coercivity. Such particles must be stabilised against the atmospheric oxidation by a careful passivation and various protective coatings. These induce additional limitations concerning the correlation of the magnetic macroscopic performances with the initial particle microstructure and phase composition. The magnetic orientation of the metallic particles in recording media has also a major effect on the magnetic properties and recording performances and should be carefully analysed by suitable methods. Since most of the magnetic recording media contain iron-based compounds, ^{57}Fe Mössbauer spectroscopy represents a powerful tool providing both particle morphology and spin orientation [2,3].

This paper is focused on the determination of particles morphology and out-of-plane easy axis distribution (EAD) for a group of four advanced double-coated MP tapes. Although the particles are distributed both in-plane and out-of-plane, we present here only properties related to out-of-plane component of magnetisation, the in-plane investigations being the subject of another paper [4]. The characterisation of the iron species has been analysed by Mössbauer spectroscopy, whereas the orientation of the iron magnetic moments with respect to the tape plane has been obtained from both magnetic measurements and Mössbauer spectra.

2. Experimental details

The analysed double-coated MP tapes have a magnetic layer and a non-magnetic underlayer. The magnetic layer contains ellipsoidal particles of metallic iron doped with a small amount of Co in order to increase the coercivity. The ellipsoids present an aspect ratio of 1:5 and a length of approximately 100nm. The particles have been passivated by an induced oxidation of the surface and subsequently uniformly distributed in a non-magnetic polymer matrix. The embedding process has been done in the presence of a small longitudinal applied field, in order to orient the ellipsoids with their easy axis along the tape direction. The non-magnetic underlayer is most probable an α-Fe_2O_3 iron oxide, as will be shown later. The metallic particles distribution and also the non-magnetic underlayer are very similar in the all four samples. The only difference between the studied tapes is the thickness of the magnetic layer (see Mrt values in Table 1). In-plane and out-of-plane hysteresis loops have been obtained at room temperature using a conventional vibrating sample magnetometer. The Mössbauer spectra have been acquired in transmission geometry using a spectrometer with symmetrical waveform and a ^{57}Co source in a Rh matrix. We performed measurements down to 4.2 K by inserting the sample in a variable temperature cryostat. The γ-ray direction was perpendicular to the sample plane.

3. Results and discussion

Typical longitudinal in-plane and the respective out-of-plane" hysteresis loop for the analysed samples are presented in Fig.1. Some basic hysteresis parameters are shown in TABLE I. Mrt represents the saturation remanent magnetisation on unit volume multiplied by the effective magnetic coating thickness and is obtained by dividing the overall magnetisation (expressed in memu) to the tape surface (expressed in cm^2). Sq is the squareness. The orientation ratio (OR) is given by the ratio between the remanent magnetization along the longitudinal direction and the remanent magnetisation along the transverse direction to the tape plane. The out-of-plane easy axis distributions have been obtained from the OR using the method described in [5]. Parameter b in TABLE I represents the semi-width of a lorentzian function, which is assumed to describe properly the out-of-plane EAD of the particle magnetic moments.

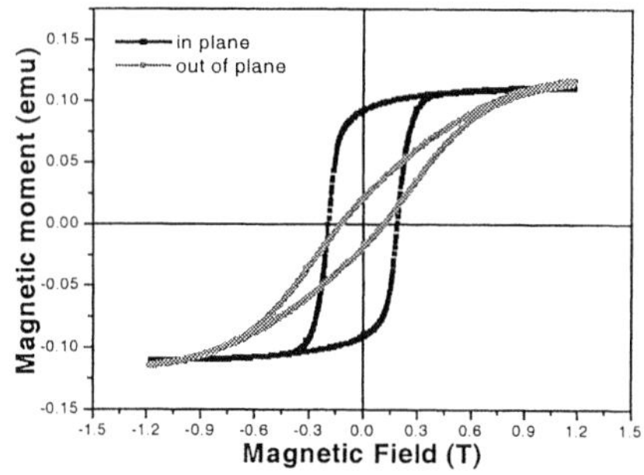

Figure 1: Longitudinal in plane and respctively out of plane hysteresis loop for sample D.

TABLE I: Samples magnetic parameters obtained from VSM measurements.

Sample	Mrt	Hc	Sq	out-of-plane	
	(memu/cm^2)	(T)		OR	b (^{0})
A	3.0	0.194	0.82	3.1	15.0
B	4.0	0.192	0.81	2.92	15.5
C	4.5	0.192	0.83	2.89	15.7
D	6.0	0.191	0.82	2.55	18.4

The Mössbauer spectra of the four analysed samples taken at room temperature (RT) are presented in Fig.2. They consist of a broad and large sextet with a hyperfine field around 49 T, a less split sextet with a hyperfine magnetic field of around 35 T and a central paramagnetic component. Excepting the less-split sextet which can be unambiguously assigned to the α-Fe(Co) metallic phase, some doubts persists to the assignment of the central pattern and the broad external sextet which may be due to either a mixture of different oxides (distributed Fe sites) or dynamical effects related to only one oxide phase. In order to solve this problem, Mössbauer spectra have been taken at 4.2 K (see Fig. 3). The low temperature spectra show clearly only two magnetic phases. The much narrower line for the external sextet stands therefore for dynamical aspects related both to the external sextet and the central pattern in the RT spectra.

TABLE II: Mössbauer hyperfine parameters, relative intensity ratio and relative areas for the two magnetic sextets in the Mössbauer spectra measured at RT and 4.2K.

Phase	T	Oxide					α-Fe(Co)				
Sample	(K)	IS/<IS> (mm/s)	QS/<QS> (mm/s)	B_{hf}/<B> (T)	R_{23}	R.A. (%)	IS (mm/s)	QS (mm/s)	B_{hf} (T)	R_{23}	R. A. (%)
A	4.2	0.38	-0.19	53.0	2.1	93.0(2)	0.06	0.00	36.8	4.0	7.0(2)
	295	0.26	-0.21	49.5	2.0	74.6	-0.06	0.00	36.0	3.6	16.8
B	4.2	0.38	-0.18	53.0	2.1	91.0(4)	0.05	0.00	36.8	4.0	9.0(4)
	295	0.26	-0.21	49.5	2.0	72.1	-0.07	0.00	35.8	3.7	18.1
C	4.2	0.38	-0.18	53.0	2.1	88.8(5)	0.06	0.00	36.8	4.1	11.2(5)
	295	0.26	-0.21	49.4	2.0	70.1	-0.07	0.00	35.8	3.5	20. 0
D	4.2	0.38	-0.19	53.0	2.1	86.2(3)	0.05	0.00	36.9	4.1	13.8(3)
	295	0.26	-0.21	49.4	2.0	64.4	-0.07	0.00	35.9	3.6	23.7

This is the reason that we fitted finally the dominant outer sextet in the RT spectra by a distribution of hyperfine fields related to relaxational aspects. The main hyperfine parameters, relative intensity ratios ($R_{23}=I_2/I_3$) and relative areas (RA) of the two magnetic sextets are presented in Table II. It may be observed that for all the analysed samples the relative area of the external sextet decreases from 4.2 K to 295 K with around 20% whereas the relative area of the inner sextet increases with around 10%. Therefore, the RT superparamagnetic central pattern could be developed only from the external sextet and consequently it has to be assigned to the oxide phase. Hence, the only trace of the metallic phase in the Mössbauer spectra belongs to the inner sextet. The increase of their relative area vs. temperature should be explained by the different temperature dependence of the Debye-Waller factor (f_{DW}) in the two implied phases, because usually f_{DW} decreases faster for an oxide than for the metallic phase. The oxide could be most probably assigned from the values of hyperfine fields and isomer shifts to an α-Fe_2O_3 or an γ-Fe_2O_3 phase [6,7]. The negative quadrupole splitting evidenced at room temperature as well as the absence of the Morin transition over the analysed temperature range (see the quadrupole split behaviour in Table II) should infer rather the presence of an γ-Fe_2O_3 phase. However, magnetic measurements suggest that the Mrt value is related primarily to the top magnetic coating and that the underlayer is essentially non-magnetic. The γ-Fe_2O_3 phase is ferrimagnetic whereas α-Fe_2O_3 is weak ferromagnetic above the Morin transition and antiferromagnetic below. Because there is no Morin transition, the negative value of the quadrupole split over the all temperature range correlated with the above magnetic result points strongly to a defected anti-ferromagnetic α-Fe_2O_3 phase in the underlayer. The higher hyperfine field in the metallic α-Fe(Co) phase is due to both the presence of the Co ions and interfacial interaction with the oxide layer. Taking into account the above Mössbauer parameters as well as the preparing procedure we may assume the particle morphology as consisting in a metallic soft magnetic α-Fe(Co) core surrounded by an oxide layer. It is worth noticing that only the low temperature spectra, where the two f_{DW} factors are comparable, give valuable information about the relative amount of the two iron phases. The results indicate (see TABLE II), as we mentioned above, only two iron phases. These are distributed in around 10% for the metallic α-Fe(Co) core and the rest of approximately 90% for the oxide present in both passivation layer and non-magnetic underlayer. There is magnetic evidence that the passivation layer on the α-Fe(Co) particles, expected to be around 2nm thick, does not contribute to the particle moment and is also non-magnetic. The volume fraction corresponding to a such thin passivation layer represents much below 10% from the metallic phase, that means below 1% from the total

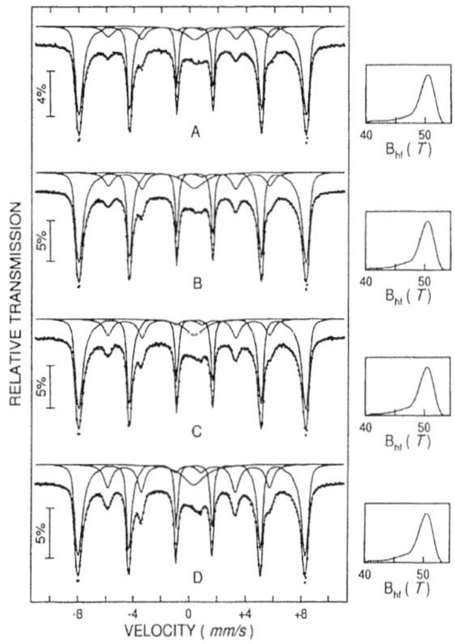

Figure 2: Room temperature Mössbauer spectra of the analysed samples.

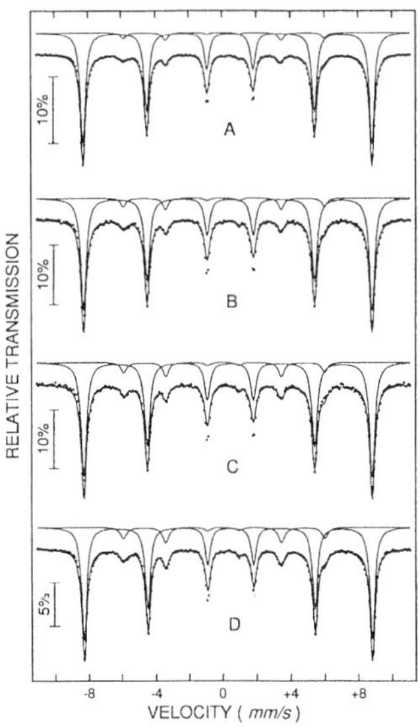

Figure 3: Mössbauer spectra taken at 4.2 K for the analysed samples

Mössbauer absorption area. Therefore, the Mössbauer data, which show only two distinct iron phases, give reliable information only about the metallic phase and the iron oxide in the non-magnetic underlayer. The linewidth behaviour of the outer sextet stands for a distribution of oxide nano-particles that behave in the regime of collective excitation. The small fraction of superparamagnetic phase identified by the central doublets in the RT spectra corresponds to the smallest particles from this distribution.

Finally, we obtained information about the out-of-plane distribution of the particle magnetic moments from the intensity ratios R_{23}. In TABLE II it can be observed that the iron spins in the metallic core at 4.2 K are all in the tape plane ($R_{23} \approx 4$). The lower values obtained at RT show that the out-of-plane spin distribution is related exclusively to dynamical aspects, the spins starting to oscillate along the longitudinal direction. The mean angle, Θ, between the "dispersed" spins and the sample plane is simply deduced by the relation $\Theta = 90° - \Psi$, where Ψ is the angle between the Fe spin and the direction of the γ ray. The angle Ψ is obtained using the well-known relation:

$$R_{23}=4\cdot\sin^2\Psi / (1+\cos^2\Psi) \qquad (1)$$

From the experimental R_{23} values at RT, the following angles Θ have been deduced: 13°, 14°, 12° and 13° for sample A, B, C respectively D (errors $\pm 1^0$). These results are in good agreement with the b values achieved from magnetic measurements (see Table I).

4. Conclusions

Magnetic properties of four advanced metal particle tapes have been studied in correlation with the particle microstructure using Mössbauer spectroscopy and magnetic measurements. The low temperature Mössbauer spectra evidence the presence of only two magnetic phases: a metallic α-Fe(Co) (the core of the metallic particles) and an oxide phase which is the main component of the non-magnetic underlayer. The metallic iron represents a fraction of ≈10% from the total amount of iron in the sample. At room temperature both the oxide and the metallic phase behave in the regime of collective excitations, excepting a low amount of very small oxide nanoparticles that behave superparamagnetically. A different temperature dependence of the f_{DW} factor in the two present phases is suggested by the evolution of their relative areas vs. temperature. Out-of-plane spin distribution deduced from Mössbauer spectroscopy is in acceptable agreement with the out-of-plane easy axis distribution derived from the magnetic data.

Acknowledgements:

The financial support of Alexander von Humboldt Stiftung is highly acknowledged by one of the co-authors (V.Kuncser). The authors wish also to thank Ulrich von Hörsten for valuable technical assistance and M.F. Thomas form Liverpool University for helping with the preliminary Mössbauer test measurements.

References:

[1] R.J. Veitch, A. Ilmer, W. Lenz, V. Richter JMMM, 193 (1999) 279-283.

[2] C.E. Johnson, J.Appl.Phys. D 29 (1996) 2266

[3] M.P.Morales, S.A.Walton, L.S.Prichard, C.J.Serna, D.P.E.Dickson and K. O'Grady, J.Magn.Magn. Mater. 190 3 (1998) 357

[4]. M. Vopsaroiu, P.R. Bissell, R. Cookson, Phys. Stat. Sol., 2001, in press.

[5]. F. Schmidllin, P.R. Bissell and J.A. Gotaas, J. Appl. Phys. vol.79 (8), 1996.

[6] N.N.Greenwood and T.C.Gibb, Mössbauer Spectroscopy, Chapmann and Hall Ltd-London (1971) 241-248

[7] J.Danon in Chemical Applications of Mössbauer spectroscopy ed. By V.I.Goldanskii and R.H.Herber, Academic Press (1968) 211-214

Mössbauer spectroscopy on ferromagnetic Fe_2VAl

D. Baabe[a], I. Maksimov[a], J. Kreitlow[a], H.-H. Klauß[a], F.J. Litterst[a], A. Matsushita[b] and S. Süllow[a]
[a] *Institut für Metallphysik und Nukleare Festkörperphysik, TU Braunschweig, D-38106 Braunschweig, Germany.*
[b] *National Research Institute for Metals, Tsukuba, Ibaraki 305-0047, Japan.*

Abstract. We present a detailed investigation of ferromagnetically ordered Fe_2VAl by means of Mössbauer spectroscopy. For this compound it has been demonstrated previously that ferromagnetism is induced by heat treatment procedures. The ferromagnetic state is attributed to the presence of 4.2(8) % Fe ions carrying a magnetic moment, randomly distributed in the sample. It has been proposed that because of Fe/V sites exchanges a magnetic moment is carried by these Fe ions. Our Mössbauer data disprove this proposal, as it is inconsistent with the upper limit of 3 % Fe/V site exchanges in sc–Fe_2VAl determined in a previous neutron diffraction study. We discuss alternative scenarios to account for the observed behavior.

Keywords: Heusler compound, Mössbauer spectroscopy, Fe-Al alloys, ferromagnetism, shell model, crystallographic disorder

The magnetic phase diagram of the alloying series $(Fe_{1-x}V_x)_3Al$ has attracted close attention in recent years because of reports about unusual ground state properties of the one member of the alloying series, Fe_2VAl. For this Heusler-ordered compound (cubic *Fm3m*) it has been argued that the observation of a semiconductor-like resistivity close to a magnetic instability represents evidence for Kondo-insulating behaviour [1, 2] analogous to the system FeSi [3]. This view was contrasted by optical conductivity studies taken as evidence for the formation of a pseudo gap in the density of states of Fe_2VAl of (0.1-0.2) eV [4]. This view is supported by various band structure calculations [5].

The problem of the ground state of Fe_2VAl grew even more complicated when Matsushita and Yamada [6] found that the bulk properties of Fe_2VAl exhibit a very strong dependence on the applied heat treatment. They demonstrated that by way of different cooling procedures after an annealing stage the nature of the ground state of Fe_2VAl can be tuned; while quenched material exhibits the previously reported semiconducting-like, non-magnetic behaviour [1], a sample slowly cooled down to room temperature after a heat treatment shows almost metallic transport and a ferromagnetic transition. Recently, we could show that such variations of the ground state properties of Fe_2VAl correspond to variations of the chemical composition of the matrix of the samples: the quenched samples are Al deficient by up to 6 %, while the composition of the slowly cooled samples is stoichiometric 2:1:1 within experimental error [7]. Similar results on the sample dependence of the properties of Fe_2VAl have been reported simultaneously by other groups [8]. On basis of neutron diffraction experiments [7], we excluded that Fe/V site exchange - as previously speculated [5] - accounts for the observation of ferromagnetism in the slowly cooled sample Fe_2VAl. In this contribution we will discuss alternative causes for the occurence of ferromagnetism in this compound, based on Mössbauer spectroscopy experiments on slowly cooled Fe_2VAl.

For our experiments a slowly cooled sample, sc–Fe_2VAl, has been prepared from stoichiometric mixtures of the constituent elements Fe (4N), V (3N) and Al (5N) arc-melted in a copper crucible under argon (5N) atmosphere and simultaneous Ti-gettering [6]. The weight loss during this process was less then 0.5 %. The specimen was sealed in an evacuated quartz ampoule for the heat treatment, homogenised at 1273 K for 24 h, slowly cooled to 553 K at a rate of -6 K/h, and subsequently furnace cooled. The sample has been characterised via resistivity, susceptibility and neutron diffraction and exhibits the previously reported ferromagnetism below T_C = 13 K [6, 7].

The ^{57}Fe Mössbauer spectroscopy experiments have been executed on this sample sc–Fe_2VAl in a standard low–temperature Mössbauer set–up at temperatures ranging from 8 K to 300 K (*source*: ^{57}Co-in-Rh matrix at room temperature; *full width at half maximum* (FWHM): 0.226 mm/s). In Figure 1 we present the Mössbauer spectra of sc–Fe_2VAl in the paramagnetic state measured at 50 K and 300 K, and below T_C measured at 8 K. The evaluation of the spectra was done using the Mössbauer fitting program NORMOS-90 [9] in the thin absorber approximation. The results of a fitting procedure with one

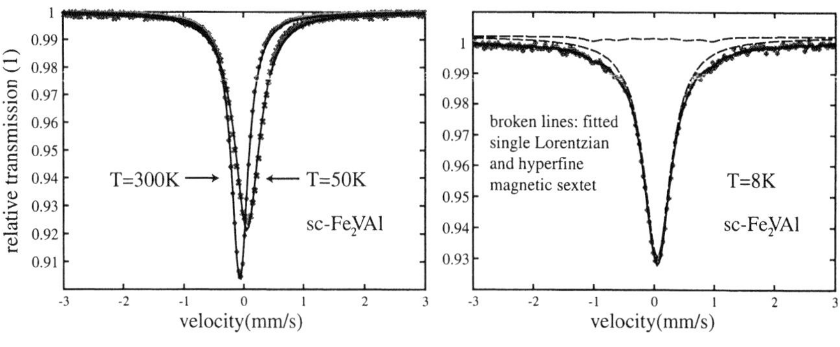

Figure 1. Mössbauer spectra measured on sc–Fe_2VAl. **Left:** Single absorption lines in the paramagnetic state at 50 K and 300 K. **Right:** A clearly broadened absorption line below T_C measured at 8 K.

single Lorentzian line in the paramagnetic state and an additional hyperfine magnetic sextet below T_C are listed in Table 1. At 300 K a broadened FWHM of 0.334(1) mm/s is observed, which is associated with a distribution of chemical shifts, probably reflecting strain or vacancy disorder in this compound. Below T_C, the further broadening of the single line and the slight change in the line shape (Figure 1, right) denote the transition into the ferromagnetic

Table I. Collection of fit results obtained by chi-square minimisation to a single Lorentzian line above and to an additional sextet below T_C. The SHIFT is constituted from the chemical shift relative to the ^{57}Co-in-Rh source and the 2nd order Doppler shift. BHF is the hyperfine magnetic field.

T (K)	SHIFT (mm/s)	FWHM (mm/s)	AREA (mm/s)	BHF (T)
300	- 0.0648 ± 0.0002	0.334 ± 0.001	0.0509 ± 0.0001	—
50	0.0554 ± 0.0006	0.498 ± 0.002	0.0616 ± 0.0002	—
8	0.053 ± 0.001	0.552 ± 0.005	0.0618 ± 0.0005	—
8	- 0.03 ± 0.03	0.31 ± 0.08	0.0027 ± 0.0005	6.1 ± 0.2

phase of sc–Fe_2VAl. A fit of the data employing a single line and a hyperfine magnetic sextet reveals a moderate hyperfine field of 6.1(2) T (Table 1). Thereby, the amount of magnetic ions is determined to 4.2(8) %.

For sc-Fe_2VAl, the observed behavior cannot be explained assuming Fe/V site exchanges and applying a *shell model* [4, 5, 10], *i.e.*, the local hyperfine field at a given Mössbauer probe is constituted by a superposition of the transferred fields of magnetic ions sited in the nearest neighbor, next-nearest neighbor etc. shell. From a neutron diffraction study on sc-Fe_2VAl we established an upper limit of 3 % Fe/V site exchanges, a value too small to account for the observed Mössbauer data [7]. In consequence, we have to consider different sources rendering locally Fe carrying a magnetic moment. We note that in our neutron diffraction study we utilised that V atoms scatters neutrons almost completely incoherently, with the effect of a very bright contrast between Fe, V and Al in that experiment. The downside of this situation is that we are not able to distinguish between vacancies and V ions. We speculate that the too large number of moment-carrying Fe ions reflects local disorder in form of vacancies on the V site. It has been demonstrated that vacancies can strongly affect the material properties in Fe–Al–alloys [11]. The situation in Fe_2VAl is complicated by a pseudo gap structure building up at the Fermi level E_F, as evidenced by the optical conductivity studies [4]. Under these semi–metallic like conditions, sample dependencies are typically much enhanced, as states within the gap might form, drastically affecting the observed properties. Further, it has been suggested that the actual position of E_F (*i.e.*, does E_F lie in, at or close to the gap) might be strongly dependend on the density of Al vacancies or (chemical) pressure effects in Fe_2VAl.

None of our observations are consistent with a Kondo–insulating ground state of Fe_2VAl [1, 12]. Instead, the strong sample dependencies are in broad agreement with a view of a pseudo gap forming in $(Fe_{1-x}V_x)_3Al$ with x approaching 0.33. In detail, however, it is very difficult to compare theoretical predictions and experimental findings, as the ground state properties of

Fe_2VAl are very much hidden behind metallurgical problems. While quenched q–Fe_2VAl is off–stoichimetric in its Al–content, even for slowly cooled samples the observation of diluted magnetism indicates that crystallographic disorder plays an important role. We exclude atomic site exchange as source of this disorder, and alternatively attribute it to the presence of vacancies in this compound.

Work at the TU Braunschweig was supported by the Deutsche Forschungsgemeinschaft DFG, under Grant No. SU 229/4-1.

References

1. Y. Nishino, M. Kato, S. Asano, K. Soda, M. Hayasaki, U. Mizutani, Phys. Rev. Lett. **79** (1997) 1909.
2. K. Endo, H. Matsuda, K. Ooiwa, M. Iijima, T. Goto, K. Sato, I. Umehara, J. Magn. Magn. Mat. **177–181** (1998) 1437.
3. Z. Schlesinger, Z. Fisk, H.-T. Zhang, M.B. Maple, J.F. DiTusa, G. Aeppli, Phys. Rev. Lett. **71** (1993) 1748; Z. Schlesinger, Z. Fisk, H.-T. Zhang, M.B. Maple, Physica B **237-38** (1997) 460.
4. H. Okamura, J. Kawahara, T. Namba, S. Kimura, K. Soda, U. Mizutani, Y. Nishino, M. Kato, I. Shimoyama, H. Miura, K. Fukui, K. Nakagawa, H. Nakagawa, T. Kinoshita, Phys. Rev. Lett. **84** (2000) 3674.
5. G.Y. Guo, G.A. Botton, Y. Nishino, J. Phys.: Condens. Matter **10** (1998) L119; D.J. Singh and I.I. Mazin, Phys. Rev. B **57** (1998) 14352; R. Weht and W.E. Pickett, Phys. Rev. B **58** (1998) 6855
6. A. Matsushita and Y. Yamada, J. Magn. Magn. Mat. **196-197** (1999) 669.
7. I. Maksimov, D. Baabe, H.H. Klauss, F.J. Litterst, R. Feyerherm, D.M. Többens, A. Matsushita, S. Süllow, J. Phys.: Condens. Matter **13** (2001) 5487.
8. Y. Feng, J.Y. Rhee, T.A. Wiener, D.W. Lynch, B.E. Hubbard, A.J. Sievers, D.L. Schlagel, T.A. Lograsso, L.L. Miller, Phys. Rev. B **63** (2001) 165109.
9. R.A. Brand, *Normos-90 Mössbauer Fitting Program Package* (Wissenschaftliche Elektronik GmbH, Starnberg, 1992).
10. H. Matsuda, K. Endo, K. Ooiwa, M. Iijima, Y. Takano, H. Mitamura, T. Goto, M. Tokiyama and J. Arai, J. Phys. Soc. Japan **69** (2000) 1004.
11. H. Domke, L.K. Thomas, J. Magn. Magn. Mat. **45** (1984) 305.
12. G. Aeppli, Z. Fisk, Comment Cond. Mat. Phys. **16** (1992) 155.

Address for Offprints: Dirk Baabe, Institut für Metallphysik und Nukleare Festkörperphysik, Mendelssohnstr. 3, D-38106 Braunschweig, Germany.

CEMS STUDY OF ^{57}Fe IMPLANTED IN DIAMOND

K. Bharuth-Ram[1], *D. Naidoo*[1], *G. Klingelhöfer*[2], *J.E. Butler*[3]

[1]Physics Department, University of Durban-Westville, Durban 4000, South Africa.
[2]Institut für Anorganische Chemis, Universität Mainz, D-55099 Mainz, Germany.
[3]Naval Research Laboratory, Washington DC 20375, U.S.A.

Conversion electron Mössbauer (CEMS) measurements have been made on a diamond sample implanted with 70 keV ^{57}Fe to a dose of 5 x 10^{14} cm^{-2}. CEMS spectra were collected in the as-implanted state and after annealing the diamond up to 1470 K. The lower temperature spectra were consistently fitted with three symmetric doublets. With annealing the areal fractions of two of the doublets increased to 44% and 48 %, respectively, while that of the third doublet decreased to be replaced by a single line component with isomer shift δ = 0.0 mm/s. Comparison of the observed isomer shifts with recent in-beam Mössbauer measurements and theoretical calculations allow us to attribute the singlet to interstitial Fe and one of the doublets to Fe close to perturbed substitutional sites.

1. Introduction

There exist several Mössbauer investigations of ^{57}Co and ^{57}Fe atoms implanted in diamond. In an early study, Sawicka *et al.* [1-3] performed a series of Mössbauer studies of ^{57}Co and ^{57}Fe implanted into natural diamond and resolved their spectra into a weak singlet and a broad doublet, suggesting that the implanted Fe ions were in highly damaged amorphous regions in the host matrix. They also performed temperature dependent measurements which showed an irreversible growth of the single line component from 3% to 20% at T>600 K, suggesting a transition of implanted ions from the low symmetry site to the high symmetry one. De Potter and Langouche [4] carried out dose dependent measurements of ^{57}Co implanted in diamond and observed that above a critical implantation dose of 10^{15} cm^{-2}, the diamond lattice was amorphized, and showed no healing with annealing to 1173 K. Recently, in-beam Mössbauer spectroscopy (IBMS) measurements have been performed on ^{57}Fe nuclei Coulomb excited and recoil implanted into single crystal diamonds [5]. CEMS spectra, collected up to sample temperatures of 800 K, were characterised by two quadrupole split doublets and a weak singlet line, and showed little change with temperature. Advantages of IBMS are the large range over which the recoils are implanted and the extremely low doses required for spectra pf acceptable statistics. A major disadvantage, however, is that the measurements are done at the sample temperature and any thermal treatment is effective only for the duration of the lifetime of the ^{57}Fe Mössbauer state (100 ns). The present work attempts to overcome this shortcoming with the ^{57}Fe ion implanted into the diamond sample at a dose below the amorphization threshold, and the CEMS measurements made after annealing the sample in vacuum up to 1470 K.

2. Experimental

To ensure good quality diamond, the samples used in this study were synthesized in the high temperature high pressure (HTHP) process with synthesis parameters controlled to ensure that nitrogen impurities were kept below 50 ppm. The diamond sample was implanted with isotopically separated ^{57}Fe of energy 70 keV and to a dose of 5x10^{14} cm^{-2}, i.e. below the amorphization threshold [4]. TRIM [6] calculations gave a mean implantation range 310Å with a straggle of 90Å. CEMS spectra were collected at room temperature, on the as-implanted sample and after annealing the sample for 30 min. in vacuum at annealing temperatures of T_A = 600 K, 950 K and 1470 K. A natural Fe foil absorber was used for the calibration of all spectra.

3. Results and Discussion

CEMS spectra for the as-implanted sample and after annealing at T_A= 600 K, 950 K and 1470 K are shown in Fig. 1.

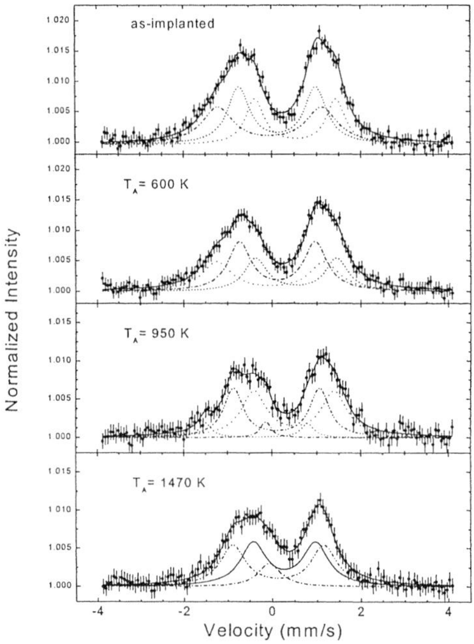

Fig. 1 : CEMS spectra of ^{57}Fe in HTHP diamond, as implanted and after annealing at 600 K, 950 K, and 1470 K

To obtain consistent fits the spectra were analysed in several different ways. The most consistent fits to the spectra were obtained in an analysis in which the lower temperature spectra were fitted with three symmetric doublets. Acceptable fits were also obtained in an analysis using two doublets and two singlets. The behaviour of two doublets (labelled D1 and D2, below) and the single line that appears after annealing at T ≥ 950 K showed no significant difference in the two analyses.

TABLE 1.

Linewidth Γ (FWHM), isomer shift, quadrupole splitting and line area A of the main doublets (D_1 and D_2) and the singlet (L_1), determined at the different annealing temperatures.

	D_1 (Γ =1.2 mm/s)			D_2 (Γ = 1.0 mm/s)			L_1 (Γ = 0.6 mm/s)	
T_A (K)	δ (mm/s)	ΔE_Q (mm/s)	A (%)	δ (mm/s)	ΔE_Q (mm/s)	A (%)	δ (mm/s)	A (%)
300	-0.13(4)	1.71(8)	38(4)	-0.53(4)	1.80(8)	26(4)		
600	-0.12(4)	1.70(8)	41(4)	-0.55(5)	1.81(8)	27(4)		
950	-0.09(3)	1.90(8)	41(4)	0.49(5)	1.80(8)	48(4)	0.13(5)	3(2)
1470	-0.13 (4)	2.00(5)	44(4)	-0.27(5)	1.40(8)	46(4)	0.06(3)	10(3)

The parameters determined from the fits are listed in Table 1, while Fig. 2 displays plots of the areal fractions of the components as a function of annealing temperature.

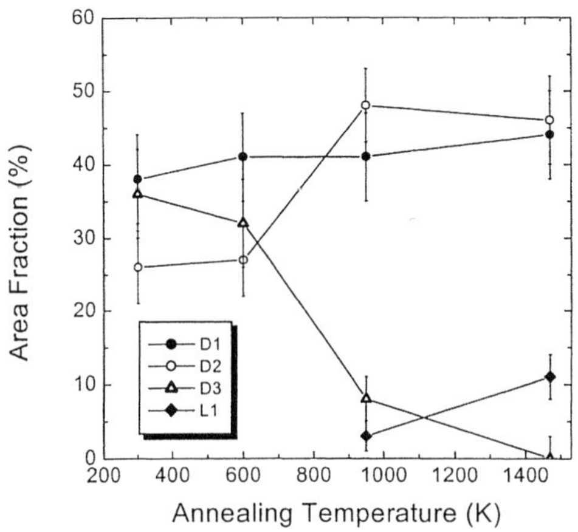

Fig. 2 Fractional population of the different components as function of annealing temperature.

The annealing behaviour of components D_1 and D_2 are in good agreement with that reported in ref.4. It is noteworthy too that only after the anneal at 1400 K are the spectral parameters in agreement with those obtained in the IBMS measurement at 300 – 800 K.

4. Discussion

In Table II we compare the isomer shifts and quadrupole splittings determined in our measurements with results obtained from IBMS measurements [5] and those of ref. 4.

TABLE II.

Isomer shifts and quadrupole splittings of the doublets determined in present measurements compared with IBMS measurements and those of ref.4

Component	Present	IBMS [5]	Ref.4
D_1 : δ (mm/s)	-0.12(4)	+0.04(3)	-0.16(5)
ΔE_Q (mm/s)	2.00(5)	2.10(3)	2.47(5)
D_1 : δ (mm/s)	-0.27 (5)	-0.51(3)	-0.70(5)
ΔE_Q (mm/s)	1.40(10)	2.33(3)	2.60(5)
L_1 : δ (mm/s)	0.06(3)	0.16(5)	1.61(5)
L_2 : δ (mm/s)			-0.79(5)

The isomer shifts of the doublets are in reasonable agreement with the IBMS results and those of Ref.4. The quadrupole splittings are smaller, perhaps reflecting the reduced damage in the vicinity of the probes in the present measurements where the dose rate was kept below the amorphization threshold. The singlet component required for a good fit at 1470 K has an isomer shift close to zero, which is quite different to that observed in ref.4. This may be due to the fact that in the latter measurements ^{57}Co were implanted in the diamond samples, and after effects of the $^{57}Co \rightarrow {}^{57}Fe$ electron capture decay may make some contribution to the electron charge distributions around the probe nuclei.

To identify the implantation sites of the Fe nuclei we compare the observed isomer shift with theoretical calculations which were carried out within the local density approximation of spin-density functional theory (LSDA) [5,7]. The calculations give δ values ranging from -0.20 mm/s for Fe at an unrelaxed substitutional site to +0.093 mm/s allowing for a small relaxation of a nearest neighbour C atom atom away from the substitutional Fe, and +0.22 mm/s for Fe at interstitial sites. The isomer shift of doublet D_1 (-0.12(4) mm/s) is in good agreement with that calculated for substitutional Fe with a relaxed Fe-C bond length. The atomic radii of the carbon and Fe atoms are 0.77 and 1.26A respectively, and so the environment of substitutional Fe should be strongly perturbed and result in large quadrupole splitting, as is observed. The singlet component shows that about 10% of the Fe atoms are located at high symmetry sites after annealing at 1470 K. The isomer shift is consistent with either interstitial or substitutional sites. However comparison with emission channelling measurements on heavy ions implanted in diamond [8,9] indicates a substitutional site as the more likely.

5. Conclusions

We have performed CEMS measurements on diamond implanted with ^{57}Fe to a dose below the amorphization threshold. The spectra show some changes with annealing, and 10% of the Fe are located at high symmetry sites after annealing at 1470 K. The isomer shift of another 45% of the Fe ions are consistent with theoretical calculations for substitutional Fe. Emission channelling measurements on As, In, Ce and Gd [8-10] ions implanted in diamond indicate that up to 50% of the implanted atoms are located near substitutional sites with small off-axis displacements. This is consistent with the present results where the large quadrupole splittings point to considerable lattice distortion in the vicinity of the probes.

References

[1] B.D. Sawicka, J.A. Sawicki, H. de Waard, Phys. Lett. 85 A (1981) 303.
[2] J.A. Sawicki and B.D. Sawicka, Nucl. Instr. Meth. 194 (1982) 465.
[3] J.A. Sawicki and B.D. Sawicka, Nucl. Instr. Meth. B46 (1990) 38.
[4] M. de Potter and G. Langouche, Z. Phys. B53 (1983) 89.
[5] K. Bharuth-Ram, M. Hartick, C. Dorn, P. Held, E. Kankeleit, R. Sielemann, L. Wende, J. Kübler and J.P.F. Sellschop, Phys. Rev. B58 (1998) 8955
[6] J.F. Ziegler, J.P. Biersack, U. Littmark, in "The Stopping and Range of Ions in Solids", (Pergamon, New York, 1985).
[7] J. Kübler, A.E. Kumm, H. Ovehof, P. Schwalbach, M. Hartick, E. Kankeleit, B. Keck, L. Wende, R. Sielemann, Z. Phys. B92 (1993) 155.
[8] K. Bharuth-Ram, H. Quintel, M. Restle, C. Ronning, H. Hofsäss and S.G. Jahn, J. Appl. Phys. 78 (1995) 5180.
[9] K. Bharuth-Ram, A. Burchard, M. Deicher, H. Quintel, M. Restle, H. Hofsäss and C. Ronning, Phys. Rev. B. (in print).
[10] K. Bharuth-Ram, U. Vetter, H. Hofsäss, C. Ronning and M. Dietrich, Proc. Intl. Conf. on Ion Beam Analysis, Cairns, Australia, July 2001.

Phonon Density of States of FeNi-Invar Determined by Nuclear Inelastic Scattering

H. Bremers (h.bremers@tu-bs.de)
Institut für Metallphysik und Nukleare Festkörperphysik, TU Braunschweig, Mendelssohnstr. 3, D-38106 Braunschweig, Germany

R. Rüffer, O. Leupold and A.I. Chumakov
ESRF, BP220, F-38043 Grenoble Cedex, France

Abstract. Since more than 100 years the Invar effect is known. It is established, that the anomalous low expansion coefficient is strongly correlated to the magnetic properties (keyword: moment volume effect). The weakening of lattice vibrations is supposed to be responsible for the high spin - low spin transition. Due to the capabilities of nuclear inelastic scattering at 3rd generation synchrotron radiation sources it is possible to determine the local phonon density of states of iron. We have investigated a $Fe_{65}Ni_{35}$ Invar foil enriched in ^{57}Fe in a range between 293 K and 540 K. Beside the local density of states the analysis delivers local thermodynamic properties such as Lamb-Mössbauer-factor and mean force constants which show anomalies near the Curie temperature.

Keywords: Invar, nuclear inelastic scattering

The lattice dynamical properties are discussed to have a strong correlation with the ferromagnetic long range order in Invar alloys. Nuclear inelastic scattering, discovered in 1995 by Seto et al.[1], is an established method to derive the density of phonon states [2]. So the aim of this paper is to study the temperature dependence of the local density of phonon states in the $Fe_{65}Ni_{35}$ Invar alloy.

The experiment was perfomed at the Nuclear Resonance Beamline, ID18 [3] at the ESRF. During the experiments the storage ring was run in 16 bunch and single bunch mode. Using high resolution monochromators a full width at half maximum (FWHM) of 0.7 meV and 4.2 meV was achieved. For the first resolution we have used a monochromator consisting of a Ge (331) and two Si (975) reflecting crystal planes and for the second we have used a nested setup with two channelcuts (Si (422) and Si (975)). The setup for the measurements consists of a forward detector to determine the instrumental function and a 4π detector close to the sample to measure the inelastic spectrum. An Invar alloy of composition $Fe_{65}Ni_{35}$, enriched to 95% in ^{57}Fe, was used as sample material. After preparation it was rolled to a thickness of 6 μm.

The local phonon density of states $g(E)$ can be derived from the normalized spectra $S(E)$ measured with the 4π detector using the following recursion formular. For this purpose the measured and normalized

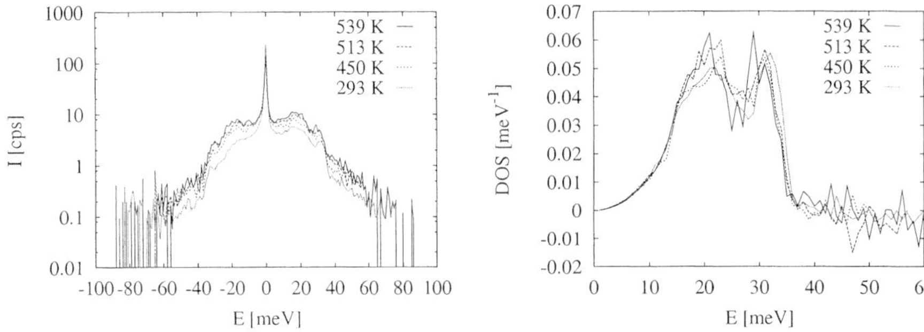

Figure 1. Left: Countrate in the 4π detector measured with the 0.7 meV bandwidth monochromator and 1 meV binning. Right: Local density of phonon states.

spectrum is decomposed in a multiphonon expansion

$$S(E) = f_{LM} \sum_{n=0}^{\infty} S_n(E) \tag{1}$$

$$\text{with } S_0(E) = \delta(E), \quad S_1(E) = \frac{E_R \cdot g(|E|)}{E\left(1 - e^{-\frac{E}{kT}}\right)} \quad \text{and}$$

$$S_n(E) = \frac{1}{n} \int_{-\infty}^{+\infty} S_1(E') S_{n-1}(E - E') dE'$$

In this multiphonon expansion S_0 represents the elastic part of absorbtion (zero-phonon term), S_1 denotes the one-phonon term and S_n may be found using recursion (f_{LM} = Lamb Mössbauer factor, E_R = recoil energy). The function $g(E)$ is the normalized density of phonon states (DOS). For a detailed theory of nuclear inelastic scattering see for example [4, 5]. Using this DOS we can derive the Lamb-Mössbauer-factor

$$f_{LM} = \exp\left[-E_R \int_0^{\infty} \frac{g(E)}{E} \frac{1 + e^{-\frac{E}{kT}}}{1 - e^{-\frac{E}{kT}}} dE \right] \tag{2}$$

or the mean force constant

$$F_{const} = \frac{M}{\hbar^2} \int_0^{\infty} g(E) E^2 dE \quad (M = \text{mass of nucleus}) \tag{3}$$

On the left hand side of figure 1 the measurements performed with the 0.7 meV monochromator during 16 bunch mode are shown. The spectra are normalized to the mean incoming flux and are binned to a grid of 1 meV. With increasing temperature the intensity of the zero-phonon peak $I(E = 0)$ decreases whereas the multiphonon part of the spectrum gains intensity. On the right the density of phonon states is presented. The softening of phonon modes espacially of the $TA_1[\xi\xi 0]$-mode as reported by Endoh et al.[6] could not be confirmed within the experimental errors. But instead of this we see a shift of the whole distribution towards lower values with increasing temperature and an increasing intensity of the peak around 20 meV.

Because of the lower intensity during single bunch mode and the lack of fine structures in the DOS we decided to change the monochromator from 0.7 meV towards 4.2 meV bandwidth. With this resolution we were able to get a series of spectra in the range 300 K to 540 K. To our surprise the elastic peaks in the forward spectrum, used to monitor the instrumental function, and the 4π spectrum exhibit a large but different variation with temperature (see fig. 2). Starting from room

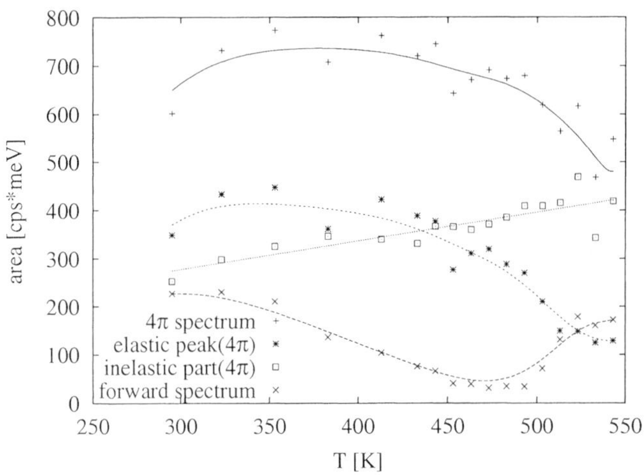

Figure 2. Area of the spectra in the forward- and 4π detector (lines are guides to the eye). The area of the elastic peak in the 4π spectrum corresponds to the Lamb-Mössbauer-factor, whereas the forward spectrum is influenced by the dephasing of the signal due to hyperfine field distributions.

temperature the area of the elastic (0-phonon) peak in the 4π spectrum decreases slightly, but in the vicinity of the Curie temperature there is a stronger decrease. The elastic peak in the forward spectrum experiences a much more pronounced decrease in intensity (approximatly a factor of 10 from 300 K to 500 K), but the intensity recovers when reaching the Curie temperature. The area of the inelastic part of the 4π spectrum shows a nearly linear increase with increasing temperature. The last observation can easily be explained by the increasing number of occupied phonon states with increasing temperature. The same effect should lead to a linear decrease in the elastic peaks. But the decrease in the forward elastic channel is much more pronounced and it does not explain the effect near the Curie temperature. For the explanation of this effect we have to keep in mind, that a hyperfine magnetic field distribution leads to a faster decay in nuclear forward scattering as shown by Shvyd'ko et al. [7]. It is a well known fact, that in samples with hyperfine magnetic field distribution the width increases with increasing temperature. This leads to the pronounced decrease of counting rate in the foward channel. But when the temperature is close to the Curie temperature this distribution collapses into a single line. So the dephasing in time vanishes and the intensity recovers.

To explain the further decrease in the elastic channel of the 4π detector in the vicinity of the Curie temperature we have to investigate the spectra in more detail. These spectra with a little poorer statistic than the previous ones were used to analyze the Lamb-Mössbauer-factor and the mean force constant using equations (1-3) versus temperature. Figure 3 shows the result of this investigation. With increasing temperature the Lamb-Mössbauer-factor decreases linear till 420 K. Above this temperature it drops down much faster, indicating a weakening of the lattice. This can be seen in the mean force constant as well (fig. 3 right).

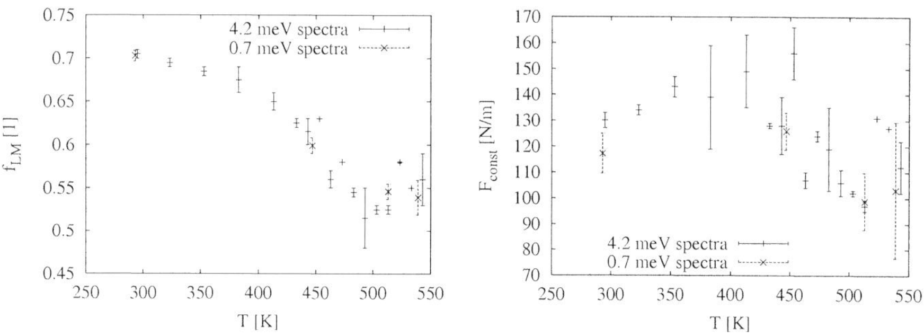

Figure 3. Left: Lamb-Mössbauer-factor. Right: Mean force constant.

References

1. M. Seto, Y. Yoda, S. Kikuta, X.W. Zhang and M. Ando, *Phys. Rev. Lett.* **74** (1995) 3828
2. For a latest review see A.I. Chumakov and W. Sturhahn, *Hyp. Int.* **123/124** (1999) 781
3. R. Rüffer and A.I. Chumakov, *Hyp. Int.* **97/98**(1996) pp. 589-604
4. V.G. Kohn, A.I. Chumakov and R. Rüffer, *Phys. Rev. B* **58** (1998) 8437
5. K.S. Singwi and A. Sjölander, *Phys. Rev.* **120** (1960) 1093
6. Y. Endoh, Y. Noda and Y. Ishikawa, *Solid state Comm.* **23** (1977) 951
7. Y. Shvyd'ko, U. van Bürk, W. Potzel, P. Schindelmann, E. Gerdau, O. Leupold, J. Metge, H.D. Rüter and G.V.Smirnov, *Phys. Rev. B* **57** 8 (1998) 3552

Spin reorientation in $HoCo_{12}B_6$

J.M. Cadogan, S.J. Campbell*, X.L. Zhao*, H.S. Li and P.W. Thompson*
School of Physics, University of New South Wales, Sydney, NSW 2052, Australia
* *School of Physics, University College, University of New South Wales, Australian Defence Force Academy, Canberra, ACT 2600, Australia*

August 31, 2001

Abstract. Mössbauer and AC susceptibility experiments on $HoCo_{12}B_6$ (^{57}Fe) show that the Fe atoms exhibit a strong preferential occupation of the *18h* site. $HoCo_{12}B_6$ undergoes an axis to plane spin reorientation at T_{sr} = 75(2) K.

Keywords: Spin reorientation, rare-earth intermetallics, crystal-fields

1. Introduction

The ternary $RCo_{12}B_6$ (R=rare-earth) intermetallic compounds crystallise in the rhombohedral $SrNi_{12}B_6$ structure [1] with the space group $R\overline{3}m$. The R atoms occupy the *3a* crystallographic sites, the Co atoms occupy the *18g* and *18h* sites, and the B atoms occupy *18h* sites. The magnetic ordering temperatures of the $RCo_{12}B_6$ compounds are in the range 134–162 K [2, 3]. Previous ^{57}Fe Mössbauer studies of $RCo_{12}B_6$ doped with enriched ^{57}Fe [4–6] indicated that the Fe atoms show a strong preferential occupancy of the *18h* site. Gubbens et al. [7] derived a set of crystal field parameters for the $RCo_{12}B_6$ series, from ^{166}Er Mössbauer data, but these parameters are unable to explain our ^{57}Fe Mössbauer observations on $NdCo_{12}B_6$ [5] and $ErCo_{12}B_6$ [8]. Here, we present our study of $HoCo_{12}B_6$ (^{57}Fe) using Mössbauer and AC susceptibility measurements.

2. Experimental methods, Results and Discussion

The $HoCo_{12}B_6$ (^{57}Fe) sample was prepared from high purity metals (99.99% Ho; 99.998% Co and 99.7% B) by arc melting in an atmosphere of purified argon. The doping with 97.5% enriched ^{57}Fe was 0.5 wt%. The arc-melted sample was wrapped in Ta foil and annealed in a sealed evacuated quartz tube at 900 °C for three weeks. The sample was characterized by powder X-ray diffraction using CuKα radiation. The lattice parameters of our sample are a=9.44(2) Å and c=7.44(2) Å. Single-phase samples could not be produced; ~10 wt% of unreacted

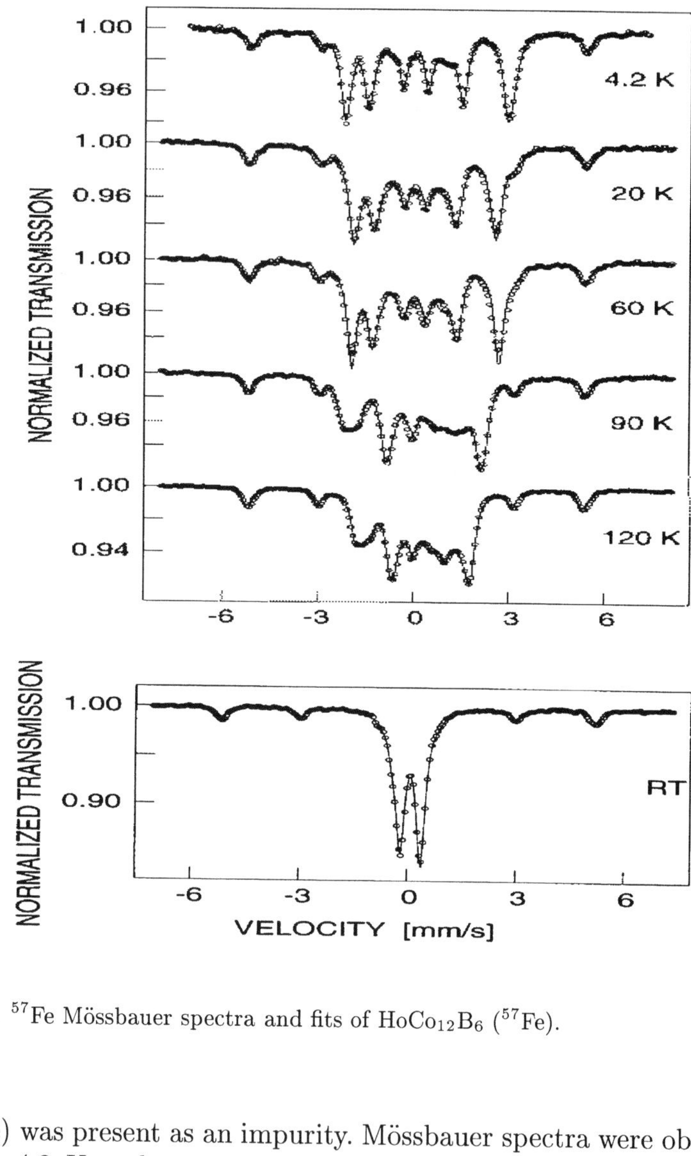

Figure 1. ^{57}Fe Mössbauer spectra and fits of $HoCo_{12}B_6$ (^{57}Fe).

Co(^{57}Fe) was present as an impurity. Mössbauer spectra were obtained between 4.2 K and room temperature (293(1) K) using a standard transmission spectrometer and a ^{57}Co**Rh** source. The spectrometer was calibrated using an α-Fe foil at room temperature and all spectra are presented relative to the centre of the α-Fe spectrum. The Mössbauer spectra of $HoCo_{12}B_6$ (^{57}Fe) are shown in Figure 1. The room temperature spectrum comprises a dominant doublet reflecting the paramagnetic state of the $HoCo_{12}B_6$ phase at room temperature, together with a magnetically-split sextet (B_{hf} = 31(1) T) due to ^{57}Fe in the Co impurity. The impurity subspectrum accounts for ~16% of the total area and was included in all fitting. As in $NdCo_{12}B_6$ (^{57}Fe) [5], the 4.2 K spectrum of $HoCo_{12}B_6$ (^{57}Fe) was fitted with two subspectra with a B_{hf} ratio of ~2.3. The relative subspectral areas of the *18h* and *18g* sites are ~88 % vs ~12 % , respectively, confirming the strong preferential site occupancy of the *18h* site by the Fe atoms, which is a feature of the $RCo_{12}B_6$ (^{57}Fe) series. The respective B_{hf} values are 15.5(5) T and 6.6(7) T. Here, we will concentrate on the dominant *18h* subspectrum.

The spectra at T ≤ 60 K are well fitted with two sextets, whereas the spectra at 90 K and 120 K require at least four sextets to obtain a reasonable fit and details of the fitting procedure are in our previous report on $NdCo_{12}B_6$ (^{57}Fe) [5]. Above 80 K the *18h* subspectrum is split into three equal-area subspectra, with B_{hf} values differing by ~1 T. The combined area of these three sextets is ~85 % of the total $HoCo_{12}B_6$ (^{57}Fe) spectral area, in agreement with the *18h* proportion of 88 % observed in the lower temperature spectra. Differences in the quadrupole splittings (QS) are also observed between the three *18h* sextets. This splitting of the *18h* subspectrum has been discussed for $NdCo_{12}B_6$ (^{57}Fe) where it was attributed to a spin reorientation from the crystal's c-axis to the basal plane. The AC susceptibility (χ_{ac}) of $HoCo_{12}B_6$, recorded in the temperature range 4.2–174 K [1] shows two sharp peaks at 146(2) K and 75(2) K. The former marks the Néel temperature of $HoCo_{12}B_6$ and the latter indicates a rearrangement of its magnetic structure. A sharp peak in χ_{ac} indicates a spin reorientation taking place with a 90° change in the magnetization orientation over a narrow temperature range. Our Mössbauer and χ_{ac} results thus provide firm evidence for a spin reorientation occurring in $HoCo_{12}B_6$ at 75(2) K, from the c-axis to the basal plane with increasing temperature.

We have carried out standard calculations over a sphere of radius 60Å to estimate the anisotropic dipolar contribution to the ^{57}Fe B_{hf} [9]. Using an average dipole value of 0.44 μ_B [10] we find that the dipolar contribution to B_{hf} is the same for all Co atoms in the *18h* and *18g* sites when the magnetization is parallel to the c-axis. When the magnetization is in the basal plane, the B_{hf} at the two Co(Fe) sites, *18h* and *18g*, differ by ~0.07 T, whereas within the same site there are three sub-groups with even smaller differences in B_{hf}, consistent with calculations by Erdmann et al. [10]. We conclude that the splitting of B_{hf}, for planar magnetization, is mainly due to orbital effects, since the calculated dipolar interaction is far too small to account for the observed splitting.

---

[1] with 10 A/m (rms) and 187 Hz

We also calculated the electric field gradient (EFG) at the ^{57}Fe sites using a similar approach to the dipolar calculations, Z_{EFG} is perpendicular to the mirror plane and X_{EFG} and Y_{EFG} lie in the mirror plane; X_{EFG} makes an angle of $\sim$-47.5° with the basal plane. The calculated asymmetry parameter is 0.75. The measured QS at a particular ^{57}Fe site depends on the orientation of B_{hf} (polar angles θ and ϕ) within the EFG principal axis frame and the subspectrum splits when the magnetization deviates from the crystal c-axis. C-axis magnetization yields a common QS for all *18h* atoms, in support of our claim of c-axis anisotropy in $HoCo_{12}B_6$ below its spin reorientation temperature. When the magnetization is along the basal plane principal directions [100], [210] or [110], two groups of *18h* atoms with different QS and an intensity ratio of 1:1 are obtained. Three groups of atoms with different QS and intensities in the ratio 1:1:1 only arise when the basal magnetization lies along [410] or a more asymmetric direction. Details of our calculations for the entire $RCo_{12}B_6$ series will be presented elsewhere.

3. Conclusions

$HoCo_{12}B_6$ has a Néel temperature of 146(2) K and an axis-to-plane spin reorientation at 75(2) K. The splitting of the ^{57}Fe magnetic hyperfine field, for basal plane magnetization, is attributed to orbital effects, while the corresponding splitting in the quadrupole interaction can be accounted for by EFG calculations.

We thank Dr A.V.J. Edge for assistance with sample preparation. This work is supported by grants from the Australian Research Council.

References

1. K. Niihara and S. Yajima, *Chem. Lett.* 5 (1972) 875
2. M. Jurczyk, A.T. Pedziwiatr, W.E. Wallace, *J. Mag. Mag. Mat.* 67 (1987) L1
3. M. Mittag, M. Rosenberg, K.H.J. Buschow, *J. Mag. Mag. Mat.* 82 (1989) 109
4. M. Rosenberg, T. Sinnemann, M. Mittag, K.H.J. Buschow, *J. Alloys and Compounds* 182 (1992) 145
5. J.M. Cadogan, S.J. Campbell, X.L. Zhao, E. Wu, *Aust. J. Phys.* 46 (1993) 679
6. E. Wu, J.M. Cadogan, S.J. Campbell, X.L. Zhao, *Hyp. Int.* 94 (1994) 1903
7. P.C.M. Gubbens, A.M. van der Kraan, K.H.J. Buschow, *J. Mag. Mag. Mat.* 87 (1990) 276
8. X.L. Zhao, J.M. Cadogan, S.J. Campbell, *J. Mag. Mag. Mat.* 140-4 (1995) 959
9. W. Steiner and R. Haferl, *Phys. Stat. Sol. (a)* 42 (1977) 739
10. K. Erdmann, M. Rosenberg, K.H.J. Buschow, *J. Appl. Phys.* 63 (1988) 4113

On the magnetic ordering of $R_6Fe_{13}X$ intermetallic compounds: (R = Pr and Nd; X = Si, Ge and Sn)

J.M. Cadogan and Suharyana
School of Physics, University of New South Wales, Sydney, NSW 2052, Australia

October 15, 2001

Abstract. We present ^{57}Fe Mössbauer spectra of $Nd_6Fe_{13}Ge$ and $Pr_6Fe_{13}Si$ obtained in the temperature range 295–450 K to show that the magnetic ordering process in $R_6Fe_{13}X$ takes place in stages, related to the layered structure.

Keywords: Magnetic ordering, rare-earth intermetallics, $R_6Fe_{13}X$

1. Introduction

Controversy over the magnetism of the $R_6Fe_{13}X$ intermetallics (R = light rare-earth and X = Cu, Ag, Au, Ga, In, Sb, Pb, Si, Ge and Sn) remains, with various claims of ferromagnetism, ferrimagnetism, antiferromagnetism, metamagnetism and spin-reorientations in the literature [1–4]. This tetragonal structure has the space group $I4/mcm$ with two R sites $8f$ and $16l$ and four Fe sites $4d, 16k, 16l_1$ and $16l_2$, forming a naturally layered atomic arrangement [1]. In a ^{57}Fe Mössbauer study of $Nd_6Fe_{13}Au$ de Groot et al. [2] observed the development of a paramagnetic doublet in this compound's spectrum *below* the Néel temperature (T_N). Curiously, the relative area of this subspectrum increases with increasing temperature as T_N is approached from below. Furthermore, there appear to be plateaux in these data. The origin of this doublet was unidentified.

Here, we present our ^{57}Fe Mössbauer study of six $R_6Fe_{13}X$ compounds (R = Pr and Nd; X = Si, Ge and Sn) above room temperature. We show that this peculiar paramagnetic doublet is, in fact, a feature of the $R_6Fe_{13}X$ series and we interpret it in terms of the magnetic structure of $R_6Fe_{13}X$. We will use $Nd_6Fe_{13}Ge$ and $Pr_6Fe_{13}Si$ to illustrate our arguments.

2. Experimental methods

The $R_6Fe_{13}X$ samples were prepared from high purity metals by arc-melting in an atmosphere of purified argon. The arc-melted samples were wrapped in Ta foil and annealed in sealed quartz tubes under a partial pressure of purified Argon at 700 °C for one week. The tubes were quenched in water at the end of the annealing period. The samples were characterized by CuKα X-ray powder diffraction. Mössbauer

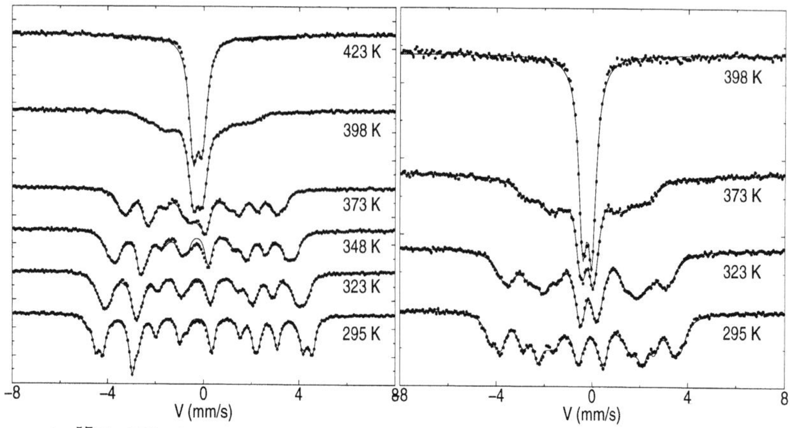

Figure 1. ^{57}Fe Mössbauer spectra and fits of (a) $Nd_6Fe_{13}Ge$ and (b) $Pr_6Fe_{13}Si$

spectra were obtained at temperatures in the range 295–500 K using a standard transmission spectrometer, calibrated with an α-Fe foil at 295 K, and equipped with an Austin Science VF-1000 vacuum furnace. The source was ^{57}Co**Rh**.

Thermomagnetic analysis was carried out above 295 K a Perkin-Elmer TGA-7 thermogravimetric analyser in a small magnetic field gradient. Néel temperatures were determined by differential scanning calorimetry on a Perkin-Elmer DSC-7, using the heat capacity peak at T_N to signal ordering. AC-susceptibility was measured in a LakeShore Closed-Cycle susceptometer at 137 Hz. Magnetically aligned samples for XRD and Mössbauer measurements were prepared by mixing fine powder with epoxy resin and setting the mixture in a magnetic field of 1.9 T.

3. Results and Discussion

XRD on $Nd_6Fe_{13}Ge$ shows that the sample is virtually single-phase with the $R_6Fe_{13}X$ *I4/mcm* structure. We used Rietveld analysis of the pattern to estimate the amount of the $Nd_2(Fe, Ge)_{17}$ impurity to be no more than ~2 wt.%. In figure 1a we show the Mössbauer spectra of $Nd_6Fe_{13}Ge$. TGA shows a very weak magnetic transition at 357(5) K and while it would be natural to assume that this transition is due to the impurity $Nd_2(Fe, Ge)_{17}$ we see that a dramatic change in the Mössbauer spectra accompanies this 'magnetic event' in the TGA. DSC shows a T_N of 429(5) K for $Nd_6Fe_{13}Ge$. The easy direction of magnetization of $Nd_6Fe_{13}Ge$ is the tetragonal c-axis. This is confirmed by our XRD and Mössbauer work on a magnetically aligned sample (at 295 K). Thus, we used 4 sextets to fit the magnetic spectra of $Nd_6Fe_{13}Ge$, in the area ratio 4:16:16:16, in agreement with other workers [3].

The hyperfine fields (B_{hf}) at 295 K are 28.6, 28.0, 25.9 and 18.8 T for the $4d$, $16k$, $16l_1$ and $16l_2$ sites, respectively. The isomer shifts are -0.14, -0.04, -0.04 and +0.02 mm/s, relative to α-Fe. At 373 K, our spectrum has one 16-type subspectrum (area = 31 %) with a B_{hf} of only 5(2) T, much smaller than the fields of the other three Fe sites (22 T, 20 T and 16 T). This site also has the most positive isomer shift, from which we identify it as the $16l_2$ site using arguments based on the Wigner-Seitz (WS) cell volume proposed by Hautot et al. [4] *vide infra*. A dramatic change in our spectra occurs between 373 K and 398 K, still significantly below the Néel temperature of 429 K, although slightly higher than the TGA result of 357 K. At 398 K, the 'paramagnetic' component of our spectrum accounts for 63 % of the total spectral area, clearly far too large to be due to an impurity. The spectrum at 423 K shows that the sample is paramagnetic at this temperature, in agreement with our DSC result.

We calculated the WS cells around the 4 Fe sites in $Nd_6Fe_{13}Ge$ using BLOKJE [5]. The number of R and Fe nearest-neighbours (nn) at the $4d$, $16k$, $16l_1$ and $16l_2$ Fe sites are (0, 12), (2, 10), (3, 9) and (5, 7), respectively. Based on the 'rule of thumb' that R nn reduce B_{hf} whereas Fe nn increase B_{hf}, it is clear that the order of the B_{hf} in the $R_6Fe_{13}X$ compounds is $B_{hf}(4d) > B_{hf}(16k) > B_{hf}(16l_1) > B_{hf}(16l_2)$. The WS volumes (in $Å^3$) for these four sites are 10.53, 11.37, 11.47 and 12.79 and our isomer shifts show the linear dependence on WS-volume previously discussed by Hautot et al. [4].

Neutron diffraction on $R_6Fe_{13}X$ compounds (Pr-Fe-Si, Nd-Fe-Au, Pr-Fe-Sn and Nd-Fe-Sn), by Schobinger-Papamantellos et al. [6, 7] shows that the $16l$ R is weakly magnetic and its refined magnetic moment at RT is almost zero. In $Pr_6Fe_{13}Si$ Schobinger-Papamantellos et al. used the temperature dependence of the integrated intensity of the (001) magnetic peak to deduce a T_N of 385(5) K, in agreement with our DSC value of 380(5) K. Unlike the (001) peak, whose intensity increases monotonically with decreasing temperature, the (003) peak increases from 385 K down to ~300 K, and then decreases with further decreases in temperature. This maximum in the (003) intensity was interpreted in terms of a "gradual ordering" of the Pr $16l$ site, occurring well below T_N. Our ac-susceptibility of $Pr_6Fe_{13}Si$ shows a rapid increase at 300(4) K, consistent with the neutron diffraction data. Furthermore, our Mössbauer spectra of $Pr_6Fe_{13}Si$ in figure 1b show a dramatic change occurring between 295 K and 323 K. The 323 K spectrum contains ~24 % of the paramagnetic doublet which increases to ~37 % by 373 K. (our XRD and Mössbauer work on a magnetically-aligned sample shows that $Pr_6Fe_{13}Si$ orders in the basal plane, in agreement with other workers [6] so the simple 4-subspectral area ratio 4:16:16:16 is inapplicable.

We propose that the magnetic ordering in $R_6Fe_{13}X$ occurs in stages. Heating $Nd_6Fe_{13}Ge$ from 373 K to 398 K results in the sudden growth of a paramagnetic component whose area is 63 % of the total spectral area. We propose that two of the 16-type Fe sites in $Nd_6Fe_{13}Ge$ lose their long-range magnetic order around 380 K; this would correspond to 61.5 % of the total spectral area, assuming the same recoil-free fractions for all Fe sites. The $16l_2$ site has only 7 Fe *nn* and 5 R *nn*, 3 of which are the Nd $16l$ site, already shown to be weakly magnetic, losing its magnetic moment well below the T_N of the $R_6Fe_{13}X$ compound. The Fe $16l_1$ site has 9 Fe *nn* and 3 R *nn*, 1 of which is the R $16l$ site. We therefore propose that the Fe $16l_1$ and $16l_2$ sites are effectively paramagnetic significantly below the T_N of the $R_6Fe_{13}X$ compound, defined by the magnetic order of the R $8f$ and the Fe $4d$ and $16k$ sites. This behaviour is most likely a consequence of the highly layered structure and its role in determining the net exchange 'seen' by a particular site.

Finally, de Groot et al. [2] also observed a small discontinuous decrease in average B_{hf} in $Nd_6Fe_{13}Au$ upon cooling through a temperature coinciding with the appearance of the Mössbauer paramagnetic doublet, once again well below T_N . We propose that this downturn in B_{hf} is not due to an impurity's magnetic ordering; one would expect an upturn in B_{hf} if this were the case. Rather, it reflects the ordering of some of the sites and the resulting effect on B_{hf} .

4. Conclusions

^{57}Fe Mössbauer spectroscopy has been used to show that the magnetic ordering in the $R_6Fe_{13}X$ of intermetallic compounds is not fully cooperative but occurs in stages related to the naturally layered structure.

This work was supported by the Australian Research Council.

References

1. J. Allemand, A. Letant, J.M. Moreau, J.P. Nozieres and R. Perrier de la Bathie, *J. Alloys and Compounds* 166 (1990) 73
2. C.H. de Groot, F.R. de Boer, K.H.J. Buschow, D. Hautot, G.J. Long and F. Grandjean, *J. Alloys and Compounds* 233 (1996) 161
3. B.P. Hu, J.M.D. Coey, H. Klesnar and P. Rogl, *J. Magn. Magn. Mater.* 117 (1992) 225
4. D. Hautot, G.J. Long, F. Grandjean, C.H. de Groot and K.H.J. Buschow, *J. Appl. Phys.* 81 (1997) 5435
5. L. Gelato, *J. Appl. Cryst.* 14 (1981) 151
6. P. Schobinger-Papamantellos, K.H.J. Buschow, C.H. de Groot, F.R. de Boer, G. Böttger and C. Ritter, *J. Phys. Condens. Matter* 11 (1999) 4469
7. P. Schobinger-Papamantellos, K.H.J. Buschow, C.H. de Groot, F.R. de Boer and C. Ritter, *J. Magn. Magn. Mater.* 218 (2000) 31

Mössbauer Hyperfine Parameters of Fe-Al Alloys

J. Deniszczyk and J. E. Frackowiak
Institute of Physics and Chemistry of Metals, Silesian University,
40-007 Katowice, Bankowa 12, Poland

August 29, 2001

Abstract. The Fermi contact magnetic hyperfine field (B_F) and isomer shift (IS) at the Fe site in bcc Fe-Al alloy for all possible configurations of the nearest neighbor (NN) atoms were calculated from first principles within the super-cell approach. The resulted B_F and IS were averaged with the use of the probabilities of specific NN configurations for disordered and ordered alloys with B2 and DO_3 superstructure taken from computer modeling. The calculated results agree very well with experimental data.

Keywords: electronic structure, intermetallic alloys, hyperfine parameters, structure simulations

1. Introduction

The magnetic properties of $Fe_{1-x}Al_x$ alloys with bcc structure have been extensively investigated experimentally by Mössbauer spectroscopy [1–8]. It has been shown that the average magnetic moment μ (extrapolated to 0 K) decreases with the Al concentration (C_{Al}) following a simple dilution model for C_{Al} up to 25 at. % [2]. For 25—30 at. % of Al it shows a sudden fall followed by a slow variation until it becomes zero at 40 at. %. The mean hyperfine magnetic field B_{hf} decreases with the Al concentration but much more slowly then magnetic moment [1, 3]. The value of isomer shift (IS) is positive for Fe-Al alloys and increases continuously with the Al concentration for disordered samples [4, 5]. For ordered samples the significant increase in value of (IS) is observed near the stoichiometric concentrations $Fe_{0.75}Al_{0.25}$ (DO_3 superstructure) and $Fe_{0.5}Al_{0.5}$ (B2 superstructure) [6, 7]. All these studies indicate that the magnetic moment as well as the hyperfine parameters (B_F and IS) of a given Fe atom in $Fe_{1-x}Al_x$ are determined primarily by its nearest neighbor (NN) configuration.

2. Method of calculations

In this work the Fermi contact magnetic hyperfine field (B_F) and isomer shift (IS) as well as magnetic moment at the Fe site in bcc Fe-Al alloys were calculated from first principles. For this purpose the super-cell approach was applied and the electronic structure calculations were performed with the use of the Tight-Binding Linear Muffin Tin Orbital (TB-LMTO) method in the Atomic Sphere Approximation (ASA) [9]. (The details of the computational setup and discussion of the structural and magnetic results of these calculations were presented in previous papers [10].) The presented values of hyperfine field were calculated from the relativistic formula given by Blügel *et al.*[13]. For isomer shift the formula $IS = -0.24 \cdot (\rho^{Fe} - \rho^{Fe-bcc})$, where ρ^{Fe} and ρ^{Fe-bcc} stand respectively for the total charge density on the surface of ^{57}Fe nucleus in alloy and elemental bcc-Fe, was used. The hyperfine parameters were calculated for a complete set of clusters $\{C_i^m\}$ of B2-type (FeAl) structure consisting of the Fe central atom (Fe-β) with the number i of neighboring Fe atoms at Al sites (i=0,1...7) allocated according to the mth crystallographically non-equivalent configuration.

To get alloy-averaged values of the hyperfine parameters the probabilities P_i^m of the occurrence of C_i^m clusters as a function of C_{Al} and atomic order must be known. With this aim we have performed the computer simulation of local atomic configurations for binary substitutional alloys by Monte Carlo method based on the modification of the Gehlen & Cohen [14] concept. Within the approach a model with 40×40×40 simple cubic unit cells was used to form four independent bcc lattices. The initial structure was taken of B2-type or DO_3-type (disordered or ordered) and four Warren short-range order parameters α_i [14] (with maximal values appropriate for a given concentration C_{Al} in the case of ordered alloys) were taken as the designed values. The convergence of the simulation procedure was achieved if simulated alphas were close to the designed values. The parameters P_i^m for all C_i^m clusters were determined and in the next step the average values of B_F and IS were calculated for ordered and disordered $Fe_{1-x}Al_x$ $(0 < x \leq 0.5)$ alloys.

3. Results and discussion

The dominant contribution to the hyperfine magnetic field originates from the Fermi contact term (B_F) proportional to the electronic spin density at the Fe nucleus. In this work the contributions to the hyperfine field from the core (1*s*, 2*s* and 3*s*) electrons — B^{cor} and from valence 4*s* electrons — B^{val} were calculated separately. The knowledge of the calculated partial (μ_{3d} and μ_{4s}) contributions to the local magnetic moments of iron allowed to do a detailed analysis of the relation between hyperfine field and magnetic subsystem in Fe-Al alloys. The results of this analysis we reported in [12] and can be summarized as follows. The B^{cor} contribution depends linearly (with negative slope) on the μ_{3d} partial self-magnetic-moment of Fe independently of its atomic position. Such dependence could be expected since the spatial spin polarization of the inner-shell electronic wave function is mainly due to the exchange-correlation coupling to the spin-polarized cloud of 3*d* shell electrons. The contribution B^{val} shows the dependence on both partial, μ_{3d}

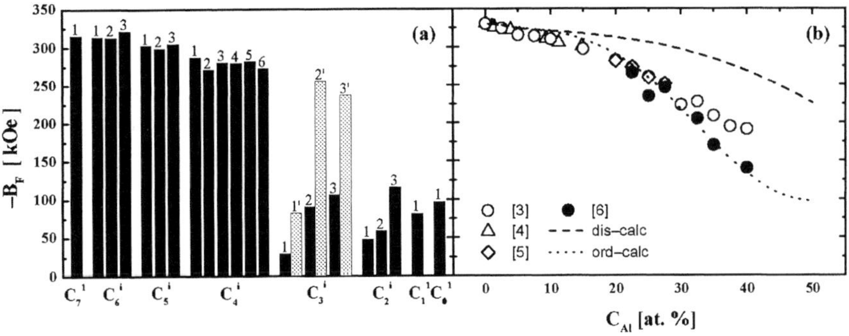

Figure 1. Fermi contact hyperfine field for calculated clusters —(a) and alloy-averaged results (lines) together with experimental data (symbols) versus Al-concentration —(b). (*The bars $C_3^{i(i')}$ represent solutions which differ in magnitudes and/or orientation of local moments. The open and solid symbols depict the disordered (dis) and ordered (ord) samples, respectively*).

and μ_{4s}, self-magnetic-moments of Fe and can be split into *local* (B^{val}_{loc}) and *transferred* (B^{val}_{tr}) parts. Because of the long-range character of interactions responsible for the spin-polarization of 4*s*-band electrons it was impossible to relate the B^{val}_{tr} part of B^{val} to the magnetic moments of surrounding atoms.

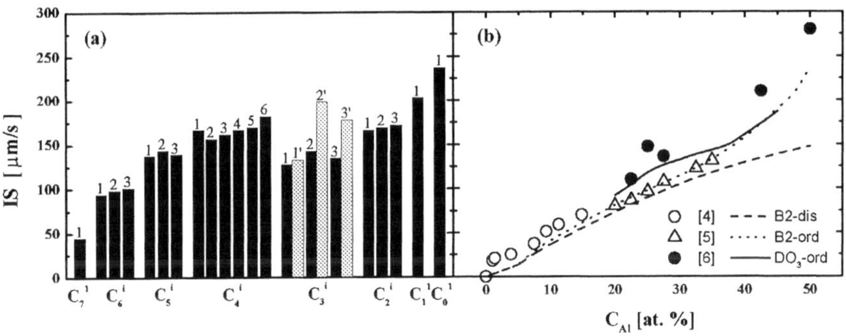

Figure 2. Calculated and measured isomer shift (*Description of the elements follows that given for figure 1*)

Figures 1 and 2 display the results of our investigations. Figure 1a shows the total Fermi contact hyperfine field ($B_F = B^{cor} + B^{val}$) at Fe-β atoms calculated for different cluster configurations C^m_i with increasing number n_{Al} of Al atoms ($n_{Al} = 8 - i$). The calculated values of B_F show well defined separation of the high-magnetic (for $n_{Al} \leq 4$) and low-magnetic (for $n_{Al} \geq 6$) states. In this regard the results may serve justification for empirical *discrete magnetic-state* models used in interpretations of the magnetic properties of Fe-Al alloy [2, 8]. In figure 1b the alloy-averaged B_F and experimental data are displayed. The agreement is satisfactory. The observed tendency that experimental points (for both ordered and disordered samples) follow the curve calculated for the ordered alloys is probably due to the too small size of the super-cell considered. Figure 2a presents the isomer shift calculated for Fe-β atoms for increasing number n_{Al} of NN Al atoms in C_i^m clusters. The variation of IS versus increasing n_{Al} differs qualitatively from that observed for B_F (fig. 1a). The value of IS increases continuously for clusters with $n_{Al} \leq 4$. After a small collapse for n_{Al} =5 the IS continues the increase up to n_{Al}=8. In figure 2b the alloy-averaged IS is compared with experimental data. For both types of atomic order (B2 and DO_3) as well as for the case of disordered B2-type structure the calculated values of IS reproduce the experimental results almost quantitatively. Detailed analysis of the results revealed that the increase of the IS on the Fe-β atoms observed for increasing Al concentration is due to the increasing Al→Fe-β electronic charge transfer with simultaneous strong intraatomic 4s→3d conversion on Fe-β atoms [11]. In effect the population of 4s(3d) states of Fe-β atoms decreases(increases) resulting in the increase of the positively valued IS.

4. Conclusions

In order to investigate the hyperfine magnetic field in $Fe_{1-x}Al_x$ alloys different parts of the Fermi contact contribution to the hyperfine field were calculated and discussed. It was found that high-magnetic states and low-magnetic states of Fe atoms are well separated for different (NN) configurations of given Fe atoms. The calculated average values of B_F agree well with experiment. The slight discrepancy suggests that in the *ab initio* calculations the next nearest-neighbor configurations have to be taken into account. For both disordered and ordered alloys the dependence of the alloy-averaged IS on C_{Al} follows the experimental results nearly quantitatively.

References

1. G. P. Haufman and R. M. Fisher, J. Appl. Phys. 38 (1967) 735-742.
2. I. Vincze, phys. stat. sol. (a) 7 (1971) K43–47.
3. G.A. Pẽrez Alcazar and E. Galvão da Silva, J. Phys. F 17 (1987) 2323–2335.
4. S.M. Dubiel and W. Zinn, Phys. Rev. B 26 (1982) 1574–1589.
5. J.E. Frąckowiak, Hyp. Interact. 28 (1986) 1067–1070.
6. J.E. Frąckowiak, Hyp. Interact. 54 (1990) 793–798.
7. Z.Q. Gao and B. Fultz, Phil. Mag. B 67 (1993) 787–800.
8. S.A. Makhlouf, M. Shiga and K. Sumiyama, J. Phys. Soc. Jap. 60 (1991) 3537–3542.
9. O.K. Andersen, O. Jepsen and D. Glötzel, in: *Highlights of Condensed Matter Theory*, eds. F. Bassani, F. Tosi and M. P. Tosi (North-Holland, Amsterdam, 1985) pp. 59–176.
10. J. Deniszczyk, Acta Phys, Pol. A 97 (2000) 583-586; *ibid* 98 (2000) 543-546.
11. J. Deniszczyk and J.E. Frąckowiak, Acta Phys, Pol. A 91 (1997) 455–458.
12. J. Deniszczyk and J.E. Frąckowiak, Mol. Phys. Rep. 30 (2000) 33–37.
13. S. Blügel, H. Akai, R. Zeller and P.H. Dederichs, Phys. Rev. B 35 (1987) 3271–3283.
14. P.C. Gehlen and J.B. Cohen, Phys. Rev. 139 (1965) A844–A855.

IRON VALENCE IN DOUBLE PEROVSKITES

A.P. Douvalis, M. Venkatesan and J.M.D. Coey
Physics Department, Trinity College, Dublin 2, Ireland

The valence state of iron in the compounds Ca_2FeMoO_6, Sr_2FeMoO_6, Ba_2FeMoO_6, Ca_2FeReO_6, Sr_2FeReO_6, $Sr_2FeMo_{1-x}W_xO_6$ (x=0.2 and 0.4), and $(Sr_{2-x}La_x)FeMoO_6$ (x=0.5 and 0.7), which have been prepared using ceramic and combustion techniques, has been studied using ^{57}Fe Mössbauer spectroscopy. Judging by the isomer shifts and hyperfine fields, intermediate iron oxidation states are found for all these compounds, represented by a $3d^{5.x}$ configuration, where 1<x<5. Here 0.x represents the number of ↓ electrons of Fe(3d) character in the Fe/Mo,W or Fe/Re conduction band.

1. Introduction

Recently the iron-based ferrimagnetic half metallic double perovskites of general formula A_2FeMO_6 (A=Ca, Sr, Ba; M=Mo, W, Re) have attracted much interest [1, 2]. The reason for this is that some of these compounds show rather good magnetoresistive properties at room temperature (RT), and additionally their Curie temperatures (T_C) have values higher than the ferromagnetic mixed valence manganites [3]. The above features make these compounds suitable candidate materials for future spin electronic devices [4]. The half metallic properties depend on the kind and valence state of the B-site cations (Fe, M), and also on the corresponding characteristics of the A site cations [3, 5, 6]. Iron ions in these compounds occupy the B-sites alternately with Mo, W or Re ions [7]. Information on the valence state of iron ions, and also on the atomic defects occurring in the B-sites, are important issues to understand their properties. Considering this importance, ^{57}Fe Mössbauer Spectroscopy offers a unique tool for the characterization of the valence state of iron ions and defects formed in these oxides.

The magnetoresistive properties of these materials in polycrystalline form has an extrinsic character [3, 8], correlated with the electronic spin scattering at the boundaries of grains, which have different relative orientations of their magnetization. Different preparation routes for these polycrystalline materials can result in different forms and sizes of grains, and also in different forms and amounts of defects (antisite defects, antiphase boundaries) in the crystal structure, affecting their magnetoresistive properties accordingly.

2. Experimental

Polycrystalline samples of Ca_2FeMoO_6, Sr_2FeMoO_6, Ba_2FeMoO_6, $Sr_2FeMo_{[1-x]}W_xO_6$ (x=0.2 and 0.4), and $(Sr_{2-x}La_x)FeMoO_6$ (x=0.3 and 0.7) were prepared using ceramic and combustion techniques. Sr_2FeReO_6 and Ca_2FeReO_6 were prepared by solid-state reaction. Mössbauer spectra (MS) were collected at RT and 16 K using spectrometers operated in the constant acceleration mode, and equipped with ^{57}Co(Rh) source kept at RT. A closed loop He cryostat (APD DMX-20) was used for measurements at low temperatures. The velocity calibration of the spectrometers was done using α-Fe and all the isomer shift (δ) values are given relative to α-Fe at RT.

3. Results and discussion

MS of Sr_2FeMoO_6, Ca_2FeMoO_6, and Ba_2FeMoO_6 at 16 K are shown in fig. 1a. Figures 1b and 1c display the corresponding spectra of $Sr_2FeMo_{1-x}W_xO_6$ (for x=0, 0.2, and 0.4), and $(Sr_{2-x}La_x)FeMoO_6$ (for x=0, 0.5 and 0.7) respectively, and in fig. 2 the corresponding MS of Sr_2FeReO_6 and Ca_2FeReO_6 are shown. The RT spectra for all compounds have rather broad and complicated absorption lines, reflecting the effect of multiple metal neighbour environments on the magnetic hyperfine field (B_{hf}) of iron ions, as also the effect of temperature in the vicinity of T_C.

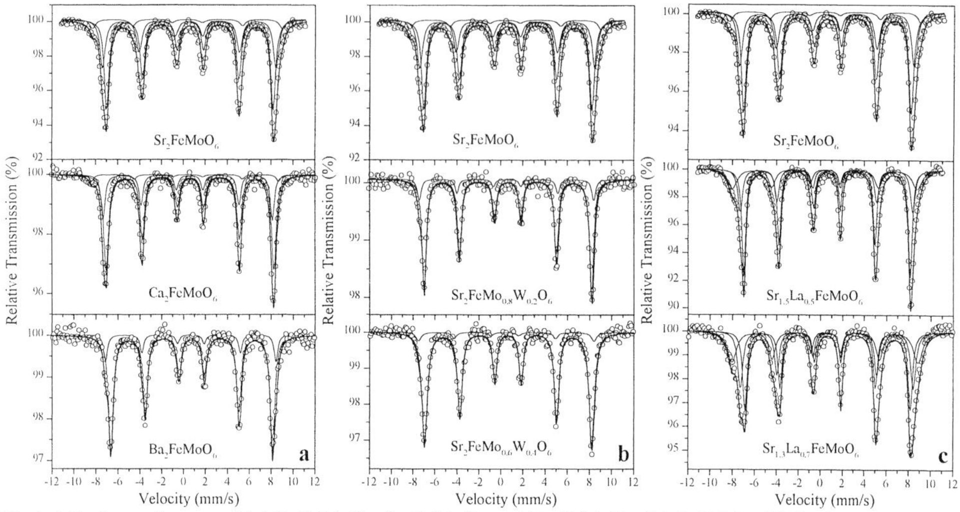

Fig 1. Mössbauer Spectra of (a) Sr_2FeMoO_6, Ca_2FeMoO_6, and Ba_2FeMoO_6; (b) $Sr_2FeMo_{1-x}W_xO_6$ for x=0, 0.2, and 0.4; and (c) $(Sr_{2-x}La_x)FeMoO_6$ x=0, 0.5 and 0.7, all recorded at 16 K.

On the contrary the low temperature spectra have better resolution. A set of two or three theoretical components was needed to describe properly the low temperature spectra for each compound. For all compounds the component with the larger absorption area (main component), is attributed to iron ions in the B-sites of the ideal, ordered double perovskite structure, that has perfect NaCl-type ordering of Fe with Mo, Re or W ions; that is 6 Mo, Re or W first cation neighbours for each Fe ion [5, 9]. The component with the smaller absorption area (secondary component) has lower δ and larger B_{hf} values than the main component for each compound. This component is attributed to iron ions in the B-sites that have a modified first cation neighbour environment, due to the existence of antisite defects [5]. These defects result in an Fe ion having 5 Mo, W or Re, and 1 Fe cation neighbours. Whenever a small third component with the lowest δ and the highest B_{hf} values of all, must be used to describe the experimental data, as in the case of Sr_2FeMoO_6, Ca_2FeMoO_6, $Sr_{1.5}La_{0.5}FeMoO_6$, and $Sr_{1.3}La_{0.7}FeMoO_6$, it is attributed to the Fe ions with a total of six other Fe ions as neighbours, also as a consequence of the antisite defects [5]. The quadrupole shift values for all components in all compounds are close to zero at 16 K. This result is indicative of a nearly cubic arrangement of the oxygen ions in the FeO_6

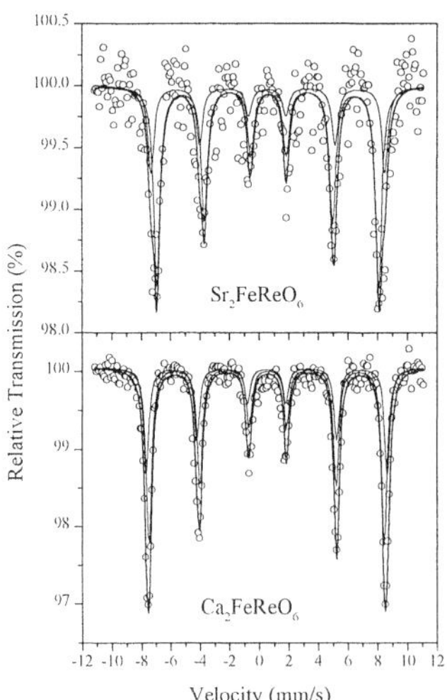

Fig 2. Mössbauer Spectra of Sr_2FeReO_6 and Ca_2FeReO_6, recorded at 16 K.

Further evidence for an intermediate electronic configuration of iron is provided by the plot of B_{hf} vs δ for all components in Fig. 3, where it can be seen that B_{hf} decreases with increasing δ. We include in this plot, not only the main components which have the greatest admixture, but also the secondary and minor components as well. There is an unmistakable correlation of isomer shift and hyperfine field. The extrapolated fields for the end member configurations Fe^{3+} (δ=0.52 mm/s) and Fe^{2+} (δ=1.20 mm/s) are 52 T and 39 T respectively, which are completely in line with values expected for oxides.

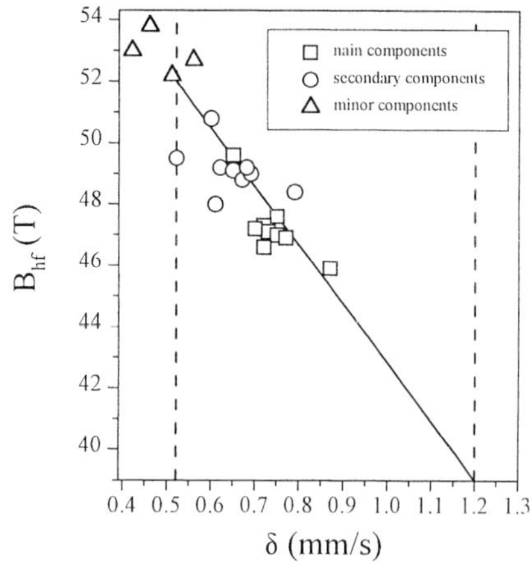

Fig. 3. Plot of B_{hf} vs δ values of all components The straight line is a guide to the eye.

4. Conclusions

The low-temperature ^{57}Fe Mössbauer spectrometry study of Ca_2FeMoO_6, Sr_2FeMoO_6, Ba_2FeMoO_6, Sr_2FeReO_6, Ca_2FeReO_6, $Sr_2FeMo_{1-x}W_xO_6$ (x=0.2 and 0.4), and $(Sr_{2-x}La_x)FeMoO_6$ (x=0.5 and 0.7) revealed intermediate valence states for the iron ions that lie in the ideal B-site NaCl type environment for every one of the compounds. In Ba_2FeMoO_6 iron is in the highest ($3d^{5.5}$) electronic configuration, while in Ca_2FeReO_6 is in the lowest ($3d^{5.2}$). In particular, iron in Sr_2FeMoO_6 cannot be considered to be in a pure Fe^{3+} state [16].

Acknowledgments

This work was supported by the AMORE project of the EU Growth program. We are grateful to Dr. J.M. Greneche for helpful advice and to Dr. R. Suryanarayanan for providing some of the samples.

References

[1] K.I. Kobayashi, T Kimura, H. Sawada, K. Terakura and Y. Tokura, Nature **395** (1998) 677.
[2] K.I. Kobayashi, T. Kimura, Y. Tomioka, H. Sawada and K. Terakura, Phys. Rev. B **59** (1999) 11159.
[3] R.P. Borges, R.M. Thomas, C. Cullinan, J.M.D. Coey, R. Suryanarayanan, L. Ben-Dor, L. Pinsard-Gaudart and A. Revcolevschi, J. Phys.: Condens. Matter **11** (1999) L445.
[4] J.M.D. Coey in : Spin Electronics, eds. M. Ziese and M.J. Thornton (Springer-Verlag Berlin Heidelberg 2001).
[5] J.M. Greneche, M. Venkatesan, R. Suryanarayanan and J.M.D. Coey, Phys. Rev. B **63** (2001) 174403.
[6] A.S. Ogale, S.B. Ogale, R. Ramesh and T. Venkatesan, Appl. Phys. Lett. **75** (1999) 537.
[7] A.W. Sleight and J.F. Weiher, J. Phys. Chem. Solids **33** (1972) 679.
[8] A. Maignan, C. Martin, M. Hervieu and B. Raveau, J. Magn. Magn. Mater. **211** (2000) 173.
[9] L. Pinsard-Gaudart, R. Surynarayanan, A. Revcolevschi, J. Rodriguez-Carvajal, J.M. Greneche, P.A.I. Smith, R.M. Thomas, R.P. Borges and J.M.D. Coey, J. Appl. Phys. **87** (2000) 7118.
[10] O. Chmaissem, R. Kruk, B. Dabrowski, D.E. Brown, X. Xiong, S. Kolesnik, J.D. Jorgensen and C.W. Kimball, Phys. Rev. B **62** (2000) 14197.
[11] N.N. Greenwood and T.C. Gibb in : Mössbauer Spectroscopy (Chapman and Hall, London 1971).
[12] H. Kawanaka, I. Hase, S. Toyama and Y. Nishihara, Physica B **281-282** (2000) 518.
[13] D.P.E. Dickson in: Mossbauer Spectroscopy (Cambridge University Press, 1986).
[14] T. Saha-Dasgupta and D.D. Sarma, Phys. Rev. B **64** (2001) 064408.
[15] M. Abe, T. Nakagawa and S. Nomura, J. Phys. Soc. Jpn. **35** (1973) 1360.
[16] S. Ray, A. Kumar, D.D. Sarma, R. Cimino, S. Turchini, S. Zennaro and N. Zema, Phys. Rev. Letters **87** (2001) 097204.

A Study of the Spin Dynamics and Mössbauer Relaxation in Sr_2FeRuO_6 by Monte Carlo Analysis

T. C. Gibb
School of Chemistry, University of Leeds, Leeds, LS2 9JT, UK

Abstract *Monte Carlo techniques have been used to gain an understanding of the spin frustration observed in Sr_2FeRuO_6 as well as the Mössbauer relaxation and spin freezing seen below 60 K. The assumption of a small antiferromagnetic coupling of next-nearest-neighbour spins can simulate the observed effects.*

1. Introduction

Magnetically concentrated oxides with two magnetic ions, for which the superexchange interactions can result in a degree of frustration, are now well known. The case of Sr_2FeRuO_6 is particularly interesting [1]: the perovskite structure is only slightly distorted and the Fe and Ru cations are randomized, while the magnetic behaviour resembles that of a spin glass with a transition temperature in the region of only 60 K. The Mössbauer spectra show that the Fe spins freeze below that temperature, although neutron diffraction is unable to detect long-range ordering of the spins.

The perovskite structure is a relatively simple structure for analysis of spin interactions, as the linear M—O—M interaction is dominant. A Monte Carlo analysis of magnetically diluted antiferromagnetic perovskites has been carried out recently [2]. Calculations using the uniaxial Ising and isotropic Heisenberg models have enabled a reconciliation of the spin dynamics of these models with the observed relaxation in the Mössbauer spectrum at temperatures well below the apparent ordering temperature in compounds typified by $SrLaFeSnO_6$.

The present work uses similar Monte Carlo techniques to probe the more subtle problem of the spin frustration in Sr_2FeRuO_6. The isotropic nature of the cations, Fe^{3+} (d^5) and Ru^{5+} (d^3), and the small crystallographic distortion of the crystal lattice would suggest that the isotropic Heisenberg model is an appropriate approximation. The presence of two metals results in many more near-neighbour (nn) and next-nearest-neighbour (nnn) interactions to consider. However, using reasonable estimated values for the ratios of the more important interactions, it has been possible to obtain a deeper understanding of the properties of Sr_2FeRuO_6.

2. Simulation method

In the Heisenberg model each spin is represented [2] as a unit vector in space using s_x, s_y and s_z as component vectors. For a cooling cycle calculation these vectors are generated to give random orientations of the spins. The energy of spin i can be written as a summation of the interactions with all neighbouring spins j

$$E_i = -\sum_j J_{i,j}\left(s_{ix}s_{jx} + s_{iy}s_{jy} + s_{iz}s_{jz}\right)$$

The choice of the values for $J_{i,j}$ are discussed below. Data sets of 40^3 magnetic atoms assigned as Fe and Ru at random were created, and allowed to evolve, including periodic boundary conditions to reduce edge effects. At each attempt at spin reversal, a prospective new spin vector was calculated: the three spin vectors were allowed to change randomly by a small increment up to a pre-set limit, and renormalized to a unit spin vector. A decision was then made to accept the new spin vector, or retain the original vector, by using the familiar Metropolis method. The spin system inevitably evolves slowly, but a cycle of 10,000 Monte Carlo cycles (MCS) to achieve equilibration and 10,000 MCS for data collection at each temperature (from kT/J_{aa} = 0.800 to 0.025 in increments of 0.025) was found to give adequate data. This necessarily limits the size of the spin system which can be used, but produces better averages over all spins for $\langle S_x \rangle$, $\langle S_y \rangle$ and $\langle S_z \rangle$ and the average spin value $\langle S \rangle$ than can be obtained for fewer MCS.

3. The exchange interactions

There are 6 nn interactions in the perovskite lattice. The exchange constants were expressed as ratios related to the J_{aa} (linear Fe—O—Fe) antiferromagnetic interaction as unity. The J_{bb} (linear Ru—O—Ru) interaction is expected to be weaker, and a value of 0.25 was adopted from a comparison of ordering temperatures of iron and chromium lanthanide-perovskites, and assuming that the stronger covalence for Ru^{5+} and the weaker 4*d* overlap would mutually cancel. The J_{ab} (linear Fe—O—Ru) value is unknown, but values in the region of –0.1 (ferromagnetic) and 0.1 (antiferromagnetic) were used. There is some uncertainty as to the likely sign of the nn d^3—O—d^5 interaction. In either event, it is not the J_{ab} coupling which leads to magnetic frustration [1], as confirmed by appropriate simulations.

There are 24 nnn interactions. They can be viewed as a non-linear Fe—O—(Fe)—O—Fe interaction, J_{aaa}, in which the central metal atom may be involved perhaps by its polarising influence on the two oxygens. The J_{aaa} and J_{bbb} interactions would be weak and antiferromagnetic, whereas the nn interactions alone should result in parallel alignment. However, in the former case J_{aaa} does not cause significant spin confusion in $SrLaFeSnO_6$. A number of diluted Ru^{5+} perovskites [3-5] have ordering temperatures in the region of 30 K, which suggests that J_{bab} could be in the region of +0.05. The earlier work [1] has suggested that J_{aba} on the other hand is small when the non-magnetic a-cation has a low oxidation state (*e.g.* Sn^{4+}), but is much larger for high oxidation state (Sb^{5+}, Nb^{5+}, Ta^{5+}) such that spin-glass behaviour is created. In the present instance, the high oxidation state of Ru^{5+} may behave similarly, so that J_{aba} can also be significant. In either case a negative value would not result in frustration. In the simulations, the other asymmetric terms such as J_{abb} have been ignored and it is suggested that they are less important.

Simulations have been carried out for a range of values of J_{ab}, J_{aba}, and J_{bab}. Only a few examples are described here to illustrate the main conclusions.

4. The generation of frustration

The final values of the three spin vectors were used [2] as red-green-blue values for a coloured graphical display of the spin system, with the appropriate staggered correlations of the antiferromagnetic order and the Fe—O—Ru coupling included. The colour display can reveal clearly groups of spins with different orientations, and therefore gives an indication of the overall coherence of the spin system.
Selected examples of the final arrangement of the cooled spin system are shown in Figure 1, but in shades of grey which are less effective than in the original. The values

of the coupling constants, $<S_a>$ and $<S_b>$ are shown in each case (with $J_{bb} = 0.25$, $J_{ab} = -0.10$). When J_{aba} is 0.1, there is significant spin misalignment, but complete spin confusion does not occur until nearer 0.2. On the other hand, if $J_{bab} = 0.05$ there is a large change, and for $J_{aba} = 0.15$ $J_{bab} = 0.05$ the spins show no long-range order. As these matrices represent a cube edge of some 160 Å, it is clear that neutron diffraction would not show any long-range order. If one were to allow the inclusion of more distant neighbours, this effect may become real for smaller values of the exchange constants.

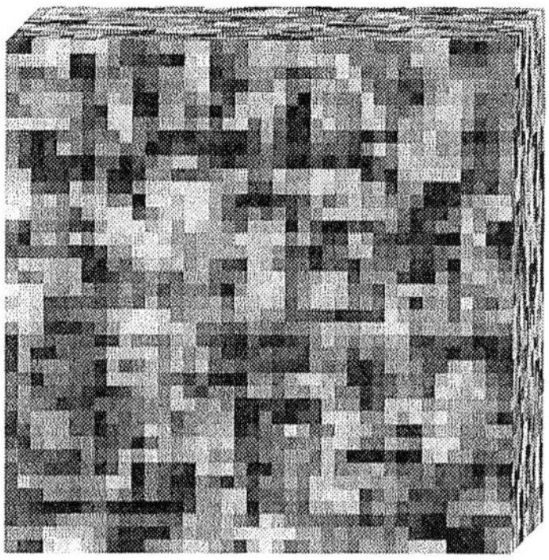

$J_{aba} = 0.20$, $J_{bab} = 0.10$, $S_a = 0.02$, $S_b = 0.02$

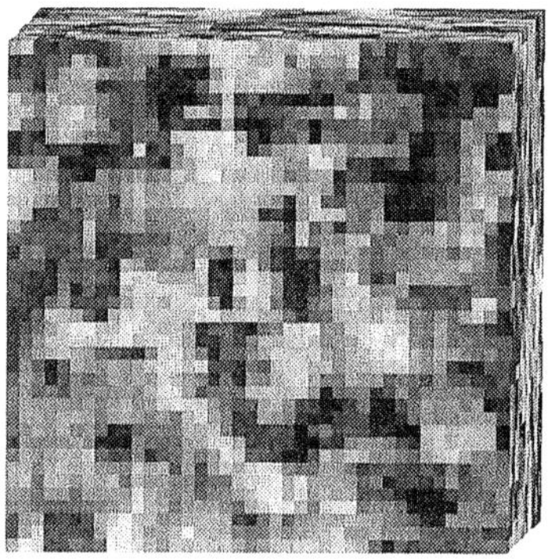

$J_{aba} = 0.15$, $J_{bab} = 0.05$, $S_a = 0.04$, $S_b = 0.05$

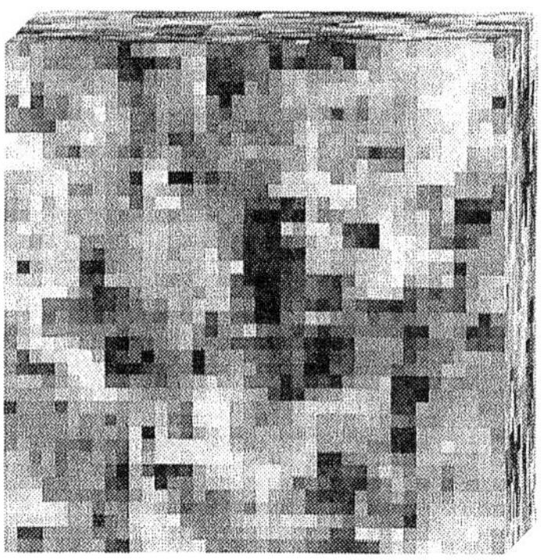

$J_{aba} = 0.10$, $J_{bab} = 0.05$, $S_a = 0.33$, $S_b = 0.33$

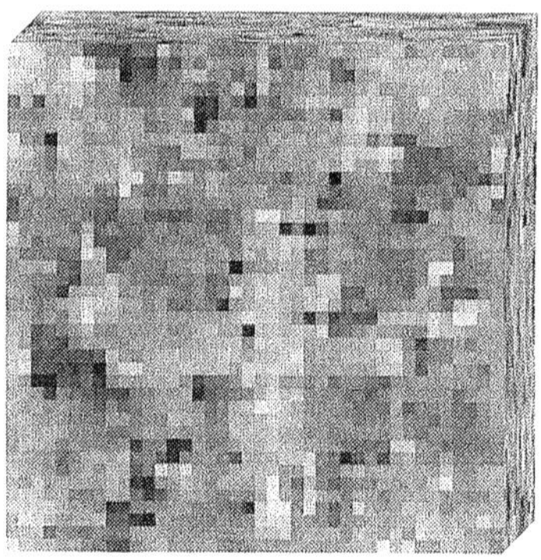

$J_{aba} = 0.10$, $J_{bab} = 0$, $S_a = 0.69$, $S_b = 0.81$

Figure 1. Pictorial representations of 40^3 spin arrays with $kT/J_{aa} = 0.025$.

5. Simulation of Mössbauer relaxation

In an attempt to demonstrate the effect of the spin dynamics on relaxation of the Mössbauer spectrum, the calculations described above included the calculation of a histogram of the $<S_i>$ values from 0 to 1 in units of 0.01. A series of 100 Mössbauer spectra using the same increments and a maximum flux density of 50 T were calculated, and then summed with weightings from the histogram. Some 10,000 MCS were used to prevent unrealistic line broadening above T_N because of inadequate averaging of the spins (the residual broadening above the ordering temperature can be substantially decreased by increasing to 50,000 MCS per step). For $J_{aba} = 0$ and $J_{bab} = 0$

(see Figure 2), the magnetic hyperfine splitting develops at a much higher temperature than for $J_{aba} = 0.10$ $J_{bab} = 0.05$, but the spin system always freezes at very low temperatures.

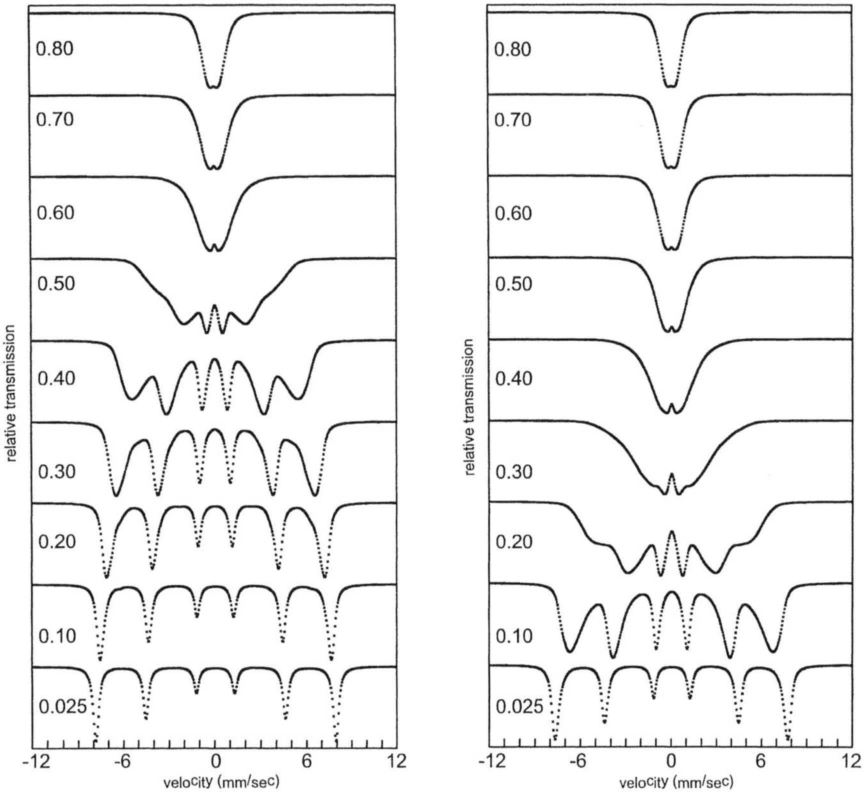

Figure 2. Simulated Mössbauer spectra for $J_{aba} = 0.0$, $J_{bab} = 0.0$ (left) and $J_{aba} = 0.10$, $J_{bab} = 0.05$ (right) for the range of values of kT/J_{aa} indicated in the Figure.

6. Conclusions

The behaviour observed in the simulations clearly reproduces the main features observed in Sr_2FeRuO_6 [1]. Despite the presence of strong spin interactions, the randomness of the site occupation and the significance of nnn spin interactions involving the Ru^{5+} cations prevent the development of a system with long-range order, although the spin system does freeze at low temperatures.

References

[1] P.D. Battle, T.C. Gibb, C.W. Jones, and F. Studer, J. Solid State Chem. 78 (1989) 281.
[2] T.C. Gibb, J. Mater. Chem. 11 (2001) 456.
[3] P.D. Battle, J.B. Goodenough, and R. Price, J. Solid State Chem. 46 (1983) 234.
[4] P.D. Battle and W.J. Macklin, J. Solid State Chem. 52 (1984) 138.
[5] P.D. Battle and C.W. Jones, J. Solid State Chem. 78 (1989) 108.

DYNAMICS OF Fe CATIONS IN AN H_2SO_4 SOLUTION BY NUCLEAR RESONANT QUASIELASTIC SCATTERING

R. Haruki, M. Seto, S. Kitao, Y. Kobayashi, Y. Yoda[a], T. Mitsui[b] and Yu. Maeda[c]

Research Reactor Institute, Kyoto University, Kumatori-cho, Sennan-gun, Osaka 590-0494, Japan
[a] *Japan Synchrotron Radiation Research Institute, 1-1-1 Kouto, Mikazuki-cho, Sayo-gun, Hyogo 679-5198, Japan*
[b] *Japan Atomic Energy Research Institute, 1-1-1 Kouto, Mikazuki-cho, Sayo-gun, Hyogo 679-5148, Japan*
[c] *Kansai Gaidai University, Hirakata, Osaka 573-1001, Japan*

Abstract: We have measured the dynamics of Fe^{3+} ions in an H_2SO_4 solution by nuclear resonant quasielastic scattering. The several sub peaks were found on the experimental spectrum with the broadening of the line width which is reduced to the diffusive motion. The form previously obtained for the diffusive motion was applied. The sub peaks were discussed comparing with the vibrational modes of the infrared data.

Keywords: iron, aqueous solution, nuclear resonant scattering, vibrational spectroscopy, synchrotron radiation, Mössbauer effect

1. Introduction

The nuclear resonant scattering by synchrotron radiation is useful to select the information on the dynamics of the specific element in a material which is not only a solid material but also a liquid state sample[1–5]. In the spectra of the Fe ions in a solution, the broadening of the line width was observed as compared to the width of the resolution function. The broadening was caused by the effect of diffusive motion of the cations. The effect of the diffusive motion in a short time range on the nuclear resonant quasielastic scattering spectrum was presented[5]. The features to measure the experimental data show remarkable progress, so the statistics of the data are also improved. Paulsen *et al* treated the nuclear resonant scattering spectra from the molecular crystals, and discussed the molecular vibrations[6]. Vankó *et al* have seen peaks in the spectrum of the Fe complexes in a solution, and the energy of the peaks were in good agreement with vibrational modes from infrared results[3].

2. Experimental

The experiment was performed at BL11XU beamline of SPring-8 Japan. The ring was operated in a 21×14 bunch-train mode where the interval of each bunch subset is 228 ns. The electron beam current was 70-40 mA at 8 GeV. The radiation was monochromatized to 3.2 meV around 14.413 keV, the excitation energy of ^{57}Fe, through a double-crystal Si(111) premonochromator and a high-resolution monochromator consisting of asymmetric Si(511) and asymmetric Si(975) channel-cut crystals. The signal from the sample was detected by a Si-avalanche photodiode detector with an area of 25 mm^2[7]. The experiment was performed at room temperature.

The sample was prepared by dissolving Fe_2O_3 in H_2SO_4 aq. and packed in a polypropylene cell.

3. Results and Discussions

The experimental results are shown in Fig.1. First we would apply the diffusive motion

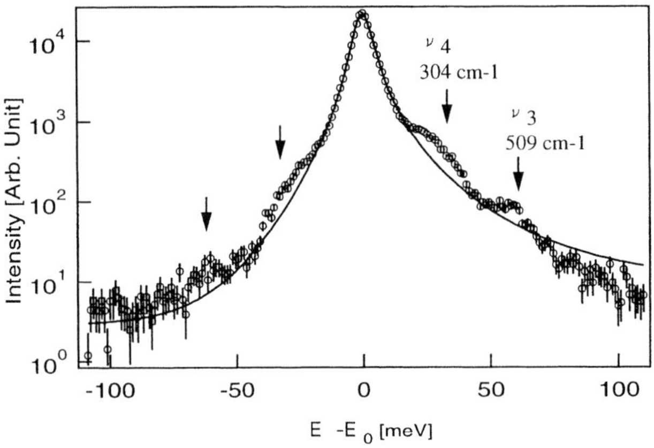

Figure 1. The closed circles with error bars show the experimental data from Fe(III) cations in an H_2SO_4 solution and the solid line shows the theoretical fitting line calculated by eq.(1) with $M = 57$. Pointers shows the energy of ν_3 and ν_4 from infrared data[11].

analysis. We have discussed that we can not ignore the effect of the diffusive motion in a short time range on the nuclear resonant quasielastic scattering spectrum[5]. As a result, we have obtained the nuclear γ ray absorption cross-section in the following form ($\alpha = (MD^2Q^2)/kT < 1$)

$$\sigma(E) = \frac{\sigma_0\Gamma}{4}\mathrm{e}^{-2W}\exp\left(\frac{E-E_0}{2kT}\right)\sum_{\nu=0}^{\infty}\frac{(-1)^\nu}{\nu!}\alpha^\nu H_\nu(E), \tag{1}$$

$$H_\nu(E) = \frac{\Gamma + 2\hbar\left(DQ^2 + \dfrac{\nu kT}{MD}\right)}{(E-E_0)^2 + \dfrac{1}{4}\left[\Gamma + 2\hbar\left(DQ^2 + \dfrac{\nu kT}{MD}\right)\right]^2}. \tag{2}$$

where D is the diffusion coefficient, σ_0 is the resonance absorption cross-section, Γ is the natural width of the excited state, $\hbar$ is the reduced Planck coefficient, E_0 is the excitation energy for the nucleus, k is the Boltzmann constant, T is temperature, Q is the wave number, M is the mass of the diffusing particle, and $2W$ is so called the Debye-Waller factor.

Now we apply eq.(1) to the experimental data from the Fe cations in an H_2SO_4 solution. The solid line in Fig.1 shows the theoretical fitting. Although the central part of the spectrum is sufficiently fitted, several peaks are obviously deviated from the line and the tails are not traced by the theoretical curve. These peaks are assumed to be caused by molecular vibrations[3]. The diffusive constant obtained here is 9.3×10^{-6} cm^2/s which is somewhat smaller than 1.2×10^{-5} cm^2/s of Fe ions in an HCl solution and 1.4×10^{-5} cm^2/s of Fe ions in Nafion membranes. Although this value might change when the molecular vibrations were considered in the fitting, it gives an outline of the diffusive motion.

Molecular vibrational analysis should be applied. The iron aquacomplex is known to be coordinated by six water molecules. We have used molecular orbital method to optimize the structure of the iron complex[6, 9]. Figure 2(a) shows the structure of $[Fe(H_2O)_6]^{3+}$.

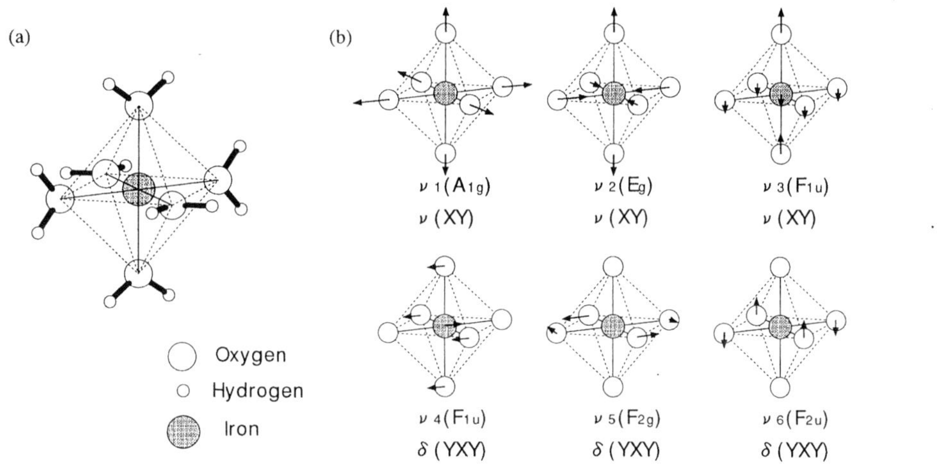

Figure 2. (a) $[Fe(H_2O)_6]^{3+}$ optimized by UHF/3-21G* MOLCAT[8]. (b) Normal modes of vibration of octahedral XY_6 molecules.

Unrestricted Hartree-Fock (UHF) method implemented in Gaussian 98 program system with 3-21G* basis set was used in the calculation. Figure 2(b) shows the normal modes vibrations of octahedral molecules which are the reduced structure of Fe hexaquacomplex[10]. In these modes, ν_1, ν_2 and ν_5 are Raman active and ν_3 and ν_4 are infrared active. Speaking of nuclear resonant scattering, the modes with iron atoms active are measurable. Those are ν_3 and ν_4. In this case, those are also infrared active, but it does not mean that infrared active modes are nuclear resonant active modes. It depends on the structure of the molecule. Mode ν_3 is exhibited at 505 cm^{-1} and mode ν_4 is exhibited at 304 cm^{-1} in infrared data[11]. Those are both in good agreement with the energy of the peaks in Fig.1.

4. Conclusion

Diffusive motion analysis was applied to nuclear resonant scattering data of Fe cations in an H_2SO_4 solution. There exists deviation which can be attributed to molecular vibrations. These peaks are in agreement with the data of the infrared experiment. These results would help us to learn more about the couplings of the diffusive motion and the molecular vibrational modes.

Acknowledgements

We are indebted to all of the staff of SPring-8 for their support and assistance.

References

[1] M. Seto, Y. Yoda, X. W. Zhang and M. Ando: Phys. Rev. Lett. **74** (1995) 3828.

[2] X. W. Zhang, Y. Yoda, M. Seto, Y. Maeda, M. Ando and S. Kikuta: Jpn. J. Appl. Phys. **34** (1995) 330.

[3] Gy. Vankó, L. Bottyán, D. L. Nagy, E. Szilágyi and A. Vertés: J. Radioanal. Nucl. Chem. **246** (2000) 97.

[4] R. Haruki, M. Seto, S. Kitao, Y. Yoda and Yu. Maeda: J. Phys. Soc. Jpn. **69**(2000) 4049.

[5] R. Haruki, M. Seto, S. Kitao, Y. Kobayashi, Y. Yoda, T. Mitsui and Yu. Maeda: J. Phys. Soc. Jpn. **70** (2001) 445.

[6] H. Paulsen, H. Winkler, A. X. Trautwein, H. Grünsteudel, V. Rusanov, H. Toftlund: Phys. Rev. B **59** (1999) 975.

[7] S. Kishimoto: Nucl. Instrum. Methods Phys. Res., Sect. A **309** (1991) 603.

[8] M. J. Frisch *et al.*: Gaussian 98, Revision A.9, Gaussian Inc., Pittsburgh, PA, 1998.

[9] Y. Tsutsui, H. Wasada: Chemistry Letters (1995) 517.

[10] K. Nakamoto: Infrared and Raman Spectra of Inorganic and Coordination Compounds 4th ed. Wiley 1986.

[11] S. P. Best, R. S. Armstrong and J. K. Beattie: Inorg. Chem. **19** (1980) 1958.

^{151}Eu Mössbauer spectroscopy of the filled skutterdite-type compound $EuRu_4P_{12}$

Kentaro Indoh*, Hideya Onodera, Chihiro Sekine[a], Ichimin Shirotani[a] and Yasuo Yamaguchi
Institute for Materials Research, Tohoku University, Sendai 980-8577, Japan
[a] *Department of Electrical and Electric Engineering, Muroran Institute of Technology, Muroran, Hokkaido 050-8585, Japan*

September 3, 2001

Abstract. $EuRu_4P_{12}$ with the filled skutterudite-type crystal structure undergoes a ferromagnetic transition at T_C= 17.8 K. ^{151}Eu Mössbauer spectroscopy was carried out on $EuRu_4P_{12}$. A magnetic hyperfine field is −398 kOe at 4.2 K and a transferred hyperfine field of −58 kOe is estimated. The transferred hyperfine field of −330 kOe in $EuFe_4P_{12}$ which is ferromagnetic below 100 K is about six times larger than that of $EuRu_4P_{12}$. This ratio of the transferred hyperfine fields correspond well to that of the Curie temperatures.

Keywords: ^{151}Eu Mössbauer spectroscopy, filled skutterudite, $EuRu_4P_{12}$, hyperfine field, transferred hyperfine field

1. Introduction

Ternary intermetallic compounds RT_4X_{12} (R=rare earth, T=Fe, Ru, Os, X=P, As, Sb) with the filled skutterudite-type ($Im\bar{3}$) crystal structure (Grandjean et al., 1984) attract our interest since these show variable physical properties: superconductivity, metal-insulator transition, magnetic transition, heavy fermion behavior and so on. Some phenomena are now open questions.

$EuRu_4P_{12}$ undergoes ferromagnetic ordering below 18 K (Grandjean et al., 1983; Sekine et al., 1997) and exhibits relatively soft ferromagnetic behavior compared to other localized-moment magnets (Sekine et al., 1997). Electrical resistivity shows metallic behavior. The present paper reports magnetic behavior of $EuRu_4P_{12}$ from a microscopic point of view by ^{151}Eu Mössbauer spectroscopic experiment.

2. Experimental

The sample of $EuRu_4P_{12}$ was prepared at high temperatures and high pressures using a wedge type cubic-anvil high pressure apparatus. More

* Corresponding author. e-mail address: indoh@imr.edu.

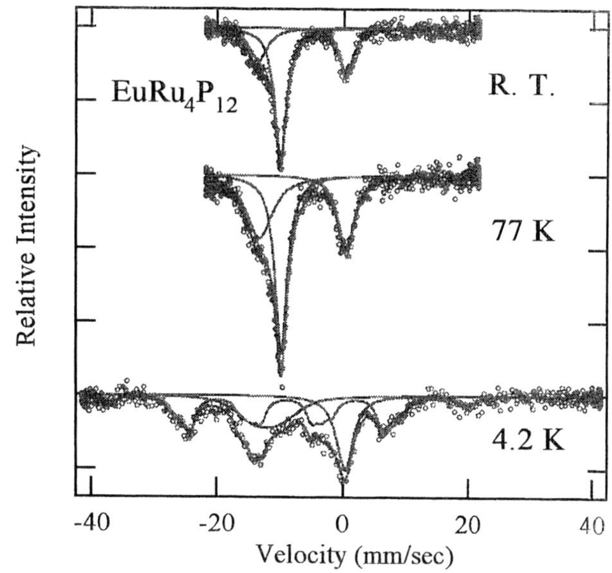

Figure 1. ^{151}Eu Mössbauer spectra at room temperature, 77 K and 4.2 K, where the observed and calculated spectra are represented by open circles and solid lines, respectively.

details of sample preparation are described elsewhere (Sekine et al., 2000). To characterize the sample, we measured magnetizations, and obtained $T_C = 17.8$ K which is consistent with the reported value (Sekine et al., 1997). The Mössbauer powder sample was prepared in a thickness of about 77 mg/cm^2. A standard radioactive source $^{151}SmF_3$ of 3.7 GBq was used at room temperature. The ^{151}Eu Mössbauer effect measurement was carried out in a temperature range from 4.2 K to room temperature with a sinusoidal velocity wave in a ordinary transmission mode.

3. Results and discussion

Figure 1 shows ^{151}Eu Mössbauer spectra at room temperature, 77 K and 4.2 K, where the observed and calculated spectra are represented by open circles and solid lines, respectively. In the paramagnetic phase at room temperature and 77 K, three absorption lines with isomer shifts of 0.3, -9.9 and -13.2 mm/sec were observed. The absorption lines at -9.9 and -13.2 mm/sec stand for two kinds of divalent Eu ions. The relative intensity of the absorption lines with the isomer shifts of 0.3,

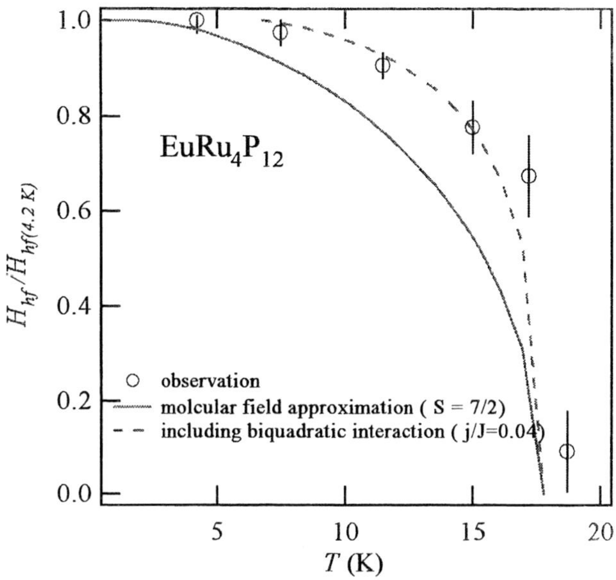

Figure 2. Temperature dependence of the magnetic hyperfine field in $EuRu_4P_{12}$.

−9.9 and −13.2 mm/sec is about 2.0 : 5.6 : 2.4. The spectrum at 4.2 K in the ferromagnetic phase was interpreted by three subspectra. The analysis reveals that the Eu^{2+} with the isomer shift of −9.9 mm/sec has a magnetic hyperfine field of −398 kOe. This absolute value is smaller than the reported value of −415 kOe (Grandjean et al., 1983). Sekine et al. reported that Eu^{2+} of 90% and Eu^{3+} of 10% were estimated from a saturation magnetization (Sekine et al., 1997). In the present sample, a saturated magnetization at 5.0 K reveals that about 60% of the Eu atoms participate the ferromagnetic order as Eu^{2+}, while a effective paramagnetic moment above 30 K shows that the Eu atoms of about 80% are divalent. These facts mean that 20% of Eu^{2+} does not participate the magnetic order below T_C. The result agrees with the result of the Mössbauer observation. Thus, the present sample contains a considerable amount of an impurity phase. However, we consider that this impurity phase will have no severe influence on the intrinsic properties of $EuRu_4P_{12}$ because the transition temperature of the sample agrees so well with that in the previous report (Sekine et al., 1997).

Temperature dependence of the magnetic hyperfine field is shown in Fig.2, where the solid and the dashed lines represent the Brillouin function for $S = 7/2$ and the molecular field approximation including a biquadratic interaction, respectively. The temperature dependence of hyperfine field seems to agree better with the molecular field approximation including a biquadratic interaction (D. S. Rodbell et al., 1963) than with the Brillouin function like the case of $EuFe_4P_{12}$ (Gérard et al., 1983). A hyperfine field of about -38 kOe was observed at 18.7 K which is higher than $T_C = 17.8$ K. This may mean that magnetic short range order exists above T_C.

In $EuFe_4P_{12}$ which is ferromagnetic below 100 K, the hyperfine field is -670 kOe at 4.2 K (Gérard et al., 1983). Since the Eu^{2+} ion with an ordered moment of 7 μ_B ($S = 7/2$) produces a core polarization field of -340 kOe, the transferred hyperfine field of -330 kOe is attributable to a conduction electron polarization (Grandjean et al., 1983). In the same way, the transferred hyperfine field can be estimated to be -58 kOe in $EuRu_4P_{12}$. In general, a transferred hyperfine field mediated by conduction electrons with s-character is positive. However, the transferred hyperfine fields of these compounds are negative. Furthermore, the ratio of magnetic transition temperatures (T_C) of $EuFe_4P_{12}$ and $EuRu_4P_{12}$ is consistent with the ratio of the transferred hyperfine fields. It is suggested that the RKKY interaction is mediated by the conduction electrons with d-character in these compounds.

References

Grandjean F., A. Gérard, D. J. Braun and W. Jeitschko. *J. Phys. Chem. Solids*, 45 (1984), 877–886.

Grandjean F., A. Gérard, J. Hodges, D. J. Braun and W. Jeitschko. *Hyperfine Interact.*, 15-16, 765-766 1983.

Sekine C., M. Inoue, T. Inaba and I. Shirotani. *Physica B*, 281&282, 308-310 2000.

Sekine C., M. Inoue, I. Shirotani and T. Yagi. *Phys. Rev. Lett.*, 79, 3218–3221 1997.

Gérard A., F. Grandjean, J. A. Hodges, D. J. Braun and W. Jeitschko. *J. Phys. C*, 16, 2797–2801 1983.

D. S. Rodbell, I. S. Jacobs, J. Owen and E. A. Harris. *Phys. Rev. Lett.*, 11, 10–12 1963.

MÖSSBAUER STUDY OF Ni-Zn FERRITES OBTAINED BY THE SELF-PROPAGATED METHOD

S.Jacobo[1], *S.Duhalde*[1], *G.Quintana*[1], *R.Rotelo*[1] *and A. F.Pasquevich*[2]
[1]Facultad de Ingeniería, Universidad de Buenos Aires. Paseo Colón 850, 1063 Buenos Aires, Argentina.
[2]Facultad de Ciencias Exactas, Universidad Nacional de La Plata, CC 67, 1900 La Plata, Argentina.

$Ni_{1-x}Zn_xFe_2O_4$ ferrites (x = 0, 0.5, 0.6 and 0.7) were prepared from self-propagated powders. Their structural and magnetic properties were characterised with x-ray diffraction, Mössbauer spectroscopy and ac susceptibility. Mössbauer analysis was performed at room temperature and at 15K, which was expected to yield useful information on the microstructure. The sintered samples present high initial permeability that increases as the concentration of Zn increases. The x= 0.7 Mössbauer spectrum exhibits superparamagnetism at room temperature and magnetic order at low temperature. The roles of cation-site occupation and magnetic coupling are discussed.

1. Introduction

In the elementary cell of a spinel structure, the cations are distributed among tetrahedral [A] and octahedral [B] sites. Zn prefers the tetrahedral sites and Ni shows a marked preference for octahedral sites. Mixed Ni-Zn ferrites could be considered as the most versatile mixed ferrites due to their high resistivity, mechanical hardness, low eddy current losses, high Curie temperature (Tc) and chemical stability [1,2]. Although this ferrite can be easily prepared by ceramic standard technology, the main problem is its high sintering temperature that greatly influences the microstructure and electromagnetic properties [3]. Recently, special attention has been given to research in nanocrystalline materials because they exhibit certain properties that are different from their bulk counterparts. It is well known that low temperature sintering on ferrites can be achieved by using active ultra fine powders synthesised *via* wet-chemical method. Particularly, the self-combustion method provides nanoparticles at very low temperatures [4]. The aim of this work is to study morphological and structural properties of $Ni_{1-x}Zn_xFe_2O_4$ ferrites prepared by a combustion method, for technological applications. In that sense, Mössbauer spectroscopy is a useful technique to study the site occupancy in Ni-Zn ferrites that greatly influences their electromagnetic behaviour [5].

2. Experimental

In order to prepare $Ni_{1-x}Zn_xFe_2O_4$, with x = 0, 0.5, 0.6 and 0.7, different proportions of iron nitrates and nickel and zinc oxalates were weighed according to the required stoichiometric proportion and diluted in water (in all preparations, [Fe(III)] + [Ni(II)] + [Zn(II)] = 1 M). A 3 M citric acid solution (50 ml) was added to each metal solution (50 ml) and heated at 40°C for approximately 30 min with continuous stirring. The final mixture was slowly evaporated until a highly viscous gel was formed. The resulting gel was heated at T~200°C, when it ignited in a self-propagated process. The final residue was calcined at 1000 °C for two hours.

The x-ray diffraction patterns of the samples were obtained under θ-2θ geometry using Cu *K*α radiation. A ^{57}Co source (Rh matrix) in constant acceleration mode at room temperature (RT) was used to record the Mössbauer spectra of the samples at RT and at 15K. The measured isomer shifts (IS) are referred relative to α-iron at RT. Magnetisation measurements were performed at room temperature in a vibrating sample magnetometer. The Curie Temperature (Tc) of each sample was determined from the permeability (μ) vs temperature curves. These complex permeabilities (not shown here) were measured on a HP4291A Material Analyzer from 1 to 1800 MHz.

3. Results and Discussion

X-ray diffraction (XRD) patterns revealed that all the samples were single phase and that the spinel structure was successfully obtained (inset of Fig.1). A reduced part of the x-ray diffractograms of $Ni_{1-x}Zn_xFe_2O_4$ ferrites are shown in Fig.1.

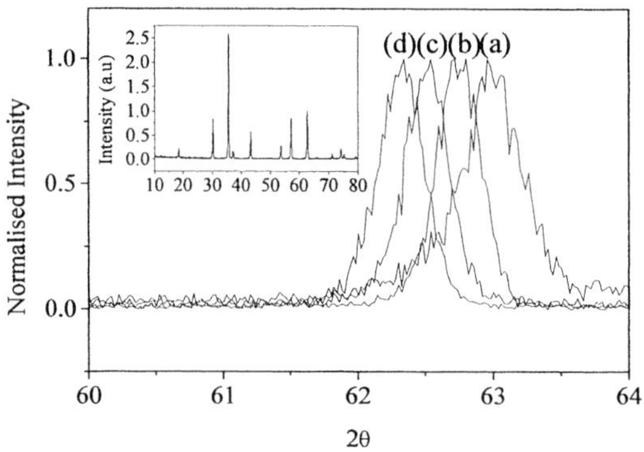

***Fig.1**: Section of the XRD patterns of Ni-Zn ferrites. (a) x=0; (b) x=0.5; (c) x=0.6; (d) x=0.7. Inset: complete XRD diffractogram for x=0.5*

***Table 1**: Lattice constant of $Ni_{1-x}Zn_xFe_2O_4$*

x	a_o(Å)
0	8.338
0.5	8.395
0.6	8.403
0.7	8.414

Cell parameters of the ferrite phase, were calculated from 21 lines, using the TREOR method [6]. The shift in the angular position of the peaks reflects the change of the lattice constant *a* due to Zn incorporation in the structure. The variation of the lattice constant *a* with x, displayed in Table 1, is in agreement with previously reported results [7,8]. Crystallite sizes were calculated from the width of the (311) lines using the Scherrer formula corrected for instrumental broadening [9]. The average particle size decreases from 50 to 40 nm as the Zn concentration increases from 0 to 0.7. The inspection by scanning electron microscopy (SEM) revealed that powder grains of about 1 μm, consisted of agglomeration of smaller particles, possibly due to magnetic attraction.

The distribution of cations in the two sites is given by $[Zn^{2+}{}_xFe^{3+}{}_{1-x}]_A[Ni^{2+}{}_{1-x}Fe^{3+}{}_{1+x}]_B$. It could be expected that the twelve magnetic ions at the [B] sites contribute to the hyperfine field at a nucleus on the [A] site. Taking into account that the contributions from Fe^{3+} and Ni^{2+} should be different, a dependence on the hyperfine magnetic field of iron ions at the [A] site with x could be expected. However, previous studies have shown that the hyperfine field of iron ions at the [A] site is insensitive to the kinds and number of cations on the [B] sites [3]. Hence, the Mössbauer spectra can be fitted with one sextet corresponding to Fe in the [A] site and a number of sextets corresponding to

[B] sites having *n* nearest-neighbour Zn atoms. So the spectrum for x=0 was fitted with two sextets while the ones at RT and at 15 K for x=0.5 sample were fitted with 4 sextets for the [B] sites, in addition to one sextet for the [A] site. The fitted results for all the samples are shown in Table 2 and agree with earlier results [3,8,10]. Fig. 2 and Fig. 3 shows the Mössbauer spectra taken at RT and 15 K respectively.

For the x=0.6 sample another sextet for the [B] sites was added to the fitting, corresponding to a smaller magnetic hyperfine field due to the increase of diamagnetic Zn nearest-neighbours. The RT Mössbauer spectrum of the x=0.7 sample clearly displayed the collapse of magnetic ordering due to superparamagnetic relaxation. This behaviour may be understood as a consequence of the decrease of particle size and the weakening of magnetic interaction due to the increase of Zn concentration. However, the 15 K spectrum clearly shows the sextets patterns. Hence, room temperature is higher than blocking temperature (T_B), that is the temperature above which superparamagnetism sets in. In agreement with that, the Tc value is much lower (≈ 100 °C) and the permeability μ vs T profile decreases almost linearly as the temperature is increased, as expected for nanosized ferromagnetic particles. So, the x=0.7 spectrum was fitted with a central doublet and a distribution of hyperfine magnetic fields. Average values of the hyperfine magnetic fields are shown also in Table 2. It should be noted that the IS values for all the sites do not show a substantial variation with x. Hence, the presence of Zn ions do not affect significantly the d-electron density which in turn affects the s-electron density. Relative areas corresponding to Fe^{3+} occupying octahedral and tetrahedral sites are shown in fig.4. For low Zinc concentrations, that information can be related to magnetisation results, which are shown in Table 3.

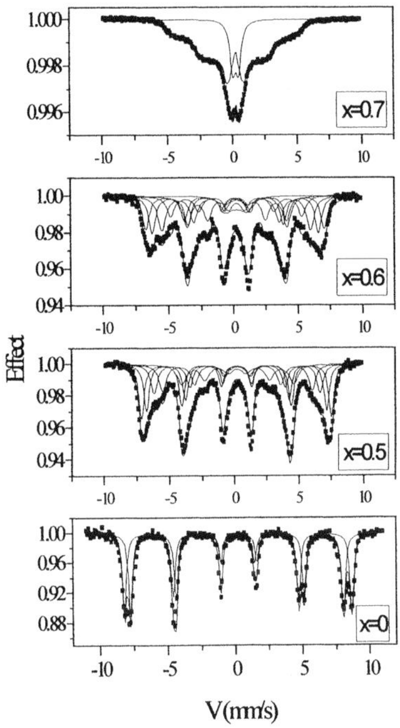

***Fig.2:** Room temperature Mössbauer spectra of $Ni_{1-x}Zn_xFe_2O_4$*

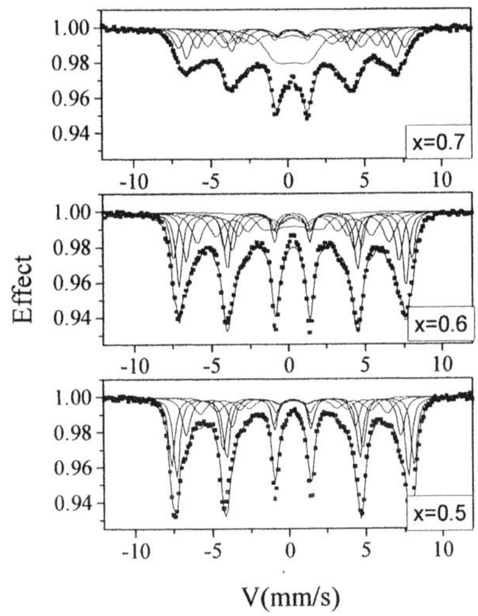

***Fig.3:** 15K-Mössbauer spectra of $Ni_{1-x}Zn_x Fe_2O_4$.*

In general, the areas corresponding to the Fe sites for x=0 calculated from Mössbauer spectra coincide, within the error of the fitting, with the distribution of Fe in both sites,

***Table 2**: Mössbauer parameters for $Ni_{1-x}Zn_xFe_2O_4$ at RT and at 15K. B: magnetic hyperfine field (T), IS: isomer shift (mm/s), B_M: value of the most probable hyperfine field over the distribution (T)*

RT X	A		B_0		B_1		B_2		B_3		B_4		B_5	
	B	IS	B	IS	B	IS	B	IS	B	IS	B	IS	B	IS
0	48.8	0.15	51.5	0.24										
0.5	45.4	0.16			44.2	0.20	39.7	0.23	35.8	0.24	26.9	0.23		
0.6	40.0	0.17			42.5	0.19	35.3	0.23	31.4	0.23	24.1	0.25	6.6	0.24
	QS	IS		B_{M2}			B_{M2}			B_{M3}			B_{M4}	
0.7	0.48	0.24		30			20			12			7	
15K X	A		B_0		B_1		B_2		B_3		B_4		B_5	
	B	IS	B	IS	B	IS	B	IS	B	IS	B	IS	B	IS
0.5	46.5	0.23			48.7	0.24	42.9	0.28	37.6	0.29	30.8	0.28		
0.6	45.3	0.24			47.8	0.27	41.5	0.27	36.4	0.30	29.4	0.30	11.9	0.29
0.7	41.9	0.25			42.3	0.27	38.1	0.29	39.9	0.31	27.7	0.32	8.1	0.32

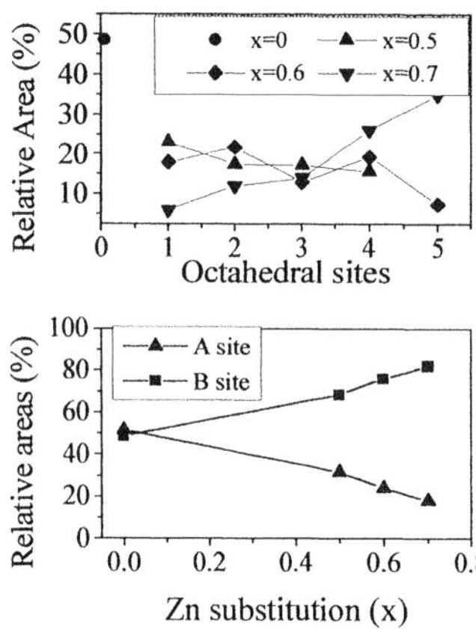

Fig.4: Relative areas calculated from Mössbauer spectra. Top panel: Octahedral relative areas vs. Octahedral sites. Botton panel: Tetrahedral and Octahedral areas vs. Zn substitution

***Table 3**: Saturation magnetisation M_s, Curie Temperature T_c and Bohr magnetons μ_B per formula unit.*

X	M_s (emu/g)	Tc (°C)	$\mu_{B/f.u}$
0	50.4	580	2.13
0.5	83.2	311	3.54
0.6	74.1	210	3.16
0.7	52.8	100	2.27

obtained from magnetisation measurements. As expected, for x=0.5 the octahedral areas calculated from Mössbauer spectrum are larger by 27%. The last results could be justified by the weakness of the magnetic coupling between Fe ions due to Zn incorporation and assuming a small amount of Ni^{2+} occupying the tetrahedral site.

4. Conclusions

Nanosized $Ni_{1-x}Zn_xFe_2O_4$ ferrites have been successfully prepared by a low-temperature-combustion method, with excellent magnetic and electrical properties. An optimal saturation magnetisation value has been found for the x = 0.5 Zn substitution. Their small size, crystalline structure and magnetic ordering features make them attractive for technological application.

Acknowledgements
Authors are thankful to Dr.H.Bertorello for magnetic measurements, to Dr. E.Sileo for diffractograms and to Dr. C.Saragovi and Mr.Angel Carnero for Mössbauer measurements at 15K. UBACyT financially supported this research.

References

1. J.Smit, H.P.J.Wijn, Ferrites, Philips Technical Library, (1959)
2. A.M.Abdem, J. Magn. Magn. Mater. 185 (1998) 199.
3. L.K.Leung, B.J.Evans, A.H.Morrish, Phys. Rev. B 8 (1973) 29.
4. S.E.Jacobo, R.Rotelo and E.Sileo, Physica B 2001 (in press)
5. M.Arshed, M.Siddique, M.Anwar-ul-Islam, N.M.Butt, Solid State Commun. 93 (1995) 599.
6. P. Werner, Z. Kristallogr.120 (1964) 375.
7. L.Wang, F.S.Li, J.Magn. Magn. Mater. 223 (2001) 233.
8. J.M.Daniels, A.Rosencwaig, Can. J. Phys. 48 (1970) 381.
9. H. Klug and L. Alexander in *X-ray Diffraction Procedures*, 2nd. ed., (John Wiley & Sons, 1973) p.656.
10. M.Arshed, M.Siddique, M.Anwar-ul-Islam, N.M.Butt, T.Abbas, M.Amhed, Solid State Commun. 93, 7 (1995) 599.

^{57}Fe Mössbauer study of novel series of intermetallic compounds $R_3(Fe_{1-x}Co_x)_{29-y}T_y$ (R=Nd, Tb, Dy; T=Ti, V)

O. Kalogirou[1], C. Sarafidis[1], T. Bakas[2] and M. Gjoka[2]

[1]*Dept. of Physics, Aristotle University of Thessaloniki, 54006 Thessaloniki, Greece*
[2]*Dept. of Physics, University of Ioannina, 45110 Ioannina, Greece*

$R_3(Fe_{1-x}Co_x)_{29-y}T_y$ (R=Nd, Tb, Dy; T=Ti, V) intermetallic compounds have been studied by ^{57}Fe Mössbauer spectroscopy at 85 and 293 K. The hyperfine interaction parameters depend on the amount of Co atoms. It is shown that Co atoms are equally distributed in all Fe sites but strongly avoid the Fe_2, Fe_3 and Fe_6 sites which are the dumb-bell sites.

Introduction

The large increase in the Curie temperatures of the $R_3(Fe,T)_{29}$ intermetallic compounds upon Co substitution and the effect of this procedure on saturation magnetisation, anisotropy field and magnetocrystalline anisotropy has recently attracted much attention. The $R_3(Fe,T)_{29}$ compounds crystallise in the monoclinic structure with space group of A2/m [1]. In this description the rare earth ions occupy two crystallographic sites (2*a* and 4*i*), the Fe atoms occupy eleven sites (4*i*, 4*g*, 4*e* and 8*j*) and the T atoms reside only in the Fe_2 (4*i*), Fe_3 (4*i*) and Fe_6 (4*g*) where the dumb-bell Fe atoms are found [1,2]. The introduction of Co in rare earth – iron intermetallic compounds considerably affects their magnetic properties. As a matter of fact, such a behaviour has been recently reported for $R_3(Fe_{1-x}Co_x)_{27.5}Ti_{1.5}$ (R=Pr [3], Gd [4]). In a previous work, we have reported the synthesis of $Nd_3(Fe_{1-x}Co_x)_{27.7}Ti_{1.3}$ ($0 \leq x \leq 0.4$) [5] and we have shown that at room temperature for $x \geq 0.2$ the compounds present an easy-magnetisation direction along the [4 0 -2] direction, showing a tilted magnetic structure for $x=0$ and 0.1. It has been shown that the Curie temperature, the saturation magnetisation, and the anisotropy field of the compounds are strongly affected by the Co content. Similar results have been obtained in the case of $Tb_3(Fe_{1-x}Co_x)_{27.4}V_{1.6}$ and $Dy_3(Fe_{1-x}Co_x)_{27.8}V_{1.2}$ [6]. In the present work we present a detailed study of the hyperfine interactions and crystal site occupancies in $R_3(Fe_{1-x}Co_x)_{29-y}T_y$ (R=Nd,Tb,Dy; T=Ti,V) compounds.

Experimental

$Nd_3(Fe_{1-x}Co_x)_{27.7}Ti_{1.3}$, $Tb_3(Fe_{1-x}Co_x)_{27.4}V_{1.6}$ ($0 \leq x \leq 0.4$) and $Dy_3(Fe_{1-x}Co_x)_{27.8}V_{1.2}$ ($0 \leq x \leq 0.2$) samples with the $Nd_3(Fe,Ti)_{29}$-type structure were prepared and characterised as described in our previous works [5,6]. ^{57}Fe Mössbauer spectra were obtained at 85 and 293 K on a conventional constant acceleration spectrometer with a $^{57}Co(Rh)$ source moving at room temperature while the absorber was at the desired temperature. The spectra were analysed as a convolution of four independent sub-spectra based on the model described in Ref. 7 for $Pr_3(Fe,Ti)_{29}$.

Results and discussion

Fig. 1 shows typical ^{57}Fe Mössbauer spectra, together with the theoretical fitting curves (solid lines), for $Tb_3(Fe_{1-x}Co_x)_{27.4}V_{1.6}$ ($0 \leq x \leq 0.4$) at 85 and 293 K. The fitting procedure was based on the model suggested in Ref. 7 for $Pr_3(Fe,Ti)_{29}$. L.C.C.M. Nagamine et al. have also used our model for the Mössbauer study of $R_3Fe_{29-x}V_x$ [8] and $R_3Fe_{29-x}Cr_x$ and $R_3Fe_{29-x}Cr_xH_y$ [9] 3:29 compounds. In the A2/m space group the $Nd_3(Fe,Ti)_{29}$-type structure is described by eleven crystallographically non-equivalent iron sites [1]. Applying a Wigner-Seitz-cell analysis [10] these sites can be divided into four groups according to the number of nearest neighbours and the bond length values. According to that analysis there are three main groups with respect to the nominal number of the nearest neighbours (nn), i.e., 13, 10, 9 nn [7]. In order to determine the actual number of the nearest neighbours the occupancy of the iron sites by Co and Ti(V) was taken into account. In the case of Co atoms the neutron diffraction results on $Pr_3(Fe_{1-x}Co_x)_{27.5}Ti_{1.5}$ (x=0.1 and 0.3) of Harris et al. [11] have shown that the Co atoms occupy those Fe sites not shared with the Ti atoms, i.e., the dumb-bell sites. On the other hand, J.J. Bara et al. have shown by Mössbauer spectroscopy that in the $RFe_{11-x}Co_xTi$ (R=Y, Dy, Er; x=0, 3, 6, 8) compounds Co atoms strongly avoid the 8i sublattice (dumb-bell) and strongly prefer the 8f and 8j sites [12]. Based on neutron diffraction data for $Nd_3(Fe,Ti)_{29}$ Hu et al. [2] have reported that in the P21/c description of the 3:29 structure the iron sites 4e_3, 4e_4, and 4e_{14} are occupied by Ti 30.45, 21.4% and 10.3% (total site occupancy), respectively. In our description these sites correspond to Fe_2, Fe_3, and Fe_6, which are, in both descriptions, the dumb-bell sites. Therefore, it was assumed that Fe_2, Fe_3 and Fe_6 are partially occupied only by Ti(V) and that the Co atoms are equally shared to all the other iron sites without preference. The so calculated numbers of nearest neighbours are given in Table 1. Thus, the iron sites were divided in three groups according to the average number of Fe nearest neighbours. Considering the bond length values too [7], the iron sites were finally divided in four groups: Fe_2, Fe_3, Fe_6 (group A, dumb-bell), Fe_1, Fe_8, Fe_{11} (group B), Fe_9, Fe_{10} (group C) and Fe_4, Fe_5, Fe_7 (group D), using the notation given in Ref. 10.

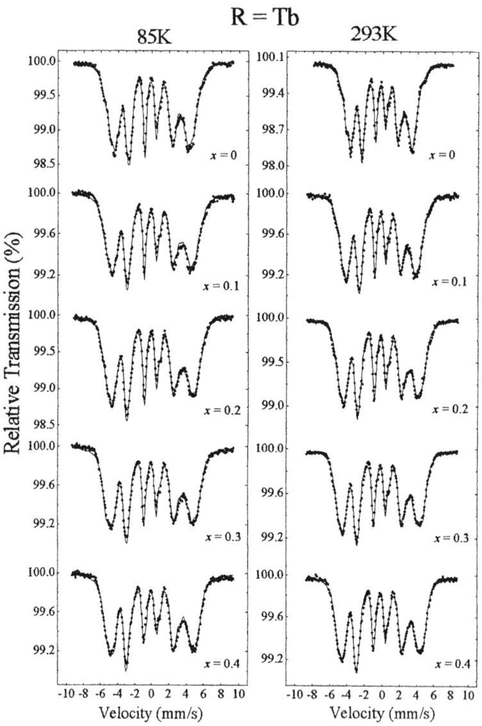

Fig. 1. The typical ^{57}Fe Mössbauer spectra at 293 and 85K of $Tb_3(Fe_{1-x}Co_x)_{27.4}V_{1.6}$ compounds.

Table 1
The number of nearest neighbours in the compounds under study, as determined taking into account the Co and Ti(V) occupancy.

x	Group A Fe_2, Fe_3, Fe_6	Group B Fe_1, Fe_8, Fe_{11}	Group C Fe_9, Fe_{10}	Group D Fe_4, Fe_5, Fe_7
0	12.3	9.4	8.7	9.4
0.1	11.2	8.5	7.8	8.5
0.2	10.1	7.7	6.8	7.7
0.3	9.0	6.9	5.9	6.9
0.4	7.9	6.0	4.9	6.0

Hence, for the compounds under study the fit was done with four components. As found from the X-ray powder diffraction patterns [5,6] some samples contained a small amount of unreacted α-Fe. Thus, in the case of R=Nd and x=0.2, 0.3 a fifth component was introduced for α-Fe (4-5 wt.%). The area ratio was constrained in order to correspond to the populations of the selected iron site groups taking into account the Ti(V) and Co occupancy of the different sites. The area ratios of the iron groups A:B:C:D, resulting from these calculation are: 1:1.5:1.3:2.1 for x=0, 1:1.3:1.1:1.9 for x=0.1, 1:1.1:1:1.6 for x=0.2, 1:0.9:0.8:1.4 for x=0.3 and 1:0.8:0.7:1.1 for x=0.4. A small amount of line broadening was allowed for each component to simulate the distribution of the environments within each component.

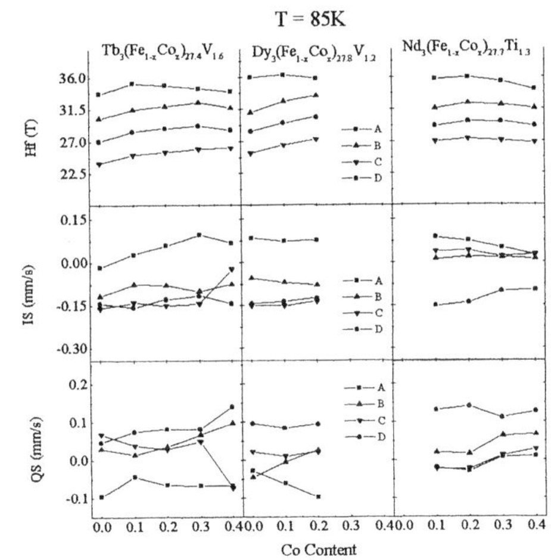

Fig. 2. The dependence of the hyperfine field, the isomer shift and the quadrupole interaction on Co concentration for $Nd_3(Fe_{1-x}Co_x)_{27.7}Ti_{1.3}$, $Tb_3(Fe_{1-x}Co_x)_{27.4}V_{1.6}$ ($0 \leq x \leq 0.4$) and $Dy_3(Fe_{1-x}Co_x)_{27.8}V_{1.2}$ ($0 \leq x \leq 0.3$) at 85 K.

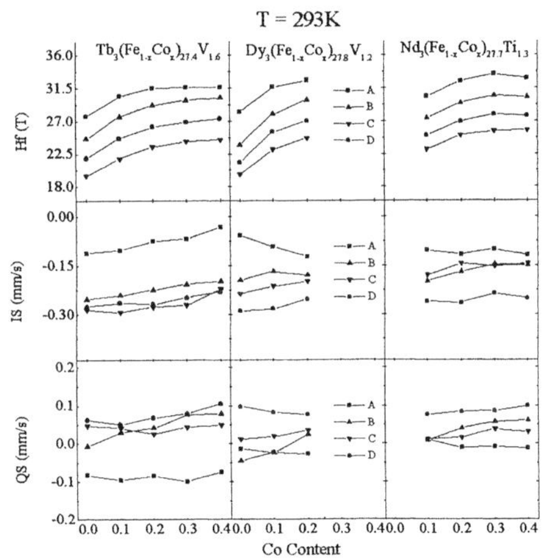

Fig. 3. The dependence of the hyperfine field, isomer shift and quadrupole interaction on Co concentration for $Nd_3(Fe_{1-x}Co_x)_{27.7}Ti_{1.3}$, $Tb_3(Fe_{1-x}Co_x)_{27.4}V_{1.6}$ ($0 \leq x \leq 0.4$) and $Dy_3(Fe_{1-x}Co_x)_{27.8}V_{1.2}$ ($0 \leq x \leq 0.3$) at 293 K.

The fitting was carried out under the assumption that H(A)>H(B)>H(D)>H(C) based on the nearest neighbours argumentation. The largest field was assigned to group A, which has the largest number of the Fe nearest neighbours. Groups B and D have the same number of the Fe nearest neighbours but the average Fe-Fe distances in $R_3(Fe,T)_{29}$ are 2.505 and 2.584 Å, respectively [7]. Therefore, we may expect a larger hyperfine field in group B than in group D. Finally, the smallest field was assigned to group C, which has the smallest number of the Fe nearest neighbours.

As shown in Fig. 2 and 3, the introduction of Co into the Fe sublattice affects the hyperfine field values for all group-sites. In the case of Nd the values of the average ^{57}Fe hyperfine field at 293 K yielded by the fit increase with the Co content up to x=0.3 and then decreases for x=0.4 in accordance to the behaviour of the corresponding saturation magnetisation values [5]. For R=Tb and Dy the dependence of the average ^{57}Fe hyperfine field is also similar to that of the saturation magnetisation values [6]. The substitution of Co for Fe has not a considerable influence on IS and QS as shown in Fig. 2 and 3. A similar behaviour has been reported for the $RFe_{11-x}Co_xTi$ (R=Y, Dy, Er; x=0, 3, 6, 8) series [12].

Acknowledgements

This work was partially supported by the European Social Fund and the General Secretariat of Research and Technology, GREECE (project 99ED 535).

References

[1] O. Kalogirou, V. Psycharis, L. Saettas and D. Niarchos, J. Magn. Magn. Mater. **146** (1995) 335.

[2] Z. Hu and W.B. Yelon, Solid State Commun. **91**(3) (1994) 223.

[3] V.R. Shah, G. Markandeyulu, K.V.S. Rama Rao, M.Q. Huang, K. Sirisha and M.E. McHenry, J. Magn. Magn. Mater. **190** (1998) 233.

[4] Guoyan Huo, Zhiyu Qiao, Guanghui Rao, Jingkui Liang and Baogen Shen, J. Alloys Comp. **285** (1999) 216.

[5] O. Kalogirou, C. Sarafidis, M. Gjoka, T. Bakas and M. Giannouri, J. Alloys Comp. **325** (2001) 59.

[6] O. Kalogirou, M. Gjoka, G. Litsardakis, C. Sarafidis, and A.C. Stergiou, presented at the 1st Joint European Magnetic Symposia, 27.8.-1.9.2001 Grenoble; J. Magn. Magn. Mater., submitted for publication.

[7] V. Psycharis, O. Kalogirou, E. Devlin, M. Gjoka, A. Simopoulos and D. Niarchos, J. Magn. Magn. Mater. **153** (1996) 75.

[8] L.C.C.M. Nagamine, E. Baggio-Saitovitch, I. Souza Azevedo, Xiu-Feng Han, Lan-Ying Lin and H.R. Rechenberg, Hyperfine Interactions **120/121** (1999) 273.

[9] L.C.C.M. Nagamine, H.R. Rechenberg, Xiu-Feng Han, E. Baggio-Saitovitch, I. Souza Azevedo and Lan-Ying Lin, Phys. Stat. Sol. **221** (2000) 767.

[10] V. Psycharis, O. Kalogirou, D. Niarchos and M. Gjoka, Mater. Sci. Forum **278-281** (1998) 526.

[11] V.G. Harris, V.R. Shah, G. Markandeyulu, K.V.S. Rama Rao, M.Q. Huang, K. Sirisha and M.E. McHenry, IEEE Trans. Magn. **35** (1999) 3286.

[12] J.J. Bara, B.F. Bogacz, A.T. Pedziwiatr and R. Wielgosz, J. Alloys Comp. **307** (2000) 45.

Crystallization and Anisotropic Hyperfine Field Fluctuation in Double Perovskite A_2FeMoO_6 (A=Ba, Sr, Ca)

Chul Sung Kim and Sung Baek Kim
Department of Physics, Kookmin University, Seoul 136-702, Korea

Abstract. The double perovskite oxides A_2FeMoO_6 (A=Ba, Sr. Ca) have been studied by Mössbauer technique, neutron and x-ray diffraction. The single phases of the polycrystalline A_2FeMoO_6 powders have been prepared by a solid-state reaction method. Their crystal structures are cubic for A=Ba, tetragonal for A=Sr, and monoclinic for A=Ca, respectively. In Sr_2FeMoO_6, the unit-cell parameters increase linearly with increasing the temperature, and the crystal symmetry changes into cubic in the paramagnetic phase (above T_C). Mössbauer spectra measurements of the A_2FeMoO_6 powder samples have been taken at various temperatures ranging from 18 to 450 K. The Curie temperatures are 345 K for Ba_2FeMoO_6, 425 K for Sr_2FeMoO_6, and 350 K for Ca_2FeMoO_6, respectively. As the temperature increases toward the Curie temperature, Mössbauer spectra show the line broadening and the difference between 1, 6 and 3, 4 line-width because of anisotropic hyperfine field fluctuation. In Sr_2FeMoO_6, the anisotropic field fluctuation of $+H(P_+=0.85)$ was great than $-H(P_-=0.15)$. We also calculated frequency factor and anisotropy energy with values of 9.8 Πh and 149.6 erg/㎤ for Sr_2FeMoO_6, respectively, using the relatively accurate data for T=260 K which is associated with the large line broadening. The anisotropy energy increase with decreasing ions radius of A-site atom (Ba, Sr, Ca).

1. Introduction

The early studies by Patterson [1] and Galasso [2] on A_2FeMO_6 (A=Ba, Sr, Ca) system examined the significance of bond distance and superexchange interaction. Recently, spin polarized giant tunneling magnetoresistance was confirmed at room temperature in powder sample of ordered double perovskite Sr_2FeMoO_6 by Kobayashi [3]. He claimed that a ferrimagnetic half metallic state is expected for this ordered perovskite with localized up spin of $Fe^{3+}(3d^5 : t^3_{2g}\, e^2_g)$ and itinerant down spin electron of $Mo^{5+}(4d^1 : t^1_{2g})$, and electronic structure calculations confirm this. The compounds of the formula A_2FeMO_6 (A=Ba, Sr, or Ca and M=Mo or Re) are of particular interest since they exhibit spin polarized giant tunneling magnetoresistance at room temperature [4, 5]. So many researchers have reported the lattice constants, magnetization and resistivity of the double perovskite transition metal oxides, especially for the Fe-based compounds.

Here, we present a detailed neutron and Mössbauer studies of the A_2FeMoO_6 (A=Ca, Sr, Ba) compounds. It is notable that line broadening and the difference between 1, 6 and 3, 4 line-width due to anisotropic hyperfine field fluctuation in Mössbauer hyperfine spectra [6, 7]. We have systematically analyzed the line broadening of Mössbauer spectra and also calculated temperature dependence of anisotropy energy from the relaxation rate.

2. Experimental

The A_2FeMoO_6 samples were prepared by standard solid-state reaction. Powders of high purity (99.99 % or better) of $BaCO_3$, $SrCO_3$, $CaCO_3$, Fe_2O_3, and MoO_3 were mixed, ground, and calcined at 1,100 ℃ for 6 h in an atmosphere of 1 % H_2 in Ar. The calcined mixtures were pressed into cylindrical bars and sintered at 1,200 ℃ for 4 h in the atmosphere. Colors of the final products were blue–black. Chemical compositions of the samples were confirmed to be stoichiometric by induction coupled plasma (ICP) and Rutherford backscattering spectrometer (RBS) analysis. Neutron diffraction data for the powder samples were collected using the High Resolution Powder Diffractometer (HRPD) at the Korea atomic energy research institute HANARO reactor. Mössbauer spectra were investigated in the temperature range from 15 to 425 K using a conventional spectrometer with a ^{57}Co source in Rh matrix [8].

3. Results and discussion

Neutron powder diffraction (NPD) patterns of the A_2FeMoO_6 were obtained at various temperatures between 10 and 473 K. Neutrons with wavelength 1.8346 Å were obtained by the (331) reflection of Ge monochromater, and 2θ ranging from 5 to 160 ° degrees. Figure 1~3 show the NPD patterns at various temperature for Ba_2FeMoO_6, Sr_2FeMoO_6, and Ca_2FeMoO_6, respectively. All samples were single phase, and exhibited a series of superstructure reflections due to cation order of Fe and Mo on alternate B-sites of the perovskite structure [9]. The structural parameters at room temperature are listed in Table I, which is the same result obtained from XRD diffraction. The unit-cell parameters for the all samples increase linearly with increasing the temperature. In Sr_2FeMoO_6, the crystal symmetry is tetragonal (*I4/mmm* ; *Z*=2) in the ferrimagnetic phase (below T_C = 425 K), but changes into cubic (*Fm3m* : *Z*=4) in the paramagnetic phase (above T_C) [10].

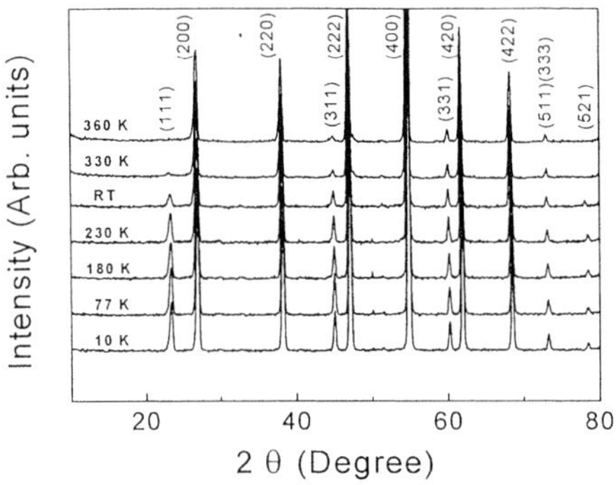

Figure 1. Neutron powder diffraction patterns at various temperature for Ba_2FeMoO_6.

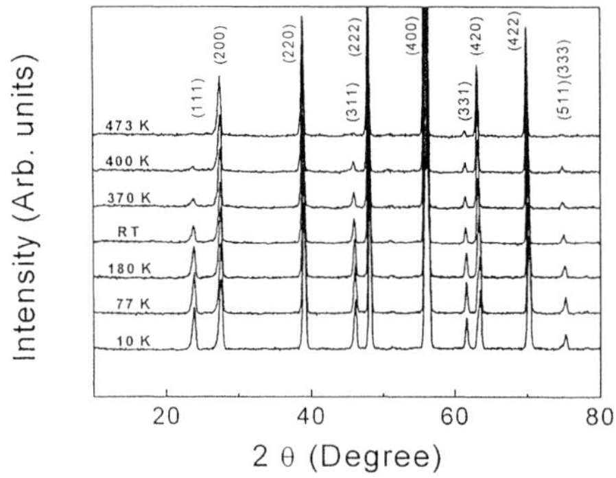

Figure 2. Neutron powder diffraction patterns at various temperature for Sr_2FeMoO_6.

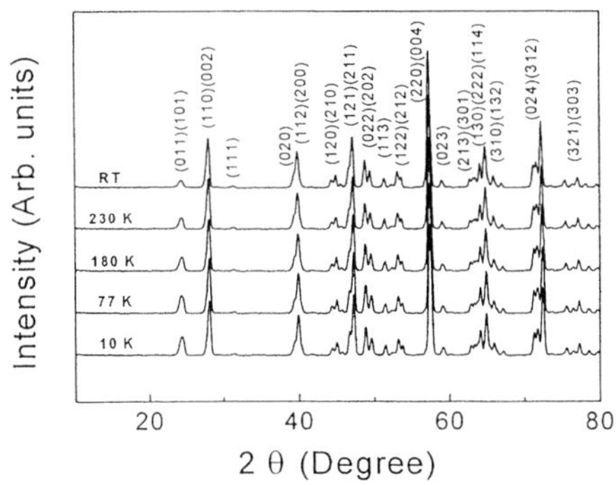

Figure 3. Neutron powder diffraction patterns at various temperature for Ca_2FeMoO_6.

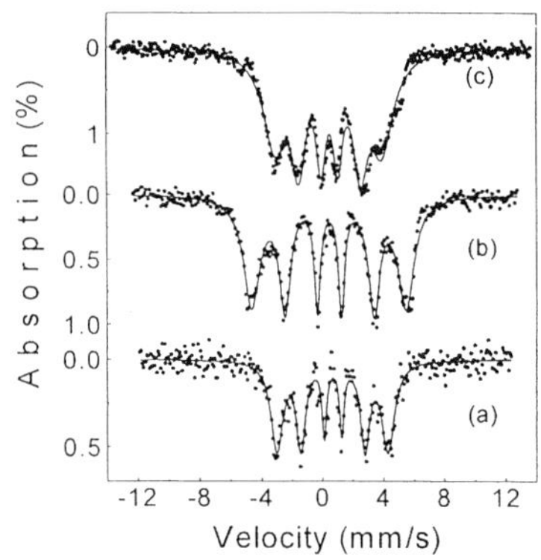

Figure 4. Mössbauer spectra at room temperature for (a) Ba_2FeMoO_6, (b) Sr_2FeMoO_6, and (c) Ca_2FeMoO_6.

Mössbauer spectra at room temperature obtained on Ba_2FeMoO_6, Sr_2FeMoO_6 and Ca_2FeMoO_6 are compared in Fig. 4. And Fig. 5 and 6 show that the Mössbauer spectra of the Ba_2FeMoO_6 and Sr_2FeMoO_6 at various temperatures. We determined the T_C = 345 K for Ba_2FeMoO_6, T_C = 425 K for Sr_2FeMoO_6, and T_C = 350 K for Ca_2FeMoO_6, respectively. Mössbauer absorption lines become broader with increasing temperature, as shown in Fig. 5 and 6. In order to explain this line broadening of Mössbauer spectra and the difference between 1, 6 and 3, 4 line-width, we used the following expression derived by M. Blume and J. A. Tjon [11] for the line shape of the Mössbauer absorption spectra in the presence of a fluctuating magnetic field, which jumps between the values $+H$ and $-H$ along the z axis with a frequency f:

$$W(k)=\frac{2}{\Gamma}\mathrm{Re}\sum_{m_0,m_1}\frac{1}{4}\left|\langle I_0m_0|H^{(+)}|I_1m_1\rangle\right|^2\times\sum_{i,j}\frac{1}{2}\langle j|(P-W-i\alpha F)^{-1}|i\rangle \tag{1}$$

where

$$P=-i(\omega-\omega_0)+\frac{1}{2}\Gamma,\quad \alpha=(g_0m_0-g_1m_1)\mu H,\quad W=\begin{pmatrix}-f & f\\ f & -f\end{pmatrix},\quad F=\begin{pmatrix}1 & 0\\ 0 & -1\end{pmatrix}.$$

Here $H^{(+)}$ is the operator for emission of gamma ray by the nucleus. Γ is the natural line-width, ω and k are the frequency and wave vector of the gamma ray photon. $|I_0m_0\rangle$ and $|I_1m_1\rangle$ $(m_0=\pm\frac{1}{2},\ m_1=\pm\frac{1}{2},\pm\frac{3}{2})$ represent the ground and the first excited states of ^{57}Fe, respectively. g_0 and g_1 are the g factors of the ground and the first excited states of ^{57}Fe, μ is the magneton and ω_0 is the frequency of the unsplit line for ^{57}Fe. Since the above expression shows only the absorption probability, it is necessary to carry out the interaction

$$a(v)=\int_{-\infty}^{\infty}W(\omega)F[\omega-\omega_0-(\omega_0/c)v]d\omega \tag{2}$$

in order to fit the theoretical curve to the Mössbauer spectra. $F(x)$ is a single line source spectrum, c is the velocity of light, and v is the Doppler velocity of the source. The line shape function $a(v)$ was calculated for each Doppler velocity v and the results are shown as solid lines through the data points in Fig. 5 and 6 with corresponding value f. The Mössbauer spectra indicate that an anisotropic field fluctuation of dwelling time probability for $+H$ ($P_+=0.85$) was greater than $-H$ ($P_-=0.15$).

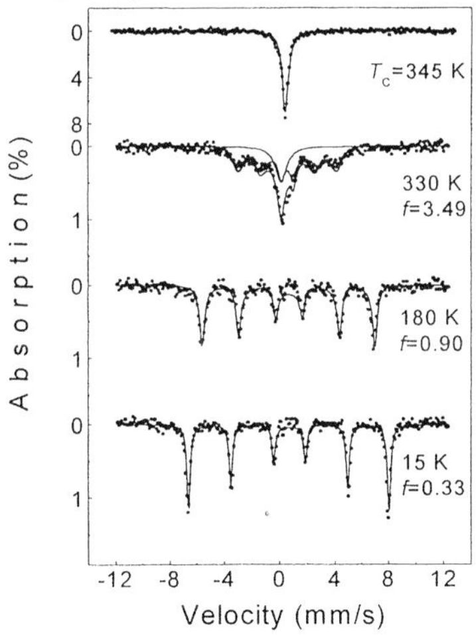

Figure 5. Mössbauer spectra at various temperature for Ba_2FeMoO_6. The relaxation frequency f is in units of $\Gamma/\hbar$.

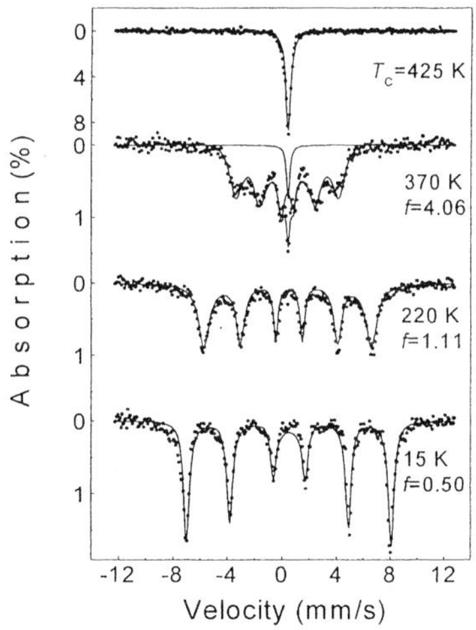

Figure 6. Mössbauer spectra at various temperature for Sr_2FeMoO_6. The relaxation frequency f is in units of $\Gamma/\hbar$.

Figure 7 (a) and (b) show temperature dependence of the fluctuation rate f. It is noted that the relaxation rate f increases rapidly as the temperature approaches the T_C. Then, the anisotropy energy can be estimated from the f values using the expression

$$f=f_0\exp(-KV/k_BT), \tag{3}$$

where f_0 is a frequency factor , K is the effective anisotropy energy per unit volume, and V is the particle volume. The volume V of the average particle for the both samples were measured to be 10^9 Å^3 using particle size analyzer made by Brookhaven Instruments Co., and frequency factors f_0 were obtained the extrapolated rate of 5.2 $\Gamma/\hbar$ for Ba_2FeMoO_6, and 9.8 $\Gamma/\hbar$ for Sr_2FeMoO_6 at T_C, respectively. Figure 8 shows temperature dependence of the anisotropy energy for Ba_2FeMoO_6 and Sr_2FeMoO_6. We calculated that the anisotropy energy was 87.8 erg/㎤ for T=230 K which is associated with the large line broadening in Ba_2FeMoO_6, and 149.6 erg/㎤ for T=260 K in Sr_2FeMoO_6. It is noticeable that calculated anisotropy energy for Sr_2FeMoO_6 at 260 K is larger than that of Ba_2FeMoO_6. We interpret it as originating from crystallographic structure and tolerance factor. Since tolerance factor for Sr_2FeMoO_6 are smaller than that of Ba_2FeMoO_6, tilting tendency of Sr ions to <110> direction are easily understood finally resulted in tetragonal phase. If crystalline anisotropy of Sr_2FeMoO_6 are larger than that of Ba_2FeMoO_6, than Sr_2FeMoO_6 should give larger anisotropy energy compared to Ba_2FeMoO_6. These results accord with present Mössbauer analyses.

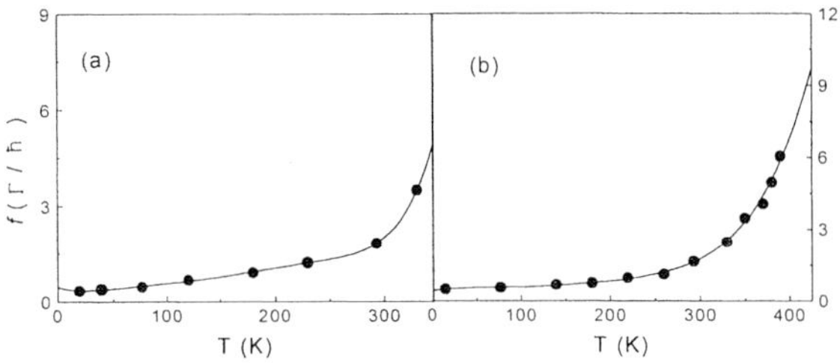

Figure 7. Temperature dependence of the relaxation frequency f for Ba_2FeMoO_6 and Sr_2FeMoO_6.

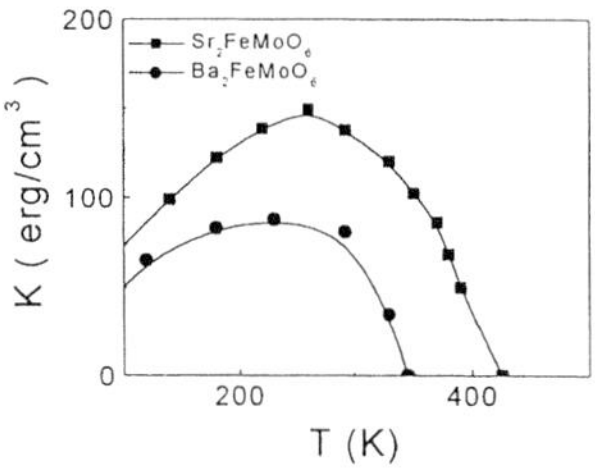

Figure 8. Temperature dependence of the anisotropy energy K for Ba_2FeMoO_6 and Sr_2FeMoO_6.

Table I. The structural parameters at room temperature for A_2FeMoO_6.

	Ba_2FeMoO_6	Sr_2FeMoO_6	Ca_2FeMoO_6
Tolerance Factor	0.975	0.918	0.884
Space Group	*Fm-3m*	*I4/mmm*	*P2₁/n*
Lattice Constants	a_0=8.0747 Å	a_0=5.5729 Å c_0=7.9077 Å	a_0=5.4056 Å b_0=5.5170 Å c_0=7.6971 Å β=89.92 °
T_C	345 K	425 K	350 K
MR ($\Delta\rho/\rho_0$)	18.8 % – 77 K 2.9 % – 300 K	19.5 % – 77 K 6.25 % – 300 K	2.6 % – 77 K –
Magnetization	3.7 μ_B – 77 K 2.2 μ_B – 300 K	3.2 μ_B – 77 K 2.3 μ_B – 300 K	3.0 μ_B – 77 K 1.8 μ_B – 300 K

Acknowledgments

This work has been supported by the Korea Science and Engineering Foundation (R01-1997-00109).

References

[1] F. K. Patterson, C. W. Moeller, and R. Waed, Inog. Chem. **2** (1963) 196.
[2] F. S. Galasso, F. C. Douglas, and R. J. Kaper, J. Chem. Phys. **44** (1966) 1672.
[3] K. -I. Kobayashi, T. Kimura, H. Sawada, K. Terakura, and Y. Tokura, Nature **395** (1998) 677.
[4] J. Gopalakrishnan, A. Chattopadhyay, S. B. Ogale, T. Venkatesan, R. L. Greene, A. J. Millis, K. Ramesha, B. Hannoyer, G. Hannoyer, and G. Marest, Phys. Rev. B **62** (2000) 9538.
[5] J. M. Greneche, M. Venkatesan, R. Suryanarayanan, and J. M. D. Coey, Phys. Rev. B **63** (2001) 174403.
[6] D. G. Rancourt, S. R. Julian, and J. M. Daniels, J. Magn. Magn. Magn. **51** (1985) 83.
[7] K. S. Baek, H. N. Ok, and J. C. Sur, Phys. Rev. B **39** (1989) 2800.
[8] C. S. Kim, S. B. Kim, J. S. Lee, and T. H. Noh, J. Appl. Phys. **79** (1996) 5459.
[9] D.D. Sarma, E. V. Sampathkumaran, S. Ray, R. Nagarajan, S. Majundar, A. Kumar, G. Nalini, and T. N. G. Row, Solid State Commun. **114** (2000) 465.
[10] Yutaka Moritomo, Sheng Xu, Akihiko Machida, Takumi Akimoto, Eiji Nishibori, Masaki Takata, Makoto Sakata, and Kenji Ohoyama, J. Phys. Soc. Japan **69** (2000) 1723.
[11] M. Blume and J. A. Tjon, Phys. Rev. **165** (1968) 4469.

Mössbauer study and electron transport properties in $Co_{0.1}Fe_{0.9}Cr_2S_4$

Sam Jin Kim[a], Woo Chul Kim[a], Jung Chul Sur[b], and Chul Sung Kim[a]

[a]*Dept. of Physics, Kookmin University, Seoul 136-702, Korea*
[b]*Dept. of Physics, Wonkwang University, Jeonbuk 570-749, Korea*

A sample of $Co_{0.1}Fe_{0.9}Cr_2S_4$ has been studied with Mössbauer spectroscopy, X-ray, SQUID magnetometer and magnetoresistance (MR). The crystal structure was determined to be cubic spinel with its lattice constant a_0=9.9696 Å. The Mössbauer spectra were recorded from 12 K to room temperature. The asymmetric line broadening is observed and considered to be dynamic, Jahn-Teller distortion. The unusual reduction of magnetic hyperfine field below 100 K could be explained in terms of cancellation effect between the mutually opposite orbital current field (H_L) and Fermi contact field (H_C). The isomer shift value of the sample at room temperature was 0.58 mm/s, which means that the charge state of Fe ions was ferrous in character. The MR peak was observed at 192 K about 10 % at an applied field of 16 kOe. The conduction mechanism in this sample is different from the double exchange mechanism in a point that there were no mixed iron charge valences.

1. Introduction

Recently the colossal magnetoresistance (CMR) effects in sulphur spinel compounds ($Fe_{1-x}Cu_xCr_2S_4$; x=0, 0.5) have been reported, and it is suggested that the conduction mechanism in these materials may not be caused by the double exchange (DE) of the carriers [1]. This is because there is no valence variation in stoichiometry as in the Mn perovskite structure. Also the crystal and magnetic structures are different from Mn perovskites, so renewed interest has been focused on conduction mechanism of suphur spinel compounds. Z. Yang's group [2] claimed the validity of the magnetic polaron model in $FeCr_2S_4$ by electron spin resonance study and C. Greaves' group [3] reported unusual reduction of Fe magnetic moment by neutron diffraction study. Also B. I. Min's group [4] interpreted its conduction mechanism by half-metallic properties. Very recently some of us have published on a $Cu_{0.5}Fe_{0.5}Cr_2S_4$ valence band photoemission study and reported mono-valent copper and ferric character of iron ions, respectively [5]. Mössbauer studies on $FeCr_2S_4$ have been reported already by many previous workers [6-8]. but the Mössbauer spectra on $FeCr_2S_4$ were found to be dependent on the sample. Furthermore some of the paper reported the line broadening originating from the excess of B site Fe^{2+} [9]. Therefore it is essential to determine the valence state of iron ions in various sulphur spinel compounds to understand the underlying physics properly. Here we present the results of temperature dependence of Mössbauer experiments and compare them with those of x-ray, magnetoresistance, and SQUID magnetometer for $Co_{0.1}Fe_{0.9}Cr_2S_4$.

2. Experiments

$Co_{0.1}Fe_{0.9}Cr_2S_4$ was prepared by usual direct reaction of the high-purity elements Fe (99.999 %), Cr (99.999 %), Co (99.999 %), and S (99.9999 %) in an evacuated quartz tube. The crystallographic structure of $Co_{0.1}Fe_{0.9}Cr_2S_4$ was analyzed by Rietveld refinement of a typical X-ray diffraction. The Mössbauer spectra were recorded using a conventional spectrometer with a ^{57}Co source in a rhodium matrix. Magnetoresistance and magnetization were measured with Van Der Pauw method and SQUID magnetometer respectively.

3. Results and Discussion

In Figure 1 we have presented the observed and calculated peak profile, and Bragg position of a typical X-ray diffraction pattern. The crystal structure of the sample at room temperature is determined to be normal cubic spinel structure *Fd3m* (Fe, Co(8*a*); Cr(16*d*); S(32*e*)(*u,u,u*)) with its lattice constant a_0=9.9696(2) Å. The final oxygen parameter *u*, Bragg factor R_B, and R_F were 0.3846 (2), 8.12 , and 6.38 %, respectively.

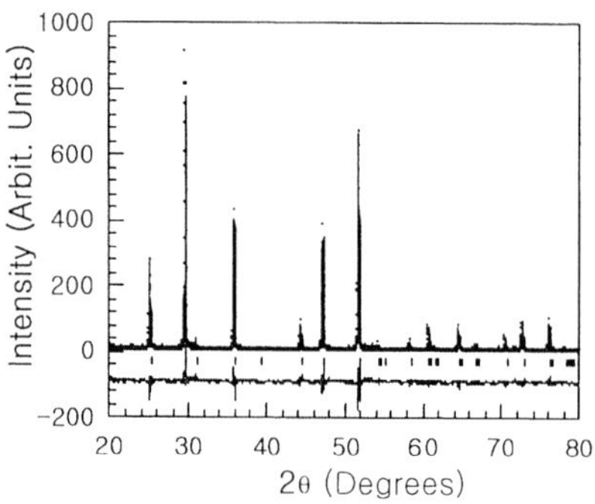

Figure 1. Refined powder X-ray diffraction pattern at room temperature. Solid circle represents the observed pattern; continuous lines represent calculated and difference (obs-calc) patterns. Tick markers correspond to the position of the allowed Bragg-reflections

Figure 2(a) shows the temperature dependence of zero-field resistance, and the resistance measured under a field of 16 kOe. It shows metallic behavior in the temperature range between 150 K and 195 K, while semiconducting behavior is seen at T>195 K and T<150 K. Figure 2(b) shows the temperature dependence of magnetization for the sample under an external field of 50 kOe. We note that the saturation magnetic moment at 80 K is about 1.44 μ_B, which means that the magnetic behavior is ferrimagnetic. Also the inset of Figure 2(b) shows the enlarged shape between 4 and 20 K. This cusp-like anomaly at 9 K is explained by changing from static to dynamic Jahn-Teller stabilization [10]. As shown in Figure 2(a) the maximum magnetoreistance temperature(T_M) was 192 K with its *MR* ratio 10 %. Assuming a simple exponential temperature dependence of resistance $R(T)= R_0 exp(E_\rho /k_B T)$, a fit to the data of Figure 2(a) above 195 K yields activation energy E_ρ = 50 meV [1].

In order to study the conduction mechanism and the local properties of iron ions, Mössbauer spectra of the sample were measured at various absorber temperatures from 12 K to room temperature. Some of the representative spectra are shown in Figure 3. Mössbauer spectra were analyzed using the eight Lorentzians fitting method [11]. The results of computer analysis are shown in Figure 4. The isomer shift at room temperature is 0.58 mm/s relative to Fe metal, which means that valence state of Fe ions is ferrous in character. The magnetic hyperfine field at 12 K is 173 kOe. This value corresponds to the typical ferrous ion. This is an important different point in comparison to Mn perovskite structure, in which the conducting mechanism is accompanied by heterogeneous Mn^{3+} and Mn^{4+} with ferromagnetic double exchange coupling. But in this sample magnetic structure is ferrimagnetic and there are no iron species other than Fe^{2+}.

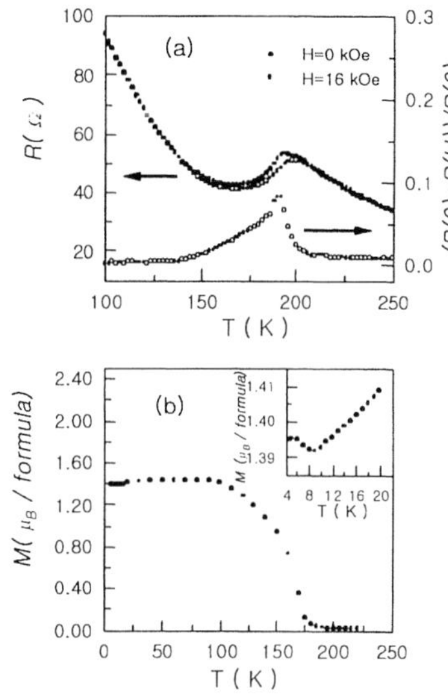

Figure 2. (a) Temperature dependence of resistance R and magnetoresistance $MR_{//}$ for $Co_{0.1}Fe_{0.9}Cr_2S_4$, where $MR_{//}=(R(0)-R(1.6T))/R(0)$. (b)Temperature dependence of magnetization M (μ_B) at 50 kOe external field. Inset shows the enlarged data between 4 and 20 K.

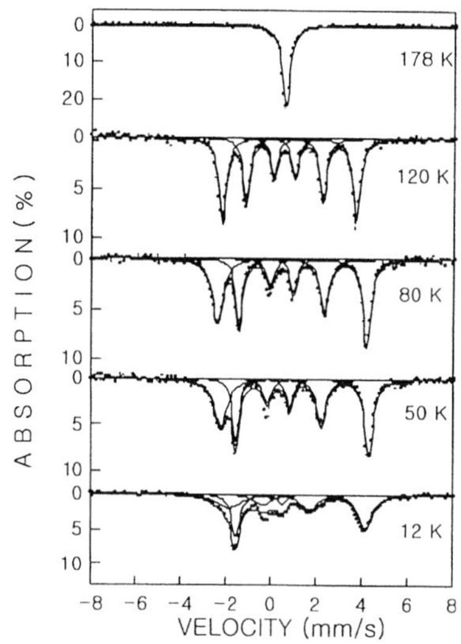

Figure 3. Mössbauer spectra for $Co_{0.1}Fe_{0.9}Cr_2S_4$ at various temperatures.

Also we note in Figure 4 that the quadrupole splitting E_Q rapidly decreases with increasing temperature up to 120 K and then approaches zero with further increase of temperature. In the tetragonal site, the ground state of Fe^{2+} is in a degenerate orbital doublet ^5E_g. This situation is unstable and must be resolved via a Jahn-Teller distortion, which may explain the observed appearance of quadrupole splitting below T_C. However, a low temperature x-ray photograph for $FeCr_2S_4$ showed that there is no static Jahn-Teller distortion above 12 K [12, 13]. Even though no static Jahn-Teller distortion is observed, dynamic Jahn-Teller distortion can cause a quadrupole shift if the distortion relaxes between a number of equivalent crystal axes [14]. This kind of relaxation is consistent with the asymmetrical line broadening apparent in Figure 3. It is noticeable that the temperature at which quadruplole splitting approaches 0 is 120 K, the same as Z. Yang *et al.* [2] claimed magnetic polarons are formed.

The unusual reduction of magnetic hyperfine field below 100 K may be explained in terms of cancellation effect between the mutually opposite H_L and H_C. Here H_L and H_C

are the orbital current field and the Fermi contact field, respectively [15].

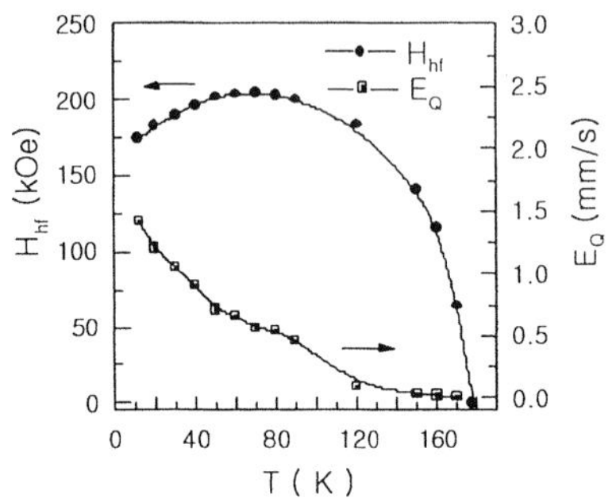

Figure 4. Temperature dependence of the quadrupole splitting and magnetic hyperfine fields for $Co_{0.1}Fe_{0.9}Cr_2S_4$.

In summary, we have studied magnetic properties and conduction mechanism in $Co_{0.1}Fe_{0.9}Cr_2S_4$. The peculiarities of Mössbauer spectra have been explained in terms of dynamic Jahn-Teller effect. The evidence that iron ions are ferrous in character led to the conclusion that conduction mechanism in this sample is different from double exchange mechanism.

The present study was supported by the Korea Science and Engineering Foundation (R01-1997-00109).

References

[1] A. P. Ramirez, R. J. Cava, and J. Krajewski, Nature 386 (1997) 156.
[2] Z. Yang, S. Tan, Z. Chen, and Y. Zang, Phys. Rev. B 62 (2000) 13872.
[3] H. M. Palmer, C. Greaves, Physica B 276-278 (2000) 568.
[4] M. S. Park, S. K. Kwon, S. J. Youn, and B. I. Min, Phys. Rev. B 59 (1999) 10018.
[5] J. S. Kang, S. J. Kim, C. S. Kim, C. G. Olson, and B. I. Min, Phys. Rev. B 63 (2001) 144412.
[6] A. M. Van Diepen, R. P. Van Stapele. Solid State Commun. 13 (1973) 1651.
[7] L. Brossard, J. L. Dormann, L. Goldstein, P. Gibart, and P. Renaudin, Phys. Rev. B 20 (1979) 2933.
[8] H. N. Ok, K. S. Baek, H. S. Lee, and C. S. Kim, Phys. Rev. B 41 (1990) 62.
[9] M. R. Spender, A. H. Morrish, Can J. Phys. 50 (1972) 1125.
[10] F. K. Lotgering, A. M. van Diepen, and J. F. Olijhohoek, Solid State Commun. 11 (1972) 1417.
[11] C. S. Kim, I. B. Shim, M. Y. Ha, C. S. Kim, and J. Y. Park, J. Appl. Phys. 73 (1993) 5707.
[12] P. Gibart and A. Beguoen-Demeaux, C. R. Acad. Sci (France) 268C (1969) 816.
[13] Z. Chen, S. T. Tan, Z. Yang, and Y. Zhang, Phys. Rev. B 59 (1999) 11172.
[14] J. A. Tjon and M. Blume, Phys. Rev. 165 (1968) 456.
[15] H. N. Ok and J. G. Mullen, Phys. Rev. 168 (1968) 563.

Electronic Properties of $CuFeS_2$ under Pressure Studied by Mössbauer Spectroscopy

Hisao Kobayashi*, Hideya Onodera[a], and Takashi Kamimura
Department of Physics, Tohoku University, Sendai 980-8578, Japan
[a] *Institute for Materials Research, Tohoku University, Sendai 980-8577, Japan*

February 28, 2002

Abstract. The ^{57}Fe Mössbauer spectra of satisfactory quality were measured with the sample $CuFeS_2$ not-enriched in ^{57}Fe under pressure up to 13 GPa. The vanishing of the magnetic hyperfine field in the spectra was observed around 6.5 GPa and room temperature where the structural phase transition occurs. It was found by the analysis of observed spectra that there exists the significant anomaly in the pressure dependence of the center shift around 6.5 GPa. These results show a substantial change in the electronic structure of $CuFeS_2$ at 6.5 GPa.

Keywords: Mössbauer spectroscopy, $CuFeS_2$, pressure, phase transition

1. Introduction

Chalcopyrite, $CuFeS_2$, is an antiferromagnetic semiconductor with extremely high Néel temperature, T_N = 853 K, at ambient pressure (Donnay et al., 1958). The ordered magnetic moment of the Fe atom is 3.8 μ_B, which is much smaller than that of the free Fe^{3+} ion. The above results indicate that the Fe 3*d* states in $CuFeS_2$ are strongly hybridized with the S 3*p* states. Only a few compounds with tetrahedrally coordinated Fe^{3+} ions are known, which could reflect on inherent instability in such compounds. The high-pressure electrical and X-ray diffraction measurements (Pitt and Vyas, 1974; Tinoco et al., 1994) indicated the first-order phase transition at 6.5 GPa and room temperature with about 10% volume reduction. There was another small anomaly at 2.8 GPa in the pressure dependence of electrical resistance.

Zaanen, Sawatzky, and Allen (ZSA) have developed a general framework for 3d transition-metal compounds where an electron correlation plays an important role (Zaanen et al., 1985). According to the ZSA scheme, these compounds can be classified into four regimes based on the relative magnitudes of the ligand-to-metal charge-transfer energy, the on-site Coulomb energy, and the bandwidth. Since pressure can change a bond length as well as a bond angle, these parameters can be varied more by pressure than by temperature and magnetic field. Thus

* kobayash@nuclear.phys.tohoku.ac.jp

it is of grate interest to study electronic states of these compounds under pressure.

Mössbauer spectroscopy is a useful method to study electronic properties of materials under pressure. However, it is not easy to measure Mössbauer effect under high pressure using a diamond anvil cell (DAC) since there is a small sample cavity on a gasket and the pair of diamonds in DAC transmit about 30% at the 14.4 keV γ-ray. Then one desires high flux density of γ-ray and samples enriched in the ^{57}Fe Mössbauer isotope to get spectra of satisfactory quality. Usually the Mössbauer absorption spectra have been measured using a ^{57}Co in Rh source with a high-space-activity area.

2. Experimental

Polycrystalline $CuFeS_2$ samples were prepared by a solid state reaction using starting materials Cu (99.99% purity), Fe (99.99%), and S (99.999%). The powdered samples were loaded into a clamp-type DAC with ruby crystals into a sample cavity of a 0.4 mm diameter in a 0.3 mm thick inconel 625 alloy gasket. Pressure was calibrated by measuring the wavelength shift of R_1 luminescence line of the ruby crystal. Fluorinert as pressure transmitting medium ensured quasihydrostatic conditions. The Mössbauer absorption spectra were measured at room temperature under pressure up to 13.5 GPa. The DAC was mounted on a translation stage with the sample at distance of approximately 3 mm from a ^{57}Co in Rh source with a high-space-activity area of 500 mCi/cm^2. The velocity scale was calibrated by α-Fe at room temperature.

3. Results and Discussion

Typical observed ^{57}Fe Mössbauer absorption spectra of $CuFeS_2$ are shown in Fig. 1. As seen in Fig. 1, spectra of satisfactory quality were obtained in 30 $\sim$ 60 hours using the samples not-enriched in ^{57}Fe. It was found that the spectra of $CuFeS_2$ below 5 GPa mainly consist of a magnetic hyperfine field with which the electric quadrupole interaction is almost zero, whereas the spectra above 7 GPa are typical of paramagnetic ones with a quadrupole splitting. The disappearance of magnetic hyperfine fields around 6.5 GPa has a good correspondence with the structural phase transition. The observed spectra were fitted well with one subspectrum in the two phases as shown in Fig. 1. Hence all Fe sites seem to be magnetically as well as crystallographically equivalent in each phase under pressure.

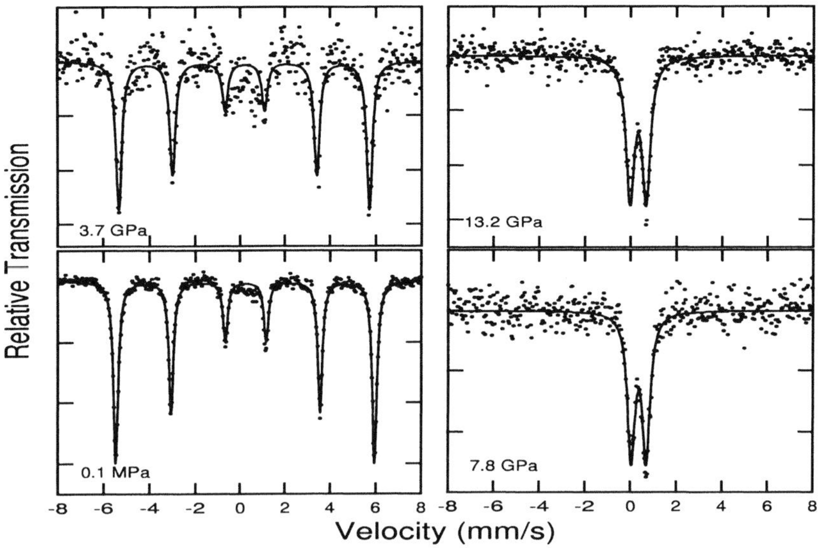

Figure 1. Typical observed ^{57}Fe Mössbauer absorption spectra of $CuFeS_2$ under pressure at room temperature. The closed circles and solid lines represent the observed and calculated spectra, respectively.

All refined hyperfine parameters show the distinct anomalies around 6.5 GPa as shown in Fig. 2. Within the accuracy of the data, we observed no anomaly at 2.8 GPa in the pressure dependence of refined hyperfine interaction. The value of center shift (δ_{CS}) decreases with pressure up to 5 GPa which values are located at the lower limit of the high-spin Fe(III) state. This result indicates the highly covalent Fe–S bond, which is consistent with the previous result (Donnay et al., 1958). In general the pressure dependence of δ_{CS} at constant temperature includes both the variation of electron density at nucleus and the change of the second-order Doppler shift (SOD). Since the pressure effect on SOD may be small, the main effect in $[\partial\delta_{CS}/\partial P]_T$ must be caused by an increase of the electron density at nucleus. Thus the volume reduction under pressure causes an increase of the *p*-*d* hybridization.

It was found that the magnetic hyperfine field (H_{hf}) decreases slightly under pressure up to 5 GPa. The H_{hf} value at finite temperature depends on the Fe magnetic moment, strength of exchange interactions and degree of covalence. As discussed in the pressure dependence of δ_{CS}, the degree of covalence increases with increasing pressure. Generally, increase in covalence contributes to decrease of H_{hf} and changes a conversion coefficient between H_{hf} and the Fe magnetic moment. In the cases of $CuFeS_2$, the observed pressure dependence of H_{hf} is most likely cause by the covalence contribution to H_{hf} because of the extremely high Néel temperature.

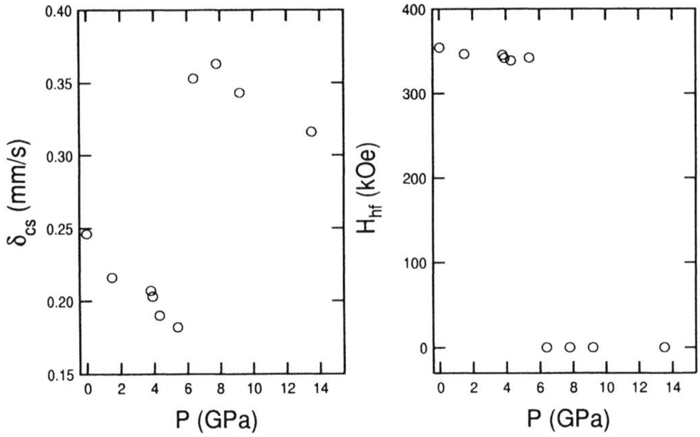

Figure 2. Pressure dependences of refined center shift (δ_{CS}) and magnetic hyperfine field (H_{hf}) in $CuFeS_2$.

A large enlargement in δ_{CS} of about 0.18 mm/s is observed around 6.5 GPa, which shows a substantial change in the electronic structure of $CuFeS_2$. The crystal structure above 6.5 GPa was determined to be a disordered rocksalt one (Tinoco et al., 1994) and then the coordination of Fe site changes from tetrahedral to octahedral ones. The δ_{CS} value and the volume of FeS_6 octahedra in this phase are comparable with those in FeS above 6.5 GPa where the Fe state with octahedral coordination has a low-spin Fe(II) state (Kobayashi et al., 1997; Kobayashi et al., 2001). Although the vanishing of H_{hf} only implies that there is no magnetic order at room temperature, these facts suggest that the spin splitting may be absent. Therefore, it is proposed that the Fe state in the high-pressure phase has the low-spin Fe(II) like state.

In summary, we have measured the ^{57}Fe Mössbauer spectra of satisfactory quality under pressure with the sample $CuFeS_2$ not-enriched in ^{57}Fe. It was found by the analysis of observed spectra that there exist the significant anomaly in δ_{CS} and disappearance of H_{hf} at about 6.5 GPa. Then it is deduced from the pressure dependence of hyperfine interaction that the electronic state of Fe ion changes from the high-spin Fe(III) to the low-spin Fe(II) like states at 6.5 GPa.

References

Donnay G., et al. *Phys. Rev.*, 112, 1917–1923 1958.
Kobayashi H., et al. *J. Phys. Condens. Matter.*, 9, 515–527, 1997.
Kobayashi H., et al. *Phys. Rev.*, B63, 115203-1–6, 2001.
Pitt G. M. and M. K. R. Vyas. *Solid State Commun.*, 15, 899–902, 1974.
Tinoco T., et al. *J. de Phys.*, C9, 151–154, 1994.
Zaanen J., et al. *Phys. Rev. Lett.*, 55, 418–421, 1985.

Defects and Disorder in Magnetoresistive Double Perovskite A_2FeMoO_6 $(A = Sr, Ba)$

R. Kruk[1,2], D.E. Brown[1], O. Chmaissem[1], R. Feng[1], B. Dabrowski[1], C.W. Kimball[1], S. Kolesnik[1]., C.E Johnson[3], E.E. Alp[3] J.D. Jorgensen[4]

[1]Physics Department, Northern Illinois University, Dekalb, IL
[2]Institute of Nuclear Physics, Krakow, Poland
[3]Advanced Photon Source, Argonne National Lab, Argonne, IL
[4]Materials Science Division, Argonne National Lab, Argonne, IL

Abstract A_2FeMoO_6 (A = Sr, Ba) is a double perovskite having an ordered ferrimagnetic structure of Fe and Mo magnetic superlattices. The Fe and Mo atoms are ordered on alternating, corner-shared octahedral sites, however, the ordered array can have imperfections that are dependent upon synthesis conditions. Mössbauer spectroscopy was used to examine the local magnetic and structural environment around the Fe atoms giving insight into what type of imperfections were present in the Fe magnetic superlattice. Transmission Mössbauer spectroscopy was performed on polycrystalline samples with varying degrees of disorder using magnetic fields of 0 and 70 kOe. We were able to clearly resolve the iron nuclei with flipped magnetic moments (iron having a magnetic dipole moment parallel to the Mo magnetic dipoles), and associate the disorder in the structure with the different sites observed in the Mössbauer spectrum. The amount of disorder induced by various synthesis conditions was consistent with magnetization measurements that consistently show a lower than expected average magnetic dipole moment.

1. Introduction

Transition metal oxides with a double perovskite structure A_2FeMoO_6 and (A = Ca, Ba, Sr) has attracted a great deal of attention owing to their high magnetic transition temperatures and spin dependent transport properties. Electronic structure calculations and experimental results show that these materials are half-metallic ferrimagnets with localized up-spin electrons on the Fe ions and itinerant down-spin electrons shared between Fe and Mo [1,2]. In this model the anticipated saturation magnetization has a value of $4\mu_B$. The reported values, however, are regularly lower than expected for a fully ordered phase. Recent experimental results along with Monte Carlo simulations show that this discrepancy is due to a partial disorder of the Mo and Fe ions (mis-site defects) [3,4,5,6]. The observed large negative magneto resistance (MR) effect is mainly attributed to the grain-boundary tunneling, although a small MR measured for a single crystal of $SrFe_{0.5}Mo_{0.5}O_3$ shows the presence of a spin dependent carrier scattering on the mis-site Fe\Mo defects [2]. In order to determine the actual defect configurations and their effect on the magnetic properties, a series of samples with varying amounts of disorder were synthesized. Mössbauer spectroscopy was used to characterize the electronic state and coordinating environment of iron.

Results and discussion

Precursors of polycrystalline samples of $SrFe_{0.5}Mo_{0.5}O_3$ and $BaFe_{0.5}Mo_{0.5}O_3$ were made from stoichiometric amounts of $SrCO_3$, $BaCO_3$, Fe_2O_3, and MoO_3 fired in air and argon at 800-1200C with periodic grinding. Then several thermal conditions and reducing atmospheres were tried to get samples with differing amounts of disorder. The samples of $SrFe_{0.5}Mo_{0.5}O_3$ were annealed at 1300C, 1100C, 960C and $BaFe_{0.5}Mo_{0.5}O_3$ at 1300C (labeled SFM1300, SFM1100, SFM960, and BFM1300 respectively) in a flowing gas mixture of 0.4-1% H_2/Ar. X-ray powder diffraction showed single-phase patterns corresponding to a tetragonal structure for the $SrFe_{0.5}Mo_{0.5}O_3$ samples and a cubic structure for $BaFe_{0.5}Mo_{0.5}O_3$. It was found that the intensity of superstructure peaks, which reflects the extent of Fe/Mo ordering, decreases upon lowering the annealing temperature. The magnetization data also shows a significant reduction of the saturation magnetic moment with decreasing annealing temperature (see Fig. 1). These results indicate that there is a correlation between the defect concentration and the magnetic properties.

The Mössbauer spectra were collected at 10K with and without an external magnetic field. The magnetic field of 70 kOe was applied perpendicular to the direction of propagation of the gamma ray, allowing determination of the arrangement of the iron spins. The zero field spectra could be fitted with three overlapping magnetically split patterns, I_1, I_2 and I_3 (see Figs. 1,2). It is interesting to note a correlation between the magnitude of magnetic hyperfine field, H_{hf}, and the isomer shift; the larger H_{hf} the lower the isomer shift observed for each component (Table 1). For the two subspectra with lower magnetic hyperfine fields, I_1 and I_2, application of an external magnetic field resulted in a 70kOe decrease in their effective fields and a change of the absorption line intensities from a 3:2:1 to a 3:4:1 ratio. This indicates that the spins of the Fe ions contributing to both components orients parallel to the applied magnetic field. The component with the largest H_{hf}, I_3, behaved differently. Its response to the application of a

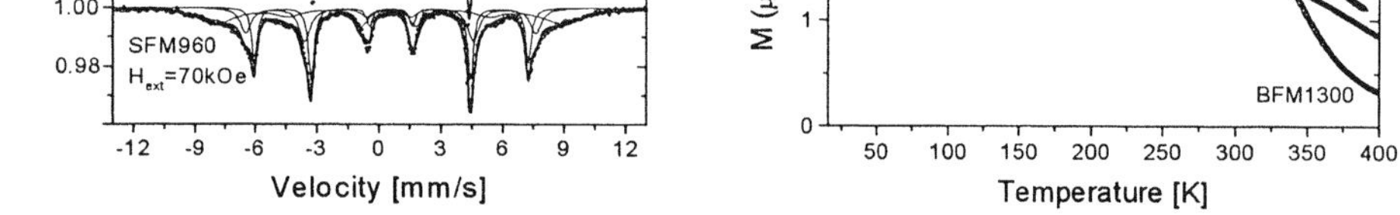

Figure 1. Mössbauer spectra and magnetization measured at 70 kOe for the samples with a different degree of the mis-site disorder.

magnetic field depended apparently on the extent of Fe/Mo disordering indicated by magnetization and X-ray data. For the relatively well ordered samples, SFM1300 and BFM1300, the third component became well resolved with an increase of the effective magnetic field of about 70kOe. Thus, for these samples the component I_3 can be identified with the Fe spins aligned antiparallel to the external magnetic field. From the magnitudes of H_{hf}, the isomer shifts, and the behavior in an external magnetic field, it was possible to assign each component to an appropriate Fe crystallographic site. The first subspectrum, I_1, with the lowest H_{hf} and narrowest lines was ascribed to the Fe ions occupying sites proper for an ideal ordered double perovskite. The mean hyperfine field and the isomer shift were found to be intermediate between high spin configuration values of Fe^{+3} and Fe^{+2} reflecting the half-metallic state predicted and observed in these compounds [1,2].

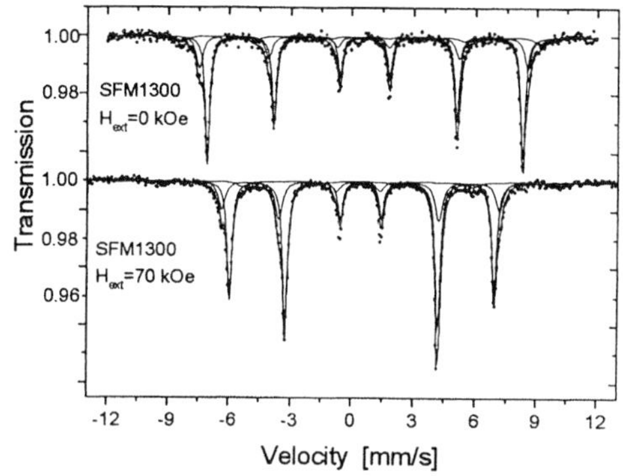

Figure 2. Mössbauer spectra of $SrFe_{0.5}Mo_{0.5}O_3$ (SFM1300) collected at 10K with and without an external magnetic field

The components I_2 and I_3 have a smaller isomer shift and a larger hyperfine field, hence they are associated with a higher charge state of iron, which may be due to the presence of a local disruption of the Fe/Mo ordering. In such a case the reduced electron density at the mis-site defect or in its vicinity can cause an enhancement of the magnetic hyperfine field, and also reduce the isomer shift. For the SFM1300 and BFM1300 Mössbauer results can be explained in terms of a random distribution of well-separated Fe/Mo mis-site defects.

Table 1. Mössbauer and magnetization data for the samples of $SrFe_{0.5}Mo_{0.5}O_3$ and $BaFe_{0.5}Mo_{0.5}O_3$. H_{eff} and H_{hf} represent hyperfine fields measured with and without an external magnetic field respectively, δ_{is} is the isomer shift (relative to metallic iron), PI_x is the relative intensity for the component I_x, M_{cal} is the magnetic moment calculated with the formula given in text, M_S is the saturated magnetization per formula unit.

	H_{hf1} H_{eff1} [kOe]	H_{hf2} H_{eff2} [kOe]	H_{hf3} H_{ef3} [kOe]	δ_{is1} [mm/s]	δ_{is2} [mm/s]	δ_{is3} [mm/s]	PI_2 [%]	PI_3 [%]	M_{cal} [μ_B]	M_S [μ_B]
BFM1300	457(1)	489(1)	520(4)							
	386(1)	417(1)	606(7)	0.88(1)	0.65(1)	0.63(1)	26	5	3.6	3.6
SFM1300	477(1)	500(1)	542(2)							
	411(1)	431(1)	620(6)	0.71(1)	0.60(1)	0.51(2)	22	5	3.6	3.6
SFM1100	466(1)	498(1)	520(4)							
	407(1)	429(1)	504*	0.71(1)	0.59(1)	0.52(3)	25	25	-	2.7
SFM960	480(1)	499(2)	522(6)							
	408(1)	429(2)	519*	0.72(1)	0.59(1)	0.56(3)	23	35	-	2.2

The major conclusion of this model is that when a Mo ion is replaced by an Fe ion the magnetic moment of Fe is coupled antiferromagnetically to its nearest neighbors. These iron atoms with flipped magnetic moments align antiparallel to the applied field thus giving rise to the component I3. Furthermore, since each Fe occupying a Mo site disturbs 5 iron nearest neighbors an additional component, I_2, is observed. The above model can be verified through magnetization measurements. When Fe and Mo swap their sites their spins flip, causing a net change of 8 μ_B. Thus the dependence of the saturation magnetization on mis-site defect concentration P is given by the formula (4-8*P) μ_B . For the samples with a low degree of disorder the net magnetic moments calculated by this formula correlate quite well with the magnetization data (Table 1).

That simple model breaks down for the more disordered samples. The component I_3 showed a remarkable broadening of the resonance lines with the mean effective field almost unchanged in an external magnetic field (Fig 1). Also the intensity ratio that best fits these components is 3:2:1 rather than expected 3:4:1. Thus, the spins of the iron ions contributing to the component I_3 are distributed randomly with respect to the external field. This could be attributed to the mis-site defects forming clusters with a very small net magnetic moment. These regions would have a higher concentration of mis-site defects where the antiferromagnetic Fe-O-Fe interactions are dominant and compete with the ferrimagnetic coupling to the host lattice on the cluster fringes.

In summary, samples of $SrFe_{0.5}Mo_{0.5}O_3$ and $BaFe_{0.5}Mo_{0.5}O_3$ with varying degree of disorder were synthesized. Mössbauer spectroscopy measurements at high magnetic fields explored the local magnetic environments around the iron. Experimental results were interpreted in terms of a mis-site defect model. It was found that the mis-site defects have a tendency to cluster for higher defect concentrations.

References

[1] K.-I Kobayashi, T. Kimura, H. Sawada, K. Terakura, and Y. Tokura, Nature 395, (1998) 677.

[2] Y. Tomioka, T. Okuda, Y.Okiomoto, R. Kumai, and K.-I. Kobayashi, Phys. Rev. B 61, (2000) 422.

[3] J. M. Greneche, M. Venkatesan, R. Suryanarayanan, and J. M. D. Coey, Phys. Rev. B 63, (2001) 174403.

[4] J. Navarro, M. Bibes, A. Roig, B. Martinez, and J. Fontcuberata, Appl. Phys. Lett. 78, (2001) 781

[5] D.D. Sarma, E.V. Sampathkumaran, Sugata Ray, R. Nagarajan, S. Majumdar, A. Kumar, G. Nalini, T.N. Guru Row, Solid State Communications 114, (2000) 465

[6] A.S. Ogale, S.B. Ogale, R. Ramesh, and T. Venkatesan, Appl. Phys. Lett. 75, (1999) 537

Acknowledgments Work at NIU was supported by the ARPA/ONR and by the State of Illinois under HECA. Work at Argonne National Lab was supported by US-DOE-BES Material Science under contract No. W-31-109-ENG-38

^{170}Yb Mössbauer study of magnetic ordering in $Yb_2Cu_2O_5$

D.H. Ryan, Saleema N.M. and R. Gagnon
Physics Department and Centre for the Physics of Materials, McGill University, 3600 University Street, Montreal, Quebec H3A 2T8, Canada

J. van Lierop
Materials and Chemical Sciences Division, Energy Science and Technology Dept., Brookhaven National Lab., Upton NY 11973-5000

J.M. Cadogan
School of Physics, The University of New South Wales, Sydney, NSW 2052, Australia

August 15, 2001

Abstract. ^{170}Yb Mössbauer shows that the Yb and Cu moments order at 15 K in $Yb_2Cu_2O_5$ and rule out models of the order that invoke distinct ordering temperatures.

Keywords: Ytterbium, Copper oxides, magnetic order

1. Introduction

$R_2Cu_2O_5$ compounds have recently been the subject of intense study, largely as a result of their relationship to the high-T_c $RBa_2Cu_3O_{7-\delta}$ materials. The blue-phase $R_2Cu_2O_5$ compounds crystallise in the orthorhombic $Pna2_1$ $Ho_2Cu_2O_5$-type structure [1] having two inequivalent rare-earth sites with equal occupancies. Susceptibility [2, 3] and neutron diffraction studies [4] show antiferromagnetic ordering of the Cu^{2+} moments accompanied by a more complex ordering of the rare-earth ion moments.

$Yb_2Cu_2O_5$ exhibits unusual magnetic behaviour. Susceptibility [2] shows two features, and heat capacity [5] yields a sharp peak well below the λ-type anomaly associated with the ordering of the Cu^{2+} ions. These results led to the suggestion of two independent orderings, with the copper ordering at 15 K and the ytterbium ordering independently at 8 K. In addition, two distinct models for the magnetic structure of $Yb_2Cu_2O_5$ have been proposed on the basis of neutron powder diffraction data. The first [4] attributes the upper transition to ordering of the copper moments, with the ytterbium moments ordering via a first-order process at the lower transition. By contrast, the other study [6] found no evidence for magnetic scattering until the lower transition was reached, and it was argued that the two magnetic species order together at the lower event.

Given the divergent views concerning magnetic order in $Yb_2Cu_2O_5$, it is clear that a direct study of the ytterbium moments in this system is needed. We present here a ^{170}Yb Mössbauer study of $Yb_2Cu_2O_5$ at temperatures from 2 K to 18 K.

2. Experimental methods

The sample was prepared by co-grinding stoichiometric amounts of Yb_2O_3 (99.9%) and CuO (99.9999%). The mixture was then pressed into a pellet and fired in air at 950°C for 24 hours. The pellet was then re-ground, pressed and re-fired for a further three days. Cu–K_α x-ray diffraction confirmed that the blue/green sample had the expected structure and no unreacted reagents were detected. ac susceptibility measurements showed the two features expected for this material [2] at 15 K and 8 K.

The 15 mCi ^{170}Yb Mössbauer source was prepared by neutron activation of a 10 wt.% Tm in Al alloy. Both source and absorber were cooled in a conventional helium-flow cryostat. Velocity calibration was achieved using a ^{57}Co source mounted on the back of the drive, with an α–Fe foil as a standard. All of the spectra were fitted using a non-linear least-squares minimisation routine. Line positions and intensities were derived from an exact solution to the full Hamiltonian combining magnetic and electrostatic contributions.

3. Results and Discussion

While there are two inequivalent Yb sites in the $Yb_2Cu_2O_5$ structure, at 18 K (above both ordering temperatures) the ^{170}Yb Mössbauer spectrum consists of a single, quadrupole-split triplet (Fig. 1). Similar behaviour was observed in $Tm_2Cu_2O_5$, where only a single component was observed by ^{169}Tm Mössbauer spectroscopy [7]. The pattern is fully consistent with Yb^{3+} and could be fitted with eQV_{zz}=-11.2(1) mm/s and an asymmetry parameter (η) of 0.64(3). Point charge calculations yield η=0.97 and 0.62 for the two Yb sites, while perturbed angular correlation (PAC) measurements [8] found η ~0.82. We find no evidence for the presence of a divalent ytterbium contribution to the spectrum, contrary to the suggestion that non Curie-Weiss behaviour of the susceptibility may be due to Yb^{2+} in this material [2].

A substantial magnetic splitting develops immediately on cooling through the transition at 15 K, and the field grows steadily as T→0 K. At 13 K, the average hyperfine field is already 50 Tesla, and by 2 K it

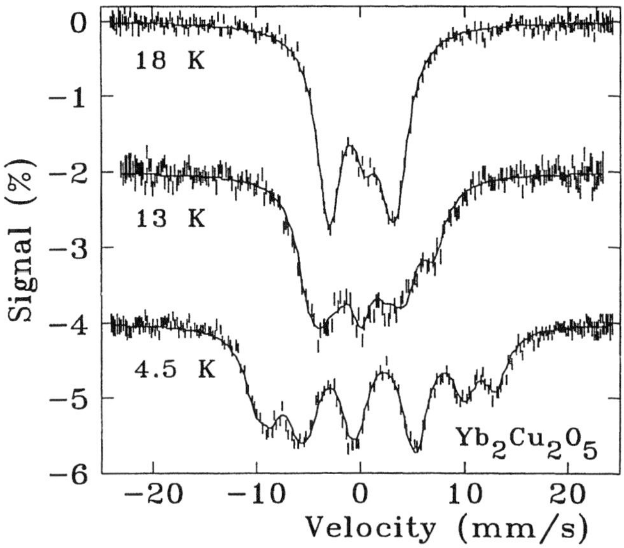

Figure 1. [170]Yb Mössbauer spectra of $Yb_2Cu_2O_5$. From top to bottom: 18 K, above the ordering temperature, showing a single paramagnetic triplet. 13 K, in the ordered state showing that the Yb is indeed magnetically ordered at this temperature. 4.5 K, showing two magnetically split quintets corresponding to the two crystallographically inequivalent Yb sites.

has reached 135 Tesla. While this field is below the 413 Tesla expected for a free-ion Yb^{3+} moment of 4 μ_B, it is far too large to simply be the result of transferred fields from the Cu^{2+} moments. Furthermore, the observed field is consistent with the reduced Yb-moments deduced from neutron scattering (1.4 μ_B [4] and 1.15 μ_B [6]) as simple scaling of the field suggests an average Yb moment of ~1.3 μ_B. Our results thus show that the Yb moments order at the 15 K event and allow us to rule out any model of the magnetic structure that requires separate ordering of the Yb and Cu moments. Furthermore, it is clear that the onset of magnetic order occurs at the upper event, and is not delayed until the feature at 8 K.

Analysis of the data obtained at 4.5 K reveals further insight into the ordering. Unlike the ^{169}Tm spectrum of $Tm_2Cu_2O_5$ [7] where only a single component was observed both above and below the ordering temperature, we clearly see two magnetic quintets of equal area in the spectrum, consistent with the two crystallographically inequivalent Yb sites in the structure. The angle (θ) between the hyperfine field and the principal axis of the electric field gradient was observed to be close to 90° at both sites and at all temperatures, however the low point symmetry (1) at the Yb sites precludes identification of the principal axis of the electric field gradient with a particular crystal axis. The Yb sites exhibit distinct hyperfine fields (147(2) T and 119(2) T) and also different values for η (see below).

The spectral changes at 8 K are quite limited, and clearly do not reflect a major re-organisation of the magnetic order. The hyperfine field at the lower field site increases by $\sim$30 T while that at the higher field site shows at most a $\sim$10 T step. More significantly, η drops to zero at the higher field site on cooling through 8 K (the other site shows no change). As η is determined solely by electrostatic effects, it is not affected by the magnetic order. Since therefore, no magnetic structure change could cause the observed drop in η, we are led to the conclusion that the feature at 8 K is associated primarily with a distortion of the crystal structure, and that the magnetic effects may only be secondary.

4. Conclusions

^{170}Yb Mössbauer spectroscopy of $Yb_2Cu_2O_5$ shows that magnetic order is established at the Yb sites at 15 K and provides no evidence to support models invoking separate Cu and Yb ordering events. Changes observed at 8 K are more consistent with a distortion of the crystal structure than a major change in the magnetic order.

This work was supported by grants from: the Natural Sciences and Engineering Research Council of Canada, Fonds pour la formation de chercheurs et l'aide à la recherche, Québec, and the Australian Research Council.

References

1. H.-R. Freund and Hk. Müller-Buschbaum, *Z. Naturforsch.*, **32b** (1977), 609.
2. R. Tróc, J. Klamut, Z. Bukowski, R. Horyn and J. Stepién-Damm, *Physica B* , **228-231** (1997), 895.
3. Z.A. Kazei, N.P. Kolmakova, R.Z. Levitin, B.V. Mill, V.V. Moshchalkov, V.N. Orlov, V.V. Snegriev and Ja. Zoubkova, *J. Magn. Magn. Mater.*, **86** (1990), 124.
4. J.L. García-Muñoz, X. Obradors and J. Rodríguez-Carvajal, *Phys. Rev.*, **B51** (1995), 6594.
5. V.V. Moshchalkov, N.A. Samarin, I.O. Grishchenko, B.V. Mill and Y. Zoubkova, *J. Magn. Magn. Mater.*, **90** & **91** (1990), 533.
6. A. Murasik, P. Fischer, H. Rundlöf, R. Tellgren and R. Tróc, *Mater. Sci. Forum* , **228-231** (1996), 895.
7. G.A. Stewart and J.M. Cadogan, *J. Magn. Magn. Mater.*, **118** (1993), 322.
8. A. Bartos, M. Uhrmacher, L. Ziegler and K.P. Lieb, *J. Alloys and Comp.*, **179** (1992), 307.

Mössbauer Study of Ba Doped Cubic Perovskite $SrFeO_3$

S. Morimoto, K. Kuzushita and S. Nasu

Division of Materials Physics, Graduate School of Engineering Science, Osaka Univ., Toyonaka, Osaka 560-8531, Japan

Ba doped cubic perovskite $SrFeO_3$ i.e. $Ba_{1-x}Sr_xFeO_{3-y}$ ($0.5 \leq x \leq 0.9$) was synthesized by two different way under high oxygen pressure; (1) 100 MPa of O_2 ($0.7 \leq x \leq 0.9$) and (2) 6.5 GPa with oxidizer ($0.5 \leq x \leq 0.8$). X-ray diffraction revealed that cubic structure was held in all composition range up to 50%Ba substitution and lattice constants increased almost linearly with Ba content. Mössbauer spectra revealed that charge disproportionation ($2Fe^{4+}$ -> $Fe^{3+}(Fe^{4-\delta})$ + $Fe^{5+}(Fe^{4+\delta})$) occurred and oxygen deficiency (nonstoichiometry) is also enhanced in highly Ba substituted specimens. Dependence of Mössbauer parameters on lattice parameter is discussed.

keywords: Fe^{4+}, perovskite oxides, charge disproportionation

Introduction

Some perovskites and related oxides with Fe^{4+} show "charge disproportionation (CD)" expressed as $2Fe^{4+}$ -> $Fe^{3+}(Fe^{4-\delta})$ + $Fe^{5+}(Fe^{4+\delta})$ under certain conditions, which is revealed clearly by ^{57}Fe Mössbauer effect: e.g. orthorhombic perovskite, $CaFeO_3$ below T_{CD} = 290 K[1], $Sr_3Fe_2O_7$ with double layer of FeO_6 octahedra below T_{CD} = 343 K[2], La doped $SrFeO_3$ of $Sr_{2/3}La_{1/3}FeO_3$ below T_{CD} = 207 K[3, 4]. In the perovskite series, $SrFeO_3$ shows no CD and metallic conductivity down to 4.2 K[5, 6], while $CaFeO_3$ shows CD and metal-semiconductor transition below T_{CD} and hexagonal $BaFeO_3$ with face-shared FeO_6 octahedra also displays CD and anomaly in conductivity below T_{CD} = 230 K. Ca substitution for Sr in $SrFeO_3$ results in showing CD gradually as increase of Ca content.

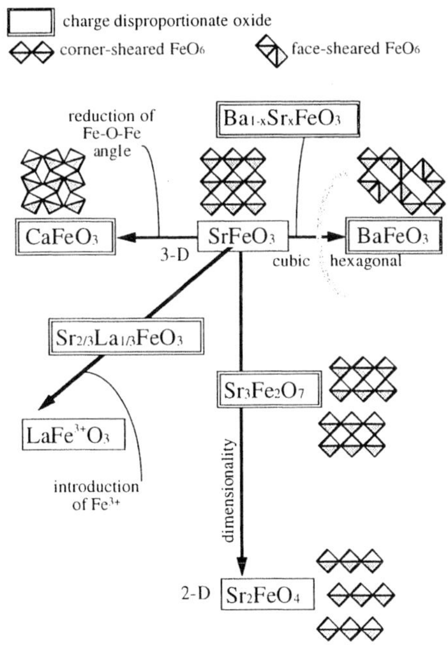

Figure 1. Charge disproportionation and structurral features of perovskites and related iron oxides.

The effect of Ba substitution on CD,

however, does not established, although earlier works is made on Mössbauser study at room temperature, electrical and magnetic properties for specimens in wide range of composition and oxygen deficiency synthesized under up to 20 MPa (200 atm) O_2[7, 8, 9]. The reaction under higher oxygen pressure easily stabilized high valence state like Fe^{4+} especially in oxides; e.g. the synthesis of a typical charge disproportionate oxide, $CaFeO_3$, was failed below 150MPa (1500 atm) O_2[10], however, it succeeds above 2GPa. The aim of this study is to investigate the perovskite system of Ba doped $SrFeO_3$ i.e. $Ba_{1-x}Sr_xFeO_{3-y}$ (referred as BSFO(x) hereafter) synthesized under applying higher oxygen pressure than previous works. In this paper we report the ^{57}Fe Mössbauer spectroscopic study and X-ray diffraction study on the effect of high oxygen pressure on the BSFO(x) system.

Experimental

BSFO(x) specimens of x = 0.5-0.9 was synthesized by conventional solid state reaction and two different way under high oxygen partial pressure as follows:

(1) under c.a. 100 MPa (1000 atm) of O_2 at 673 K for 2 days ($0.7 \leq x \leq 0.9$)

(2) under 6.5GPa with oxidizer ($KClO_4$) at 873 K for 30 min. ($0.5 \leq x \leq 0.8$)

For second annealing condition, six-anvil type high pressure apparatus was used. No chemical analysis on Ba, Sr and O contents is performed. Powder X-ray diffraction patterns were collected by RAD-1A system (Rigaku) and M21X system (Mac Science) using Co Kα and Cu Kα radiation. Lattice constants evaluated by the WPPD (Whole Powder Pattern Decomposition) method developed by Toraya[11]. ^{57}Fe Mössbauer spectra were measured using ^{57}Co in Rh as a γ-ray source; Center shifts were evaluated relative to α-Fe at room temperature.

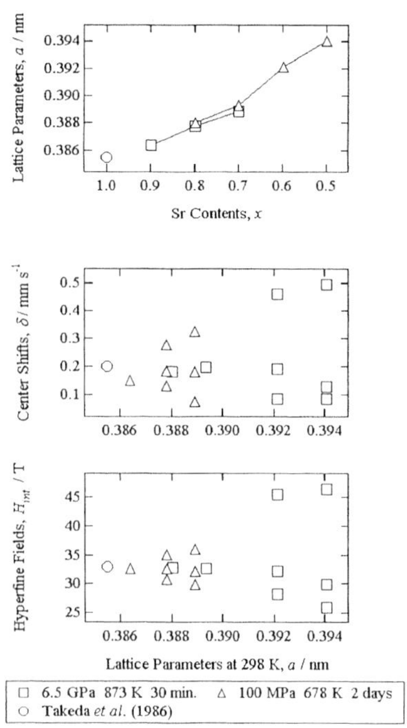

Figure 2. Lattice parameters, center shifts and hyperfine fields of BSFO(x).

Results and Discussion

Powder X-ray diffraction patterns are well indexed assuming cubic lattice for all composition range; however, 100 reflection of BSFO(0.6) (synthesized under 6.5GPa) of somewhat large width eliminating from the evaluation of the lattice parameters. Cubic structure is thus held at least up to 50%Ba substitution. Lattice parameters increase almost linearly in good accordance with Ba content for both annealing conditions (Figuer. 2). ^{57}Fe Mössbauer

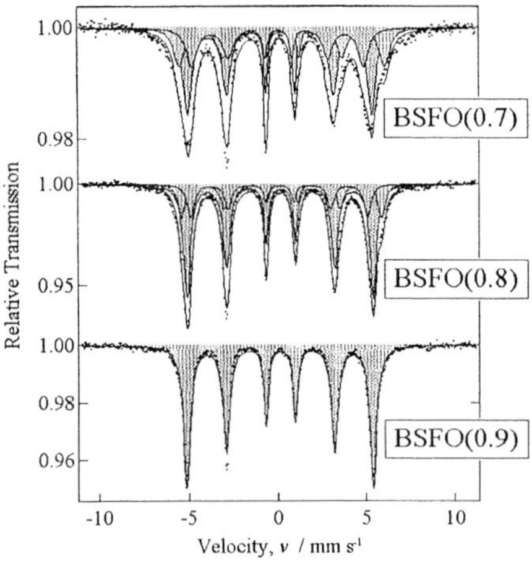

Figure 3. Mössbauer spectra at 5 K of BSFO(x) synthesized under 100 MPa O_2.

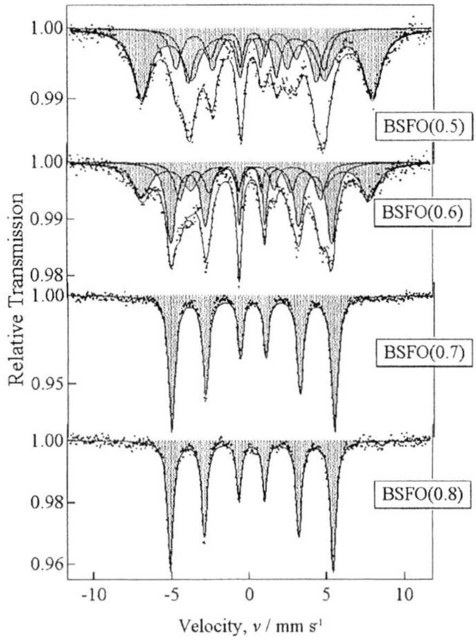

Figure 4. Mössbauer spectra at 5 K of BSFO(x) synthesized under 6.5 GPa.

spectrum of BSFO(0.8) and BSFO(0.7) at 5 K synthesized under 100 MPa O_2 is reproduced three hyperfine sextets with different center shifts and hyperfine fields, which are assigned to $Fe^{4+\delta}$ and $Fe^{4-\delta}$ components with equal intensity plus Fe^{4+} component. The difference in the magnitude of hyperfine fields between $Fe^{4+\delta}$ and $Fe^{4-\delta}$ components is a good measure of degree of CD, and is proportional to Ba content (i.e. lattice parameters) shown in Figure 2.

Specimens synthesized under 6.5 GPa show very different behavior in their Mössbauer spectra from those obtained under 100 MPa O_2. BSFO(0.8) and BSFO(0.7) shows no CD even at 5 K although having similar lattice parameters of samples of same Ba/Sr ratio synthesized under 100 MPa O_2. Spectra of BSFO(0.6) and BSFO(0.5), however, is interpreted by CD at 5 K; The difference in hyperfine fields between $Fe^{4+\delta}$ and $Fe^{4-\delta}$ still larger than those of BSFO(0.7) under 100 MPa O_2 in accordance with their lattice parameters (Figure. 2). The area ratios of decomposed spectra as follows: ($Fe^{4+\delta}$, Fe^{4+}, $Fe^{4-\delta}$) = (25%, 42%, 33%) for BSFO(0.8) and (29%, 35%, 36%) for BSFO(0.7) obtained under 100MPa; (35%, 43%, 22%) for BSFO(0.6) and (56%, 23%, 21%) for BSFO(0.5) obtained under 6.5GPa. The difference between the ratio of $Fe^{4+\delta}$ component and that of $Fe^{4-\delta}$ one indicates the specimens with high Ba content synthesized under 6.5GPa have high oxygen deficiency due to unknown reason. This is consistent with larger lattice parameters than the extrapolation of those for the specimens obtained under 100MPa. It still should be noted that the $Fe^{4-\delta}$ component is confirmed even in the spectra of the specimens with large oxygen deficency.

Fe^{4+} has $(t_{2g})^3(e_g)^1$ electronic configuration which is same for Mn^{3+}. Fe^{4+} in $SrFeO_3$, which is cubic and metallic, causes no distortion on FeO_6 octahedra in contrast to distorted MnO_6 due to

Jahn-Teller(JT) effect in $LaMnO_3$, which is orthorhombic and insulator. Metallic conductivity, no JT distortion and no CD in $SrFeO_3$ is expressed by the collective electron in narrow σ^*-band formed by the mixture of t_{2g} (Fe) and 2p (O) in similar energy level[8]. Form cubic $SrFeO_3$ without CD to orthorhombic $CaFeO_3$ with CD, small Ca cation make Fe-O-Fe angle decreasing from 180°. to ca. 160°[9]; band width decreases and itinerant character of t_{2g} electron is suppressed; thus $CaFeO_3$ shows CD (not JT) to stabilized localized t_{2g} electron. From $SrFeO_3$ with three-dimensional FeO_6 network to charge disproportionate $Sr_3Fe_2O_7$ with double layered FeO_6, two-dimensionality i.e. the introduction of disconnection to FeO_6 network[10] causes to decrease band width and occur CD. Ba doped $SrFeO_3$ synthesized in this study increases lattice parameters with Ba content keeping cubic structure thus Fe-O atomic distance is expected to increase in accordance with expansion of FeO_6 octahedra; CD is induced in Ba doped $SrFeO_3$ with large lattice parameters due to narrowing band width caused by large Fe-O distances, although the question still exists for no CD in the specimens obtained under 6.5 GPa with composition range of $0.7 \leq x \leq 0.9$.

Acknowledgment

The authors express their gratitude to Prof. Takano, Dr. Azuma and Dr. Kawasaki of Institute for Chemical Research in Kyoto Univ. for their help in sample preparation under high pressure.

References

[1] S. Nasu, T. Abe, K. Yamamoto, S. Endo, M. Takano and Y. Takeda, *Hyperfine Interactions*, **70** (1992) 1063.
[2] K. Kuzushita, S. Morimoto, S. Nasu and S. Nakamura, *J. Phys. Soc. Jpn.*, **69** (2000) 2767.
[3] P.D. Battle, T.C. Gibb and S. Nixon, *J. Solid State Chem.*, **77** (1988) 124.,
[4] S. Nasu, unpublished (1999).
[5] J.B. MacChesney, R.C. Sherwood and J.F. Potter, *J. Chem. Phys.*, **43** (1965) 1907.
[6] T. Takeda, R. Kanno, Y. Kawamoto, M. Takano, S. Kawasaki, T. Kamiyama and F. Izumi, *Solid State Sci.*, **2** (2000) 673.
[7] C. Gleitzer, X. Getbaux, M. M. Pradhan, R. Thomas, M. Zanne, J. Aubry and A. Hadni, *J. Chim. Phys. (France)*, **70** (1973) 276-81.
[8] M. Zanne and C. Gleitzer, *J. Solid State Chem.*, **6** (1973) 163.
[9] S. Udagawa, S. Shin and Y. Fumoto, *Yogyo-Kyokai-Shi*, **82** (1974) 75.
[10] S. Shin, *Mater. Res. Bull.*, **16**, 299 (1981).
[11] H. Toraya, *J. Appl. Crystallogr.*, **19**, (1986) 440.
[12] Y. Takeda, R. Kanno, T. Takeda, O. Yamamoto, M. Takano, N. Nakayama and Y. Bando, *J. Solid State Chem.*, **63** (1986) 237.
[13] H. Adachi and M. Takano, *J. Solid State Chem.*, **93**, 556-66 (1991).
[14] S. Morimoto, T. Yamanaka and M. Tanaka, *Physica B*, **237-238** (1996) 66.
[15] P. M. Woodward, D. E. Cox, E. Moshopoulou, A. W. Sleight, and S. Morimoto, *Phys. Rev. B*, **62** (2000) 844.
[16] S.E. Dann, M.T. Weller, D.B. Currie, M.F. Thomas and A.D. Al-Rawwas, *J. Mater. Chem.*, **3** (1993) 1231.

MÖSSBAUER AND XRD STUDY OF (Zn, Cd)$_x$Co$_{1-x}$Fe$_{2-x}$Al$_x$O$_4$ SPINEL FERRITES

J. Z. Msomi, T. Moyo, and K. Bharuth-Ram***
School of Pure and Applied Physics, University of Natal, Durban 4041, South Africa
***Physics Department, University of Durban-Westville, Durban 4000, South Africa*
*E-mail: moyo@nu.ac.za, Fax: +27 (031) 2616550

We report results of Mössbauer and x-ray diffraction measurements on powdered samples of (Zn, Cd)$_x$Co$_{1-x}$Fe$_{2-x}$Al$_x$O$_4$ spinel oxides (where x = 0 to 1.0) obtained at about 300 K. The spectra for both Zn- and Cd-substituted spinel oxides transform with increase in x from ordered magnetic state to disordered spin-state. Differences in the evolution of Fe atoms on tetrahedral (A) and octahedral (B) sites between Zn- and Cd- based compounds are observed. Lattice parameters, density and porosity of the samples are also reported.

1. Introduction

Several reports have appeared in the literature recently [1-4], associated with substitutions by non-magnetic ions on tetrahedral (A) and octahedral (B) sites of spinel structure. These ions significantly affect the properties of the spinel oxides. Substitutions by Zn ions are common and generally assume that the ions prefer tetrahedral sites. Thanki et al [4] have presented the magnetic properties of the spinel system Zn$_x$Co$_{1-x}$Fe$_{2-x}$Al$_x$O$_4$. Simultaneous site dilutions by Zn and Al ions on A and B sites respectively were investigated. The general cation distribution was taken to be $\left(Zn_x^{2+}\ Fe_{1-x}^{3+}\right)^A \left(Co_{1-x}^{2+}Fe_{1.0}^{3+}Al_x^{3+}\right)^B O_4^{2-}$. In a preliminary study [5], it was observed that Zn ions appeared to occupy both A and B sites. We have therefore investigated the magnetic properties of a complete series of Cd or Zn and Al substituted CoFe$_2$O$_4$ ferrites. The Cd^{2+} ion has similar electronic structure to Zn^{2+} but differs in ionic size. It is also important to investigate the effect of this size difference on magnetic properties. In the present work we have made a systematic study of the composition dependence of magnetic properties of (Cd, Zn)$_x$Co$_{1-x}$Fe$_{2-x}$Al$_x$O$_4$ spinel for $0.0 \le x \le 1.0$ in order to ascertain the site occupation of Cd and Zn ions and the validity of simultaneous site dilution in this series of compounds.

2. Experimental

The samples investigated here were prepared by the standard ceramic technique from high purity starting oxides mixed by grinding in methanol. The dry powders were formed into pellets in a 12 mm diameter die under a pressure of about 2 tons/cm^2 applied for 2 minutes. The pellets were pre-sintered at 950 °C for about 30 hours. All samples were finally sintered at about 1050 °C for another 30 hours except for Zn-based samples in the composition range $x \ge 0.5$, which were sintered at 1100 °C in order to ensure single-phase formation. All heat treatments were carried out in flowing oxygen gas. This is the main difference with other samples investigated [4,5]. Bulk densities of the pellets were obtained from measurement of mass and dimensions of the pellets. The quality and structure of the samples produced for this study was confirmed by x-ray diffraction. ^{57}Fe Mössbauer measurements were recorded at about 300 K on powdered samples of the spinel oxide absorbers using a spectrometer operated in constant acceleration mode.

3. Results

Sharp x-ray diffraction lines were obtained for all samples except for $x \ge 0.8$ in Zn- based samples as shown in figs.1. The most significant lines were successfully indexed to cubic spinel structure confirming single-phase formation. The lattice parameters, a derived from x-ray results are shown in fig. 2. The data are consistent with Vergad's law. Cd-substituted

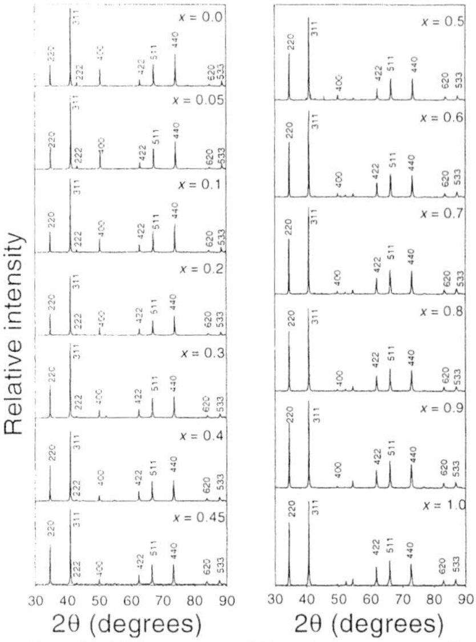

Fig. 1(a) The x-ray diffraction patterns for $Cd_xCo_{1-x}Fe_{2-x}Al_xO_4$ spinel oxides.

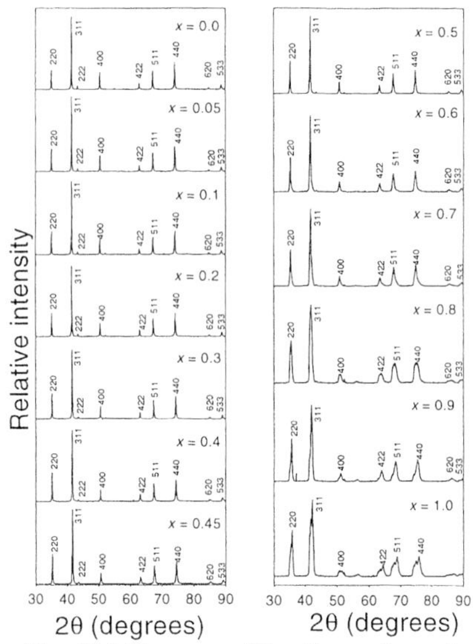

Fig. 1(b) The x-ray diffraction patterns for $Zn_xCo_{1-x}Fe_{2-x}Al_xO_4$ spinel oxides.

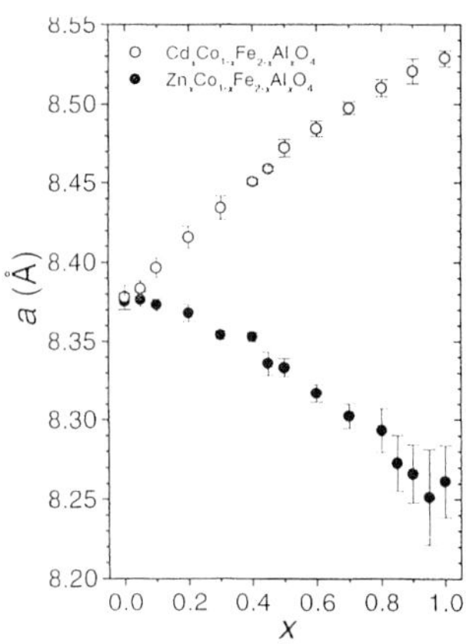

Fig. 2 Variation of lattice parameter *a* with Cd or Zn content.

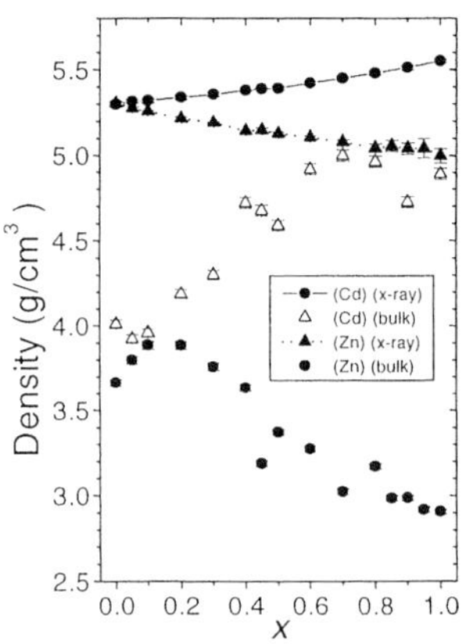

Fig. 3 Bulk and x-ray densities of $(Cd, Zn)_xCo_{1-x}Fe_{2-x}Al_xO_4$ spinel oxides.

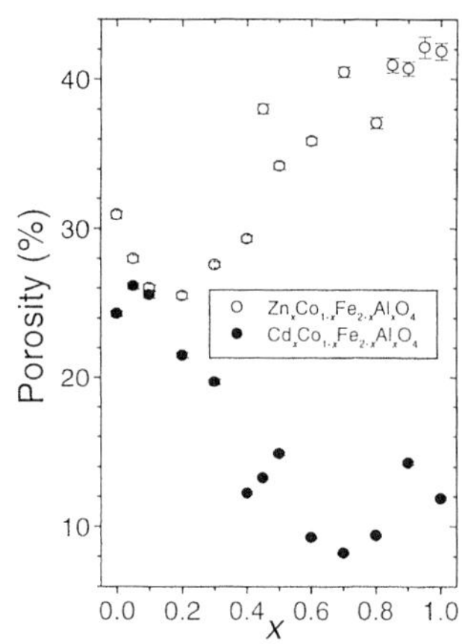

Fig. 4 Porosity as a function of composition for Cd and Zn based spinel oxides.

compounds show an increase in a with x. This is attributed to larger Cd^{2+} ion (0.97 Å) that is replacing smaller Fe^{3+} (0.64 Å) ions on A sites. For Zn substituted compounds, a decreases with increase in x. A decrease of the lattice parameter with increase in Zn substitution has been observed in Fe_2MoO_4 spinel oxides [6]. In fig. 3 we show results of densities of the samples derived from direct measurements (bulk density) and from lattice parameters (x-ray density). The x-ray density was deduced using the formula $\rho_x = \frac{8M}{N_A a^3}$ where 8 is the number of molecules in a unit cell of the spinel lattice, M is the molecular mass of the sample and N_A is the Avogadro's number. The percentage porosity (P) has been calculated

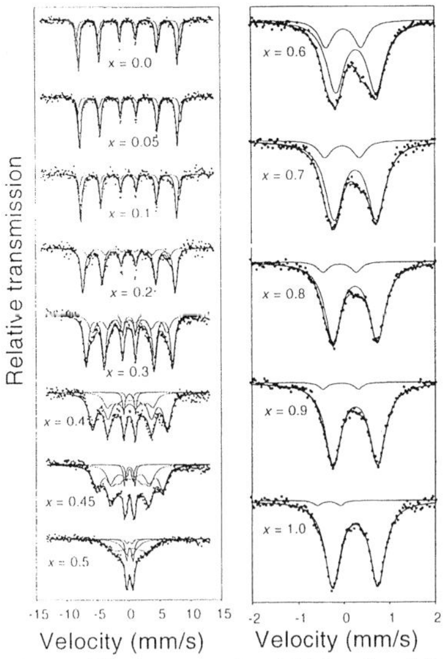

Fig. 5(a) Mössbauer spectra of $Cd_xCo_{1-x}Fe_{2-x}Al_xO_4$ based compounds measured at room temperature.

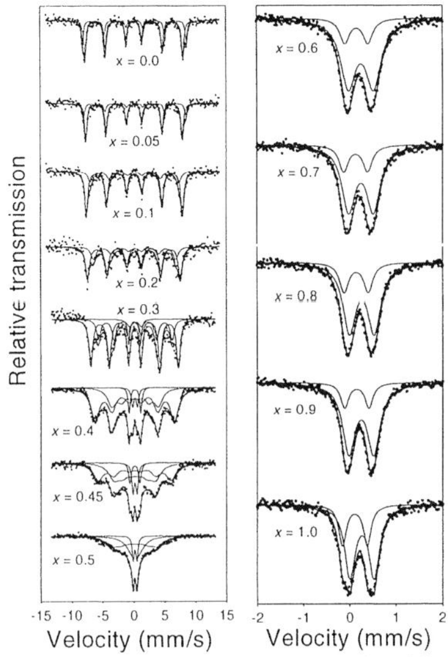

Fig. 5(b) Mössbauer spectra of $Zn_xCo_{1-x}Fe_{2-x}Al_xO_4$ based compounds measured at room temperature.

using the formula [3] $P(\%) = \left(1 - \frac{\rho_b}{\rho_x}\right) \times 100$ where ρ_b is bulk density of the pellet. Fig. 4 shows that our samples are highly porous. This is a result probably related to the heat treatment of the samples in flowing oxygen.

The compositional dependence of Mössbauer spectra is shown in figs. 5. Two sextets and symmetric doublets with equal line width are used to fit the spectra. The quadrupole splitting was found to be negligible for sextets. The criterion used to allocate sextets and doublets to A or B sites is based on fitted results of the quadrupole splitting and isomer shifts which are expected to be lower for A sites [7]. Tetrahedral sites are known to have higher symmetry and hence lower value of quadrupole splitting than octahedral sites. No significant variation of isomer shift (IS) and quadrupole splitting (QS) are observed as shown in table 1.

Table 1

Average values of isomer shifts (IS) and quadrupole splitting (QS) of sextets or doublets.

Zn-based spinels	(mm/s)	x	Cd-based spinels	(mm/s)	x
$< IS >_A$	0.15 ± 0.03	$(0.0 \le x \le 1.0)$	$< IS >_A$	0.18 ± 0.03 -0.04 ± 0.04	$(0.0 \le x < 0.5)$ $(x \ge 0.6)$
$< IS >_B$	0.25 ± 0.03	$(0.0 \le x \le 1.0)$	$< IS >_B$	0.25 ± 0.04	$(0.0 \le x \le 1.0)$
$< QS >_A$	0.52 ± 0.02	$(x \ge 0.6)$	$< QS >_A$	0.75 ± 0.02	$(x \ge 0.6)$
$< QS >_B$	0.534 ± 0.006	$(x \ge 0.6)$	$< QS >_B$	0.94 ± 0.04	$(x \ge 0.6)$

The hyperfine fields deduced from the fits are plotted in fig. 6. The decrease in the fields on both A and B sites occur due to dilution of magnetic ions by diamagnetic ions. A cross over effect between hyperfine fields on A and B sites is observed in both Cd- and Zn-based compounds at low Cd and Zn content. The compositional dependence of area fractions of sextets and doublets corresponding to A and B sites are shown in fig. 7. The trends are similar for both series of compounds. A maximum in the Fe fraction on A site at low values of x is observed. This can be attributed to occupation of both A and B sites by Cd and Zn atoms. The areas reflect the relative fraction of Fe atoms on A and B sites, and hence, inversely, the occupation of these sites by the Cd or Zn and Al ions. In Cd- based spinels for $x > 0.4$, there is a gradual decrease of Fe atoms on A site and an increase on B site. This is the expected trend if Cd and Al ions have preference for A and B sites respectively. In Zn-based spinels, the situation appears to be more complicated and contradicts the cation distribution assumed earlier [4]. A broad peak in the Fe fraction on A site reappears in the range $0.6 \leq x \leq 0.95$ suggesting again that Zn atoms go to both sites. A systematic reduction in the proportion of Fe atoms on A site to zero does not occur for Zn- based samples.

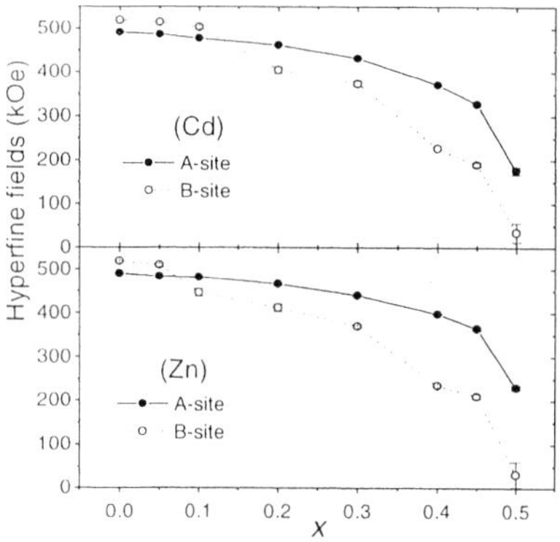

Fig. 6 Hyperfine fields for Cd and Zn based spinel oxides.

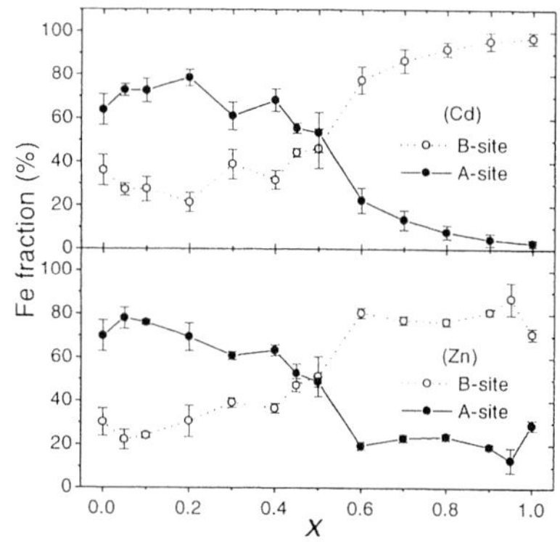

Fig. 7 Fe fraction on A and B sites in Cd and Zn based spinel.

4. Conclusion

No significant impurity phases were detected in the samples studied in this work showing that simultaneous site dilution by Cd or Zn and Al is obtained. The magnetic properties of Cd and Zn- based samples investigated appear to be similar as expected. The results of our analysis show that Zn ions appear to have a stronger preference for both tetrahedral and octahedral sites than Cd ions. Work is now in progress to investigate temperature dependence and effect of changes in sample preparation conditions on magnetic properties.

Acknowledgements

The authors wish to thank the University of Natal and the National Research Foundation of South Africa for supporting this work.

References

[1] S. A. Jadhav, J. Magn. Magn. Mat. 224 (2001) 167.
[2] O. M. Hemeda and M. M. Barakat, J. Magn. Magn. Mat. 223 (2001) 127.
[3] D. Ravinder, T. Alivelumanga, Material Lett. 49 (2001) 1.
[4] V. T. Thanki, K. H. Jani, B. S. Trivedi, K. B. Modi, and H. H. Joshi, Materials Lett. 37 (1998) 236.
[5] T. Moyo, B. Giordanengo, M. A. C. de Melo, A. Y. Takeuchi, P. R. J. Silva, H. Saitovitch and E. Baggio-Saitovitch, Hyperfine Interactions, 120/121 (1999) 285.
[6] A. Roy, J. Ghose, A. Ray, R. Rangathan, J. Magn. Magn. Mat. 202 (1999) 359.
[7] S. C. Bhargawa, A. H. Morrish, H. Kunkel and Z. W. Li, J. Phys.: Condens. Matter. 12 (2000) 9667.

Change of the Nuclear Charge Radius for the ^{197}Au Mössbauer Resonance

P. Palade [a] F.E. Wagner [b] G. Filoti [a]

[a] *Department of Solid State Magnetism, National Institute of Physics of Materials, 76900 Bucharest-Magurele, Romania*

[b] *Physics Department, Technical University of Munich, 85747 Garching, Germany*

The electron density at the gold nuclei in metallic gold and in the cubic intermetallic compounds CsAu, RbAu, AuMg, AuZn, AuCd, AuSc, $AuAl_2$, $AuGa_2$, $AuIn_2$, $AuSn_2$ and $AuSb_2$ have been calculated using the Wien97 computer code. The electron densities thus obtained and the experimental values of the isomer shifts define a linear correlation spanning nearly the whole range of isomer shifts observed for the 77 keV transition of ^{197}Au. From the slope of this correlation one obtains $\Delta\langle r^2\rangle = (4.6 \pm 0.2) \cdot 10^{-3}$ fm^2 for the change of the mean square nuclear charge radius accompanying 77 keV transition in ^{197}Au.

Keywords: Gold, intermetallic compounds, nuclear charge radius, Wien97

1. Introduction

The 77 keV Mössbauer resonance ^{197}Au has been used in many studies of the electronic state of gold in gold-containing materials, with isomer shifts and quadrupole splittings being the main primary information obtained from the Mössbauer spectra. The chemical applications have, for instance, been reviewed by Parish [1]. For many applications only relative values of the isomer shift are relevant and it is not important to obtain absolute values of the differences $\Delta\rho(0)$ of the electron density $\rho(0)$ at the gold nuclei giving rise to the observed isomer shifts (see, e.g., [2])

$$IS = \frac{c}{E_\gamma} \cdot \frac{Ze^2}{6\epsilon_0} \cdot \Delta\langle r^2\rangle \cdot \Delta\rho(0),$$

where c is the velocity of light, E_γ is the gamma ray energy, Z the atomic number, e the elementary charge, ϵ_0 the electric field constant, and $\Delta\langle r^2\rangle$ is change of the mean square nuclear charge radius accompanying Mössbauer transition.

In the past, crude estimates of $\Delta\langle r^2\rangle$ were mainly based on Hartree-Fock or Dirac-Fock calculations of the electron charge density at the nucleus in free atoms with certain configurations of outer electrons and on isomer shift differences between suitable chemical compounds, to which approximate idealized valence electron configurations were attributed on the basis of qualitative chemical arguments. After an early estimate of $\Delta\langle r^2\rangle \approx 18 \cdot 10^{-3}$ fm^2 by Shirley [3], Faltens and Shirley [4] estimated $\Delta\langle r^2\rangle$ to lie between 12 and $18 \cdot 10^{-3}$ fm^2. Kaindl et al. [5] obtained $\Delta\langle r^2\rangle = 8.6 \cdot 10^{-3}$ fm^2 from the isomer shift difference between fluorides of trivalent and pentavalent gold and Dirac-

Fock calculations for free gold ions with $5d^8$ and $5d^6$ electron configurations. Kalvius and Shenoy [6] arrived at the same value by combining the isomer shift difference between AuCl and $KAuCl_4$ [7] with Dirac-Fock-Slater calculations for free ions with $5d^{10}$ and $5d^8$ electron configurations. The value of $8.6 \cdot 10^{-3}$ fm^2 has also been quoted in several compilations of $\Delta\langle r^2\rangle$ values [8] [9]. In a somewhat different approach, the increase of the isomer shift of metallic gold under high pressure [10] was used to estimate $\Delta\langle r^2\rangle$, yielding a value of $8.9 \cdot 10^{-3}$ fm^2 [10,11].

Recently, the full-potential linearised augmented plane wave (LAPW) method has become capable of calculating the electronic properties of solids with sufficient accuracy to give meaningful values for the electron densities at the nuclei even of heavy elements. Such calculations can therefore be used to derive reliable values of $\Delta\langle r^2\rangle$. We have used the Wien97 computer code [12] to calculate the electron densities at the gold nuclei in metallic gold and in a number of its intermetallic compounds in order to derive the value of $\Delta\langle r^2\rangle$ for the 77 keV transition in ^{197}Au.

2. LAPW Calculations

The WIEN97 code uses the LAPW method [13] to calculate the electronic properties of crystalline solids. The electron charge density is calculated by the density functional theory (DFT) in the generalised gradient approximation (GGA), using the parametrisation of Perdue at al. [14] for the local exchange correlation energy. The core states are assumed to be fully occupied and the calculations are fully relativistic for them. The valence states are, in a first step, analyzed in the scalar relativistic approximation (without the spin-orbit term in the Dirac equation) for both spin up and spin down electrons. In a second step the spin-orbit coupling is introduced and the matrix elements are calculated using the full Dirac hamiltonian and the restricted wave function basis obtained by the scalar relativistic approximation. The potential between the muffin-tin spheres is expanded into plane waves and calculated self-consistently from the interstitial charge density. For the wave functions inside the muffin-tin spheres an expansion up to an orbital quantum number of $l = 12$ was used in the present calculations. The charge density and the potential inside the muffin tin spheres were expanded into lattice harmonics up to $L = 6$ for all atoms. The Fermi level was calculated by the tetrahedron method. We used about 100 plane waves for each atom in the unit cell and a k-mesh with 40-50 k-points in the irreducible wedge of the first Brillouin zone. A muffin tin radius of 2.6 a.u. was used for gold in all compounds in order to facilitate a comparison between the electron transfer processes which affect the number of electrons in the muffin-tin sphere of gold in the different compounds. The muffin-tin radii of the other metals were adjusted according to the specific crystal structures [15], avoiding overlapping spheres.

3. Results and Discussion

The intermetallic compounds for which calculations were performed (Table 1) are all cubic and hence do not show an electric quadrupole splitting. The LAPW formalism allows the valence electron charge density inside the muffin-tin spheres to be decomposed according to angular momenta, thus giving information on the electron transfer processes

Table 1
Experimental values of the isomer shifts with respect to Au:**Pt** and valence charge counts inside the muffin-tin sphere of gold.

Material	*IS* (mm/s)	s	*p*	*d*	*f*
Au metal	−1.23	0.552	6.366	8.796	0.044
AuCd	0.77	0.627	6.336	8.775	0.019
AuZn	1.37	0.637	6.400	8.741	0.025
$AuSb_2$	2.18	0.671	6.505	8.665	0.029
$AuSn_2$	2.74	0.694	6.488	8.695	0.026
$AuIn_2$	3.41	0.712	6.471	8.671	0.018
AuMg	3.74	0.719	6.347	8.718	0.010
$AuGa_2$	4.33	0.739	6.553	8.645	0.026
AuSc	5.31	0.771	6.418	8.640	0.009
$AuAl_2$	6.03	0.787	6.702	8.582	0.016
AuRb	6.81	0.838	6.010	8.841	0.010
AuCs	7.09	0.846	6.007	8.837	0.001

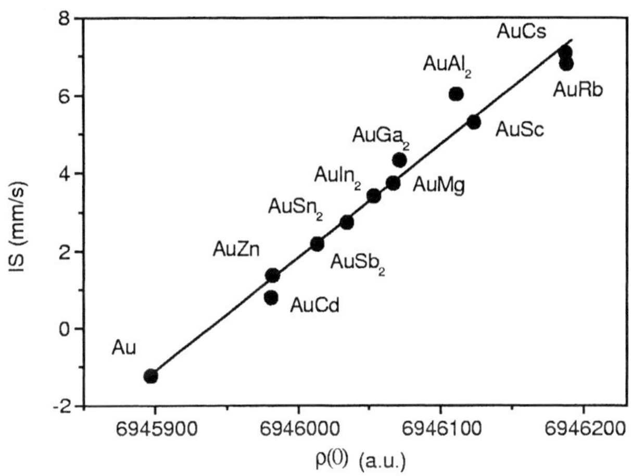

Figure 1. Comparison of the experimental values for the isomer shifts, *IS*, and the calculated electron densities, $\rho(0)$. The electron densities are given in atomic units, i.e., in electrons per a_0^3, where a_0 = 0.0529 nm is Bohr's radius

inside the muffin-tin sphere of gold. The charge densities of the valence electrons with s, p, d, and f character within the muffin tin sphere of gold given in Table I refer to the 6s electrons, the 5p plus 6p electrons, and the 5d electrons; as expected the contribution of f electrons is negligible. The values given in Table I show that there is a monotonic but moderate increase of the 6s electron charge with the isomer shift, and a slight decrease of the 5d charge relative to pure gold in all intermetallics except AuCs and AuRb, which are often considered as ionic compounds of a positive alkali metal and negatively charged gold, which would ideally have a $5d^{10}6s^2$ electron configuration.

In Fig. 1 the calculated electron densities at the gold nuclei are compared with the experimental isomer shifts (Table 1), which were taken from the literature [16-22] except for the *bcc* compound AuSc, for which the isomer shift was measured in the course of this work. The data points define a linear correlation, with but minor deviations attributable to inaccuracies of the calculations rather than the errors of the experimental values, which are no larger than a few times 0.01 mm/s. From the slope of the best fit of a straight line to the data points, one derives $\Delta\langle r^2\rangle = (4.6 \pm 0.2) \cdot 10^{-3}$ fm^2. The error given for this value reflects only the standatd deviation of $\Delta\langle r^2\rangle$ due to the scatter of the data points, but not any systematic errors of the LAPW calculations.

The value derived in this way confirms the positive sign of $\Delta\langle r^2\rangle$, but is more than a factor of two smaller than any of those given previously [3-11]. This means that the electron density differences at the gold nuclei in compounds and alloys of gold are considerably larger than the previous estimates of $\Delta\langle r^2\rangle$ would have suggested.

References

[1] R.V. Parish, in: *Mössbauer Spectroscopy Applied to Inorganic Chemistry*, Vol. 1, ed. G. J. Long, Plenum Press, New York and London, 1984, pp. 577-615.
[2] B.D.Dunlap and G.M. Kalvius, in: *Mössbauer Isomer Shifts*, eds. G.K. Shenoy and F.E. Wagner, North Holland, Amsterdam, 1978, pp. 15-48.
[3] D.A. Shirley, Rev. Mod. Phys. 36 (1964) 329.
[4] M.O. Faltens and D.A. Shirley, J. Chem. Phys. 53 (1970) 4249.
[5] G. Kaindl, K. Leary and N. Bartlett, J. Chem. Phys. 59 (1973) 5050–5054.
[6] G.K. Shenoy and B. D. Dunlap, in: *Mössbauer Isomer Shifts*, eds. G.K. Shenoy and F.E. Wagner, North Holland, Amsterdam, 1978, pp. 869-875.
[7] H.D. Bartunik, W. Potzel, R.L. Mössbauer and G. Kaindl, Z. Physik 240 (1970) 1.
[8] G.M. Kalvius and G.K. Shenoy, Atomic Data and Nuclear Data Tables 14 (1970) 639.
[9] G.K. Shenoy and B. D. Dunlap, in: *CRC Handbook of Spectroscopy*, Vol. 3, ed. J.W. Robinson, CRC Press Incorporated, 1983, pp. 490-495.
[10] L.D. Roberts, D.O. Patterson, J.O. Thomson, and R.P. Levy, Phys. Rev. 179 (1969) 656
[11] D.L. Williamson, in: *Mössbauer Isomer Shifts*, eds. G.K. Shenoy and F.E. Wagner, North Holland, Amsterdam, 1978, pp. 371-360.
[12] P. Blaha, K. Schwartz and J. Luitz, WIEN97, Vienna University of Technology, 1997. (Improved and updated version of the original copyrighted WIEN-code, which was published by P. Blaha, K. Schwartz, P. Sorantin, and S. B. Trickey in Comp. Phys. Commun. 59 (1990) 399.)
[13] O.K. Andersen, Phys. Rev. B 12 (1975) 3060.
[14] J.P. Perdue,K. Burke, and M. Ernzerhof, Phys. Rev. B 77 (1996) 3865.
[15] Th.B Massalski, *Binary Alloy Phase Diagrams*, ASM International, Materials Park, Ohio, 1990.
[16] J.O. Thomson, F.E. Obenshain, P.G. Huray, J.C. Love, and J. Burton, Phys. Rev. B 11 (1975) 1835.
[17] T.K. Sham, M.L. Perlman, and R.E. Watson, Phys. Rev. B 19 (1979) 539.
[18] G.K. Wertheim, R.L. Cohen, G. Crecelius, K.W. West, and J.H. Wernick, Phys. Rev. B 20 (1979) 860.
[19] J. Batchelor, T. Birchall, and R.C. Burns, Inorg. Chem. 25 (1986) 2009.
[20] J. Friedl, F.E. Wagner, J.A. Sawicki, D.C. Harris, J.A. Mandarino, and Ph. Marion, Hyp. Int. 70 (1992) 943.
[21] A. Kyek, F.E. Wagner, P. Palade, A. Jianu, D. Macovei, R. Popescu, R. Manaila, and G. Filoti, Journal of Alloys and Compounds 313 (2000) 15.
[22] A.-V. Mudring, *Darstellung, Struktur und Eigenschaften von Auriden, Auraten und Auridauraten*, Thesis, University of Bonn, 2001.

The Magnetism of FeOCl Revisited: Observation of Low Field Spin–Flop Behavior for the Polycrystalline Form

By

W. M. Reiff * and P. D. Stupik, Dept. of Chemistry, Northeastern University, Boston, Ma 02115, USA, J. L Manson, Materials Science Div., Argonne National Laboratory, Argonne, Il 60439-4821, USA, and C.C. Torardi, Central Research and Development Dept., E. I. Dupont Co., Wilmington, De 19880-0356, USA.

Abstract- Applied field Mössbauer spectroscopy is used in the initial characterization of the low field part of the H versus T phase diagram of FeOCl in its polycrystalline form. These experiments indicate that FeOCl has largely "spin-flopped" for H_0~3 Tesla at 4.2K.

Introduction-FeOCl is an important two-dimensional magnetic material that has been extensively investigated in the context of its reductive topotactic intercalation chemistry [1], [2]. There have also been several studies of this material from the point of view of its basic magnetism [1], [3], [4]. FeOCl exhibits intrinsic moisture sensitivity that has not always been properly taken into account during sample handling by various investigators. Thus, the literature concerning its magnetism is somewhat confused. In any event, it is generally agreed that pristine FeOCl exhibits very strong 2D magnetic behavior with $T\chi_{max\text{-}2D}$ occurring well above ambient temperature and that $T_{Néel}$ corresponding to its long range 3D order ~ 85K. The strong 2-dimensionality is consistent with its (orthorhombic) structure, namely **double layers** of distorted edge sharing cis-$[Fe^{III}Cl_2O_4]$ octahedra where the primary inter double layer magnetic exchange interactions are articulated by weak van der Waals contacts between chlorine atoms.

Mössbauer spectroscopy study of the long-range (3D) ordering of FeOCl was reported initially by Grant [3] using single crystal samples. He reported the onset of magnetic hyperfine interactions at ~92K. His crude susceptibility data indicated that FeOCl exhibits antiferromagnetic interactions. A spectrum determined at 4.2K shows the presence of two (inequivalent) iron sites. No hyperfine-split spectra were reported at any other temperature.

Adam and Buisson [4] subsequently used neutron diffraction to confirm the presence of multiple sites, and in fact discovered that while along the a and b axes there was indeed an antiferromagnetic alignment of spins, the detailed magnetic structure is much more complex than initially suggested by Grant. Specifically, the spins rotate by 99° from one unit cell to the next along the c direction. These authors report Néel temperature of 89K from their single crystal neutron diffraction measurements.

Results-We now present high-resolution Mössbauer spectra of a high purity polycrystalline sample of FeOCl at temperatures ≥ 4.2K. Sample spectra in this temperature range (figure 1) clearly show two hyperfine patterns, which essentially coalesce into a single pattern, as the temperature is lowered (top of figure 2). Further, we find a T_N = 84K ± 3K for our powder sample. The hyperfine field at 4.2K is 44T in reasonable agreement with Grant's [3] single crystal result of 43T.

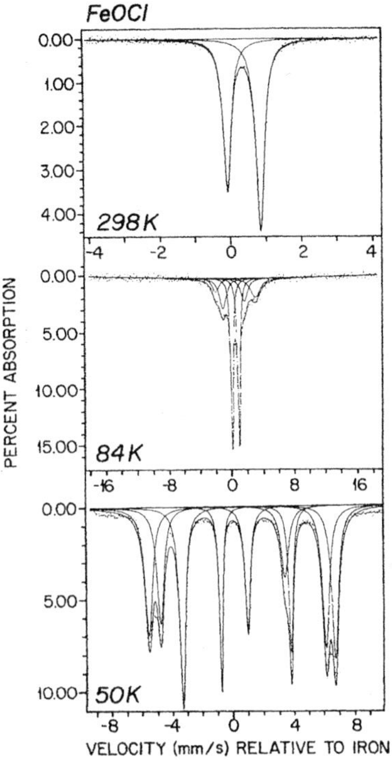

With such well-resolved powder spectra, and the knowledge of the antiferromagnetic nature of the 3D ground state of the system, it was hoped that magnetically perturbed Mössbauer spectra would show evidence of spin-flop behavior. That is, at some critical field, H_{sf}, the opposing moments of the antiferromagnetically coupled sublattices orient themselves perpendicular to the applied field direction in a first order field induced phase transition. At even higher fields, H_p, one can gradually align the moments parallel to the applied field.

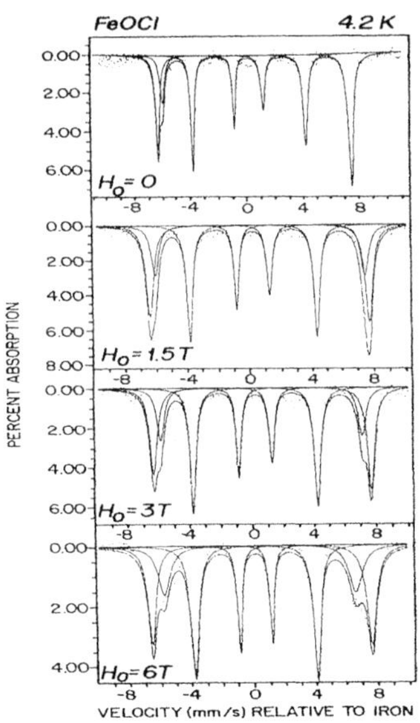

For the case of a longitudinally applied field ($H_0 \parallel E_\gamma$), a spin-flop transition is observed in Fe^{57} Mössbauer spectra as the <u>enhancement</u> of transitions 2 and 5 of the Zeeman pattern assuming that the intensity of the $\Delta m_I = \pm 2$ transitions is zero. This arises from the $\sin^2\theta$ intensity dependence of the $\Delta m_I = 0$ spectral transitions where θ is the angle between the direction of gamma-ray propagation and the internal field (easy axis of magnetization).Some high field spectra are shown in Figure 2 where one can see the evidence of a spin-flop transition as the gamma ray transitions at ~±4 mm/sec grow in intensity with increasing field strength. The fitted line areas do indeed reflect this. The zero field spectra yield a ratio of areas of the $\Delta m_I = 0$ transitions to the innermost lines of 1.7 : 1. This ratio steadily increases until it reaches a maximum of 2.2 : 1 at an applied field of 5T. The fact that the transition is not sharp as expected for first order behavior is undoubtedly due in part to the polycrystalline nature of the sample.It is interesting to note that at 6 T the intensity ratio has decreased to approximately the same value as at 4.5 T (2.0: 1). This may indicate that we are now entering the paramagnetic phase at a relatively low field. Wether this is the case or not, it seems clear that FeOCl is an ideal candidate for further study at higher fields and in particular using single crystal samples in view of its convenient orthorhombic unit cell. Further detailed study is also warranted in the context of the zero point spin reduction possible for a highly 2D magnet of high spin ironIII [5][6].

* Author to whom correspondence should be addressed.

[1] S. M. Kauzlarich, J. F. Ellena, P.D. Stupik, W. M. Reiff, and B. A. Averill, J. Amer. Chem Soc. 109 (1987) 4561.

[2] R.H. Herber, Acc. Chem. Res., 15 (1982) 216.

[3] R. W. Grant, J. App. Phys., 42 (1971) 1619.

[4] A. Adam and G. Buisson, Phys.Stat. Sol. A, 30 (1973) 323.

[5] C. E. Johnson, “Proceedings of the International Conference on the Applications of the Mössbauer Effect”, Jaipur, India, 1981, p 72.

[6] G. P. Gupta, D. P. E. Dickson, C. E. Johnson, and B. M. Wanklyn, J. Phys C: Solid State Phys., 10 (1977) L459.

High field Mössbauer investigations of $CoGa_{2-x}Fe_xO_4$ spinels

M.Reissner[a], W.Steiner[a], Z.Seidov[b], G.Sultanov[c]

[a]*Institut für Angewandte und Technische Physik, Technische Universität Wien, A-1040 Wien, Austria*
[b]*Experimentalphysik V, Universität Augsburg, D-86135 Augsburg, Germany*
[c]*Institute of Physics, Academy of Sciences of Azerbaijan, 370143 Baku, Azerbaijan*

From ^{57}Fe Mössbauer spectroscopy in external fields up to 13.5 T on $CoGa_{2-x}Fe_xO_4$ spinels with $x = 1.0$, 0.8, 0.3, and 0.2 a preferred substitution of the Co-atoms at the A-sites by Fe with increasing x is concluded. Whereas the internal fields are found to point into nearly the same direction as the applied field for Fe at the A-sites, they point – with deviations of up to 40° – into the reverse direction for Fe at the B-sites, indicating an antiferromagnetic coupling between Fe atoms on the two different sites. The values of the internal fields increase slightly with x and are higher for iron at the B-sites.

Keywords: High field Mössbauer spectroscopy, partial inverse spinel, $Co(Ga,Fe)_2O_4$

In spinels of general formula AB_2O_4 the distribution of magnetic ions over the octahedral B and the tetrahedral A sites is one of the keypoints for the magnetic behaviour. In case of the gallium-substituted cobalt ferrites $CoGa_{2-x}Fe_xO_4$ no clear picture concerning the site distribution is given in literature. X-ray investigations [1] have shown that, although the Fe to Ga ratio is different for both A- and B-sites, for the whole series Co mainly occupies only the B-sites, thus forming an inverse spinel. In contrast, a neutron-diffraction investigation [2] of the boundary phase $CoGa_2O_4$ indicated that only 60% of the Co atoms at the A-site are substituted by Ga. Therefore this sample was characterised to be a partial inverse spinel.

Preceding magnetic and ^{57}Fe Mössbauer investigations on polycrystalline samples of $CoGa_{2-x}Fe_xO_4$ point for high iron content to ferromagnetic correlations and for low iron concentrations to spin glass like behaviour [3]. To get more information about the magnetic structure of these compounds we have continued these investigations by ^{57}Fe Mössbauer experiments at 5 K in external fields up to 13.5 T selecting two samples from the iron rich side (x = 1.0, 0.8) and two from the spin glass regime (x = 0.3, 0.2).

In the experiments the γ-ray direction was parallel to the applied field B_a. The ^{57}Co**Rh** source was placed in a field compensated area to exclude line broadening caused by the external field. The velocity drive was run in sinusoidal mode. Calibration was performed by simultaneous measurement of an α-iron foil. The obtained spectra were analysed by solving the complete Hamiltonian with magnetic and quadrupole splitting in the fast-relaxation limit. Sample thickness was taken into account by solving the transmission integral.

Typical spectra are shown in Fig.1. There exist two components: one which increases and one which decreases with B_a, indicating that the compounds cannot be characterized as regular spinels and that Fe is distributed over both the A- and the B-sites. A simple analysis by means of two spectra with relative broad lines shows that the area ratio for both compo-

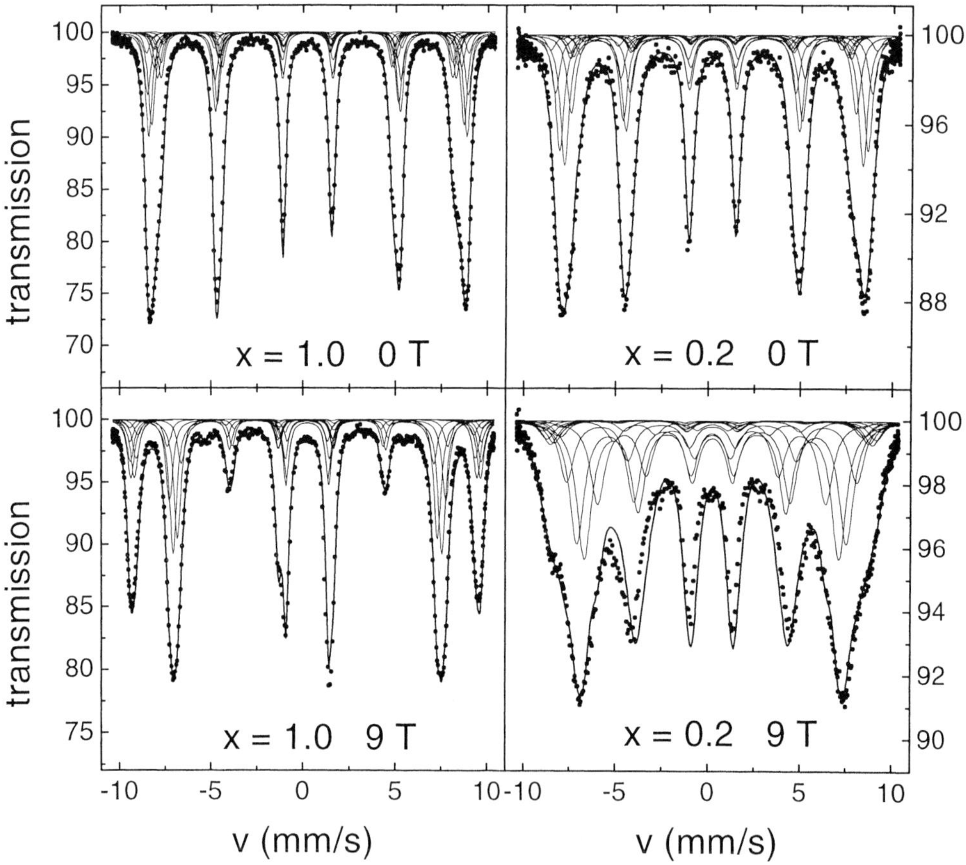

Figure 1: ^{57}Fe Mössbauer transmission spectra of $Co_1Ga_{2-x}Fe_xO_4$ recorded at 5 K with B_a = 0 and 9 T for x = 1.0 and 0.2.

nents deviate from unity. Therefore the samples also do not form a pure inverse spinel, where all the A atoms are replaced by 50% of the B ions, but a partial inverse spinel of the general formula $(Co_{\lambda}Ga_{1-\lambda-y}Fe_y)_A(Co_{1-\lambda}Ga_{1+\lambda-z}Fe_z)_BO_4$ with $y + z = x$.

In the final fit the spectra were analysed by a superposition of subspectra taking into account only nearest neighbour (*nn*) interactions. The intensity ratios of the subspectra were determined by means of binomial distributions and only contributions with intensities higher than 1% were taken into account as individual subspectra. Therefore for iron atoms at the A-site, which have 12 *nn* B atoms, between 7 and 8 and for the iron atoms at the B-site, which have 6 nn B atoms, between 5 and 6 subspectra were used. The relative intensity of the sum of all subspectra for A-site atoms to the one for the B-site atoms was always determined from the 9 T spectrum (Fig. 1). No difference in quadrupole splitting $eQV_{zz}/4$, centre shift *CS*, and angle Θ between applied field B_a and hyperfine field B_{hf} was allowed for subspectra according to the same site. The line width Γ was kept constant for all subspectra. Therefore only B_{hf} varied between the different subspectra belonging to A or B-sites. In the analysis of the spectra it was assumed that the hyperfine field increases with increasing number of Fe atoms in the *nn* shell.

Excellent results are obtained with this model for the Fe-rich samples with $\Gamma/2 = 0.18 \pm 0.01$ mm/s (Fig.1 left). For the Fe-poor samples the spectra could only be described satisfactorily well with strongly increased $\Gamma/2$ (e.g. 0.2 and 0.25 mm/s at 0 T for $x = 0.3$ and 0.2, respectively). For the spectra with highest external field even higher values up to 0.40 mm/s are necessary (Fig.1 right). This increase in $\Gamma/2$ with B_a for the Fe poor samples point to increasing influence of relaxation effects. Such effects are characteristic of spin glasses and are not unexpected because magnetic measurements [3] have characterised those samples as spin glass-like with a characteristic temperature $T_A = 26$ and 32 K for $x = 0.2$ and 0.5, respectively. For all spectra $eQV_{zz}/4$ was 0 ± 0.03 mm/s. Whereas $CS = +0.24 \pm 0.01$ mm/s (rel. to Fe**Rh** at 30 K) was obtained for Fe on B-sites independent of concentration x, *CS* is lower and decreases with x for Fe on A-sites ($+0.22 \pm 0.02$ and 0.14 ± 0.04 mm/s for $x =$ 0.3 and 1.0). The ratio for Fe on A-sites (y) to Fe on B-sites (z) was found to be 0.25, 0.33, 0.54, and 0.61 for $x = 0.2$, 0.3, 0.8, and 1.0, respectively. The same tendency, but much more pronounced was found by Otera Arean et al [1].

The change in hyperfine field B_{hf} is only determined by the change in applied field. B_{hf} increases linearly with B_a for Fe on A-sites, whereas it decreases by the same amount for Fe on B-sites. The angle Θ between B_a and B_{hf} decreases with field and also mildly with x (Table I). From B_{hf}, Θ and B_a the internal field B_{int} at the Fe atom was calculated according to $\boldsymbol{B}_{hf} = \boldsymbol{B}_{int} + \boldsymbol{B}_a$. Within the experimental error the obtained values for B_{int} are independent of B_a (Table I). Probably due to the demagnetizing field the values for B_{int} are lower (higher) for Fe on the A-sites (B-sites) by about 1 T compared to the zero field values. Whereas B_{int} for the Fe-atoms at the A-sites points in the same direction as the applied field B_a, with only small (max. 18°) deviations from the B_a-direction, B_{int} for the Fe-atoms at the B-sites points in opposite direction, indicating that the moments on both sites

Table I. Mean values of measured hyperfine field B_{hf}, angle Θ between B_{hf} and applied field B_a and calculated internal field B_{int} for both A- and B-sites for different B_a and Fe concentration x.

x		1.0			0.8			0.3			0.2		
	B_a [T]	B_{hf} [T]	Θ [°]	B_{int} [T]	B_{hf} [T]	Θ [°]	B_{int} [T]	B_{hf} [T]	Θ [°]	B_{int} [T]	B_{hf} [T]	Θ [°]	B_{int} [T]
A	0	50.1			49.0			46.8			46.2		
	4	53.2	12	49.1	51.7	18	47.9	49.2	18	45.4	49.2	18	48.0
	9	58.4	12	49.2	56.9	18	48.4	53.7	18	45.2	54.1	18	45.6
	13.5	62.9	5	49.6	61.4	18	48.7	58.4	18	45.8			
B	0	53.3			52.0			51.6			50.5		
	4	49.8	210	53.5	49.5	230	52.2	49.0	232	51.6	48.1	237	50.4
	9	45.0	205	53.3	44.8	218	52.2	44.7	230	51.0	43.0	230	49.3
	13.5	40.4	202	53.2	40.4	214	52.1	40.2	225	50.7			

couple antiferromagnetically. Deviations from antiparallel alignment are larger for smaller iron content (e.g. at 4 T 39° and 19° for $x = 0.2$ and 1.0, respectively). Also the value of B_{int} increases mildly with x. Both facts point to a weakening of the coupling with decreasing iron content. Assuming a direct proportionality of B_{int} to the magnetic moment, one can conclude, that the moments at the B-sites are slightly larger than at the A-sites and are only very little dependent on x. Together with the fact that for Fe the moments are usually oriented antiparallel to the internal fields, a resultant moment in direction of the external field is present. This explains the ferromagnetic-like behaviour found in the magnetic investigations [3] for the Fe rich samples.

In conclusion, the different field dependence of the two components found in the spectra give clear indication of an occupation of both A- and B-sites by iron atoms. The different intensities of these two components prove that the samples are partial inverse spinels. Antiferromagnetic coupling is found between Fe on A- and B-sites. The values of the internal fields are around 10% higher for the iron atoms at the B-sites and are only mildly dependent on x.

References

[1] C. Otero Arean, J.L. Rodriguez Blanco, J.M. Rubio Gonzalez and M.C. Trobajo Fernandez, J.Mat.Sci.Letts. 9 (1990) 229.

[2] J.L. Soubeyroux, D. Fiorani and E. Agostinelli, J.Magn.Magn.Mat. 54-57 (1986) 83.

[3] Z. Seidov, G. Sultanov, S. Ibragimov and A. Kaplienko, J.Phys. IV France 7 (1997) C1-265.

^{57}Fe Mössbauer studies on $BaFe_2X_3$ compounds with X = S, Se

M. Reissner[a], W. Steiner[a], H. Boller[b]

[a]*Institut für Angewandte und Technische Physik, Technische Universität Wien, Austria*
[b]*Institut für Allgemeine und Anorganische Chemie, Universität Linz, Austria*

^{57}Fe Mössbauer investigations are reported for $BaFe_2X_3$ compounds with X = S, Se in the temperature interval $4 < T < 293$ K. According to temperature dependence of quadrupole splitting and value of centre shift both samples can be characterised as mixed valence compounds. Whereas at 4.2 K for the sulphur compound only one hyperfine split spectrum is present, a field distribution with approximately a factor of two larger mean hyperfine field is observed for the Se sample.

Keywords: Iron sulphides, iron selenides, $BaFe_2S_3$, $BaFe_2Se_3$, Mössbauer spectroscopy

The crystal structures of $BaFe_2S_3$ and $BaFe_2Se_3$ [1] are composed of polyanionic infinite $[Fe_2X_3]^{2-}$ double chains (X = S, Se) formed by edge-sharing $[FeX_4]$ tetrahedra. Ba^{2+} cations are interspersed between these chains which are separated by roughly 6 Å. Although the structures of the sulphide and the selenide are basically the same and only one Fe site is present, they differ in symmetry having space groups Cmcm and Pnma, respectively. In $BaFe_2Se_3$ the double tetrahedral $[Fe_2Se_3]$ chains are twisted with respect to each other, and the environment of Ba and Fe is less regular than in $BaFe_2S_3$. Mössbauer investigations on these compounds have been performed by Reiff et al. [2], however, only down to 78 K. Recently Gönen et al. [3] concluded from magnetic measurements spin glass like behaviour for the sulphur compound, which should be caused by strong interchain antiferromagnetic coupling. The temperature for the susceptibility maximum was observed around 25 K. Preliminary magnetic measurements on the Se sample revealed a smeared maximum around 250 K [4] and no magnetic order could be observed for the sulphur compound in unpolarised neutron diffraction measurements at 4 K [5]. Concerning to these neutron investigations the crystallographic structure remains unchanged with decreasing temperature. To obtain further information on the magnetic behaviour of these compounds, we performed a Mössbauer study down to 4.2 K.

The samples were prepared from stoichiometric mixtures of powdered BaS (BaSe), Fe, and S (Se). Since barium selenide is very moisture sensitive, it had to be handled under inert conditions (glove box). The educts were slowly heated up to 900°C in evacuated quartz tubes and then cooled to 300°C at a cooling rate of 20°C/h. Products were checked by X-rays (Guinier photographs).

The spectra were observed using a ^{57}Co**Rh** source, relative to which all centre shift data, CS, are given. During the measurements the samples were kept in helium atmosphere. The temperature stability was ±0.5 K. The spectra were analysed solving the full Hamiltonian with magnetic and quadrupole splitting in the fast relaxation limit. Sample thickness was taken into account using the approximations given by [6].

Although for both compounds crystallographically only one Fe site is present, the spectra observed at room temperature could only be fitted by a superposition of several subspectra. The additional two components for the sulphur sample with centre shift CS = 0.25 and 0.60 mm/s and $eQV_{zz}/2$ = 0.70 and 0.80 mm/s, and the one for the selenium sample with CS = 0.30 mm/s, $eQV_{zz}/2$ =1.00 mm/s point to the presence of small amounts of secondary phases. The quadrupole splitting and centre shift of these compounds are close to the ones reported for $Ba_9Fe_{16}S_{32}$ and $Ba_6Fe_8S_{15}$ [2]. The contribution of these phases are shown in all spectra in Fig. 1, but will be ignored in the following discussion.

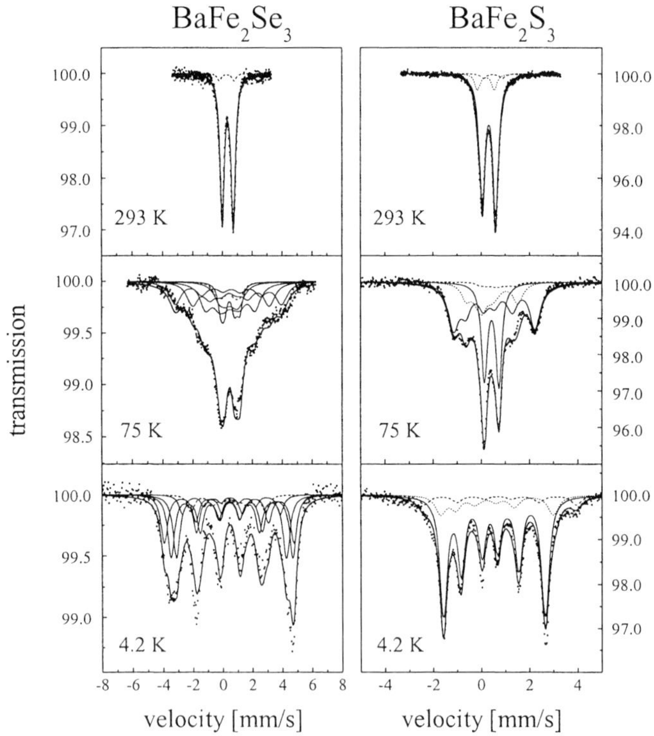

Figure 1: ^{57}Fe Mössbauer spectra of $BaFe_2X_3$, X = S, Se at various temperatures. Spectra concerning to impurity phases are indicated by dashed lines.

The spectrum for $BaFe_2S_3$ exhibit an asymmetry which can be explained by a very small degree of texture (Fig. 1). At room temperature quadrupole splitting (0.56 and 0.74 mm/s) and centre shift (0.33 and 0.37 mm/s) for both samples (S and Se) are in excellent agreement with the values reported by [2]. Compared to the S compound, the line width Γ for the Se sample is 18% lager.

The dependence of quadrupole splitting on temperature is larger for Se than for S (e.g. at 4.2 K 0.61 and 0.96 mm/s, respectively) but remains generally small. CS exhibits the expected temperature dependence due to second order Doppler shift (Fig. 2). The Debye temperatures derived from fits to a Debye model are 320 and 435 K for S and Se, respectively.

At 100 K the spectrum observed for the Se compound could be fitted in the same manner as for higher temperatures, however, an increase of Γ by 38% results. Below this temperature a magnetic hyperfine splitting is observable. The spectra were analysed by an

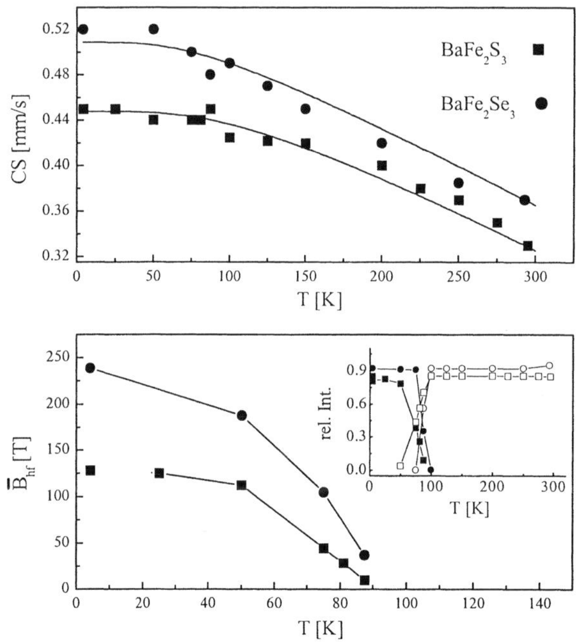

Figure 2: Temperature dependence of centre shift CS (full lines analysis according to the Debye model) and mean hyperfine field $\overline{B}_{hf}$ (lines are only guides for the eyes) for $BaFe_2X_3$, X = S, Se. Inset: relative intensities of the paramagnetic (open symbols) and magnetic (full symbols) component in the spectra.

arbitrary superposition of 4 magnetically split subspectra in addition to a spectrum exhibiting only a quadrupole splitting (Fig.1). The latter component decreases in intensity and for 75 K only magnetically split components are present (inset Fig. 2). At 4.2 K the spectrum could be analysed by means of a superposition of at least 4 subspectra with extremely broad lines (Γ = 0.52 mm/s, Fig. 1). Quadrupole splitting and centre shift are constraint to be the same for all spectra. The remaining discrepancies indicate, that these 4 subspectra can be only seen as a crude approximation in spite of the condition that crystallographically only one single Fe position is present.

For the S compound the magnetic splitting appears around 90 K and the paramagnetic component remains down to 50 K (inset Fig. 2). Only one magnetically split spectrum is necessary to analyse the low temperature measurements. The line widths get broad for the spectra in the transition region (100 to 50 K), but stay around 0.34 mm/s for the lowest temperatures. At 4.2 K the hyperfine field is nearly half the one obtained for the Se compound (Fig. 2).

Although a Fe^{2+} ionic state can be expected, both the value of CS and the small dependence of the quadrupole splitting on temperature are indications, that these samples should be characterised as mixed valence compounds. A similar interpretation was given by [2].

The well resolved spectrum with small line width for the S compound below 50 K indicates a rather simple type of magnetic structure and is in contradiction to the proposed spin glass like behaviour by [3], because due to different magnetic hyperfine interactions of the individual Fe sites the overall shape of the spectra for spin glasses is generally more complex (e.g. [7]). Interesting to mention is the change in temperature dependence of the susceptibility around 90 K reported by [3] which would roughly correspond with the appearance of magnetic hyperfine splitting in the present investigations.

The local symmetry around the Fe atom is different in the two structure types. In $BaFe_2S_3$ two different S atoms whereas in $BaFe_2Se_3$ three different Se atoms built up the Fe environment. This might explain the larger quadrupole splitting for the Se sample. The mean bond lengths Fe - Fe and Fe - S/Se are approximately 3% and 10% larger in the Se compound. It has to be proved by further high field Mössbauer investigations and band structure calculations, how these changes in local symmetry cause the surprisingly large difference of nearly a factor of two for the mean magnetic hyperfine field at 4.2 K.

References

[1] H. Y. Hong, H. Steinfink, J. Solid State Chem. 5 (1972) 93.
[2] W. M. Reiff, I. E. Grey, A. Fan, Z. Eliezer and H. Steinfink, J. Solid State Chem. 13 (1975) 32.
[3] Z. S. Gönen, P. Fournier, V. Smolyaninova, R. Greene, F. M. Araujo-Moreira and B. Eichhorn, Chem. Mater. 12 (2000) 3331.
[4] K. Hiebl, H. Boller, unpublished.
[5] H. Boller, G. André, unpublished.
[6] S. Morup, E. Both, Nucl. Instr. and Methods 124 (1975) 445.
[7] A. Pösinger, H. Winkler, W. Steiner, A.X. Trautwein, M. Reissner, J. Phys.: Condens. Matter 5 (1993) 3653.

Mössbauer Analysis of ^{57}Fe Substituted for Cu in 1212-Phase $(Pb,Cu)Sr_2(Y,Ca)Cu_2O_z$ Superconductor

Masafumi Taniwaki[1], Masanori Takeuchi[2] and Toshihiko Maeda[3]

[1]*Kochi University of Technology, Tosayamada-cho, Kochi, 782-8502, Japan*
[2]*TFT Liquid Crystal Display Group , Sharp Corporation, Taki-cho, Mie, 519-2192, Japan*
[3]*Superconductivity Research Laboratory, ISTEC, Shinonome, Tokyo, 135-0062, Japan*
E-mail: taniwaki.masafumi@kochi-tech.ac.jp

^{57}Fe Mössbauer effect in the 1212-phase superconductor is studied by comparison with that in the 123-phase. Spectra at room temperature consist of three doublets (D1 with Δ = 0.56 mm/s, D2 with 1.01 mm/s, D3 with 1.80mm/s), and two antiferromagnetically split sextets (S1with H_i = 472 kOe, S2 with 381 kOe) are observed at low temperatures. D3(= S1) attributes to Fe^{4+} (S = 2) at Cu(1) site and D1(= S2) do to Fe^{3+} (S = 5/2) at Cu(2).

1. Introduction

There are two crystallographical sites for Cu in a high Tc copper oxide superconductor $(Pb,Cu)Sr_2(Y,Ca)Cu_2O_z$ (z∼7) with Tc = 65K which has a crystal structure similar to the 123 phase [1]. One is the Cu (1) site which is octahedrally coordinated by six oxygen atoms and the other is the Cu(2) site pyramidally coordinated by five oxygen. Substituted cations are expected to occupy the cation sites with the similar ion radius in copper oxide superconductors, therefore Fe ions with radii of 0.64-0.74Å are considered to occupy Cu or Pb sites. In this work, we prepared the samples substituted Fe for Cu and clarified Fe sites from the dependence of ^{57}Fe Mössbauer spectrum on Fe concentration and temperature.

2. Experiments

The powders were mixed to nominal compositions of $(Pb_{0.5},(Cu_{1-x}Fe_x)_{0.5}Sr_2(Y_{1-2.5x},Ca_{2.5x})$ $(Cu_{1-x}Fe_x)_2O_z$ (x=0.05, 0.10, 0.15, 0.20), where the composition ratio of Y and Ca was determined to balance the total valences. The powders were calcined at 850℃ in air for 10h, and then fired at 1050℃ for 1h and cooled at 60 deg/h in O_2 flow. It was shown by x-ray powder diffractometry that the crystal structures of all samples were single 1212-phase. The superconductivity, however, was not observed in any sample. ^{57}Fe Mössbauer spectra of all samples were measured at room temperature in order to obtain the dependence of spectrum parameters on Fe concentration, and the temperature dependence of the spectrum was taken for the sample of x=0.2 at temperatures from 11.6 K to room temperature. The obtained spectra were standardized relative to Fe at room temperature.

3. Results

All obtained spectra at room temperature were successfully analyzed by non-linear least square method assuming three doublets (D1, D2, D3) as shown in Fig. 1. This is consistent with the result of ^{57}Fe Mössbauer study by Maeda et al. [2]. Spectrum parameters are plotted against x (Fe concentration) in Fig. 2, in which those in 123 phase-$YBa_2Cu_3O_7$ are shown for comparison. The isomer shift (δ), quadrupole splitting (Δ) and line width (Γ) of each dou-

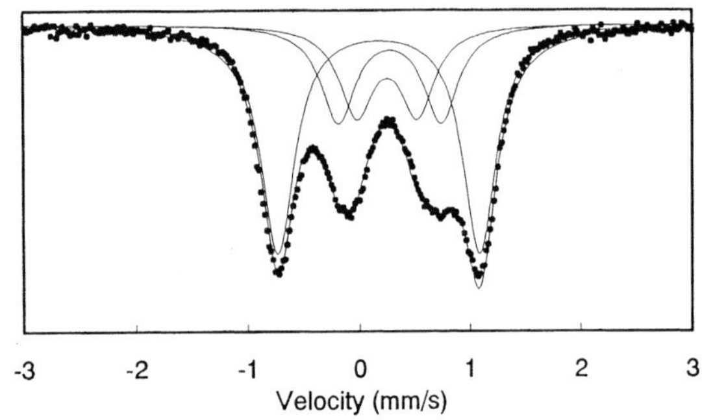

Fig. 1. ^{57}Fe Mössbauer spectrum of 1212-phase $(Pb_{0.5},(Cu_{1-x}Fe_x)_{0.5}Sr_2(Y_{1-2.5x},Ca_{2.5x})(Cu_{1-x}Fe_x)_2O_z$ (x = 0.05) at room temperature.

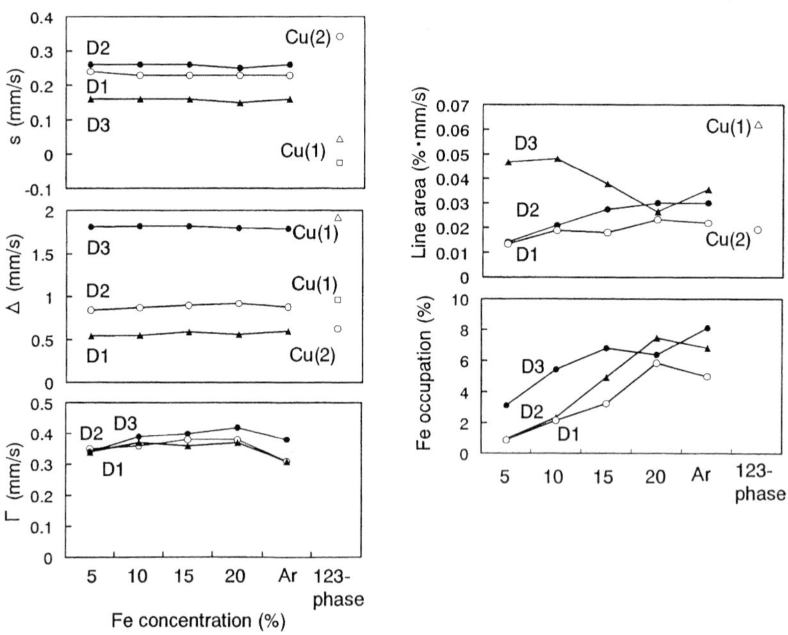

Fig. 2. Dependence of Mössbauer parameters on Fe concentration.

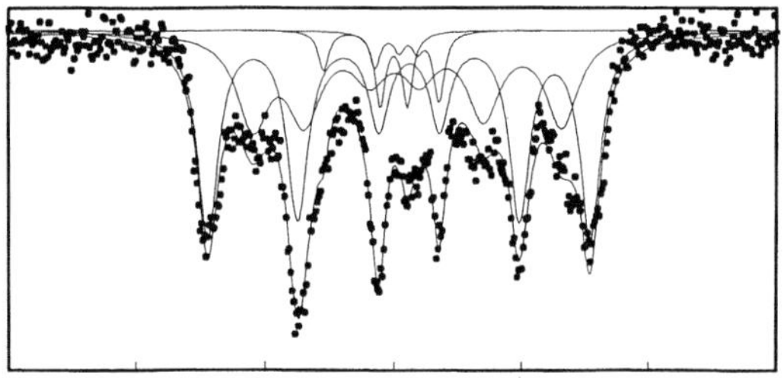

Fig. 3. ^{57}Fe Mössbauer spectrum of 1212-phase $(Pb_{0.5},(Cu_{1-x}Fe_x)_{0.5}Sr_2(Y_{1-2.5x},Ca_{2.5x})(Cu_{1-x}Fe_x)_2O_z$ (x = 0.2) at 11.6 K.

blet are almost constant in all samples, which shows that this analysis is suitable. The δ is nearly equal in D1 and D2 (δ = 0.23-0.24 mm/s in D1, δ = 0.25-0.26 mm/s in D2), and that of D3 is 0.15-0.16 mm/s which is considerably small as compared with D1 and D2. The values of Δ are 0.54-0.59 mm/s in D1, 0.91 - 1.01 mm/s and 1.80-1.82 mm/s. The dependence of the peak area (A) is similar in D1 and D2, which increases monotonously with increasing Fe concentration, while the A in D3 decreases at x $\geqq$ 0.1. The spectrum of the sample of x = 0.2 was magnetically split below 60K as shown in Fig. 3. Two major sextets (S1, S2) are observed as split peak components. The sextet S2 with H_i = 380.7 kOe disappeared at 40K, and S2 with H_i = 471.5 kOe did above 60K. Discussion on other two components (a doublet and a sextet) is impossible because of weakness in their intensity.

4. Discussion

The major components of ^{57}Fe Mössbauer spectrum of 123-phase at room temperature are three doublets, A(δ = 0.05 mm/s, Δ = 1.9 mm/s), B(δ = 0.0 mm/s, Δ = 1.0 mm/s) and C (δ = 0.3 mm/s, Δ = 0.6 mm/s) [3]. They are generally explained to correspond to two or four folded Fe(1) at Cu(1) site, five folded Fe(1) and pyramidally O-coordinated Fe(2) at Cu(2) site(Fig. 4.). At low temperatures magnetically split lines appear [4].

The spectrum parameters of D1(δ = 0.23-0.24 mm/s, Δ = 0.54-0.59 mm/s) are very similar to the doublet, C, and furthermore, Cu-O distances are nearly equal in the 1212-phase and the 123-phase. Therefore D1 is assigned to Fe(2) pyramidally O-coordinated at Cu(2) site. The quadrupole splitting of D3 (1.79-1.82 mm/s), which is the largest in three, is close to that of the doublet A (1.9mm/s), therefore it is considered that the coordination of Fe is the same as that of the doublet A. The A corresponds to two or four folded Fe(1) at Cu(1) site. However, there remain two questions; the isomer shift (0.15-0.16 mm/s) is significantly larger than that of A (0.05 mm/s), and there are not the two or four folded Cu sites in the 1212-structure. The magnetic split components, S1 with H_i = 472 kOe and S2 with H_i = 381 kOe have line areas of 53.5 % and 37.4 % in total absorption line area, respectively. It was confirmed by SQUID that these ordering are anti-ferromagnetic, and T_N is 51

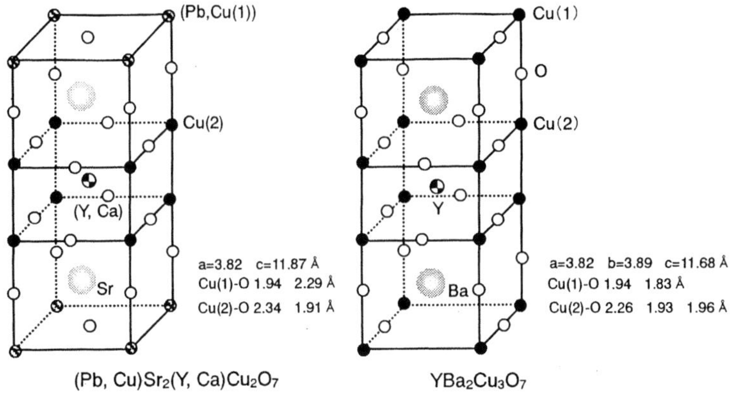

Fig. 4. Crystal structure of 1212 phase and 123 phase.

K for S1 and 29 K for S2. The spin state of Fe, S is estimated to be 5/2 for S1 and 2 for S2 using the relation of $H = 220\ S$. Only Fe^{4+} has a spin state, $S = 4$, therefore S2 is assigned to D3 (δ = 0.15-0.16 mm/s). The peak areas of D3 and S2 are nearly equal, which is not contradictory to this assignment. The sextet S1 is similar to the outer magnetic component of Fe in 123-phase corresponding to the doublet C, namely Fe(2). From this, it is considered that S1 corresponds to D1. However there is a contradiction in their peak areas; the area of S1 is almost 2 times as large as D1. Probably not only D1 but also D2 will have turned to S1 with antiferromagnetic transition. The isomer shifts of D1 and D2 are nearly equal and their peak areas change similarly with Fe concentration, from which we consider that both doublets correspond to two kinds of Fe at Cu(2) site. Those have the same number of O atoms but their crystallographical sites are differently displaced due to the doped Pb. Two kinds of Fe are magnetically ordered by the antiferromagnetism of Cu(2)O plane and the split sextets are overlapped.

Ms. Noriko Nitta, a graduate student in Kochi University of Technology, helped us with all her might for completion of this work. We would sincerely like to express our gratitude to her.

References

[1] T. Maeda, K. Sakuyama, S. Koriyama and S. Tanaka, Phys. Rev. **B43** (1991)7866.

[2] T. Maeda, M. Taniwaki, K. Isawa, K. Sakuyama and H. Yamauchi, in:Proc. 5th Int. Symp. on Superconductivity, Kobe 1992, Y. Bando and H.Yamauchi, eds. Springer-Verlag, Tokyo, 1993, p353.

[3] J.Degler, G. Ritter, G.Saemann-Iscenko, B.Roas, J.Schultz, B. Mlnar, D.L. Nagy and I.S.Szücs, Hyp. Int. **55** (1990)1267.

[4] L. Buttyán, B. Molnár, D. L. Nagy, I. S. Szücs, J. Tóth. J. Dengler, G. Ritter and J. Schober, Phys. Rev. **B38** (1988)11373

Influence of the Statistical Fluctuation of Fe^{2+} Localization upon the Tilt of Fe^{2+} Spin in $Fe_xNi_{1-x}Cl_2$

T. Tamaki

Department of General Studies II, Faculty of Engineering, Gunma University, 4-2, Aramaki, Maebashi, Gunma 371-8150, Japan

The Fe^{2+} spin behavior in $Fe_xNi_{1-x}Cl_2$ of $x<0.15$ containing the oblique antiferromagnetic (OAF) phase has been investigated especially of the tilt of the Fe^{2+} spin to the c-axis. From the analyses of Mössbauer spectra, we have found that two types of Fe^{2+} spins as type I and type II exist in this system. The Fe^{2+} spin of type I makes a small tilting angle from the c-axis, while that of type II has a larger angle than type I. We consider that the tilting angle is strongly connected to the statistical fluctuation of the localization of minor Fe^{2+} spins in the Ni^{2+} host. From the results of the computer calculation, we got the simulated localization of the Fe^{2+} spins in a lot of triangular lattice points. It is considered that the nearest neighbor and next nearest neighbor Fe-Fe couplings originate the Fe^{2+} spins of type I.

Keywords : $Fe_xNi_{1-x}Cl_2$, Mössbauer spectroscopy, the statistical fluctuation of Fe^{2+} localization

1. Introduction

$Fe_xNi_{1-x}Cl_2$ is a random mixture with competing orthogonal spin anisotropies. In pure systems, Fe^{2+} has a strong Ising anisotropy towards the hexagonal c-axis and Ni^{2+} has a weak XY anisotropy in the c-plane. In the early study of this system, we found from the neutron scattering measurements that the samples of $0.03<x<0.12$ form the $NiCl_2$-type antiferromagnetic (AF) order at the Néel temperature T_N and undergo the transition to the oblique antiferromagnetic (OAF) phase at lower temperature T_L [1]. Namely, the OAF phase in $Fe_xNi_{1-x}Cl_2$ system appears in the concentration region $0.03<x<0.12$ below T_L. We have investigated the average spin behavior of Fe^{2+} spins and Ni^{2+} spins by the neutron scattering measurements. In order to investigate the local spin behavior of minor Fe^{2+} spins, we carried out the systematic ^{57}Fe Mössbauer measurements. Mössbauer spectra are composed of two kinds of sub-spectra: the spectrum I and the spectrum II[2-4]. The spectrum I is originated from the Fe^{2+} spin of type I which makes a small tilting angle from the c-axis, while the spectrum II is originated from the Fe^{2+} spin of type II which has a larger angle than that of type I. At 4.2 K, the spectrum I already appears for the sample of $x=0.020$ with the relative fraction of the spectrum I (W_I) = 0.4, although this sample is still in the $NiCl_2$-type AF phase. The W_I increases rapidly with increasing the Fe concentration, and approaches 1.0 near $x=0.068$, though the sample is in the OAF phase and still far from the boundary between the OAF phase and the $FeCl_2$-type AF one. We guess that the Fe-Fe interaction reflects strongly on the behavior of W_I. Therefore, we simulated the Fe localization in the Ni host by the computer calculation. We show the results of the computer simulations.

2. Experimental results

With Mössbauer technique, we have investigated the $Fe_xNi_{1-x}Cl_2$ samples of many kinds of the Fe concentration x. Those are the samples of $x=0.010$, 0.020, 0.034, 0.05, 0.06, 0.068, 0.079, 0.15, and of even more higher concentrations. We have already published the temperature variations of the observed Mössbauer spectra of the samples of $x=0.034$, 0.068, 0.079[2,4] and those of the samples of $x=0.010$, 0.020, 0.034, 0.05, 0.06, 0.068, 0.079, 0.15

observed at 4.2 K[3] in the referred papers. In this report, only the vector models of two types of Fe^{2+} spins analyzed for the Mössbauer spectra of the samples of x=0.034, 0.068, 0.079 are shown in Fig.1, for our convenience. Fig.1 is an enlarged part of the concentration (x)-temperature (T) phase diagram. The vector of a thick line shows the Fe^{2+} spin of type I, and that of a thin line does the Fe^{2+} spin of type II. A length of a vector is drawn in proportion to the value of W_I or W_{II}. Namely, it represents the number of Fe^{2+} spins. The angle of the vector from the temperature axis coincides with the tilting angle of Fe^{2+} spins, θ_I or θ_{II}.

On the figure, it is easily seen that the numbers of the Fe^{2+} spins of both types change systematically with the temperature and concentration variations. The number of the Fe^{2+} spins of type I increases as the temperature decreases and the Fe concentration increases. The tilting angle (θ_I) of those spins does not change with the temperature and concentration variations, and is always θ_I=20°. On the other hand, both of the number of the Fe^{2+} spins of type II and the tilting angle (θ_{II}) of those spins decrease as the temperature decreases and the Fe concentration increases. It seems that the temperature variations of $W_{I,II}$ or $\theta_{I,II}$ are smooth for each sample and have no correlation with the lower transition temperature T_L. However, it is suggested that the magnetic behavior of the minor Fe^{2+} spins closely affects on the average magnetic behavior of this system.

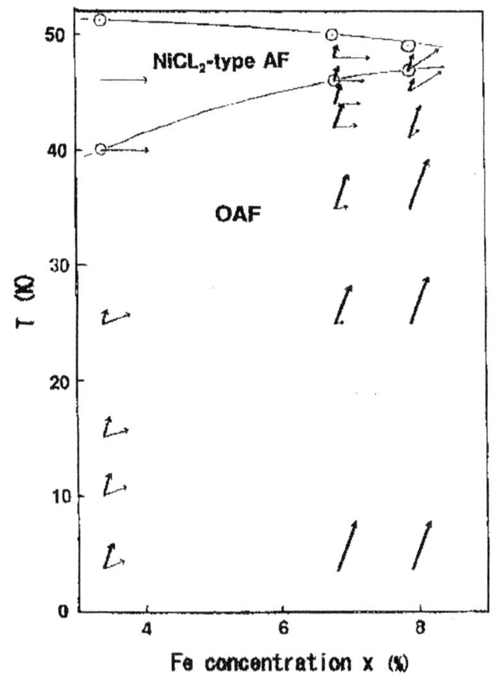

Fig.1 The vector models of two types of Fe^{2+} spins are shown on the concentration (x)-temperature (T) phase diagram. A vector of a thick line shows the Fe^{2+} spin of type I, and that of a thin line shows the Fe^{2+} spin of type II. The length of a vector is drawn in proportion to the value of W_I or W_{II}. The tilting angle of the vector from the temperature axis coincides with θ_I or θ_{II}.

3. Computer simulations

In Figs.2 (a)~(f), we show patterns of the Fe localization in Ni host. Those are figures of the results of computer calculation in which the random distribution of Fe(x)Ni(1-x) is calculated. Total number of the triangular lattice points is 1170. In the

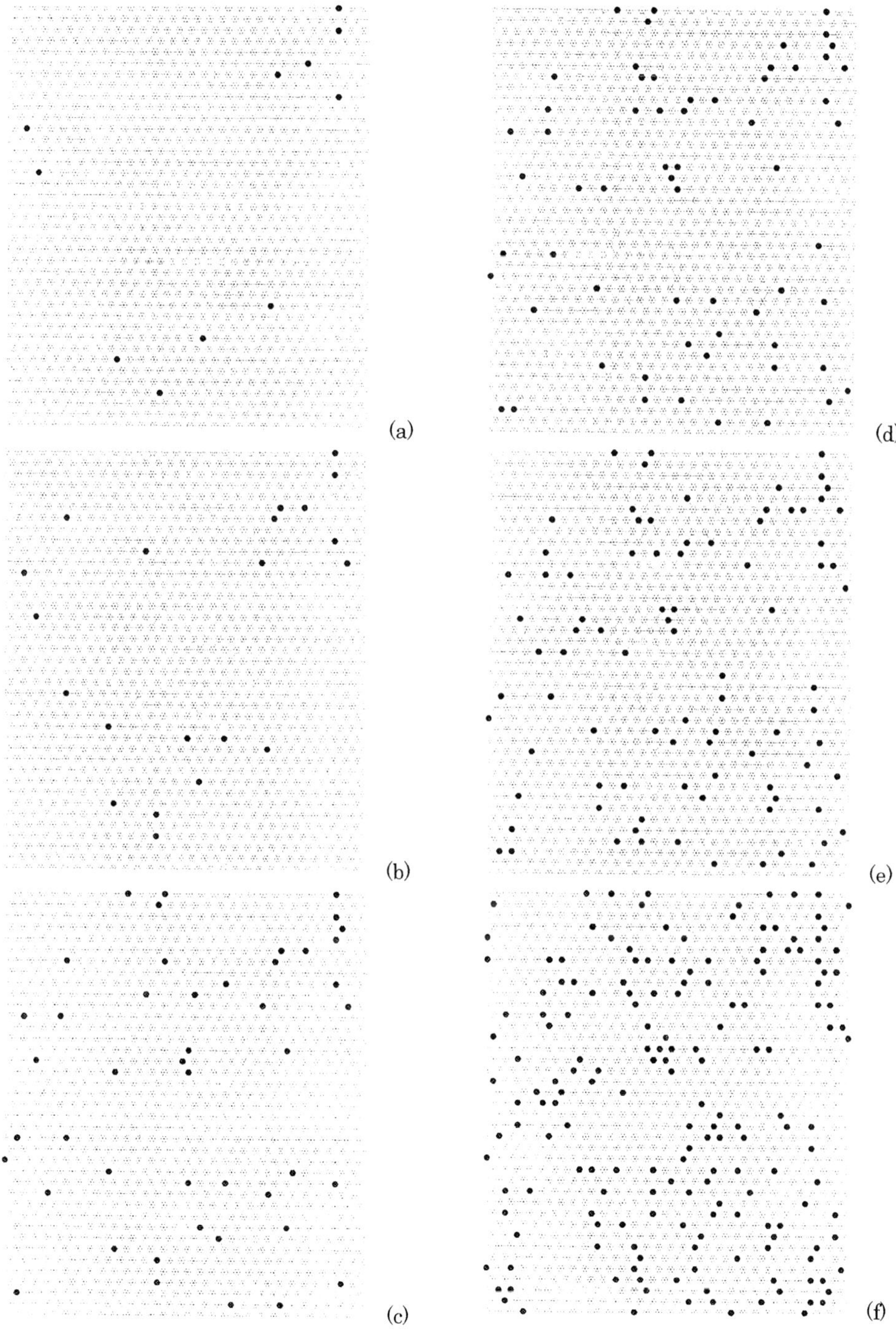

Fig.2 The simulated patterns of Fe (●) localization in the Fe(x)Ni(1-x) triangular lattice points are shown. Those are the cases of x=0.01(a), 0.02(b), 0.04(c), 0.06(d), 0.08(e) and 0.15(f).

figures, the cases of x=0.01, 0.02, 0.04, 0.06, 0.08, 0.15 are shown. A transparent point indicates the Ni location and a black point does that of Fe. At first sight, we can recognize that the Fe-Fe coupling increases as the Fe concentration increases.

4. Discussion

The variation of W_I with Fe concentration x is shown in Fig.3. The values of W_I were published in the referred paper[3] as the results analyzed for the Mössbauer spectra observed at 4.2 K. At the Fe concentration x=0.010 the value of W_I is zero, increases rapidly to 0.4 at x=0.020 and reaches 1.0 at x=0.068. It increases drastically in two concentration regions $0.01<x<0.02$ and $0.06<x<0.07$.

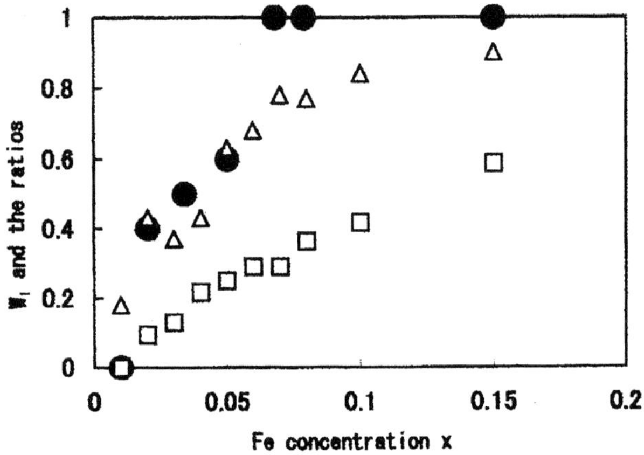

Fig.3 The concentration variations of W_I (●) determined at 4.2 K and the ratios (□, △) of the number of Fe's manually counted to the hole Fe's in the simulated distributions are shown. The symbol □ shows the ratio of Fe's located in the nearest neighbor lattice points each other. Another symbol △ shows that of Fe's located in the nearest neighbor and next nearest neighbor lattice points.

On the other hand, for the simulated Fe(x)Ni(1-x) distribution, we counted manually the number of Fe's those coupled with Fe in the nearest neighbor each other. The ratio (□) of the number of Fe's thus counted to the total Fe's is shown in Fig.3. It increases gradually with increasing the Fe concentration, however does not reach 1.0 but ~0.6 even in the higher concentration x=0.15. As the second trial, we counted the number of Fe's to extend those located in the next nearest neighbor lattice points. The ratio (△) of those Fe's determined as well is also shown in Fig.3. It shows the similar concentration dependence as W_I though fluctuates around x~0.04 and does not reach 1.0 at x>0.07 but 0.9 even at x=0.15. In the concentration region of x>0.07, it is considered that a group of Fe's glows in the c-plane and puts the average anisotropy of Ni host around the group under the influence of the strong anisotropy of Fe^{2+} spins. As the result, a lone-paired Fe also behaves like the Fe^{2+} spin of type I. Moreover, the inter-plane couplings should be considered. In anyway, Fig.3 suggests us that the nearest neighbor and next nearest neighbor Fe-Fe couplings originate the Fe^{2+} spin of type I. The effect of those couplings on the temperature variation of W_I for each concentration is another problem to be next picked up.

References

1) A.Ito, T.Tamaki, Y.Someya and H.Ikeda, *Physica* **120B** (1983) 207.
2) T.Tamaki and A.Ito, *Hyperfine Interactions* **68** (1991) 365.
3) T.Tamaki and A.Ito, *Proc. Conf. 50 "ICAME-95"*, I.Ortalli (Ed.), SIF, Bologna (1996) 223.
4) T.Tamaki and A.Ito, *J. Radioanal. Nucl. Chemi.* **246** (2000) 101.

EVIDENCE OF INTERMEDIATE MAGNETIC STATES IN NON-STOICHIOMETRIC HEMATITE

R.E. Vandenberghe, E. Van San, E. De Grave
NUMAT, Subatomic and Radiation Physics, Ghent University,
Proeftuinstraat 86, B-9000 Ghent, Belgium

Abstract A small-particle hematite and a bulk-sized Al-substituted hematite sample have been reinvestigated with high-resolution Mössbauer spectroscopy. By using a distribution analysing method, based on a linear $\cos^2\theta$ relationship for the quadrupole shift and hyperfine field, it is shown that the Morin transition is accompanied with two intermediate magnetic states. These states probably originate from an ordering of "defect moments".

1. Introduction

In hematite, the magnetic transition from an antiferromagnetic (AF) state at low temperature to a weakly ferromagnetic (WF) state at high temperature is a well-known phenomenon. In pure single-crystal hematite this so-called Morin transition is sharp and occurs at about 265K. However, for non-stoichiometric hematite, i.e. hematite featuring a lot of defects, incorporated water or OH- and/or cation substitution (e.g. Al for Fe), this magnetic changeover behaves quite differently. In most cases the transition is characterised by a temperature region in which both phases coexist and wherein the contribution of the WF phase augments at the expense of that of the AF phase with increasing temperature [1-5]. This has been clearly observed through the presence of two apparently discerned sextets in the Mössbauer spectra with varying relative intensities. On the other hand, a particular pure and rather well-crystallised commercial hematite powder sample showed a narrow transition region in which the Mössbauer spectra did not exhibit such a simple two-phase behaviour [6,7]. In the transition region, the spectra could then be better described by using a correlated QS– HF distribution, revealing the presence of two additional intermediate states [7]. It was further shown that the extensiveness of these intermediate states over temperature diminished strongly after annealing the sample at 900°C, inferring lattice defects as the main origin of their existence.

Questions arise now whether these two apparently different Morin transition behaviours are consistent with one another. In other words, one can wonder whether these intermediate magnetic states are also present in non-stoichiometric hematite, because they cannot be resolved from a conventional fitting procedure. For that purpose, two previously well-characterised hematite samples have been reinvestigated by Mössbauer spectroscopy and their spectra have also been analysed using the aforementioned distribution method.

2. Experimental

Two specific hematite samples have been chosen for the present study. The first sample HL121 is picked out of a series of non-substituted hematites, prepared by decomposition of lepidocrocite [3,5]. Hematite with a large particle size (d ≈ 140 nm), has been selected in order to avoid asymmetric line shapes in the spectra measured at high temperatures. The second sample RD2C consists of a typical bulk-sized (> 1μm) 2at.% Al-substituted hematite, prepared by homogeneous precipitation and subsequent combustion of oxinates [4,5]. Spectra of both samples were recorded between 80 and 300K over 2048 channels (unfolded), the latter to improve spectral resolution. Analyses of the spectra were carried out using the conventional two-sextet fitting procedure and further by the distribution method outlined in ref. [7]. The latter procedure assumes both the quadrupole shift (2ε) and the hyperfine field (H_{hf}) to be linearly correlated with $\cos^2\theta$. The angle θ between the *c*-axis (EFG principal axis) and the spins is allowed to vary from 0° (AF) to 90° (WF).

3. Results and discussion

3.1. The two-state transition model

In the two-state transition model the transition region is described by two sextets which are usually attributed to AF and WF phase. Outside this region the spectra can only be significantly fitted with one sextet. The transition behaviour is best monitored by the quadrupole shift having opposite signs in the extreme cases. As an example, the temperature dependence of the quadrupole shifts for the sample RD2C is represented in Fig. 1.

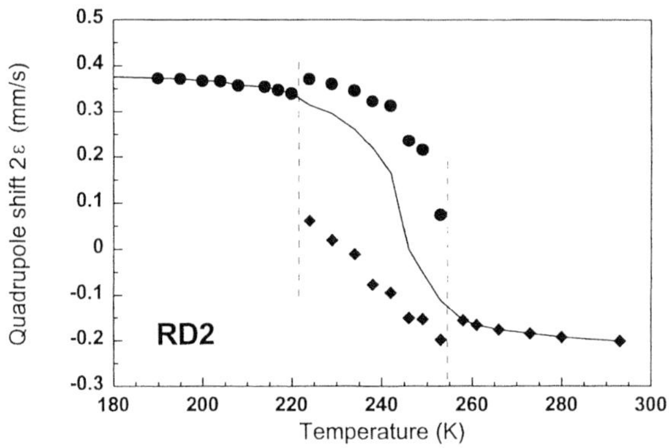

Fig. 1. Quadrupole shift 2ε vs. temperature for the sample RD2C from the analyses with the two-state model for the transition region: ● AF state; ■ WF state; the solid line represents the weighted average.

From such an observation a model has been proposed in which the AF spins move gradually away from the [111] direction with increasing temperature, whereas some of them further flip towards the basal plane with a maximum canting angle of 30 degrees with the latter. Finally they reorient themselves until the stable WF state in the basal plane is reached [2]. However, from the analyses of these spectra, the quadrupole-shifts at the extremities of the two-sextet fitting region clearly show discontinuities, which are here much more pronounced than in the case of the analyses of the formerly measured spectra over 512 channels (folded) [4,5]. Moreover, a detailed inspection at the 6th lines of the spectra in the transition region shows that a two-sextet model does not adequately fit the spectra in the transition region (Fig. 2.). The central part between the two lines seems to be more filled up indicating a possible existence of additional lines.

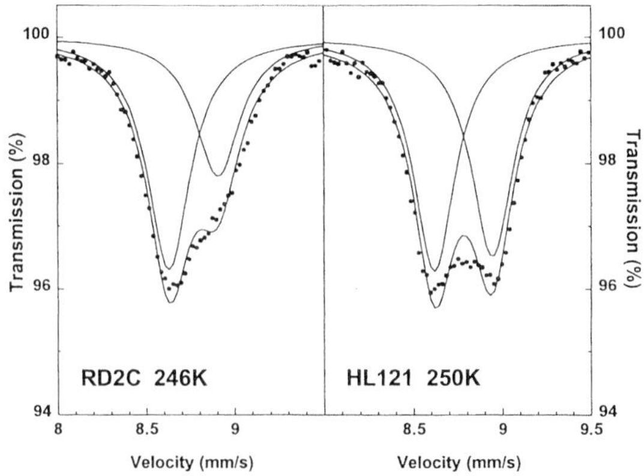

Fig. 2. The sixth lines of some spectra in the transition region fitted according to the two-state model.

3.2. The four-state transition model

The spectra were alternatively fitted with the aforementioned distribution method. The fitting routine is based on that of Le Caër and Dubois [8] constraining the probability values to be positive. In our case a linear line broadening in the elemental spectra has been included in order to account for small local magnetic hyperfine field distributions. Additionally, the possibility of a linear relationship for the intensities of the second and fifth line has been introduced to account for small possible texture effects. The program has beforehand been fully tested on simulated spectra.

The $\cos^2\theta$ distribution profiles of the two samples at various temperatures in the transition region are displayed in Fig. 3. They clearly show the presence of two intermediate states: an AF-like state in the vicinity of $\theta = 33$ degrees and a WF-like state around $\theta = 53$ degrees.

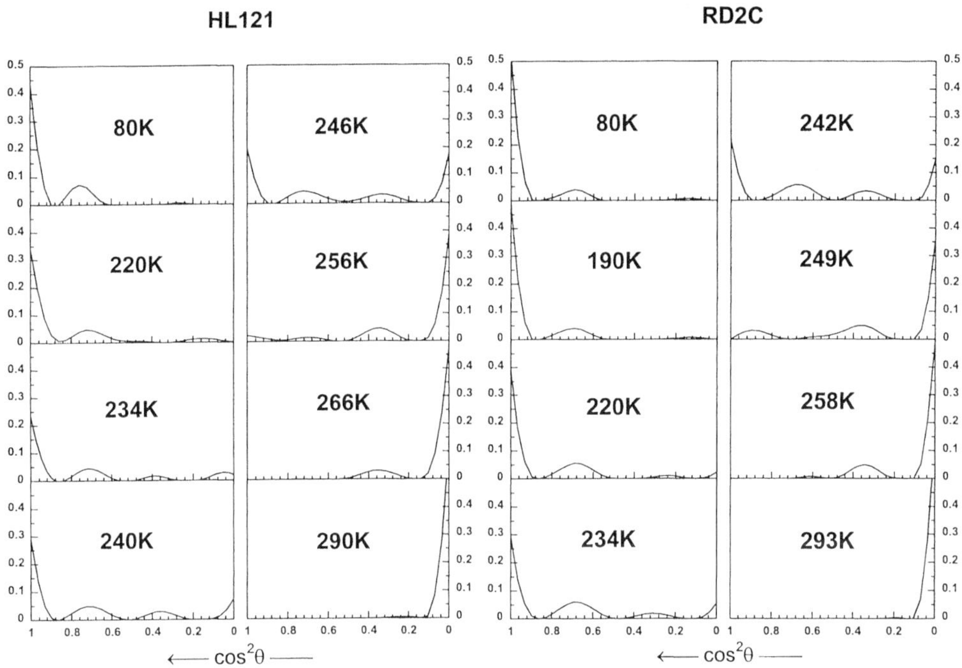

Fig. 3. Obtained $\cos^2\theta$ distribution profiles at various temperatures for both samples LH121 and RD2C.

From each part in the distribution the average quadrupole shift has been calculated. The results are shown in Fig. 4 in which the four states can be clearly distinguished. It is worth remarking that the different states undergo some changes at their extremities. Whether these are due to magnetic interactions with other states remains unclear for the moment. They might also result from an artefact of the fitting procedure because of the very low intensities of some phases involved at these regions.

4. Conclusions

The distribution method reveals that the two intermediate magnetic states which have been previously established in relatively well-crystallised and pure hematite are also present in non-stoichiometric or Al-substituted hematite. In a two-state fitting model they are usually masked by the simultaneous presence of the predominant AF and WF states, which in these samples extend over a relatively broad temperature range. The gradual change of the quadrupole shift in the transition region obtained from the two-sextet fitting can now be explained by a gradual increase of the population of the intermediate states, probably due to various kinds of distributive effects in the lattice. The persistence of the AF-like phase, down to low temperatures also explains why the quadrupole shift at low temperature as obtained from a one-sextet fitting is always smaller than twice the absolute value of that at high temperature.

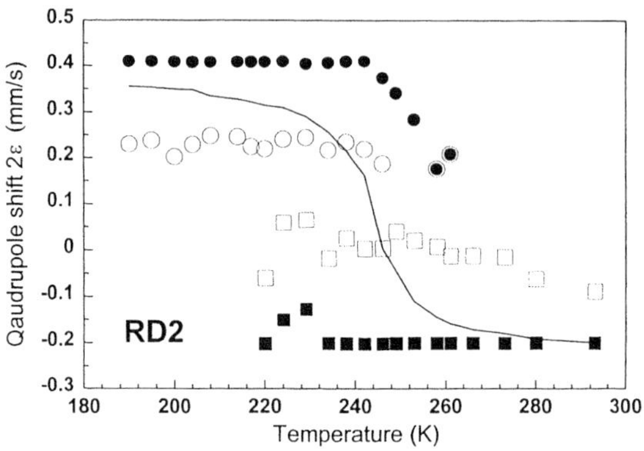

Fig. 4. Quadrupole shift 2ε vs temperature for the sample RD2C derived from the distribution (four-state) model for the transition region: ● AF state; O AF-like intermediate state; ■ WF state; □ WF-like intermediate state; the solid line represents the weighted average.

The origin of these intermediate states is not quite understood yet. As these states seem to be relatively stable, it cannot be expected that they are a mere result of isolated spins that are loosely bound due to the missing neighbouring Fe atoms, caused by lattice defects, non-stoichiometry (proto- or hydrohematite) or Al-substitution. These energetically stable states might therefore be caused by some kind of order in these "defect moments". Unfortunately, it will remain difficult to support these findings by theoretical calculations because of the complexity of the problem and, particularly, the lack of values of the magnetic energy terms that play a key role in the Morin-transition behaviour.

References

[1] R.C. Nininger and D. Schroeer, Phys. Chem. Solids 39 (1978) 137

[2] E. De Grave, D. Chambaere and L.H. Bowen, J. Magn. Magn. Mater. 30 (1983) 349

[3] E. Van San, E. De Grave, R.E. Vandenberghe, L. Datas, V. Barrón and A. Rousset, Phys. Chem. Minerals, 28 (2001) 488

[4] G.M. da Costa, E. Van San, E. De Grave, R.E. Vandenberghe, V. Barrón and L. Datas, Phys. Chem. Minerals, (2001) *in press*

[5] R.E. Vandenberghe, E. Van San, G.M. da Costa and E. De Grave, Czech. J. Phys. (2001) *in press*

[6] N.-H.J. Gangas, T. Bakas and A. Moukarika, Hyp. Int. 23 (1985) 245

[7] R.E. Vandenberghe, T. Becze-Déak and E. De Grave, ICAME95, Conference Proceedings Vol 50, Ed. I Ortalli, SIF, Bologna, 1996, p207

[8] G. Le Caër and J.M. Dubois, J. Phys. E: Sci. Instrum. 12 (1979) 1083

Mössbauer and magnetic evidence for spin frustration in hetero-trinuclear clusters

M. Vopsaroiu[1*], V. Kuncser[1], J. Bartolome[2], V.M. Meriacre[3], K.I. Turta[3], I. Prisecaru[1] and G. Filoti[1]
1. National Institute for Materials Physics, Bucharest-Magurele, Romania
2. ICMA – CSIC – University of Zaragoza, 50009, Spain
3. Institute of Chemistry – Moldavian Academy of Science, Chisinev
* *Correspondence:* University of Central Lancashire, Magnetic Materials Research, Preston, PR1 2HE, UK

Four new trinuclear complexes of the type $[Fe_2M^{2+}O(C_9H_{19}COO)_6(H_2O)_3]$ where M^{2+}= Fe, Co, Mn, Ni, have been synthesised and investigated using Mössbauer Spectroscopy and magnetic measurements. The Mössbauer data point to a high spin state (S=5/2) for trivalent iron and a variation of the magnetic and hyperfine parameters with the substituted bivalent metal in the trinuclear cluster. The results obtained from magnetic measurements and Mössbauer spectroscopy suggest a strong frustration of the spin configuration in all investigated compounds.

Keywords: trinuclear cluster, magnetic and Mössbauer measurements, spin frustration

1. Introduction

Transition metal carboxylates represent a class of compounds with both fundamental and applied interest [1,2]. New trinuclear complexes with the composition $[Fe_2M^{2+}O(C_9H_{19}COO)_6(H_2O)_3]$, where M^{2+} = Fe, Mn, Co, Ni have been prepared in order to study the influence of the bivalent metal on the magnetic and hyperfine properties when substitutes one Fe^{3+} in the previously reported compound, $[Fe_3O(C_9H_{19}COO)_6(H_2O)_3]NO_3$ [3,4]. The magnetic susceptibilities as well as the Mössbauer spectra have shown a paramagnetic behaviour even at low temperatures (1.8K). In the most cases the Mössbauer spectra of the homo-trinuclear carboxylates present asymmetrical doublets over a large temperature range [5,6] and the X-ray investigations evidence equilateral or isosceles triangular arrangement of metal atoms around the central oxygen [7,8].

Even paramagnetic, due to the inter-cluster relaxation, these compounds do not obey the Curie-Weiss law and generally present a spin-cluster temperature dependent. The Mössbauer spectroscopy has been used as a complementary technique to provide information about the local spin configuration as well as about spin relaxation phenomena.

2. Experimental details

The investigated compounds are presented in Table 1. The iron complexes with the main formula $[Fe_2M^{2+}O(C_9H_{19}COO)_6R_3]$, where M^{2+} = Mn, Co, Ni and R = H_2O have been obtained with the following procedure: 5.16 g of $C_9H_{19}COOH$ and 10 ml hexane was added to a suspension of 3 g of $[Fe_2M^{2+}O(CH_3COO)_6\,(H_2O)_3]\cdot 2H_2O$. The mixture was warmed up to 60^0C and agitated at the same temperature during 2 h.

Table 1. The analysed compounds and the total spin values as experimentally obtained or theoretically expected. The codes defined in the column 2 will be used throughout the text for quick referring to the samples.

Chemical formula	Code	S' (RT)	S' – ground state		
			Exp. from S'(T)	Exp. from m(H)	Theory
$[Fe_2Fe^{2+}O(C_9H_{19}\text{-}COO)_6(H_2O)_3]$	Fe^{2+} - C_9	2.8	0.8	0.9	0
$[Fe_2MnO(C_9H_{19}\text{-}COO)_6(H_2O)_3]$	Mn- C_9	3.25	1.5	1.4	0.5
$[Fe_2CoO(C_9H_{19}\text{-}COO)_6(H_2O)_3]$	Co - C_9	2.5	1.25	1.3	0.5
$[Fe_2NiO(C_9H_{19}\text{-}COO)_6(H_2O)_3]$	Ni - C_9	2.17	1.25	0.8	0

The solution was dissolved in ether, filtered and evaporated in air until became a powder. The resulting powder was washed by decantation several times using hexane or acetonitryl until the infrared (IR) spectral band (1700 – 1740 cm^{-1}) corresponding to the free fatty acids had vanished. The compound with $M^{2+} = Fe^{2+}$ was obtained following the same technique but using as suspension the mixed valence acetate: $[Fe^{3+}{}_2Fe^{2+}O(CH_3COO)_6(H_2O)_3]$ [9].

Magnetic measurements between 1.8 K and 300 K have been performed using a SQUID magnetometer. The Mössbauer spectra have been acquired at 80K and room temperature (RT) using a constant acceleration spectrometer with a ^{57}Co source in rhodium matrix. The thickness of the samples was around 7 mg natural iron per cm^2.

3. Results and discussion

The Mössbauer spectra obtained at RT and 80K are asymmetrical quadrupolar doublets (Fig. 1) showing relaxation effects [4], i.e. a temperature dependent change of the intensity ratio between the left and the right line of the doublet. The specific Mössbauer parameters, namely isomer shift (IS) and quadrupole splitting (QS), suggest a high spin configuration for the Fe^{3+} ions (Table 2). Only the Fe^{2+}- C_9 compound presents a second doublet corresponding to Fe^{2+}. The large QS value of this doublet suggests also a high spin state configuration for the Fe^{2+}. Therefore, we will consider all three metallic ions in the cluster in a high spin state. The asymmetry of Mössbauer doublets was analysed in terms of the well-known Blume relaxation model. The line shape is obtained from time dependent nuclear perturbed Hamiltonian [5,10] with the hyperfine field jumping randomly between well-specified values:

$$H=H_0 + H_{IS} + k(3I^2_z - I^2) - g\,\mu_N\,h\,I\,f(t) \tag{1}$$

where: H_0 is the unperturbed iron nuclear Hamiltonian, H_{IS} corresponds to the isomer shift interaction, k is a proportionality factor related to the quadrupol interaction Hamiltonian, I is the nuclear spin (I_z his projection along z-axis, i.e the axis along the main component (V_{zz}) of the electric field gradient) and h represents the reference magnetic field (related to the electronic spin) making the angle θ with z-axis. It is important to mention that the time random function $f(t)$ takes only values of +/- 1, namely with spins jumping only between the two h projections.

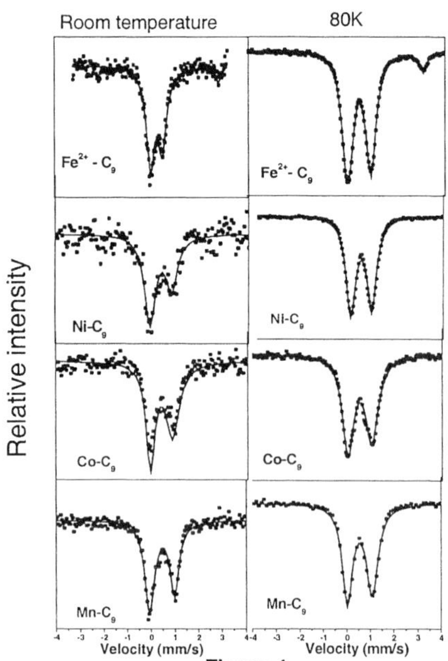

Figure 1.
The Mossbauer transmision spectra measured at RT and 80K.

Let's define the asymmetry factor of a relaxed doublet spectrum as $F = I_2 / I_1$ where I_2 is the intensity of the right line of the doublet and I_1 the intensity of the left line. Previous simulations performed with h = 55T and for different angles θ [4,6] evidenced two distinct situations: (i) θ close to 0° (the magnetic field jumps randomly along Vzz) gives rise to an asymmetrical doublet with F < 1 (higher intensity at smaller velocity) and (ii) θ close to 90° (the magnetic field jumps randomly perpendicular to Vzz) gives rise to an asymmetrical doublet with F > 1. In both cases, the spectra became much symmetrical (F ≈ 1) by increasing the jumping rate, i.e by increasing the temperature. For θ around 35° the spectra are almost symmetrical over a large domain of jumping rates.

Table 2. Hyperfine Mössbauer parameters and the asymmetry factors for the spectra measured at RT and 80K.

Substance	T(K)	Hyperfine Mössbauer parameters (mm/s) IS	QS	Asymmetry factor (F)
Fe^{2+} - C_9	80	1.33 / 0.52	2.65 / 0.77	0.99
Fe^{2+} / Fe^{3+}	300	1.78 / 0.31	3.48 / 0.76	0.96
Mn- C_9	80	0.53	0.88	0.91
	300	0.48	0.86	0.83
Co - C_9	80	0.53	0.84	0.89
	300	0.38	0.75	0.71
Ni - C_9	80	0.48	0.75	0.95
	300	0.44	0.74	0.62

Absolute error for IS, QS: ±0,02 mm/s.

The experimental spectra and respectively the F values (Table 2) suggest that for the all studied compounds the relaxation takes place close to the Vzz direction (F < 1). The spectra of the present compunds show a decrease of the F factor with temperature pointing to reduced jumping rates at higher (RT) temperatures. This peculiar behaviour may be the result of much closer spin energy states at low temperatures leading to a faster jumping rate and could be also a first indication for spin frustration phenomena.

Assuming an effective spin per cluster, S', and also considering no inter-cluster interaction, we can estimate the temperature dependence of the total spin per cluster (Fig. 2). The experimental susceptibility data were analysed in terms of the well-known relation:

$$\chi_m(T) = \frac{N_A g^2 \mu_B^2}{3K_B T} S'(S'+1) \qquad (2)$$

where: N_A – Avogadro number, k_B – Boltzman constant, g – Lande factor (g = 2,0023), S´- the quantum number of the total spin per cluster considered as an effective spin per cluster and χ_m – the magnetic molar susceptibility. The temperature variation of the effective spin per cluster stands for temperature dependent interactions between spins inside the cluster. Effective spins at RT and in the ground state were estimated from S'(T) dependences. Moreover, the effective spin per cluster in the ground state has been also experimentally obtained from the low temperature magnetisation curves, m(B), with B the induction of the applied magnetic field (more details in a forthcoming paper). The effective spins in the ground state derived from both S'(T) and m(B) curves are rather closed (see Table 1), but they are far different from the theoretical total spins derived in the frame of the Heisenberg-Dirac-van-Vleck (HDVV) model. Higher spin states density around the ground level in the real compounds, could explain again the higher experimental obtained spin values. That means a certain degeneracy of the ground level, typical for spin-frustrated

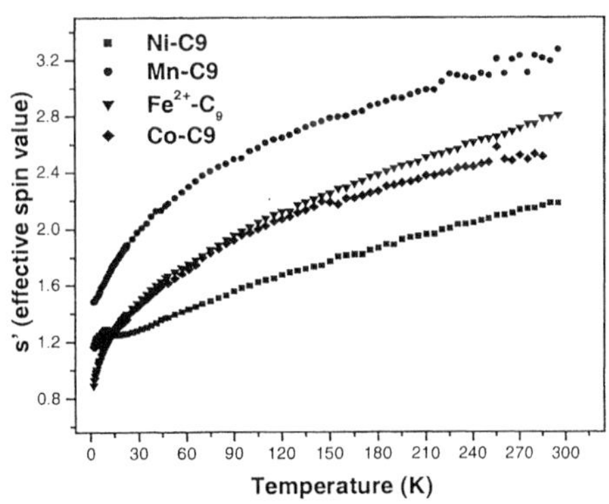

Figure 2. The effective spin per cluster vs. temperature.

systems. Another argument, which supports the presence of the spin frustration, comes from the linear contribution of the M^{2+} spin quantum number, S_3, to the total spin per cluster at RT (see Figure 3). In this temperature range, there are populated higher spin energy levels. An additional theoretical increase of the total spin per cluster in steps of 0.5 is expected due to M^{2+} ion substitution in the order Ni, Co, Fe and Mn. From the linear mentioned dependence, only a 0.3 experimental contribution to the total spin for each M^{2+} ion has been obtained. This smaller contribution may be related to spin frustration mechanisms involving the M^{2+} spins, too.

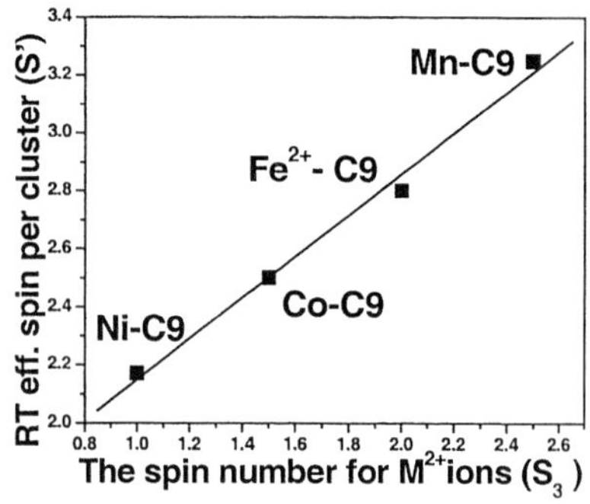

Figure 3. The contribution of M^{2+} spins to the total spin per cluster at room temperature

4. Conclusion

The new prepared compounds with a bivalent ion as the third member of the trinuclear cluster show relaxation type Mössbauer spectra. Consequently, they have been analysed within Blume's model. The unusual decreases of the asymmetrical factors with the temperature have been correlated with the previous results [4,6], and finally related to spin frustration phenomena. The effective spin per cluster presents strong variation vs. temperature. The experimental effective spins corresponding to the ground state are far from the theoretical estimations, pleading also for intra-cluster spin frustration phenomena.

References

[1] J. Catterick and P. Thornton, Adv. Inorg. Chem. Radiochem. 20 (1977) 291.
[2] C. Oldham, Progress in Inorg. Chem., 10 (1968) 223.
[3] K.I.Turta, V.M.Meriacre, V.E.Zubareva, I.V.Dranca Russ. J. Coord. Chem. 22 (1996) 134. (rus.)
[4] G. Filoti, V.M. Meriacre, E. Mateescu, V. Kuncser and K.I. Turta, Hyp. Inter. 116 (1998) 127-136.
[5] J.F. Duncan, C.R. Kanekar, K.F. Mok, J. Chem. Soc. (1969) 480.
[6] G. Filoti, J. Bartolome, D.P.E. Dickson, C. Rillo, I. Prisecaru, T. Jovmir, V. Kuncser, C.Turta, JMMM 196-197 (1999) 561.
[7] K. Anzenhofer and J.J. De Boer, Recuel (Trav. Chim. Pay-Bas) 88 (1969) 286.
[8] K.I. Turta, S.G. Shova, E.A. Spataru, M.D. Mazus, T.I. Malinovskii, Zh. Struct. Khimii 35 (1994) 112, (rus)
[9] Chretien A., Lous E., Bull Soc Chim. Franse. 1944. -V.11, Nr.9-10. - P.446-453.
[10] Blume M., Tjon J., Phys. Rev. – 1968.–V. 165.–Nr.2 P.446-456.

^{119}Sn Mössbauer study of the giant magnetocaloric compound: Gd_5Sn_4

D.H. Ryan, H.B. Wang and Z. Altounian
Physics Department and Centre for the Physics of Materials, McGill University, 3600 University Street, Montreal, Quebec H3A 2T8, Canada

October 9, 2001

Abstract. Gd_5Sn_4 exhibits remarkably large transfered hyperfine fields at the two tin-sites (36.6 T and 29.4 T) at 12 K. The alloy undergoes a first-order structural change from a ferromagnetic to non-magnetic form at 82 K. The transition can be reversed by the application of a magnetic field indicating that the structural and magnetic properties of this system are intimately related.

Keywords: Tin, Gadolinium, Structure, Magnetocaloric effect

1. Introduction

The recent discovery of a giant magnetocaloric response in $Gd_5Si_2Ge_2$ [1] has revived interest in rare-earth/Si-group alloys. The structural [2] and zero-field magnetic [3] phase diagrams for $Gd_5Si_xGe_{4-x}$ have been published and $R_5Si_xGe_{4-x}$ ternary compounds with other rare-earths are being investigated [4].

In the $Gd_5Si_xGe_{4-x}$ giant magnetocaloric materials, the magnetic transition is accompanied by, or possibly driven by, a first order structural phase transition from orthorhombic to monoclinic [5]. It is crucial, therefore, to determine the structural as well as magnetic phase diagrams for these compounds. Unfortunately, the high absorption cross section of natural gadolinium presents a severe barrier to neutron diffraction studies. We have therefore turned to Mössbauer spectroscopy as providing both magnetic and structural information on a local scale.

In this paper, we report the results of a combined high-field magnetisation and ^{119}Sn Mössbauer study of Gd_5Sn_4. Our motivation was twofold. First, if the Sn/Si-based ternary 5:4 compounds exhibit the same giant magnetocaloric effect as the Si/Ge containing alloys, then they will do so at a lower temperature. This may permit a wider operating temperature range by combining several alloys. Second, one can use ^{119}Sn Mössbauer spectroscopy to investigate the magnetic and chemical environment of each crystallographically inequivalent Sn atom.

2. Experimental Methods

The sample was prepared in a tri-arc furnace using stoichiometric amounts of Gd (99.9%) and Sn (99.99%). The alloy was air-sensitive, tending to segregate into metallic tin and gadolinium oxide on exposure to air for a few hours. All sample handling was therefore performed under argon in a glove box.

Powder x-ray diffraction measurements were made using Cu K_α radiation. Magnetisation measurements were carried out on a commercial susceptometer/magnetometer in fields of up to 9 Tesla.

Mössbauer measurements were made on a constant-acceleration spectrometer using a 74 MBq ^{119m}Sn $BaSnO_3$ source and a vibration-isolated closed-cycle refrigerator. Spectra were fitted using a conventional non-linear least-squares minimisation routine. The spectrometer was calibrated against a 99.99% $\alpha-$Fe foil.

3. Results and Discussion

Cu K_α x-ray diffraction showed that the sample was single-phased, and that the alloy adopted the Pnma orthorhombic Sm_5Sn_4-type structure. AC susceptibility showed evidence for two magnetic transitions in Gd_5Sn_4. The largest change is associated with the event at 82 K, however there is also a weaker pedestal that disappears at 137 K. Such behaviour might suggest an impurity phase is present, however none was detected by x-ray diffraction. High-field magnetisation showed the alloy to be ferromagnetic at 5 K with an average Gd moment of $6.5\mu_B$.

The room temperature Mössbauer spectrum of Gd_5Sn_4 consists of a single sharp line at +1.9 mm/s, consistent with the alloy being single-phased. By contrast, the spectrum at 12 K is far more complex. Fig. 1) shows that the 12 K pattern is dominated by two well-split magnetic sextets that account for about 70% of the total absorbed area. In addition, there is a singlet with an isomer shift of +2.4 mm/s. With these primary features subtracted, an additional component is visible. This could be fitted as the sum of two Gaussian distributions of hyperfine fields, and accounted for about 25% of the total area. The hyperfine fields at the tin sites in the two sharp sextets are remarkably large, being 29.4 T and 36.6 T. These fields are both larger than the 28 T reported for Gd_5Sn_3 [6]. The area ratio for these sites is 2.5:1, somewhat lower than the 3:1 ratio expected from the ($4c_2$ + 8d):($4c_1$) pairing derived from the structural analysis. However the field ratio (1.24:1) is very close to the 5:4 expected when the nearest Gd neighbours were counted.

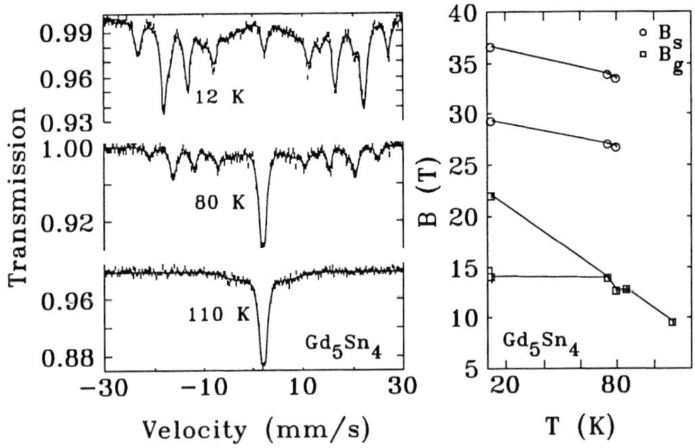

Figure 1. Left: ^{119}Sn Mössbauer spectra of Gd_5Sn_4. There is an abrupt transition to a non-magnetic state above 80 K as the sharp sextets disappear.
Right: Temperature dependence of the hyperfine fields for the sharp (B_s) and Gaussian-broadened (B_g) components. The sharp sextets are clearly lost above 80 K.

The temperature dependence of the spectra (Fig. 1) provides clear evidence for a thermally driven change in structure. While the hyperfine field at each of the two sharp sextets changes only slowly on heating, their area declines rapidly, at the expense of the central feature. By 80 K, the two fields are 34 T and 27 T, however the area fraction of this component is down to 45%, and at 85 K it is gone entirely. The loss of the sharp magnetic component clearly corresponds to the main feature in the AC susceptibility data. Inspection of the spectrum at 110 K shows that a broad magnetic component with an average field of ~13 T is still present and it is this contribution and its ultimate disappearance at 140 K that provides the pedestal in the AC susceptibility data. The progressive structural change evident on cooling explains why the sample can appear single-phased in RT x-ray diffraction, while exhibiting mixed-phase behaviour at 12 K.

The abrupt transition from a magnetically ordered form to a non-magnetic structure on heating is precisely the behaviour required for a giant magnetocaloric effect. In order to demonstrate that the magnetic and structural changes are coupled, we have made magnetisation measurements at temperatures just above the structural change to the non-magnetic form. As Fig. 2 shows, fields of a few Tesla are sufficient to reverse the change and restore a highly magnetic form. The field required to reverse the magnetic→non-magnetic change increases rapidly with temperature, and by 110 K (~30 K above the zero-field transition temperature) a field of 8 Tesla is required to restore the magnetic state.

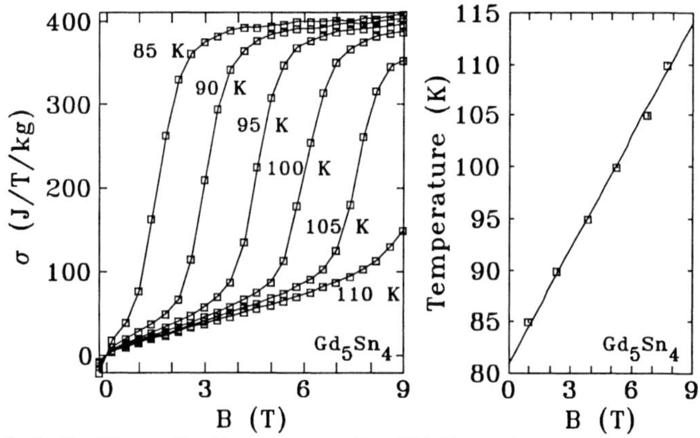

Figure 2. Left: Magnetisation curves for Gd_5Sn_4 at several temperatures above the structural transformation. The sample exhibits an abrupt transition back to a high-magnetisation state in an applied field. Right: Field dependence of the non-magnetic to magnetic transition temperature for Gd_5Sn_4.

4. Conclusions

^{119}Sn Mössbauer spectroscopy shows remarkably large transfered hyperfine fields at the Sn sites in Gd_5Sn_4 at 12 K. The alloy exhibits a first-order structural change from a ferromagnetic to non-magnetic form at 82 K. The transition can be suppressed by the application of a magnetic field indicating that the structural and magnetic properties of this system are intimately related. The transition is expected to be associated with a giant magnetocaloric effect.

This work was supported by grants from the Natural Sciences and Engineering Research Council of Canada and Fonds pour la formation de chercheurs et l'aide à la recherche, Québec.

References

1. V.K. Pecharsky and K.A. Gschneidner Jr. *Phys. Rev. Lett.*, **78** (1997), 4494.
2. V.K. Pecharsky and K.A. Gschneidner Jr. *J. Alloys Compds.* , **260**, (1997) 98.
3. V.K. Pecharsky and K.A. Gschneidner Jr. *Appl. Phys. Lett.*, **70** (1997), 3299.
4. K.A. Gschneidner, V.K. Pecharsky, A.O. Pecharsky, V.V. Ivtchenko and E.M. Levin, *J. Alloys Compds.* , **303**, (2000) 214.
5. L. Morellon, P.A. Algarabel, M.R. Ibarra, J. Blasco, B. Garcia-Landa, Z. Arnold and F. Albertini, *Phys. Rev.* , **B58** (1998), R14721.
6. D. Ravot, A. Percheron-Guegan, J. Jove, J.L. Dormann and O. Gorochov, *Proceedings of the International Conference on the Applications of the Mossbauer Effect Soc. Italiana di Fisica.*, **50** (1996), 363.

The Magnetic Structure of Doped Gamma Iron Oxide

A.A. Yousif[1], A.M. Gismelseed[1], A. D. Al-Rawas[1], M. E. Elzain[1], I. Ayub[2], H.M. Widatallah[3], F.J. Berry[2]

[1]Physics Department, Sultan Qaboos University, Box 36-123, Alkhod, Oman
[2]Chemistry Department, The Open University, Milton Keynes, MK7 6AA, UK
[3]Physics Department, Faculty of Education, Khartoum University, Sudan.

Microcrystals of doped gamma iron oxide samples with copper, magnesium, aluminium, chromium, titanium and tin have been prepared and studied by electron microscopy, x-ray powder diffraction and Mössbauer spectra recorded between 80 and 300K. The Mössbauer spectra at room temperature reveal different degrees of superparamagnetism. The hyperfine interaction parameters have been calculated using a distribution of hyperfine fields and the magnetic anisotropy constant for each sample has been deduce using either superparamagnetic relaxation or collective magnetic excitations models. The influence of the dopant on the magnetic structure of the iron oxide has been investigated.

1. Introduction

The magnetic properties of small particles of γ-Fe_2O_3 has been the subject of numerous investigations. Superparamagnetism, collective magnetic excitations, spin canting, surface effects, have all been associated with particle size effects [1-3]. The nonstoichiometric spinel has Fe(III) occupying the 8 tetrahedral sites and 13 1/3 of the 16 octahedral sites of the unit cell, leaving the rest of the octahedral sites vacant. The doping of pure γ-Fe_2O_3 with cations of different charge states and site preference would change the chemical environment of Fe(III) and may affect the magnetic properties.

2. Experimental

The compounds of nominally 8% metal-doped iron oxide, where the metal-dopant is either magnesium, copper, chromium, aluminium, tin or titanium, were prepared using hydrothermal and reflux methods by adding aqueous ammonia to aqueous mixtures of iron(III) chloride hexahydrate, iron(II) chloride tetrahydrate (Fe(III):Fe(II) = 2:1), and the metal-dopant chloride until pH7. X-ray powder diffraction and transmission electron microscopy (TEM) data were recorded to verify the structure and determine particle size. Mössbauer measurements in the temperature range 80-300 K were recorded using a spectrometer in transmission mode and powdered samples placed in a continuous flow cryostat. Metallic iron was used for calibration and least-square fitting programmes were used for the data analysis.

3. Results and Discussion

The metal-doped samples were found by x-ray powder diffraction to be monophasic cubic spinels. Electron microscopy showed a distribution of particle size, with an average diameter, D, of each sample given in Table 1. The Table also shows the nominal chemical formula of each specimen, although the real metal dopant content could be less. The mctal–dopants are either divalent (Mg, Cu), trivalent (Cr, Al) or tetravalent (Sn,Ti) and to preserve charge neutrality, every 4 ferric ions are replaced by 6 divalent ions, 4 trivalent ions or 3 tetravalent ions. Thus, the environment of iron atoms will depend on the site preference of the metal- dopant.

The room temperature Mössbauer spectra of all samples show broadened Zeeman patterns with different concentrations of a superimposed paramagnetic component. The intensity of the paramagnetic component was reduced in spectra measured at lower temperatures and almost disappeared in samples measured at 80 K. The Mössbauer spectrum of well crystallized γ-Fe_2O_3 consists of a broadened sextet that can be fitted with one sextet with a hyperfine field of 49.9 T for the unresolved tetrahedral and octahedral iron sites, and a small quadrupole splitting [4]. The introduction of metal-dopants would create neighbouring configurations around the tetrahedral or octahedral sites which would necessitate fitting the spectra using a large number of discrete Lorentzian subspectra. We have, instead, used a distribution of hyperfine fields and one paramagnetic component, using the NORMOS fitting program. In some spectra which experience marked relaxation, the fitting was improved by introducing an additional very broad singlet of several mm/s width. Representative spectra are shown in figure 1, for the experimental data and theoretical fit of the room temperature and liquid nitrogen measurements. The hyperfine field distribution profile of each spectrum is also plotted in the same figure. An average hyperfine field for each spectrum, B, and standard deviation, $\Delta\Gamma_B$, have been deduced from the fit, the latter gives a measure of the distribution width. As expected, the distribution of the hyperfine field, especially at room temperature, produces a tail at smaller fields in al! samples reflecting the particle size distribution and/or contributions due to the doping. The distribution width, $\Delta\Gamma_B$, narrows as the temperature is lowered, which is an indication of the slowing down of relaxation times. The average hyperfine field at room temperature and at 80K, with the width of each distribution, are presented in Table 1.

The table shows no evidence of noticeable reduction in the hyperfine field at liquid nitrogen temperature of Sn, Ti, Cr, Al, or Mg-doped samples from that of pure γ-Fe_2O_3 of comparable particle size [2]. Moreover, the Fe nucleus senses similar hyperfine interactions as the result of doping with either nonmagnetic Sn^{4+} or Ti^{4+}. Both cations have octahedral site preference. There is no evidence of formation of any Fe^{2+}, indicating that more octahedral Fe^{3+} vacancies would be produced at those sites [5]. Berry *et al* have found that, at room temperature, the hyperfine field of γ-$Fe_{1.9}Sn_{.1}O_3$ is reduced slightly from the pure sample (48.8 vs 49.9 T), while the current measurements show a pronounced reduction in the hyperfine field due to particle size effects. Moreover, the slightly higher hyperfine field in the Sn-doped sample compared to the Ti-doped sample is due to particle size difference, which is also consistent with the intensity of the paramagnetic component developed in each sample. Both Cr^{3+} and Al^{3+} have octahedral site preference, and may thus substitute for Fe^{3+}. As expected, although the two samples have similar particle sizes, the average hyperfine field of Cr- doped sample (Cr^{3+} is magnetic) is consistently higher than that of the Al-doped sample (Al^{3+} is diamagnetic), with the latter showing pronounced relaxation effects. Similar discussion on the hyperfine field could be pursued for the Mg-doped sample, where Mg^{2+} has octahedral site preference to substitute for Fe^{3+} and occupy some vacancies.

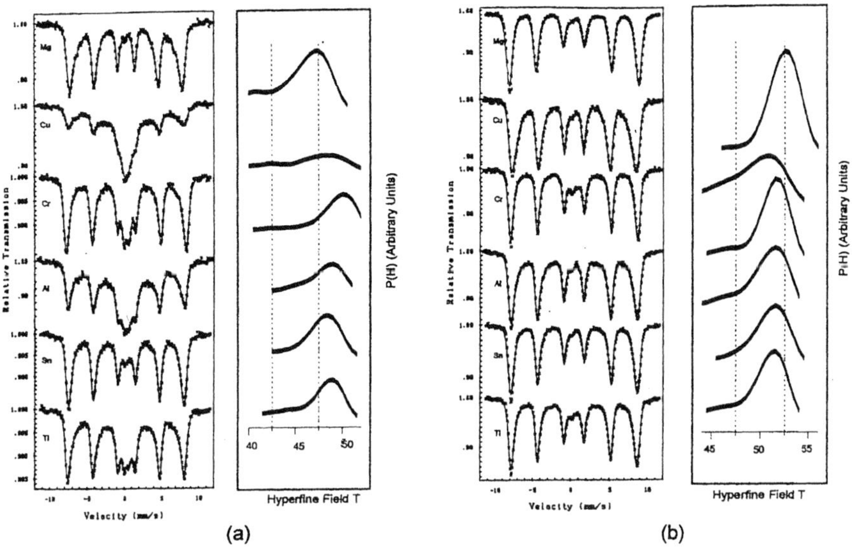

Figure 1: Mössbauer spectra calculated distribution P(H) at (a) RT and (b) 78 K

Table 1: Parameters Deduced from TEM and Mössbauer Measurements

Dopant	D nm	$B(\Delta\Gamma_B)$, 300K T	$B(\Delta\Gamma_B)$, 80K T	$\Delta B/B(0)\Delta T$ K^{-1} $X10^4$	K J/m^3 $X10^{-4}$
Mg	13	44.7(3.9)	52.5(2.3)	6.1	1.0
Cu	9	44.1(5.7)	48.3(4.1)	-	1.7[a]
Cr	18	48.5(3.2)	50.6(2.8)	2.1	1.1
Al	16	46.7(3.6)	49.7(3.6)	-	0.4[a]
Sn	33	47.0(3.0)	50.4(2.9)	2.6	1.4
Ti	12	46.6(3.6)	50.1(2.9)	3.0	2.6

[a]K calculated from superparamagnetic relaxation.

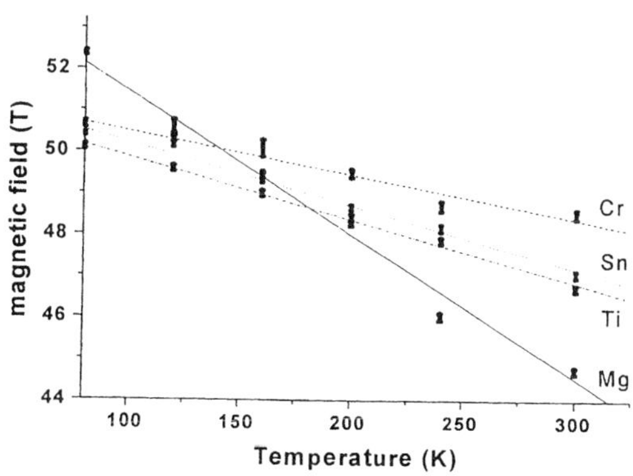

Figure 2: Temperature dependence of the average hyperfine fields of selected dopants

The Cu-doped sample, where Cu^{2+} has tetrahedral site preference [6], however, shows obvious relaxation and the lowest hyperfine field in all samples for the whole temperature range. The reduction in the average hyperfine field, which resulted from the weighted mean of the reduced hyperfine fields at the octahedral sites and the already lower hyperfine field at the tetrahedral sites, reflects the influence of A-site substitution.

Superparamagnetic relaxation at different levels has been observed in all samples (with a very small contribution in the Mg-doped sample), but more pronounced in Cu and Al-doped samples. A blocking temperature, T_B, of about 160 and 200K may be estimated for the latter samples respectively. We have estimated the magnetic anisotropy, K, for these 2 samples (Table 1) from the paramagnetic fraction and using the exponential dependence of the superparamagnetic relaxation frequency, υ, with temperature [7], $\upsilon = \upsilon_0 \exp(-KV/k_BT)$, where V is the volume of the particle.

Below the blocking temperature, as in the cases of the rest of the samples, the hyperfine field has been reduced by 4-15% in the measured temperature range. The Neel temperature of small particle size γ-Fe_2O_3 is not found to differ appreciably from the bulk material [7]. In addition, recent measurements of Berry *et al* on Sn-doped γ-Fe_2O_3 have shown that the Neel temperature is reduced from the pure material (~ 790K compared 850K), though is still much higher than the measured temperature range. Thus the reduction in the hyperfine field cannot be attributed to variation of magnetization with temperature. It is proposed that collective magnetic excitations may be responsible for the reduction of the average hyperfine field in Mg, Cr, Ti or Sn-doped samples. A plot of the average hyperfine field with temperature is shown in Figure 2 and the error bars give the statistical error calculated from the distribution and do not represent the width of distribution. A good linear dependence according to the relation $B(T) = B(0)(1 - k_BT/2KV)$ has been derived. The extrapolated value of the hyperfine field at T=0, B(0), is taken as the saturation average hyperfine field in this temperature range which is used instead of the bulk value used by Mørup *et al* in their study of Fe_3O_4 [7].The slope was to used to calculate the magnetic anisotropy constant (table 1).

Although the magnetic anisotropy constant of the metal doped samples is smaller than that of the pure material of smaller particle size [6], the large difference of 1 to 2 orders of magnitude cannot be attributed to particle size variation only and suggests a more significant role of the environment of iron. A more detailed systematic study will follow.

References

[1] J.M.D. Coey and D. Khalafalla, Phys. Stat. Sol. (a) 11(1972)229.
[2] K. Haneda and A.H. Morrish, Phys. Lett. 64A(2)(1977)259.
[3] P.J. Picone, K. Haneda and A.H. Morrish, J. Phys. C: Solid State Phys. 15(1982)317.
[4] R.J. Pollard and A. H. Morrish, IEEE Trans. Magn. MAG-23, 42 (1987).
[5] F.J. Berry and Ö. Helgason, Hyp. Int. 126(2000)269
[6] R.E. Vandenberghe and E De Grave, in: *Mössbauer Spectroscopy Applied to Inorganic Chemistry Vol. 3*, ed. G.J. Long, New York: Plenum, (1989), p59.
[7] S. Mørup, J.A. Dumesic and H. Topsoe, in: *Applications of Mössbauer Spectroscopy V II*", ed. R.L. Cohen, Academic Press (1980), pp 1.

MÖSSBAUER AND NEUTRON DIFFRACTION STUDIES ON THE MAGNETIC PHASE TRANSITION IN ANTIFERROMAGNETIC $FePt_{0.5}Ir_{0.5}$ ALLOY

T. Goto[a], J. Takahashi[a], Y. Itoh, K. Ohyama[b], Y. Yamaguchi[b] and H. Yoshida[b]
[a]*Faculty of Engineering, Tohoku Gakuin University, Tagajo 985-8537, Japan*
[b]*IMR, Tohoku University, Sendai 980-8577, Japan*

1. Introduction

The ferromagnetic FePt alloy with a tetragonal CuAuI-type structure has a large uniaxial anisotropy constant K_u=7x10^7 erg/cc [1]. The large magnetocrystalline anisotropy energy may be due to the spin-orbit coupling of Pt atoms [2]. It is expected that the substitution of Ir for Pt gives rise to considerable change in magnetic properties of FePt alloys, because Ir atoms have no magnetic moment. In fact, it has been recently found that in the Ir substituted system $FePt_{1-x}Ir_x$ ($0 \leq x \leq 0.6$) with CuAuI-type structure, the spontaneous magnetization decreases with increasing Ir concentration x and the alloys become antiferromagnetic for x>0.35 [3].

In this paper, the relation between the magnetic structure and the ^{57}Fe hyperfine structure of $FePt_{1-x}Ir_x$ are reported. Moreover, the phase transition observed for $FePt_{0.5}Ir_{0.5}$ by neutron diffraction (ND) measurement is reported and the characteristic behavior of the ^{57}Fe hyperfine field due to the peculiar spin arrangement is also reported.

2. Experimental

Samples of $FePt_{1-x}Ir_x$ ($0 \leq x \leq 0.6$) alloys were prepared by arc melting. The ingots were annealed at 1373K for 48h, and then annealed at 1053K for 48h. The ^{57}Fe Mössbauer spectra and the ND patterns of $FePt_{0.5}Ir_{0.5}$ were taken at various temperatures for the powders.

3. Results and discussion

In the $FePt_{1-x}Ir_x$ system, the magnetic phase change from ferromagnetic to antiferromagnetic state with Ir concentration x gives rise to the obvious change in ^{57}Fe Mössbauer spectrum. Some examples of ^{57}Fe Mössbauer spectra at 295 and 80K are shown in figure 1. From the spectrum analysis, it was found that three kinds of ^{57}Fe hyperfine structures, which are different each other in hyperfine field, exist in the spectra of this system. The individual subspectra obtained by spectral fitting are indicated by dotted lines in figure 1. The highest field is in the range of 274~285kOe at 80K and corresponds to that of ferromagnetic FePt. The lowest field is in the range of 184~189kOe at 80K and corresponds to that of antiferromagnetic $FePt_{0.45}Ir_{0.55}$. The field with intermediate value of about 230kOe at 80K is apparently seen in the samples with small spontaneous magnetizations. This field disappears at low temperatures for x=0.4. In this alloy, the small spontaneous magnetization disappears with decreasing temperature.

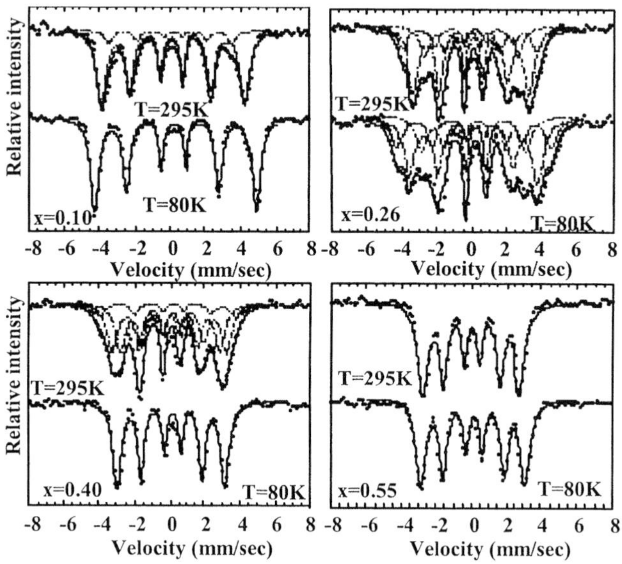

Figure 1. ^{57}Fe Mössbauer spectra of $FePt_{1-x}Ir_x$ at 295 and 80K.

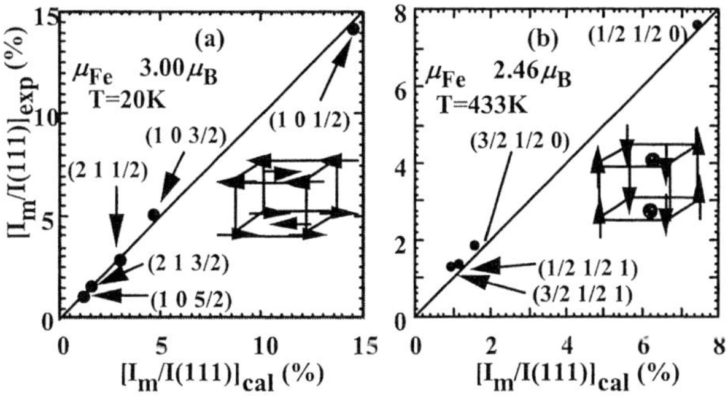

Figure.2. Relation between the experimental and calculated intensities of magnetic diffraction lines of $FePt_{0.5}Ir_{0.5}$ at (a) 20K and (b) 433K.

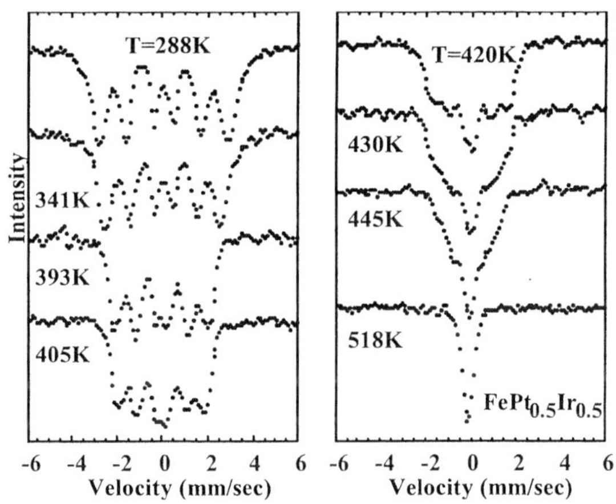

Figure 3. Temperature dependence of Mössbauer spectrum of $FePt_{0.5}Ir_{0.5}$.

Neutron diffraction (ND) measurements have been made at various temperatures on the antiferromagnetic $FePt_{0.5}Ir_{0.5}$ with a Neel temperature (T_N) of 505K, in order to investigate the spin structure and the Fe moment. The results revealed the existence of antiferro-antiferromagnetic phase transition with temperature. The reasonable spin structures in the low, and high temperature antiferromagnetic phases determined from the intensity analysis, which are shown in figure 2, are as follows. The directions of Fe-spins are along the a-axis in the temperature below 280K and the adjacent Fe-spins are antiferromagnetically arranged in the basal planes of Fe, and between the planes separated by (Pt,Ir)-planes. On the other hand, above 370K the Fe-spins are arranged along the c-axis, and form the antiferromagnetic (π π 0) arrangement. In this spin structure, spin frustration may possibly appear, because the exchange fields may be canceled for half of the Fe atoms.

Some of Mössbauer spectra of $FePt_{0.5}Ir_{0.5}$ taken at various temperatures are shown in figure 3. A characteristic feature of the spectra in the temperature range between 370K and T_N supports the existence of spin frustration, because a relaxation spectrum appears near the center of spectrum in addition to a sextet. Moreover, the occurrence of the relaxation spectrum gives rise to destruction of clear split of the remaining sextet. Details of spectral analysis will be published separately. The relatively small hyperfine field of $FePt_{0.5}Ir_{0.5}$, as compared with the value of FePt, does not correspond to the large Fe moment of 3.0μB obtained by ND measurement Therefore, it is satisfactory to consider that the ^{57}Fe hyperfine field of $FePt_{1-x}Ir_x$ is influenced considerably by the transferred hyperfine field which depends strongly on the spin structure. In the magnetic phase transition observed here, the Fe-Fe exchange interaction along c-axis changes its sign from antiferromagnetic to ferromagnetic with increasing temperature and at the same time the spin direction changes from in- plane to c-axis. Therefore, it is considered that the anisotropic exchange interactions between Fe atoms and the dependence of the exchange interactions on atomic distance play an important role for the antiferro-antiferromagnetic phase transition of $FePt_{0.5}Ir_{0.5}$.

References

[1] O.A. Ivanov, L.V. Solina, V.A. Denshina and L.M. Magat, Phys. Met. Metallogr. 35 (1973) 81.

[2] I. V.Solovyev, P. H.Dederichs and I. Mertig, Phys. Rev. B 52 (1995) 13419.

[3] T. Goto, J. Takahashi, M. Nakamura, T. Hirose, K.Watanabe, and H. Yoshida, J. Magn. Magn. Mater. 226-230 (2001) 1656.

High-Pressure Mössbauer Study of Perovskite Iron Oxides

Takateru Kawakami[1], Saburo Nasu[1], Tetsuya Sasaki[1], Kaori Kuzushita[1], Shotaro Morimoto[1], Shoichi Endo[2], Shuji Kawasaki[3], Miko Takano[3]
[1]*Division of Materials Physics, Department of Physical Science Graduate School of Engineering Science, Osaka 560-8531, Japan*
[2]*Research Center for Materials Science at Extreme Conditions, Osaka University, Toyonaka, Osaka 560-8531, Japan*
[3]*Institute for Chemical Research, Kyoto University, Uji, Kyoto 611-0011, Japan*

Abstract: $CaFeO_3$, $La_{1/3}Sr_{2/3}FeO_3$ and $SrFeO_3$ in the perovskite oxides have been investigated by high-pressure ^{57}Fe Mössbauer spectroscopy. In $CaFeO_3$ the charge disproportionation (CD: $2Fe^{4+} \rightarrow Fe^{3+} + Fe^{5+}$) occurs about 290 K in the pressure range of 0-17 GPa. Above 20 GPa, the CD is suppressed. The magnetic order (MO) temperature of 125 K at an ambient pressure rises to room temperature above 34 GPa. In $La_{1/3}Sr_{2/3}FeO_3$ the CD ($3Fe^{11/3+} \rightarrow 2Fe^{3+} + Fe^{5+}$) and the MO occur at the same temperature up to 21 GPa, which decrease from 207 K to 165 K with increasing pressure. Further increase of pressure above 25 GPa, the CD reaction is suppressed and the MO temperature rises above 400 K.

Key words: High-pressure ^{57}Fe Mössbauer spectroscopy, Diamond anvil cell (DAC), Charge disproportionation, $CaFeO_3$, $La_{1/3}Sr_{2/3}FeO_3$, $SrFeO_3$

INTRODUCTION

One of the most interesting problems in the condensed-matter systems is metal-insulator transitions driven by the correlation effects associated with electron-electron interaction [1]. A charge order in perovskite iron compounds of $CaFeO_3$ and $La_{1/3}Sr_{2/3}FeO_3$ often accompany their metal-insulator transitions. A high valent state of Fe in $SrFeO_3$, $CaFeO_3$ and $La_{1/3}Sr_{2/3}FeO_3$ of the perovskite iron oxides is quite important to study in a charge disproportionation (CD) and a magnetic order (MO).

$SrFeO_3$ is one of the typical Fe^{4+} oxide having a cubic perovskite structure. Mössbauer spectrum at 4.2 K consists of a single magnetically-split sextet with hyperfine magnetic field of 33.1 T (Fe^{4+}, $t_{2g}^3\ e_g^1$) [2]. $SrFeO_3$ shows a metallic conductivity maintaining the valence state of Fe^{4+} down to 4 K and an antiferromagnetic long-lange order with a helical spin structure below 134 K [3,4,5,6]. Photoemission spectroscopic study of $SrFeO_3$ has suggested that the $(FeO_6)^{8-}$ octahedron has a large negative effective charge-transfer energy of about -3 eV [7].

The crystal structure of $CaFeO_3$ is $GaFeO_3$-type [8,9,10,11]. $CaFeO_3$ shows a metal-insulator transition at 290 K. At the same temperature the Mössbauer spectrum becomes to split gradually into two paramagnetic absorption lines suggesting the charge disproportionation (CD : $2Fe^{4+} \rightarrow Fe^{3+} + Fe^{5+}$) with two different valent states with decreasing temperature in a second-order transition. $CaFeO_3$ below 125 K shows an antiferromagnetic long-range order with a helical spin structure.

$La_{1/3}Sr_{2/3}FeO_3$ in the solid solution system shows a metal-insulator transition with a charge order and antiferromagnetic order at the same temperature of 207 K with narrow thermal hysteresis suggesting the occurrence of a first-order phase transition. The charge disproportionation (CD : $3Fe^{11/3+} \rightarrow 2Fe^{3+} + Fe^{5+}$) in $La_{1/3}Sr_{2/3}FeO_3$ is detected by Mössbauer spectroscopy [12]. From the analysis of the observed Mössbauer spectra, two kind of Fe-site species with valences of Fe^{3+} and Fe^{5+} are identified with a number ratio of 2:1.

In this investigation, we examined the pressure effects on the CD, the MO and the crystal structure of $SrFeO_3$, $CaFeO_3$ and $La_{1/3}Sr_{2/3}FeO_3$ by ^{57}Fe Mössbauer spectroscopy. Results obtained are summarized as a *P-T* magnetic phase diagram of $SrFeO_3$, $CaFeO_3$ and $La_{1/3}Sr_{2/3}FeO_3$.

MATERIALS AND METHODS

^{57}Fe Mössbauer measurements have been performed under high pressure using a diamond anvil cell (DAC). ^{57}Fe enriched (50 %) $SrFeO_3$, $CaFeO_3$ and $La_{1/3}Sr_{2/3}FeO_3$ powder are synthesized under high oxygen pressure and used as an absorber in transmission geometry. DAC used in this experiment is of Basset type [13]. The powder sample and a small ruby chip are enclosed in a hole of diameter 200 μ m in rhenium gasket. The pressure-transmitting media used are: a 4:1 methanole:ethanol solution. The actual pressure is determined from the wavelength shift of the ruby line using the non-linear pressure calculation curve. All of the Mössbauer spectra have been measured by a combination of a specimen powder in DAC and a velocity transducer attached with 370 MBq ^{57}Co in Rh as a high density γ -ray source. The velocity scale of the spectrum is relative to α -Fe at room temperature.

RESULTS AND DISCUSSION

The high pressure ^{57}Fe Mössbauer spectra of $La_{1/3}Sr_{2/3}FeO_3$ are shown in FIG. 1. At an ambient pressure and 298 K and 210 K the spectra are a paramagnetic singlet of an average valence $Fe^{11/3+}$, as shown in FIG. 1(a) and (b). With a decrease of the temperature below 200 K the CD and the MO occur simultaneously. At 200 K, the spectrum consists of three components, those are one pramagnetic singlet of $Fe^{11/3+}$ and two magnetically-split sextets of Fe^{3+} and of Fe^{5+} as shown in FIG. 1(c). Their hyperfine magnetic fields are determined to be 38.4 T for Fe^{3+} and 23.8 T for Fe^{5+}. At 5 K the hyperfine magnetic fields are determined to be 46.3 T for Fe^{3+} and 26.5 T for Fe^{5+} whose values correspond to Fe of high spin configurations, as

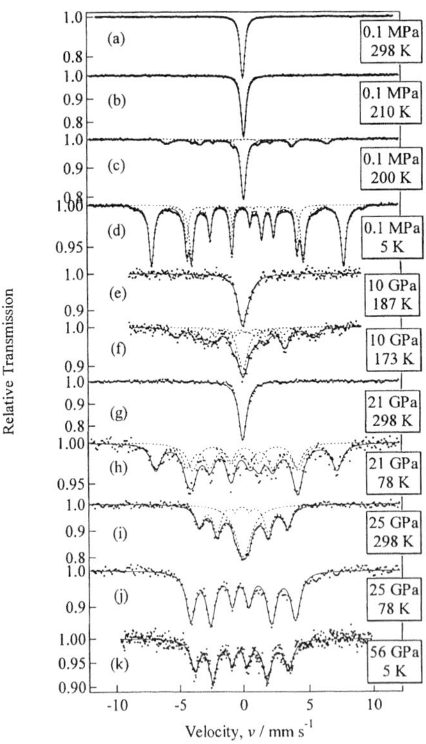

FIG. 1. Typical pressure-dependent Mössbauer spectra obtained from $La_{1/3}Sr_{2/3}FeO_3$ under various temperatures.

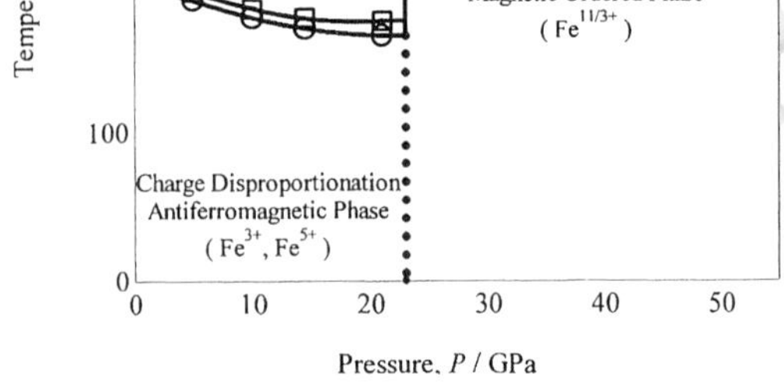

FIG. 2. *P-T* phase diagram of $La_{1/3}Sr_{2/3}FeO_3$. The critical temperature of the charge dispropotionation (CD) and the magnetic order (MO) temperature are shown as a function of the pressure.

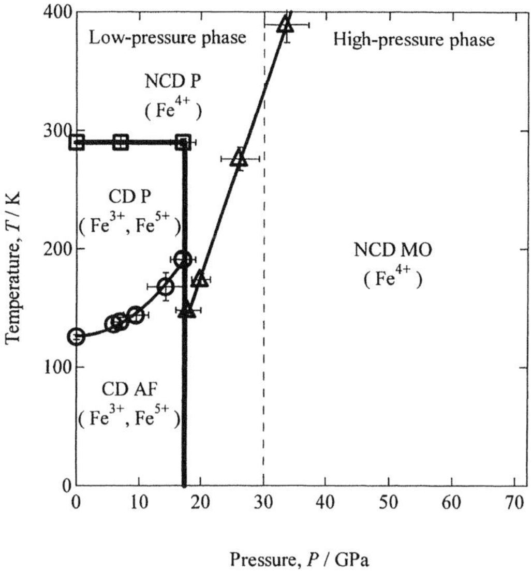

FIG. 3. *P-T* phase diagram of $CaFeO_3$. The critical temperature of the charge disproportionation (CD) and the magnetic order (MO) temperature are shown as a function of the pressure.

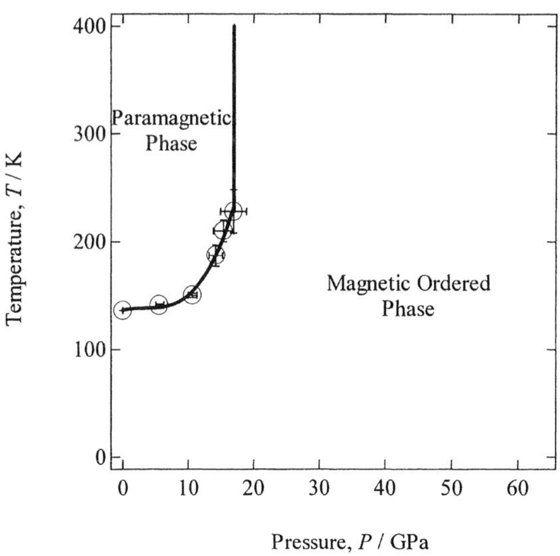

FIG. 4. *P-T* phase diagram of $SrFeO_3$. The critical temperatures of the magnetic order are shown as a function of the pressure.

shown in FIG. 1(d). $La_{1/3}Sr_{2/3}FeO_3$ shows the first-order phase transition which is a simultaneous CD and MO transition at an ambient pressure. At 10 GPa and 187 K, the spectrum is paramagnetic singlet of $Fe^{11/3+}$. At 173 K just below T_{MO}, the spectrum shows the occurrence of both the CD and the MO, as shown in FIG. 1(e) and (f). It is concluded the simultaneous CD and MO transition is maintained under high pressure up to 21 GPa.

The pressure between 21 GPa and 25 GPa the Mössbauer spectra are drastically changed and the CD is suppressed although the magnetic order temperature (T_{MO}) suddenly increases above 400 K at 25 GPa. At 21 GPa and 298 K the spectrum is still a paramagnetic singlet. At 21 GPa and 78 K, the spectrum is two magnetically-split sextets showing the CD. At 25 GPa and 298 K, the spectrum shows an extra magnetically-split sextet. At 78K, the spectrum is a only one magnetically-split sextet, from which the CD is suppressed by the high pressure, as shown in FIG 1(g), (h), (i) and (j). According to the inclination of the pressure, the critical pressure of the CD is deduced to be about 23 GPa.

The critical temperature of the charge disproportionation (T_{CD}) and the T_{MO} as a function of pressure of $La_{1/3}Sr_{2/3}FeO_3$, $CaFeO_3$ and $SrFeO_3$ are shown in FIG. 2, FIG. 3 and FIG. 4 as a *P-T* phase diagram. The T_{CD} and the T_{MO} are shown up to 65 GPa in the temperature range of 6-400 K [14].

CONCLUSION

The *p-d* hybridization between

O 2*p* and Fe 3*d* is enhanced with a reduction in Fe-O distance by external pressure. In $CaFeO_3$ the magnetic ordering temperature increases with an increase of pressure, but the T_{CD} does not show any significant change up to 17 GPa and is suddenly suppressed above 20 GPa. Different pressure dependences on the charge and spin order in $CaFeO_3$ might be interpreted as follows. The CD is controlled mainly by holes and the magnetic order is enhanced by the electron transfer between oxygen and iron.

In $La_{1/3}Sr_{2/3}FeO_3$ the appearance of the CD state below 21 GPa suggests that the localization of the charge at the Fe sites occurs and the resultant Mössbauer spectra consist of two components rising from Fe^{3+} and Fe^{5+} at low temperatures. The spectra above 25 GPa consist of a single component with an average valence state of $Fe^{11/3+}$, since the *p-d* hybridization is enhanced by the pressure and the collective motion of holes can occur between Fe sites. As a result of their collective motion, $La_{1/3}Sr_{2/3}FeO_3$ might show a good electrical conductivity, for which an appropriate experimental evidence has not yet obtained.

Oxygen-hole character in $SrFeO_3$, $CaFeO_3$ and $La_{1/3}Sr_{2/3}FeO_3$ with the strong *p-d* hybridization under high pressure may play an important role for the understanding of its transport and magnetic properties.

ACKNOWLEDGEMENTS

The authors thank Prof. A. Fujimori of University of Tokyo for his helpful discussion. This work has been supported by Core Research for Evolutional Science and Technology (CREST) of the Japan Science and Technology Corporation (JST) and a Grant-in-Aid for COE Research (10CE2004) and for Scientific Research (A, 12304018) from the Ministry of Education, Science, Sports and Culture.

REFERENCES

[1] M. Imada, A. Fujimori, and Y. Tokura, Rev. Mod. Phys. **70**, 1211 (1998).

[2] P. K. Gallagher, J. B. MacChesney, and D. N. E. Buchanan, J. Chem. Phys. **41**, 2429 (1964).

[3] J. B. MacChesney, R. C. Sherwood, amd J. F. Potter, J. Chem. Phys. **43**, 1907 (1965).

[4] T. Takeda, Y. Yamaguchi, and H. Watanabe, J. Phys. Soc. Jpn. **33**, 967 (1972).

[5] H. Oda, Y. Yamaguchi, H. Takei, and H. Watanabe, J. Phys. Soc. Jpn. **42**, 101 (1977).

[6] S. Kawasaki, M. Takano, R. Kanno, T. Takeda, and A. Fujimori, J. Phys. Soc. Jpn. **67**, 1529 (1998).

[7] A. E. Bocquet, A. Fujimori, T. Mizokawa, T. Sitoh, H. Namatame, S. Suga, N. Kimizuka, Y. Takeda, and M. Takano, Phys. Rev. B **45**, 1561 (1992).

[8] M. Takano, N. Nakanishi, Y. Takeda, S. Naka, and T. Takeda, Mater. Res. Bull. **9**, 923 (1977).

[9] S. Morimoto, T. Yamanaka, and M. Tanaka, Physica B **237-238**, 66 (1997).

[10] P. M. Woodward, D. E. Cox, E. Moshopoulou, A. W. Sleight, and S. Morimoto, Phys. Rev. B **62**, 844 (2000).

[11] T. Takeda, et al., Solid State Sciences 2, 673-687 (2001).

[12] M. Takano, and Y. Takeda, Bull. Inst. Chem. Res. Kyoto Univ. **61**, 406 (1983).

[13] W. A. Bassett, T. Takahashi and P. W. Stook: Rev. Sci. Instrum. **38**, 37 (1967).

[14] T. Kawakami, S. Nasu, T. Sasaki, S. Morimoto, S. Endo, S. Kwasaki, and M. Takano, J. Phys. Soc. Jpn. **70**, 1491(2001).

APH INDUCED RED CELL OXIDATION IN HEALTHY PERSONS

S. Croci, I. Ortalli, G. Pedrazzi, R. Delsignore[1], G. Passeri[1]

Istituto di Scienze Fisiche, University of Parma, INFM Udr Parma,43100 Parma Italy
1 Dipartimento di Medicina Interna e Scienze Biomediche, University of Parma, Italy

In a previous work we have shown that the red blood cells of patients suffering from breast cancer (P) seem more sensitive to the denaturing action of acetylphenylhydrazine (APH) than those from healthy donors (H). The amount and the formation time of the hemin, one of the last oxidation products, are significantly different in the patients. Therefore, the spectral contribution of hemin after the same incubation time could be a factor of discrimination between the samples from the patients and healthy population. The aim of the present work was to gain a deeper understanding of the oxidation processes and establish a standard setup for the measurements.

Key words: Haemoglobin oxidation, free radicals, Mössbauer Spectroscopy

1. Introduction

The production of oxidative substances like free radicals (FRs) in tumoral pathologies is well know [1, 2]. Erythrocyte physico-chemical properties can be modified by FRs presence and may lead to lipid peroxidation and haemoglobin oxidation [3]. A previous work performed by our group on erythrocytes collected from healthy volunteers and women suffering from breast cancer showed a very different oxidation pattern after 50 minutes of incubation with acetylphenylhydrazine (APH) [4]. These results supported the hypothesis of a pre-existent erythrocytes liability in the blood samples of the patients with respect to those of healthy people. The aim of the present work is to set a procedure for sample preparation and measurement in order to establish a range of "healthy behaviour". The final goal is to arrive to discriminate the blood from healthy persons from that of the patients.

2. Materials and Methods

The sample preparation procedure used in this work has been slightly modified with respect to the previous one [4]. The degree of erythrocytes damage is dependent on the amount of oxidant and antioxidant substances present in the blood [5, 6] and the level of Hb oxidation is connected with the presence of free radicals and deoxyHb content. In order to eliminate antioxidants from the blood and to convert all the amount of deoxyHb to oxyHb red cells have been washed before the treatment with APH. Compared to [4] this operation resulted in a more uniform behaviour under the APH oxidant action.

An aliquot of 15 ml of venous blood was collected from each female healthy volunteer and then divided in three tubes with heparin. Each tube was then washed in saline solution and centrifuged (3000 rpm) for 10 minutes. The supernatant was then removed. One of the tube was used as control (C) and the packed red cells at this stage were collected in a Mössbauer sample holder and immediately frozen in liquid nitrogen. The red cells of the other two tubes were diluted in 100 ml of a 30 mM APH solution at pH = 7.6 (Sörensen buffer), and incubated at 37 °C for 50 (APH50) and 120 (APH120) minutes, respectively. The final samples (APH50, APH120) were again washed in saline solution, centrifuged, and the packed cells were finally adjusted in a suitable sample holder and frozen in liquid nitrogen.

The detection of hemin by spectrophotometric measurements is quite a long and difficult procedure if the sample, as in present case, is a mixture of several components like met-Hb and other oxidation products ; on the contrary the Mössbauer parameters isomer shifts and quadrupole splittings, characteristic of met-Hb and hemin, are quite different and can be readily recognised. For these reasons we carried out the experiments with Mössbauer spectroscopy and the measurements were performed at 77K with a conventional spectrometer.

3. Results and Discussions

The packed red cells have been analysed by Mössbauer spectroscopy after incubation with APH for 50 and 120 minutes, respectively. In figure 1 the spectra obtained from three healthy subjects (E,T,G) are shown and the relevant spectral areas are reported in table I .

For each person the "Control", "APH50" and "APH120" samples are presented and compared. The clinical data of the donors have been also recorded and showed that the hemograms, as well as the number of erythrocytes per μl of blood, and average Hb concentration per cell, were all within the normal physiological range.

Analysing the results we have found that control spectra (figure 1C), after the washing procedure, show the presence of Oxy-Hb only without any deoxy-Hb left. On the contrary, for the spectra of red cells incubated for 50 minutes and 120 minutes with APH (figure 1: APH50, APH120) two components are present, one characteristic of Oxy-Hb and the other of the envelope of the various oxidation products. This latter asymmetric and broad peak is due to the superposition of methemoglobin, hemichromes and hemin [4, 5].

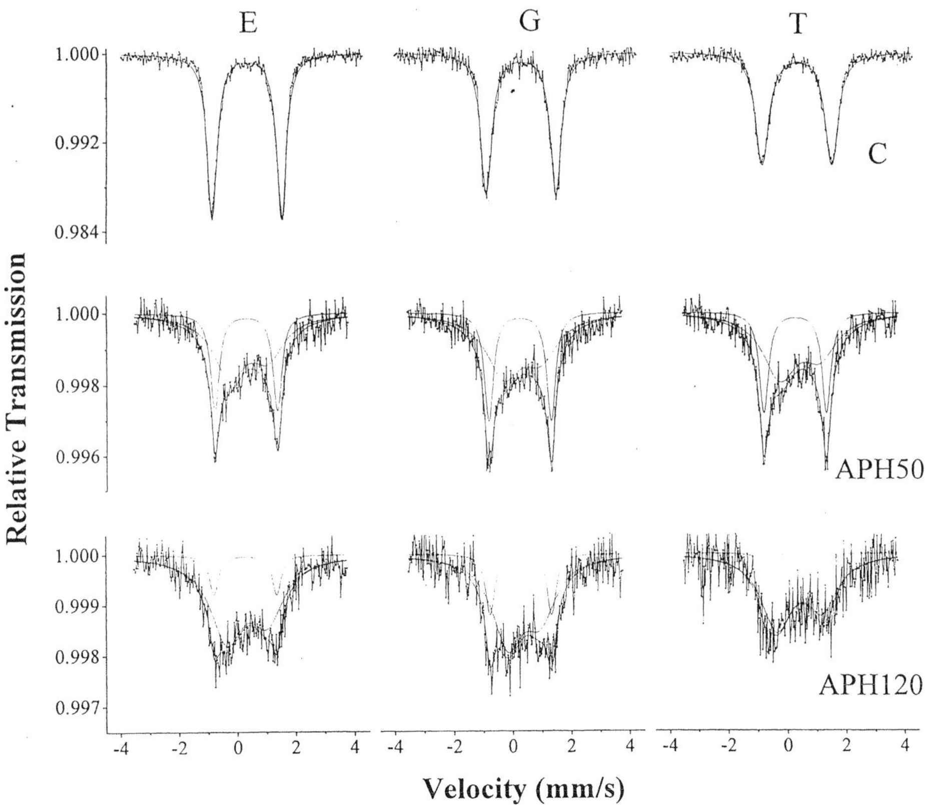

Figure1. Erythrocyte spectra of three healthy women (G,E,T) : C - control, APH50 and APH120.

	Control			APH - 50			APH - 120		
	E	*G*	*T*	*E*	*G*	*T*	*E*	*G*	*T*
Area (%) HbO2	100	100	100	35	35	34	13	20	8
Area (%) ox. products	-	-	-	65	65	66	87	80	92

Table I. Mössbauer spectral area relative to E, G, and T samples after different time of incubation with APH. Errors can be estimated to be in the order of some percent.

It can be noted that in the case of APH50 samples, the spectral pattern are very similar each other and the areas are comparable, while the spectra obtained after 120 minutes of incubation (APH120 samples) present quite a different spectra, clear signs of different oxidation states.

After 50 minutes of APH incubation about 65% of the total oxyHb is transformed in oxidation products, with about the same oxidation level for all the samples.

In contrast, if the red cell are incubated for a longer time, as in the case of 120 minutes, Hb oxidation together with lipid peroxidation lead to stiffness of the membrane and release of hemin. This advanced oxidation causes a drastic modification of the red cell and Hb and prevents a clear comparison between the spectra.

In conclusion we believe that an APH incubation time equal to 50 minutes may be a good choice for a standard procedure and for comparisons. In fact, after this time interval we have observed the largest difference between the samples collected from healthy subjects and patients and, moreover, the data obtained on healthy volunteers seem stable with comparable oxidation products among all the samples.

References

[1] B. Halliwell, Nutr. Rev. 52 (1994) 253-265;

[2] M.R.Clemens, Klin. Wochenschr 69 (1991): 1123-1134

[3] W.H. Reinhart, L.A. Sung, S. Chien, Blood 68 (1986) 1376-1383

[4] S.Croci, I.Ortalli, G.Pedrazzi, G.Passeri and P.Piccolo, Hyp Int 126 (2000): 47-52;

[5] G. Lelkes, I. Fodor, G. Lelkes, S. Hollán and A.J. Verkleij Biochim Biophys Acta 945 (1988) 105-110;

[6] S.Turi, I.Németh,A.Torkos, L.Saghy, I.Varga, B.Matkovics, J.Nagy, Free Radical Biology & Medicine 22 (1997)161-168;

Structural Relaxation from X- ray irradiated metastable Hbmet investigated by Mössbauer Spectroscopy

S.Croci[1,3], A.Dilg[1], N.Engler[1],V. Prusakov[2], O. Iakovleva[1], I. Ortalli[3], F.Parak[1]

[1]Physik-Department E17, Technische Universität München, 85747 Garching, Germany;
[2]Institute of Chemical Physics, Russian Academy of Science, Moscow, Russia;
[3]Istituto di Scienze Fisiche INFM - Udr Parma, University of Parma, 43100 Parma.

Protein functions are strictly connected to their structural dynamics. Actually, the study of haemoglobin metastable states is a useful way to have insight in to structural changes that lead the protein from one stable state to another. Mössbauer spectroscopy has been used to follow the kinetics from the irradiated aquo-haemoglobin to carboxy-haemoglobin (HbCO), Hbdeoxy and Hbmet pre-irradiated states. Thermal cycles from 140 K to 200 K allow a study of the kinetics of relaxation from the low spin metastable state of the irradiated aquo-haemoglobin to equilibrium states. The population exchange of the different conformations during the relaxation is reported as a function of time and temperature.

Introduction

In order to have an insight into haemoglobin (Hb) function, a study on Hb conformational changes has been performed with Mössbauer spectroscopy. The importance of structural dynamics connected with the change of different conformations is well know in the study of the protein activity. In the case of haemoglobin that has more conformations depending on its ligands and bonding, the production of a metastable state, as in case of myoglobin [1], is a good trick to study the protein kinetics. In the liganded state, the heme is planar, the iron has spin 0, and the quaternary haemoglobin structure assumes the relaxed (R) configuration, while in the unliganded state the iron has spin 2, the heme move from the mean heme plane reaching the stable tense configuration (T). Haemoglobin is characterised by assuming a large number of conformational substates, these different substates have the same overall structure but differ in details[2].

In particular the relaxation from a metastable FeII state, to two stable states, deoxyhemoglobin (deoxyHb) and carboxyhemoglobin (HbCO) has been followed. The metastable FeII state has been produced by X-ray irradiation of aquohemoglobin (Hbmet) at nitrogen temperature, while the relaxation has been obtained performing thermal cycles in the 140 K – 200 K range.

Material and Method

Hbmet has been enriched with Fe^{57}, then optical measurements [3] have been performed in order to check the sample preparatio. Enriched Hbmet has been concentrated and 50% of glycerol has been added to the sample, to prevent free radical formation. The sample has been quickly frozen in condensed propane to have less relaxation between high spin and low spin, and placed into a Mössbauer sample holder at T=77K. The sample has been irradiated with an X-ray beam at 90K for about 70h with a 30 mA cathode current and 40KV peak voltage. The exposure has been done on both sides of the sample to obtain a uniform irradiation and Mössbauer measurements have been performed at T=6K and T=77K before and just after irradiation. Thermal cycles consist of raising the sample temperature for a different time and cooling again to nitrogen temperature, keeping the sample in the Mössbauer cryostat. The temperature in thermal cycles takes values from 140K to T=200K while all the measurements were performed after cooling to T= 77K. The measurements are done with a conventional Mössbauer spectrometer in transmission mode.

Results and discussion

Measurements done before irradiation of the sample exhibit the presence of FeIII high spin and low spin components [4] as shown in figure1(a) while after irradiation almost all the high and low spin FeIII component is reduced to FEII low spin as shown in figure 1(b).

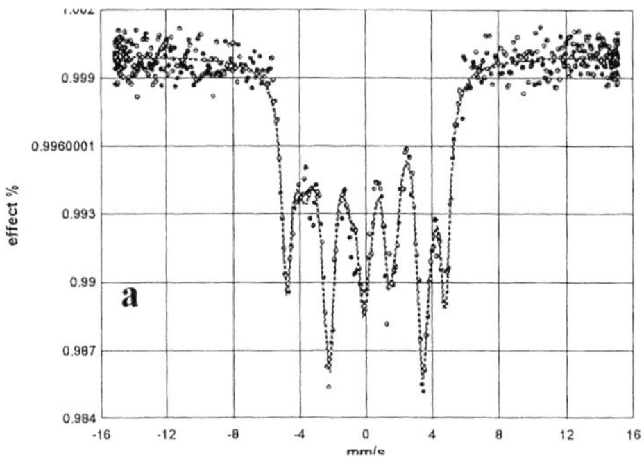

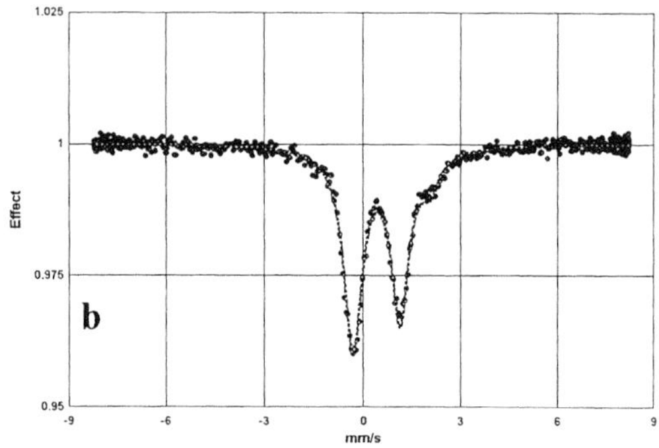

Figure 1. Mössbauer spectra before (a) and after (b) irradiation at T=5K

The spectra of the freshly irradiated sample is fitted with 84% of FeII low spin component, 10% of FeII high spin similar to deoxyHb component and about 6% of Hbmet (FeIII) not reduced component. During thermal cycles, the (Fe(II)HbH$_2$0) that is in a metastable state, relax to deoxy and HbCO stable states. In particular the water molecule can escape, heme can bind a CO or move to the deoxy position with a consequent FeII changing from low spin to high configuration. The Mössbauer parameters of the three new species, FeII low spin , FeII high spin (deoxyHb) and HbCO, are reported in table 1.

Species	Excursion temperature	QS (mm/s)	IS (mm/s)
FeII LS	**140 K <T <180 K**	1.40 ± 0.08	0.52 ± 0.02
FeIIHS (deoxyHb)		0.41 ± 0.02	0.25 ± 0.01
HbCO		2.43 ± 0.04	0.92 ± 0.02
FeII LS	**190 K < T < 200 K**	1.31 ± 0.01	0.55 ± 0.01
FeIIHS (deoxyHb)		0.39 ± 0.02	0.26 ± 0.02
HbCO		2.30 ± 0.02	0.92 ± 0.01

Table 1. Mössbauer parameters of the components present after irradiation at T=77K

The Mössbauer spectra obtained after irradiation at 77K, with different excursion temperatures, has to be fitted with four components as reported in figure 2, in which are shown the spectra of the sample from the heating cycles at T=170K and T=180K .

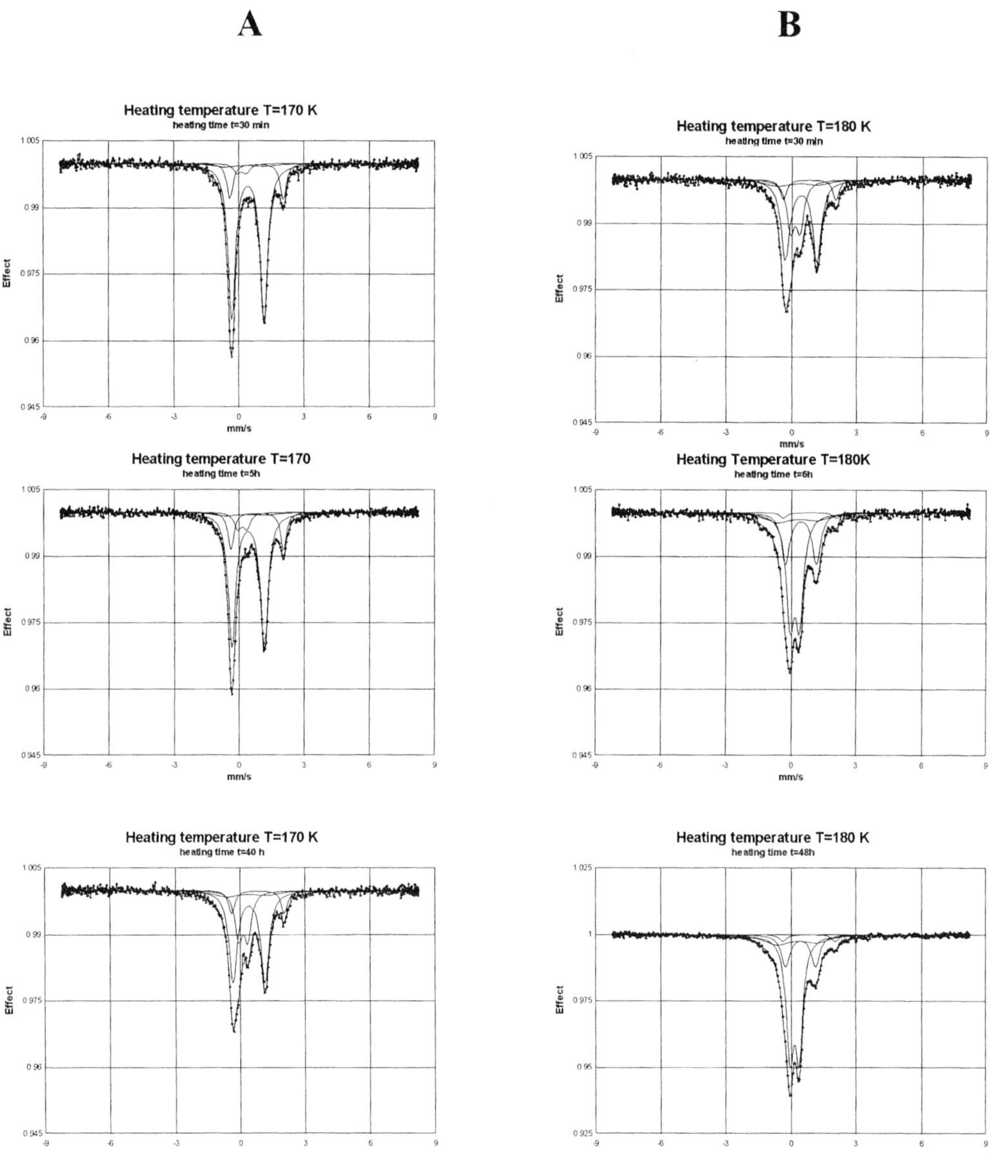

Figure2. Mössbauer spectra of irradiated samples after thermal cycles at T=170 K (A) and at T=180 K (B). In A are shown spectra after 0.5h, 5h, and 40h of heating time and in B are shown the spectra after 0.5, 6h, and 48h.

Electrons produced by irradiation at 90 K convert FeIII into FeII. Nevertheless at 80 K the water molecule is still bound as a sixth ligand, (Fe(II)HbH$_2$0). The relaxation of the metastable state FeII low spin occurs mainly after the heating beyond T=160K as it is shown in figure 3 where the relative amount of the species against heating time is reported.

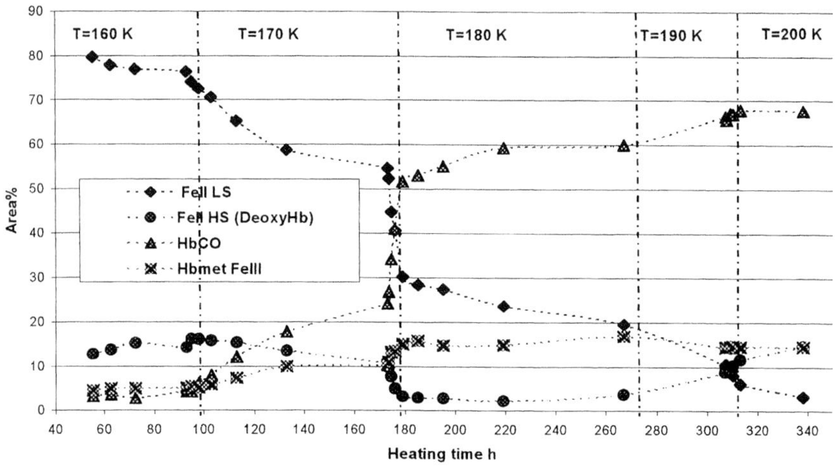

Figure3. Relative fractions of the different components against heating time.

At T=180K there is a dramatic decrease in the FeII low spin component, together with an increase of the HbCO species. Moreover the conversion of deoxyHb state into Hbmet FeII component mainly above 180K is probably due to interaction of free radicals produced by irradiation.
The different conformational states[5], FeII low spin and FeII high spin (deoxyHb), are allready present after irradiation, showing different enthalpy barriers of the protein. Nevertheless up to 160 K only 4% of the FeII low spin state is relaxed. The different relaxing behaviour of FeII low spin at 170 K and 180 K reflects the structural change of the protein above 180K .

References:

[1] V.E.Prusakov, J.Steyer, F.Parak ,Biophys. J. 68,2524-30,1995;
[2] A.Ansari, J.Berendzen, S.F.Bowne, H.Fraunfelder, I.E.T.Iben, T.B.Sauke,E.Shyamsunder, R.D.Young , Proc. Natl. Acad. Sci. USA 82, 5000-4, Aug 1985
[3 E. Antonini & G. Brunori (1971) *Hemoglobin and Myoglobin in their Reaction with ligand* North-Holland Amsterdam.]
[4] U.F. Thomanken, F.Parak, B. Wintergerst ,Z.Naturforschu.32c,11-19;
[5] N. Engler, A. Osterman , A. Gassmann, Don C.Lamb, V.E.Prusakov, J.Schott, R.Schweizer-Stenner, F.Parak Biophys. J. 78 ,2081-92,2000;

POTASSIUM ASCORBATE AS PROTECTIVE AGENT IN THE OXIDATION OF THE RED BLOOD CELLS

S. Croci[1], G. Pedrazzi[1], G. Paoli[2], D. Monetti[2], I. Ortalli[1]

1 - Istituto di Scienze Fisiche, University of Parma, INFM-Udr Parma, Via Volturno 39 Parma, Italy
2 - Fondazione Internazionale Valsè Pantellini per la Ricerca e lo Studio di Malattie Degenerative, Firenze

Free radicals and oxidative substances are involved in many degenerative diseases modifying cell physiological properties. Many mechanisms can be disrupted, including the K-Na channels, with a consequent potassium loss. Free radicals are also able to modify erythrocytes stability leading to peroxidation of the lipid membrane, oxidation of haemoglobin (Hb) and finally to the formation of Heinz Bodies. Such a process can also be induced or enhanced by strong oxidants like acetylphenylhydrazine (APH). In the present work the antioxidant properties of potassium ascorbate acting on the red cells treated with APH are investigated.

1. Introduction

Oxidative processes are involved in many degenerative diseases leading to various cell modifications. Recognised facts that deserve particular attention are the oxidant action exhibited by free radicals (FRs), that is involved in promoting and developing cancer [1-3], and the potassium loss from the cell with the consequent alteration of Na-K pumps. The loss of potassium is strictly connected with apoptosis ([4] and references therein). At normal intracellular K+ concentration, about 140-150 mM, the apoptotic mechanisms are dormant due to the inability to activate key protease and nuclease enzymes involved in the apoptotic process. The ubiquitous occurrence of such processes in living cells and tissues makes reasonable to extend the previous considerations to the red blood cells. Due to their systemic character the erythrocytes circulating in the blood stream can reach all the organs and undergo FRs action. The erythrocytes stability can be reduced by peroxidation of the lipid membrane and oxidation of hemoglobin (Hb). The Hb oxidation process starts with the formation of methemoglobin, continue until endoerythrocytes inclusion called Heinz Body are formed inside the red cell, and ends with the release of hemin [2, 5]. This oxidation pathway can also be induced or enhanced by a strong oxidant like acetylphenylhydrazine (APH).

In the present work we are interested in examining the effects of potassium ascorbate (K-asc) on APH treated erythrocytes collected from healthy volunteers using Mössbauer spectroscopy.

2. Materials and methods

The samples were prepared from venous blood taken from healthy volunteers. Five heparinized tubes (5 ml of blood each one) were collected from each person. One of the five tubes was used as control (C), the other four were incubated in a 30mM APH solution in Sörensen buffer (pH=7.6), with the following procedures : in the first sample (APH50) 5 ml of whole blood were incubated for 50 minutes with APH; in the second (ASC50), third (K50) and fourth sample (KASC50), 5 ml of whole blood were added to APH and mixed with a solution of ascorbic acid, potassium bicarbonate and potassium ascorbate (K-asc), respectively.

K-asc is a salt resulting from the salification of ascorbic acid by potassium bicarbonate, and has antioxidant properties. In this preliminary work the K-asc concentration is based on the potassium concentration present in the cell at physiological condition. Ascorbic acid, potassium bicarbonate and potassium ascorbate were diluted in distilled water to prepare the solutions.

At the end of the incubation time, the samples were centrifuged for 10 minutes at 3000 RPM (~ 1500 g). After removal of the supernatant the cells were resuspended in 0.9% NaCl solution and centrifuged again at 3000 RPM for 10 minutes. The procedure was repeated twice. The packed red cells obtained from the last centrifugation were collected in a Mössbauer sample holder and immediately frozen in liquid nitrogen. The measurements were performed at 77K with a conventional Mössbauer spectrometer (Takes inc.) working at constant acceleration (drive from Wissel Intruments) in transmission mode. The source was 57Co/Rh with an activity of 3.7 GBq (100 mCi).

3. Results and discussion

The antioxidant action of potassium ascorbate was followed during the incubation with the strong oxidant APH in various environments. Since K-Asc is the reaction product of ascorbic acid (Asc) and potassium bicarbonate (K), the K50 and ASC50 samples were also prepared, as reported in the materials and methods section, in order to observe the action of the two substances separately.

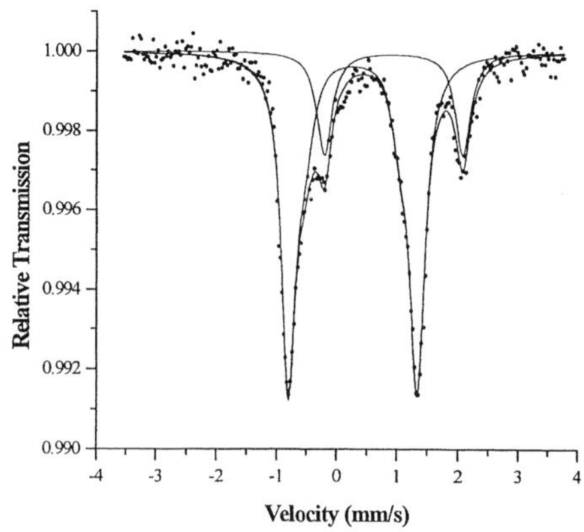

Figure1. Untreated erythrocytes (Control): oxyHb and deoxyHb subspectra are indicated.

Mössbauer spectra revealed that only oxy-Hb and deoxy-Hb, as expected, are present in the spectrum of the control sample (C - figure 1). The spectra of the treated samples (figure 2) show the disappearance of the Deoxy-Hb subspectra and the formation of a new component characteristic of oxidation products like methaemoglobin and haemichromes.

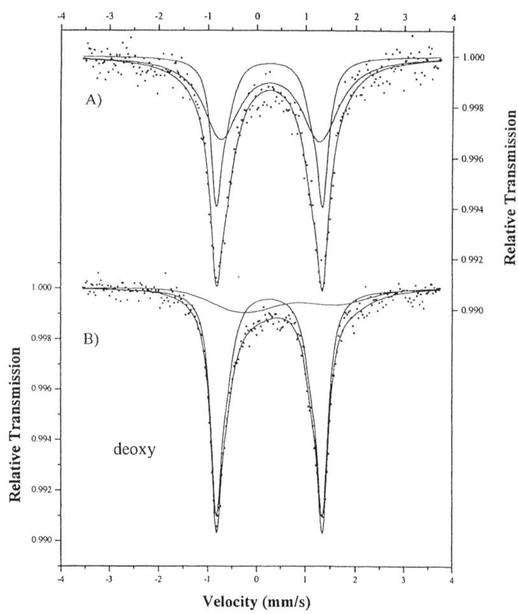

Figure2. Red cells incubated with APH for 50' (A); red cells incubated with APH and potassium ascorbate for 50' (B).

The increase of this latter oxidation component corresponds to a decrease of the oxy-Hb component (APH50, figure 2A). All the Mössbauer spectra were fitted with two subspectra obtained from the characteristic parameters of oxy-Hb and deoxy-Hb (figure 1) and from oxy-Hb and the envelope of all the oxidation products (figure 2) [5].

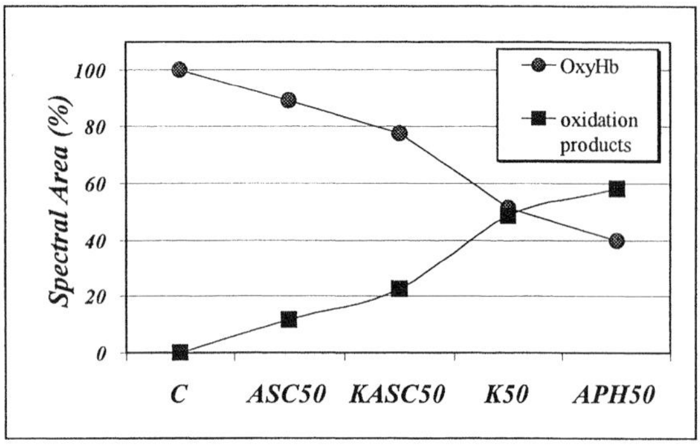

Figure3. Oxidation products in the erythrocyte samples after 50 minutes incubation with APH alone (APH50) and together with ascorbic acid (ASC50), potassium ascorbate (KASC50), potassium bicarbonate (K50).

As it can be noted from figure 2B, that represent the spectrum of KASC50, the oxidation pattern is almost completely absent when K-Asc is mixed with the APH added blood samples.

This fact indicates a strong antioxidant behaviours of this compound.

it can be calculated that the relative amount of oxidation components is about 60% after 50 minutes of incubation with APH, while in the KASC50 sample the action of potassium ascorbate reduces the oxidation to about 20%. The percent area of the oxidation product of all the samples (figure 3) confirms once more the antioxidant behaviour of ascorbic acid (Vitamin C). Our results show that potassium ascorbate too has antioxidant behaviour, together with the ability to carry potassium inside the red cells. As reported in [6], in fact, ascorbic acid bound to K+ can be transported across the membrane maintaining the physiological potassium concentration.

References

[1] M.R.Clemens, Klin. Wochenschr 69 (1991): 1123-1134;

[2] P. Jarolim, M.Lahav, S.C. Liu, J. Palek, Blood 76 (1990) : 2125-2131

[3] B. Halliwell, Nutr. Rev. 52 (1994) 253-265;

[4] F.M. Hughes Jr, C.D. Bortner, G.D. Purdy, J.A. Cidlowski, J. Biol. Chem. 272 (1997) 30567-30576

[5] S.Croci, I.Ortalli, G.Pedrazzi, G.Passeri and P.Piccolo, Hyperfine Interaction 126 (2000): 47-52

[6] W. Lohman, Biophys-Struct-Mech. 184; 10(4):205-10

Dynamic phenomena in deoxy- and oxymyoglobin investigated by nuclear resonant forward scattering of synchrotron radiation

C. Herta, H. Winkler, R. Benda, A.X. Trautwein and M. Haas[1]

Institut für Physik, Medizinische Universität zu Lübeck, D-23538 Lübeck, Germany
[1]*Institute of Physics, University of Tartu, EE-2400 Tartu, Estonia*

Myoglobin is probably the best studied iron protein because of its simple globular structure and its vital importance as oxygen storage protein in muscles. Nevertheless, there are still a series of questions open, e.g., about its phonon density of states governing the paramagnetic relaxation of the iron center and details of the iron-oxygen bond. Nuclear resonant forward scattering (NFS) of synchrotron radiation is expected to provide additional information on these topics complementary to conventional Mössbauer spectroscopy.

1. Introduction

Myoglobin (Mb) is still an object of intense studies because of its important physiological role as an oxygen storage protein. Information about the phonon density of states (DOS) can be deduced from the temperature dependence of the spin-lattice relaxation rates in deoxy Mb where the iron is in the ferrous high-spin state. Application of a magnetic field produces spin expectation values which are different for the various spin sublevels and which interact with the nuclear spin via the magnetic hyperfine coupling tensor. At temperatures above about 10 K phonon-driven transitions between these sublevels take place, which manifest themselves in the time-delayed nuclear resonant forward scattering (NFS) of synchrotron radiation.

The purpose of oxygen storage can only be fulfilled if the oxygen molecule is bound reversibly to the heme iron. X-ray and conventional Mössbauer studies of FeO_2 "picked-fence" porphyrins [1] provide strong evidence that some dynamic structural disorder of the FeO_2 moiety exists. Such a disorder would explain the temperature dependence of the quadrupole splitting observed also in oxy Mb. Further proof should come from a faster decay of the intensity of the time-delayed NFS at higher temperatures, because stochastic fluctuations of hyperfine parameters destroy the coherence within the nuclear exciton [2].

2. Materials and Methods

The samples of Mb were prepared as described in [3]. The experiments were performed at the beamline BW4 of the Hamburg synchrotron laboratory (HASYLAB). The measurements were carried out in the "2-" and "5-bunch" mode with time spacing between two bunches of about 500 ns and 200 ns, respectively. The simulations of the spectra were performed with the program package SYNFOS which allows the simulation of NFS spectra including powder averaging as well as paramagnetic relaxation and fluctuation of the efg [4,5].

3. Results and Discussion

3.1 *Deoxymyoglobin*

The temperature-dependent NFS spectra of the deoxy Mb sample in applied fields (Fig. 1) are fitted under the assumption that spin-lattice relaxation is present with transition probabilities, which for $\Delta E_{ij} = E_j - E_i$ (with $E_i < E_j$) are taken as [6]

$$w_{i\to j} = w_0 \cdot \frac{|\Delta E_{ij}|^{d_s}}{\exp\{\Delta E_{ij}/(k_B T) - 1\}} \qquad w_{j\to i} = w_{i\to j} \cdot \exp\{\Delta E_{ij}/(k_B T)\}.$$

The results of the fits are shown in Fig. 1 by solid lines. The spectroscopic dimension of Mb has been determined as $d_s = 2.26$, which is appreciably lower than the value of 3 expected for bulk solids.

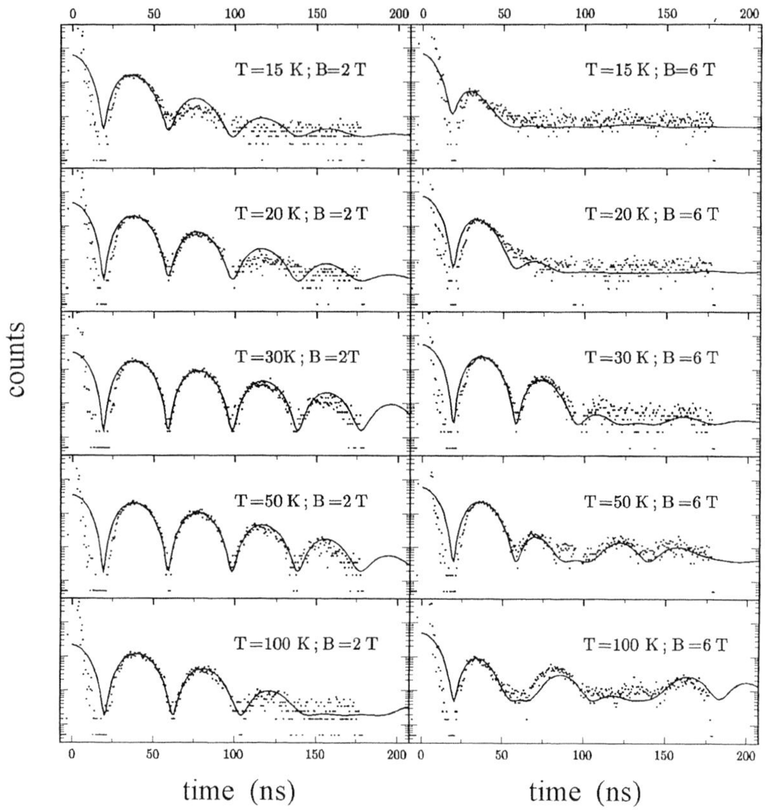

Figure 1. Measured NFS spectra of deoxy Mb in applied fields and at temperatures as indicated.

3.2 *Oxymyoglobin*

The NFS spectra of the MbO_2 sample taken in the temperature range 3.2 - 200 K are reproduced in Fig. 2. Analysis by simply allowing ΔE_Q and the effective thickness t_{eff} to be fitted yields an unphysical increase of the effective thickness with increasing temperature.

Studies of the FeO_2 moiety in picket-fence porphyrin have already suggested that relaxation due to dynamic structural disorder of O_2 takes place. This structural disorder was also detected by X-ray diffraction, which shows two or three positions of the terminal O-atom depending on the kind of picked-fence used. Different positions of the terminal O atom with respect to the heme iron cause different electric field gradients (efgs). The observed temperature-dependent quadrupole splitting is in this way easily explained as caused by fluctuations of the efg tensor [1]. The X-ray structure of MbO_2 also exhibits a structural disorder of FeO_2 with two different positions of the terminal O-atom [7]. Conventional Mössbauer studies on single crystals of MbO_2 have shown that the principal component V_{zz} of the efg tensor is oriented along the projection of the O-O bond onto heme plane [8]. If the terminal O-atom is located in the position 2 the respective V_{zz} should be of the same magnitude as in position 1, but the efg tensor is rotated accordingly. Therefore the physical picture which we have to envisage is that the efg fluctuates between two orientations with a rate that depends on temperature.

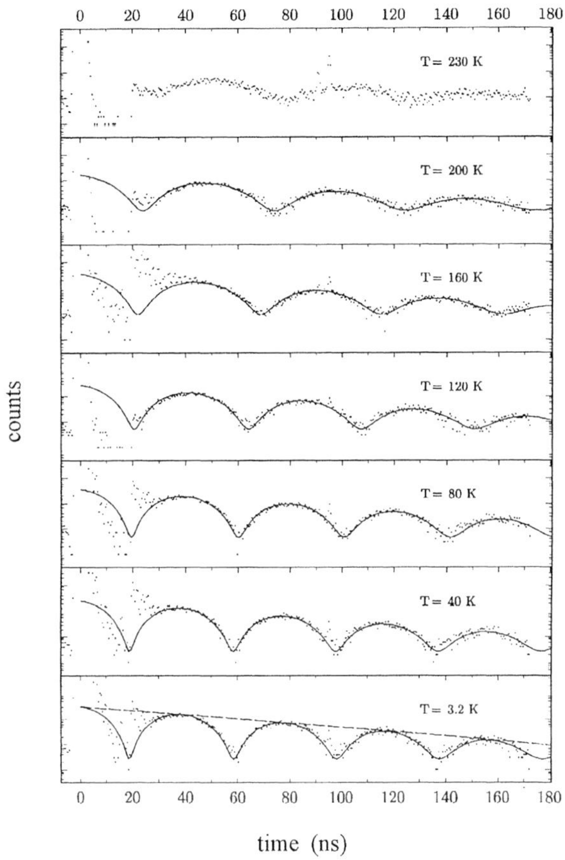

Figure 2. NFS of MbO_2 taken at the temperatures as indicated. The solid lines are fits as described in the text. The dashed line demonstrates the decay at 3.2 K due to the life time and effective thickness alone.

Assuming that the asymmetry parameter η of the efg in both positions is zero as found in the case of the model compounds only the rotation angle β between the two efg tensors has to be considered as structural parameter. The ratio of the occupation probabilities p_1 and p_2 of the two positions of the terminal O-atom is given by the Boltzmann factor with ΔE being the energy difference between these positions:

$$p_2 : p_2 = e^{-E_A / k_B T} .$$

In the final approach we have fitted all NFS spectra of the MbO_2 sample in the temperature range 3.2 - 200 K simultaneously by using the Debye model with Θ_D = 215 K for the temperature dependence of the effective thickness t_{eff}. For a chosen rotation angle β in the range 30 - 55° the energy difference ΔE and the temperature-dependent relaxation rate $w(T)$ were taken as free parameters in the fit. The lowest value of χ^2 is achieved around $\beta = 40°$ corresponding to $\Delta E = 111\ \mathrm{cm}^{-1}$. Calculated curves, which correspond to the best global fit to all NFS spectra of MbO_2, are entered as solid lines in Fig. 2. The resulting relaxation rates could be fitted by the expression

$$w(T) = w_\mathrm{T} + w_0 \cdot \mathrm{e}^{-E_\mathrm{A}/k_\mathrm{B}T}$$

with $w_\mathrm{T} = 10^7\ \mathrm{s}^{-1}$, $w_0 = 10^9\ \mathrm{s}^{-1}$ and $E_\mathrm{A} = (260 \pm 35)\ \mathrm{cm}^{-1}$, which indicates that thermal activation plus a low-temperature tunneling process are responsible for the dynamics of the O_2 ligand.

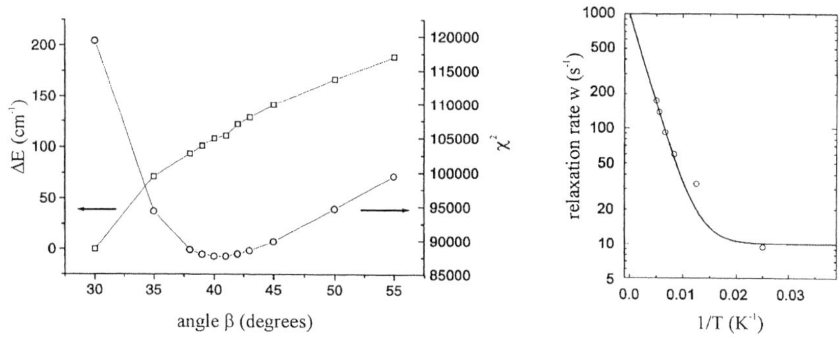

Figure 3 left. Sum of squares χ^2 and resulting energy difference ΔE between position 1 and 2 of the terminal O-atom in dependence of β.
Figure 3 right. Arrhenius plot of the relaxation rates $w(T)$.

4. Conclusion

We have successfully applied NFS using synchrotron radiation to determine the spectroscopic dimension of deoxy Mb and to demonstrate that the FeO_2 moiety in MbO_2 undergoes thermally activated structural disorder, i.e. the terminal O-atom in FeO_2 jumps stochastically between two possible positions which causes relaxation of the NFS.

Acknowledgements

This work was supported by the Federal Ministry of Education and Research (BMBF) under grant no. 05 SK8FLA8.

References

[1] M. Schappacher, L. Ricard, J. Fischer, R. Weiss, E. Bill, R. Montiel-Montoya, H. Winkler and A.X. Trautwein, Eur. J. Biochem. 168(1987)419-429
[2] G.V. Smirnov and V.G. Kohn, Phys. Rev. B 52(1995)3356-3365
[3] C. Keppler, K. Achterhold, A. Ostermann, U. van Bürck, A.I. Chumakov, R. Rüffer, W. Sturhahn, E.E. Alp and F.G. Parak, Eur. Biophys. J. 29(2000)146-152
[4] M. Haas, E. Realo, H. Winkler, W. Meyer-Klaucke, A.X. Trautwein, O. Leupold and H.D. Rüter, Phys. Rev. B 56(1997)14082-14088
[5] M. Haas, E. Realo, H. Winker, M. Meyer-Klaucke, A.X. Trautwein and O. Leupold, Phys. Rev. B 61(2000)4155-4159
[6] H. Winkler, C. Schulz and P.G. Debrunner, Phys. Letters 69A(1979)360-363
[7] S. E. V. Phillips, J. Mol. Biol. 142(1980)531-554
[8] Y. Maeda, T. Harami, Y. Morita, A.X. Trautwein and U. Gonser, J. Phys. Colloq. C2(1979)500-5001

Characterisation of Cytochrome c-552 from *Nitrosomonas europaea* with EPR and Mössbauer spectroscopy

T. Teschner, R. Benda, V. Schünemann, A. X. Trautwein, D. Arciero[1], A. Hooper[1], and K. K. Andersson[2]

Institute of Physics, Med. University, 23538 Lübeck, Germany
[1]Department of Biochemistry and Mol. Biol., University of Minnesota St.Paul, Minnesota 55108, USA
[2]Department of Biochemistry, University of Oslo, N-0316 Oslo, Norway

Abstract

The mono-heme protein cytochrome c-552 from *Nitrosomonas europaea* has been studied by EPR as well as low temperature Mössbauer spectroscopy in fields up to 7 T. The obtained data have been analyzed within the model of Oosterhuis and Lang in the reference frame of Taylor for ferric low spin hemes. Our analysis yields $\delta = 0.26$ mm/s, $\Delta E_Q = 2.06$ mm/s, $\eta = 0.9$, $\alpha = 90°$, $\mathbf{A}/g_N\mu_N = (-33, 28, 76)$ T and $\mathbf{g} = (1.31, 1.77, 3.34)$ consistent with an electronic $(d_{xy})^2\,(d_{xz})^2\,(d_{yz})^1$ ground state typical for Type I model hemes

Introduction

Cytochrome (cyt.) c-552 from *Nitrosomonas europaea* is a 9.1 kDa mono-heme protein and belongs to the bacterial cyt. c family. It is an electron-transfer protein, which is responsible for shuttling electrons from the mitochondrial protein-bc_1-complex to cyt. oxidase. The heme group in cyt. c-552 is axially coordinated to a histidine and a methionine residues, and from previous EPR and MCD studies [1] it is known that the heme iron is in the ferric low- spin ($S = 1/2$) state. The information obtained from EPR, however, is limited since only two g-values are resolved and only the absolute values of the electronic g-tensor elements are known. Therefore we have undertaken a complementary Mössbauer study in order to determine the electronic g-tensor and thereby the electronic ground state of cyt. c-552. For this purpose Mössbauer data obtained at 4.2 K in various fields up to 7 T have been analyzed by the method of Oosterhuis and Lang [2] in the reference frame of Taylor [3].

Materials and Methods

Cyt. c-552 from *Nitrosomonas europaea* was isolated as previously described [1]. ^{57}Fe enrichment was achieved by growing *Nitrosomonas europaea* in a medium enriched with ^{57}Fe, with ^{56}Fe removed by Chelex. EPR spectra were recorded with a conventional X-Band spectrometer (Bruker ER 200 D SRC) equipped with a He-flow cryostat (ESR 900, Oxford Instruments). Mössbauer spectra were recorded using a conventional spectrometer in the constant- acceleration mode. Isomer shifts are given relative to α-Fe at room temperature. The spectra obtained at 8 and 20 mT were measured in a He bath cryostat (Oxford MD 306) equipped with a pair of permanent magnets. For the high-field spectra a cryostat equipped with a superconducting magnet was used (Oxford Instruments). Magnetically split spectra of paramagnetic samples were simulated in the spin-Hamiltonian approximation [4] described below, otherwise spectra were analyzed by least-square fits using Lorentzian line shape.

Results and Discussion

Fig. 1 shows a Mössbauer spectrum of cyt. c-552 obtained at T = 210 K. The main component (96 % relative area) is a doublet with an isomer shift δ = 0.23 mm/s and a quadrupole splitting of ΔE_Q = 2.03mm/s. These values are identical to those reported for the structural related cyt. c-551 from *Pseudomonas aeruginosa* [5]

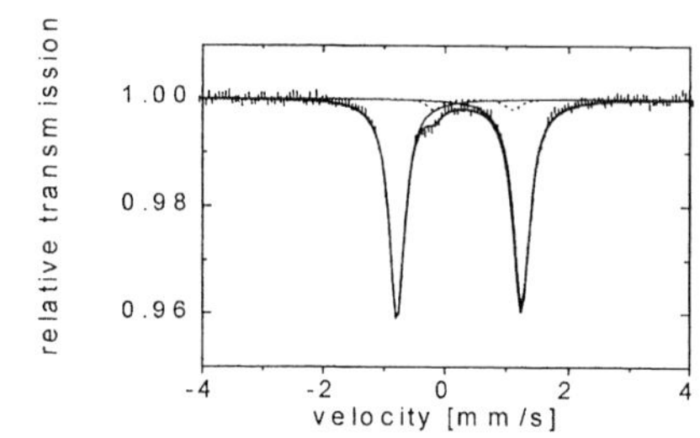

Figure 1: Mössbauer spectrum of cyt. c-552 obtained at T = 210 K. The solid line is a Lorentzian fit with an asymmetric doublet yielding δ = 0.23 mm/s and ΔE_Q = 2.03 mm/s (Γ_{left} = 0.23 mm /s, Γ_{right} = 0.32 mm/s). A minor doublet (dashed line, 8.6 % relative area) has δ = 0.46 mm/s, ΔE_Q = 1.29 mm/s and Γ = 0.30 mm/s.

The EPR spectrum of cyt. c-552 shows a highly axial low spin (HALS) type signal (Fig. 2). Such resonances are characteristic for so called Type I -heme centers with an Fe $(d_{xy})^2 (d_{xz})^2 (d_{yz})^1$ electronic configuration and a ferric low spin (S = 1/2) state [6]. Type I centers exhibit strong g-tensor anisotropy and often only the g_z-value with the z-axis being perpendicular to the heme plane is seen in EPR. Here we observe g_z = 3.34 (see arrow in Fig. 2). Unfortunately the region around g ~ 2 is dominated by a Cu(II) impurity (app. 10 % of the total spin intensity). From a previous EPR study g_y has been determined to be 1.77, but g_x has not been observed.

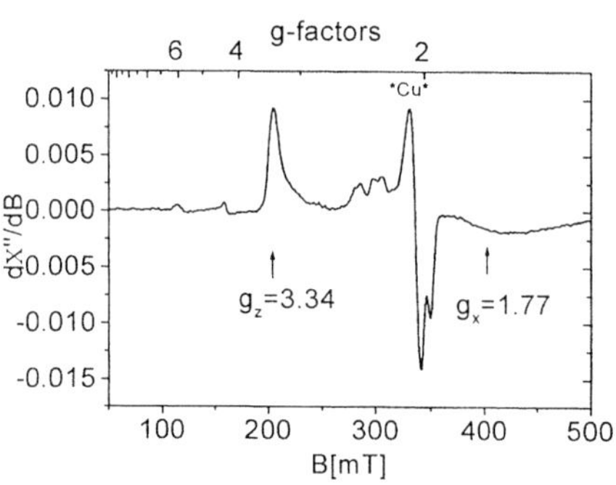

Figure 2: EPR-Spectrum of ^{57}Fe enriched cyt. c-552. The arrow indicates g_z = 3.34. The signal around g ~ 2 is due to a Cu(II) impurity (10 % relative spin-density). Experimental parameters: T = 10 K; ν = 9.6422 GHz; microwave power: 80 μW; modulation frequency: 100 kHz ; modulation amplitude: 5 mT.

In order to determine the full g-tensor and thereby the electronic ground state of cyt. C-552 Mössbauer data obtained at 4.2 K in fields up to 7 T (Fig. 3 and 4) have been collected. Adopting the coordinate system of Taylor [2] the ground state S = 1/2 Kramers-doublet is written as a linear combination of the three t_{2g} orbitals, and the coefficients of the ground state wave function (a, b and c) can be calculated from the g-tensor elements. According to Oosterhuis and Lang [2] it is possible to calculate from the coefficients a, b and c spectral parameters like the A-tensor.In our analysis we made use of g_z = 3.34 and g_y = 1.77. In order to determine g_x we have additionally assumed $g_x^2 + g_y^2 + g_z^2 = 16$ [7,8] and have yielded g_x = 1.31.

With the full g-tensor we were able to calculate the hyper fine coupling tensor elements (Tab. 1) which were used as starting parameters for the simulations shown in Fig. 3 and 4. We want to emphasize that the parameter set given in Tab. 1 fits the field-dependent low-temperature Mössbauer data over such a broad range of conditions. It is clearly seen from table 1 that g_{zz} and A_{zz} are the dominating fine and hyperfine tensor quantities, hole being strongly influenced by large orbital contributions This causes the magnetic hyperfine splitting being dominated by the product $<S_z^{eff}>A_{zz}$ and thus confirms that cyt. c-552 belongs to the class of Type I heme centers [5,9].

In conclusion we note the relatively simple crystal-field model of Oosterhuis and Lang [2] successfully describes the magnetic hyperfine interactions in ferric-low spin heme centers and especially confirms that the heme center of cyt. c-552 from *Nitrosomonas europaea* is of Type I.

Table 1: Crystal field and hyperfine parameters of the cyt. c-552 from *Nitrosomonas europaea* obtained from the simulations shown in Fig.3 and 4.

	g	Δ/λ	V/λ	δ (mm/s)	ΔE_Q (mm/s)	η [a)]	$A/g_N\mu_N$ (T)	Γ (mm/s)
Experiment	(- ,1.77, 3.34)			0.26	2.06	-0.9 ± 0.6	(-35±14, 27±15, 76±2)	0.3
Theory	(1.31, 1.77, 3.34)	1.13	7.0			-2.2	(-33.9 ,3.6, 75.3)	

a) Here η is given in the molecular frame with the z – axis perpendicular to the heme – plane. However, the coordinate system of the efg tensor is commonly chosen in such a way that $|V_{zz}| \geq |V_{yy}| \geq |V_{xx}|$; in order to reach this convention in the present case the coordinate system of the efg tensor has to be turned around the z – axis of the molecular frame by the Euler angle $\alpha = 90°$, so we obtain for the experimental value η'=0.9.
Concerning the theoretical value the coordinate system of the efg tensor has to be turned around the y-axis of the molecular frame by the Euler angle $\beta = 90°$, and instead of $\eta = -2.2$ and and $\Delta E_Q = -2.06$ mm/s the values $\eta' = 0.25$ and $\Delta E_Q = -2.06$ mm/s.

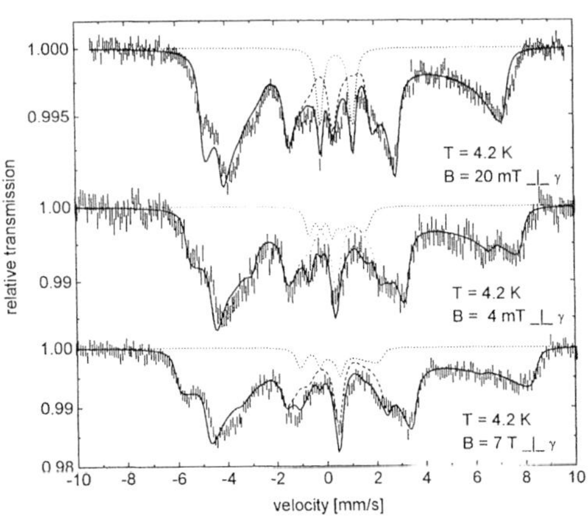

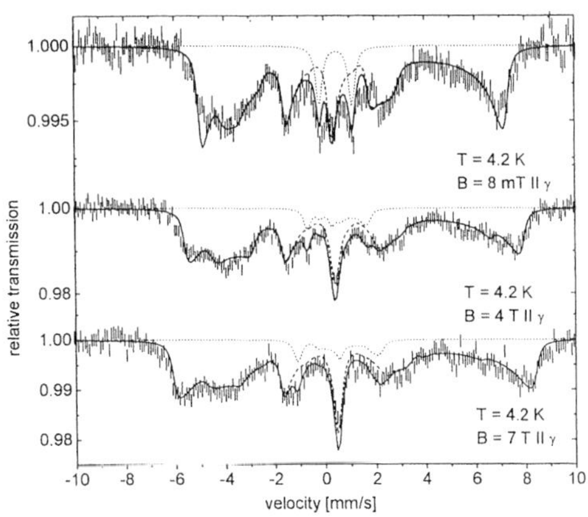

Fig. 3 (left): Field-dependant Mössbauer spectra of cyt. c-552 obtained at T = 4.2 K in fields perpendicular to the γ-beam. The solid line in each spectrum is a sum of two contributions: (1) the ferric low-spin heme center with parameters given in Tab. 1 (91.4 % relative contribution; dashed line) and (2) a diamagnetic component with $\delta = 0.48$ mm/s and $\Delta E_Q = 1.29$ mm/s which is attributed to reduced cytochrome c-552 with the heme iron in the low-spin ferrous (S = 0) state (8.6 % relative contribution; dotted line)

Fig. 4 (right) Field-dependant Mössbauer spectra of cyt. c-552 obtained at T = 4.2 K in fields parallel to the γ-beam. The solid line in each spectrum is again a sum of the two components as described in Fig. 3 with parameters given in Tab. 1.

References

[1] Arciero D. M., Peng Q., Peterson J., Hooper A. B., (1994) FEBS Letters **342**: 217-220

[2] Oosterhuis WT, Lang G (1969) Phys Rev **178** (2): 439-456

[3] Taylor CPS (1977) Biochim Biophys Acta **491**:137-149

[4] Walker FA (1999) Coord Chem Rev **185-186**: 471-534

[5] Schünemann V. and Winkler H., Rep Prog.Phys.**63** (2000) 263-353

[6] Timkovich R., Bergmann D., Arciero D. M., and Hooper A. B. (1998) Biophysical J. 75 1964-1972

[7] Debrunner P.G. (1993) in Biological Magnetic Resoannce, Volume **13**:EMR of Paramagnetic Molecules (Berliner, L. J., and Reuben, J. Eds.) pp 59 –101, Plenum Press, New York.

[8] Griffith J.S. (1956) Proc. R. Soc. London, Ser. **A 235**, 23-36

[9] Benda R., Schünemann V., Trautwein A.X., Cai S., Polam J.R., Watson C.T., Shokhireva T.Kh. and Walker F.A. to be published

How a blood sucking insect gets its meal: The ferriheme proteins Nitrophorin 2 and 4 studied by Mössbauer Spectroscopy

Patrick Wegner[a], Rüdiger Benda[a], Volker Schünemann[a], A. X. Trautwein[a], Robert E. Berry[b], Celia A. Balfour[b], David Wert[b], F. Ann Walker[b]

[a] *Institut für Physik, Med. Universität Lübeck, Ratzeburger Alle 160, 23538 Lübeck, Germany*

[b] *Department of Chemistry, University of Arizona, Tucson, Arizona 85721,* USA

E-Mail: wegner@physik.mu-luebeck.de

Abstract. Nitrophorins are ferriheme proteins contained in the salivary glands of the blood-sucking insect *Rhodnius prolixus*, also known as the kissing bug. The heme iron binds nitric oxide (NO). These proteins liberate NO after injection into the victim's tissues; NO can diffuse to the capillaries to dilate them and bring more blood to the site of the bite. The open binding-site at the iron is used by the protein to bind histamine, which has been produced by the victim as a response to the injury. Nitrophorin 2 and 4 (NP2, NP4) in the NO- and histamine-ligated form have been studied by low temperature Mössbauer spectroscopy in magnetic fields up to 7T. At 4.2 K the NO complex of NP2 exhibits $\delta = 0.00$ mm/s and $\Delta E_Q = 1.84$ mm/s. High-field measurements indicate that the Fe-NO adduct is diamagnetic, which is also the case for the Fe-NO adduct of NP4. The histamine-ligated form of NP2, on the other hand, shows a broad magnetic pattern at 4.2 K in applied fields from 10 mT up to 7T, characteristic for a ferric low-spin heme-iron, which is consistent with a recent EPR investigation performed on a Nitrophorin 1-histamine complex [2].

1. Introduction

The proteins Nitrophorin 2 and 4 (NP2, NP4) are found in the salivary glands of the Amazon river-based kissing bug *Rhodnius prolixus*. For obtaining a sufficient blood-meal, these proteins are injected into the victim by the insect prior to feeding. When the pH value suddenly rises from 5 in the salivary glands of the kissing bug to pH 7.5 in the tissues of the victim, the binding affinity is changed and thus NO, which is bound to the iron in the active heme-center of the protein, is released. The NO can migrate through the victim's tissue to the capillaries to dilate them to allow more blood to flow to the site of the bite.
Histamine, which is produced by the victims immune-system as a response to the bite, is bound at the open binding site of the iron. The release of NO enhances the blood flow towards the insect's mouth-part and the binding of the histamine prevents the insect being detected before it has finished its meal. The *Rhodnius* NP's have a molecular weight of 20 kDa. Crystallographic data show that the tertiary structure of the NP's exhibit a β-barrel with a histidine residue (His-59) that serves as the proximal ligand to the heme [3]. The NP's represent the first examples of proteins with stable Fe(III)-NO complexes, where the NO can be stored for a long period of time.

2. Material and methods

The NP's were prepared as described previously [4].
Mössbauer spectra were recorded using a conventional spectrometer in the constant-acceleration mode. Isomer shifts are given relative to α-Fe at room temperature. The spectra obtained at 10 and 20 mT (perpendicular to the γ-beam) were measured in a He-bath cryostat (Oxford MD 306), equipped with a pair of permanent magnets. For the high-field spectra (7 T) a cryostat equipped with a superconducting magnet was used (Oxford Instruments). The obtained spectra were analyzed by least square fits using Lorentzian line shape, while the magnetically split spectra of the paramagnetic and diamagnetic samples were simulated within the spin Hamiltonian formalism [5]

$$\hat{H} = \frac{eQV_{zz}}{4I(2I-1)}\left[3\hat{I}_z - I(I+1) + \eta(\hat{I}_x^2 - \hat{I}_y^2)\right] - g_N\mu_N \vec{I}\cdot\vec{B} + \langle\vec{S}\rangle\cdot\vec{A}\cdot\vec{I}.$$

I denotes the spin of the nuclear ground or excited state, Q the nuclear quadrupole moment of the nuclear excited state, V_{zz} the z-component of the electric-field gradient (efg), η the asymmetry-parameter of the efg and g_N the nuclear g factor.

3. Results and discussion

Figure 1a shows the Mössbauer spectra of the sample NP2-NO at pH 5 obtained at 4.2K. The main doublet (91±2% relative contribution) exhibits an isomer shift of $\delta = 0.00$ mm/s, the value of which is, within experimental error, identical to that recently found for the diamagnetic Fe(III)-NO heme model complex ($[OEP^{57}Fe(NO)(N\text{-}MeIm)]^+Cl^-$ ($\delta = 0.02$ mm/s) [1]. The quadrupole splitting of $\Delta E_Q = 1.84$ mm/s is slightly larger as that reported for the model complex ($\Delta E_Q = 1.64$ mm/s). Figure 1b shows the Mössbauer

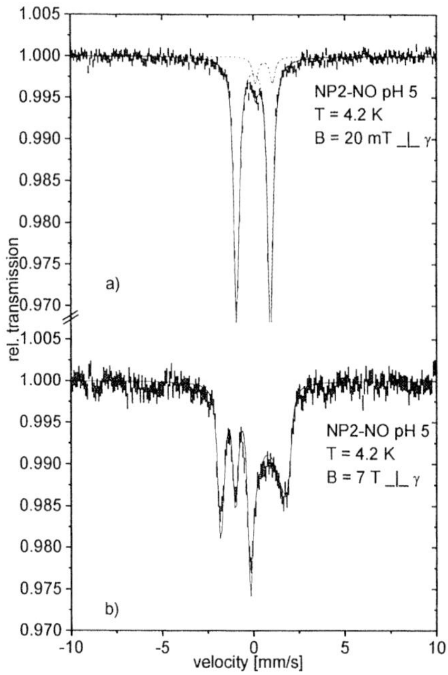

Figure 1: Mössbauer spectra of NP2-NO at pH 5 taken at 4.2 K in an applied field of 20 mT (a) and of 7 T (b) perpendicular to the γ-beam. The solid line in (a) represents a Lorentzan fit with two components: (i) $\delta = 0.00$ mm/s, $\Delta E_Q = 1.84$ mm/s, $\Gamma = 0.34$ mm/s (ii) $\delta = 0.58$ mm/s, $\Delta E_Q = 0.92$ mm/s, $\Gamma = 0.33$ mm/s. The solid line in (b) is a spin Hamilton simulation for component (i) with parameters given in Table 1. Component (ii) splits in the high-field measurements to a six-line pattern and therefore vanishes in the statistical background of the spectrum.

spectrum of NP2-NO at pH 5 obtained in an external field of 7T. The observed magnetic splitting is only due to the external field, which confirms the diamagnetic state of the iron-NO complex. The Mössbauer study of NP4-NO at pH 5 has yielded $\delta = 0.03$ mm/s and $\Delta E_Q = 1.64$ mm/s and a diamagnetic iron-NO complex as well. In both isoforms there is also a minority species (9±2% relative contribution in NP2-NO at pH 5 and 20±2% relative contribution in NP4-NO at pH 7.5) that exhibits $\delta = 0.58$ mm/s and $\Delta E_Q = 0.92$ mm/s for NP2-NO and $\delta = 0.45$ mm/s and $\Delta E_Q = 0.75$ mm/s for NP4-NO.

These parameters, obtained in the low-field measurements, strongly suggest a Fe(III) high-spin component, which splits magnetically into a six-line pattern and therefore vanishes in the statistical background due to the low signal to noise ratio.

When the protein is injected into the victim the pH-value is changed from 5 to 7.5. A Mössbauer study of NP2-NO at pH 7.5 yields identical Mössbauer parameters as found for NP2-NO at pH 5 (not shown here).

Changing the pH value of the environment of the protein to pH 7.5, the binding affinity towards NO is changed as well: The NO is released more easily and can migrate into the victim's tissues. In turn, the open binding site is occupied by a histamine molecule produced by the victim in response to the bite of the bug. The Mössbauer spectrum of NP2 at pH 7.5, with histamine bound to the iron (NP2-His), obtained at 150 K is shown in Fig. 2a. The main component is an asymmetric doublet (84±2% relative contribution) and exhibits $\delta = 0.22$ mm/s and $\Delta E_Q = 2.14$ mm/s. Since these parameters are in the range of parameters observed for ferric low-spin (S=1/2) heme centers [5], and since it is expected that binding of a histamine to the open iron binding site would produce a 6-coordinated ferric low-spin heme iron, we attribute the main doublet of Figure 2a to the iron-histamine complex.

As in the case for the NP2-NO, there is a second doublet (16±2% relative contribution) which has $\delta = 0.48$ mm/s and $\Delta E_Q = 0.56$ mm/s characteristic for a ferric high-spin iron.

The presence of a major ferric low-spin and a minor ferric high-spin species has been confirmed by an accompanying EPR study of NP-2-His (data not shown) which shows a rhombic signal with $\ddot{g}$ = (1.35, 2.25, 3.02), characteristic for ferric low-spin heme iron (87±4 % relative contributions), and an axial signal with $\ddot{g}$ = (6.0, 6.0, 2.0) characteristic for the ferric high-spin iron center (13±4% relative contribution), in accordance with the results obtained from Mössbauer spectroscopy.

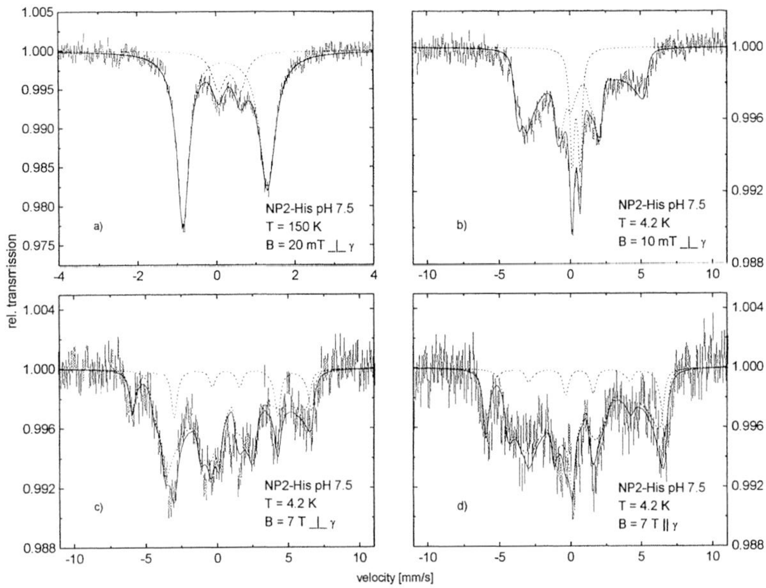

Figure 2: Mössbauer spectra of NP2-His at pH 7.5 obtained at conditions indicated in the plots. (a) The dashed line (i) is a Lorentzian fit assuming an asymmetric doublet with $\delta = 0.22$ mm/s, $\Delta E_Q = 2.14$ mm/s, and linewidths $\Gamma_1 = 0.38$ mm/s, $\Gamma_2 = 0.49$ mm/s but identical relative spectral area (84± 4% relative contribution). The dot-dashed line (ii) exhibits $\delta = 0.48$ mm/s, $\Delta E_Q = 0.56$ mm/s, $\Gamma = 0.45$ mm/s (16±4% relative contribution). (b-d) The dashed line is a spin Hamiltonian simulation for the iron-histamine-complex with the parameters given in Table 1. Component (ii) has been assumed to be a fast relaxing ferric high-spin species in (b), which in (c,d) splits magnetically in high fields to a six line pattern, which has been simulated with $\delta =$ 0.48 mm/s, $\Delta E_Q = 0.56$ mm/s, $\Gamma = 0.45$ mm/s $A_{xx}/g_N\mu_N = A_{yy}/g_N\mu_N = A_{zz}/g_N\mu_N = -20.7$ T, $g_x = g_y = g_z = 2$, E/D = 0, D = 8 cm^{-1}.

NP2-His at pH 7.5 shows a broad paramagnetic spectrum at 4.2K in a field of 20mT (Fig. 2b). The contribution of the ferric low-spin iron-histamine complex to the field-dependant Mössbauer spectra shown in Fig. 2b-d has been successfully simulated with the parameter set given in Table 1. Also listed are theoretical values of the asymmetry parameter η and the magnetic hyperfine coupling tensor $\overleftrightarrow{A}$ obtained by the model of Oosterhuis and Lang [6] in the reference frame of Taylor [7]. Given the simple crystal-field approach, the agreement of the experimentally determined value $\overleftrightarrow{A}_{exp}$ = (-37.9, -5.8, 53.9) and the calculated one $\overleftrightarrow{A}_{theo}$ = (-40.5, 16.7, 50.0) is reasonable, and we attribute the deviations to covalency effects not included in the model. The minor ferric high-spin component shows an unusual behavior: It relaxes rapidly within the Mössbauer time window of about 10^{-7} sec at 4.2 K in a field of 20 mT, which leads to the appearance of a doublet in Fig. 2b, and it relaxes slowly in large applied fields of 7 T (Fig. 2c,d). Nevertheless its Mössbauer signature under high-field conditions could be successfully simulated with spin Hamiltonian parameters characteristic for ferric high-spin heme centers (Fig. 2c,d) [5] in the slow relaxation limit. We tentatively assign it to a 5-coordinated high-spin heme iron(III) center of NP2 that, at pH 7.5, has released its NO-ligand.

In conclusion, our Mössbauer study of the NP-NO samples reveals a diamagnetic iron-NO complex, consistent with the electron configurations of either a ferrous low-spin Fe(II)-NO^+- or an antiferromagnetically coupled ferric low-spin Fe(III)-NO-complex, similar to that recently found for the diamagnetic Fe(III)-NO heme model complex ($[OEP^{57}Fe(NO)(N\text{-}MeIm)]^+Cl^-$.

Table 1: Mössbauer parameters of the main component shown in Figs. 1 – 2. Calculated values according to [6,7] are given in italics.

Sample	spin	δ [mm/s]	ΔE_Q [mm/s]	η [a)]	$\overleftrightarrow{A}/g_N\mu_N$ [T]	Γ [mm/s]
NP2-NO pH 5	0	0.00	1.84			0.3
NP4-NO pH 5	0	0.03	1.64			0.34
NP2-NO pH 7.5	0	0.00	1.83			0.35
NP2-His pH 7.5	1/2	0.22	2.14	-1.6 *-2.6*	(-37.9, -5.8, 53.9) *(-40.5, 16.7, 50)*	0.35

[a)] Here η is given in the molecular frame with the z-axis perpendicular to the heme-plane. However, the coordinate system of the efg tensor is commonly chosen in such a way that $|V_{zz}| \geq |V_{yy}| \geq |V_{xx}|$; in order to reach this convention in the present case the coordinate system of the efg tensor has to be turned around the y – axis of the molecular frame by the Euler angle $\beta = 90°$ and around the new z′- axis by $\gamma = 90°$. In this new coordinate system $\eta' = 0.1$ and $\Delta E_Q' = -2.14$ mm/s.

[1] Schünemann, V., Benda, R. Trauwein, A. X., Walker, F. A., (2000) *Israel J. Chem.*, **40**, 9.
[2] Astashkin, V. A., Raitsimring, A.M., Walker, F. A., (1999) *Chem. Phys. Lett.* , **306**, 9.
[3] Andersen, J. F., Montfort, W. R., (2000) *J. Biol. Chem.*, **275**, 30496.
[4] Andersen, J. F., Weichsel, A., Balfour, C. A., Champagne, D. E., Montfort, W. R., (1998) *Structure*, **6**, 1315.
[5] Debrunner, P. G., In *Iron Porphyrins*, Part III; Lever, A.B.P.; Gray, H.B., Eds.; VCH Publishing: Weinheim, 1989; pp 139 – 234.
[6] Oosterhuis, W. T., Lang, G., (1969) *J. Chem. Phys.*, **50**, 4381 – 4387.
[7] Taylor, C. P. S., (1977) *Biochim. Biophys. Acta*, **491**, 137.

A STUDY ON RED BLOOD CELLS OF THE PATIENTS WITH CYANOTIC CONGENITAL HEART DISEASE

X.F. Zhang[1 *], *H.P. Gong*[1], *Y.D. Gong*[1], *N.M. Zhao*[1], *Y.W. Liu*[2] *and L. Han*[2]
[1]Department of Biological Sciences and Biotechnology, State Key Laboratory of Biomembrane and Membrane Biotechnology, Tsinghua University, Beijing 100084, China;
[2]Anzhen Hospital, Beijing 100029, China

Abstract The red blood cells (RBCs) of the patients with Cyanotic Congenital Heart Disease (CCHD) have been studied by using the Mossbauer Spectroscopy (MS), Circular Dichroism (CD) and Fluorescence Depolarization (FD) measurements, comparing with normal person's. The results may interpret the clinical behavior of the patients.

1. Introduction

Cyanotic Congenital Heart Disease (CCHD) is a general term that includes many kinds of congenital heart diseases. The common characteristics of CCHD patients are low saturation degree of oxygen in blood, elevation of the number of RBCs and high viscosity of blood. On the other hand, the RBCs of the patients are different from those of normal persons [1-3]. We studied hemoglobin (Hb) and blood ghosts of the patients and normal persons by using Circular Dichroism, Mossbauer Spectroscopy [4], and Fluorescence Depolarization [5,6], respectively. All the results confirmed the defect of the patients' RBCs.

2. Experiments and Results

The blood samples were collected from eight patients (6 males and 2 females) in Anzhen Hospital of Beijing. Among the eight patients, four were Tetralogy of Fallot, two Monoventricle and two Double Right Ventricular Outflow. The clinical data on the patients are listed in Table 1.

Table 1. The clinical data on blood samples from the patients with CCHD

No.	1	2	3	4	5	6	7	8
Age	15	8	15	13	4	21	3	11
Sex a)	M	F	M	M	M	M	M	F
RBCs(10^{12} /L)b)	6.67	7.86	7.17	7.37	7.59	8.17	8.99	7.40
Hb(g/L)c)	188	227	214	232	200	247	188	222
Hct(L/L)d)	0.574	0.684	0.624	0.685	0.600	0.752	0.602	0.651

a) M- male, F-female. b) RBCs - concentration of RBCs in blood. c) Hb - concentration of Hb in blood. d) Hct –hematocrit, is the ratio: the volume of RBCs / the volume of blood.

* To whom correspondence will be addressed

Other data of the patients were as follows: age average = 11.25; RBCs concentration > 6.6×10^{12} L^{-1}; Hb concentration>180g/L; Hct>0.57L/L. The blood samples of control group were from normal persons. 1ml anticoagulant and physiological salt solution (PSS) was added into every sample, respectively. RBCs were separated from the blood samples by centrifugation (3000 rpm, 5 minutes) and washed 3 times with PSS to remove leukocytes and platelets. These RBCs were respectively prepared as samples for MS, CD and FD measurements later.

(1) Fluorescence Depolarization experiments.

Washed RBCs were added with 5 mM hypotonic Tris-HCl buffer (volume of RBCs / volume of buffer = 1/50), then were kept at 4°C for 2 hours to rupture the membrane. The ghosts obtained by centrifugation (9000 rpm, 15 minutes) were washed with 5mM Tris-HCl buffer for 2 times to remove Hb. After centrifugation (15000 rpm, 15 minutes), the purified ghosts were washed with 8 mM PBS (isotonic) buffer. Then the solution was centrifuged (15000 rpm, 15 minutes) to remove the supernatant and the ghost was suspended with PBS(phosphate buffer solution)(2-4ml). The concentration of membrane protein was determined by Bradford method. Then suspension was diluted with PBS till the concentration of membrane proteins reached about 400 μg/ml. 2ml of this suspension was added with 2 ml DPH (diphenyl-hexatrienyl, 2μM), kept at 30°C for 30 minutes, washed with PBS and then centrifuged (15000 rpm, 15 minutes) to remove to unbound DPH in supernatant. The ghosts bound with DPH were suspended with 4 ml PBS, and then were used to measure fluorescence depolarization P on HITACHI-850 Fluorescence Spectrophotometer. The P value was listed in Table 2 and Table 3.

Table 2. The results obtained from FD and CD experiment on the patients' ghost and Hb.

No	1	2	3	4	5	6	7	8
$P^{a)}$	0.338	0.329	0.348	0.344	0.347	0.322	0.331	0.332
F_h (%)$^{b)}$	63.5	67.5	65.0	64.0	67.5	65.5	65.5	62.0
F_b (%)$^{c)}$	0	0	0	0	0	0	0	0
Ft (%)$^{d)}$	0	0	0	0	0	0	0	0
F_r (%)$^{e)}$	36.5	32.5	35.0	36.0	32.5	35.0	35.0	38.0

a) P- polarization degree. b) F_h- percentage of α-helix. c) F_b- percentage of β strand. d) F_t - percentage of turn. e) F_r - percentage of random coil

Table 3. The calculated results of FD and CD experiments on normal persons and the patients [a)]

	n	P	F_h(%)	F_b(%)	F_t(%)	F_r(%)
patients group	8	0.336±0.010	65±1	0	0	35±2
control group	7	0.350±0.010	56±1	0	0	44±1

a) n - number of persons in the group. The P and F_h values of the patients group are significantly different with control group ($P < 0.05$).

(2) Mossbauer experiments.

The washed RBCs were encapsulated in perspex containers (1.5cm Diameter, 0.7cm Thickness) and then were stored in liquid nitrogen until measurements. Samples were maintained at 80K during measurements. Two samples (No.1 and No.5) weren't measured for some reasons. The Mossbauer parameters were obtained by least-square fitting using Lorentzian lines. The "a" components corresponded to the contribution of oxy-hemoglobin, and the "b" components to deoxy-hemoglobin'. The data were listed in table 4.

Table 4. The Mossbauer parameters of the patients' RBCs at 80K

No.	component	I.S.(mm/s)[a)]	Q.S.(mm/s)[b)]	Γ(mm/s)[c)]	I(%)[d)]
2	a	0.267±0.011	2.016±0.023	0.371±0.013	92.35
	c	0.196±0.046	0.502±0.051	0.246±0.073	7.65
3	a	0.257±0.012	2.014±0.023	0.402±0.016	90.27
	b	0.917±0.058	2.131±0.068	0.322±0.110	9.73
4	a	0.257±0.014	1.925±0.023	0.438±0.021	71.13
	b	0.898±0.018	2.145±0.026	0.273±0.025	28.87
6	a	0.260±0.012	2.074±0.023	0.362±0.016	61.31
	b	0.922±0.016	2.284±0.026	0.324±0.021	38.69
7	a	0.266±0.011	2.051±0.023	0.401±0.014	83.23
	b	0.935±0.022	2.227±0.031	0.252±0.034	16.77
8	a	0.262±0.011	2.055±0.022	0.357±0.013	72.22
	b	0.921±0.020	2.258±0.029	0.333±0.029	27.78

a) I.S.-isomer shifts relative to α-Fe at room temperature. b) Q.S.- quadrupole splitting. c) Γ- line width. d) I- percentage of the component.

(3) Circular Dichroism experiments.

Following MS measurement, the RBCs were regenerated at room temperature and were added with water to rupture cell membrane to release Hb. Hb were obtained after centrifugation (13000 rpm; 15 minutes) to remove the ghosts. All the purified Hb were diluted with water till A_{415}= 0.9 and measured on J-500C Spectropolarimeter. The obtained CD spectra were analyzed by computer program according to the method of Chen et al [7] to calculate the second structures of Hb. The data were listed in Table 2 and Table 3.

3. Discussion

The MS experiment results showed that the spectra of all six patients except one and all normal persons were composed of two components which were quadrupole splitting peaks marked as "a" and "b" respectively. The "a" component is corresponding to low spin Fe^{2+}(S=0), that is, it is the contribution of Fe in oxyhemoglobin. While "b" component is generated by high spin Fe^{2+}(S=2). It is the contribution from deoxyhemoglobin [4]. Because blood sample was exposed to air when washed, the ratio of oxy-Hb to deoxy-Hb reflects the optimum capacity of Hb to carry oxygen. This ratio of normal person is very high (average value: 13.6±2.1), suggesting the strong ability of RBCs to carry oxygen in normal persons. On the contrary, the ratio of components "a" to "b" is lower than 9.3 for the patients, which showed that the patients' RBCs could not provide human body with the necessary oxygen. The results corresponded to lower saturation degree of oxygen in blood of the patients. Elevation of the number of RBCs and Hb were an indemnity for lower saturation of oxygen in the blood of the patients. The high viscosity of blood of the patients came from high numbers of RBCs in the blood. The No.2 patient has different components from other patients in that there was anomalous "c" component instead of "b" component. The abnormal "c" component may be denatured hemoglobin, but it hasn't been explained clearly so far.

From the results of CD experiment, we found that the secondary structures of Hb from CCHD patients were different from those of normal persons in that their percentages of α-helix in Hb were significantly more than of normal persons. More α-helix in Hb may result in less flexibility of Hb structure, which may account for the low ability of patients' RBC to carry oxygen. This agreed with the results of MS experiments.

Reference

[1] Willem A. Helbing, Sidney A. Rebergen, Chris Maliepaard, Bettina Hansen, *et al.*, *American Heart Journal* **130**(1995): 828.

[2] Leitch CA; Karn CA; Ensing GJ; Denne SC, J Pediatr (UNITED STATES) **137**(2000):381.

[3] McManus A, Leung M, *Sports Med (NEW ZEALAND)* **29**(2000): 229.

[4] Zhang Xiufang, Shen Linming, Bao Hongxia, Ding Xiaolan, *et al.*, *Hyperfine Interactions* **71**(1992): 1251.

[5] Xue Quan-fu, Chai Dan, Miao Hui, Lin Yuo-hua, *et al.*, *Progress in Microcirculation Research*, (1994) p70.

[6] Chai Dan, Xue Quan-fu and Si Qin, *Proceedings of Beijing Satellite Symposium of the Eighth International Congress of Biorheology*, Beijing, 1992.

[7] Chen Y. H., *Biochemistry* **13**(1974):3350.

A High Pressure Mössbauer Spectroscopy Study of [^{57}FeOEPCl]

S. Anandan[1], J. Silver[1], G. R. Fern[1], P. J. Titler[1], C. A. McCammon[2].

[1]School of Chemical and Life Sciences, University of Greenwich, Wellington Street, Woolwich, London, SE18 6PF. E-mail s.anandan@gre.ac.uk.

[2]Bayerishes Geoinstitut, Universität Bayreuth, D-95440, Bayreuth, Germany.

Introduction

Iron porphyrins are widespread in nature, being found as the active sites in hemoglobin, myoglobin and a wide range of cytochromes[1]. The most common porphyrin is Protoporphyrin IXiron(II)/(III) [PPIXFe(II)/(III)] (haem), however this complex is difficult to crystallise and its chemistry is often modelled using man made porphyrins. We have previously studied [Fe^{III}PPIXCl][2] by high pressure Mössbauer spectroscopy and found evidence of a change of spin from high spin state to a spin admixed state S = +/- $^{3}/_{2}$, +/- $^{5}/_{2}$.

To extend our work in this area we present here the results of a high pressure Mössbauer spectroscopic study produced 2,3,7,8,12,13,17,18-octaethylporphyrin ^{57}iron(III) chloride [$^{57}Fe^{III}$OEPCl]. The reason for choosing this porphyrin is that it has ethyl side groups substituted in the same positions on the macrocycle ring as the side groups in PPIX and thus a better model for the latter than tetraphenylporphyrins.

Experimental

The preparation of [^{57}FeOEPCl] was prepared according to a literature method[3]. Only anhydrous and purified solvents were used for the synthesis. ^{57}Fe foil (0.01g, 0.18 mmol) was dissolved in deoxygenated conc. HCl (2ml) under nitrogen. Two drops of 2, 6-lutidine were added to a solution of H_2OEP[4] (5mg, 0.09 mmol) in $CHCl_3$ (3ml) and anhydrous $^{57}FeCl_2$ in CH_3Cl. The reaction mixture was stirred at room temperature for 5 hours. Identification of the product - [^{57}FeOEPCl] was achieved by a comparison of the electronic absorption spectrum of a purchased reference sample.

High pressure Mössbauer spectra were recorded using a Merill Basset diamond anvil cell (DAC) under hydrostatic pressure. 4:1 Methanol:Ethanol was used as the hydrostatic pressure fluid[5]. The ruby fluorescence method was used to measure the pressure in the DAC[6].

Results

Table 1 Mössbauer Spectroscopic Parameters for [^{57}FeOEPCl]

Pressure GPa	δ mms^{-1}.	$ΔE_Q$ mms^{-1}	Γ mms^{-1}	% Area	δ mms^{-1}.	$ΔE_Q$ mms^{-1}	Γ mms^{-1}	% Area
1.38(5)	0.28(1)	0.82(2)	0.64(2)	100(2)				
1.88(5)	0.34(3)	0.98(4)	0.77(5)	100(4)				
2.05(21)	0.30(2)	0.91(4)	0.78(3)	100(3)				
2.57(31)	0.314(3)	0.937(4)	0.85(1)	100(1)				
2.75(20)	0.304(5)	0.898(9)	0.78(1)	100(1)				
3.50(23)	0.318(4)	0.968(7)	0.73(1)	100(1)				
4.37(24)	0.327(2)	0.783(4)	0.61(1)	100(1)				
5.40(33)	0.310(4)	0.95(1)	0.65(2)	76(2)	0.31(1)	2.30(4)	0.76(6)	24(2)
6.30(41)	0.315(1)	0.95(1)	0.65(2)	70(2)	0.31*	2.29*	0.81(6)	30(1)

* values held in fitting routine.

Figure 1

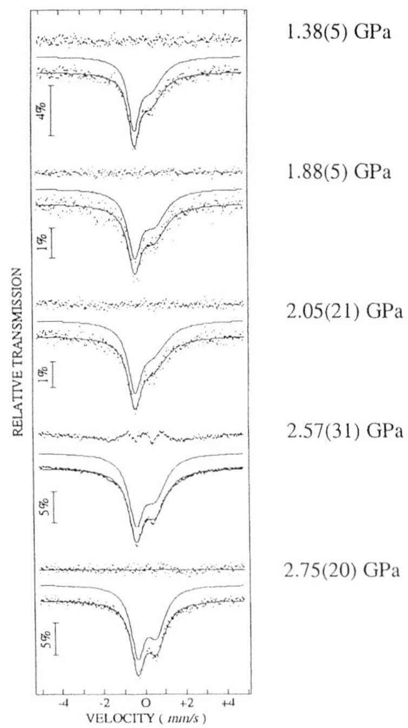

Figure 2

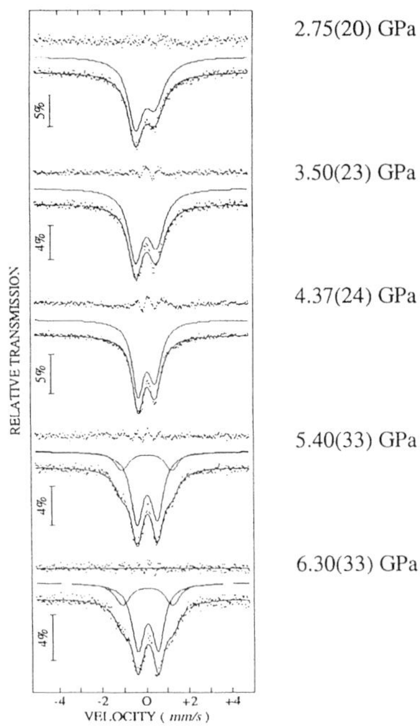

Figure 1 and 2 of ^{57}Fe Mössbauer spectra under high pressure of [^{57}FeOEPCl], with increasing pressure from 1.38(5) GPa to 6.30(41) GPa.

Figure 3

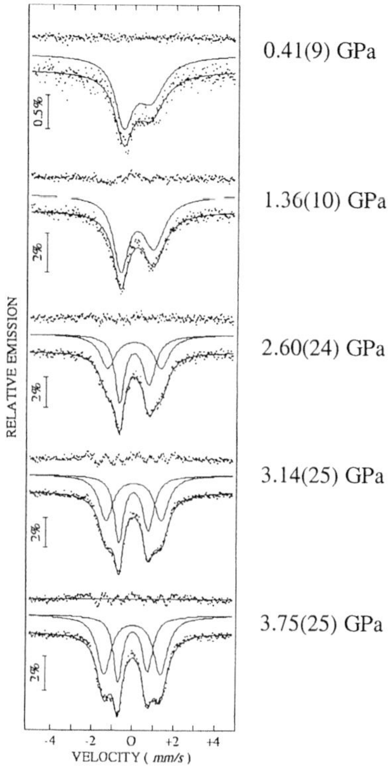

Figure 4

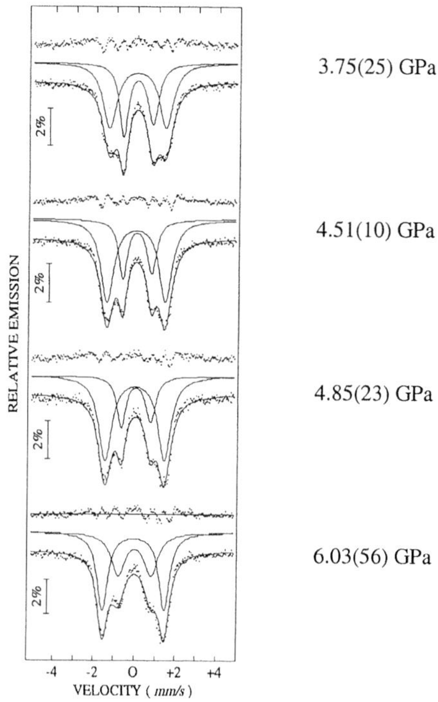

Figure 3 and 4 of ^{57}Fe Mössbauer spectra under high pressure of [^{57}FePPIXCl], with increasing pressure from 0.41(9) GPa to 6.03(56) GPa.

Discussion

The Mössbauer spectroscopic data for [^{57}FeOEPCl] are presented in Table 1 and in figs. 1 and 2. The room temperature Mössbauer spectroscopic data are in agreement with literature values[7] and indicate that the compound contains high spin Fe^{III}. The first effect of applying pressure was to decrease the broadness of the background and as the pressure increased this process steadily continued. These spectra were fitted as doublets with a height variation allowing for asymmetry and for the ratio of the widths of the two lines to vary. The given linewidth is for the low velocity line. The quadrupole splitting (ΔE_Q) for the high spin Fe^{III} site becomes more obviously resolved as the pressure increased to 4.37(24) GPa. It then increased at higher pressures. This differed for the ΔE_Q of the high spin Fe^{III} in [^{57}FePPIXCl](fig. 3 and 4) which initially increased with increasing pressure and then levelled off. This was in keeping with our findings for [^{57}FePPIXCl] (figs. 3 and 4).

At 5.40(33) GPa there is evidence for the development of a second site in the spectrum, which persists at 6.30(41) GPa. The parameters for this site are compatible with those of an S = +/- $^{3}/_{2}$, +/- $^{5}/_{2}$ spin admixed state. Again this is in agreement with our findings for [^{57}FePPIXCl] (figs. 3 and 4). However for the [^{57}FeOEPCl] the spin admixed site only occurs at higher pressures and the proportion of this site is much less than in [^{57}FePPIXCl].

It should be noticed that in both figs. 1 and 2 and figs. 3 and 4 the asymmetry of the high spin Fe^{III} site decreased dramatically as the pressure increased. The room pressure asymmetry is due to the presence of Kramer's doublets. This suggests that at higher pressures as the lattice is compressed, the higher energy Kramer's doublet becomes less populated due to pressure increasing the energy differences between these energy levels.

At low temperature (4.2K) and normal pressure, symmetrical Mössbauer spectra are observed, as only the S_z = +/- ½ states are populated and - as Blume[8] has explained, the spin-spin relaxation is fast. As the temperature increases, the upper states populate (+/- $^{3}/_{2}$ +/- $^{5}/_{2}$) with longer relaxation times, leading to asymmetric Mössbauer spectra - like those at higher pressures in figs. 2 and 4. This means that the application of pressure has a similar effect on the Mössbauer spectra, to that of lowering temperature, as we have reported previously[9-10].

Acknowledgements

High-pressure experiments were performed at the Bayerisches Geoinstitut under the E. C. 'Human Capital and Mobility' - Access to large scale facilities programme. (Contract No. ERBCHGET940053 to D. C. RUBIE).

References

[1} B.P Berezin, "Coordination compounds of Porphyrins and Phthalocyanines", John Wiley & Sons Ltd., (1981).

[2] P.J. Titler, D.A. Davies, J. Silver and C.A. McCammon - unpublished results, presented as a poster at ICAME 1999.

[3] V.V. Borovkov, J. M. Lintuluoto andY. Inuone, Synlett., 1(1999) 61.

[4] J. Sessler, A. Mozaffari, M. Johnson, Org. Synth., 70(1991) 68.

[5] A. Jayaraman, Rev. Mod. Phys., 55(1983) 1.

[6] J. D. Barnett, S. Block and G. J. Piermarinni, Rev. Sc. Instrum, 44(1973) .

[7] B. W. Fitzsimmons, J. R. Sams and T. B. Tsin, Chem. Phys. Lett., 119(1997) 2563.

[8] M. Blume, Phys. Rev. Lett., 18(1967) 305.

[9] J. Silver, R. M. G. R. Roberts, D. A. Davies and C. A. McCammon, Chem. Commun., (1996) 11.

[10] J. Silver, J.R. Miller, D.A. Davies, and C.A. McCammon, Inorg. Chem., 36(1997) 4017.

In Situ ^{57}Fe Mössbauer Spectroscopy Investigation of the Thermal Treatment of Metal-Doped Fe_3O_4

Ibrar Ayub[1], Frank J Berry[1], Eleanor Crabb[1] and Örn Helgason[2]

[1]*Department of Chemistry, The Open University, Walton Hall, Milton Keynes MK7 6AA, United Kingdom*

[2]*Science Institute, University of Iceland, Reykjavik, Iceland.*

Abstract

^{57}Fe Mössbauer spectra recorded *in situ* as a function of increasing temperature show that oxidation of spinel-related titanium- doped Fe_3O_4 to structurally-related titanium-doped γ–Fe_2O_3 occurs at *ca.* 300 °C and its subsequent conversion to corundum-related titanium-doped α-Fe_2O_3 begins at *ca.* 550 °C but is only completed after thermal treatment at higher temperatures. The results contrast with those recorded from the thermal treatment of magnesium- doped Fe_3O_4 which showed oxidation to magnesium- doped γ–Fe_2O_3 and the simultaneous onset of conversion to magnesium-doped α-Fe_2O_3 at *ca.* 300 °C.

Introduction

The doping of M^{4+} ions into Fe_3O_4 and γ–Fe_2O_3 with the inverse spinel-related structure and into α-Fe_2O_3 with the corundum-related structure has been an area of activity for some time because of interest in the thermal, magnetic, electrical and other physical properties of the systems. Although it is possible, by selecting appropriate heating regimes, to convert Fe_3O_4 to γ–Fe_2O_3 and subsequently to α-Fe_2O_3, an understanding of the transformation of titanium-doped Fe_3O_4 to γ–Fe_2O_3 to α-Fe_2O_3 appears not to have received significant attention. We report here on the *in situ* investigation of these transformations by ^{57}Fe Mössbauer spectroscopy and we compare the results with those recorded from magnesium-doped Fe_3O_4.

Experimental

Magnesium- and titanium-doped Fe_3O_4 were prepared by reducing magnesium-doped α-Fe_2O_3 and titanium-doped γ-Fe_2O_3 in a flow of 10 % hydrogen – 90% nitrogen at 400 °C. The single phase nature of the materials was verified by X-ray powder diffraction and ^{57}Fe Mössbauer spectroscopy. The ^{57}Fe Mössbauer spectra were recorded from powdered samples with a constant acceleration spectrometer and a *ca.* 400 MBq ^{57}Co/Rh source. The furnace design for the *in situ* study of phase transformations has been described in elsewhere [1]. The sample thickness was 50-80 mgcm^{-2}. The isomer shift data are quoted relative to α-iron.

Results and Discussion

Magnesium-doped Fe_3O_4

The ^{57}Fe Mössbauer spectra recorded from magnesium-doped Fe_3O_4 (shown by EDX analysis to contain 2.32% magnesium) following heating in an atmosphere with a partial oxygen pressure of 0.2 Pa are collected in Figure 1. The magnesium-doped Fe_3O_4 gave a characteristic ^{57}Fe Mössbauer spectrum (Figure 1a). Heating at 250 °C (Figure 1b) caused a decrease in the magnitude of the magnetic hyperfine field, as observed previously in tin- doped magnetite [2], but without the detection of a phase transformation. However, during treatment at 300 °C (Figure 1c) the oxidation of magnesium-doped Fe_3O_4 (Mg/Fe_3O_4) to magnesium-doped γ–Fe_2O_3 (Mg/γ–Fe_2O_3) and the conversion to magnesium-doped α-Fe_2O_3 (Mg/α-Fe_2O_3) was observed in the spectra which were recorded every two hours over a 40 hour period. Similar behaviour has previously been observed for tin-doped magnetite [3]. The process continued during treatment at 350, 400 and 450 °C (Figure 1d, 1e and 1f). The spectrum recorded after cooling to 127 °C (Figure 1g) showed that about one-third of the $Mg/\gamma-Fe_2O_3$ is converted to $Mg/\alpha-Fe_2O_3$ at 450 °C.

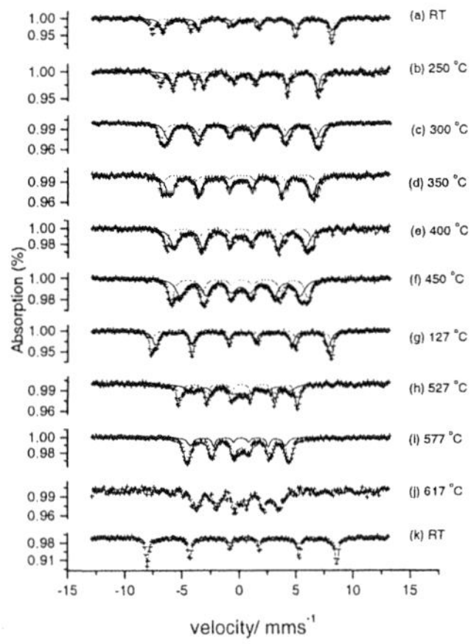

Figure 1 ^{57}Fe Mössbauer spectra recorded *in situ* from (a) magnesium-doped Fe_3O_4 and following heating at (b) 250 °C; (c) 300 °C; (d) 350 °C; (e) 400 °C; (f) 450 °C; (g) 127 °C; (h) 527 °C; (i) 577 °C; (j) 617 °C; (k) 617 °C and cooled to room temperature.

Following heating at 577 and 617 °C the broad lined spectra (Figure 1i and j) demonstrated total conversion to $Mg/\alpha-Fe_2O_3$. X-ray powder diffraction recorded from the material cooled to room temperature confirmed that magnesium had not segregated from the material during the heat treatment and the Mössbauer spectrum (Figure 1k) (δ 0.38 mms^{-1}, H 51 T) confirmed only the presence of $\alpha-Fe_2O_3$.

Titanium-doped Fe_3O_4

The ^{57}Fe Mössbauer spectrum recorded from titanium-doped Fe_3O_4 (shown by EDX analysis to contain 4.89% titanium) following heating in an atmosphere with a partial pressure of oxygen of 1 Pa are collected in Figure 2. The ^{57}Fe Mössbauer spectrum recorded from the titanium-doped Fe_3O_4 (Ti/Fe_3O_4) (Figure 2a) showed the characteristic sextet pattern together with a doublet characteristic of small particle material. Materials heated between 300 and 400 °C (Figure 2b, 2c and 2d) showed broad lined spectra best fitted to a distribution of sextet patterns and demonstrating conversion to titanium-doped $\gamma-Fe_2O_3$ ($Ti/\gamma-Fe_2O_3$) together with a doublet characteristic of small particle iron oxide. The spectrum recorded from the material following cooling to room temperature (Figure 2e) was best fitted to a distribution of hyperfine parameters characteristic of conversion to Ti/γ–Fe_2O_3 and the absence of Ti/α-Fe_2O_3. The appearance of a component typical of Ti/α-Fe_2O_3 was only observed after heating at 550 °C (Figure 2h) and was confirmed by cooling to room temperature (Figure 2i). X-ray powder diffraction confirmed that titanium had not segregated from the iron oxides during the thermal treatment. The higher temperature conversion of titanium- doped γ–Fe_2O_3 to titanium- doped α-Fe_2O_3 contrasts with the results recorded from magnesium- doped γ–Fe_2O_3 and demonstrates that titanium is capable of stabilising the spinel-related γ–Fe_2O_3 structure with respect to transformation to the corundum-related α-Fe_2O_3 structure.

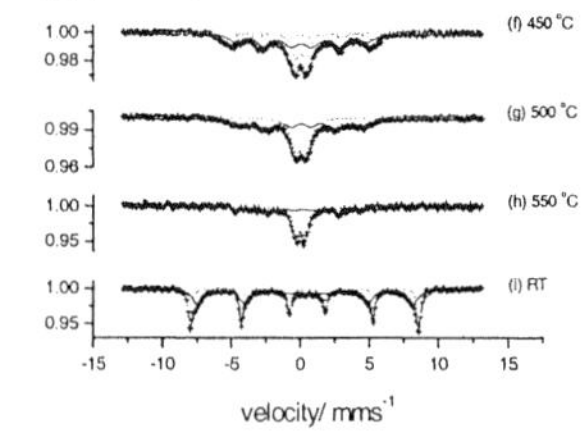

Figure 2 ^{57}Fe Mössbauer spectra recorded *in situ* from (a) titanium-doped Fe_3O_4 and following heating at (b) 300 °C; (c) 350 °C; (d) 400 °C; (e) 400 °C and cooled to room temperature; (f) 450 °C; (g) 500 °C; (h) 550 °C; (i) 550 °C and cooled to room temperature.

We thank Johnson Matthey plc for financial support.

References

[1] Ö. Helgason, H.P. Gunnlaugsson, K. Jonsson and S. Steinthorsson, Hyp. Int. 91(1994) 595.
[2] F.J. Berry, Ö Helgason and S.J. Skinner, J. Solid State Chem. 122(1996) 353.
[3] Ö. Helgason, F.J. Berry, K. Jonsson and S.J. Skinner, in: Proceedings of the International Conference on the Application of Mössbauer Effect, Ed. I. Orteilli, S.I.F. Bologna 1996, p59.

Nuclear resonant forward scattering using synchrotron radiation applied to study the low-spin ferriheme complex [TPPFe(NH_2PzH)$_2$]Cl

R. Benda, C. Herta, V. Schünemann, H. Winkler, A. X. Trautwein, Y. Shvyd'ko[1], and F. A. Walker[2]

Institute of Physics, Med. University, 23538 Lübeck, Germany
[1]Institute of Exp. Physics, University of Hamburg, 22761 Hamburg, Germany
[2]Department of Chemistry, University of Arizona, AZ 85721 Tucson, USA

Abstract

We have characterised the low–spin ferriheme complex bis(3-aminopyrazole)tetraphenylporphyrinatoiron(III) chloride ([TPPFe(NH_2PzH)$_2$]Cl) by nuclear resonant forward scattering (NFS) using synchrotron radiation at 4.2K in fields up to 6T. The analysis of the field dependant NFS spectra obtained with the **g**-tensor determined by previous pulsed EPR measurements [1] (**g**=(1.87, 2.28, 2.39)) yields the hyperfine parameters η = 0.0; β = 90°; γ = 90°; ΔE_Q = -2.56 mm/s and $\mathbf{A}/\mu_n g_n$ = (-46.9, 9.5, 17.6) T. This investigation shows that NFS provides higher accuracy in determining hyperfine parameters than conventional Mössbauer spectroscopy, if the delay time is long enough and the effective thickness is relatively high (in the present case t $\approx$ 150 ns and t_{eff} $\approx$ 10).

Introduction

The biological function of low-spin ferriheme complexes ranges from electron transfer (e.g. in the membrane protein complex cytochrome *bf)* to enzymatic catalysis. Examples of the latter are the cytochromes P450, which are essential for detoxification in various living organisms. In order to obtain a better understanding of the biologically active iron site of these proteins, it is useful to compare their electronic properties with those of chemical model compounds by means of various spectroscopic methods like electron paramagnetic resonance (EPR), electron spin echo envelope modulation (ESEEM), magnetic circular dichroism (MCD), nuclear magnetic resonance (NMR) and Mössbauer spectroscopy [2]. We have applied Mössbauer spectroscopy in the time domain, i.e. nuclear resonant forward scattering (NFS) of synchrotron radiation to prove the electronic properties of the iron centre in the low-spin ferriheme complex bis(3-aminopyrazole)tetraphenylporphyrinatoiron(III) chloride ([TPPFe(NH_2PzH)$_2$]Cl) and we have compared these results with those obtained from a previous pulsed EPR and Mössbauer study [1]. This model complex exhibits conventional Mössbauer spectra which are very similar to those of cytrochrome $P450_{cam}$ from *Pseudomonas putida* [3] and to Chloroperoxidase from *Caldariomyces fumago* [4].

Material and Methods

[TPP^{57}Fe(NH_2PzH)$_2$]Cl was prepared by dissolving TPP^{57}FeCl synthesised by literature methods [5] in a 3-aminopyrazole/DMF solution. The iron concentration of the sample used for conventional Mössbauer spectroscopy was approximately 10 mM and that for the NFS samples was 11 mM (ESRF) and 17 mM (HASYLAB), respectively.

Conventional Mössbauer spectra were recorded using a spectrometer in the constant acceleration mode. Isomer shifts are given relative to α - Fe at room temperature. The spectrum obtained at 20 mT was measured in a He-bath cryostat (Oxford MD 306) equipped with a pair of permanent magnets. The measurement at 1 T was performed in a self-constructed bath cryostat equipped with a superconducting magnet. For the high-field spectrum (7 T) a cryostat also equipped with a superconducting magnet was used (Oxford Instruments, Spectromag 4000). The magnetically split spectra of the paramagnetic samples were simulated with the spin-Hamiltonian formalism [6].

The NFS measurements were performed at HASYLAB station BW4 in Hamburg and at ESRF station ID22N in Grenoble. The experimental conditions of BW4 are described in [7] and the experimental conditions of ID22N in [8]. All NFS spectra were simulated with the program package SYNFOS [9].

Results and Discussion

NFS spectra of $[TPP^{57}Fe(NH_2PzH)_2]Cl$ obtained at 4.2K in fields from 61 mT up to 6 T are shown in Figure 1. For comparison the conventional Mössbauer spectra obtained at 4.2 K in fields of 20 mT up to 7 T are shown in Figure 2. Using the **g**-tensor from pulsed EPR measurements (**g**=1.87, 2.28, 2.39) [1] we performed simultaneous fits of the NFS spectra[1] and compared these results with those of the conventional spectra. The results of our analysis are listed in Tables 1 and 2. The solid lines in Figures 1 and 2 represent the parameters (Table 1) obtained from the best fit of the NFS spectra. In contrast the dashed lines in Figures 1 and 2 represent the parameters (Table 2) obtained solely from the conventional Mössbauer spectra.

Table 1: Spin – Hamiltonian and Mössbauer parameters obtained from simulations of the NFS spectra (solid lines in Figure 1).

g – tensor [a)]	ΔE_Q[mm/s]	η	$A/g_N\mu_N$ [T]
$g_{xx} = 1.87$			$A_{xx} = -46.9 \pm 2.0$ [d)]
$g_{yy} = 2.28$	+2.56[c)]	-2.2 >-2.9 [b)] >-3.8 [d)]	$A_{yy} = 9.5 \pm 2.0$ [d)]
$g_{zz} = 2.39$			$A_{zz} = 17.6 \pm 1.5$ [d)]

a) Derived from EPR [1]

b) Here η is given in the molecular frame with the z – axis perpendicular to the heme – plane. However, the coordinate system of the efg tensor is commonly chosen in such a way that $|V_{zz}| \geq |V_{yy}| \geq |V_{xx}|$; in order to reach this convention in the present case the coordinate system of the efg tensor has to be turned around the y – axis of the molecular frame by the Euler angle $\beta = 90°$ and around the new z´- axis by $\gamma = 90°$. In this new coordinate system $\eta' = 0.0$ and $\Delta E_Q' = -2.56$ mm/s.

c) Sign of quadrupole splitting does not influence the NFS spectra

d) Error margins result from the simulation of the NFS spectra

1 The different concentrations and real thickness' of the samples for the experiments at ESRF and HASYLAB were taken into account by using the following values for the effective thickness: a) $t_{eff} = 13.6$ and b) – f) $t_{eff} = 12$.

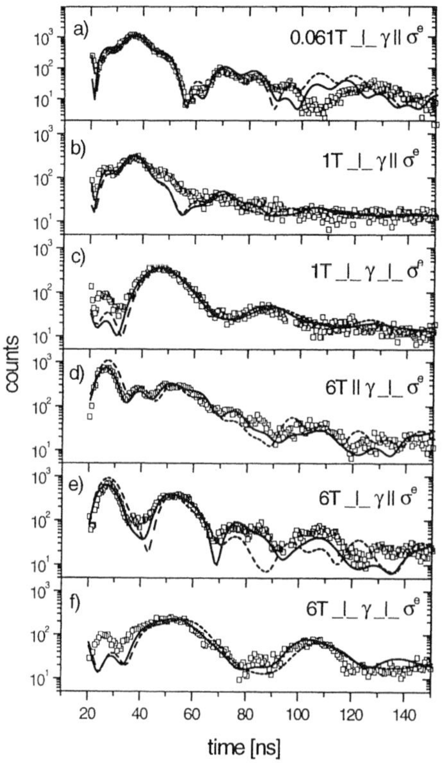

Figure 1: NFS spectra of [TPPFe $(NH_2PzH)_2$]Cl. Spectra were recorded at 4.2 K and with magnetic fields as indicated. Parameters for solid lines, which represent the best fit of NFS spectra, are summarised in Table 1. Parameters for the dashed lines, which represent the best fit of conventional Mössbauer spectra, are summarised in Table 2. NFS measurements were performed a) at ID22N (ESRF) and b) – f) at BW4 (HASYLAB).

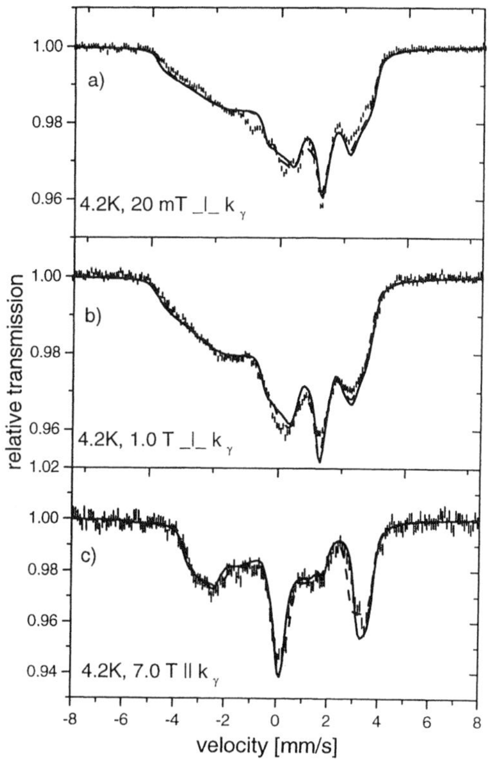

Figure 2: Field-dependent conventional Mössbauer spectra of [TPPFe$(NH_2PzH)_2$]Cl. Spectra were recorded at 4.2 K and with magnetic fields as indicated. Dashed lines result from simultaneous simulation of conventional Mössbauer spectra (without having seen NFS spectra before); corresponding parameters are summarised in Table 2. Solid lines represent simulations using the parameters obtained from the NFS analysis (Table 1).

Comparing solid and dashed lines in Figures 1 and 2 it is obvious that small variations of hyperfine parameters (in the present case the asymmetry parameter η and the magnetic hyperfine coupling tensor A) are more sensitively reflected in NFS spectra than in conventional Mössbauer spectra. This is especially the case if NFS spectra are recorded to long enough delay times; the smaller the variations the longer the delay time has to be in order to make these variations detectable. Due to the experimental decrease of count rate with increasing delay time (at least neglecting dynamical beat structure) it is necessary to provide high enough effective thickness t_{eff} in the sample to make the quantum beat structure visible at long delay times. In our case recording and analysing NFS spectra up to 150 ns required $t_{eff} > 10$.

Table 2: Spin – Hamiltonian and Mössbauer parameters obtained from simulations of the conventional Mössbauer spectra (dashed lines in Figure 2).

g – tensor [a]	ΔE_Q[mm/s]	η	$A/g_N\mu_N$ [T]
$g_{xx} = 1.87$	+2.56	-1.7> -2.3 [b]>-3.5 [c]	$A_{xx} = -45.6 \pm 3.5$ [c]
$g_{yy} = 2.28$			$A_{yy} = 6.0 \pm 5.5$ [c]
$g_{zz} = 2.39$			$A_{zz} = 16.9 \pm 2.5$ [c]

a) Derived from EPR [1]

b) Here η is given in the molecular frame with the z – axis perpendicular to the heme – plane. However, the coordinate system of the efg tensor is commonly chosen in such a way that $|V_{zz}| \geq |V_{yy}| \geq |V_{xx}|$; in order to reach this convention in the present case the coordinate system of the efg tensor has to be turned around the y – axis of the molecular frame by the Euler angle $\beta = 90°$ and around the new z´- axis by $\gamma = 90°$. In this new coordinate system $\eta' = 0.2$ and $\Delta E_Q' = -2.56$ mm/s.

c) Error margins result from the simulation of the conventional Mössbauer spectra

In summary we note that the hyperfine parameters of $[TPP^{57}Fe(NH_2PzH)_2]Cl$ are determined with better accuracy with NFS than with conventional Mössbauer spectroscopy; necessary condition is that appropriate experimental conditions (high effective thickness and long delay time) are provided. An effective thickness $t_{eff} > 10$ is easily obtainable with ^{57}Fe – enriched model compounds like the one under study here, in contrast to ^{57}Fe – enriched proteins like hemoglobin or cytochrome P450. Nevertheless this investigation motivates for future studies of enzymes by NFS, since this method requires much less sample volume than conventional Mössbauer spectroscopy (< 50 µl at 3rd generation synchrotron sources).

This work was supported by the BMBF – grant 05SK8FLA8.

References

[1] Schünemann, V., Raitsimring, A. M., Benda, R., Trautwein, A. X., Shokireva, T. K., Walker, F. A., (1999) *JBIC* **4**,708.

[2] Walker, F. A., (1999) *Coord. Chem. Rev.* **471**, 185.

[3] Sharrock, M., Debrunner, P. G., Schulz, C., Lipscomb, J. D., Marshall, V., Gunsalus, I. C., (1976) *Biochim. Biophys. Acta* **420**, 8.

[4] Champion, P. M. , Münck, E., Debrunner, P. G., Hollenberg, P. F., Hager, L. P., (1973) *Biochemistry* **12**, 426.

[5] Walker, F. A., Huynh, B. H ., Scheidt, W. R., Osvath, S. R., (1986), *J. Am. Chem. Soc.* **108**, 5288.

[6] Debrunner, P. G., In *Iron porphyrins*, **part III.**; Lever, A. B. P., Gray, H. B., Eds.; VCH: Weinheim, 1989; pp 139 –234.

[7] Leupold, O., Winkler, H., (1999) *Hyp. Interact.* **123/124**, 571.

[8] Rüffer, R., and Chumakov, A. I., (1996) *Hyp. Interact.* **97/98**, 589.

[9] Haas, M., Realo, E., Winkler, H., Meyer-Klaucke, W., Trautwein, A. X., Leupold O., and Rüter, H. D., (1997) *Phys. Rev. B* **56**, 14082.

^{57}Fe Mössbauer Spectroscopy Investigation of Cobalt-Substituted Lanthanum Orthoferrites

Frank J. Berry[a], Ramón Gancedo[b], José F Marco[b] and Xiaolin Ren[a]

[a]*Department of Chemistry, The Open University, Walton Hall, Milton Keynes MK6 7AA, United Kingdom*
[b]*Instituto de Química Física 'Rocasolano', Consejo Superior de Investigaciones Científicas, c/Serrano 119, 28006 Madrid, Spain.*

The products derived from the temperature programmed reduction of cobalt substituted lanthanum ortoferrites have been examined by ^{57}Fe Mössbauer spectroscopy. The results show that $LaFeO_3$ is partially reduced in 10% hydrogen – 90% nitrogen to metallic iron. The material resisted complete reduction even at 1200°C. In contrast, iron and cobalt in the material $LaFe_{0.5}Co_{0.5}O_3$ segregated under treatment in hydrogen to form an iron-cobalt alloy and metallic iron.

1. Introduction

The rare earth orthoferrites of composition $REFeO_3$ contain only trivalent metal ions in the structure and are suitable systems for isovalent substitution. Materials of the type $LaFe_{1-x}Co_xO_3$ have attracted interest in recent years [1-3] because of their potential in electroceramic and catalytic oxidation applications.

We report here on the characterisation of the material $LaFe_{0.5}Co_{0.5}O_3$ and its reduction properties by ^{57}Fe Mössbauer Spectroscopy.

2. Experimental

Compounds of the type $LaFe_{1-x}Co_xO_3$ (x=0, 0.5) were prepared by grinding appropriate quantities of lanthanum(III) oxide, α-iron(III) oxide and cobalt(II) oxide in an agate pestle and mortar and heating in air at 1200 °C(12h).

^{57}Fe Mössbauer spectra were recorded at 298K with a constant acceleration spectrometer in transmission geometry using a ^{57}Co/Rh source. The drive velocity was calibrated with the ^{57}Co/Rh source and a natural iron foil. All the isomer shift data are reported relative to that of metallic iron at room temperature. Temperature programmed reduction profiles were recorded from *ca.* 150 mg samples in flowing 10% H_2/ 90% N_2 (15-20ml/min) with the temperature being increased by 5°C/min.

3. Results and Discussion

Materials of composition $LaFe_{1-x}Co_xO_3$ were shown by X-ray powder diffraction to be single phase. $LaFeO_3$ was found to be orthorhombic whilst $LaFe_{0.5}Co_{0.5}O_3$ was hexagonal as reported previously [1,2].

The tpr profile recorded from the compound $LaFeO_3$ (Figure 1) showed two broad peaks at ca 475° and 1150°C. The tpr profile recorded from $LaFe_{0.5}Co_{0.5}O_3$ (Figure 1) also showed two peaks which were at lower temperature.

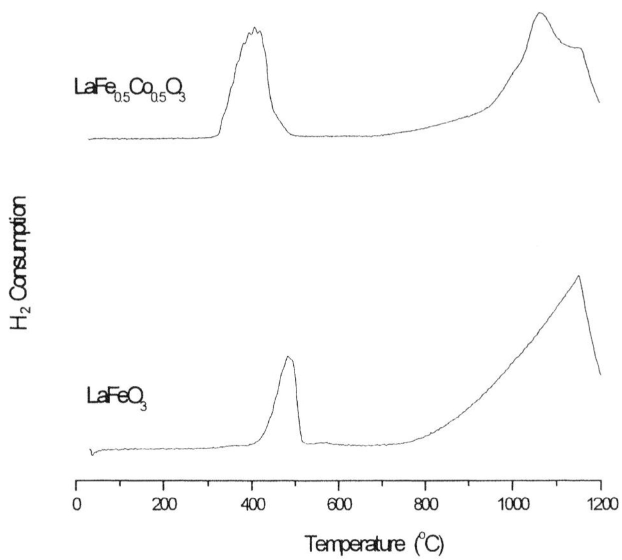

Figure 1. Temperature programmed reduction profiles recorded from $LaFeO_3$ and $LaFe_{0.5}Co_{0.5}O_3$

The ^{57}Fe Mössbauer spectrum recorded from $LaFeO_3$ (Figure 2a) was similar to that reported earlier [4] and consisted of a magnetic sextet with parameters δ=0.36 mms^{-1}, 2ε= -0.09 mms^{-1} and H=52.7 T. The spectrum recorded following treatment in 10% hydrogen / 90% nitrogen at 600°C (after the first peak in the tpr profile) showed the partial reduction of the perovskite-related $LaFeO_3$ phase to metallic iron (Figure 2b). The relative intensity of the sextet characteristic of metallic iron amounted to 16%. The spectrum recorded following treatment in the hydrogen / nitrogen gas mixture at 1200°C (after the second peak in the tpr profile) (Figure 2c) showed an enhanced amount of metallic iron (22%) but, and significantly, without complete reduction of lanthanum orthoferrite.

The ^{57}Fe Mössbauer spectrum recorded from $LaFe_{0.5}Co_{0.5}O_3$ (Figure 3a) was very different. The spectrum showed a distribution of magnetic fields with an average hyperfine magnetic field of 26.3 T. The origin of the effect of cobalt on the ordering temperature of $LaFeO_3$ is to be subject of further investigation. The spectrum (Figure 3b) recorded after the first reduction peak in the tpr profile (Figure 1) also showed a distribution of magnetic fields although the average hyperfine magnetic field was 36.7 T, i.e. 10.4 T higher than that shown by pure $LaFe_{0.5}Co_{0.5}O_3$. The maximum of the hyperfine magnetic field distribution was located at *ca.* 47.0 T, a significantly lower value than that characteristic of $LaFeO_3$. The broadness of the spectral lines and poor statistics precluded the identification of a spectrum corresponding to metallic iron. Whether or not iron and/or cobalt segregate

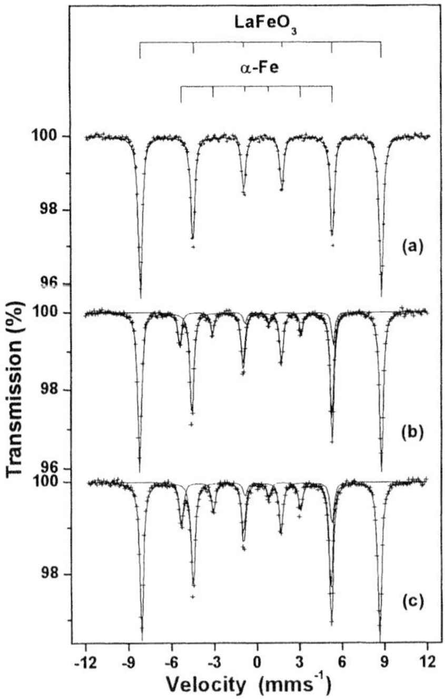

Figure 2. [57]Fe Mössbauer spectra recorded from (a) $LaFeO_3$ (b) $LaFeO_3$ following treatment at 600°C and (c) $LaFeO_3$ following treatment at 1200°C

from the material during the initial treatment in hydrogen is therefore unclear and we have yet to ascertain from other techniques wether cobalt or lanthanum are involved in the reduction step indicated by the first peak in the tpr profile. The Mössbauer spectrum recorded from the material following reduction at 1100°C (Figure 3c) showed a sextet pattern with parameters similar to those characteristic of $LaFeO_3$ (with relative spectral area of 53%) together with another broad sextet which was deconvoluted into two sextets: one characteristic of metallic iron (6%) and a second more intense component (41%) with Mössbauer parameters δ=0.04 mms^{-1}, 2ε= -0.04 mms^{-1} and H=34.8 T which are compatible with the presence of an iron-cobalt alloy [5]. The result indicates that treatment under reducing conditions at 1100°C induces the partial segregation of iron and cobalt from the perovskite related oxide structure and the formation of metallic and alloy phases. The spectrum recorded from the material following further treatment in hydrogen at 1200°C did not show any sextet which could be associated to any oxide species. It consisted of a sextet containing contributions from both metallic iron (33%) and iron-cobalt alloy. The hyperfine magnetic field obtained from the fit of the spectrum for the iron-cobalt alloy (35.7 T) was slightly higher than that observed in the case of the material formed at 1100°C indicating that the composition of this phase could be slightly different. The result contrasts with that recorded from $LaFeO_3$ under reducing conditions in that the presence of cobalt appears to induce the segregation of both iron and cobalt and the formation of metallic iron and iron-cobalt alloy phases.

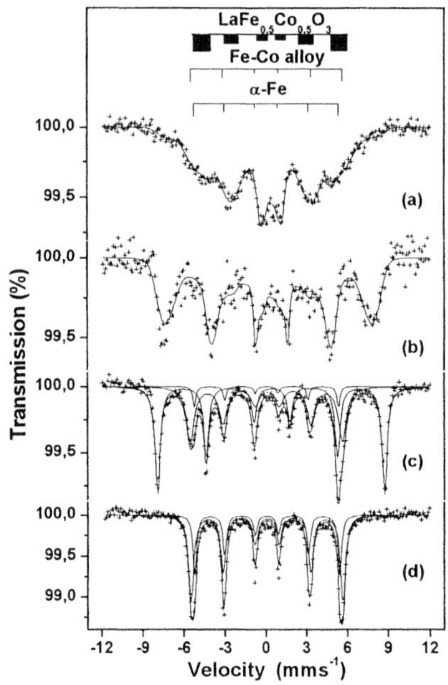

Figure 3. [57] Fe Mössbauer spectra recorded from (a) $LaFe_{0.5}Co_{0.5}O_3$, (b) $LaFe_{0.5}Co_{0.5}O_3$ following treatment at 500°C (c)) $LaFe_{0.5}Co_{0.5}O_3$ following treatment at 1100°C and (d) $LaFe_{0.5}Co_{0.5}O_3$ following treatment at 1200°C.

Acknowledgement

We thank The China Scholarship Council for the award of a scholarship (XR)

References

[1] Y.Sadaoka,E.Traversa,P. Nunziante, M.Sakamoto, *J. Alloy Compounds,* 1997,261, 182
[2] E.Traversa, P.Nunziante, M. Sakamoto, Y.Sodoaka and R. Montanar, *Mater Res.Bull.,* 1998,33, 673
[3] V. R.Choudhary, B.S.Uphade, S.G. Pataskar, *Fuel,* 1999,78, 919
[4] D. Treves, *J. Appl. Phys.* 1965, 36, 1033
[5] V. Carles, C. Laurent, M. Brieu and A. Rousset, *J. Mater. Chem*, 1999, 9, 1003

X-RAY AND MÖSSBAUER STUDY OF MIXED POTASSIUM-SODIUM FERRATE (VI)

S. K. Dedushenko[a*], Yu. D. Perfiliev[b], A. M. Golubev[a], P. Melnikov[c], P. P. Corbi[d], A. A. Saprykin[b]

[a]*Bauman Moscow State Technical University, 5, 2-nd Baumanskaya Street, Moscow 107005, Russia*
[b]*Department of Chemistry, Moscow State University, Lenin Hills, Moscow 119899, Russia*
[c]*Department of Physics, Federal University of Mato Grosso do Sul-UFMS, Cidade Universitária s/n, 79070-900, Campo Grande, MS, Brazil*
[d]*Institute of Chemistry, University of the State of São Paulo-UNESP, Rua Professor Francisco Degni s/n, 14801-970, Araraquara, SP, Brazil*

Mixed potassium-sodium ferrate(VI), $K_3Na(FeO_4)_2$ has been synthesised by precipitation from alkaline solution. It crystallise in a hexagonal system with lattice parameters a=5.8273±0.0018 Å, c=7.5415±0.0032 Å and belongs to glaserite-like structures (space group P/3m1). Iron is present in oxidation state +6: room temperature Mössbauer spectrum is a doublet with isomer shift δ=-0.89±0.01 $mm \cdot s^{-1}$ (relative to α-iron), quadrupole splitting Δ=0.21±0.01 $mm \cdot s^{-1}$ and full width at half-height Γ=0.22±0.01 $mm \cdot s^{-1}$.

1. Introduction

Although wine-coloured FeO_4^{2-} ion has been known since 18th century and is characterised in some detail in solutions and solids, no reliable information is so far available on Fe^{6+}-derivatives with different coordination polyhedra, except tetrahedron, or other ligands, besides oxygen. Crystalline ferrates(VI) of all alkali metals and barium have been thoroughly described [1-3]. Mixed potassium-strontium ferrate $K_2Sr(FeO_4)_2$ [4] and derivatives with quaternary ammonium cations [3] are also known. The existence of other ferrates(VI) is still to be verified [1,3].

The present work deals with the data concerning a new mixed ferrate(VI) of $K_3Na(FeO_4)_2$ composition.

2. Experimental

The compound was obtained by adding solid potassium hydroxide to 40% NaOH solution containing *ca.* 0.1 M FeO_4^{2-}. Then the precipitated tiny crystals of $K_3Na(FeO_4)_2$ were separated, washed with cold alcohol and dried under vacuum.

X-ray diffractogram was obtained using a Siemens Kristalloflex diffractometer, CuKα radiation, Ni-filter in the range 2ϑ 4÷150°.

Mössbauer absorption spectra were measured on a Perseus spectrometer working at constant velocities, the control and the adjustment of the spectrometer-vibrator rate being performed by a laser interferometer. A standard γ-source of ^{57}Co in metallic rhodium matrix with the activity of 1.2 GBk (a product of Cyclotron, Co., Ltd., Obninsk, Russia) was employed.

Atomic absorption analysis for K and Na was performed using A Analyst 300 Perkin Elmer spectrometer.

3. Results and discussion

Atomic absorption technique proved that potassium to sodium ratio correlates with the suggested chemical formula.

X-ray powder diffraction pattern of the salt (Table 1) has been indexed in a hexagonal system with the following lattice parameters: a=5.8273±0.0018 Å, c=7.5415±0.0032 Å. $K_3Na(FeO_4)_2$ is isostructural to the double potassium-sodium chromate $K_3Na(CrO_4)_2$ [5] and is very close to the latter as to the lattice parameters, obviously because of the almost identical of Fe^{6+} and Cr^{6+} ionic radii (according to Shenon [6] their values are 0.25 and 0.26 Å respectively). Both compounds are intrinsically related to glaserite, $K_3Na(SO_4)_2$, space group P/3m1.

---

* To whom to address correspondence.

Table 1
Indexation of $K_3Na(FeO_4)_2$ X-ray diffraction pattern.

I/I_{max}, %	h k l	d_{obs}, Å	d_{cal}, Å
6	1 0 0	5.04	5.05
32	1 0 1	4.20	4.20
10	0 0 2	3.75	3.77
48	1 0 2	3.02	3.02
100	1 1 0	2.912	2.914
3	1 1 1	2.714	2.718
21	0 0 3	2.515	2.514
4	2 0 1	2.392	2.393
1	1 1 2	2.304	2.306
10	1 0 3	2.249	2.250
60	2 0 2	2.094	2.097
3	1 1 3	1.903	1.903
7	0 0 4	1.884	1.885
4	2 1 1	1.848	1.849
4	2 0 3	1.778	1.781
10	1 0 4	1.765	1.766
18	2 1 2	1.701	1.702
18	3 0 0	1.681	1.682
1	3 0 1	1.642	1.642
5	1 1 4	1.583	1.583
1	3 0 2	1.535	1.536
17	2 0 4	1.510	1.510
18	2 2 0	1.457	1.457
4	1 0 5	1.445	1.445
1	3 1 0	1.399	1.400
1	3 1 1	1.377	1.376
3	2 2 2	1.359	1.359
5	2 1 4	1.341	1.341
5	3 1 2	1.313	1.312
3	2 0 5	1.295	1.295
5	0 0 6	1.258	1.257
2	3 1 3	1.224	1.223
5	4 0 2	1.197	1.196
4	2 1 5	1.183	1.183

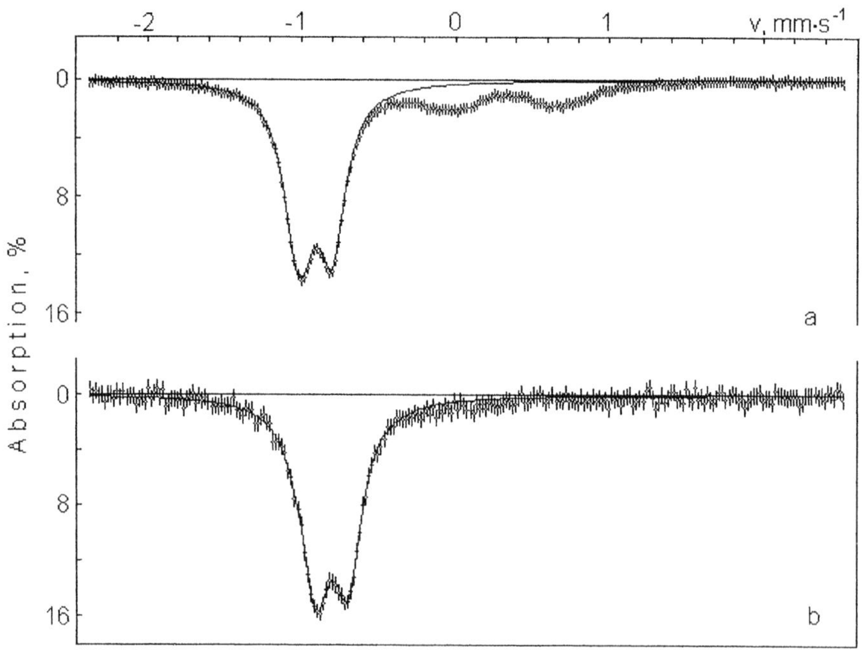

Figure 1. Mössbauer spectra of $K_3Na(FeO_4)_2$ at room temperature (a) and at 4.2 K (b).

Room temperature Mössbauer spectrum of the compound is shown in Figure 1a. It contains a doublet with isomer shift δ=-0.89±0.01 mm·s^{-1} (relative to α-iron), quadrupole splitting Δ=0.21±0.01 mm·s^{-1} and full width at half-height Γ=0.22±0.01 mm·s^{-1}. The measured value of the isomer shift correlates well with the data for other iron(VI) compounds [4,7,8]. The image of spectrum is not changed down to liquid helium temperature (fig.1b). At 4.2 K the isomer shift reduces to -0.80±0.01 mm·s^{-1} that corresponds to estimated temperature drift of its value, quadrupole splitting being constant. Low difference between the intensities of doublet's lines is in our opinion due to Gol'danskii-Karyagin effect and is not the result of a texture. It should be underlined that quadrupole interaction in hexavalent iron is clearly observed for the first time.

Quadrupole splitting in this glaserite-like salt may appear due to a considerable charge asymmetry within the second coordination sphere of iron. It is well known that glaserite structure is closely related with that of β-K_2SO_4 [8]. The simple ferrates(VI) of potassium, rubidium and caesium, which crystallise in this structural type show no quadrupole splitting in their Mössbauer spectra because tetrahedral FeO_4^{2-} ion is in symmetrical crystallographic environment. On the contrary, in glaserite structure, the layers similar to those contained in the β-K_2SO_4 (see Figure 2b), are shifted in a half of translation along c-axis resulting in a distortion of bigger K-Na-containing polyhedra. We presume that this distortion is responsible for the alterations of the above charge distribution.

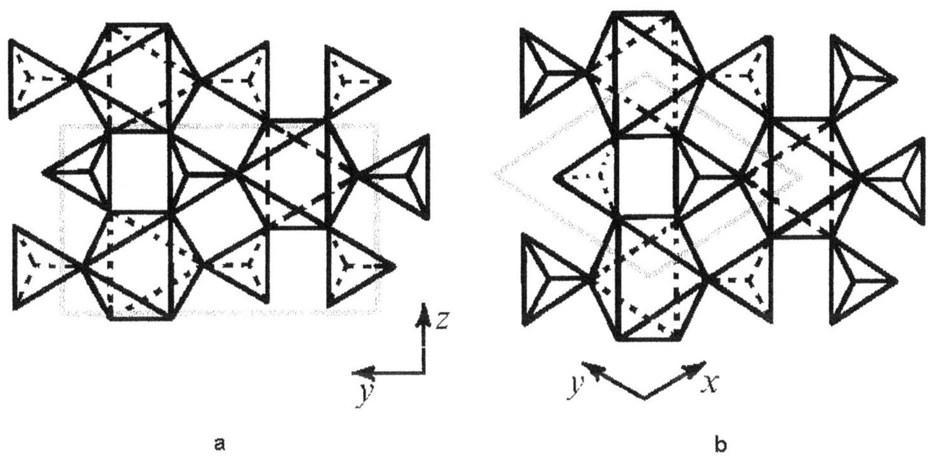

Figure 2. The crystallographic layers in β-K_2SO_4 (a) and glaserite (b) as the projections on (1 1 0) and (1 $\bar{1}$ 1) planes respectively [9]; unit cells are marked out in grey.

Ferrates(VI) prepared by wet chemistry approach are usually contaminated with amorphous, X-ray silent oxo-hydroxo derivatives of iron(III). That is why the spectra (Figure 1) contain also two broadened asymmetrical absorption lines in their right parts. These however do not interfere with determination of the isomer shift and quadrupole splitting, since the absorption lines of iron(VI) and iron(III) are located in the different range of the spectrum.

Acknowledgments

This work was supported by Russian Foundation for Basic Research (project 01-03-32407) and the Ministry of Education of the Russian Federation.

References

[1] I. G. Kokarovceva, I. N. Beliaeva and L. V. Semeniakova, Russ. Chem. Rev. 41 (1972) 929.
[2] R. J. Audette, J. W. Quail, W. H. Black and B. E. Robertson, J. Solid State Chem. 8 (1973) 43.
[3] M. Malchus and M. Jansen, Z. anorg. allg. Chem. 624 (1998) 1846.
[4] S. Ogasawara, M. Takano and Y. Bando, Bull. Inst. Chem. Res. Kyoto Univ. 66 (1988) 64.
[5] A. Goldberg *et al.* , JCPDS-ICDD Database, File No.32-841.
[6] R. D. Shannon, Acta Crysallogr., Sect. A: Fundam. Crystallogr. 32 (1976) 751.
[7] R. H. Herber and D. Johnson, Inorg. Chem. 18 (1979)2786.
[8] F. Menil, J. Phys. Chem. Solids 46 (1985) 763.
[9] B. I. Lazoryak, Russ. Chem. Rev. 65 (1996) 307.

MÖSSBAUER STUDIES OF BONDING AND TIN(II) SUBLATTICE SOFTENING IN THE HIGHLY UNUSUAL DOUBLY DISORDERED $Ba_{1-x}Sn_xCl_{1+y}F_{1-y}$ SOLID SOLUTION

Georges Dénès* and Abdualhafeed Muntasar

Laboratory of Mössbauer Spectroscopy and Solid State Chemistry, Laboratories for Inorganic Materials, Concordia University, Montréal, Québec, Canada

Two methods (wet and dry) have been designed to prepare a wide $Ba_{1-x}Sn_xCl_{1+y}F_{1-y}$ solid solution. Mössbauer spectroscopy has shown that the tin(II) recoil-free fraction becomes very small at high x and highly positive y, revealing considerable lattice softening.

INTRODUCTION AND PROCEDURES

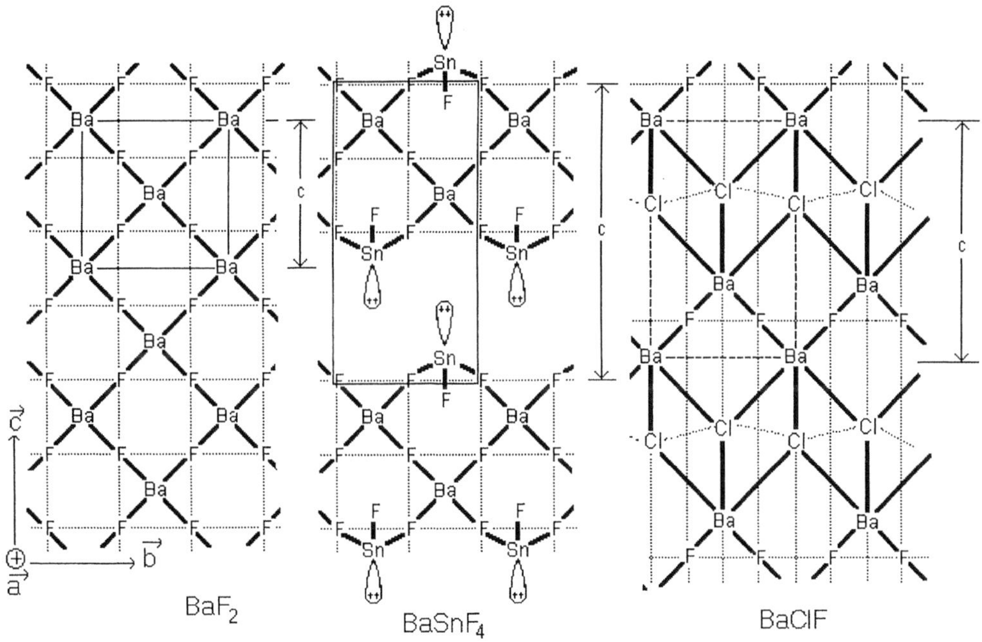

Figure 1: Projection of a slice of the structures of BaF_2, $BaSnF_4$ and BaClF

* Author to whom all correspondence should be addressed (gdenes@vax2.concordia.ca)

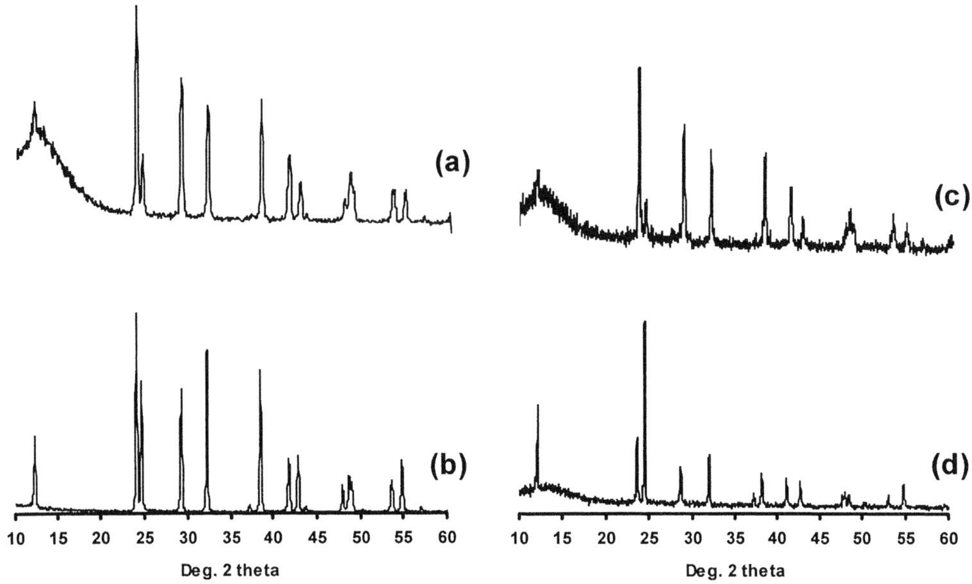

Figure2: X-ray diffraction pattern of (a) unsubstituted BaClF prepared by precipitation, (b) substituted $Ba_{1-x}Sn_xCl_{1+y}F_{1-y}$ prepared by precipitation, (c) substituted $Ba_{1-x}Sn_xCl_{1+y}F_{1-y}$ prepared by direct reaction for x = 0.225, y = -0.100 and (d) substituted $Ba_{1-x}Sn_xCl_{1+y}F_{1-y}$ prepared by direct reaction for x = 0.225 and y = 0.100

The structures of superionic $BaSnF_4$ and BaClF are both tetragonal distortions of fluorite-type BaF_2 with metal ordering in $BaSnF_4$ and anion ordering in BaClF (fig. 1). BaClF is a highly ionic structure (PbClF-type), with order between the two anions of highly difference sizes, and therefore both (i) Ba substitution by weak metal tin that usually binds covalently, and (ii) Cl/F disorder, are highly unexpected. Nevertheless, a wide $Ba_{1-x}Sn_xCl_{1+y}F_{1-y}$ solid solution has been prepared, that is doubly disordered, i.e Ba/Sn and Cl/F, while keeping the Cl layer/F layer order characteristic of the PbClF-type structure. The unexpected solid solution has rather unusual properties, two being a change of bonding type and tin(II) lattice softening as a function of the method of preparation and the compositional coefficients x and y. The discovery and study of these properties were carried out by Mössbauer spectroscopy. The samples were prepared by precipitation from solutions of SnF_2 and $BaCl_2.2H_2O$, and by direction reactions between appropriate amounts of BaF_2, SnF_2 and $BaCl_2$ in dry conditions at high temperature [1]. X-ray Diffraction and Mössbauer spectroscopy were carried out using literature methods [1].

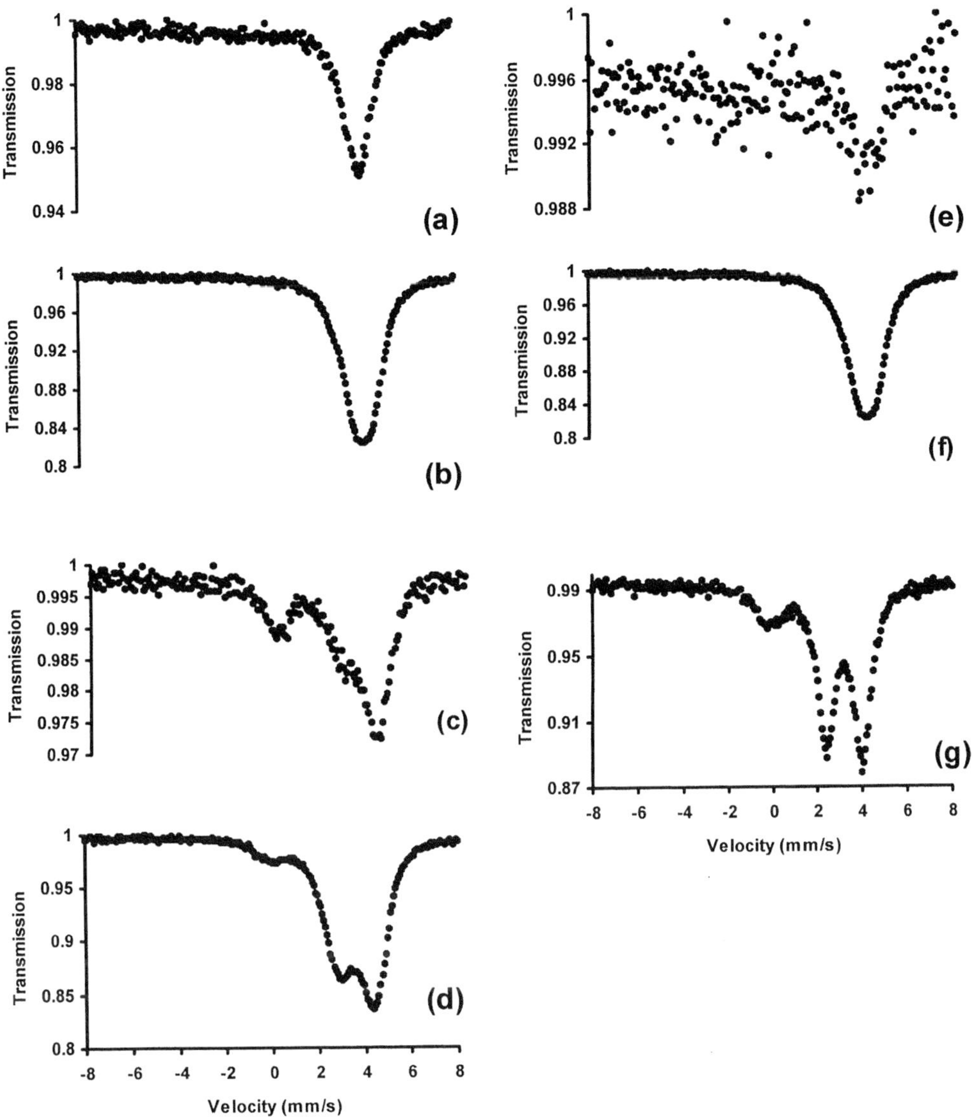

Figure 3: Mössbauer spectrum of (a) $Ba_{1-x}Sn_xCl_{1+y}F_{1-y}$ prepared by precipitation (RT), (b) $Ba_{1-x}Sn_xCl_{1+y}F_{1-y}$ prepared by precipitation (15 K), (c) $Ba_{1-x}Sn_xCl_{1+y}F_{1-y}$ prepared by direct reaction (x, y, T) (0.225, -0.15, RT), (d) $Ba_{1-x}Sn_xCl_{1+y}F_{1-y}$ prepared by direct reaction (x, y, T) (0.225, -0.15, 15 K), (e) $Ba_{1-x}Sn_xCl_{1+y}F_{1-y}$ prepared by direct reaction (x, y, T) (0.225, 0.250, RT) (f) $Ba_{1-x}Sn_xCl_{1+y}F_{1-y}$ prepared by direct reaction (x, y, T) (0.225, 0.250, 15 K) and (g) $BaSnF_4$ prepared by direct reaction (RT spectrum)

RESULTS AND DISCUSSION

X-ray diffraction (fig. 2) shows that, regardless of the method of preparation and of its chemical composition, the $Ba_{1-x}Sn_xCl_{1+y}F_{1-y}$ solid solution gives the same set of Bragg peaks as unsubstitued BaClF, and can therefore be assumed to have the same crystal structure and to be fully disordered. Wild variations in peak intensity have been shown to be due to preferred orientation. The Mössbauer spectra (fig. 3) show the following: (i) precipitated samples (figs. 3a & 3b) contain the highly unexpexted stannous Sn^{2+} ions (single line at 4.1 mm/s) in a slightly distorted environment (broadened line); (ii) samples prepared by the dry method for y<0 (Cl/F<1) (fig. 3c & 3d) contain a mixture of the expected covalently bonded tin(II) (quadrupole doublet) and unexpected Sn^{2+} ions; covalent tin(II) has the weakest lattice since it gains more on cooling; (iii) samples prepared by the dry method for y>0 (Cl/F>1) (fig. 3e & 3f) contain only the unexpected Sn^{2+} ions, and its recoil-free fraction at RT is so low that the signal is barely detectable; however, on cooling, the Sn^{2+} ion signal is strong (iv) $BaSnF_4$, for comparison, gives a large quadrupole doublet, generated by the large e.f.g., which is created by tin lone pair (fig. 1). A similar environment is present in the covalently bonded tin(II) of $Ba_{1-x}Sn_xCl_{1+y}F_{1-y}$ obtained by the dry method for y<0. This is expected in an environment enriched in fluorine, since Sn-F bonds are always covalent. The single line at high isomer shift obtained for the stannous ion is expected, since in the Sn^{2+} ion, the tin orbital are not hybridized, and therefore the lone pair is located on the 5s orbital, that is normally spherical, and therefore generates no e.f.g., although a small deviation from sphericity due to polarization by the distorted environment is expected to generate a small e.f.g., hence line broadening. The drastic lattice softening observed at high chlorine content (y>0, i.e. Cl/F>1) can be explained by the fact that since the Sn^{2+} ion is quite substantially smaller than the Ba^{2+} ion, making the stannous ion loose in an oversized cage. The problem is aggravated by the extra Cl, that can be expected to cluster around tin, giving a situation similar to that of $SnCl_2$ [2]. The small signal at ca. 0 mm/s is due to tin(IV), probably SnO_2, generated by surface oxidation of the particles [3].

REFERENCES

[1] A. Muntasar and G. Dénès, in *Mössbauer Spectroscopy in Materials Science*, ed. M. Miglierini and D. Petridis, NATO Science Series, 3. High Technology, Vol. **66**, Kluwer Academic Publishers, Dordretch (Netherlands), 1999, p. 39.
[2] J.D. Donaldson and B.J. Senior, J. Chem. Soc. (A) (1969)2358.
[3] G. Dénès and E. Laou, Hyp. Int. 92(1994) 1013.

A MÖSSBAUER SPECTROSCOPIC STUDY OF SELF-PASSIVATION IN TIN(II) CHLORIDE FLUORIDES: THE SPECIAL CASE OF THE $Ba_{1-x}Sn_xCl_{1+y}F_{1-y}$ SOLID SOLUTION

Georges Dénès* and Abdualhafeed Muntasar
Laboratory of Mössbauer Spectroscopy and Solid State Chemistry, Laboratories for Inorganic Materials, Concordia University, Montréal, Québec, Canada

A wide disordered $Ba_{1-x}Sn_xCl_{1+y}F_{1-y}$ solid solution has been prepared by wet and dry methods. Mössbauer spectroscopy has been used to study the oxidation of tin(II) to tin(IV) and the phenomenon, or lack of, self-passivation, and how it varies with the method of preparation, with the type of bonding, and with the lattice strength.

INTRODUCTION AND PROCEDURES

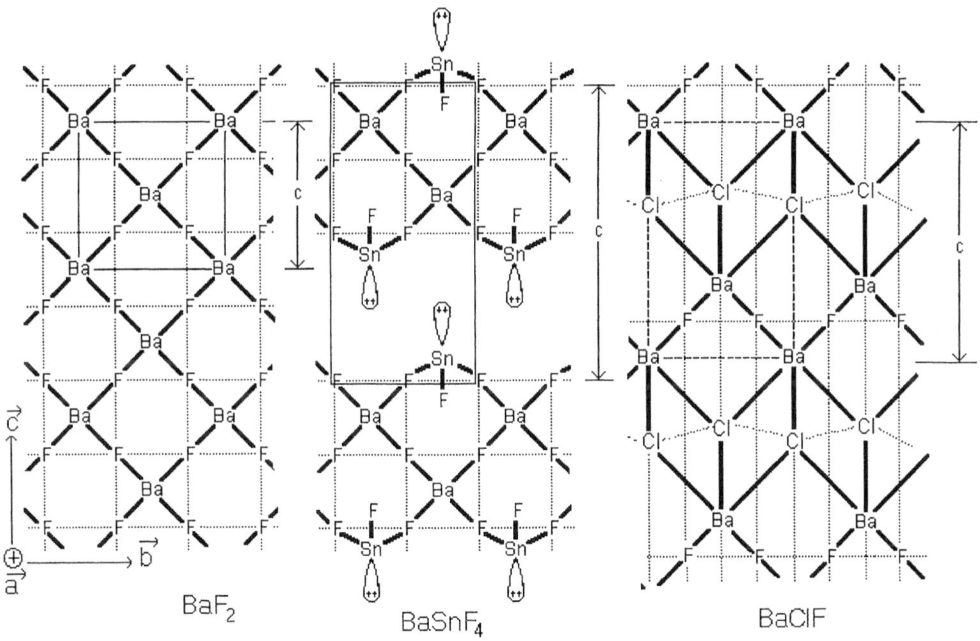

Figure 1: Projection of a slice of the structure of BaF_2, $BaSnF_4$ and BaClF on the (a,b) plane

* Author to whom all correspondence should be addressed (gdenes@vax2.concordia.ca)

The structures of BaClF is a tetragonal distortion of fluorite-type BaF_2. Metal ordering in $BaSnF_4$ (fig. 1). BaClF is a highly ionic structure (PbClF-type), with order between the two anions of highly difference sizes (fig. 1), and therefore both (i) Ba substitution by weak metal tin that usually binds covalently, and (ii) Cl/F disorder, are highly unexpected. Nevertheless, a wide $Ba_{1-x}Sn_xCl_{1+y}F_{1-y}$ solid solution has been prepared, that is doubly disordered, i.e Ba/Sn and Cl/F, while keeping the Cl layer/F layer order characteristic of the PbClF-type structure [1]. The unexpected solid solution has highly unusual properties, including a change of bonding type, tin(II) lattice softening, and surprising oxidation behavior in air. Samples were prepared by precipitation and by direction reactions in dry conditions at high temperature, and studied by X-ray diffraction and Mössbauer spectroscopy [1].

RESULTS AND DISCUSSION

X-ray diffraction (fig. 2) shows that, regardless of the method of preparation or its chemical composition, the $Ba_{1-x}Sn_xCl_{1+y}F_{1-y}$ solid solution gives the same set of Bragg peaks as unsubstitued BaClF, therefore it has the same crystal structure and is fully disordered. In addition, no change is observed after prolonged storage under air. Therefore, X-ray diffraction shows no obvious sign of tin oxidation as the samples age. The same is also observed for $BaSnF_4$. The large variations in peak intensity were shown to be due to preferred orientation. The Mössbauer spectra (fig. 3) show the following: (i) $BaSnF_4$ contains a minor amount of tin(IV), and it is well passivated (figs. 3a & 3b), (ii) $Ba_{1-x}Sn_xCl_{1+y}F_{1-y}$ from dry method for $y<0$ ($Cl/F<1$) (fig. 3c & 3d) contains a mixture of the expected covalently bonded tin(II) (quadrupole doublet) and unexpected Sn^{2+} ions, and it is also well passivated, (iii) $Ba_{1-x}Sn_xCl_{1+y}F_{1-y}$ from dry method for $y>0$ ($Cl/F>1$) (fig. 3e & 3f) contain only the unexpected Sn^{2+} ions, and the recoil-free fraction at RT is very low. They also contains a minor amount of tin(IV) when fresh, however, they do not passivate well since the intensity of the tin(IV) signal (0 mm/s) increases dramatically with aging, (iv) precipitated $Ba_{1-x}Sn_xCl_{1+y}F_{1-y}$ gives also only the unexpected Sn^{2+} ions, however its recoil-free fraction is high (fig. 3f); in addition, it shows a non negligible increase of the relative amount of tin(IV) with aging (fig. 3g). The above observations show that covalently bonded tin(II) is fairly well protected against passivation, whilst ionic tin is not, particularly when its sublattice is soft.

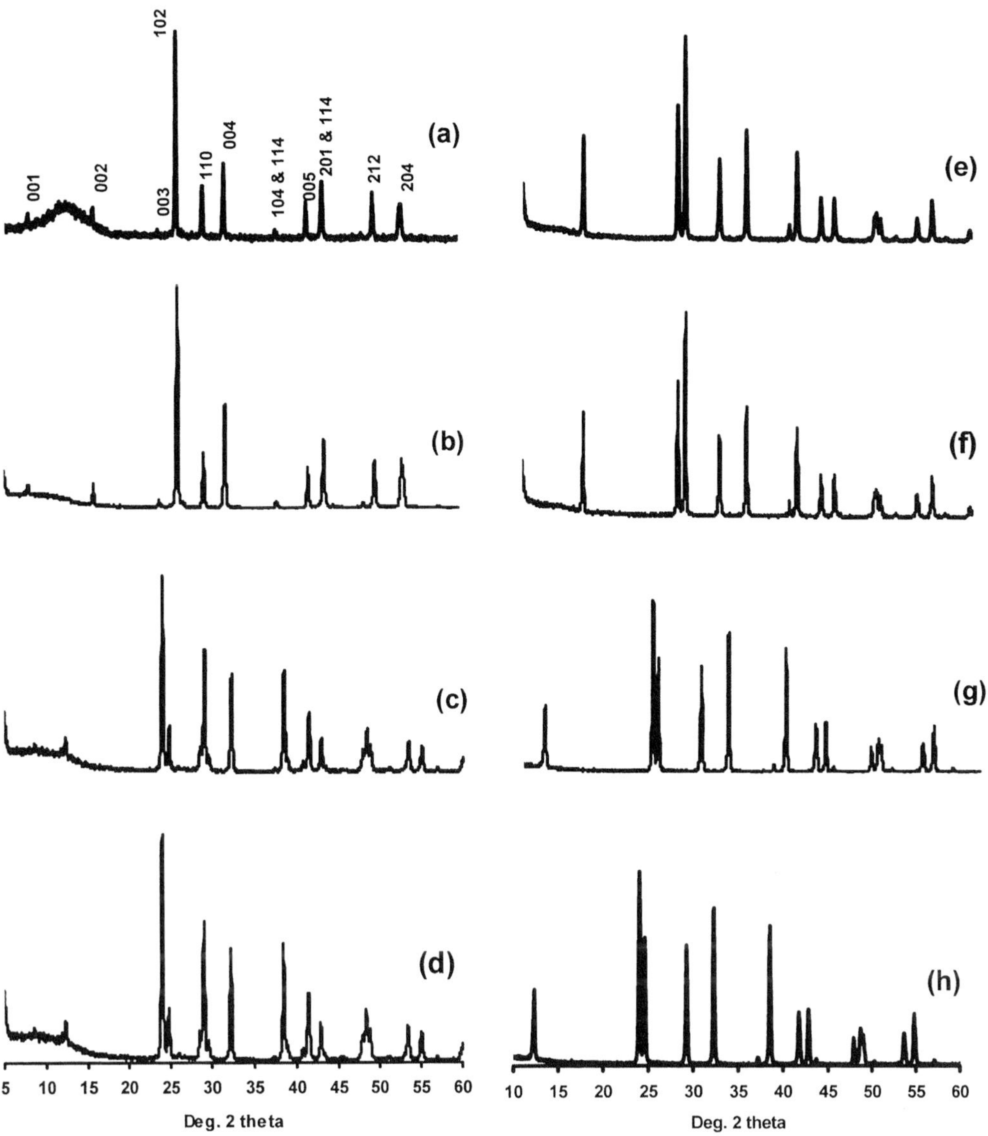

Figure 2: X-ray powder diffraction pattern of: (a & b) $BaSnF_4$ fresh and aged 69 months (c & d) $Ba_{1-x}Sn_xCl_{1+y}F_{1-y}$ from dry method (x = 0.225, y = -0.100) fresh and aged 32 months, (e & f) $Ba_{1-x}Sn_xCl_{1+y}F_{1-y}$ from dry method (x = 0.225, y = 0.225) fresh and aged 32 months, (g & h) precipitated (X = 0.860) $Ba_{1-x}Sn_xCl_{1+y}F_{1-y}$ fresh and aged 48 months

REFERENCES

[1] A. Muntasar and G. Dénès, Mater. Struct 3(1996)246.

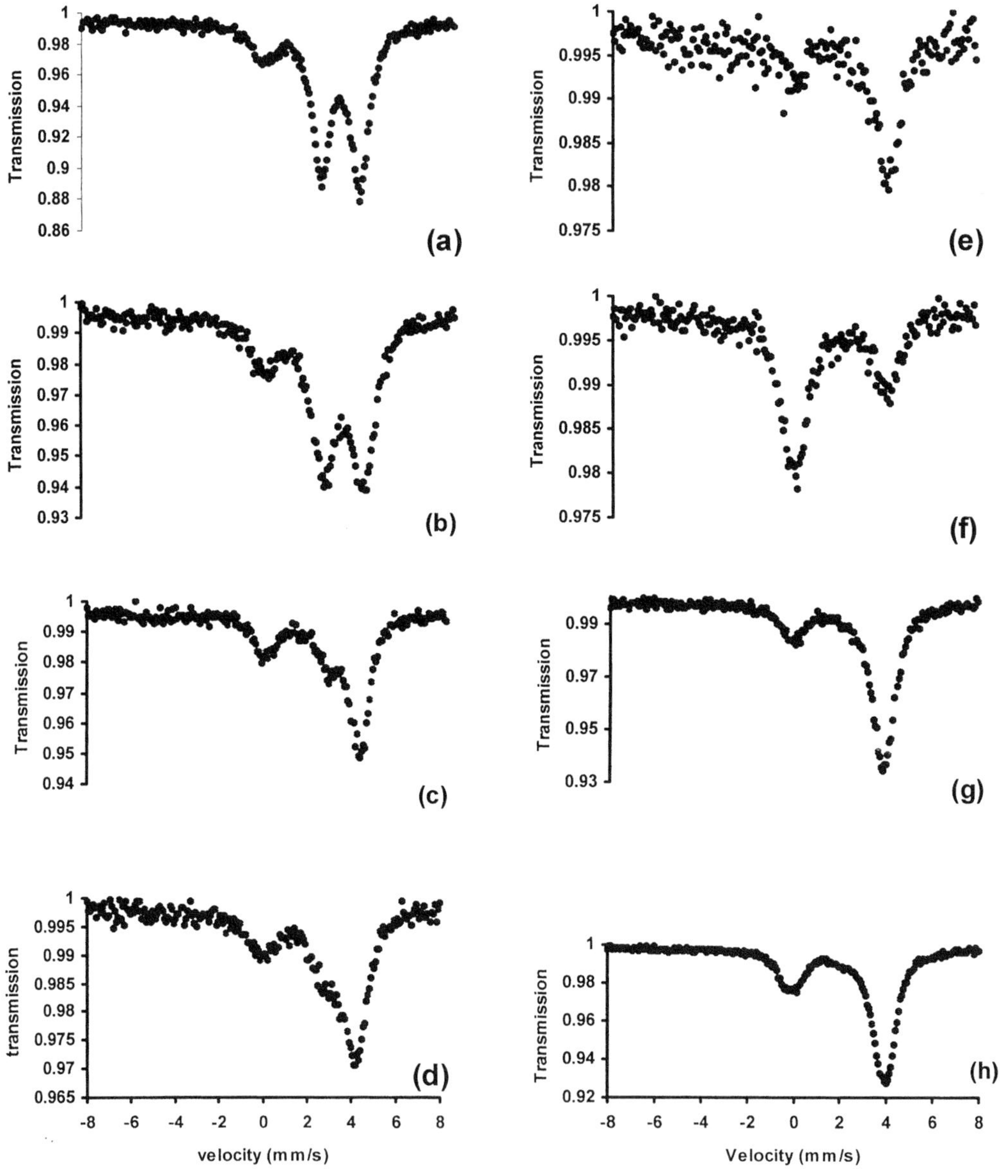

Figure 3: Ambient Mössbauer spectrum of: (a & b) $BaSnF_4$ fresh and aged 69 months (c & d) $Ba_{1-x}Sn_xCl_{1+y}F_{1-y}$ from dry method ($x = 0.225$, $y = -0.100$) fresh and aged 32 months, (e & f) $Ba_{1-x}Sn_xCl_{1+y}F_{1-y}$ from dry method ($x = 0.225$, $y = 0.225$) fresh and aged 32 months, (g & h) precipitated ($X = 0.860$) $Ba_{1-x}Sn_xCl_{1+y}F_{1-y}$ fresh and aged 48 months.

VARIATION IN SPIN DIRECTION OBSERVED FOR OXALATE-BRIDGED Mn(II)-Fe(III) SYSTEMS

S. Iijima and F. Mizutani
National Institute of Advanced Industrial Science and Technology, Higashi 1-1-1, Tsukuba, Ibaraki 305-8566, Japan

Abstract The directions of spin alignment were estimated for A[Mn(II)Fe(III)$(ox)_3$] compounds from their ^{57}Fe Mössbauer spectra. The compounds with large templating cations, $A^+ = P(C_6H_5)_4^+$ and $As(C_6H_5)_4^+$, exhibited a spin direction almost perpendicular to the honeycomb layers consisting of the alternate array of Mn(II) and Fe(III) ions through oxalate ligands. On the other hand, the parallel spin direction was observed for the compound with the small cation, NPr_4^+. The spectrum of the compound with $A^+ = PBu_4^+$ indicated the presence of two kinds of Fe(III) ions with different spin directions.

1.Introduction

^{57}Fe Mössbauer spectroscopy has provided significant information regarding the electronic and magnetic structures of the iron-containing members of the molecule-based magnets A[M(II)M'(III)$(ox)_3$], where A^+ = organic cations such as quarternary ammonium and ox^{2-} = oxalate ion [1-8]. The A[M(II)M'(III)$(ox)_3$] compounds have the same basic structure that is constructed by the honeycomb layers consisting of the alternate array of M(II) and M(III)' ions through ox^{2-} ligands, being separate by the templating counter ions A^+ [9]. The ^{57}Fe Mössbauer investigations of NBu_4[M(II)M'(III)$(ox)_3$] (NBu_4^+ = tetra(*n*-butyl)ammonium ion) [1-8] revealed that the compounds containing a metal ion with single ion anisotropy, such as NBu_4[Fe(II)Cr(III)$(ox)_3$] [1,2], NBu_4[Fe(II)Fe(III)$(ox)_3$] [1] and NBu_4[Co(II)Fe(III)$(ox)_3$] [8], exhibit a spin direction parallel to the basal honeycomb layers. On the other hand, the directions in the compounds consisting of only magnetically-isotropic metal ions, such as NBu_4[Ni(II)Fe(III)$(ox)_3$] [4] and ^{57}Fe(III)-doped NBu_4[Mn(II)Cr(III)$(ox)_3$] [6], are nearly perpendicular to the basal layers. NBu_4[Mn(II)Fe(III)$(ox)_3$] (**1**) was found to be an exception, which gives an intermediate direction of spin alignment [4]. In the present work, we have measured ^{57}Fe Mössbauer spectra for A[Mn(II)Fe(III)$(ox)_3$] compounds with $A^+ = P(C_6H_5)_4^+$ (**2**), $As(C_6H_5)_4^+$ (**3**), NPr_4^+ (**4**) and PBu_4^+ (**5**), in order to elucidate the effect of the kinds of the counter cation (A^+) on the magnetic structure of the Mn(II)-Fe(III) system, where C_6H_5 and Pr are phenyl and *n*-propyl groups, respectively.

2.Experimental

Compounds **2-5** were prepared as microcrystals according to the procedures previously reported [10,11]. The authenticity of the compounds was confirmed by microanalyses for C, H, N, Mn and Fe. These samples were further checked by the cell parameters (a_0, c_0) of hexagonal lattice that were essentially consistent with the values reported in the literature [9]. The distance between the basal layers (d) of each compound was obtained from the c_0 value [12]. The Mössbauer spectra were measured in the temperature range of 4.2–298 K with an instrument described previously [4]. The isomer shifts are reported relative to metallic iron foil.

3.Results and Discussion

All the compounds **2-5** exhibited a magnetically-relaxed or split Mössbauer spectrum below the Néel temperature (T_N = 25–27 K [11]). The Mössbauer spectra at 4.2 K of the compounds **2-4**, except for **5**, consisted of a well-resolved sextet showing an axial symmetry around the ^{57}Fe nucleus, as exemplified for **2** and **4** in Figures 1a and 1b, respectively. The angle (Θ) between the internal magnetic field ($\boldsymbol{H}_n$) and the principal axis of V_{zz} at the Mössbauer nucleus in each of the three compounds was estimated at 4.2 K by applying the following approximate equation for axial symmetry; $S_1 - S_2 = -1/2eV_{zz}Q(3\cos^2\Theta - 1)$, where S_1 and S_2 represent the interval between the lowest-energy and the second lowest-energy peaks, and that between the highest-energy and the second highest-energy peaks, respectively [13]. The quadrupole splitting (ΔE_Q) just above T_N (0.61, 0.62 and 0.66 mm/s at 30 K for **2-4**, respectively) was employed as the value of $|1/2eV_{zz}Q|$ in the equation. The principal axis of V_{zz} of iron in both the M(II) and M'(III) sites of the A[M(II)M'(III)(ox)$_3$] compounds with spherical A^+ is rationally thought to coincide with the D_3 axis which goes through the iron atoms and is perpendicular to the basal layers [2,6].

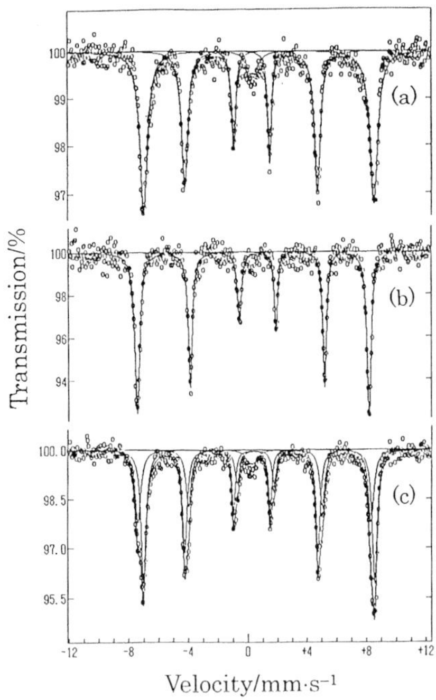

Figure 1. *^{57}Fe Mössbauer spectra at 4.2 K of (a) compound 2, $P(C_6H_5)_4[Mn(II)Fe(III)(ox)_3]$, (b) compound 4, $NPr_4[Mn(II)Fe(III)(ox)_3]$ and (c) compound 5, $PBu_4[Mn(II)Fe(III)(ox)_3]$.*

Compounds **2** and **3** having larger values of the distance between the basal layers (d = 9.52 and 9.65 Å, respectively) than that of **1** (d = 8.95 Å) exhibited Mössbauer spectra noticeable for the $S_1 < S_2$ relationship, which afforded small Θ values (21° for **2** and 20° for **3**) indicating spin directions almost perpendicular to the basal layers. The positive sign of V_{zz} was determined for both compounds; the same sign was reported for $NBu_4[M(II)Fe(III)(ox)_3]$ compounds [4,6,7]. The Θ values of **2** and **3** were different from the intermediate Θ value (52°) of **1**, and well coincided with those of $NBu_4[Ni(II)Fe(III)(ox)_3]$ [4] and ^{57}Fe(III)-doped $NBu_4[Mn(II)Cr(III)(ox)_3]$ [6]. The Mössbauer parameters obtained were listed in Table 1 along with the d value for each compound. A similar perpendicular spin alignment was also shown for $P(C_6D_5)_4[Mn(II)Fe(III)(ox)_3]$ by a neutron diffraction study [14].

In the Mössbauer spectrum of compound **4** with the shortest d (8.20 Å), on the other hand, the S_1 was observed to be fairly larger than S_2 (Figure 1b); the estimated Θ value (83°) indicates a nearly parallel direction of spin alignment. The above results suggest, at least for the A[M(II)M'(III)(ox)$_3$] compounds consisting of only magnetically-isotropic metal ions, the spin direction is very sensitive to the d value and it can be controlled by choosing the A^+ of appropriate bulkiness, although the mechanism governing the spin direction is not understood at present. To get information about the effect of A^+ in the compounds including a metal ion with single ion anisotropy, for comparison, the Θ values were estimated for $P(C_6H_5)_4[Fe(II)Cr(III)(ox)_3]$ and $P(C_6H_5)_4[M(II)Fe(III)(ox)_3]$ (M = Fe, Co) by the spectroscopy. All the compounds showed Θ values close to 90° in accord with the cases of the NBu_4^+ analogues [1,2,8].

Table 1

*The magnitude of internal field (H_n), the Θ angle and the distance between the basal layers (d) estimated for compounds **1-5**.*

Compound	**1**	**2**	**3**	**4**	**5** [a)]
H_n/kOe	502 [c)]	489	493	493	491 and 497
Θ [b)] /°	52 [c)]	20	21	83	19 and 60
d /Å	8.95	9.52	9.65	8.20	9.19

a) Two components were observed in the spectrum of **5**.
b) The Θ value of the component of **5** with H_n = 497 kOe was estimated on the assumption of $V_{zz} > 0$. For the other cases, the positive sign of V_{zz} was determined from the line-shape analysis.
c) References 4 and 7.

The d value of compound **5** (9.19 Å) was intermediate between those of **1** and **2**. This compound exhibited a complicated feature of Mössbauer spectrum at 4.2 K which can be decomposed into two sets of sextet (Figure 1c), while only a quadrupole doublet was observed above 28 K. The two components (ca. 1:2 area ratio) were characterized by different Θ values, 60° and 19°; the former and the latter values are comparable to that of **1** and that of **2** (or **3**), respectively. Some polymorphism or disorder in the lattice structure of A[M(II)M'(III)(ox)$_3$], suggested by Nuttall, et al. [14] and Ovanesyan, et al. [15], would be responsible to the presence of two kinds of Fe(III) ions in **5**.

References

[1] S. Iijima, T. Katsura, H. Tamaki, M. Mitsumi, N. Matsumoto and H. Ōkawa, Mol. Cryst. Liq. Cryst. 233 (1993) 263.
[2] N. S. Ovanesyan, G. V. Shilov, L. O. Atovmyan, R. N. Lyubovskaya, A. A. Pyalling and Yu. G. Morozov, Mol. Cryst. Liq. Cryst. 273 (1995) 175.
[3] W. M. Reiff, J. Kreisz, L. Meda and R. U. Kriss, Mol. Cryst. Liq. Cryst. 273 (1995) 181.
[4] S. Iijima, F. Mizutani, M. Mitsumi, N. Matsumoto and H. Ōkawa, Inorg. Chim. Acta 253 (1996) 47.
[5] S. Iijima and F. Mizutani, Mol. Cryst. Liq. Cryst. 306 (1997) 227.
[6] S. Iijima, S. Koner and F. Mizutani, J. Radioanal. Nucl. Chem. 239 (1999) 245.
[7] S. Iijima and F. Mizutani, Mol. Cryst. Liq. Cryst. 335 (1999) 143.
[8] S. Iijima and . Mizutani, unpublished results.
[9] R. Pellaux, H. W. Schmalle, R. Huber, P. Fisher, T. Hauss, B. Ouladdiaf and S. Decurtins, Inorg. Chem. 36 (1997) 2301, and references quoted therein.
[10] H. Tamaki, M. Mitsumi, K. Nakamura, N. Matsumoto, S. Kida, H. Ōkawa and S. Iijima, Chem. Lett. (1992) 1975.
[11] C. Mathonière, C. J. Nuttall, S. G. Carling and P. Day, Inorg. Chem. 35 (1996) 1201.
[12] G. V. Shilov, N. S.Ovanesyan, N. A. Sanina, A. A. Pyalling and L. O. Atovmyan, Russ. J. Coord. Chem. 24 (1998) 802.
[13] W. H. Reiff, Coord. Chem. Rev. 10 (1973) 37.
[14] C. J. Nuttall, S. G. Carling and P. Day, Mol. Cryst. Liq. Cryst. 334 (1999) 615.
[15] N. S. Ovanesyan, G. V. Shilov, N. A. Sanina, A. A. Pyalling, L. O. Atovmyan and L. Bottyán, Mol. Cryst. Liq. Cryst. 335 (1999) 91.

Mössbauer Spectroscopic Study on $x Fe_2O_3 \cdot (40-x)Na_2O \cdot 60P_2O_5$ (x=5,20,30) Glasses at low temperature

T. Inamura, T. Toriyama[a], H. Wakabayashi[a], H. Iijima, and K. Yoshino
General Education Center, Musashi Institute of Technology,
1-28-1 Tamazutsumi, Setagayaku, Tokyo 158-8557,Japan
[a]*Department of Energy Science and Engineering, Musashi Institute of Technology,*
1-28-1 Tamazutsumi, Setagayaku, Tokyo 158-8557, Japan

It is well known that Fe^{3+} ions in alkali iron phosphate glasses are in the octahedral site and play a role as the NWM(network modifier) from the isomer shift (I.S.) and the quadrupole splitting (Q.S.) at room temperature. Recently, Day et al. reported that the addition of Al_2O_3 or Fe_2O_3 to a sodium phosphate glass yields the glass with considerably good chemical durability and suggested that Al^{3+} and Fe^{3+} ions constitute octahedra to be combined with PO_4-chain or tetrahedra to be substituted for P^{5+} ions in phosphate glasses. On the other hand, using the criterion for the Debye temperature of Fe^{3+} ions in glasses proposed by Nishida, we can examine whether they are present in the NWM or in the NWF (network former). In this paper, Mössbauer spectroscopic study of [$x Fe_2O_3 \cdot (40-x)Na_2O \cdot 60P_2O_5$ (x=5,20,30)] was performed in the range of 6K to 300K. It was confirmed that both ions are in the octahedral site from I.S. and Q.S. values of Fe^{3+} ions. Their Debye temperature were determined to be about 350K at x=5,20,30. It was concluded that although Fe^{3+} ions in the sodium iron phosphate glass are present in the NWM, they are strongly combined with oxygen and act as NWF.

1. Introduction

Phosphate glasses are generally of practical interest because the combination of high thermal expansion and low melting temperature makes phosphate glasses potential candidates for sealing to high expansion metals such as aluminium, aluminium alloys, or stainless steels. But their relatively poor chemical durability limits their practical application to nuclear waste hosts. About this problem, Y. B. Peng and D. E. Day found that the additions of Al_2O_3 and Fe_2O_3 to alkali phosphate glasses are very effective in improving the chemical durability [1]. Then, as the iron phosphate glasses have lower chemical durability and higher crystallization tendency than borosilicate glasses, they are considered as an alternated host for vitrifying high nuclear level wastes [2].

In the Mössbauer spectroscopic study, it has been suggested by a number of the researchers that in the alkali phosphate glasses whose network former (NWF) constitute PO_4 tetrahedra, Fe^{2+} and Fe^{3+} ions are present and both ions constitute FeO_6 octahedra from Isomer shift (I.S.) and Quadrupole splitting (Q.S.) for Fe^{3+} ions at room temperature and play a role of network modifier (NWM)[3,5,6,8]. Recently, X. Yu and D. E. Day et al. reported that sodium phosphate glasses which contain more than 30 mol% of Fe_2O_3 content have characteristics such as the better water resistance and the origin comes from the fact that Fe^{3+} ions locate in the distorted octahedral sites and provide strong P-O-Fe^{3+} bonds instead of P-O-P bonds [7].

On the other hand, T. Nishida proposed the relation between NWF and NWM and Debye temperature of the Fe^{3+} ions in a variety of glasses except alkali phosphate ones, such as the NWM for the Debye temperature less than 270K and the NWF for the one more than 280K [8]. Thus, it plays a significant role in measuring I.S. and Q.S. for Fe^{3+} ions and their Debye temperature in alkali iron phosphate glasses, which indicate how strongly they are combined with the surrounding oxygen in spite of NWM.

In this paper, Mössbauer spectroscopic study on $x Fe_2O_3 \cdot (40\text{-}x)Na_2O \cdot 60P_2O_5$ (x=5,20,30) was performed to determine the coordination of the Fe^{3+} ions from I.S. and Q.S. values at room temperature and to measure Debye temperatures of those ions from temperature dependency of absorption area. The results were compared with those of references [5,6] and the strength of P-O-Fe^{3+} bonds in these alkali iron phosphate glasses were discussed.

2. Experimental

The Na_2CO_3, Fe_2O_3 and H_3PO_4 of granted reagent grade were used for preparing the glass samples. The starting mixture was sintered in an alumina crucible at about 500°C and melted at 1250°C in air for 2 hours, and then quenched on an iron plate at room temperature. Mössbauer spectra for the powder $x Fe_2O_3 \cdot (40\text{-}x) Na_2O \cdot 60P_2O_5$ (x=5,20,30) samples which are all thin to extent that the absorption area were in proportion to recoilless fraction, were measured by the constant acceleration mode at room temperature and in the low temperature range from 5K to 300K. ^{57}Co (1.11GBq) diffused into Rh matrix was used as a source. Mössbauer parameters for each spectrum were calculated with the NORMOS least-square fitting program. A pure iron foil was used for calibration of the spectrometer.

3. Results and discussion

Mössbauer spectrum of $5Fe_2O_3 \cdot 35Na_2O \cdot 60P_2O_5$ glass at room temperature is shown in Fig. 1. A and B doublets correspond to the components of Fe^{2+} ion and Fe^{3+} ion,

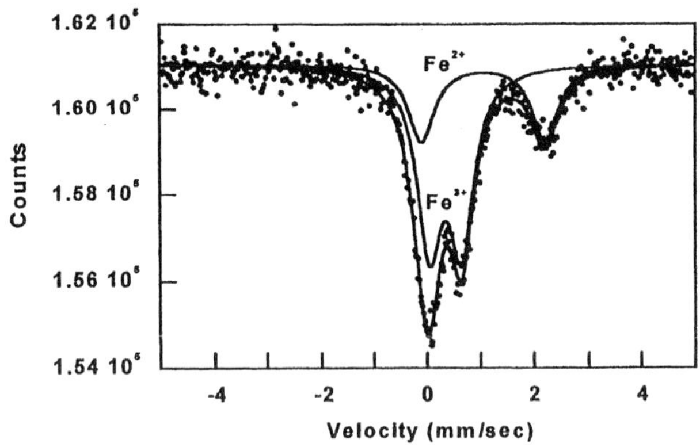

Fig. 1. Mössbauer spectrum of $5Fe_2O_3 \cdot 35Na_2O \cdot 60P_2O_5$ glass at room temperature.

respectively. Mössbauer spectra of $x\mathrm{Fe_2O_3}$-$(40-x)\mathrm{Na_2O}$-$60\mathrm{P_2O_5}$ (x=20,30) glasses at room temperature were similar to that of $5Fe_2O_3\cdot35Na_2O\cdot60P_2O_5$ glass. Table 1 shows the results of I.S. and Q.S. for Fe^{3+} ions at room temperature against Fe_2O_3 content of the samples. As shown in table 1, the values of I.S. for Fe^{3+} ions were 0.431-0.448 mm/s for the present

Table 1

I.S. of Fe^{3+} ions in $x\mathrm{Fe_2O_3}\cdot(40-x)\mathrm{Na_2O}\cdot60\mathrm{P_2O_5}$($x$=5,20,30) glasses at room temperature.

x (mol.%)	I.S. of Fe^{3+} ions (mm/s)	Q.S. of Fe^{3+} ions (mm/s)
5	0.448	0.605
20	0.435	0.640
30	0.431	0.728
Brooks et al. [4]	0.43	0.49
Taragin et al. [5]	0.41	0.80

samples. According to the literature values (0.40-0.55 mm/s) presented by M. D. Dyar [3], T. Nishida [8] and G. Tomandl et al. [9], these values correspond to NWM (O_h). Furthermore, the values of Q.S were 0.605-0.728mm/s and similarly, comparing these values with the literature values (0.36-0.88 mm/s) by the same authors mentioned above, they also correspond to NWM (O_h).

Fig. 2 shows the temperature dependency of absorption area ratio $A(T)/A(5\mathrm{K})$ normalized to absorption area at 5K for Fe^{3+} and Fe^{2+} ions of $5Fe_2O_3\cdot35Na_2O\cdot60P_2O_5$ glass and that of the recoilless fraction ratio $f(T)/f(5\mathrm{K})$ normalized to recoilless fraction calculated at 5K for the Debye model with θ_D=359 K. The recoilless fraction $f(T)$ were

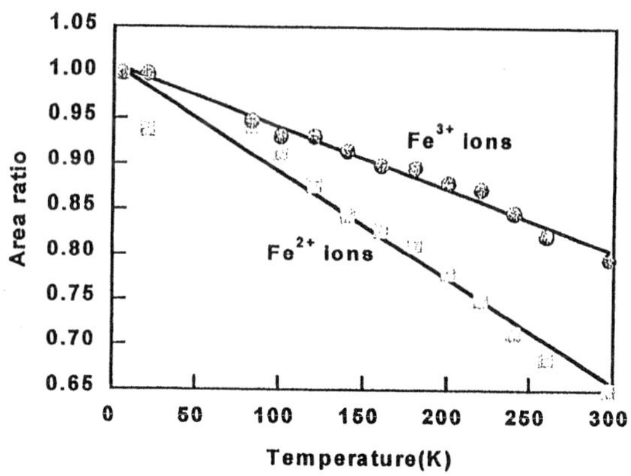

Fig. 2. Temperature dependency of the absorption area ratio $A(T)/A(5\mathrm{K})$ normalized to absorption area at 5K and the curves obtained by least square fittings for Fe^{3+} and Fe^{2+} ions in $5Fe_2O_3\cdot35Na_2O\cdot60P_2O_5$ glass.

calculated using the approximation described in ref. 10. We deduced Debye temperature θ_D of Fe^{3+} and Fe^{2+} ions for sample ($5Fe_2O_3 \cdot 35Na_2 \cdot 60P_2O_5$) glass by a least-squares fitting method. The calculated results are shown in table 2 together with those for the samples (x=20, 30). The θ_D values for Fe^{3+} and Fe^{2+} ions were determined to be 345-359K and 265-270K, respectively.

Table 2

Debye temperatures of Fe^{3+} and Fe^{2+} ions for samples.

x (mol.%)	Debye temperature of Fe^{3+} ions (K)	Debye temperature of Fe^{2+} ions (K)
5	359	270
20	347	269
30	345	265
Ref.4	325	251
Ref.5	385	230

The former are slightly higher than the θ_D value (325K) obtained by J.S. Brooks et al. [5] for alkali iron phosphate glass and lower than that (385K) obtained by M.F. Taragin et al. [6] for sodium iron aluminophoshate glass. As shown in table1, both I.S. values belong to NWM. (*Oh*). Therefore, we can conclude that Fe^{3+} ions in sodium iron phosphate glass are located in NWM (*Oh*), but they are considered to form a network with the same degree of strong binding state as NWF (*Td*) to keep the Debye temperature high.

4. Conclusion

It was concluded that Fe^{3+} ions in $x Fe_2O_3 \cdot (40-x) Na_2O \cdot 60P_2O_5$ (x=5,20,30) glasses are located in the coordination of NWM (*Oh*) from the values of I.S. and Q.S. at room temperature, but they are considered to form a network with the same degree of strong binding state as NWF (*Td*) judging from the Debye temperature as described in reference 8.

Acknowledgement

We are grateful to Prof. T. Nishida, Department of Biological and Environmental Chemistry, Kinki University, for his fruitful comments and discussion.

References

[1] Y. B. Peng and D. E. Day. Glass Technol. 32(1991)166.
[2] C. S. Ray et. al., Non-Cryst. Solids 249(1999)1.
[3] M. Darby Dyar, Amer. Min. 70(1985)304.
[4] Y. Lin et al., J. Non-Cryst. Solids 112(1986)136.
[5] J. S. Brooks et al., Phys. Chem. of Glasses 33(5)(1992)167.
[6] M. F. Taragin, J. C. Eisenstein and W.Haller, Phys. and Chem. of Glasses 13(1972)149.
[7] X. Yu, D.E.. Day, G.J. Long and R.K. Brow, J. Non-Cryst. Solids 215(1997)21.
[8] T. Nishida, J. Non-Cryst. Solids 177(1994)257.
[9] G. Tomandl, in: *Glass Science and Technology*, Vol. 4B, eds. D.R. Uhlmann and N.J. Kreidl (Academic Press, New York, 1990)p.273.
[10] J. Herberle, in: *Mössbauer Methodology*, Vol. 7, eds. I.J. Gruverman and C.S. Seidal (Plenum Press, New York, 1971)p.299.

Synthesis and ^{57}Fe and ^{151}Eu Mössbauer Spectroscopic Studies of New Lanthanoid-iron Complexes

M. Katada[a], T. Nawa[a], H. Kumagai[b], S. Kawata[c], and S. Kitagawa[d]

[a] *Graduate School of Science, Tokyo Metropolitan University, Minami-ohsawa, Hachioji, Tokyo 192-0397, Japan*
[b] *Institute for Molecular Science, Myodaiji, Okazaki, Aichi 444-8585, Japan*
[c] *Department of Chemistry, Shizuoka University, Shizuoka 422-8529, Japan*
[d] *Graduate School of Engineering, Kyoto University, Sakyo-ku, Kyoto, 606-8501, Japan*

New lanthanoid-iron 2,2'-bipyridine (bpy) and 2,2':6',2"-terpyridine (terpy) complexes have been investigated by a variable temperature Mössbauer spectroscopy and a single crystal structure analysis. The crystal structures of Ln-bpy (Ln = La, Nd and Eu) and Eu-terpy complexes were determined as follows; $\{[Ln(bpy)_2(H_2O)_4]_2[(\mu\text{-}NC)_2Fe(CN)_4]\}[Fe(CN)_6]\cdot 8H_2O$ (Ln = La and Nd), $\{[Eu(bpy)(H_2O)_4][(\mu\text{-}NC)_2Fe(CN)_4]\cdot 1.5bpy\cdot 4H_2O]\}_x$ and $\{[Eu(terpy)(H_2O)_4][(\mu\text{-}NC)_2\text{-}Fe(CN)_4]\}_x$, respectively. The structure of Eu-complexes consists of one-dimensional zig-zag chain polymer. While La- and Nd-bpy complexes consist of discrete $\{[Ln(bpy)_2(H_2O)_4]_2[(\mu\text{-}NC)_2\text{-}Fe(CN)_4]\}^{3+}$ (Ln = La and Nd) cation and $[Fe(CN)_6]^{3-}$ anion. The temperature dependences of ^{151}Eu-Mössbauer spectral area for Eu-complexes lay between those of three-dimensional polymer and monomeric molecular compounds.

1. Introduction

Although ^{151}Eu and ^{153}Eu can be used as Mössbauer nuclei, ^{151}Eu is more frequently used for Mössbauer spectroscopic studies because of a narrow line-width and large recoil-free fraction [1]. Hence, after ^{57}Fe and ^{119}Sn, ^{151}Eu is one of the most studied Mössbauer isotopes. However, there are few works studied by using both ^{151}Eu and ^{57}Fe isotopes. We have reported the variable temperature Mössbauer studies and X-ray crystal analyses for new lanthanoid-iron ($Fe^{III}(CN)_6^{3-}$) complexes with 1,8-naphtyridine (Eu-napy) [2] and 1,10-phenanthroline (Eu-phen) [3]. It was found that Eu-napy and Eu-phen complexes consist of one-dimensional zig-zag chain structure. The temperature dependence of area intensity of Mössbauer lines fell in the region between that of three-dimensional polymer and monomeric molecular complex. In the present study, some new lanthanoid-iron complexes with bipyridine (Ln-bpy) and terpyridine (Ln-terpy) were synthesized and characterized by means of ^{57}Fe and ^{151}Eu Mössbauer spectroscopies and X-ray crystal analysis.

2. Expermental

Samples. Ln-bpy complexes were prepared by mixing 5 mmole of $LnCl_3\cdot 6H_2O$ and 15 mmole of bpy in 25 ml of H_2O and 5 mmole of $K_3Fe(CN)_6$ in 25 ml of H_2O. In the preparation of Eu-terpy complex, a mixture of H_2O and ethanol (1:1) was used as a solvent instead of H_2O. The purity of these complexes was confirmed by elemental analyses for C, H and N.

Mössbauer Spectroscopic Measurement. ^{57}Fe and ^{151}Eu Mössnbauer spectra of the complexes were measured with a constant-acceleration spectrometer using $^{57}Co(Rh)$ and $^{151}SmF_3$ sources, respectively, at room temperature. The spectra were fitted with Lorentzian line shapes using a least-squares fitting procedure. The isomer shift values of ^{57}Fe and ^{151}Eu spectra are reported relative to α-Fe and EuF_3 resonances, respectively, at room temperature.

Crystallographic Data Collection and Refinement of the structure. Data collection was carried out on a Rigaku AFC7R using Mo K_α radiation. Crystallographic data are given in Table 1.

Table 1

Summary of crystal data of {La(bpy)$_2$(H$_2$O)$_4$]$_2$[(μ-NC)$_2$Fe(CN)$_4$]}$_2$[Fe(CN)$_4$]·8H$_2$O, {Nd(bpy)$_2$(H$_2$O)$_4$]$_2$[(μ-NC)$_2$Fe(CN)$_4$]}-[Fe(CN)$_4$]·8H$_2$O, {[Eu(bpy)(H$_2$O)$_4$][(μ-NC)$_2$Fe(CN)$_4$]·1.5bpy·4H$_2$O}$_x$ and {[Eu(terpy)(H$_2$O)$_4$][(μ-NC)$_2$Fe(CN)$_4$]}$_x$.

Complex	La-bpy	Nd-bpy	Eu-bpy	Eu-terpy
Formula Weight	1614.71	1625.38	898.50	669.25
Crystal System	triclinic	triclinic	triclinic	monoclinic
Space Group	$P\bar{1}$	$P\bar{1}$	$P\bar{1}$	$P2_1/c$
a, Å	11.497(3)	11.486(3)	10.644(5)	9.708(3)
b, Å	15.708(3)	15.608(2)	19.892(6)	16.700(4)
c, Å	10.214(2)	10.182(2)	9.724(3)	15.703(5)
α,°	92.28(2)	92.44(1)	96.09(3)	
β,°	113.50(2)	113.68(2)	115.66(2)	
χ,°	87.17(2)	87.20(1)	83.08(3)	99.05(3)
V Å	1689.2(7)	1669.2(5)	1838(1)	2514(1)
Z	1	1	2	4
Dc, gcm^{-3}	1.587	1.617	1.623	1.829
μ(MoKα), cm^{-1}	17.31	20.32	21.43	30.93
Crystal Size, mm	0.60x 0.40x 0.15	0.20x 0.20x 0.15	0.20x 0.20x 0.15	0.2x 0.2x 0.2
Crystal Color, Habit	orange, prismatic	orange, prismatic	yellow, prismatic	orange, prismatic
$^F000$	810.00	816.00	904.00	1356.00
No. Observations (I>3.00 δ(I))	7130	6722	7556	2613
No. Variables	418	418	469	343
R, R(w)	0.026, (0.032)	0.046, (0.055)	0.031, (0.036)	0.045, (0.046)

3. Results and Discussion

3.1 Crystal Structure

In the Ln-bpy and Ln-terpy (Ln-terpy) series, the {[La(bpy)$_2$(H$_2$O)$_4$]$_2$[(μ-NC)$_2$Fe(CN)$_4$]}-[Fe(CN)$_6$]·8H$_2$O, {[Nd(bpy)$_2$(H$_2$O)$_4$]$_2$[(μ-NC)$_2$Fe(CN)$_4$]}[Fe(CN)$_6$]·8H$_2$O, {[Eu(bpy)(H$_2$O)$_4$][(μ-NC)$_2$Fe-(CN)$_4$]·1.5bpy·4H$_2$O]}$_x$ and {[Eu(terpy)(H$_2$O)$_4$][(μ-NC)$_2$Fe(CN)$_4$]}$_x$ were prepared as single crystals and their crystal structures were determined by X-ray analysis. ORTEP drawing of La-bpy complex and structure of Nd-bpy are shown in figure 1. The structure of these complexes consist of discrete {[Ln(bpy)$_2$(H$_2$O)$_4$]$_2$[(μ-NC)$_2$Fe(CN)$_4$]}$^{3+}$ (Ln = La and Nd) cation and [Fe(CN)$_6$]$^{3-}$ anion. The structure of Nd-bpy is consistent with the result reported by Yi et al. [4].

ORTEP drawing and structure of Eu-bpy is shown in figures 2 and 3, respectively. In Eu-bpy complex, two types of uncoordinated bipyridine molecules exist: one is the bipyridine bound to hydrogen atom of water molecule coordinated to europium atom in an adjacent polymer chain. The N position in the bipyridine ligand becomes *trans*-form during the formation of the hydrogen bond. This type of uncoordinated bipyridine molecules makes stacking form with the bipyridine molecules coordinated to Eu atom. The other is the bipyridine bound to hydrogen of water molecules coordinated

Eu atoms with *cis*-form. This type of bipyridine stacks the same type of bipyridines as shown in figure 3. The structure of Ce-bpy and Pr-bpy complexes are in agreement with that of La-bpy, whereas the structure of Sm-bpy is consistent with that of Eu-bpy from the X-ray powder diffraction measurements.

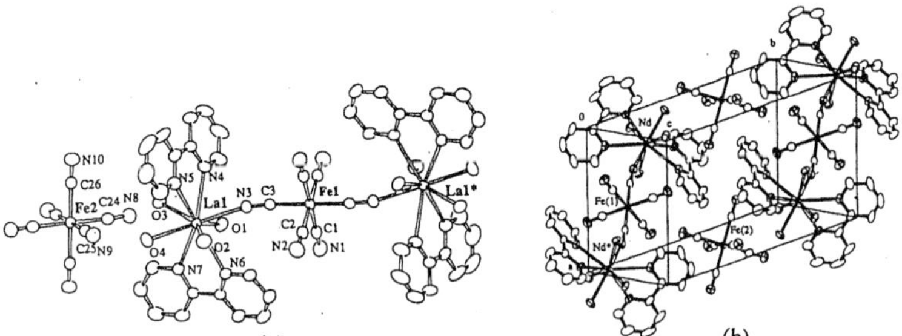

Figure 1. ORTEP drawing of of $\{La(bpy)_2(H_2O)_4\}_2[(\mu\text{-}NC)_2Fe(CN)_4]\}[Fe(CN)_6]\cdot 8H_2O$ (a) and $\{Nd(bpy)_2(H_2O)_4\}_2[(\mu\text{-}NC)_2Fe(CN)_4]\}$-$[Fe(CN)_6]\cdot 8H_2O$. Select bond angle and bond length are: N(4)-Eu-N(1)(cyano) = 74.1(3); C(1)-Fe(1)-C(4) = 178.0(6); C(5)-Fe(1)-C(2) = 177.5(6); C(6)-Fe-C(3) = 178.6(9) ° ; Fe(1)-C(1) = 1.92(1); Fe(1)-C(2) = 1.95(1); Fe(1)-C(3) = 1.96(1); Fe(1)-C(4) = 1.92(1); Fe(1)-C(5) = 1.93(1) and Fe(1)-C(6) = 1.94(1) A.

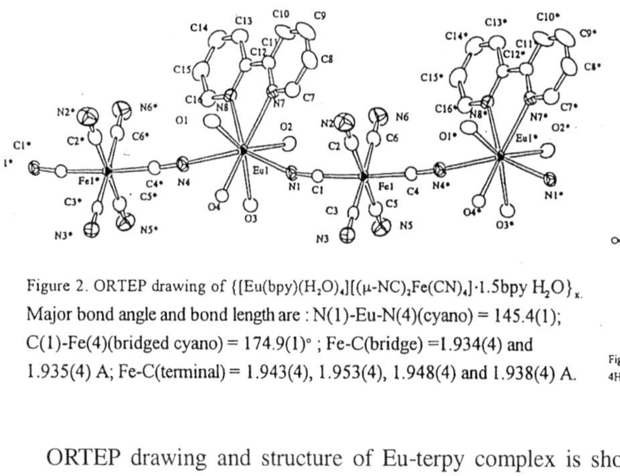

Figure 2. ORTEP drawing of $\{[Eu(bpy)(H_2O)_4][(\mu\text{-}NC)_2Fe(CN)_4]\cdot 1.5bpy\ H_2O\}_x$. Major bond angle and bond length are : N(1)-Eu-N(4)(cyano) = 145.4(1); C(1)-Fe(4)(bridged cyano) = 174.9(1)° ; Fe-C(bridge) =1.934(4) and 1.935(4) A; Fe-C(terminal) = 1.943(4), 1.953(4), 1.948(4) and 1.938(4) A.

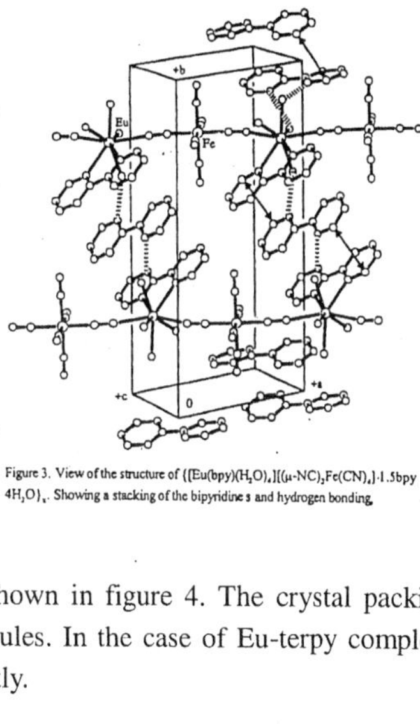

Figure 3. View of the structure of $\{[Eu(bpy)(H_2O)_4][(\mu\text{-}NC)_2Fe(CN)_4]\cdot 1.5bpy\ 4H_2O\}_x$. Showing a stacking of the bipyridines and hydrogen bonding.

ORTEP drawing and structure of Eu-terpy complex is shown in figure 4. The crystal packing structure consists of the stacking form of the terpyridine molecules. In the case of Eu-terpy complex. terpyridine ligand coordinated to Eu atom stacks each other directly.

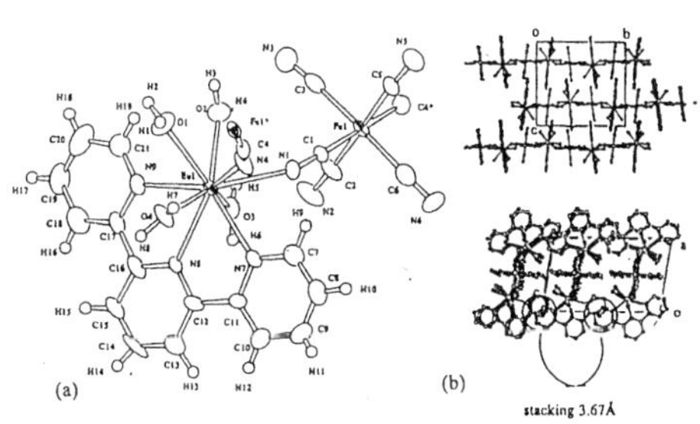

Figure 4. ORTEP drawing (a) and crystal structure (b) of $\{[Eu(terpy)(H_2O)_4][(\mu\text{-}NC)_2Fe(CN)_4]\}x$. Select bond angle and bond length are: N(4)-Eu-N(1)(cyano) = 74.1(3); C(1)-Fe(1)-C(4) = 178.0(6); C(5)-Fe(1)-C(2) = 177.5(6); C(6)-Fe-C(3) = 178.6(9)° ; Fe(1)-C(1) = 1.92(1); Fe(1)-C(2) = 1.95(1); Fe(1)-C(3) = 1.96(1); Fe(1)-C(4) = 1.92(1); Fe(1)-C(5) = 1.93(1) and Fe(1)-C(6) = 1.94(1) A.

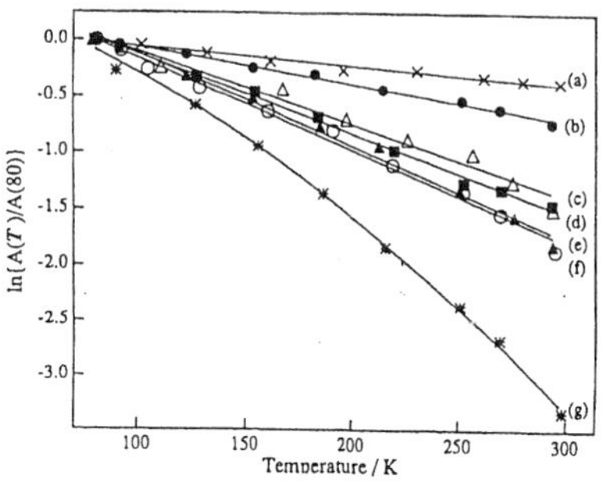

Figure 5. Temperature dependence of [151]Eu-area intensity for (a) EuF_3, (b) EuFe, (c) Eu-terpy, (d) Eu-napy, (e) Eu-bpy, (f) Eu-phen and (g) $Eu(DPM)_3$.

3.2 *Mössbauer spectra*

The ^{151}Eu-Mössbauer spectra observed exhibit a single broad line with unresolved hyperfine structure, while ^{57}Fe-Mössbauer spectra of Ln-bpy and Eu-terpy consist of a well resolved quadrupole doublet at room temperature. ^{151}Eu and ^{57}Fe Mössbauer spectra parameters of the complexes are listed in tables 2 and 3, respectively. The values of ^{151}Eu-isomer shift (*IS*) is in the range of trivalent state. The ^{57}Fe-isomer shift values showed that irons atom in these complexes are trivalent low spin state. Two kinds of doublet were observed for Ln-bpy (Ln = La, Ce, Nd, Pr) in which outside doublet corresponds to the iron in bridged FeC_6 octahedron, while inside one is attribute to that in discrete $[Fe(CN)_6]^{3-}$ ion.

Table 2
^{151}Eu Mössbauer parameters for europium-iron complexes at 78 K.

Complex	IS mms^{-1}	QS(η) mms^{-1}	Γ_{exp} mms^{-1}	Ref.
Eu-bpy	0.59	8.4(0)	2.3	This Work
Eu-terpy	0.36	4.0(0.86)	2.5	This Work
Eu-napy	0.45	4.1(0.70)	2.2	[2]
Eu-phen	0.77	4.3(0.76)	2.3	[3]
Eu-Fe	0.52	7.0(0.43)	2.3	[2]

Table 3
^{57}Fe Mössbauer parameters for Ln-bpy and Eu-terpy complexes at room temperature

Complex	IS mms^{-1}	QS(η) mms^{-1}	Γ_{exp} mms^{-1}
La-pby	-0.12	0.53	0.25
	-0.14	1.26	0.24
Ce-bpy	-0.15	0.54	0.29
	-0.16	1.23	0.22
Pr-bpy	-0.13	0.57	0.28
	-0.15	1.19	0.24
Nd-bpy	-0.15	0.64	0.25
	-0.16	1.21	0.23
Sm-bpy	-0.15	1.62	0.25, 0.25
Eu-bpy	-0.13	1.65	0.26, 0.29
Eu-terpy	-0.14	1.79	0.26, 0.26

The temperature dependence of the spectral area, ln{A(T)/A(78)}/dT of the Mössbauer lines for Eu-bpy and Eu-terpy are shown in figure 5 together with the results of related complexes reported previously [2, 3]. The slop of EuF_3 was the smallest. This indicates that the Eu atoms in this compound are bound strongly due to the formation of a three-dimensional network. The temperature dependence of the molecular complex, $Eu(DPM)_3$ (DPM = 2,2,6,6-tetramethyl-3,5-heptanedionato) is large and deviates significantly from the straight line. This suggests that intermolecular interaction is weak and the molecular motion becomes large with an increasing of temperature. The slope for Eu-bpy and Eu-terpy is smaller than that of Fe-Eu complex, $\{[Eu(H_2O)_2][(\mu\text{-}NC)_6Fe]\cdot 2H_2O\}_x$. The structure of Fe-Eu complex consist of infinite three-dimensional network of FeC_6 groups bridged by cyanide linkages to eight-coordinated $EuN_6(H_2O)_2$ groups. The temperature dependences of the Eu-bpy and Eu-terpy can be interpreted by assuming that the Eu atoms in the one-dimensional polymer complex are more labile than those in the three-dimensional one.

References

[1] F. Grandiean and G.J. Long, in: Mössbauer Spectroscopy Applied to Inorganic Chemistry, Vol. 3 eds. G.J. Long and F.Grandjean, Plenum Press, New York, 1989, p. 513.

[2] M. Katada, T. Ishikawa, S. Kawata, and S. Kitagawa, in: Conf. Proc. Vol. 50 "ICAME-95", I. Ortalli, ed. SIF, Bologna, 1996, p.111.

[3] M. Katada, T. Nawa, H. Kumagai, S. Kawata and S. Kitagawa, J. Radioanal. Nucl. Chem. 239 (1999) 227 .

[4] T. Yi. S. Gao X-W. Chen, C-H. Yan, and B-G. Li, Acta Cryst. C54 (1998) 41.

[5] N.N. Greenwood and T.C. Gibb, Mössbauer Spectroscopy, Chapman and Hall, London, 1971.

In-beam Mössbauer Study of ^{57}Fe Species Arising from ^{57}Mn Implanted into $KMnO_4$

Y. Kobayashi[a], M. K. Kubo[b], Y. Yamada[c], T. Saito[c], A. Yoshida[d], H. Ogawa[a], H. Ueno[a], and K. Asahi[a]

[a] *Applied Nuclear Physics Lab., RIKEN, Hirosawa, Wako, Saitama 351-0198, Japan.*
[b] *School of Science, Univ. of Tokyo, Hongo, Bunkyo, Tokyo 113-0033, Japan.*
[c] *Dept. of Chemistry, Science Univ. of Tokyo, Kagurazaka, Shinjuku, Tokyo 162-8601, Japan.*
[d] *RI Beam Science Lab., RIKEN, Hirosawa, Wako, Saitama 351-0198, Japan.*

We have succeeded in obtaining well-resolved in-beam Mössbauer spectra of ^{57}Fe arising from short-lived ^{57}Mn ($T_{1/2}$ = 1.45 min) implanted into potassium permanganate ($KMnO_4$) between 11 K and 155 K. The ^{57}Fe Mössbauer spectra obtained below 90 K were analyzed with two components of a doublet and a singlet. The singlet was tentatively assigned to the substitutional ^{57}Fe atoms for Mn-sites in tetrahedral $[MnO_4]^-$ with an unusual valence state of Fe(VIII), based on the experimental results of the isomer shifts of the singlet and the calculations of the molecular orbital.

Keywords: in-beam Mössbauer spectroscopy, ^{57}Mn, RI beam, implantation, ^{57}Fe, hyperfine interaction, valence state, isomer shifts, molecular orbital, Fe(VIII)

1. Introduction

Emission Mössbauer spectroscopy combined with an ion implantation technique has provided unique information on the final site positions, chemical states, and dynamical behavior of implanted atoms, and has contributed greatly to the characterization of extremely dilute impurities in materials [1]. Many emission studies using ^{57}Fe produced by EC decay of ^{57}Co have been performed in metals, semiconductors, insulators, and chemical compounds. In these investigations, however, there has not been determined chemically attractive phenomenon concerning valence states of ^{57}Fe arising from ^{57}Co, ^{57}Fe atoms decayed from implanted or doped ^{57}Co probes occur in mainly Fe^{2+} and/or Fe^{3+}, as well as in stable and ordinary chemical states in solid state chemistry.

^{57}Mn is another well known source nuclide of ^{57}Fe Mössbauer spectroscopy, as shown in Fig. 1. About 80 % of ^{57}Mn decays directly to the Mössbauer level of 14.4 keV through the β^-–decay process. The applications of ^{57}Mn to the Mössbauer spectroscopy have been very rare, especially from the viewpoint of chemistry [2-4], because its half-life ($T_{1/2}$ = 1.45 min) is rather short. Mn sits next to Fe in the periodic table, but the chemical properties of Mn are much different from those of Fe and Co. Mn takes various stable oxidation states from 0 up to 7+ in ordinary solids, while the highest oxidation state of Fe is 6+ in K_2FeO_4. Charge balance indicates that after a β–decay the oxidation state of the atom containing the daughter nuclide may remain the same as the one of the parent. If the

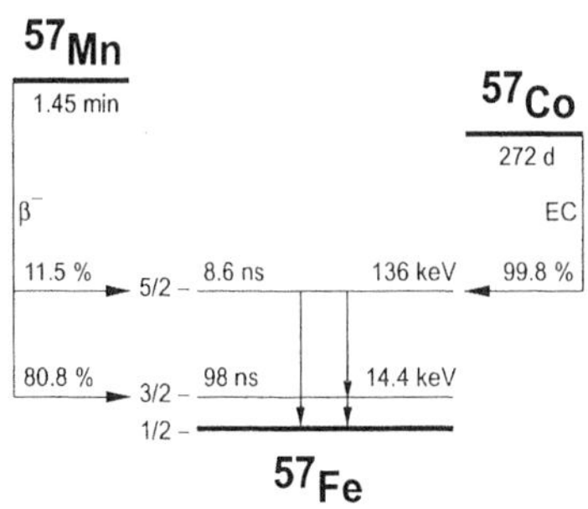

Fig. 1 Simplified level diagram of ^{57}Fe.

decay electron flies away from the parent, the oxidation number of the product atom increases by 1. Ion implantation generally produces a system far from equilibrium and the after effects caused by the β–decay around the decaying probes are not so serious compared to those by the EC-decay of ^{57}Co. Therefore, unusual valence states of ^{57}Fe atoms higher than 6 are expected to be synthesized after ^{57}Mn implantation into compounds in which Mn^{6+} or Mn^{7+} stably exists.

In this study, the in-beam Mössbauer technique was applied to study chemical states, especially concerning the valence states of ^{57}Fe arising from ^{57}Mn in $KMnO_4$ where Mn ions are in a 7+ ([Ar]$3d^0 4s^0$) state.

2. Experimental

^{57}Mn beam was produced as a secondary RI beam following a nuclear projectile-fragmentation reaction of ^{59}Co beam (E/A = 80 MeV, I = 10 pnA) with Be production target, and separated by the in-flight isotope separator using RIPS (Riken Projectile-fragment Separator). Subsequently, the ^{57}Mn was implanted into a sample of $KMnO_4$ after passing through an Al plate with a thickness of 205 μm to suppress the beam energy down to 15 MeV/nucleon. The beam intensity of ^{57}Mn, typically 1.37×10^5 particle/sec, was measured by a plastic scintillation counter located in the beam line.

The sample was a pellet of $KMnO_4$ of 40 mm × 40 mm × 0.5 mm$t$ and was placed in a Liq. He cryostat. The Mössbauer spectra were measured over the temperature range of 11 K to 155 K. Two parallel-plate avalanche counters (PPAC) with an ^{57}Fe-enriched stainless-steel foil were used for 14.4 keV γ-rays detection. The Mössbauer γ-ray of 14.4 keV was also monitored by a CdZnTe detector during all measurement. The detailed experimental layout was described in Ref. [5] and [6].

3. Results and Discussion

In-beam Mössbauer spectra of ^{57}Fe arising from ^{57}Mn in $KMnO_4$ obtained at 11, 90, and 130 K are shown in Fig. 2. Each spectrum was accumulated for more than 24 hrs. In general, analysis of Mössbauer spectra of probes implanted into chemical compounds is not straightforward compared to those for metals and semiconductors. The spectra observed in this experiment, however, were not so complicated and could be fitted tentatively by two well-defined components, which consisted of a *doublet* and a *singlet*, as shown in Fig. 2. It does tell us that the ^{57}Fe atoms produced from the implanted ^{57}Mn probes have landed finally in limited positions after about 2 min of the lifetime.

The isomer shifts of the *doublet* and the *singlet* at 11 K were obtained to be 0.57(4) mm/s and –0.38(5) mm/s, respectively. The ratio of area intensity of the *singlet* to the *doublet* at 11 K was almost 1:4. The area intensity of the *singlet* decreased with the increase of temperature, and vanished over 130 K. The Mössbauer parameters of the *doublet* at temperatures between 11 K and 155 K, for example I.S. = 0.57(4) mm/s and Q.S. = 0.80 mm/s at 11 K, revealed an Fe^{2+} in a high spin state. The Fe^{2+} doublet is ascribed to ions located in interstitial sites or substituted for the K atoms in $KMnO_4$.

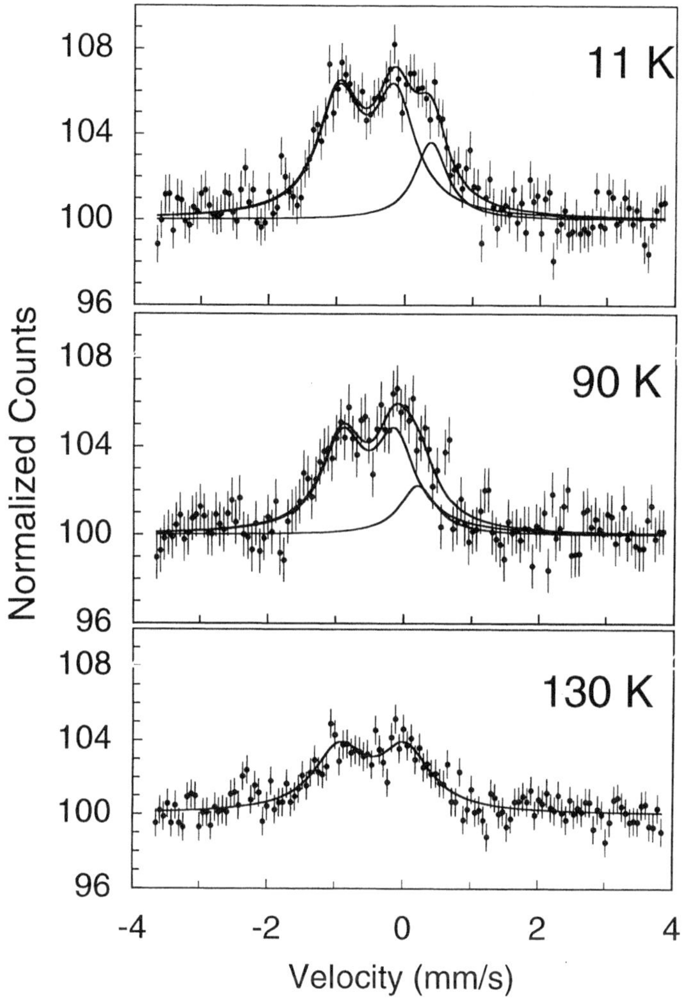

Fig. 2 The in-beam Mössbauer spectra of ^{57}Fe arising from ^{57}Mn implanted into $KMnO_4$ between 11 and 130 K. The isomer shift is given relative to α-iron at room temperature, and the sign of velocity is in the emission Mössbauer experiment.

On the other hand, the isomer shift of the *singlet* suggests an existence of exotic Fe species in unusual and higher valence state than 4+. Potassium permanganate is ionic, that forms orthorhombic (*Pnma*) crystals. A Mn atom in permanganate ion, $[MnO_4]^-$, is surrounded by four oxygen atoms at the corners of a regular tetrahedron, in which the average Mn–O distance is 0.1629 nm. Suppose that the anomalous *singlet* resulted in the Fe species substituted for regular Mn atoms in $[MnO_4]^-$. The calculations of molecular orbital using DV-Xα and Gaussian methods were performed in order to estimate the relationship between the valence states and the electron densities of Fe atoms in tetrahedral $[FeO_4]^{n-}$ with a similar distance of Mn–O. The result of Gaussian calculation indicates that two species of $[FeO_4]^{2-}$ by Fe^{6+} and $[FeO_4]^0$ by Fe^{8+} are most stabilized tetrahedrons, and that tetrahedral configurations with other valence states of Fe are geometrically unstable. The EFGs at iron nuclei are calculated to be zero for $[FeO_4]^{2-}$ and $[FeO_4]^0$, which is in agreement with the experimental peak shape of the *singlet*. With increasing the valence state of Fe, the bonding nature of Fe–O shifts from ionic to covalent; therefore, the contribution of 3*d*-electrons to bonding orbital increase, while those to antibonding orbital decrease. Though the total population number of 3*d*-electrons of Fe atom decreases as the number of antibonding electrons becomes fewer in the region from Fe^{2+} to Fe^{6+}, the population of 3*d*-electrons of Fe atom increases because of the covalent nature in the region from Fe^{6+} to Fe^{8+}. Therefore, the isomer shift of $[FeO_4]^{2-}$ corresponding to Fe^{6+} presents a maximum value, i.e., –0.85 mm/s in alkali oxoferrates [7]. The isomer shift of tetrahedral $[FeO_4]^0$ with higher valence state is considered to be less than the value of alkali ferrate.

On the basis of the isomer shift of the *singlet* observed in this experiment and the calculations of the molecular orbital described above, it is preliminarily concluded that the *singlet* might be assigned to the *substitutional* ^{57}*Fe atoms* for Mn-sites in a tetrahedral $[MnO_4]^-$ with an unusually high valence state of Fe^{8+}.

References

[1] H. de Waard and L. Niesen, "Ion Implantation Mössbauer Effect Studies", in Mössbauer Spectroscopy Applied to Inorganic Chemistry, (Plenum, New York 1987) p1-87.

[2] R. S. Preston and B. J. Zabransky: Phys. Lett. **A 55** (1975) 179.

[3] M. Nakada et al., Bull. Chem. Soc. Jpn. **65** (1992) 1.

[4] Y. Kobayashi et al., Euro. Phys. J., in press.

[5] Y. Kobayashi et al., Hyp. Int., **(C)3** (1998) 273.

[6] Y. Kobayashi et al., Hyp. Int., **126** (2000) 417.

[7] R. H. Herber and D. Johnson, Inorg. Chem., **18** (1979) 2786.

^{151}Eu Mössbauer spectroscopic and XRD study on some fluorite-type solid solution systems, $Eu_yM_{1-y}O_{2-y/2}$ (M=Zr, Hf, Ce)

N.M. Masaki[1], H. Otobe[1], A. Nakamura[1], N.R.D. Guillermo[2], Y. Izumiyama[3], D. Harada[3] and Y. Hinatsu[3]

[1]Department of Materials Science, Japan Atomic Energy Research Institute, Tokai-mura, 319-1195, Japan,

[2]Philippine Nuclear Research Institute, Quezon city, 1101, Philippine,

[3]Department of Chemistry, Hokkaido University, Sapporo, 060-0810, Japan.

Abstract ^{151}Eu Mössbauer spectroscopy and powder x-ray diffractometry (XRD) have been applied to elucidate the local structure around the trivalent europium (Eu^{3+}) in the titled three defect-fluorite systems. From this study, it has been clarified that in contrast to Eu-Ce system the stabilized cubic-fluorite phase in Eu-Zr and Eu-Hf systems has most likely a pyrochlore-type local structure (microdomain structure).

1. Introduction

The so-called defect-fluorite systems, $M'_yM_{1-y}O_{2-y/2}$ (M^{4+}=Zr, Hf, Ce, Th, etc., M'^{3+}= trivalent lanthanides(Ln), Y, etc.) are oxide-ion conductors due to the high mobility of oxygen-vacancies (V_o) introduced by these lower-valent cations substitution ($M^{4+} \rightarrow M'^{3+}$). Among these systems, the Zr- and Hf-based systems with smaller M^{4+} have the stabilized cubic-fluorite phase first realized after ~20 % M'^{3+} substitution (y~0.20). In addition, in these so-called stabilized zirconias and hafnias containing larger Ln^{3+}s (La-Gd), pyrochlore (P)-type (both cations and V_o long-range ordered) superstructure phase is formed around y~0.50 (accompanying with some off-stoichiometric ($y \neq 0.5$) stability region) (see Fig. 1). Recently, more structural evidences have been accumulating on stabilized zirconias through EXAFS [1], single-crystal XRD [2] studies, etc. that the 'disordered' defect-fluorite (DF) phase outside the pyrochlore-stability region is in fact more likely to have a pyrochlore-type local structure (P-type microdomain structure).

Mössbauer spectroscopy is a powerful tool to investigate local structure around Ln. In order to shed new light on the local structural features of these defect-fluorite systems, we have performed a systematic ^{151}Eu Mössbauer and XRD study on $Eu_yM_{1-y}O_{2-y/2}$ (M=Zr, Hf, Ce) over the whole composition range ($0 \leq y \leq 1.0$). To our knowledge, no such Ln-Mössbauer spectroscopic study has been reported on these systems, except for that on the stoichiometric (y=0.5) pyrochlore ($Ln_2M_2O_7$) by Chien et al.[3].

2. Experimental

Samples of three systems were prepared by the usual ceramic methods, using ZrO_2, CeO_2, HfO_2 and Eu_2O_3 powders (all purchased reagents, 3 to 4N). The reaction mixtures were fired at 1500 °C for 16 hrs in air, two to three times, with intermediate grinding(s). Powder XRD measurements were performed using Cu-Kα radiation (JDX-3531, JEOL) in the range, $10° \leq 2\theta \leq 110°$, at 0.02° interval. Lattice parameters were calculated using the program Cell [4]. ^{151}Eu-Mössbauer spectroscopy was conducted on powder samples lapped in thin aluminum foil (0.010 mm thick). A spectrometer (Laboratory Equip. Co.) was calibrated using ^{57}Fe Mössbauer spectra of α-Fe. A standard $^{151}SmF_3$ source (1.85 GBq) and

NaI(Tl) scintillation detector were used. The isomer shifts (IS) are given relative to EuF_3 at 300K.

3. Results and discussion

Figure 1 shows the lattice parameter data of three systems obtained in the present study, together with calculated curves based on Kim's [5] and Otobe & Nakamura's formula[6]. For Eu-Ce system, almost complete solid solution is formed over the entire $0<y<1.0$ range from MO_2 (CN=8) to $M'O_{1.5}$ (CN=6); a V_O disordered defect fluorite (DF) phase for $y<0.62$ and a V_O-ordered C-type phase for $y>0.62$. Thus, average oxygen coordination number (CN) around the cation in this system is supposed to vary according to CN = 8 - 2y.

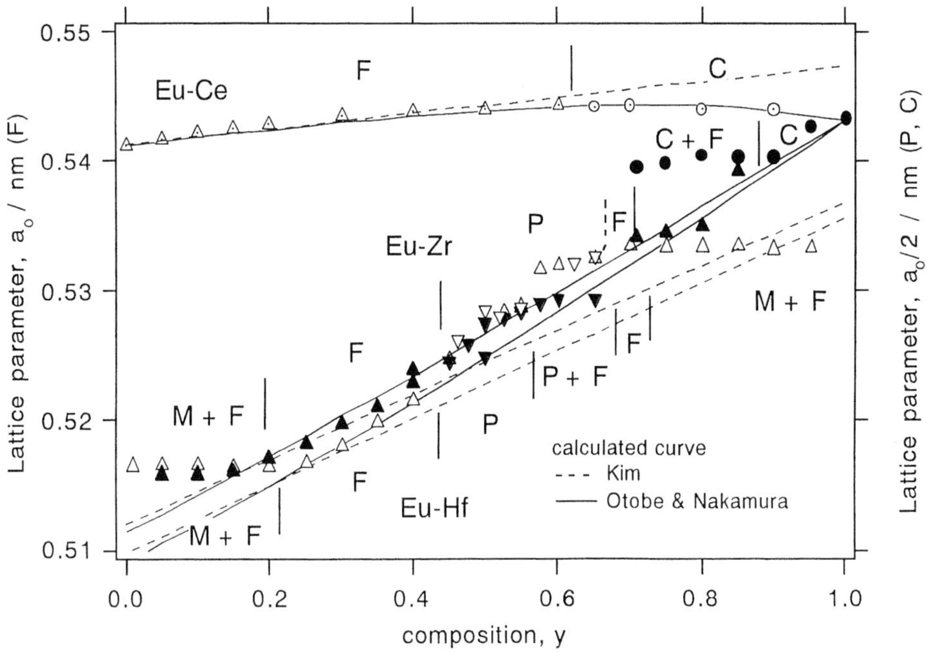

Fig. 1. Lattice parameter for cubic phases vs. composition (y).
F : Fluorite (▲ (Zr), △ (Hf), ◬ (Ce)), P : Pyrochlore (▼ (Zr), ▽ (Hf)),
C : C-type (● (Zr), ⊙ (Ce)), M : Monoclinic
For both P and C superstructure phases, one half of their lattice parameters are plotted.

In contrast, for Zr and Hf systems, the stabilized defect-fluorite (DF) based solid solution phases extend over more limited (but still wide) $\sim0.20<y<\sim0.70$ range. In both systems, the disordered DF phase changes to the pyrochlore(P)-type (both the cations (M^{4+},M'^{3+}) and V_O ordered) superstructure phase above $y\sim0.45$, which extends up to almost full $y\sim0.70$ and much narrower $y\sim0.55$ in the Zr and Hf systems, respectively. Such pyrochlore-type phase returns to the disordered DF-type phase again at higher boundary region around $y\sim0.70$, which has much narrower homogeneity range than that in the lower y side ($\sim0.20<y<\sim0.45$) for both systems, as seen there.

Mössbauer spectra of three systems consist of almost symmetrical broad single line around 0.5-1.0 mm/s. These isomer shifts (ISs) are characteristic for Eu^{3+}, and no spectra for Eu^{2+} around -12mm/s were observed. The Mössbauer spectra were fitted by one Lorentz function assuming that the quadrupole splitting equals zero. The IS of Eu^{3+} vs. composition y plots obtained by the present study are shown in Fig. 2 for the three systems. It is striking that in contrast to the Ce system for which the ISs of Eu^{3+} increase smoothly with y, those of the Zr and Hf systems exhibit a V-shaped pronounced minimum around the stoichiometric pyrochlore (y~0.5).

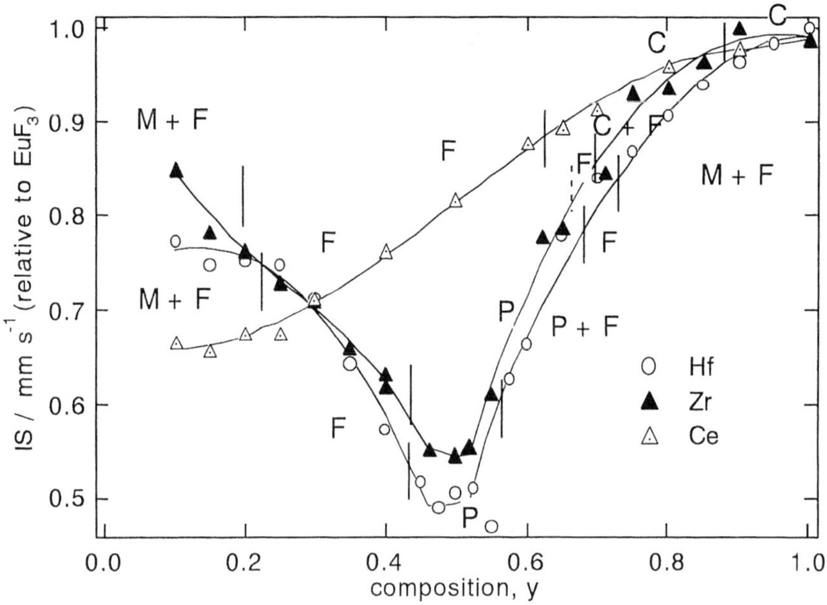

Fig. 2. Isomer shift of Eu^{3+} vs. composition(y).

It is rather well established that the IS of Eu^{3+} correlates with the average Eu-O bond length in oxide systems [7], i.e., shorter the bond length, higher the IS. This means that in the Ce system with increasing y the average Eu-O bond length of the system smoothly decreases. As discussed in [6], this is due to the oxygen-vacancy (V_O) effect, and, as mentioned formerly, one can reasonably assume that the average oxygen coordination number (CN) around the Eu^{3+} smoothly decreases according to CN=8-2y in this system. The decrease in repulsion between the coordinating oxide-ions around the Eu^{3+} with y leads to the decrease of the Eu^{3+} ionic-radius, r(Eu^{3+})(CN=8-2y)[8], and hence of the Eu-O bond length.

In contrast, in pyrochlore(P)-type superstructure phase formed at y=0.5 for the Zr and Hf systems, the Eu^{3+} (Zr^{4+}, Hf^{4+}) is oxygen 8(6) coordinated (CN=8(6)), the two V_O being adjacent to the smaller Zr^{4+} or Hf^{4+}. This increase in oxygen coordination number (CN) to 8 in pyrochlore phase in these systems from 7 in disordered DF phase in the Ce system at the same y=0.5 is considered to be the main reason for the significant decrease of the IS of the Eu^{3+} to the minimum. In pyrochlore structure, the two short Eu-O and six long Eu-O bonds exist, and the respective bond lengths are calculated by knowing the oxygen positional parameter, x of 48f site, and lattice parameter (a_0) of the system[3]. From the XRD results, $Eu_2Zr_2O_7$ and $Eu_2Hf_2O_7$ are found to have the two short (0.229 and 0.228 nm) and six long Eu-O bonds (0.250 and 0.254 nm) (x=0.343 and 0.336), respectively, and hence the respective average Eu-O bond lengths are 0.245 and 0.248 nm. Whereas, the Eu-O bond length at the same y=0.5 in the Ce system (CN=7) is calculated to be r(Eu^{3+})(CN=7)(=0.101nm) +

$r(O^{2-})$ (CN=4)(=0.138nm) = 0.239 nm[8]. Thus the order of the Eu-O bond lengths in these three systems at y=0.5, 0.239 nm (Ce) < 0.245 nm (Zr) < 0.248 nm (Hf), is in fair agreement with that of the ISs of the Eu^{3+} shown in Fig. 2.

The fact that in the stabilized DF phase in 0.20<y<0.45 region the ISs of Eu^{3+} decrease smoothly with y to the respective minimums at y~0.5 for both Zr and Hf systems, opposite to that for the Ce system, clearly indicated that this stabilized DF phase by no means has the simple parent-fluorite derived DF structure as the Ce system has, but rather has also a (short-range ordered) pyrochlore-type local and/or microdomain structure. Though not detailed here, by assuming constant 8 coordination (CN=8) for the Eu^{3+} and a smoothly varying oxygen positional parameter x from 0.342 at y=0.20 to the above value at y=0.50 for the both systems, this decrease of IS in this region can be almost quantitatively interpreted.

As for sharp upturn of the ISs in higher y>0.5 region, similar argument is possible, but here also the effects of Vo formation around Eu^{3+} site and the partial occupation of Zr site (CN=6) by the excess Eu^{3+} should be properly taken into account. The both effects decrease the CN from 8, therefore is considered to accelerate the rate of IS upturn in this region. For the Zr system, also the stabilized c-type solid solution phase (CN=8-2y) is formed at further higher y region. In the c-type Eu_2O_3 (y=1.0), the average Eu-O bond length is 0.234 nm which is much shorter than that for pyrochlore phase at y=0.5. Thus, for the Zr system the Eu-O bond length - IS correlation seems to hold well up to y=1.0. However, for the Hf system, in equilibrium with the DF phase is pure monoclinic Eu_2O_3 (CN=7), in which three kinds of the Eu^{3+} sites and a wide distribution of individual Eu-O bond lengths exist. The overall average Eu-O bond length (=0.244 nm) in monoclinic Eu_2O_3 is almost comparable to that of pyrochlore phase at y=0.5. Large IS of this monoclinic Eu_2O_3 comparable to that of the c-type Eu_2O_3 is not well understood at present.

The present ^{151}Eu Mössbauer spectroscopic study for the three defect-fluorites systems, $Eu_yM_{1-y}O_{2-y/2}$ (M=Zr, Hf, Ce), almost unambiguously have shown that the stabilized cubic fluorite phase in the Zr and Hf systems also have a pyrochlore-type local structure, different from the Ce system and consistent with the reported EXAFS study [1] on stabilized zirconias.

Acknowledgement

N. R. D. Guillermo, Y. Izumiyama and D. Harada appreciate the Japan Atomic Energy Research Institute (the JAERI) for the offer of STA Scientist Exchange Program, Graduate Student Summer Training Program, Fellowship of Advanced Science, respectively. They could participate in the present study during their stay at the JAERI through these programs.

References

[1] T. Uehara, K. Koto, S. Emura and F. Kanamura, Solid State Ionics 23(1987)331.
[2] T. Moriga, S. Emura, A. Yoshiasa, S. Kikkawa, F. Kanamura and K. Koto, Solid State Ionics 40/41(1990)357.
[3] C. L. Chien, A. W. Sleight, Phys. Rev. B, 18(1978)2031
[4] Y. Takaki, T. Taniguchi, H. Yamaguchi, K. Nakata, J. Ceram. Soc. Jpn., Int. Ed., 96(1988)13.
[5] D. J. Kim, J. Amer. Ceram. Soc.,72 (1989) 1415.
[6] H. Otobe and A. Nakamura, Solid Oxide Fuel Cell (SOFC)-VI, S. C. Singhal and M. Dokiya eds. The Electrochemical Soc., NJ, U.S.A., 1999, p. 463.
[7] S. Tanabe, K. Hirao and N. Soga, J. Non-Cryst. Solids 113 (1989) 178.
[8] R. D. Shannon, Acta Crystallogr.Sect.A. 32 (1976) 751.

Study on porphyrin complex ligated with azaferrocene derivatives

S. Nakashima [a], A. Negishi [b], J. Nakamura [b] and T. Okuda [b]
[a]Radioisotope Center, Hiroshima University, Kagamiyama, Higashi-Hiroshima 739-8526, Japan
[b]Department of Chemistry, Graduate School of Science, Hiroshima University, Kagamiyama, Higashi-Hiroshima 739-8526, Japan

[5,10,15,20-Tetrakis(pentafluorophenyl)porphyrinato]iron complex ligated with azaferrocene has two crystal forms, while the corresponding complex ligated with 3-methylazaferrocene has only one crystal form. It was revealed that an introduction of methyl substituent to the pyrrole ring makes a less stable complex.
Keywords: Mössbauer spectroscopy, porphyrin, azaferrocene, polynuclear complex, redox process, UV irradiation

1. Introduction

The chemistry of ferrocene derivatives has received considerable attention[1]. Although the chemistry of azaferrocene derivatives has been the subject of relatively few investigation, the two active sites of iron and nitrogen atoms offer a wide range of chemistry[2]. An interesting example is that the oxidation site in iodine oxidation changes by an introduction of substituent[3]. Another example is the use as ligand[4].

There has been considerable interest in the stereochemistry and electronic properties of iron porphyrins containing imidazole and pyridine ligands. The photoinduced electron transfer processes of bi- and poly-nuclear transition metal complexes are of intense current interest. The porphyrin complexes ligated with azaferrocene are known to be photosensitive, while the complexes ligated with pyridine or imidazole are not photosensitive[5,6]. In the present study, we re-synthesized the [5,10,15,20-tetrakis(pentafluorophenyl)porphyrinato]iron complex ligated with azaferrocene and newly synthesized the complex ligated with 3-methylazaferrocene. Two crystal forms were obtained in azaferrocene complex, while only one crystal form was obtained in 3-methylazaferrocene complex. The stability was compared from the point of UV irradiation.

2. Experimental

Azaferrocene(=AF) was synthesized according to the reported procedure[7]. 3-Methylazaferrocene(3-MeAF) was also synthesized with the same procedure by using 3-methylpyrrole. The products were isolated by column chromatography on alumina. Azaferrocene was purified by sublimation. The compounds were identified by ^{1}H NMR.

The porphyrin complexes were obtained according to the literature[5] by mixing the iron-5,10,15,20-tetrakis(pentafluorophenyl)porphyrin(=tfpp) chloride and azaferrocene or 3-methylazaferrocene in chloroform and then putting them in the refrigerator. Ligation was confirmed by using FAB Mass[8,9].

A $^{57}Co(Rh)$ source moving in a constant-acceleration mode was used for ^{57}Fe Mössbauer spectroscopic measurements. Variable-temperature ^{57}Fe Mössbauer spectra were obtained by using a Toyo-Research spectrometer and a continuous-flow cryostat. The ^{57}Fe Mössbauer parameters were obtained by least-squares fitting to Lorentzian peaks. The isomer-shift values are referred to metallic iron.

3. Results and discussion

Two crystal forms were obtained in azaferrocene-ligated complex. One is plate-like red-purple crystal, and the other is needle-like blue-violet one. The IR absorption peak of the out-of-plane bending vibration of C-H bond at 819 cm^{-1} in azaferrocene changes to 829 cm^{-1} by ligation for both crystals of azaferrocene. The change is similar to the case of N-methylation[3]. ^{57}Fe Mössbauer spectroscopy showed azaferrocene-like doublet and porphyrin-like doublet. The Mössbauer parameters are shown in Table 1. The intensity ratio was rationally explained by the ratio of porphyrin and azaferrocene (nearly 1:2).

Table 1 ^{57}Fe Mössbauer parameters at 80 K

Sample	Porphyrin site (A)		AF, 3-MeAF site (B)		Area ratio
	I.S.[a]/ mm s^{-1}	Q.S./ mm s^{-1}	I.S.[a]/ mm s^{-1}	Q.S./ mm s^{-1}	(B/A)
Complex of AF (Red-purple)	0.46±0.02	1.21±0.02	0.54±0.02	2.44±0.02	1.6±0.3
Complex of AF (Blue-violet)	0.45±0.02	1.25±0.02	0.55±0.02	2.42±0.02	1.5±0.3
Complex of 3-MeAF	0.50±0.01	1.16±0.01	0.53±0.02	2.44±0.02	2.1±0.2

[a]Isomer-shift data are reported with respect to iron foil.

The electronic spectra were observed in chloroform. The spectra are shown in Fig. 1. There is a difference in the electronic spectra between the two crystal forms. The red-purple crystals showed the same spectrum with the crystal already reported[5]. On the other hand, the blue-violet crystals showed a different spectrum. The ^{57}Fe Mössbauer parameters for the red-purple crystals are also similar to those of reported values[4].

Only one crystal form was obtained from the complex ligated with 3-

methylazaferrocene. The color is blue-violet. ^{57}Fe Mössbauer spectroscopy showed two doublets. The parameters are also shown in Table 1. The intensity ratio was also nearly 1:2. The electronic spectrum in chloroform is very similar to that of blue-violet one ligated with azaferrocene, which is shown in Fig. 1.

The UV-irradiation dependence in electronic spectrum with time was pursued. The change is shown in Fig. 1. Every sample changed to the same spectrum by UV irradiation. It can be seen that the spectrum of red-purple one changes to Fe^{III} type via that of blue-violet one. The spectrum of 3-methylazaferrocene-ligated sample changes more rapidly by the UV irradiation. One might think that the red-purple one is ligated with two azaferrocenes and the blue-violet one is ligated with one azaferrocene. But the ligation number of azaferrocene is thought to be the same, because the ^{57}Fe Mössbauer spectrum of blue-violet one at 80 K shows the area ratio of nearly 1:2 and the FAB Mass spectra are similar to each other. And the color of blue-violet gradually changes to red-purple in the solid state.

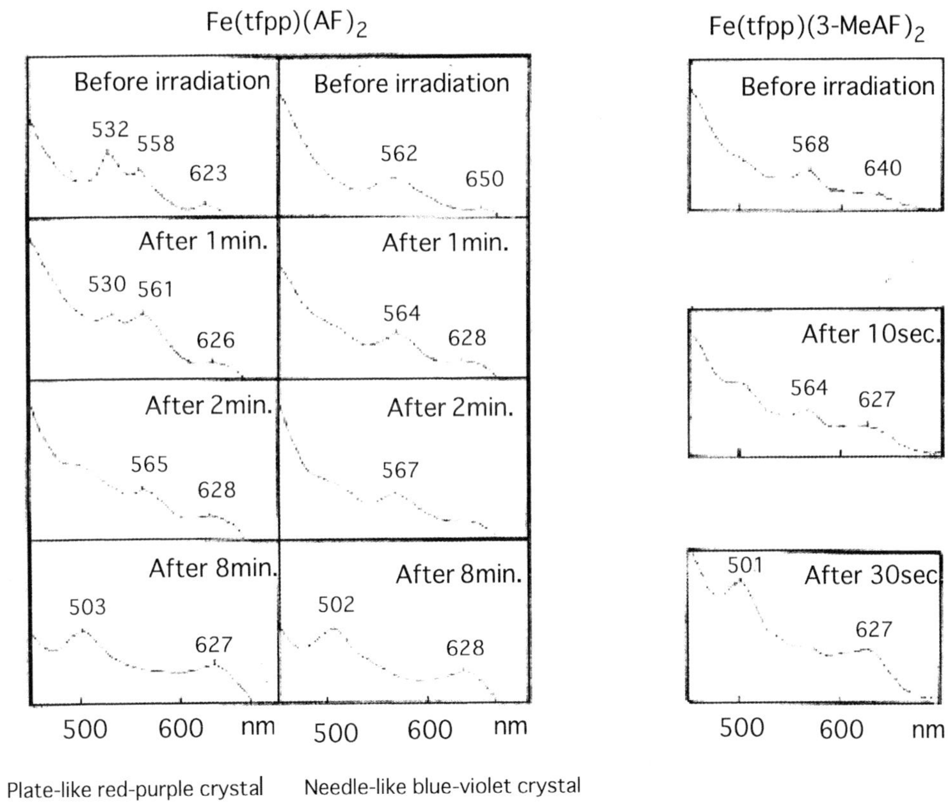

Fig. 1 Changes in electronic absorption spectrum in chloroform by irradiation with UV light.

The methyl substituent has an electron donating character. One might expect that the stable sample can be obtained by using 3-methylazaferrocene because of electron donating character of methyl substituent, which is opposite against the present result. The complex ligated with 3-methylazaferrocene is actually unstable one. This reflects the importance of the steric effect.

Moreover, we oxidized these complexes with iodine. ^{57}Fe Mössbauer spectra revealed the oxidation of porphyrin site.

In the present study we obtained stable and metastable states of azaferrocene-ligated porphyrin. It is judged from the result of 3-methylazaferrocene-ligated complex that the origin of the metastable state is maybe due to the bulkiness of the azaferrocene. The bulkiness is strongly important in these system. Actually, we could not obtain the complex ligated with 2,4-dimethylazaferrocene. The structure of the metastable state is under study.

References and Notes

[1] "Ferrocenes", A. Togni and T. Hayashi (Ed), VCH, Wheinheim (1995).

[2] J. Zakrzewski and C. Giannotti, *Coord. Chem. Rev.*, 140 (1995) 169.

[3] S. Nakashima, T. Kitao, H. Matsunaga, I. Kimura, H. Inamura, T. Okuda, and H. Sakai, *J. Radioanal. Nucl. Chem.*, 239 (1999) 279.

[4] M. Cesario, C. Giannotti, J. Guilhem, J. Silver and J. Zakrzewski, *J. Chem. Soc., Dalton Trans.*, (1997) 47.

[5] J. Zakrzewski, M. Cesario, J. Guilhem, and C. Giannotti, *J. Chem. Soc., Dalton Trans.*, (1992) 3059.

[6] J. Zakrzewski and C. Giannotti, *J. Chem. Soc., Chem. Commun.*, (1993) 1109.

[7] K. K. Joshi, P. L. Pauson, A. R. Qazi, and Stubbs, *J. Organomet. Chem.*, 1 (1964) 471.

[8] Fast atom bombardment mass spectrum (positive, 3-nitrobenzyl alcohol matrix): no parent peak (m/z 1402) but peak at m/z 1215 {Fe(tfpp){Fe(C_5H_5)(C_4H_4N)]}.

[9] Fast atom bombardment mass spectrum (positive, 3-nitrobenzyl alcohol matrix): no parent peak (m/z 1430) but peak at m/z 1229 {Fe(tfpp){Fe(C_5H_5)(C_5H_6N)]}.

Characterization of Heavy Metal Waste Glass by Mössbauer Spectroscopy

Tetsuaki NISHIDA[a] and Shiro KUBUKI[b]
[a]*Department of Biological and Environmental Chemistry, Kyushu School of Engineering, KINKI University, Kayanomori 11-6, Iizuka 820-8555, Japan*
[b]*Department of Chemical and Biological Engineering, Ube National College of Technology, Tokiwadai 2-14-1, Ube 755-8555, Japan*

Mössbauer spectroscopy is utilized for characterizing heavy metal waste glasses prepared from synthetic soda-lime silicate glass, glass bottle for commercial drinks and zeolite (mordenite). Local structure of heavy metal waste glasses is discussed with the chemical durability investigated with the acid rain simulant (pH 3.5) composed of HNO_3 and H_2SO_4.
Keywords: heavy metal waste glass, silicate glass, glass bottle, zeolite, chemical durability

1. Introduction

Mössbauer spectroscopy is powerful when investigating the local structure of new glass [1,2]. For example, Mössbauer study of semiconducting vanadate glass proved that the structural relaxation of structural units (VO_4 and VO_5) causes a distinct increase in the electrical conductivity [1]. Silicate glass is a very stable and reliable material for the fixation of hazardous heavy metals [3]. They are generally buried under the ground in the form of hydroxide or oxyhydroxide. It is expected that such heavy metals will migrate into the environment and may cause severe environmental pollution when they encountered underground water, acid rain, *etc.*

Nishida has found the following three *"Experimental Rules"* that enable the structural role of Mössbauer atoms in glass [1,2].

Experimental Rule 1 (Tg- Δ Rule); a linear relationship exists between the glass transition temperature (Tg) and the quadrupole splitting (Δ) of Fe(III) atoms. Slope of the straight line indicates the site occupation of Mössbauer atoms, *i.e.*, network-forming (NWF) sites or network-modifying (NWM) sites.

Experimental Rule 2; Debye temperature becomes larger than 280 degrees when Mössbauer atoms occupy NWF sites, whereas it is less than 270 degrees when they occupy NWM sites.

Experimental Rule 3; gamma-ray irradiation of glass results in a "reduction" of Mössbauer atoms, *e.g.*, from Fe(III) to Fe(II), when they occupy NWF sites. By contrast, an "oxidation" takes place, *e.g.*, from Fe(II) to Fe(III), when they occupy NWM sites.

In the present paper, Mössbauer study was carried out in heavy metal waste glasses prepared from glass and zeolite in order to elucidate the relationship between the local structure and the chemical durability.

2. Experimental

Suspended waste solution containing hydroxide (or oxyhydroxide) of heavy metals was treated with 20 % NaOH solution containing small amounts of $AlCl_3$, $CaCl_2$ and $FeSO_4$ as coagulant, precipitant and reducing agent for Cr(VI), respectively. Precipitate of hydroxide (or oxyhydroxide) was dried with diatomaceous earth in order to prepare yellow sludge. This procedure is often conducted in several waste water treatment facilities for a tentative fixation or a storage of heavy metals. The main component of the diatomaceous earth is SiO_2 (> 92 mass%) [3]. Soda-lime silicate glass, $17Na_2O \cdot 13CaO \cdot 65SiO_2 \cdot 2Fe_2O_3 \cdot 3Al_2O_3$ (in mass%), of light brown color was prepared for the fixation of heavy metals. It was prepared by melting the reagent mixture of Na_2CO_3, $CaCO_3$, SiO_2, Fe_2O_3 and Al_2O_3 placed in a platinum crucible at 1400 °C for 1 h in an electric muffle furnace. Ice-cold water was used for quenching the melt; outside the platinum crucible was quickly immersed into the ice-cold water. Heavy metal waste glass of light green color was prepared by melting the yellow sludge and the synthetic soda-lime silicate glass at 1350 °C for 1 h in an electric muffle furnace. Weight ratio of the yellow sludge to the synthetic glass was fixed to 1:9. Another heavy metal waste glass of light brown color was prepared in the same manner by melting the yellow sludge together with zeolite (mordenite).

Each glass sample was pulverized in an agate mortar for ^{57}Fe Mössbauer and X-ray diffraction (XRD) measurements. ^{57}Co(Pd) source of 3.7×10^8 Bq (10 mCi) was used as the Mössbauer source. A sheet of metallic iron (α-Fe) foil enriched with ^{57}Fe was used as the reference for the isomer shift (δ) and for calibrating the velocity scale of the spectrometer. Spectral analysis was made by a least-squares method. XRD pattern was recorded at a scanning rate of 0.5 ° min^{-1} with Cu-K$_\alpha$ rays.

3. Results and discussion

X-ray spectrometry proved that the average composition of the yellow sludge in mass% was Si (75.0), Al (8.35), Fe (2.65), K (0.30), Ca (6.76), Cr (0.39), Ni (0.94), Cu (1.34), Zn (0.68) and Pb (3.48) [3]. XRD study of the yellow sludge shows a broad halo pattern ascribed to amorphous structure. Several weak peaks due to crystalline SiO_2 are observed at 2θ = 20.10, 21.57, 22.45, 26.50 and 27.64 °, which are superimposed on the halo peak. Besides these peaks, weak peaks are observed at 7.34, 14.55 and 29.61 °. They are ascribed to crystalline γ-FeOOH: *i.e.*, lepidocrocite [2-7].

Mössbauer spectrum of the yellow sludge is shown in figure 1(a), which shows one paramagnetic quadrupole doublet due to Fe(III) with δ and Δ values of 0.33 and 0.84 mm s^{-1}, respectively. These parameters are comparable to those reported for γ-FeOOH [2-7], *i.e.*, δ = 0.30—0.36 mm s^{-1} and Δ = 0.55—0.74 mm s^{-1}. Slightly larger value of Δ obtained in this study (0.84 mm s^{-1}) reflects distortion at the octahedral Fe(III) sites. γ-FeOOH is the main component of the passive film, corrosion product and rust of iron [2-7]. The present XRD and Mössbauer studies (figure 1(a)) indicate that most of the heavy metals were coprecipitated with $Fe(OH)_3$, and are stored as amorphous oxyhydroxide together with small amount of γ-FeOOH.

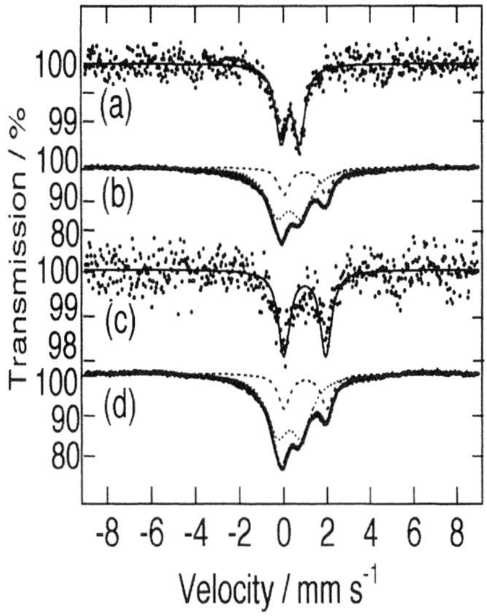

Figure 1. Mössbauer spectra of (a) yellow sludge, (b) synthetic silicate glass, (c) heavy metal waste glass and (d) γ-irradiated synthetic silicate glass.

Figure 1(b) shows the Mössbauer spectrum of synthetic silicate glass ($17Na_2O \cdot 13CaO \cdot 65SiO_2 \cdot 2Fe_2O_3 \cdot 3Al_2O_3$), which shows a doublet due to Fe(III) superimposed on a doublet due to Fe(II). The Fe(III) has δ and Δ values of 0.23 and 1.14 mm s^{-1}, respectively, reflecting distorted tetrahedral symmetry. Values of δ (0.97 mm s^{-1}) and Δ (1.92 mm s^{-1}) for Fe(II) also reflect distorted tetrahedral symmetry. These δ values suggest that both Fe(III) and Fe(II) occupy substitutional sites of Si(IV), playing a role of NWF.

Mössbauer spectrum of heavy metal waste glass is illustrated in figure 1(c). The paramagnetic doublet with δ of 0.97 mms^{-1} and Δ of 1.92 mm s^{-1} indicates that all the Fe(III) ions in the yellow sludge are reduced to Fe(II) in the heavy metal waste glass when prepared at 1350 °C. These Mössbauer parameters indicate that Fe(II) atoms occupy distorted tetrahedral sites as do Si(IV), playing a role of NWF. It is known that γ-FeOOH is easily modified to γ-Fe_2O_3 when heated at temperature higher than 450 °C [2]. It is therefore considered that Fe(III) is reduced to Fe(II) after γ-FeOOH was modified to γ-Fe_2O_3 when melted at 1350 °C.

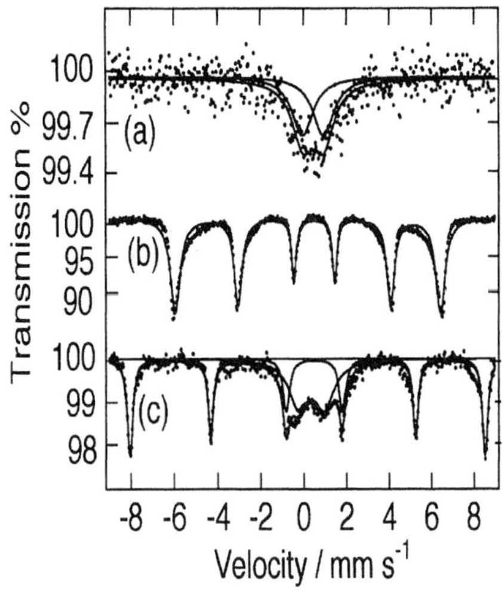

Figure 2. ^{57}Fe Mössbauer spectra of (a) heavy metal waste glass prepared from mordenite, (b) α-FeOOH and (c) "mordenite glass" prepared by melting mordenite and α-FeOOH at 1400 °C.

Figure 1(d) is the Mössbauer spectrum of synthetic silicate glass ($17Na_2O \cdot 13CaO \cdot 65SiO_2 \cdot 2Fe_2O_3 \cdot 3Al_2O_3$) irradiated with ^{60}Co-γ rays of 1.1×10^4 Gy. Two doublets due to Fe(III) and Fe(II) are observed in the irradiated silicate glass, as observed in the spectrum of non-irradiated silicate glass (see figure 1(b)). A slight increase in the relative absorption area of Fe(II) from 25.7 (figure 1(b)) to 26.3 % (figure 1(d)) proves that the Fe(II) and Fe(III) occupy NWF sites, as described above (see *Experimental Rule 3*). Large values of linewidth (Γ) obtained in these samples, 0.61 − 1.19 mm s^{-1}, indicate that distorted tetrahedra of Fe(III) and Fe(II) constitute amorphous structure.

Additional heavy metal waste glass prepared from mordenite is illustrated in figure 2(a); it can be analyzed into one quadrupole doublet due to octahedral Fe(III) with δ, Δ and Γ values of 0.44, 0.99 and 1.20 mm s^{-1}, respectively. It is well known that Fe_2O_3 is a so-called intermediate oxide that can occupy both NWF and NWM sites in glasses, depending on the composition. Low content of Na_2O in the mordenite ($8Na_2O \cdot 13Al_2O_3 \cdot 79SiO_2$) will be responsible for the preferential occupation of Fe(III) for NWM sites. From the δ value of 0.44 mm s^{-1}, we can easily deduce that Fe(III) atoms occupy NWM sites [1,2,8,9].

Figure 2(b) is the Mössbauer spectrum of commercially available α-FeOOH (geothite). δ value of 0.36 mm s^{-1} and Δ value of -0.27 mm s^{-1} are comparable to the corresponding Mössbauer parameters for octahedral Fe(III) reported in the literature [1,2].

Figure 2(c) is the Mössbauer spectrum of "mordenite glass" prepared for comparison by quenching the melt composed of α-FeOOH and mordenite from 1400 to 0 °C. The δ value of 0.37 mm s^{-1} and Δ value of -0.27 mm s^{-1} obtained for the sextet indicate an α-Fe_2O_3 phase [1,2]. It is considered that most α-FeOOH molecules have been modified into α-Fe_2O_3 during the melting of mordenite at 1400 °C. The inner doublet of figure 2(c) has a relative absorption area of 16.7 %. The δ value of 0.36 mm s^{-1} and Δ value of 1.21 mm s^{-1} for the inner doublet are ascribed to paramagnetic Fe(III) with octahedral symmetry [1,2].

Leaching test of these heavy metal waste glasses (figure 3) proved that concentration of Zn(II) could be reduced to less than 20 ppb in every case in the acid rain simulant, which was composed of equal amount of HNO_3 (pH 3.5) and H_2SO_4 (pH 3.5). Open circles refer to sludge dried with diatomaceous earth. Solid circles (●) refer to heavy metal waste glass. Solid triangles (▲) refer to the heavy metal glass prepared from glass bottle for refreshing drinks. Additional waste glasses prepared from mordenite at 1200 (▼) and 1400 °C (■) show comparable chemical durability. Figure 3 indicates that mixing and drying of hydroxides (or oxyhydroxides) with diatomaceous earth will be effective for the *short-period storage* of heavy metals. For the long-time storage, however, diatomaceous earth will be less effective than the silicate glass, since heavy metals are physically adsorbed on the surface of the holes in the former, whereas they constitute a three-dimensional network in the latter.

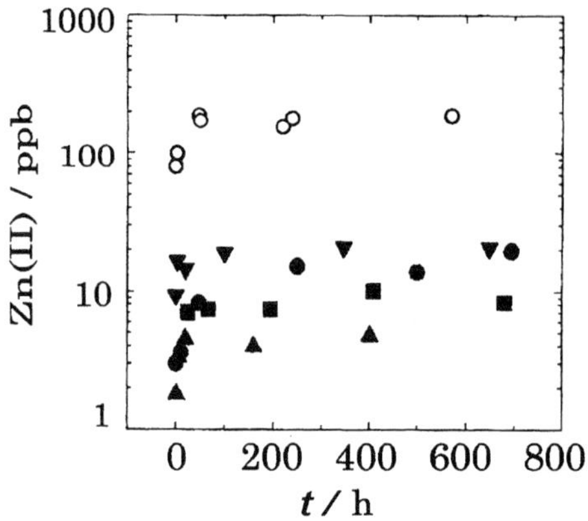

Figure 3. Concentration of Zn(II) leached into the acid rain simulant at 25 °C.

These experimental results prove that soda-lime silicate glass and mordenite are effective media for the stable fixation or solidification of hazardous heavy metals. Several types of "glass wastes" will be effectively reconstructed for the environmental purification. They will also have several potential applications in the industrial fields.

Acknowledgments

One of the authors (TN) is deeply grateful to Kyushu Environmental Evaluation Association (President Yoshimasa Takashima, Prof. Emer., Kyushu University) for financial support. They are grateful to Tosoh Cooperation for providing mordenite.

References

[1] Z. Homonnay, S. Music, T. Nishida, N.S. Kopelev and A. Vértes, in: Mössbauer Spectroscopy of Sophisticated Oxides, Akadémiai Kiadó, Budapest, 1997, pp. 27-87.

[2] F.E. Fujita, S. Nasu, T. Nishida and Y. Yoshida, in: Introduction to the Mössbauer Spectroscopy -Principles and Applications-, Agne Gijutsu Center, Tokyo, 1999, pp. 169-266 [in Japanese].

[3] T. Nishida, M. Seto, S. Kubuki, O. Miyaji, T. Ariga and Y. Matsumoto, J. Ceram. Soc. Jpn., 108 (3) (2000) 245.

[4] K. Nomura, M. Tasaka and Y. Ujihira, Corros., 44 (1988) 131.

[5] Cs. Vértes, M. Lakotos-Vársanyi, A. Vértes, W. Meisel and P. Gütlich, Hyp. Int., 93 (1994) 1817.

[6] N.N. Greenwood and T.C. Gibb, in: Mössbauer Spectroscopy, Chapman and Hall, London, 1971, pp. 254−56.

[7] A. Nakanishi, K. Fukumura and T. Kobayashi, Conf. Proc. Ital. Phys. Soc., 50 (1996) 669.

[8] T. Nishida, K. Maki and M. Seto, Radioisotopes, 48 (1999) 313.

[9] T. Nishida and K. Maki, J. Radioanal. Nucl. Chem., 242 (1999) 557.

HETEROBIMETALLIC [2- (DIMETYLAMINOMETHYL) –FERROCENYL] DERIVATIVES INVESTIGATED WITH ELEMENTS OF THE FOURTH MAINGROUP

C.Pietzsch,[1] D. Hohme,[1] K.Jacob,[2] A. Fischer [3]
[1]) Inst. Angew. Physik, TU BA Freiberg, [2]) Inst. Anorg. Chemie , Univ. Halle (Merseburg),
[3]) Chemische Institute der Otto-von-Guericke Universität, Magdeburg

Introduction

In recent years, great interest has been focused on using heterobimetallic synthesis reagents to generate bimetallic layers of a base on carbo/ nitride and nitride through thermolytic reactions. Subject of our Mössbauer investigations were heterobimetallic organic substances with the heteroelements Si, Ge, Sn, Pb and their influence on the charge transfer Fe^{II} – Fe^{III} in the corresponding complexes.

The strengths of the transition enthalpy in dependence on the heteroelement was another question of interest. Dependent on the structure of the investigated heterobimetalorganica one can distinguish two different types of Mössbauer spectra. In the first spectra type the measured spectra can be resolved into three subspectra (D_1, D_2, S). D_1 holds (due to its parameters) for Fe^{II} in the ferrocene skeleton. D_2 is assignet to Fe^{III}, the single line S represents the transition state Fe^{II}/ Fe^{III}. This transition state, a result of a chelate formation, represents an oxidation state averaged over time due to a fast electron exchange in the order of the Mössbauer window ($k > 10^7 s^{-1}$). The proposed chelate formation mechanism can be interpreted by the help of the MO – diagram of ferrocene. A further possible mechanism for electronic interaction, and the corresponding formation of the doublet D_2 (representing Fe^{III} in the Mössbauer spectra)is to fix ligands with an acceptor character on the FcN- group. This group of compounds shows another type of Mössbauer spectra for example, the spectrum of $FcNSiCl_3$ (1) /1/. The acceptor properties of trichlorsilyle ($-SiCl_3$) result in a charge transfer ferrocene (iron II) → ligand via the Cp-group and the corresponding formation of the ferricinium (iron III) component.

A charge- back transformation heterometal→ ferricinium ion as found in the case of chelate formation does not take place due to the missing chelate bond and the strong electron suction of the chlorine atoms at the ligand. The Mössbauer parameters are listed in table 1 .

Table 1:

		D1			D2			S	
Compound	T/K	δ/mms^{-1}	ε/mms^{-1}	I/%	δ/mms^{-1}	ε/mms^{-1}	I/%	δ/mms^{-1}	I/%
$(FcN)SiCl_3$ (1)	80	0.511(9)	2.30(2)	94(1)	0.16(3)	0.47(5)	6(1)		
	298	0.46(1)	2.30(1)	86(2)	0.11(3)	0.95(1)	24(3)		
$(FcN)GeCl_3$ (2)	80	0.50(5)	2.29(3)	88(1)	0.21(2)	0.95(1)	4(2)	0.32(2)	8(1)
	298	0.47(1)	2.27(1)	66(5)	0.21(2)	0.56(1)	24(2)	0.33(4)	10(1)
$(FcN)SnCl_3$ (3)	80	0.54(6)	2.44(1)	87(2)	0.25(3)	0.60(6)	10(3)	0.65(5)	3(1)
	298	0.44(1)	2.31(1)	40(3)	0.11(3)	0.43(3)	45(4)	0.39(5)	15(4)
$(FcN)_2Sn$ (5)	80	0.52(2)	2.37(4)	94(1)	0.22(5)	0.50(9)	4(1)	0.36(1)	4(1)
	298	0.45(1)	2.35(1)	79(4)	0.16(8)	0.64(8)	14(3)	0.36(5)	7(2)
$(FcN)_2Pb$ (6)	80	0.52(6)	2.39(1)	88(3)	0.18(9)	0.46(1)	6(2)	0.48(2)	2(1)
	298	0.12(1)	2.38(3)	40(3)	0.12(1)	0.18(5)	45(6)	0.32(1)	10(6)
$(FcN)_3SnCl$ (7)	80	0.53(6)	2.39(1)	77(4)	0.26(3)	0.25(7)	19(4)		
	298	0.45(1)	2.35(5)	40(6)	0.12(1)	0.42(2)	60(7)		
$(FcN)_4Sn$ (8)	80	0.53(3)	2.38(7)	95(2)	0.23(2)	0.50(0)	5(1)		
	298	0.54(6)	2.36(1)	81(4)	0.08(3)	0.20(5)	18(4)		

δ = isomer shift rel.α iron, ε = quadrupole splitting, I = intensity

The interactions found by Mössbauer spectoscopy were confirmed by X-ray single crystal investigations. On a part of the investigated complexes the transition enthalpy could be estimated using an Arrhenius plot.

Tin

Oranotin(IV)compounds act against transition metal chlorides with metal atoms of higher oxidation numbers as less reductive organyl group transmitters in methathese reactions due to their weak polar Sn- C- σ bond. This was the reason to prepare compounds of fourfold bonded tin with the general formula $(FcN)_4\ SnCl_{(4-n)}$, (n = 1-4).
The question of interest was whether the voluminous ligand FcN makes it possible to form heterobimetallic tinorganic complexes by chelate formation.
Fig. 1 shows the Mössbauer spectra from the organotin compounds $(FcN)SnCl_3$ (3), $(FcN)_3SnCl$ (7), $(FcN)_4Sn$ (8).

Fig.1:

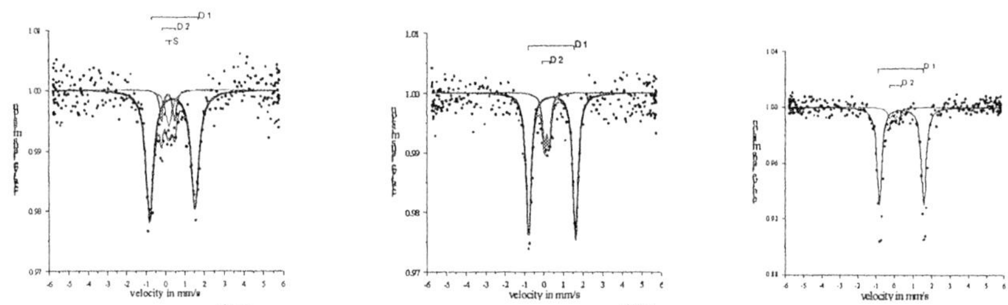

(3) (7) (8)

X-ray single crystal analyses could be made. The structures found out by these X-ray investigations confirm the Mössbauer spectroscopic results, Fig.2.

Fig.2:

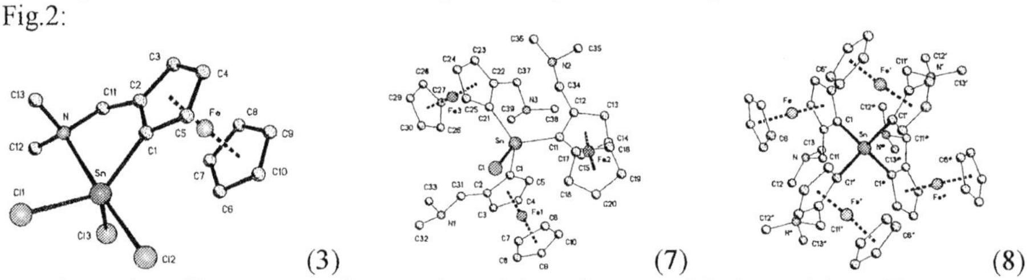

(3) (7) (8)

Another point of interest was the question of the existence of the heterobimetallic $(FcN)_2Sn$ (5) as a tin(II)organic donor molecule, since large organo groups support the kinetic stabilisation of the Sn-C-σ bond.
The Mössbauer spectrum and the X-ray picture confirm the chelate structure for (5), Fig.3.

Fig. 3:

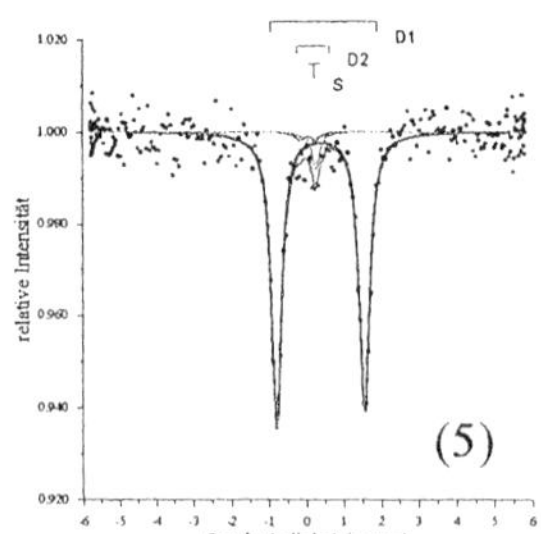

(5)

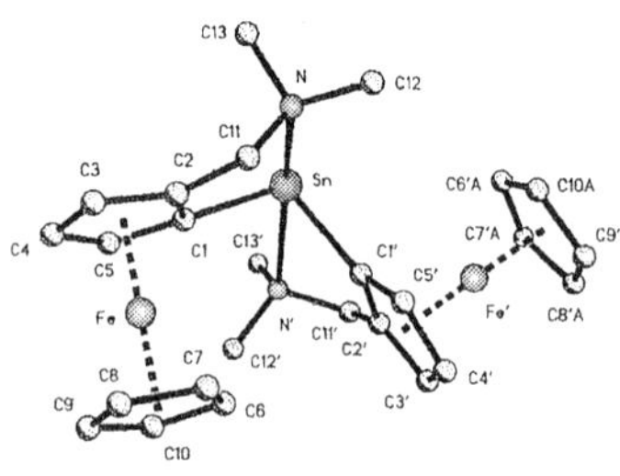

(5)

Silicon

U-V measurements on $(FcN)SiCl_3$ (1) show an $^1A_1$-$^1E_1$ transition at 443pm, which is shifted hypsochromely by 5.2 pm, compared with the free amine.
X-ray single crystal analyses show that the Si – N distance (268.5 pm) is clearly shorter than the sum of the van der Waal radii (350 pm) but distinctly longer than the dative Si-N bond.

By this the $-SiCl_3$ group acceptor properties lead to donor/ acceptor interactions in $(FcN)SiCl_3$ (1).
The X-ray result is confirmed by a possible D_1, D_2 Mössbauer fit, Fig.4.
Fig.4:

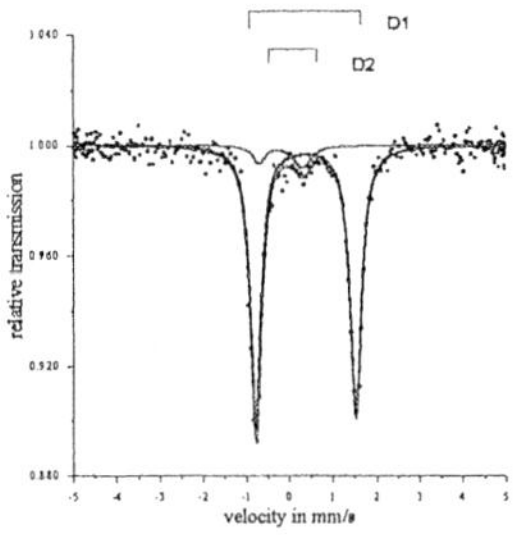

(1)

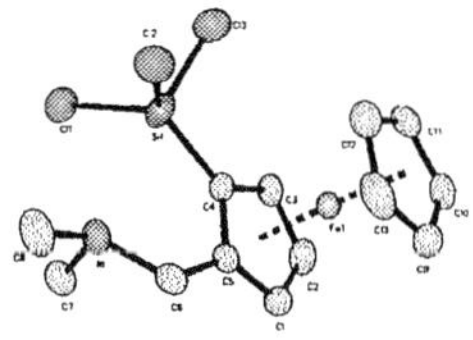

(1)

Lead
In the case of $(FcN)_2Pb$ (6) a large chemical reactivity was assumed. In this case the N-Pb interaction plays an important part in the stabilisation of this compound. The X-ray single crystal investigations of the compounds 4 and 6 confirm the chelate bond between the respective metal atom and the FcN ligand found out by Mössbauer spectroscopy, Fig.5.
Fig.5:

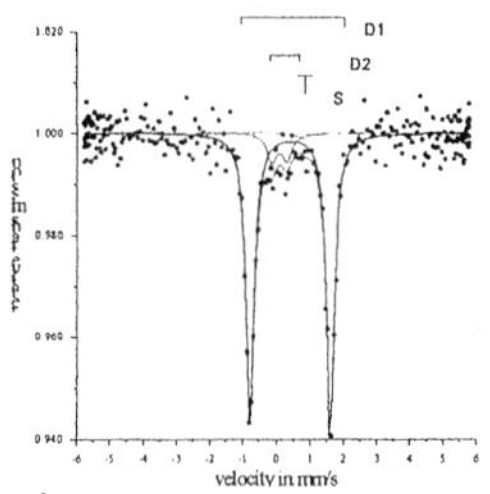

(6)

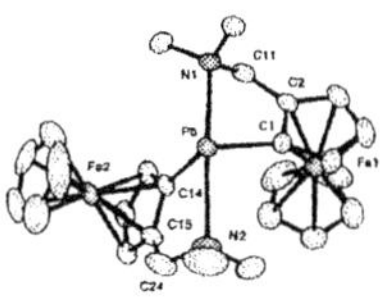

(6)

Germanium
A D_1, D_2, S fit of the Mössbauerspectrum on $(FcN)GeCl_3$ (2) and with that the interpretation as a chelate structure was possible. The X-ray single crystal measurement confirms this result, Fig.6.

Fig.6:

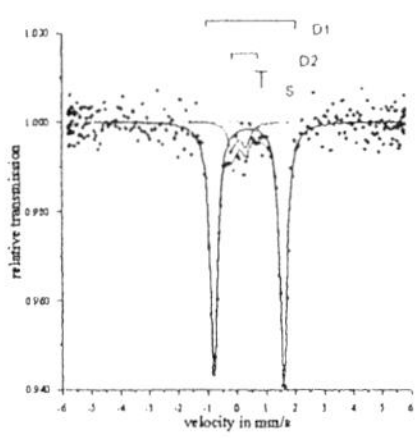

(2)

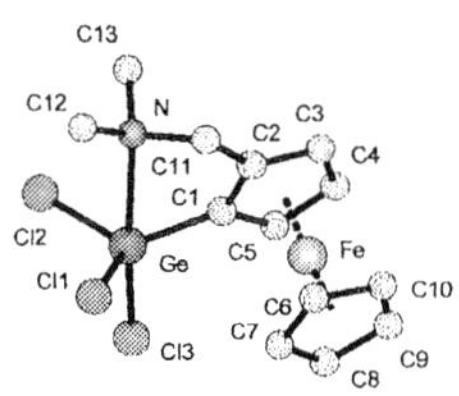

(2)

A D_1, D_2, S fit of the Mössbauer spectra and with it the existence of a chelate bond for $(FcN)_2Ge$ (4) could not by clearly confirmed. An X-ray single crystal investigation to clearify this qucstion was impossible due to missing suitable crystals.

Table 2:

Compound	Structure Chelate	D/A	Radius (heteromet.)/ Å	dt 80 K	298 K	transition enthalpy $\Delta H^{\emptyset}$ / kJ·mol^{-1}
$FcNSiCl_3$ (1)		+	1,17	6(2)	24(5)	1,13(41)
$FcNGeCl_3$ (2)	+		1,22	12(2)	35(4)	1,26(44)
$FcNSnCl_3$ (3)	+		1,41	13(3)	60(4)	1,16(41)
$(FcN)_2Sn$ (5)	+		1,41	8(1)	21(9)	1,04(61)
$(FcN)_2Pb$ (6)	+		1,54	8(2)	55(9)	n.d.
$(FcN)_3SnCl$ (7)		+	1,41	19(4)	60(4)	1,28(25)
$(FcN)_4Sn$ (8)		+	1,41	5(1)	18(4)	1,17(32)

dt = degree of transformation (ferrocene-ferricinium); n.d. = not determinated

Conclusion

At the compounds $(FcN)SiCl_3$ (1), $(FcN)GeCl_3$ (2) and $(FcN)SnCl_3$ (3) the stability of the chelate structure and with that the degree of transformation at 298K rises with increasing radii of the heterometal atom. We find the same behaviour for compounds 4, 5 and 6.
The temperature dependent Mössbauer measurements show an enlarged degree of transition ferrocene-ferricinium at a rising temperature in chelate as well as in donor/acceptor complexes (table 2).
A comparison of the tin-dimethylaminomethylferrocenyl- compounds showed that with an increase of the number of (FcN) groups the probability of chelate formation decreases. This result can by explained with the bulkynes of the (FcN) ligand.
The Arrhenius plot of the degree of transformation of all investigated compounds shows a nearly linear connection. From the slope of this line a standard transition enthalpy $\Delta H^{\ominus} = 1.22$ kJmol^{-1} can be estimated. The formulated transformations are intramolecular interactions. For this kind of interactions the transition enthalpies are in the range from 0.2 kJmol^{-1} to 12 kJmol^{-1} (melting enthalpy). A separate inspection of chelate bond and donor / acceptor ineraction showed that the chelate bond is favoured energetically ($\Delta H^{\ominus} = 1.15$ kJmol^{-1}) in contrast to the donor/acceptor interaction ($\Delta H^{\ominus} = 1.28$ KJmol^{-1}). The presence of the electron attractive chlorine ligand causes a higher degree of transformation.

References

/1 / W.Palitzsch, C.Pietzsch, K. Jacob, F.T. Edelmann, Th. Gelbrich, V. Lorenz, M. Putnatt, G.Roewer; J. Organomet. Chem. 554, (1998), 139

Acknowledgements:
The autors wish to thank the DFG for financial support.

Mössbauer Studies of Fe(II)-nitrosyl Porphyrin Model Compounds

C.E. Schulz,[a] G.R.A. Wyllie,[b] and W.R. Scheidt[b]

[a] *Department of Physics, Knox College, Galesburg, IL 61401, USA*
[b] *Department of Chemistry and Biochemistry, University of Notre Dame, Notre Dame, IN 46556, USA*

We have studied the temperature-dependent Mössbauer spectra of polycrystalline samples of several five-coordinate Fe(II)-NO model compounds: octaethylporphyrin (OEP), tetraphenylporphyrin (TPP), tetramesitylporphyrin (TMP), and tetratolylporphyrin (TTP). The OEP model shows a single symmetric quadrupole doublet, while for all the other species, the quadrupole double is asymmetric, with the lower-velocity line significantly broadened. X-ray structure determinations of Fe(TPP)NO show significant disorder of the Fe-N-O plane, and we believe the broadened Mössbauer spectra also reflect this disorder. Data on OEP and TPP at 4.2K in fields of 4T and 9T are fit well by a paramagnetic S=1/2 model. While the temperature variation of the quadrupole splitting and isomer shift values for these species is weak, the compounds exhibit an unusually rapid drop in absorption area with temperature, consistent with Debye temperatures of 81K(OEP), 64K(TMP), 55K(TPP) and 47K(TTP).

Nitric oxide (NO) has received a great deal of study as a result of its biological function in processes of neurotransmission, vasodilation, and immune function [1][2]. There has been a long history of Mössbauer studies of heme proteins and model compounds with an axial NO ligand bound to the heme iron. In 1969 Oosterhuis and Lang [3] analyzed hemoglobin and myoglobin NO complexes. They showed that high-field spectra are simulated well with a positive quadrupole splitting of axial symmetry, and an electronic effective spin S=1/2.

The complexity of the hemoglobin system leaves open many questions about the fundamental determinants of the physical characteristics of the heme-NO system. Hence the study of porphyrin-NO model compounds should be of great value. Surprisingly, there have been relatively few studies of 5-coordinate porphyrin-NO complexes. The first 5-coordinate nitrosyl porphyrin characterized was the $\{CoNO\}^8$ [4] complex Co(TPP)NO [5][6], followed by the $\{FeNO\}^7$ complex Fe(TPP)NO [7]. Since then a variety of other $\{FeNO\}^7$ systems have been studied, which can have a total electronic spin S=1/2, or S=3/2 (or have a spin equilibrium between the 1/2 and 3/2) [8].

The Fe-(TPP)NO complex has S=1/2. X-ray data show that the NO is bound at an angle to the heme normal with the Fe-N-O angle near 145°, and that there is disorder in the orientation of the NO group. Later high-resolution x-ray structures demonstrated that in a variety of 5-coordinate porphyrin-NO compounds, the Fe-N bond was tilted slightly (about 5°) from the normal to the mean plane of the porphyrin [9]. Cross-polarization magic-angle NMR studies on the Co(TPP)NO compound show that there is a dynamic effect, probably a rotation or swinging of the NO around the normal to the heme plane, which takes place only at temperatures above 220K [10]. This sort of dynamic disorder has been well documented in heme O_2 complexes [11].

We report here Mössbauer studies on four different Fe(II) porphyrin NO complexes: Fe(TPP)NO, Fe(OEP)NO, Fe(TTP)NO, and Fe(TMP)NO. Mössbauer spectra were taken on polycrystalline samples dispersed in Apiezon M grease. The room temperature spectra of these compounds are shown in Figure 1, and the corresponding quadrupole splitting, isomer shift, and linewidth parameters are given in Table 1.

Figure 1 – Room Temperature Mössbauer Spectra

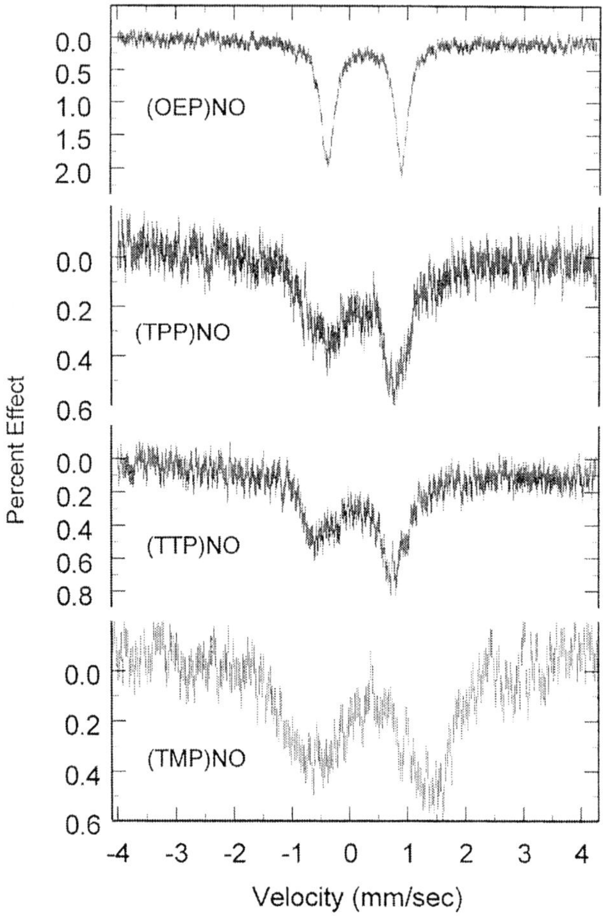

Table I – Room Temperature Parameters

	Δ_Q(mm/s)	δ_{Fe}(mm/s)	Γ_L(mm/s)	Γ_R(mm/s)
Fe(OEP)NO	1.27	0.26	0.28	0.28
Fe(TPP)NO	1.18	0.21	0.85	0.59
Fe(TTP)NO	1.24	0.18	0.80	0.66
Fe(TMP)NO	1.16	.27	0.71	0.58

The quadrupole splitting and isomer shift are very weakly temperature dependent, and are characteristic of low-spin ferrous iron systems. There are several interesting features to note: while the OEP spectrum shows a narrow, nearly-symmetric quadrupole doublet, the TPP, TTP, and TMP spectra all have very broad lines, with the left line broader than the right. The broadening of the lines appears not to be due to dynamic effects: the linewidths remain fairly constant from room temperature down to 4.2K. An obvious interpretation is that the broadening is due to a heterogeneity of iron environments, perhaps created by varied orientations of the NO group. For the low-velocity line to be broadened more than the right, there would have to be a correlated distribution of isomer shift and quadrupole splitting values, with the smaller quadrupole splitting corresponding to the larger isomer shift. Lending weight to this idea is the fact that the x-ray structure of the OEP-NO complex does not show the disordered NO direction seen in the TPP-NO complex [12], and OEP-NO has a narrow, symmetric Mössbauer spectrum.

A more detailed characterization of these molecules would include a determination of the EFG asymmetry parameter η, and the components of the magnetic hyperfine tensor A. Although the compounds studied are all paramagnetic (S=1/2), they exhibit no magnetic hyperfine splitting even in a weak applied field due to rapid fluctuations of the electron spins, presumably due to spin-spin relaxation processes. However, at 4.2K in a large applied field, spin-spin processes are quenched and magnetic hyperfine structure appears. Figure 2 shows the variable-field Mössbauer spectra of OEP and TPP compounds, respectively. For each compound, the spectra were simultaneously fit with an S=1/2 Spin Hamiltonian model using least squares minimization with a downhill simplex algorithm [13]. No attempt was made to include in the calculations a distribution of quadrupole splittings and isomer shifts for the TPP-NO complex. The resulting fit parameters are shown in Table II.

Figure 2 – OEP and TPP Applied-field spectra at 4.2K

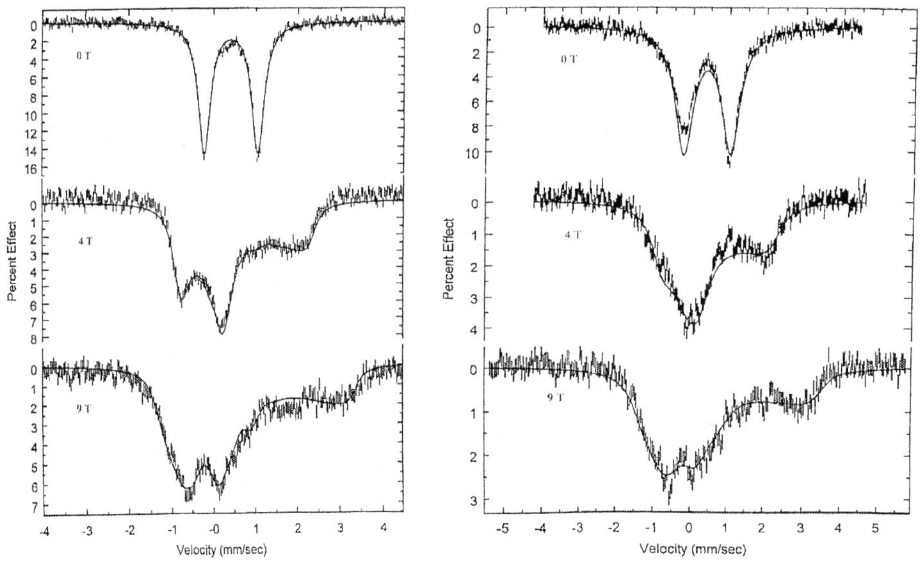

Table II – 4.2K High Field Fit

	Δ_Q(mm/s)	η	δ_{Fe}(mm/s)	A_{xx}(T)	A_{yy}(T)	A_{zz}(T)	Γ(mm/s)
Fe(TPP)NO	1.24	0	0.40	-28	-16	13	0.48
Fe(OEP)NO	1.27	0	0.43	-26	-26	13	0.27

High-field data are not yet available for the TTP and TMP compounds. It should be noted that the unpaired electron spin resides in a molecular orbital located primarily on the NO group — the EPR spectrum shows clear nitrogen hyperfine lines. Thus the A tensor given Table II measures the transferred spin polarization onto the iron. The fact that the A-tensor components are so large (~26T) is evidence of strong mixing of the NO π^* orbitals with the iron d orbitals.

While the quadrupole splitting, isomer shift and linewidth parameters of these four model complexes are only weakly temperature dependent, the area under the Mössbauer absorption curve has a very rapid temperature dependence. This area, proportional to the recoilless fraction can be fit for each of the models by a function of the form Area(T) = $A_o\exp(-T/T_o)$. Using a Debye model, and assuming $T>\theta_D/2$, we have $T_o = Mc^2k_B\theta_D^2/3E_o$ where E_o = 14.4keV for Fe^{57}, M is the mass of the recoiling unit, and θ_D is the Debye temperature [14]. The Debye model is admittedly a crude approximation to the vibrational dynamics of porphyrin model compounds in the solid state, but should give a rough characterization of the system. We assume excitation of acoustic modes for which the porphyrin recoils as a whole, (M~400Da) and calculate Debye temperatures of 81K(OEP), 62K(TMP), 55K(TPP) and 47K(TTP).

It's interesting to compare the calculated Debye temperature of 55K for Fe(TPP)NO with the vibrational density of states (VDOS) function recently measured for this compound in a nuclear resonance vibrational spectroscopy (NRVS) experiment [15]. The VDOS exhibited a set of 17 modes of energy between 74 and 540 cm^{-1}, with little evidence of a continuum of acoustic modes between the large resonances. Vibrational modes of energy lower than 74 cm^{-1} are lost in the wings of the recoilless absorption peak of the NRVS spectrum. The lowest observed mode of 74 cm^{-1} would act as an Einstein oscillator with θ_E = 106K. Thus our calculated Debye temperature of 55K certainly has a reasonable order of magnitude. This value could be compared in future work with calorimetric measurements.

References and Endnotes

[1] S.H.Snyder and D.S. Bredt, Sci. Am. 266 (1992) 68.
[2] J.S. Stamler, S.J. Singel, and J. Loscalzo, Science 258 (1992) 1898.
[3] W.T. Oosterhuis, and G. Lang, J. Chem. Phys. 50 (1969) 4381.
[4] In the $\{MNO\}^n$ notation the *n* denotes the total number of metal d electrons plus NO π^* electrons.
[5] W.R. Scheidt and J.L. Hoard, J. Am. Chem. Soc. 95 (1973) 8281.
[6] Abbreviations used: TPP = tetraphenylporphyrin, OEP = octaethylporphyrin, TMP = tetramesitylporphyrin, TTP = tetratolylporphyrin, EFG = electric field gradient.
[7] W.R. Scheidt and M.E. Frisse, J. Am. Chem. Soc. 97 (1975) 17.
[8] C. Hauser, T. Glaser, E. Bill, T. Weyhermuller, and K. Wieghardt, , J. Am. Chem. Soc. 122 (2000) 4352.
[9] W.R. Scheidt, H.F. Duval, T.J. Neal, and M. K. Ellison, J. Am. Chem. Soc. 122 (2000) 4651.
[10] C.J. Groombridge, L.F. Larkworthy, and J. Mason, Inorg. Chem. 32 (1993) 379.
[11] R. Montiel-Montoya, E. Bill, A.X. Trautwein, H. Winkler, L. Ricard, M. Schappacher, and R. Weiss, Hyperfine Interactions 29 (1986) 1411.
[12] M. K.Ellison, and W.R. Scheidt, J. Am. Chem. Soc. 119 (1997) 7404.
[13] W.H. Press, W.T. Vetterling, S.A. Teukolsky, and B.P. Flannery, *Numerical Recipes in C – The Art of Scientific Programming, 2nd Ed.* (Cambridge. Univ. Press, Cambridge, 1992).
[14] N.N. Greenwood, and T.C. Gibb, *Mossbauer Spectroscopy,* p. 11, (Chapman and Hall Ltd., London, 1971).
[15] J. T. Sage, C. Paxson, G.R.A. Wyllie, W. Sturhahn, S.M. Durbin, P.M. Champion, E.E. Alp, and W.R. Scheidt, J. Phys. Condens. Matter 13 (2001) 7707.

ANTIMONY-121 MÖSSBAUER SPECTRA FOR CROWN ETHER COMPLEXES OF ANTIMONY(III) TRIHALIDES, SbX_3(crown)

Masashi Takahashi, Atsushi Matsuura, Chikara Fujita, Kazuyuki Minowa, Masuo Takeda
Department of Chemistry, Faculty of Science, Toho University
Miyama, Fuanbashi, Chiba 274-8510, Japan

Abstract Antimony-121 Mössbauer spectra for the crown ether complexes of SbX_3 (X = Cl, Br, F) are measured at 20 K. The Mössbauer parameters indicate that the stereochemical activity of the lone pair is changed as a result of the weak interaction between SbX_3 and crown ether. A good relation between Mössbauer parameters and Sb–O bond length is observed though some intermolecular interactions are important in SbF_3 complexes.

1. Introduction

The X-ray crystal structure determinations of SbX_3(crown) have indicated that the distance between the Sb atom and the crown ether is shortend as the size of the crown ether becomes small [1 – 9]. The stereochemical role of the lone pair is one of the subjects for the structural discussion. Hough et al. and Alcock et al. have suggested the almost invariant stereochemical acitivity in the crown ether complexes of $SbCl_3$ [1, 2]. ^{121}Sb Mössbauer spectroscopy gives useful information on electronic states and have been applied to some crown ethers [3, 4, 7] but systematic study has not been done. Thus we have been carried out a systematic study to examine the stereochemical activity of lone pair using ^{121}Sb Mössbauer spectroscopy and X-ray crytallography. In this paper the following crown ethers are employed; 12-crown-4 (12C4), benzo-12-crown-4 (B12C4), 15-crown-5 (15C5), benzo-15-crown-5 (B15C5), 18-crown-6 (18C6), dibenzo-18-crown-6 (DB18C6). A preliminary result has been reported already [6].

2. Experimental

Complexes were prepared by the same way as described in the literature.

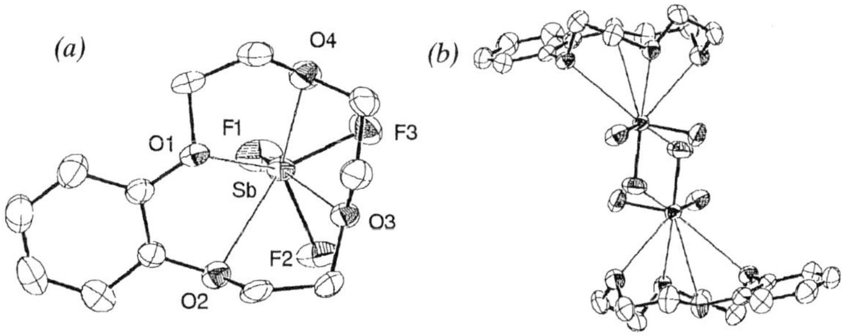

Figure 1. Crystal structure of SbF_3(B12C4). (a) top view, (b) side view showing a dimeric structure

Table 1 ^{121}Sb Mössbauer Parameters at 20 K for SbX_3(crown) and mean Sb–O bond length

Compound	δ [a] / mm s^{-1}	e^2qQ / mm s^{-1}	η	ΔN_s	ΔN_p	d(Sb—O) / Å
$SbCl_3$(12C4)	-9.10	6.3	0.66	0.22	–0.26	2.81 [c]
$SbCl_3$(B12C4)	-8.47	7.1	0.61	0.18	–0.23	2.85
$SbCl_3$(15C5)	-8.57	10.2	0.53	0.19	–0.11	2.90 [d]
$SbCl_3$(B15C5)	-7.59	10.4	0.26	0.12	–0.10	2.96
$SbCl_3$(18C6)	-6.51	12.7	0.27	0.06	–0.02	3.18 [e]
$SbCl_3$(DB18C6)	-6.32	12.6	0.18	0.04	–0.02	
$SbCl_3$ (crystal)	-5.90	13.1	0.33	0.02	0.00	
$SbCl_3$ / DCE [b]	-5.65	13.1	0.15	reference		
$SbBr_3$(12C4)	-8.89	4.9	0.95	0.21	–0.23	2.94
$SbBr_3$(15C5)	-7.75	7.5	0.59	0.14	–0.13	3.06 [f]
$SbBr_3$(B15C5)	-7.24	7.2	0.32	0.10	–0.14	
$SbBr_3$ (crystal)	-6.09	11.8	0.46	0.03	0.04	
$SbBr_3$ / DCE [b]	-5.61	10.8	0.35	reference		
SbF_3(B12C4)	-5.60	17.7	0.25	0.03	–0.03	2.96
SbF_3(15C5)	-6.72	15.6	0.02	0.11	–0.11	2.94 [g]
SbF_3(B15C5)	-6.29	15.4	0.11	0.08	–0.12	2.97
SbF_3(18C6)	-5.15	18.8	0.00	0.01	0.01	3.084 [h]
SbF_3(B18C6)	-5.06	18.5	0.00	reference		
SbF_3 (crystal)	–6.25	17.8	0.22	0.08	–0.03	

[a] relataive to InSb at 20 K, [b]frozen dichloroethane solution, [c] ref. [5, 6], [d] ref. [1], [e] ref. [2], [f] ref. [4], [g] ref [3], [h] ref. [8].

Antimony-121 Mössbauer spectra were measured at 20 K using a $Ca^{121m}SnO_3$ source on an Austin S-600 Mössbauer spectrometer. The spectra were computer-fitted using a transmission integral method and the isomer shift (δ) is given relative to InSb at 20 K.

Crystals for X-ray determination were obtained by slow evaporation of the solvent. X-ray determinations were carried out by a conventional manner. The data were collected on the Rigaku AFC5S diffractometer and the structure were solved by a direct method and refined by a full-matrix least-squares analysis using a TEXSAN program package.

3. Results and Discussion

3.1. Crystal structures

Crystal structures of $SbCl_3$(B12C4) [10], $SbCl_3$(B15C5) [11], $SbBr_3$(12C4) [12], SbF_3(B12C4) [13] and SbF_3(B15C5) [14] have been determined in this study. Each complex except SbF_3(B12C4) has a similar molecular structure. They have essentially the same structure as the known crown ether complexes. Each SbX_3 has a pyramidal geometry and the crown ether is puckered so that the oxygen atoms face to Sb atom. The lone pair of Sb atom is supposed to be directed to the crown ether. The crystal structure of SbF_3(B12C) has a dimeric structure in which one F atom of each SbF_3 unit bridges another SbF_3 unit forming a $(\mu\text{-F})_2Sb_2F_4$ unit, and B12C4 coordinates SbF_3 unit in an inclined manner (Figure 1). Details of the crystal structures will be published elsewhere.

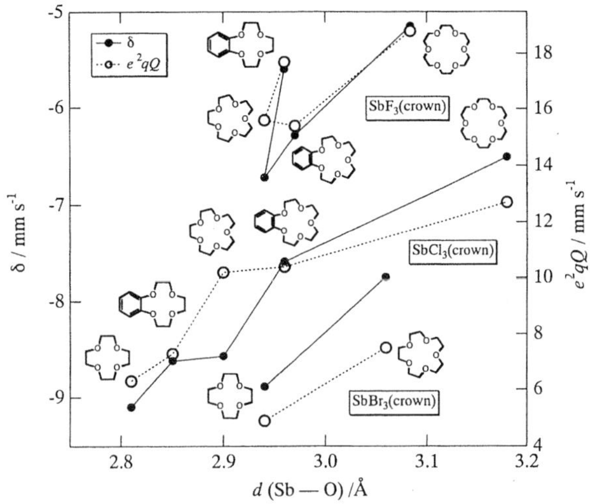

Figure 2. Relation between Sb–O bond length and Mössbauer parameters for SbX_3(crown). Solid circles: δ, open circles: e^2qQ.

3.2 ^{121}Sb Mössbauer parameters

The Mössbauer parameters are listed in Table 1. Each compound has a positive quadurapole coupling constant (e^2qQ) value, indicating that the principal component of the electric field gradient V_{zz} is negative. The negative V_{zz} is due to the lone pair at the Sb atom. The remarkable changes in δ and e^2qQ are ascribed to the change in stereochemical activity of the lone pair. The decrease in δ implies the increase in s elecron density and the reduction in e^2qQ indicates the decrease in p electron imbalance. The change in the stereochemical acitivity of the lone pair causes the change in the relative contribution of the s and p character of the lone pair. Hence the Mössbauer parameters clearly indicate that the stereochemical activity of the lone pair is much more decreased as the size of the crown ether is reduced.

We can estimate the change in s and p electron populations of Sb atom (ΔN_s and ΔN_p) using the Mössbauer parameters according to the Parish's method [16, 17]. A reference compound is necessary for the estimation and the change from the reference to the sample is calculated as ΔN_s or ΔN_p. The cumene solutions of $SbCl_3$ and $SbBr_3$ are used as references for $SbCl_3$(crown) and $SbBr_3$(crown). For SbF_3(crown), SbF_3(18C6) is used as a reference. The estimated ΔN_s and ΔN_p values are also included in Table 1. According to Parish the uncertainty of the estimation would be about 0.05 mm s^{-1}. The ΔN_s and ΔN_p values suggest that the increase in the s electron population is almost canceled out by the decrease in the p electron population, and the electron density at Sb atom is kept unchanged as a result. This clearly indicates that the stereochemical activity of the lone pair is affected by the interactions between SbX_3 and crown ether.

3.3 Relation between ^{121}Sb Mössbauer parameters and Sb–O bond lengths

The Mössbauer parameters are plotted against the Sb–O bond lengths in Figure 2. In $SbCl_3$(crown) both δ and e^2qQ decrease according to the decrease in the Sb–O bond lengths. This indicates again that the stereochemical activity of the lone pair at Sb atom is reduced as the results of Sb–crown ether interactions. Clearly the approach of the crown ether to $SbCl_3$ causes some suppression of the space for the lone pair. Figure 2 indicates also that the smaller the size in crown, the more the suppression occurs. The same size effect might operate in $SbBr_3$(crown) but the data is short for such discussion. For SbF_3(crown) other effects also exert as well as the size effect of the crown ether. SbF_3 complex with B12C violates the relation. The Sb–O bond lengths are longer than that expected from the $SbCl_3$(crown) and the Mössbauer parameters are also larger than the expectation. This is apparently caused by the nature of the SF_3 molecule that the molecules are subject to associate through the F atoms. Such an aggregation effect expected to operate effectively in the small crown ethers, and the effect is exactly observed in SbF_3(B12C4). And the polymeric structure for $(SbF_3)_2$(12C4) [7] would be the extreme example.

The shortest mean Sb–O bond length is 281 pm for $SbCl_3$(12C4). This length is apparently longer than the sum of covalent radii (216 pm) [17]. According to bond valence (BV) calculation proposed by Brown and Wu [18], the BV value for Sb–O bond in $SbCl_3$(12C4) is only 0.18. Thus the coordination of the crown ether is quite weak but still affects the stereochemical activity of the lone pair as mentioned above. The same has been also demonstrated in the antimony-molybdenum double oxides, in which the weak Sb–O interaction cannot be negligible [19].

References

[1] E. Hough, D.G. Nicholson and A.K. Vasudevan, J. Chem. Soc., Dalton Trans. (1987) 427.

[2] N.W. Alcock, M. Ravindran, S.M. Roe and G.R. Willey, Inorg. Chim. Acta 167 (1986) 115.

[3] M. Schäfer, J. Pebler, B. Borgsen, F. Weller and K. Dehnicke, Z. Naturforsch. 45b (1990) 1243.

[4] M. Schäfer and K. Dehnicke, Z. anorg. allg. Chem. 611 (1992) 149.

[5] N.W. Alcock, M. Ravindran and G.R. Willey, Acta Cryst. B49 (1993) 507.

[6] M. Takahashi, T. Kitazawa and M. Takeda, J. Chem. Soc., Chem. Commun. (1993) 1779.

[7] I. Becker, M. Windhaus and R. Mattes, Z. Naturforsch. 49b (1994) 870.

[8] J. Lipkowski, M.S. Fonari, V.Ch. Kravtov, Y.A. Simonov, E.V. Ganin and V.O. Gelmboldt, J. Chem. Cryst. 26 (1996) 823.

[9] G.R. Willey, D.R. Aris and W. Errington, Inorg. Chim. Acta 300-302 (2000) 1004.

[10] Crystal data: $C_{12}H_{16}O_4Cl_3Sb$, M = 452.36, orthorhombic, *Pnma* (No.62), a = 15.569 (3), b = 11.371 (2), c = 9.560 (2) Å, U = 1692 (1) Å^3, Z = 4, D_x = 1.775 g cm^{-3}, μ(MoKα) = 21.24 cm^{-1}, R = 0.060, R_w = 0.080 for 1556 reflections with $I > 3\sigma$.

[11] Crystal data: $C_{14}H_{20}O_5Cl_3Sb$, M = 496.42, monoclinic, *P21/c* (No.14), a = 9.147 (2), b = 12.636 (2), c = 16.439 (2) Å, b = 99.36 (1)°, U = 1874.8 (1) Å^3, Z = 4, D_x = 1.759 g cm^{-3}, μ(MoKα) = 19.29 cm^{-1}, R = 0.035, R_w = 0.037 for 3267 reflections with $I > 3\sigma$.

[12] Crystal data: $C_{12}H_{12}O_4Br_3Sb$, M = 581.69, orthorhombic, *Pnma* (No.62), a = 30.39 (2), b = 12.499 (7), c = 8.142 (4) Å, U = 3093 (6) Å^3, Z = 8, D_x = 2.498 g cm^{-3}, μ(MoKα) = 94.81 cm^{-1}, R = 0.059, R_w = 0.060 for 1366 reflections with $I > 3\sigma$.

[13] Crystal data: $C_{12}H_{16}O_4F_3Sb$, M = 403.00, monoclinic, *P21/c* (No.14), a = 11.311 (3), b = 10.264 (3), c = 12.489 (2) Å, β = 90.79 (2)°, U = 1449.8 (6) Å^3, Z = 4, D_x = 1.846 g cm^{-3}, μ(MoKα) = 19.43 cm^{-1}, R = 0.028, R_w = 0.038 for 2825 reflections with $I > 3\sigma$.

[14] Crystal data: $C_{14}H_{20}O_5F_3Sb$, M = 447.05, monoclinic, *P21/c* (No.14), a = 13.382 (3), b = 7.702 (3), c = 15.905 (4) Å, β = 95.91 (2)°, U = 1630.6 (8) Å^3, Z = 4, D_x = 1..821 g cm^{-3}, μ(MoKα) = 17.42 cm^{-1}, R = 0.035, R_w = 0.043 for 2759 reflections with $I > 3\sigma$.

[15] R.V. Parish, Coord. Chem. Rev. 42 (1982) 1.

[16] A. Ishiguro, M. Takahashi and M. Takeda, J. Organomet. Chem., 611 (2000) 558; Eqs. (1) and (3) in this papaer should be $U_p = e^2qQ / (e^2qQ)_0$ and $\delta = aN_s + bN_p$, respectively.

[17] J.E. Huheey, Inorganic Chemistry, 3rd ed., Harper & Row, New York, 1983, p. 258.

[18] I.I. Brown and K.K. Wu, Acta Cryst. B32 (1976) 1957.

[19] Y. Kajitani, M. Takahashi and M. Takeda, Int. J. Inorg. Mater. 3 (2001) 337.

Variable Temperature 57Fe Mössbauer Spectroscopic Studies on Ferrocenyl "Push-Pull" Molecules.

P. J. Titler[1], J. Silver[1], J. Zakrzewski[2], A. Klys[2] and I. Janowska[2]

[1]School of Chemical and Life Science, University of Greenwich, Wellington Street, Woolwich, London SE18 6PF Email tp06@gre.ac.uk
[2]Department of Organic Chemistry, University of Lodz, 90-136 Lodz, Narutowicza 68, Poland

Introduction

Although there have been a number of Mössbauer spectroscopic studies on NLO properties of ferrocenyl "push-pull" molecules [1-6] there is room for more characterisation of the electronic environment of the iron atom in such molecules. With this in mind two push-pull ferrocenes (**I** and **II**) have been synthesised and characterised. Within these push-pull molecules the ferrocene moiety acts as an electron donor, with the electron directed through the conjugated *p*-bridge to a strong electron acceptor group. Molecules of this type display large second order non - linear optical (NLO) properties (quadratic hyperpolarizabilities).
These compounds are particularly well suited to ^{57}Fe Mössbauer spectroscopic studies and an investigation has been carried out to enable a deeper insight into their ground-state structures as well as charge distribution in the solid state.

Experimental

^{57}Fe Mössbauer spectra were recorded on the apparatus as previously reported[7]. In addition specific temperature measurements were carried out using a variable temperature cryostat (custom made by Prototech UK) controlled by a Lake Shore Cryotronics DRC 80C temperature controller.
Complexes **I** and **II** were prepared by the Knoevenagel condensation of ferrocenecarbaldehyde with 1,3-bis(dicyanomethylene)indane in acetic anhydride and with 3-dicyanomethyleneindan-1-one in methanol, respectively . The (*E*)-configuration of **II** has been confirmed by X-ray crystallography (details will be published elsewhere). All starting materials were purchased from Aldrich Chemicals Ltd.

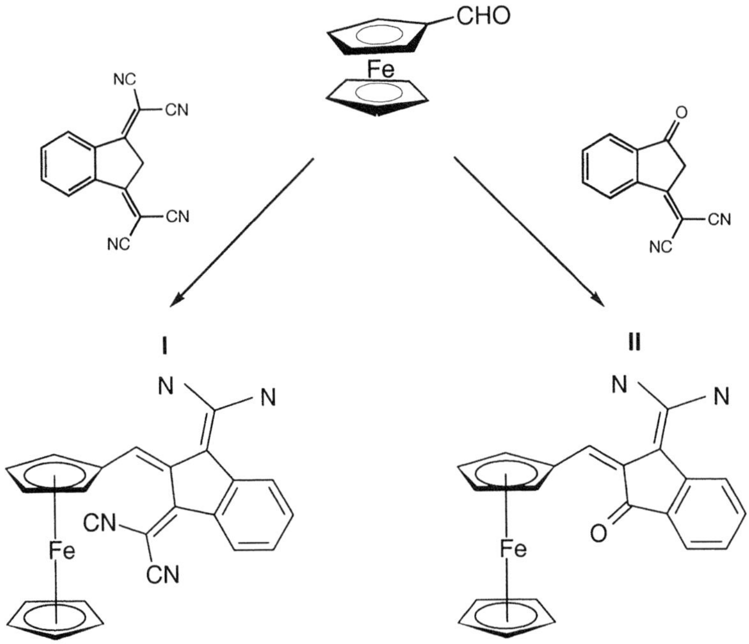

Fig 1 Preperation of compounds **I** and **II**

Compound I To a boiling solution of ferrocenecarbaldehyde (40 mg, 0.19 mmol) in methanol (5 ml) 3-dicyanomethyleneindan-1 one (110 mg, 0.57 mmol) was added, and the solution was stirred without heating for 15 min. After evaporation of the solvent, chromatography (SiO_2 / chloroform-hexane 1:1) and recrystallisation from hexane, green II was obtained in 86 % yield. ^{1}H NMR (200MHz, $CDCl_3$, d): 8.66 (d, J = 6.6 Hz, aromatic H), 8.46(s,1H,-vinylic H), 7.88 (d, J = 6.6 Hz, aromatic H),7.74 (m, 2H, aromatic H),5.32 (bs, 2H, substituted Cp) 5.05(bs 2H, substituted Cp), 4.35 (s,5H, unsubstituted Cp) VIS (l_{max}/solvent): 613 nm/ hexane, 622 nm/ benzene, 632 nm/ chloroform. Correct C, H, N analysis.

Compound II A solution of ferrocenecarbaldehyde (106 mg 0.5 mmol) and 1,3-bis(dicyanomethylene)indane (80 mg, 0.33 mmol) in acetic anhydride (12 ml) was heated to 55-60^{0}C for 1.5h. After evaporation of the solvent, chromatography (SiO_2 / chloroform-hexane 1:1) and recrystallisation from hexane, dark green I was obtained in 67 % yield. ^{1}H NMR (200MHz, acetone-d_6, d): 8.71(s,1H,-vinylic H),8.60 and 8.49(m's,1H each H-4 and H-7), 8.00 (m,2H, H-5 and H-6),5.18 (t, J = 2.1 Hz, 2H, substituted Cp) 5.00(t, J = 2.1 Hz, 2H, substituted Cp), 4.38

(s,5H, unsubstituted Cp). VIS (l_{max}/solvent): 637 nm/ hexane, 656 nm/ benzene, 659 nm/ chloroform. Correct C, H, N analysis.

Results and discussion

Table 1 ^{57}Fe Mössbauer Parameters for compounds **I** and **II**

Temperature (K)	δ (mms^{-1})	Δ (mms^{-1})	Γ† (mms^{-1})	% Area	δ (mms^{-1})	Γ† (mms^{-1})	% Area
				I			
77K	0.53(1)	2.167(2)	0.207(3)	100(1)			
140K	0.51(1)	2.17(2)	0.26(3)	81(7)	0.22(2)	0.25(8)	19(4)
298K	0.505(4)	2.200(5)	0.27(2)	38(2)	0.289(6))	0.33(2)	62(2)
77K*	0.53.(3)	2.180(6)	0.22(1)	65(3)	0.96(21)	1.05(20)	35(8)
				II			
77K	0.542(3)	2.186(6)	0.24(1)	100(3)			

* Post 298K
† Full width at half height

The results in table 1 and figs 2, 3 and 4 indicate significant changes of charge distribution in the solid state with variation of temperature. A significant change in the Mössbauer spectrum in fig 1 of compound I is shown at room temperature. A second site usually associated with 'oxidised' ferrocene is seen as a broad single line between the ferrocenyl doublet. This signal almost disappears on reverting to 77K. This is of the total switching of an electron in over 50% of the molecules between 78K and 298K from the ferrocenyl iron to an excited state molecular orbital in which the iron orbitals are not significantly involved. Compound II exhibits similar behaviour. Further studies on this are being carried out.

It has been found that these compounds display especially low lying charge transfer bands and significant solvatochromism which gives a promise of large second-order NLO properties.

Previously Mössbauer spectroscopic studies on ferrocenyl molecules manifested evidence for such total switching of an electron. The main difference in the present site is the closeness of the acceptor site to the ferrocene moiety.

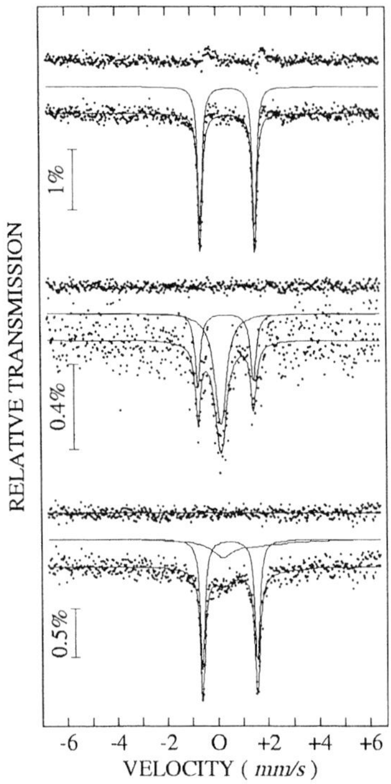

Fig 1 ^{57}Fe Mössbauer Spectrum of compound **I** at a) 77K, b) 298K and c) 77K after 298K run

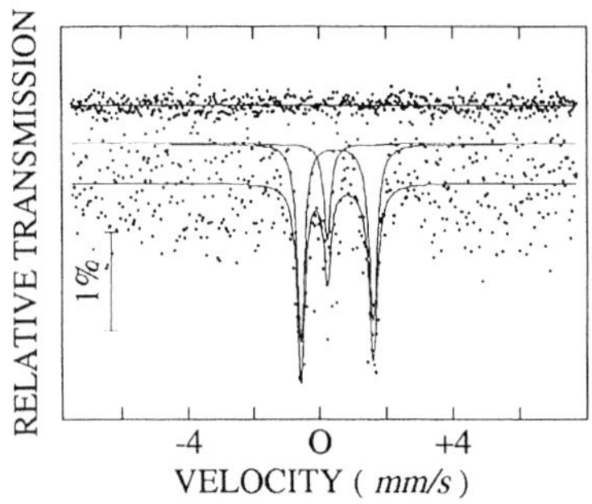

Fig 3 ^{57}Fe Mössbauer Spectrum of compound **II** at 140K

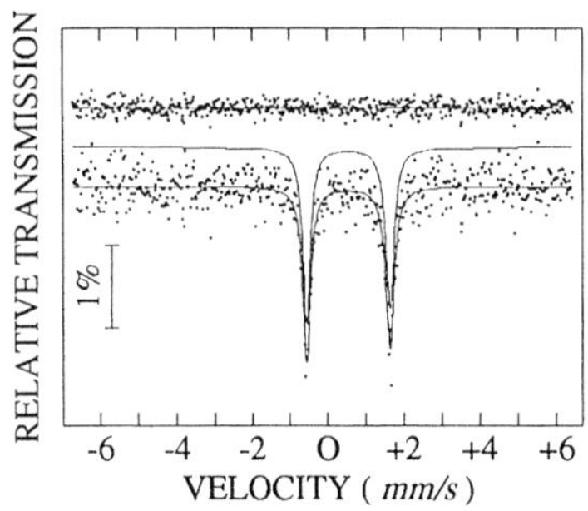

Fig 4 ^{57}Fe Mössbauer Spectrum of compound **II** at 77K

References

[1] A. Houlton, N Jassim, R.M.G. Roberts, J. Silver, D. Cunningham, P. McArdle and T. Higgins, J. Chem. Soc. Dalton Trans., (1992) 2335.

[2] C. Lopez, J. Sales, X. Solans and R. Zquiak, J. Chem. Soc. Dalton Trans., (1994) 747.

[3] A. Houlton, J.R. Miller, J. Silver, N Jassim, M. J. Ahmet, T. L. Axon, D. Boor and G. H. Cross, Inorg. Chim. Acta, 205(1993), 67.

[4] C. Lopez, R. Bosque, X. Solans, M. Font-Bardia, D. Tramuns, G. Fern and J. Silver J. Chem. Soc. Dalton Trans., (1994) 3039.

[5] J. Silver, J.R. Miller, A Houlton and M.J. Ahmet, J. Chem. Soc. Dalton Trans., (1994) 3355.

[6] D.A. Davies, J. Silver, G. Cross and P. Thomas J. Organomet. Chem., In Press 2001.

[7] Y. Hamed, R.C. Hider and J. Silver, Inorg. Chim. Acta, 66(1982), 13.

THE Ca BINDING TO THE HULL OF *NIXTAMALIZED* CORN GRAINS AS DETECTED BY ^{57}FE MÖSSBAUER SPECTROSCOPY AND RELATED TECHNIQUES

E. Reguera[1], *H. Yee-Madeira*[2,*], *R. Gonzalez*[3], *F. Sánchez-Sinencio*[4]

[1]Institute of Materials and Reagents, Havana University, Havana, Cuba
[2] Depto. Física, Esc. Sup. Física y Matemáticas-IPN, Av. IPN s/n, Edif . 9, U. Prof. "ALM" Col. Lindavista, 07738, México, D.F., México
[3] Faculty of Chemistry, Havana University, Havana, Cuba
[4] Departments of Physics, CINVESTAV-IPN, Mexico, D.F.

Abstract. Tortilla and other products elaborated from *nixtamalized* (alkaline cooking in a saturated solution of $Ca(OH)_2$) corn grains are the main source of Ca intake in Mexico, mainly in rural regions. In that cooking process Ca is mainly retained in the germ and the corn hull. In the germ Ca appears as salts of fatty acids. According to IR and ^{13}C CP/MAS NMR data, the hull binds Ca through the hemicelluloses acidic species (uronic acids). Mössbauer spectra of *nixtamalized* hull samples, where Ca^{2+} was substituted by $^{57}Fe^{2+}$, suggest the existence of two slightly different sites for Ca in the corn hull.

1. Introduction

In Mexico corn is consumed mainly as Tortillas. Tortillas are prepared from mass or flours obtained through an alkaline cooking and steeping of corn grains in a saturated solution of $Ca(OH)_2$, process that is used from prehispanic times. The alkaline treatment makes corn proteins available and incorporates Ca to the cooked grains, increasing the nutritional levels of the mass or flours elaborated with *nixtamalized* corn [1]. Approximately a 50% of Ca intake in Mexico, mainly in rural regions, is provided by Tortilla and other products elaborated from *nixtamalized* corn [2]. It is accepted that of the total Ca employed in *nixtamalization* only a small fraction is retained in the grain during the cooking and resting periods [3]. In the germ Ca is found as salts of fatty acids, due to a partial saponofication of its fats during the alkaline cooking, while in endosperm it is forming inclusion compounds of these salts within the amylose helical structure [3]. Little attention has been given to the possible retention of Ca by the corn hull (pericarp) and on its state in this part of the grain, although, as will be discussed later, the presence there of species capable of bind Ca. In this contribution we are reporting a study on the state of Ca in the hull of *nixtamalized* corn grains, using infrared (IR), ^{57}Fe Mössbauer, ^{13}C cross polarization magic angle spinning NMR (^{13}C CP/MAS NMR) and X-ray energy-dispersed spectroscopy (EDS) as sensing techniques.

2. Experimental

Samples of the different parts of the corn grain before and after their alkaline cooking for 30 min. at 85 °C in a saturated solution of $Ca(OH)_2$ were studied. Additionally, samples of the corn hull previously separated from the grain with and without the alkaline treatment were also analyzed.

The IR spectra were run in KBr pressed disks using an FTIR spectrophotometer (from Bruker). The ^{13}C CP/MAS NMR spectra were recorded in a Bruker ASX300 (300 MHz) spectrometer.
The Mössbauer spectra of *nixtamalized* hull were obtained from samples in which part of Ca^{2+} was substituted by $^{57}Fe^{2+}$. The spectra were recorded at room temperature with a ^{57}Co/Rh source using a constant acceleration spectrometer (from Wissel) and then fitted with a least-squares minimization algorithm in order to obtain the values of isomer shift (δ), quadrupole splitting (Δ), line width (Γ) and relative area (A). The values of δ are reported relative to sodium nitroprusside.

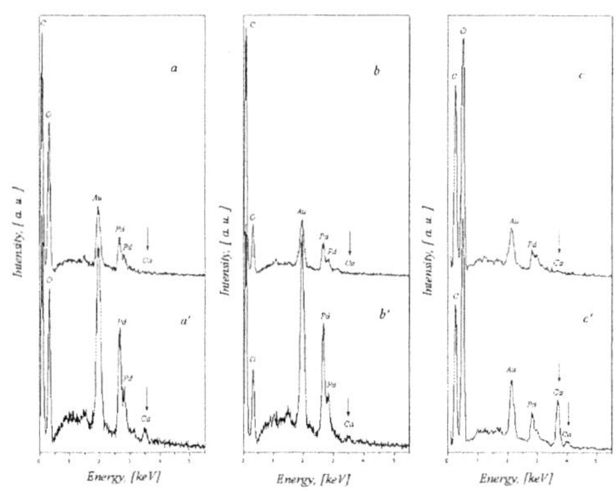

Figure 1. EDS spectra of different parts of the corn grain before (*a,b,c*) and after (*a´,b´,c´*) its alkaline cooking (nixtamalization). (*a,a´*): Germ; (*b, b´*): Endosperm; (*c,c´*): Hull. Observe that Ca is mainly retained in the hull.

* Corresponding author E-mail: yee@esfm.ipn.mx and COFAA-Fellow

3. Results and discussion.

A) Ca retention by the corn grain.

Figure 1 shows the EDS spectra of the different parts of the corn grain before and after its *nixtamalization*. The fluorescent peaks of Ca, which are practically absent in untreated grains, appear relatively intense after the alkaline cooking indicating a significant Ca retention. According to the peaks intensity, Ca is retained in the following order: hull > germ > endosperm. Since the hull only represents a 5% of the grain weight, from Figure 1 one concludes that this part of the corn grain shows a pronounced ability to bind Ca.

Hemicelluloses are the major component of the hull and they are acidic type polysaccharides with a main chain of xylose units to which single units of L-arabinose and uronic acids are attached [4]. The acid-base reaction of $Ca(OH)_2$ with these acidic groups could be responsible of the Ca retention by the corn hull. From pH titration, we found a Ca retention by the hull of approximately 4 mg/g.

B) On the nature of Ca binding in the corn hull.

Figure 2 shows the IR spectra of the corn hull before and after its alkaline treatment. In untreated hull samples strong IR absorptions at 1700 and 1736 cm^{-1} are observed, corresponding to carboxylic groups vibrations. The IR region of these acidic vibrations shows a complex structure suggesting that slightly different carboxylic species could be present. On the alkaline treatment these acid bands disappear and new bands at 1418, 1475 and 1629 cm^{-1} due to carboxylate vibrations appear. These last absorption bands result from the neutralization reaction of the acidic species and Ca salts formation. When the samples were immersed in a solution of oxalic acid the acidic bands were restored and Ca oxalate is formed, as detected by IR spectroscopy.

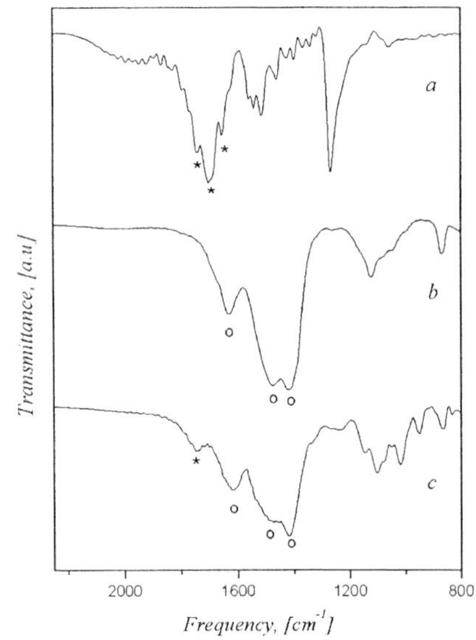

Figure 2. IR spectra (2250-800 cm^{-1} region) of the corn hull before (a) and after (b) the alkaline cooking. For comparison, the IR spectra of the Ca salt of the polygalacturonic acid was included (c). In (c) a small residual acidic signal is also observed.

Figure 3 shows the ^{13}C CP/MAS NMR spectra of hull samples before and after their alkaline cooking. These are complex spectra due to the nature of the studied material. The main signals can be assigned to the saccharide matrix [5] but the broad one at 173.53 ppm is mainly due to the carbonyl groups of uronic acids. When this signal is decomposed a contribution of carbonyl groups of esters (at 170 ppm) and from other carbonyl groups (at 177 ppm) are also observed. On the alkaline cooking, the signal at 173.53 ppm practically disappears due to the acid groups neutralization and Ca salts formation. The other changes observed in the 13C CP/MAS NMR spectra on cooking process are related with the progressive degradation of the hull hemicelluloses and lignin, which pass to the cooking liqueour.

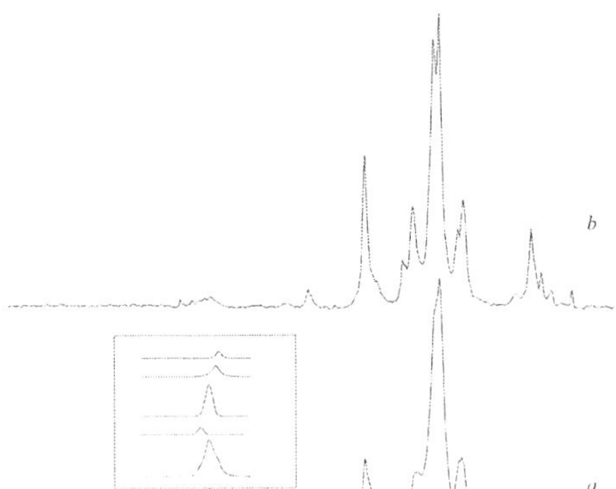

Figure. 3. ^{13}C CP/MAS NMR spectra of the corn hull before (a) and after (b) the alkaline cooking. The inset shows the deconvolution of the carboxylic signals.

C) On the state of Ca in nixtamalized corn hull as detected by ^{57}Fe Mössbauer spectra.

The Ca retained in the corn hull during the *nixtamalization* process can be exchanged by Fe in order to study the local Ca environment using Mössbauer spectroscopy. Due to the relatively low concentration of Ca in *nixtamalized* hull samples, a convenient option is to do the ionic exchange using a ^{57}Fe enriched ferrous salt. To extract Ca from the hull and place divalent iron one needs to find an anion that prefers Ca over Fe. A possibility is the use of $^{57}FeSO_4.xH_2O$. In aqueous solution the following reaction should occur:

$$\text{Hull-Ca}^{2+} + FeSO_4 \longrightarrow \text{Hull-Fe}^{2+} + CaSO_4\downarrow$$

$CaSO_4$ is a very insoluble salt, which will precipitate in the solution or remains as an independent phase occluded in the sample

Figure 4 shows the EDS spectra of *nixtamalized* hull samples before and after their reaction with $FeSO_4$. As expected, the exchange of Ca by Fe effectively takes place. The exchanged samples adopt a brownish color suggesting that, in the treatment in aqueous solution, iron is finally oxidized to Fe^{3+}, even if a glove box filled with Ar-99.99 % is used.

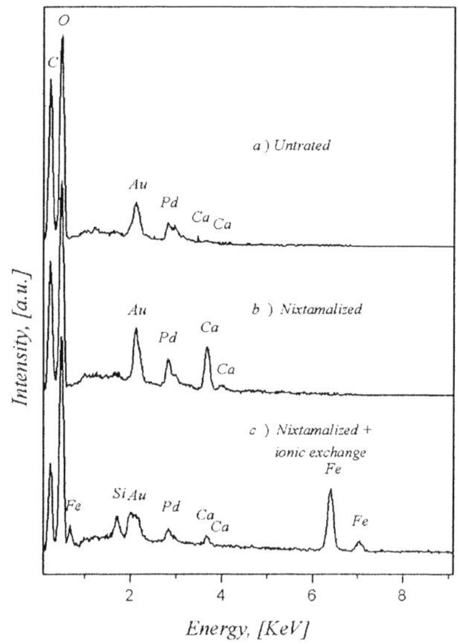

Figure 4. EDS spectra of: a) hull of untreated corn grains; b) nixtamalized hull; c) nixtamalized hull and then treated in an aqueous solution of $FeSO_4$ to exchange Ca by Fe. Observe that the ionic exchange effectively takes place

It is well known that certain oxidation-reduction processes can be retarded by carrying the reaction in the solid state [6], milling the reactants in an agate mortar. Based on this experience, and in order to minimize the oxidation of Fe^{2+} to Fe^{3+}, in the samples for Mössbauer spectroscopy the exchange reaction of Ca^{2+} by $^{57}Fe^{2+}$ was also performed milling in an agate mortar the appropriate quantities of *nixtamalized* hull, $^{57}FeSO_4.xH_2O$ and hydrazine hydrochloride.

Figure 5 shows the typical Mössbauer spectra corresponding to hull-Fe^{2+} samples oxidized and partially oxidized. In Table I are collected the corresponding Mössbauer parameters obtained by a fitting of two quadrupole doublets from Fe^{2+} and Fe^{3+} of similar isomer shift values.

From the analysis of the Mössbauer spectra of *nixtamalized* hull samples we can infer the existence of two structurally different sites for Fe, and if this element effectively substitutes Ca, then there are also two different sites for Ca. This could result by two different chemical surroundings or different geometrical position of the ligands. The values of δ observed for Fe^{2+} and Fe^{3+} suggest that the iron cations are coordinated to oxygen atoms as first neighbors. This is in accordance with above discussed results from IR and ^{13}C CPMAS NMR data, which suggest that Ca is linked to the oxygen atoms from the carboxylate groups of the hull hemicelulloses.

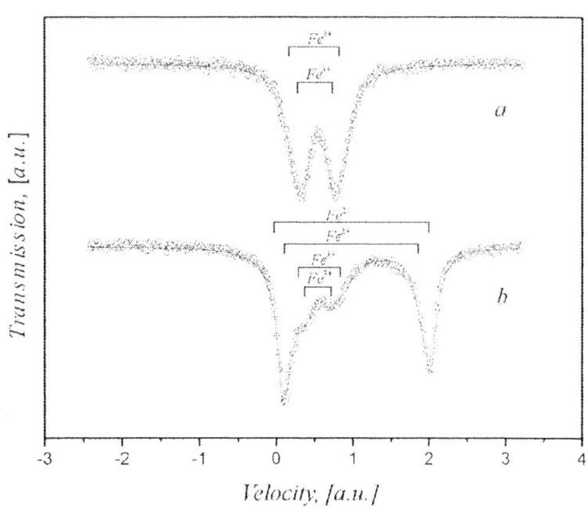

Figure 5 Mössbauer spectra of nixtamalized corn hull where Ca^{2+} was exchanged by $^{57}Fe^{2+}$: a) Ionic exchange carried out in an aqueous medium; b) Ionic exchange performed by a mechanochemical route. In both spectra two different structural sites for iron can be identified.

4. Conclusions

During the cooking of corn in an alkaline solution of $Ca(OH)_2$, Ca is incorporated to the grain. The largest Ca content is found in the corn hull, in amount of approximately 4 mg/g. According to IR and ^{13}C CP/MAS NMR data, Ca in the hull is linked to the carboxylate groups of the hemicelluloses and where slightly different carbonyl groups were detected (uronic acids). The ionic exchange of Ca for ^{57}Fe in the *nixtamalized* hull and the corresponding Mössbauer spectra suggest the existence of two slightly different structural sites for Ca. The obtained Mössbauer parameters are in accordance with the IR and ^{13}C CPMAS results.

Table I. Mössbauer parameters of ^{57}Fe in nixtamalized corn hull samples.

Sample	Temp. [K]	δ [mm/s]	Δ [mm/s]	Γ [mm/s]	A [%]	Assignment
$FeSO_4.xH_2O$	300	1.01	3.22	0.29	100	Fe^{2+}
$Hull-Fe^{2+}$; Exchange in aqueous media	300	0.54	0.62	0.35	41	Fe^{3+}
		0.55	0.37	0.31	59	Fe^{3+}
$Hull-Fe^{2+}$; Exchange in a solid state reaction	300	0.52	0.66	0.32	14	Fe^{3+}
		0.57	0.32	0.33	22	Fe^{3+}
		1.06	1.93	0.29	43	Fe^{2+}
		1.05	1.79	0.34	21	Fe^{2+}
$Hull-Fe^{2+}$; Exchange in aqueous media	12	0.71	0.65	0.35	67	Fe^{3+}
		0.73	0.89	0.31	33	Fe^{3+}

The values of δ are reported relative to sodium nitroprusside; Errors in δ, Δ and Γ are smaller than 0.01 mm/s.

References

[1] A. Trejo-González, A. Morales and C. Wild-Altamirano, Adv. Chem. Series, 198 (1982) 145.
[2] S.O. Serna-Saldivar, L.W. Roney and L.W. Greene, Cereal Chem. 68 (1991) 565; 69 (1992) 78.
[3] E. Reguera, H.Yee-Madeira, J. Fernández and F. Sánchez-Sinencio, *Topics in Contemporary Physics*, (Edited by H. Heras and R.V. Jímenez, Monash Litho, Mexico City) p. 221.
[4] L. Saulnier, C. Mestres, J.L. Doublier, P. Roger and J.J. Thibault, J. Cer. Sci. 17(1993) 267.
[5] J. Fernandez, E. Reguera and P. Ortiz, Spectrochim, Acta A, (2001), Accepted.
[6] E. Reguera, L. Núñez and J. Fernandez, Z. Naturforsch. 50B (1995)1067.

Acknowledgments The authors thank Dr. G. Cabañas-Moreno for EDS spectra acquisition.

Mössbauer Spectroscopic Study of Iron-Containing Sodium Borosilicate Glasses

K. Yoshino[a], T. Toriyama[b], T. Inamura[a], H. Wakabayashi[b] and T. Nishida[c]

[a] *General Education Center, Musashi Institute of Technology, 1-28-1 Tamazutsumi, Setagaya-ku, Tokyo 158-8557, Japan*

[b] *Department of Energy Science and Engineering, Musashi Institute of Technology, 1-28-1 Tamazutsumi, Setagaya-ku, Tokyo 158-8557, Japan*

[c] *Department of Biological and Environmental Chemistry, Kyushu School of Engineering, Kinki University, Kayanomori 11-6, Iizuka 820-8555, Japan*

Mössbauer study of xNa$_2$O•5Fe$_2$O$_3$•(95-x)(B$_2$O$_3$•SiO$_2$) glasses (x=5-40) was carried out in order to evaluate the chemical bond strength of Fe-O-B and Fe-O-Si bonds in sodium borosilicate glasses having different numbers of nonbridging oxygen (NBO). From the isomer shift values (0.22-0.36 mm/s) obtained for Fe(III) at 300K, we can deduce that Fe(III) atoms occupy substitutional sites of B(III) and Si(IV) and play a role of network former. A marked decrease of isomer shift is observed with an increase in the Na$_2$O content when x>10. This result indicates a formation of NBO in B(III)O$_4$ and Si(IV)O$_4$ units. From the temperature dependence of the absorption area for Fe(III), Debye Temperature (θ_D) of 280 K was estimated for 35Na$_2$O•5Fe$_2$O$_3$• 60(B$_2$O$_3$•SiO$_2$) glass having NBO, while that of 5Na$_2$O•5Fe$_2$O$_3$•90(B$_2$O$_3$•SiO$_2$) glass was 380 K since it has no NBO. It was found that the composition-dependent decrease of θ_D in borosilicate glasses, caused by the formation of NBO, is larger than that observed in borate glasses.

1. Introduction

It is well known that alkali borosilicate glasses have high chemical durability, high thermal and mechanical stability and high radiation durability [1]. Nishida *et al.* proved that borosilicate glasses are more resistant to thermal neutron than borate glasses [2]. When a large amount of alkali oxide is introduced into glass matrix, chemical durability, mechanical stability, *etc.* are generally lowered owing to the formation of nonbridging oxygen (NBO). Formation of NBO in glass network has been investigated in the Mössbauer study of several glasses containing a small amount of Fe$_2$O$_3$ and different amounts of alkali oxides [3,4]. It is interesting to investigate how alkali oxide affects the local structure, Mössbauer parameters and Debye temperature (θ_D) of alkali borosilicate glasses.

In this paper, Mössbauer study of sodium iron borosilicate glasses, xNa$_2$O•5Fe$_2$O$_3$•(95-x)(B$_2$O$_3$•SiO$_2$), was carried out in order to elucidate the effect of NBO on the chemical and physical properties. Absorption area for Fe(III) was estimated at lower temperatures in order to evaluate the θ_D value of 5Na$_2$O•5Fe$_2$O$_3$•90(B$_2$O$_3$•SiO$_2$) and 35Na$_2$O•5Fe$_2$O$_3$•60(B$_2$O$_3$•SiO$_2$) glasses, since the latter has a large number of NBO, while the former not.

2. Experimental

Na$_2$CO$_3$, Fe$_2$O$_3$ and H$_3$BO$_3$ of guaranteed reagent grade were used for preparing the glass samples. Starting mixture was sintered in an alumina crucible at about 500 °C and was melted at 1250 °C in air for 2h. The mixture was quenched with a twin roller. Mössbauer spectra of powder samples were recorded by the constant acceleration mode at room temperature. For low temperature Mössbauer measurement, a commercial cryostat filled with liq. N_2 was used in the range between 78 and 295 K. ^{57}Co (1.11 GBq) diffused into Rh matrix was used as a Mössbauer source. Mössbauer parameters for each spectrum were calculated with a least-squares fitting program NORMOS. A pure iron foil was used for calibrating the velocity scale of the spectrometer.

3. Results anddiscussion

Mössbauer spectra of $x N_2O \cdot 5Fe_2O_3 \cdot (95-x)(B_2O_3 \cdot SiO_2)$ glasses (x=5-40) measured at room temperature are shown in Fig. 1. All the Mössbauer spectra consist of two

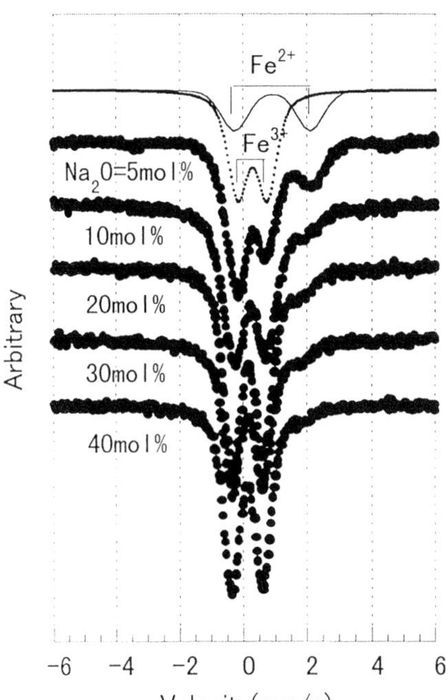

Fig. 1 Mössbauer spectra of $x Na_2O \cdot 5Fe_2O_3 \cdot (95-x)(B_2O_3 \cdot SiO_2)$ glasses (x=5-40) measured at room temperatuer.

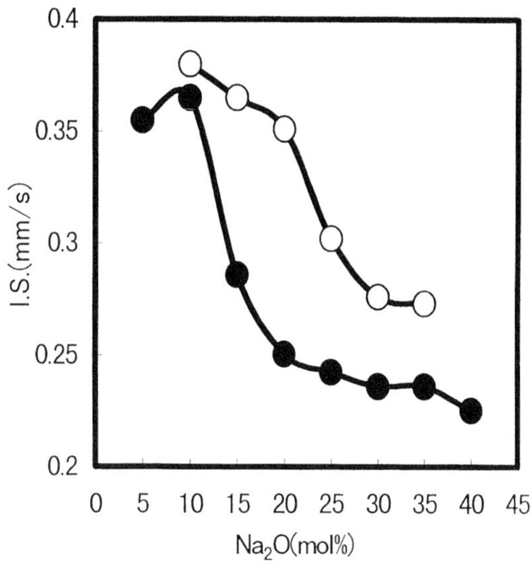

Fig.2 Composition dependency of the isomer shift values for Fe(III) atoms in $x Na_2O \cdot 5Fe_2O_3 \cdot (90-x)(B_2O_3 \cdot SiO_2)$ glasses (●) and $x Na_2O \cdot 5Fe_2O_3 \cdot (90-x)B_2O_3$ glasses (○).[4] measured at room temperature.

quadrupole oublets due to Fe(III) and Fe(II) species. The component of Fe(II) species decreases with an increase in the Na_2O content. From the values of IS (1.03 mm/s) obtained for Fe(II) species, we can consider that Fe(II) atoms occupy octahedral sites and play a role of network modifier.

Figure 2 shows composition dependency of the isomer shift (IS) values for Fe(III) in $xN_2O{\bullet}5Fe_2O_3{\bullet}(95\text{-}x)(B_2O_3{\bullet}SiO_2)$ glasses (x=5-40) obtained at room temperature. Corresponding composition dependency of IS values for $xN_2O{\bullet}5Fe_2O_3{\bullet}(95\text{-}x)B_2O_3$ glasses (x=10-35) [4] is also illustrated for comparison. From the values of IS (0.22-0.36 mm/s) obtained for Fe(III), we can deduce that Fe(III) atoms occupy substitutional sites of tetrahedral B(III) atoms and play a role of network former. A marked decrease in the IS values of Fe(III) was observed with an increase in the Na_2O content when $x \geqq 10$. From the comparison of the IS curve of $xN_2O{\bullet}5Fe_2O_3{\bullet}(95\text{-}x)(B_2O_3{\bullet}SiO_2)$ glasses with that of $xN_2O{\bullet}5Fe_2O_3{\bullet}(95\text{-}x)B_2O_3$ glasses [4], we can speculate that NBO is primarily formed in $B(Fe)O_4$ tetrahedra when x=10-15, while it also appears in $Si(Fe)O_4$ tetrahedra when $x \geqq 15$. The decrease of IS values can be explained by a decrease in the distance between Fe(III) atoms and oxygen, giving rise to an increased overlap of 4s-orbital of Fe(III) with p_π orbital of oxygen [3-5]. Namely, s-electron density at iron nucleus will increase, and as a result, the value of IS decreases. The decrease in IS observed in borosilicate glasses is ascribed to the formation of NBO in both $B(Fe)O_4$ and $Si(Fe)O_4$ units. Formation of NBO in $Si(Fe)O_4$ units will be associated with the smaller values of IS observed in borosilicate glasses, which are obviously smaller than those of borate glasses [4].

A marked decrease is observed in quadrupole splitting (QS) for Fe(III), when $x \geqq 10$, in the similar manner as IS decreases. Very similar results were previously reported by Nishida *et al.* [5] in $xK_2O{\bullet}7Fe_2O_3{\bullet}(100\text{-}x)(B_2O_3{\bullet}2SiO_2)$ glasses (x=10-40), in which a remarkable decrease was observed in both IS and QS values in the same composition region, owing to the formation of NBO.

Mössbauer spectra of $5Na_2O{\bullet}5Fe_2O_3{\bullet}90(B_2O_3{\bullet}SiO_2)$ and $35Na_2O{\bullet}5Fe_2O_3{\bullet}60(B_2O_3{\bullet}SiO_2)$ glasses were measured between 78 and 295 K. Each absorption area for Fe(III) in both samples is normalized by the value obtained at 78K, as shown in Fig. 3. Solid and open circles respectively correspond to the normalized absorption area for Fe(III) in $5Na_2O{\bullet}5Fe_2O_3{\bullet}90(B_2O_3{\bullet}SiO_2)$ and $35Na_2O{\bullet}5Fe_2O_3{\bullet}60(B_2O_3{\bullet}SiO_2)$ glasses. The absorption area was utilized for estimating the θ_D value, since d ln$f(T)$/dT can be approximated by d ln$A(T)$/dT. The recoil-free fraction, $f(T)$, is expressed by the following equation [6], *i.e.*

$$f(T) = \exp\left\{-\frac{3E_R}{2k\Theta_D}\left[1+4\left(\frac{T}{\Theta_D}\right)^2\int_0^{\Theta_D/T}\frac{x}{e^x-1}\mathrm{d}x\right]\right\}.$$

Solid lines in Fig. 3 were obtained by a least squares fitting. From the slope of the straight lines shown in Fig. 3, θ_D values of 380 and 280 K were estimated for $5Na_2O{\bullet}5Fe_2O_3{\bullet}90(B_2O_3{\bullet}SiO_2)$ and $35Na_2O{\bullet}5Fe_2O_3{\bullet}60(B_2O_3{\bullet}SiO_2)$ glasses, respectively.

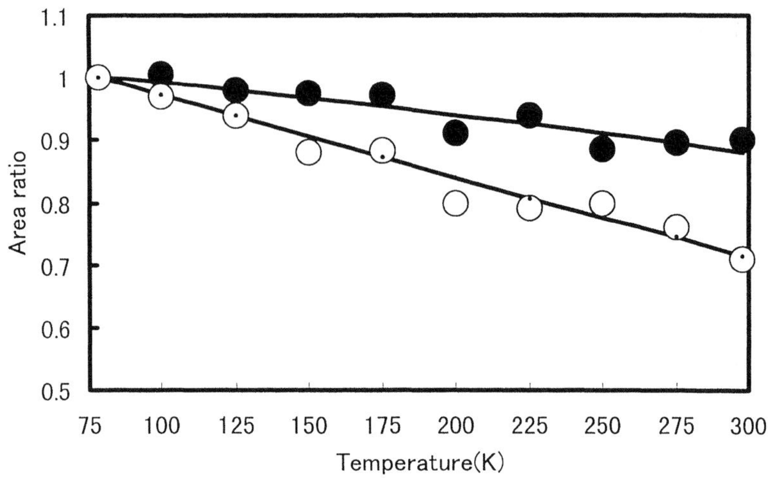

Fig. 3. Temperature dependency of the absorption area in the Mössbauer spectra of $5Na_2O{\bullet}5Fe_2O_3{\bullet}90(B_2O_3{\bullet}SiO_2)$ (●) and $35Na_2O{\bullet}5Fe_2O_3{\bullet}60(B_2O_3{\bullet}SiO_2)$ glasses (○).

The decrease of θ_D value from 380 to 280 K is ascribed to the formation of NBO atoms that means breaking of Fe-O-B and Fe-O-Si bonds in the network. The decrease of θ_D values obtained for Fe(III) in borosilicate glasses is larger than that found in borate glasses. The experimental result that the θ_D value (380K) obtained for Fe(III) in borosilicate glass is larger than that (320K) obtained in borate glass without NBO [4] indicates that B-O-Fe-O-Si bonds in borosilicate glass is more intense than B-O-Fe-O-B bonds in borate glass.

References

[1] S. Music, J. Mater. Sci. Lett., **8** (1989) 537.
[2] T. Nishida, T. Hirai and Y. Takashima, Radiochem.Radioanal.Lett., **49** (1981) 307.
[3] T. Nishida, in: *Mössbauer Spectroscopy of Sophisticated Oxides*, eds. A. Vértes and Z. Homonnay (Akadémiai Kiadó, Budapest, 1997) p. 31.
[4] K. Yoshino, T. Toriyama, T. Inamura, H. Wakabayashi and H. Iijima, J. Radioanal. Nucl. Chem., **246** (2000) 39.
[5] T. Nishida, T. Hirai and Y. Takashima, Phys. Chem. Glasses, **22** (1981) 94.
[6] D. Barb, in: *Grundlagen und Anwendungen der Mössbauer Spectroscopie*, (Akademie, Berlin, 1980) p. 20.

Investigation of the Fe^{3+} luminescence centre in $LiAlO_2$ phosphors activated with iron.

D. A. Davies, J. Silver, P. J. Titler and R. Haywood.
Centre for Phosphor and Display Materials, Department of chemical and life sciences,
University of Greenwich, Woolwich Campus,
Wellington Street, Woolwich, SE18 6PF.

1. Abstract

A Mössbauer spectroscopic study of $LiAlO_2$ doped with ^{57}Fe (0.1 mol.%) has shown that the Fe in the material undergoes paramagnetic relaxation. The spectra recorded during the stages of the synthesis have shown a change in the paramagnetic relaxation rate which allows some conclusions to be drawn on the effect of molten fluxes.

2. Introduction

Lithium aluminium oxide ($LiAlO_2$) has been studied as a material for:- molten carbonate fuel cells, nuclear reactors, and as a host lattice for phosphors.[1] $LiAlO_2$ can and has been made by a variety of methods including:- Sol Gel chemistry, hydrothermal synthesis, and direct combination of oxides.[1] $LiAlO_2$ exists in at least three polymorphs (α, β and γ) with the γ polymorph being stable at high temperatures.[2] In 1974 Lehman patented the synthesis of $LiAlO_2$ activated with iron ($LiAlO_2$:Fe) as a visible red/infra red emitting phosphor for use in fluorescent lamps.[3] Synthesis of phosphor powders usually involves a high temperature heat treatment and thus for phosphor materials it is usually γ-$LiAlO_2$ which is prepared. This polymorph has a $P4_12_12$ space group and consists of an array of edge sharing tetrahedral of Li and Al co-ordinated to O.[2] However the ESR data of the Fe in γ-$LiAlO_2$:Fe is consistent with octahedral co-ordination, not tetrahedral co-ordination.[4] To understand more about the Fe environment a sample of γ-$LiAlO_2$:Fe made with ^{57}Fe enriched iron was prepared for Mössbauer spectroscopic studies, our initial findings are reported herein. They have enabled a significant advance in the understanding of molten fluxes during high temperature crystallisation.

Fluxes are widely used during high temperature crystallisation of materials, specifically in phosphor chemistry they are used to promote growth and to improve luminescent properties. Although they have been the subject for much discussion in the literature on how they affect crystallisation, there is little, if any, direct evidence on how they improve luminescence in phosphor materials.[5]

Mössbauer spectroscopy offers an excellent method for examining specific atoms. When the activator in a phosphor material is Mössbauer active it may be studied without the lattice material interfering, however the level of activator doping in phosphor materials is much lower than those usually required for Mössbauer spectroscopic studies. If, as in the case of this work, the activator concentration is very low (0.1 mol %) then isotopic enrichment of the phosphor is desirable.

3. Experimental.

$LiAlO_2$:Fe phosphors (0.1 mol % Fe) was prepared by the method reported in the literature.[2] Characterisational and spectroscopic methods used are also reported in the literature.[6] Starting materials were of AnalaR grade or higher, the Fe used was 99.00 % ^{57}Fe.

4. Results and discussion.

In the documented synthesis a complex firing process is followed, with the final firing involving the use of Li_2CO_3 as a flux.[1] The ^{57}Fe enriched sample, showed a significant improvement in PL emission intensity after the final firing. It must be realised that the majority of the emission spectra of $LiAlO_2$ is in the IR region of the electromagnetic spectrum and is thus imperceptible to the human eye, see figure 1. High temperature firing using a flux can result in an increase in particle size compared to materials fired without added Li_2CO_3. Luminescence is known to be a phenomena which is affected by particle size and the screens tested were of comparable thickness and not necessarily optimal for the sample. Thus the improvement in luminescence could be said to be caused by the increase in particle size and/or by the associated changes in the prepared phosphor screen. However the Mössbauer spectra of the ^{57}Fe enriched sample, before and after the final

Figure 1. Emission spectrum of $LiAlO_2$:Fe.

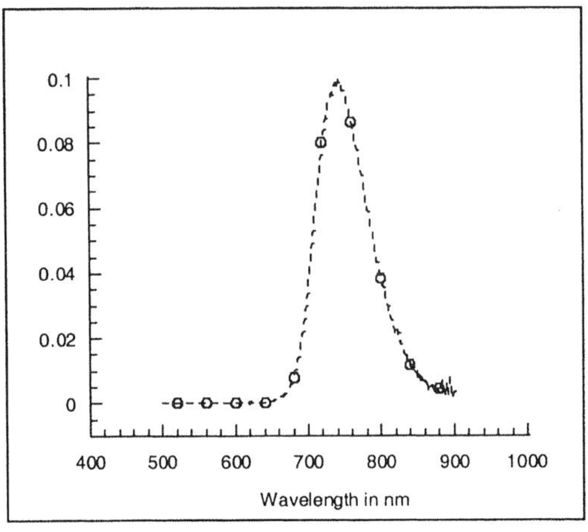

thermal processing with added Li_2CO_3, show that another important phenomena has also occurred.

Concentration quenching occurs in phosphors when the activator concentration exceeds and optimum level.[7] Though concentration quenching is not fully understood it occurs when activator to activator distances become shorter due to increasing activator concentration. Thus even when below the concentration at which concentration quenching begins (0.1 mol % Fe for $LiAlO_2$:Fe)[2] homogeneous distribution of the activator through the host lattice is desired to achieve the highest emission intensity. Another phenomena associated with low level dopants in host lattices is paramagnetism.[8] The distribution of the dopant through the host lattice affects the spin relaxation rate and close contacts raise the rate of relaxation.

Figure 2. Spectra recorded in this work Pre and Post final firing.

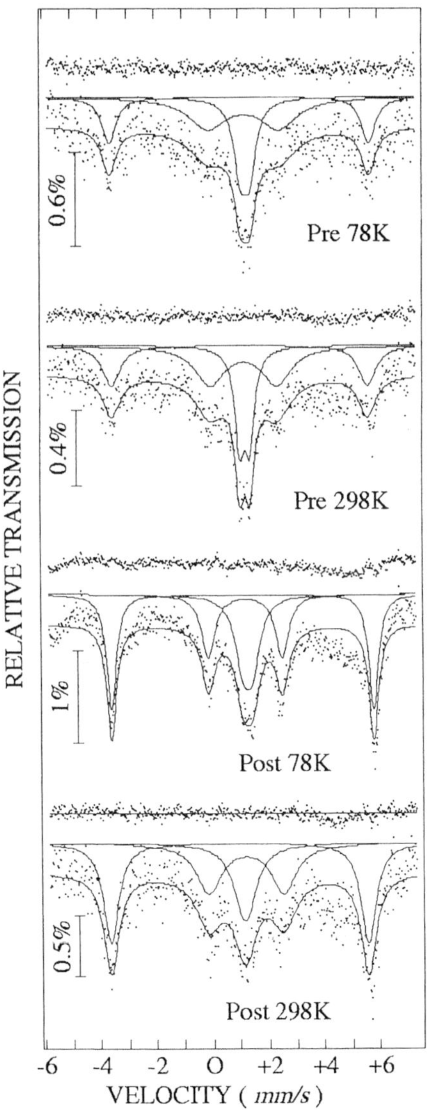

This phenomena can be clearly observed using Mössbauer spectroscopy. The Mössbauer spectrum of Fe doped in a host lattice undergoing slow paramagnetic relaxation consists of a one Fe site which gives a magnetic sextet with a second site yielding a narrow quadrupole splitting in the centre of the sextet. This is consistent with the spectra recorded for the $LiAlO_2$:^{57}Fe sample before and after the final thermal processing. The rate of relaxation can be determined from the ratio of the areas of the magnetic sextet and the quadrupole doublet, The more dominant the area of the quadrupole doublet, the higher the rate of relaxation. It can be seen from the spectra in figure 2 that the relative area of the doublet has decreased with the final firing showing a decrease in the paramagnetic relaxation rate. A decrease in paramagnetic relaxation rate observed for samples with the same concentration of Fe after the final firing therefore suggests that homogeneous distribution of the iron activator had not been achieved before the final firing. Indeed it may not have been achieved by the final firing, however it is improved compared to the sample before the final firing.

5. Conclusions.

The observations of an improvement in $LiAlO_2$ materials undergoing the final firing with added Li_2CO_3 show that the Li_2CO_3 acts as a flux and that an action of a flux is to promote homogeneous distribution of the activator through the host lattice. We believe that this is the first time that this has been demonstrated directly for a phosphor material rather than inferred from the results of luminescence characterisation.

6. References.

1, M.A. Valenzuela, J. Jimenez-Becerril, P. Bosch, S. Bulbulian and V.H. Lara, *J. Amer. Ceram. Soc.*, 1996, **79**, 455.

2, R. Dronskowski, *J. Inorg. Chem.*, **32**, 1993, 1.

3, W. Lehmann, US Patent number US3857054, 1974.

4, T.R.N. Kutty and M. Nyak, *Mat. Res. Bull.*, 1999, **34**, 249.

5, E. Nakazawa in Phosphor Handbook, 1st edition, Editors: S. Shionoya, and W.M. Yen, CRC Press, London, 1999, 85.

6, Y. Hamed R. C. Hider, J. Silver, *Inorg. Chem. Acta.,* 1982, **66**, 13.

7, S. Inaho and T. Hase in Phosphor Handbook, 1st edition, Editors: S. Shionoya, and W.M. Yen, CRC Press, London, 1999, 459.

8, N.N. Greenwood and T.C. Gibb, Mössbauer Spectroscopy, Chapman Hall Ltd, London, 1971.

Nickel- and Cobalt-doped magnetite as catalysts on the oxidation of CO

Maria F. F. Lelis*, Rachel V. R. A. Rios, Rochel M. Lago, José D. Fabris and Wagner N. Mussel
Departamento de Química, Universidade Federal de Minas Gerais, Campus - Pampulha 31270-901 Belo Horizonte, MG, Brazil
**Permanent address: Departamento de Química, Universidade Federal do Espírito Santo, 29060-900 Vitória, ES, Brazil*

Waldemar A. A. Macedo and Maxmiliano D. Martins
Laboratório de Física Aplicada, CDTN, 31270-901, Belo Horizonte, MG, Brazil

August 16, 2001

Abstract. The catalytic activity of nickel-, $Ni_xFe_{(3-x)}O_4$, (x = 0; 0.10; 0.28; 0.54) and cobalt-, $Co_xFe_{(3-x)}O_4$ (x = 0; 0.19; 0.38; 0.75) doped magnetites on the oxidation of carbon monoxide was investigated. Magnetite samples were produced by heating mixed hydrate ferric oxyhydroxide (IHA) at 430 ^oC and characterized by conventional chemical analyses, saturation magnetization measurements, XRD and room temperature Mössbauer spectroscopy. The catalytic experiments showed the influence of structural Ni^{2+} or Co^{2+} in promoting the oxidation of CO at lower temperatures, compared to the pure magnetite. XRD and Mössbauer analyses after the oxidation of CO indicate that pure magnetite is fully converted to hematite. The doped magnetites were converted to a mixture of maghemite and hematite. These results suggest that the spinel structure is thermally stabilized if it contains Ni^{2+} or Co^{2+}, occurring its complete conversion to hematite at significantly higher temperatures than that of pure magnetite.

Keywords: Ni and Co doped-Ferrites, Mössbauer, Catalysis

1. Introduction

Iron oxides such as magnetite (Fe_3O_4) and goethite (α-Fe_2O_3) have been used as catalyst in several processes (Andreev et al., 1992), (Zhang et al., 2000) and (Wang and Davis, 1999). These oxides present semi-conducting characteristics, which are suitable to be used in oxi-redox reactions. Magnetites can be obtained in a large range of chemical composition by adding a third cation ($M_xFe_{3-x}O_4$, where M = metal and $0 \leq x \leq 1$) to the spinel structure. The resulting materials have some particular properties which could be selected for the reaction of interest. In the present work, doped magnetites with Ni and Co were synthesized, characterized (Table I) and their use on the carbon monoxide oxidation has been evaluated.

Table I. Saturation magnetization measurement and chemical composition for nickel- and cobalt-doped magnetites ($\Diamond$ = cation vacancy).

Sample	Formula	Saturation Magnetization, σ/J $T^{-1}kg^{-1}$
Mag00	$Fe^{3+}_{2.03}Fe^{2+}_{0.95}\Diamond_{0.02}O_4$	92.5
Co019	$Fe^{3+}_{2.07}Fe^{2+}_{0.71}Co^{2+}_{0.19}\Diamond_{0.03}O_4$	66.6
Co038	$Fe^{3+}_{2.04}Fe^{2+}_{0.56}Co^{2+}_{0.38}\Diamond_{0.02}O_4$	50.0
Co075	$Fe^{3+}_{2.02}Fe^{2+}_{0.22}Co^{2+}_{0.75}\Diamond_{0.01}O_4$	39.9
Ni010	$Fe^{3+}_{2.08}Fe^{2+}_{0.77}Ni^{2+}_{0.10}\Diamond_{0.04}O_4$	76.6
Ni028	$Fe^{3+}_{2.05}Fe^{2+}_{0.64}Ni^{2+}_{0.28}\Diamond_{0.03}O_4$	70.1
Ni054	$Fe^{3+}_{2.03}Fe^{2+}_{0.41}Ni^{2+}_{0.54}\Diamond_{0.01}O_4$	66.3

2. Material and methods

The samples were obtained by the method proposed by Abreu Filho and Galembeck (1987), based on a precursor synthesis and its conversion to magnetite by heating it in N_2 atmosphere at 430 ^oC and characterized by conventional chemical analysis, atomic absorption (Carls Zeiss Jena AAS), saturation magnetization measurements, X-ray diffraction in a Rigaku Geigerflex 3064 diffractometer, equipped with a graphite diffracted beam monochromator, using Cu($K\alpha$) radiation and room temperature Mössbauer spectroscopy. The catalytic reactions were carried out in a tubular quartz reactor by using 20 mg of magnetite in a CO/O_2 (2:1) mixture flux of 55 mL/min. The reacting products were analyzed with a Shimatdzu 17A gas chromatograph.

3. Results

The saturation magnetization decreases as the doping cation increases in the crystal structure (Table I). The Fe^{3+} amount remains constant but the Fe^{2+} content decreases as Ni^{2+} and Co^{2+} increases in the structure. Ni and Co tend to occupy the octahedral site of the spinel structure. The catalytic tests show that the presence of Ni^{2+} and Co^{2+} in the magnetite structure has strong influence on the catalytic activity, by promoting the CO oxidation at much lower temperatures, if compared to the pure magnetite. The sample with the highest nickel amount (Ni054) has catalytic activity starting at 100 ^oC, converting up to 55% of CO at 200 ^oC. On the other hand, samples Mag00, Ni010 and

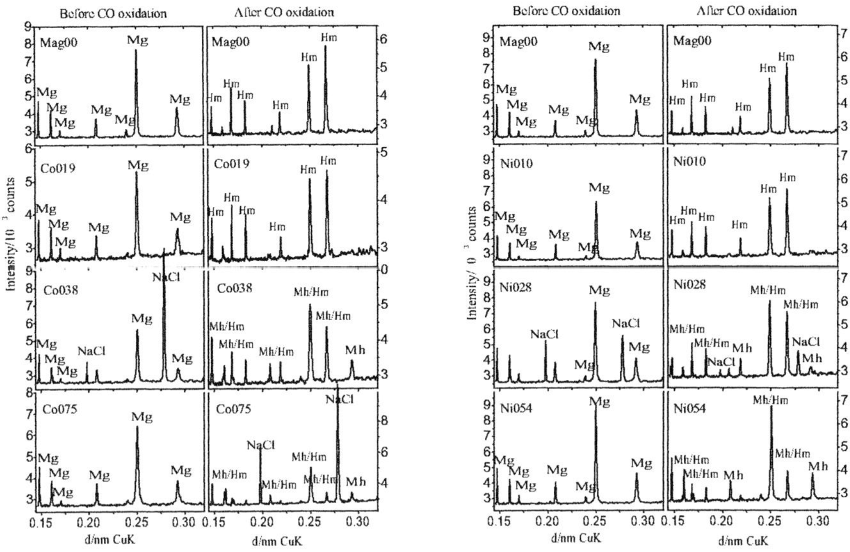

(a) XRD for Co samples (b) XRD for Ni samples

Figure 1. XRD for magnetites samples, before and after the CO oxidation (NaCl = internal standard, Mg = magnetite, Mh = maghemite and Hm = hematite.)

Ni028 start the catalytic activity only above 200 °C. For the highest substituted sample (Co075), it is observed a significant increase in the CO oxidation at temperatures above 100 °C, whereas for samples Co019 and Co038, the corresponding conversion only starts at temperatures higher than 200 °C. For all samples before oxidation, the XRD experiments showed characteristics reflections of a well-crystallized spinel phase (Figure 1). Mössbauer spectra obtained at room temperature before and after the catalysis experiments are shown in Figure 2. The intensity of the external sextet due to tetrahedral site of the spinel structure decreases according to the nickel increase in the lattice. Mossbauer parameters $\delta/\mathrm{Fe} = 0.66$ mm s^{-1} and H = 46.2 T were ascribed to ions in octahedrally coordinated sites whereas $\delta/\mathrm{Fe} = 0.28$ mm s^{-1} and H = 49.2 T were due to iron in tetrahedral sites. The XRD patterns of samples Mag00 after CO oxidation show characteristic reflections of hematite, which suggest that magnetite is (partially) converted to hematite in the chemical process; patterns for samples Ni010, Ni028 and Ni054 show the presence of both hematite and maghemite. The

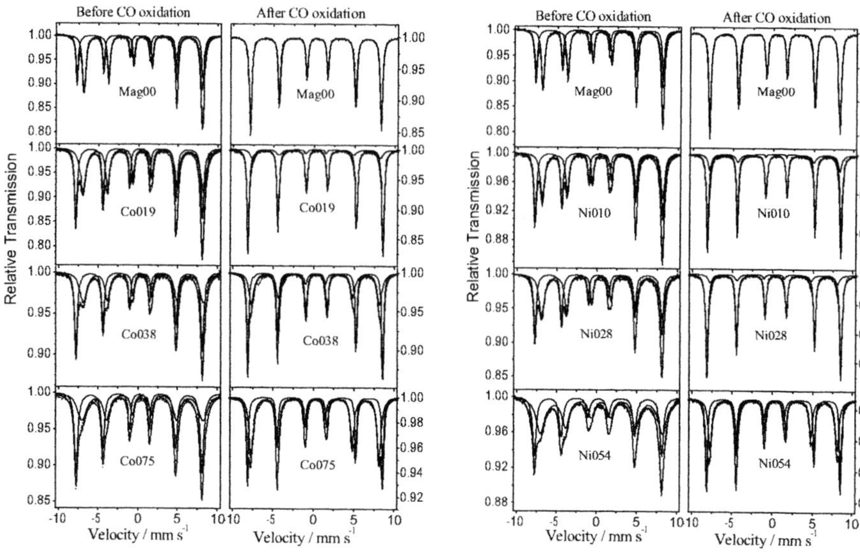

(a) Mössbauer spectra for Co samples (b) Mössbauer spectra for Ni samples

Figure 2. Room temperature Mössbauer spectra for the series of before and after the CO oxidation.

Mössbauer spectra after the CO oxidation show the complete conversion of the pure magnetite sample to hematite. The Ni- doped samples Ni010, Ni028 and Ni054 are converted to a mixture of hematite and maghemite at ratios Hm:Mh of 89:11, 79:21 and 48:52, respectively. Co-doped samples Co019, Co038 and Co075 are correspondingly converted to Hm:Mh ratios of 87:13, 59:41 and 40:60, respectively.

4. Conclusions

The main observed effects of nickel or cobalt substitution for iron in the spinel structure of magnetite during CO oxidation are: (i) it increases the catalyst activity and (ii) the thermal stability of the resulting doped-oxides during the thermally induced reaction, relatively to the pure oxide, and (iii) (Ni or Co)-doped magnetites are converted to a mixture of maghemite and hematite, instead of only hematite as it was observed for pure magnetite.

5. Acknowledgements

Work financially supported by CNPq, FAPEMIG and CAPES/PICDT (Brazil).

References

A. Andreev; I. Mitov; V. Idakiev; T. Tmomov; S. Asenov The Crystal Structure of Chalcopyrite Determined by X Rays Proceeding of the International Congress on Catalysis **Budapest, Hungary (1992)19-24**.

C. Zhang, S. Li, L. Wang, T. Wu, S. Peng Materials Chemistry and Physics **(2000) 62:1:52**.

Y. Wang; B. H. Davis Applied catalysis A. General **180: (1999) 1-2** pp. 277.

P.P. De Abreu Filho; E.A. Pinheiro; F. Galembeck Reactivity of solids **(1987) 3** pp. 241.

Hydrogenation of HoFe2

P. J. Schurer, B. Bennett and M. Evans
Royal Military College, Kingston On, Canada

X-ray diffraction and Mossbauer effect spectroscopy investigations show that pure homogeneous $HoFe_2H_x$ hydrides exist for $x \geq 1.5$. The Mossbauer spectrum for $HoFe_2H_2$ can be analyzed by assuming that the hydrogen has occupied interstitial sites in tetrahedra that consist of two iron and two holmium atoms and also in tetrahedra formed by three iron and one holmium atom. A minimum hydrogen-hydrogen distance of 2.05 Å is required in the hydrides. The isomer shift values increase monotonically with the increase in the number of hydrogen nearest neighbors.

I. Introduction

The cubic C15 Laves intermetallic RFe_2 compounds (R = rare earth) absorb large quantities of hydrogen at room temperature. RFe_2H_x compounds up to x = 4.2 have been reported for R between Ce and Er [1-4]. The absorption of hydrogen causes the cubic structure to expand. The absorption of hydrogen affects strongly the strength of the magnetic exchange interactions in the ferrimagnetic structure of $HoFe_2$. The Curie temperature is lowered by hundreds of degrees over the range of the hydrogenation [5]. In the present paper we present the results of a X-ray diffraction (XRD) and Mossbauer effect spectroscopy (MES) study of $HoFe_2H_2$.

II. Preparation

The alloy $HoFe_2$ was prepared by arc melting Ho (3N) and Fe (5N) under Ar atmosphere. The ingot was annealed for more than 4 days at 800 °C under 10^{-9} mbar. The hydrides were prepared following a procedure similar to the one described by Przewoznik et al. [6]. Grains (< 200 mesh) were activated at 140 °C in a vacuum of less than 10^{-4} mbar for two hours and subsequently exposed to hydrogen gas at room temperature. The total hydrogen content in the sample was determined from the drop in hydrogen pressure in the fixed volume of the apparatus. After all or most of the hydrogen was absorbed, the hydrogen distribution in the sample was homogenized by annealing at 200 °C for at least three hours. As the sample is heated above room temperature, we observed that desorption of some of the hydrogen takes place. The hydrogen that was lost is reabsorbed as the hydride is slowly cooled down over a time period of at least eight hours from 200 °C back to room temperature. This slow cool down is done in an attempt to maintain a homogeneous hydrogen distribution. The highest amount of hydrogen that was obtained with this method of preparation was $x \simeq 3.6$ hydrogen per formula unit. The same result was found for other samples that we prepared in the R = Gd to Er range.

XRD of the hydrides shows that for nominal values of $x \leq 1.5$ the samples contain two phases. A cubic hydride phase with a lattice constant of 7.57 Å grows at the expense of the parent cubic phase that has a lattice constant of 7.30 Å. Apparently, all hydrogen that is available is first used up to produce a hydride with close to 1.5 H per formula unit. It is not possible to make $HoFe_2H_x$ with $x \leq 1.5$ using the present preparation method. For $x \geq 1.5$ the lattice constant of the single-phase cubic hydride samples increases with hydrogen content.

III. Results and discussion.

In the parent $HoFe_2$ all iron atoms in the crystal structure are equivalent with the local threefold axis lying along one of the <111> directions. The iron atoms and rare earth atoms form corner sharing tetrahedra in the crystal structure. Three types of tetrahedra are present in the crystal structure: tetrahedra formed by four iron (B4), by one holmium and three iron (AB3), or by two holmium and two iron atoms (A2B2). By heating above about 600K, $HoFe_2H_2$ transforms into an amorphous phase. Aoki proposed that this transformation is driven thermodynamically in order to maximize the number of Ho atoms nearest neighbors (NN) for the hydrogen atoms [7]. Heating the compound makes it possible for the metallic atoms to rearrange such that the hydrogen atoms can be located in tetrahedra consisting of up to four Ho atoms. From this model it follows that for the crystalline phase one expects that the hydrogen atoms preferentially occupy the A2B2 tetrahedral interstices after absorption by $HoFe_2$. This assumption is supported qualitatively by enthalpy predictions [8].

A central iron atom is associated in principle with 12 NN A2B2 tetrahedral interstices. However, not all of these interstitial sites can be occupied at the same time. It has been proposed that repulsive interaction between NN hydrogen atoms produces a limiting H-H internuclear distance of about 2.1 Å [9]. In $HoFe_2$ a central iron atom can only have three hydrogen NN that satisfy this selection rule. The next two hydrogen atoms inserted in an A2B2 would be separated from the hydrogen NN already in place by only 1.94 Å. Without expansion, $HoFe_2$ can only accommodate three hydrogen atoms per formula unit. For the expanded crystal structure of $HoFe_2H_{1.5}$ there are possibly two more tetrahedral interstices available. Two additional hydrogen atoms at A2B2 interstices would be 2.01 Å away from hydrogen NN already in place. It is more likely that the two additional hydrogen atoms can occupy AB_3 interstices, because these hydrogen atoms would be 2.05 Å separated from another hydrogen already in place. This separation is very close to the limiting distance of about 2.1 Å. An absorption of hydrogen atoms in three A2B2 and two AB3 interstices gives a maximum of 4.3 hydrogen per formula unit. This value is in agreement with the maximum value of x = 4.2 found experimentally.

In contrast to the spectrum for the parent material [10], the spectra recorded for $HoFe_2H_2$ indicate the presence of several different iron sites (Fig. 1). The spectra have been fitted by assuming that the hydrogen atoms are distributed over three A2B2 and two AB3 interstices. This model is supported by the fact that at least five components are needed for producing a satisfactory least square fit. For the case of a completely random distribution of hydrogen atoms over the five available sites, the probabilities for the occurrence of each of the resulting six Fe NN configurations can be calculated. Table 1 shows the comparison between the relative areas of the five components in the Mossbauer spectrum (Fig. 1) and the values expected from the random distribution of hydrogen NN. A 2 % component with $H_{hf} \simeq$ 330 kG observed in the spectrum is probably due to metallic iron present at the surface of the fine powder hydride particles. A similar result was found for a hydride of FeTi [11]. An additional singlet is included to improve the fit for the inner part of the spectrum. The singlet may represent a contribution from very small superparamagnetic particles that may be formed in the sample as the result of decomposition into very fine powder during the hydrogenation.

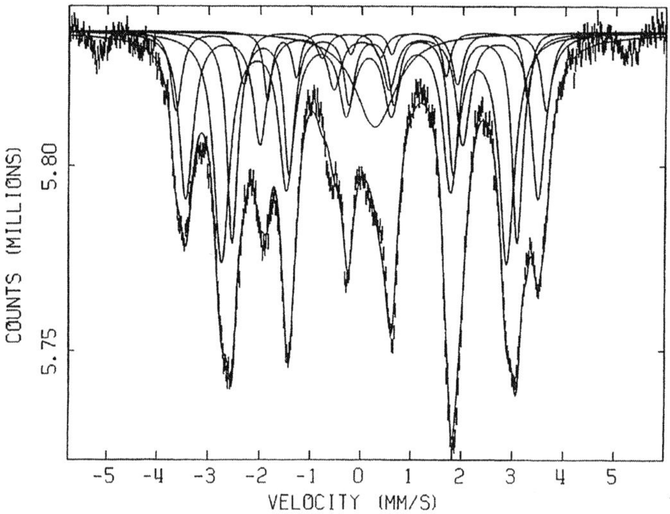

Fig 1. Mossbauer effect spectrum of $HoFe_2H_2$. The five components corresponding to Fe sites with zero to four hydrogen nearest neighbors are shown in the plot.

The isomer shift (i.s) values increase monotonically when the number of hydrogen NN is increasing (Table1). The variation is qualitatively consistent with expectations based on a model proposed by Miedema and van der Woude [12].

The value of the magnetic hyperfine fields H_{hf} (Table1) indicate that the presence of hydrogen NN and the increase of the lattice parameter affect the magnitude of the magnetic moment and the strength of magnetic interactions. Alternatively, the distribution of H_{hf} values may be due to the change in local symmetry and reflects the anisotropy of the magnetic hyperfine field as has been observed in some other RFe2 intermetallic compounds [10].

In conclusion, the Mossbauer spectrum for $HoFe_2H_2$ supports the model that H can occupy three interstices in A2B2 tetrahedra and two in AB3 tetrahedra. The fact that homogeneous hydrides $HoFe_2H_x$ only exist for $x \geq 1.5$ indicates that a minimum distance of 2.05 Å is required between hydrogen nearest neighbours in these hydrides.

Table 1

Mossbauer effect parameters of the five components in the Mossbauer spectrum of $HoFe_2H_2$. The relative areas are compared with the values expected from a random distribution of five hydrogen nearest neighbors.

N	Random Distribution	Relative Areas	i.s. (mm/s)	2ϵ (mm/s)	H_{hf} (kG)
0	0.05	0.10(2)	-0.12	0.25	226
1	0.19	0.20(2)	0.00	0.04	215
2	0.33	0.41(5)	0.11	-0.08	174
3	0.29	0.21(4)	0.11	0.08	174
4	0.12	0.07(1)	0.44	0.52	160
5	0.02				

Acknowledgement

This work was financially supported by the department of National Defence Canada.

References.

[1] H.H. van Mal, K. H. J. Buschow, A. R. Miedema, J. Less-Common Met. 49(1976)473.
[2] H. A. Kierstead, J. Less Common Meatals 70(1980)199.
[3] Y. Berthier, Th. De Saxce, D. Fruchart, P Vulliet, Physica 130B(1985)520.
[4] A. Ye. Yermakov, N. V. Mushnikov, N. K. Zajkov, V. S. Gaviko, V. A. Barinov, Phil. Mag. B68(1993)883.
[5] K. H. J. Buschow and A. M. van Diepen. Sol. St. Com. 19(1976)79
[6] J. Przewozik , V. Paul-Bonccour, M. Latrocche, A. Percheron-Guegan, J. Alloys Comp. 225(1995)436.
[7] K. Aoki in Displacive Phase Transformations and Their Applications in Materials Engineering. Ed. K. Inoue, K. Mukherjee, K. Otsuka and H. Chen. The Minerals, Metals & Materials Society. (1998)101.
[8] D. Ivey and D. Northwood, J. Less-Common Metals 115 (1986) 23.
[9] A.C. Switendick, Z. Phys, Chem. N.F. 117(1979)89.
[10] G. J. Bowden, D. St. P. Bunbyry, A. P. Guimaraes and R. E. Snyder, J. Phys. C1(1968)1376.
[11] G. K. Shenoy, B. D. Dunlap, P. J. Viccaro, and D. Niarchos, Hyp Int. 9(1981)531.
[12] A. R. Miedema and F. van der Woude, Physica. B100(1980)145.

Metal - tellurium bonds in 3d-transition metal ditellurides

J. Stanek[a], S.S. Hafner[b], P. Fornal[c]

[a] *Institute of Physics, Jagiellonian University, 30-059 Cracow, Poland*
[b] *Material Science Centre, University of Marburg, 35032 Marburg, Germany*
[c] *Institute of Physics, Technical University of Cracow, 30-083 Cracow, Poland*

The hyperfine interaction parameters of ^{57}Fe probes in $Mn_{1-x}Fe_xTe_2$ (pyrite), $FeTe_2$ (marcasite) and $Ni_{1-x}Fe_xTe_2$ (CdI_2 - type structure) show structurally induced 3d electron transfer from metal to Te increasing in order: $MnTe_2$, $FeTe_2$ and $NiTe_2$. The determined signs of V_{zz} are such as expected for lattice contributions of electric field gradients.

1. Introduction

The crystal structures of *3d* transition metals ditellurides, $MeTe_2$, Me=Mn, Fe, Co, Ni evolve from cubic pyrite type for $MnTe_2$ (space group Pa3) through orthorhombic marcasite type for $FeTe_2$ (Pnnm) to hexagonal CdI_2 (C3m) type for $NiTe_2$. Metals ions are six coordinated by Te atoms which form compressed, quasi-planar trygonal anti -prisms in $MnTe_2$ and $NiTe_2$ or compressed octahedra in $FeTe_2$ with quasi-linear coordination.

The previous ^{125}Te Mössbauer studies of MTe_2 (M=Ti, V, Cr, Fe, Co and Ni) [1,2] showed an increase of the effective charge of Te from *-1* in pyrite structure ($MnTe_2$) through *-1.5* in marcasite structure ($CoTe_2$, $FeTe_2$) to nearly *-2* in CdI_2 type structures ($TiTe_2$, VTe_2, $NiTe_2$) which points out the unusual valence states of corresponding cations, for example Fe^{+3} state in $FeTe_2$ and Ni^{+4} in $NiTe_2$. In the present work we attempted to study cation electronic states by replacing Mn and Ni cations by ^{57}Fe probes without altering the original pyrite and CdI_2 structures, respectively, and to compare them with Fe state in $FeTe_2$.

2. Experimental

Samples. Samples of $Mn_{1-x}{}^{57}Fe_xTe_2$, $Ni_{1-x}{}^{57}Fe_xTe_2$ (x=0.03, 01) and $FeTe_2$ were synthesised from high purity elements employing dry evacuated silica tube technique. All specimens were heated, quenched, reground and annealed at 400 °C for several weeks until complete, single phase, equilibrium, tested by X-ray diffraction, was attained. The $FeTe_2$ sample was prepared from natural iron.

Mössbauer measurements. For recording the spectra between 20 K and 293 K a gas flow cryostat was applied. One sample ($FeTe_2$) was measured at 4.2 K in external magnetic field of 4.5 T parallel to the gamma radiation. The paramagnetic spectra were fitted assuming Lorentzian lines, the spectra exhibiting mixed magnetic - quadrupole interactions were evaluated by a Hamiltonian solving program [3] while the spectrum of $FeTe_2$ in external magnetic field was fitted according to Gabriel - Ruby procedure.

3. Results

At room temperature all pectra were in form of quadrupole doublets, without any traces of foreign phases. The hyperfine parameters, isomer shifts, δ, quadrupole splitting, Δ, and internal magnetic fields, H, are collected in Table 1

Table 1.

Hyperfine parameters of measured spectra. δ, Δ and H are isomer shift (vs. metallic iron at room temperature), quadrupole splitting and magnetic hyperfine field, respecively. The data marked by "*" refer to the source kept at 4.2 K.

Sample	T [K]	δ [mm/s]	Δ [mm/s]	H [T]
$Mn_{0.9}Fe_{0.1}Te_2$	293	0.78	1.28	
	247	0.82	1.29	
	197	0.85	1.31	
	102	0.91	1.34	
	85	0.91	1.34	2.76
	75	0.92	1.37	4.60
	20	0.94	1.45	9.32
$Mn_{0.97}Fe_{0.03}Te_2$	293	0.78	1.28	
$FeTe_2$	293	0.47	0.52	
	4.2*	0.46	- 0.52	
$Ni_{0.9}Fe_{0.1}Te_2$	293	0.42	(+)0.33	
	150*	0.44	0.41	
	4.2*	0.40	0.44	
$Ni_{0.97}Fe_{0.03}Te_2$	293	0.42	0.31	

$Mn_{1-x}Fe_xTe_2$. Examples of the obtained spectra are shown in Fig. 1. Below 85 K, in the antiferromagnetic state, the spectra showed mixed magnetic - quadrupole interaction with H=9.3 T at 20 K. The second derivative of the potential, V_{zz}, is positive, the electric field gradient tensor, efg, is axial (asymmetry parameter $\eta \approx 0.$) and H is parallel to *z*, the main axis of electric field gradient tensor.

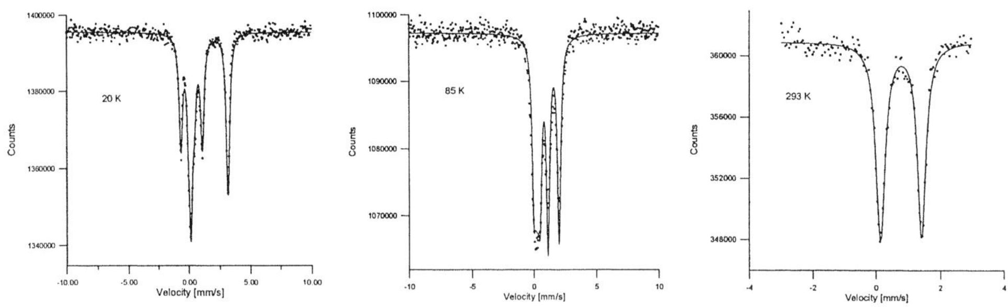

Fig. 1. [57]Fe Mössbauer spectra of $Mn_{0.9}Fe_{0.1}Te_2$ at temperatures as marked.

The temperature dependence of quadrupole splitting, Δ, is weak, it decreases only by 13% between 20 K and 373 K.

$FeTe_2$. The well known Mössbauer spectrum of $TeFe_2$ is in form of a symmetric quadrupole doublet, cf. Fig.2, with Δ invariant between 4.2 and 293 K. The spectrum recorded at 4.2 K in external magnetic field of 4.5 T, see Fig. 2, shows that electric field gradient is fairly axial ($\eta \leq 0.2.$) with negative V_{zz} .

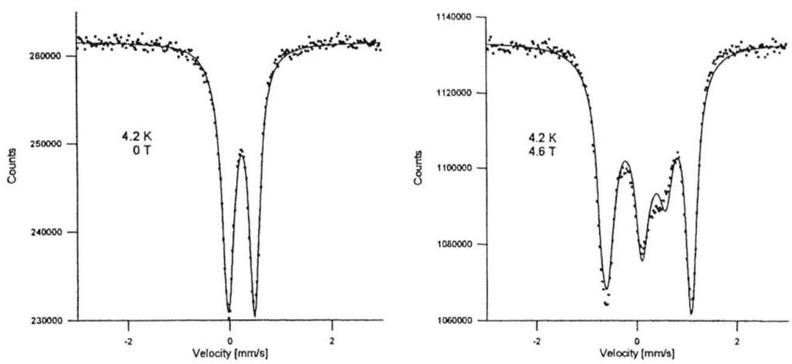

Fig. 2. ^{57}Fe Mössbauer spectrum of $FeTe_2$ at 4.2 K, left. Right: the same but measured in external magnetic field of 4.6 T.

$Ni_{1-x}Fe_xTe_2$. The asymmetric doublet of this sample, see Fig. 3, was for us a puzzle until the electron microscopy study showed that that our powder sample consisted of a small "discs" preferentially oriented into the sample plane. Indeed, the asymmetry of the spectrum

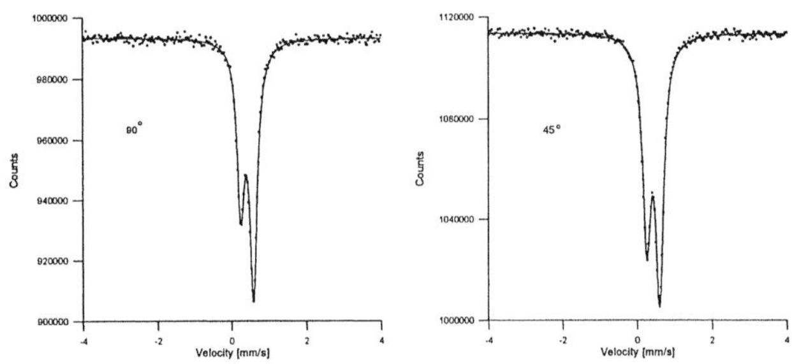

Fig. 3. Room temperature ^{57}Fe Mössbauer spectra of $Ni_{0.9}Fe_{0.1}Te_2$ measured with γ beam perpendicular to the sample plane (left) and at 45° to the sample plane (right).

strongly depends on the orientation of the sample against gamma beam, see Fig. 3. Thus, the spectrum is interpreted as a single quadrupole doublet which asymmetry arises from the texture. If one assumes that the main axis of the field gradient is perpendicular to the sample plane the more intense high energy line shows that V_{zz} is positive. Δ decreases from 0.44 mm/s to 0.33 mm/s between 4.2 and 293 K.

4. Discussion

The values of δ and Δ are rather unusual and hardly can be used allone for determination of the iron charge states. In the best known case of the low spin Fe^{II} in $FeTe_2$ they are rather typical for Fe^{+3} high spin compounds. This was interpreted by Ingals in terms of molecular orbital approach and explained as result of strong covalency effects. However, the determined charge state of $(Te_2)^{-3}$ molecule in this compound [2] suggests that the electronic levels of Fe^{II} (three lowest levels equally populated by paired electrons) may by partly depopulated by electron transfer to Te anions. Thus, the nominally low spin Fe^{II} has the charge state similar +3 cation. It is interesting to note that the sign of V_{zz} is negative as expected for lattice contribution of efg for the quasi linear coordination of compressed T_6 octahedra of Fe in $FeTe_2$.

In $Mn_{1-x}Fe_xTe_2$ the divalent Mn^{+2} and univalent Te^{-1} should force the Fe^{+2} (high spin ?) state of iron. However, the $\delta = 0.78$ mm/s at 293 K and rather small $\Delta = 1.28$ mm/s must cause some doubts. The weak temperature dependence of Δ, and the positive V_{zz}, as expected for the quasi planar coordination of Fe, suggest the lattice origin of efg.

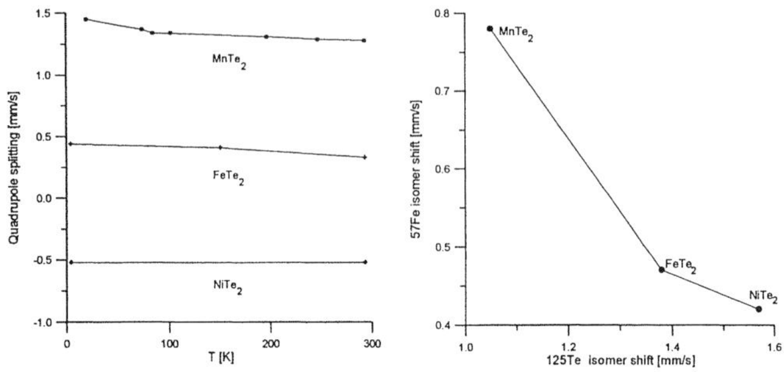

Fig. 4. The temperature dependence of quadrupole splitting of ^{57}Fe in $Mn_{1-x}Fe_xTe_2$, $FeTe_2$,and $NiTe_2$ (left). ^{125}Te vs ^{57}Fe isomer shifts in the studied samples.

In addition, in antiferromagnetic state, the internal magnetic field on ^{57}Fe is 9.3 T, only, which is very close to the value of transferred magnetic field acting on ^{125}Te, 11.4 T [4]. Thus, it may be postulated, that the measured on ^{57}Fe magnetic field is the transferred field as well. All these facts could be explained with the assumption that in this case iron is in low spin state, Fe^{II} , as in $FeTe_2$.

The Δ and δ values in $Ni_{1-x}Fe_xTe_2$ are quite characteristic for Fe^{+3} state although the Te^{-2} valence state could force rather unusual Fe^{+4} state of iron. The weak temperature dependence of Δ and positive V_{zz} point out the lattice origin of efg.

The comparison of ^{125}Te and ^{57}Fe isomer shifts is shown in Fig. 4. The increase in *5p6s* Te orbitals (increasing δ of ^{125}Te) is related to the decrease of δ of ^{57}Fe which may be interpreted as a gradual decrease in *3d* population of ^{57}Fe probe.

5. Conclusion

We were successful in introducing ^{57}Fe probes in substitutional cation sites of $MnTe_2$ and $NiTe_2$. The ^{57}Fe hyperfine parameters, together with the previous ^{125}Te data allow to propose the following scenario for the evolution of structure of the *3d* transition metal ditellurides. The pyrite structure ($MnTe_2$) consists of covalently bounded $(Te_2)^{-2}$ cubically close packed pairs hold by metal ions. With increasing number of *3d* electrons ($CoFe_2$, $FeTe_2$) charge transfer from metal to Te weakens the Te-Te bonds leading to the marcasite structure. Finally, in the $NiTe_2$ case, the electron transfer breaks the Te-Te covalent bonds and Te^{-2} ions form hexagonal, close packed arrangement with Ni^{+4} cations nested between alternative layers of anions.

The work was partly supported by Polish - German scientific collaboration (Grant 98-031)

References.

[1] H. Binczycka, S.S. Hafner, G. Moh, J. Stanek, Phys. Lett. A145(1990) 467.
[2] J. Stanek, A. M. Khassanov, S.S. Hafner, Phys. Rev. B 45(1992) 56.
[3] D.G. Rancourt, Nucl. Instrum. Methods B44(1989) 1999.
[4] M. Pasternak, A.L. Spijkervet, Phys. Rev. 181(1969) 181.

PREPARATION AND EUROPIUM-151 MÖSSBAUER SPECTROSCOPY OF $Ca_{1-x}Eu_xAl_{12}O_{19}$

T. ARAKAWA and H. FUCHIGAMI

Department of Biological and Environmental Chemistry, Kyushu School of Engineering, Kinki University, Iizuka, Fukuoka 820-8555, Japan

$Ca_{1-x}Eu_xAl_{12}O_{19}$ were prepared by the solid state reaction using $Al(OH)_3$, $Ca(OH)_2$ and Eu_2O_3 at 1373K under a mixture gas atmosphere(H_2 : 20 %, N_2:80%). Europium ions in these compounds were almost divalent state, judging from the measurement of the photoluminescence and Mössbauer spectra. The emission spectrum of $Ca_{1-x}Eu_xAl_{12}O_{19}$ (x=0.01) under 325 nm excitation consists of a broad and with maximum at 425 nm(Eu^{2+} $4f^56d^1$-$4f^7$ transition). ^{151}Eu Mössbauer spectra show that Eu^{2+} is present in sites with isomer shifts in the range from –12.2 to –11.7 mm/s.

1. INTRODUCTION

There are some important compounds between CaO and Al_2O_3. $CaAl_{12}O_{19}$ has the magnetoplumbite structure being similar to the beta-alumina, which are available for the solid electrolytes, catalysts and laser host materials. In order to understand the functional properties of these compounds, it is necessary to investigate the state of active centers. In this paper $Ca_{1-x}Eu_xAl_{12}O_{19}$ were synthesized by direct solid-phase reaction, and we have measured the photoluminescence and the ^{151}Eu Mössbauer spectra of $Ca_{1-x}Eu_xAl_{12}O_{19}$. Eu^{2+} ion among many divalent cations was chosen from the following reasons. The ionic radius of Eu^{2+} ion is close to that of Ca^{2+} ion. Also, the photoluminescence of Eu^{2+} ion is unique and is affected by the crystal field surrounding Eu^{2+} ion. On the other hand, Mössbauer spectroscopy has proved quite useful in characterizing the valence state and coordination of cations because the isomer shift is caused by small changes in the electronic charge density at nucleus. Europium is ideal in this respect.

2. EXPERIMENTAL

$Ca_{1-x}Eu_xAl_{12}O_{19}$ were prepared by the solid state reaction using $Al(OH)_3$, $Ca(OH)_2$ and Eu_2O_3 under a mixture gas atmosphere(H_2 : 20%, N_2:80%) at 1373K. The obtained phases have been checked through XRD technique. The emission spectra were measured with a HITACHI recording absolute spectrofluorophotometer (F-4500) at room temperature. Mössbauer spectra were recorded at room temperature using a Mössbauer spectrometer(VT - 6000), which was calibrated with EuF_3. A ^{151}Sm (SmF_3) source(1.85G Bq) was used.

3. RESULTS AND DISCUSSION

3.1 Photoluminescence of $Ca_{1-x}Eu_xAl_{12}O_{19}$

The photoluminescence spectra of $Ca_{1-x}Eu_xAl_{12}O_{19}$ under excitation at 325 nm are shown in figure.1. The spectra consist of broad, single peaked emission bands. These bands were assigned to the

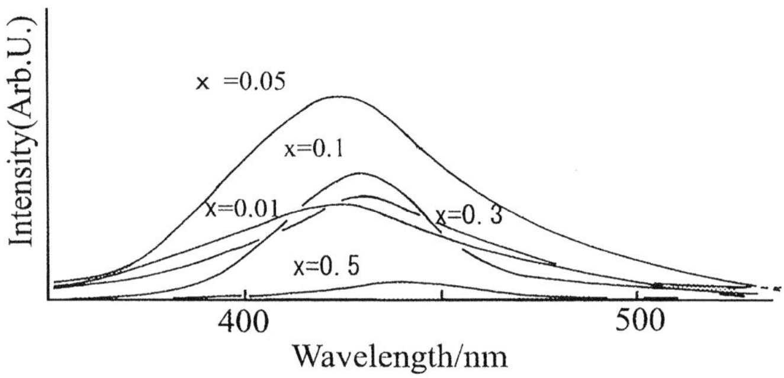

Figure 1. Emission spectrum of $Ca_{1-x}Eu_xAl_{12}O_{19}$.

Eu^{2+} ions($4f^7$ - $4f^65d^1$ transition), judging from the range of wavelength. The peak position slightly shifted higher wavelength with the increase of Eu^{2+} ions. And the effect of Eu^{2+} ion concentration was appeared. The intensity of photoluminescence reached a maximum at ca. x=0.05. The emission peak position of Eu^{2+} center is known to be easily affected by the crystal field surrounding in the Eu^{2+} ions[1].

3.2 ^{151}Eu Mössbauer spectra of $Ca_{1-x}Eu_xAl_{12}O_{19}$

Figure 2 shows the Mössbauer spectra of $Ca_{1-x}Eu_xAl_{12}O_{19}$. The isomer shift value is - 11.7

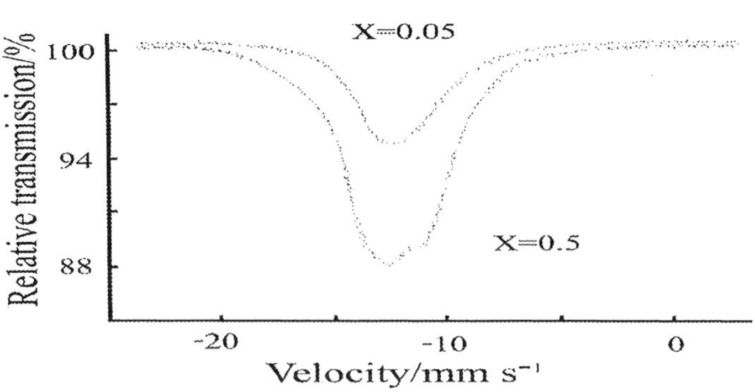

Figure 2. ^{151}Eu Mössbauer spectra of $Ca_{1-x}Eu_xAl_{12}O_{19}$ at room temperature.

mm/s for x=0.05 and –12.2 mm/s for x=0.5. Thus, it is clear that there is a difference of the isomer shift with x of $Ca_{1-x}Eu_xAl_{12}O_{19}$. ^{151}Eu Mössbauer spectra of Eu_2NbO_3, which occurred in divalent and trivalent oxidation state, show that Eu^{2+} was present in sites with isomer shift value in the range –13 to –12 mm/s[2]. Also, the isomer shift values of Eu^{2+} ions for $Eu_3V_2O_7$ and Eu_2VO_4, which demonstrates the presence of both divalent and trivalent Eu ions, were respectively observed at -12.53 mm/s and –11.86 mm/s[3]. It is evident from figure 2 that europium is present in divalent state because this is well within the range of isomer shift reported for divalent europium compounds. Whereas the Eu^{2+} resonance line of $Ca_{1-x}Eu_xAl_{12}O_{19}$ is broad compared with EuF_2 as reported by Coey et al.[4]. The reason for broadening of the Eu^{2+} line would be considered as follows:(1) a distribution of isomer shifts corresponding to Eu^{2+} in sites with different coordination numbers and different Eu-O bond lengths, (2) unresolved electric quadrupole interaction, (3) $Eu^{2+} \Leftrightarrow Eu^{3+}$ electronic charge fluctuations. (3) can be neglected because there is no corresponding Eu^{3+} line. In the case of (2), linewidth of the crystalline europium compounds are usually of the order of 2.5-3.5 mm/s as reported elsewhere[2,3]. However, since linewidth in figure 2 was slightly larger than 3.5 mm/s, the broadening of these compounds may be slightly affected by (2). The most reasonable explanation of the Eu^{2+} line is a variety of cation sites with isomer shifts in the range from –12.2to –11.7 mm/s. It is reported that the range of Eu-O bond lengths corresponding to the range of isomer shift of –11.3 to –13.5 mm/s is found to be 0.26 to 0.30 nm[5]. Therefore, it is thought that there is a difference in Eu-O bond length of $Ca_{1-x}Eu_xAl_{12}O_{19}$ with x values.

The magnetoplumbite structure of the $CaAl_{12}O_{19}$ host can be viewed as two spinel blocks separated by a layer containing large cation, one small cation and three oxygen, along the c-axis. In the case of $Ca_{1-x}Eu_xAl_{12}O_{19}$, the large cations are calcium and europium, and the small cation is aluminium. That is, the layer containing cations(Eu^{2+} or Ca^{2+} ions) and oxygens lies between the two spinel blocks. When the large cation substitutes for the calcium in $CaAl_{12}O_{19}$, the resulting structure is distorted[6]. Thus, a significant amount of strain is introduced into the crystal, which accounts for the change of ^{151}Eu Mössbauer spectra and the photoluminescence spectra of $Ca_{1-x}Eu_xAl_{12}O_{19}$.

References

[1] D.J.Simkin, J.Phys.Chem.Solids, 52(1991)175.

[2] M.Saeki, M.Nakada, M.Masaki and A.Nakamura, Conf.Proc.Ital.Phys.Soc. "ICAME'95", 50(1996)119.

[3] Z.M.Stadnik, G.Stronik and T.Arakawa, Hyferfine Interact., 68(1991)209.

[4] J.M.D.Coey, A.McEvoy and M.W.Shafer, J.Non - Crystalline Solids, 43(1981)387.

[5] A.Nakamura, N.M.Masaki, M.Nakada, M.Saeki, K.Tomimoto and J.Akimitsu, Ceram.Transact., 71(1996)295.

[6] A.L.N.Stevels, J.Luminesc., 48&49(1991)251.

SPIN CROSSOVER IN Fe(II) MOLECULAR COMPOUNDS – MÖSSBAUER and μSR INVESTIGATIONS

S.J. Campbell[1, 2], V. Ksenofontov[1], Y. Garcia[1,3], J.S. Lord[4], S. Reiman[1,5], P. Gütlich[1]
[1] *Institut für Anorganische Chemie und Analytische Chemie, Johannes Gutenberg Universität, 55099 Mainz, Germany,*
[2] *School of Physics, University College, The University of New South Wales, Australian Defence Force Academy, Canberra, ACT 2600, Australia,*
[3] *Department of Chemistry, Catholic University of Louvain, 1348 Louvain-la-Neuve, Belgium,*
[4] *ISIS, Rutherford Appleton Laboratory, Chilton, Didcot, OX11 OQX, UK,*
[5] *Institute of Nuclear Physics of Moscow State University, 119899 Moscow, Russia*

The compound $[Fe(ptz)_6](ClO_4)_2$ (ptz = 1-propyl-tetrazole) displays a complete and gradual spin crossover centred around 125 K as evidenced by magnetic and muon measurements over the temperature range ~ 4.2–300 K. Although the crystal structure reveals only one crystallographic site, line broadening is observed in the Mössbauer spectra in the vicinity of the spin transition. The muon spin relaxation behaviour of this compound indicates that a structural transformation rather than dynamic processes may account for the observed spectral features. Both the Mössbauer and muon measurements are consistent with a mixture of high and low spin Fe ions in the transition region.

1. Introduction

Iron (II) spin crossover (SC) molecular materials represent an important class of switchable transition metal compounds. In these materials the spin state can be modified from high-spin (HS, S = 2) to low-spin (LS, S = 0) by several external perturbations including temperature, pressure or light [1]. ^{57}Fe Mössbauer spectroscopy has been used extensively in probing the valence and spin states of iron SC compounds. Besides allowing the amount of HS and LS components to be determined, Mössbauer spectroscopy also provides insight to the dynamics of spin interconversion within an observation time window.

$[Fe(ptz)_6](ClO_4)_2$ (ptz = 1-propyl-tetrazole) displays a complete and gradual spin transition around 125 K [2]. As shown in Fig. 1, the central iron(II) ion of the compound is octahedrally coordinated by six tetrazole ligands [3]. Although the crystal structure reveals only one crystallographic iron site [3], line broadening was observed in the Mössbauer spectra on cooling through the spin transition [2, 4]. In addition despite the absence of thermal hysteresis, the crystals were found to crack upon cooling [4]. While this behaviour could indicate a first order structural phase transition, the line broadening could also be associated with the dynamic behaviour of the octahedral iron complexes during the SC. We have applied the μSR technique in order to obtain further microscopic information about the SC transition. In μSR spectroscopy spin-polarised positive muons are implanted into a

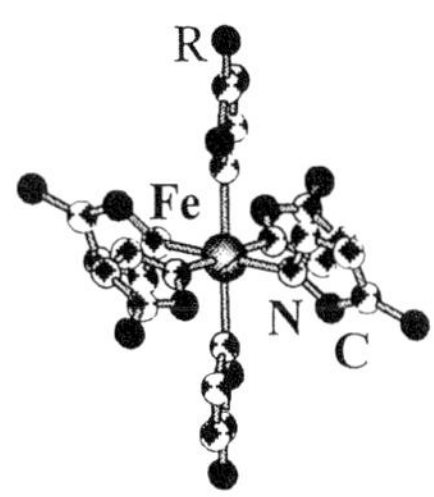

Figure 1. Molecular structure of $[Fe(ptz)_6](ClO_4)_2$. The substituent is shown as R and the perchlorate anions have been omitted for clarity.

sample, thereby acting as probes of the local magnetisation [5]. While Mössbauer spectroscopy enables the local hyperfine fields at lattice sites to be determined, positive muons mainly sense the interstitial sites and hence the dipolar fields generated by surrounding paramagnetic ions. In addition, ^{57}Fe Mössbauer spectroscopy enables dynamical properties of magnetic spins in the time scale $10^{-7} - 10^{-8}$ s to be investigated, whereas the μSR technique measures processes within a time window $10^{-5} - 10^{-9}$ s [6].

2. Experimental

The compound $[Fe(ptz)_6](ClO_4)_2$ was synthesised and characterised as described elsewhere [2, 4]. The ^{57}Fe Mössbauer measurements were carried out using a constant-acceleration conventional spectrometer and helium cryostat and the spectra fitted using Recoil 1.03a Mössbauer Analysis Software. The magnetic susceptibility measurements (4.2–300 K) were carried out using a PAR 151 Foner-type magnetometer and a helium cryostat. The μSR measurements were performed on the MuSR spectrometer at ISIS, Rutherford Appleton Laboratory, UK. The muons with energy ~ 3.2 MeV and implantation ranges of the order of 100 mg cm^{-2} thermalise within the sample on time scales much shorter than the spin relaxation times. The muons decay with mean lifetime 2.2 μs and emit positrons preferentially in the direction of the muon spin. The positrons are detected by scintillation detectors surrounding the sample. The μSR spectra were recorded over the temperature range ~ 12–300 K using a closed cycle refrigerator.

3. Results and discussion

The SC transition as observed in magnetic susceptibility and muon measurements is shown in Fig. 2. Representative Mössbauer spectra as measured on cooling through the SC transition are shown in Fig. 3. The HS state at 270 K reveals only one iron site with isomer shift and quadrupole splitting values of $\delta_{HS} = 0.96$ mm s^{-1} and $\Delta_{HS} = 1.32$ mm s^{-1} respectively. Line broadening occurs around ~200 K before the onset of the SC around ~190 K as indicated by the magnetic measurements (Fig 2). The Mössbauer spectrum at 200 K is well represented by two subspectral components with parameters $\delta_{HS1} = 1.01$ mm s^{-1}, $\Delta_{HS1} = 1.75$ mm s^{-1} and $\delta_{HS2} = 1.06$ mms^{-1}, $\Delta_{HS2} = 1.35$ mm s^{-1}. The Mössbauer spectrum below 125 K corresponds to an almost pure LS state with $\delta_{LS} = 0.44$ mm s^{-1} and $\Delta_{LS} = 0.10$ mm s^{-1}.

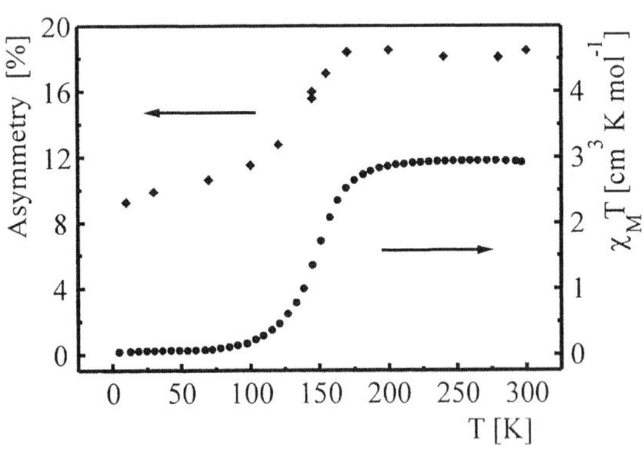

Figure 2. Temperature dependence of the asymmetry parameter (zero field) and $\chi_M T$.

The zero and longitudinal magnetic field muon relaxation spectra were found to comprise both fast and slow components with characteristic times ~ 0.5 μs^{-1} and ~ 10 μs^{-1}. As shown in Fig. 2, the temperature variation of the muon

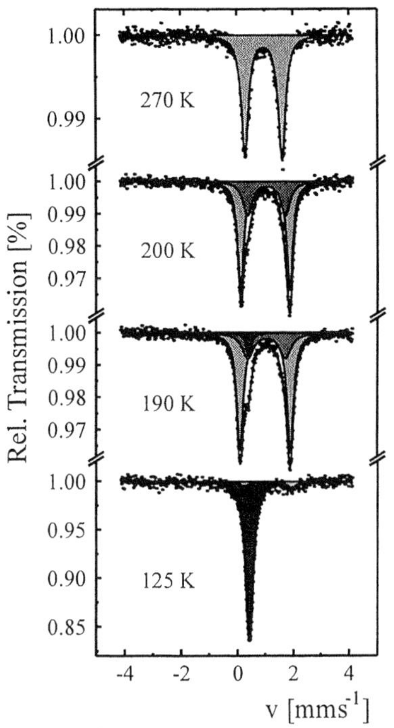

Figure 3. Mössbauer spectra over the region of the SC transition.

asymmetry parameter correlates well with the spin transition curve obtained from magnetic measurements. The dependence of the asymmetry parameter on the longitudinal magnetic field at different temperatures was measured in order to determine the manner in which the implanted muons stabilise in the $[Fe(ptz)_6](ClO_4)_2$ sample (Fig. 4). The set of experimental curves presented in Fig. 4 provides evidence for the formation of a paramagnetic state, such as muonium μ^+e^-, a light analogue of the hydrogen atom. The measured initial asymmetry of such a state increases with increasing magnetic field. Fitting the data at 10 K gives a value of the hyperfine coupling A=500 MHz which is typical of molecular radical states rather than the free muonium atom. Of special interest is the predominant diamagnetic-like muon fraction present in the HS state of the $[Fe(ptz)_6](ClO_4)_2$ compound. The temperature dependence of the asymmetry parameter provides an indication of the fraction of the muons that interacts with unpaired spins in the SC compound, and therefore the fraction of high spins present at any temperature. The similarity of the susceptibility and muon asymmetry curves (Fig. 2) also suggests that either the muon radical only interacts with one neighbouring Fe spin or that in the mixed state the high and LS Fe ions are clustered rather than being randomly mixed. On the basis of the structure of $[Fe(ptz)_6](ClO_4)_2$ (Fig. 1), the muonium is likely to be localised at the tetrazole ring, i. e. bonded to the carbon atom. This agrees with the measured hyperfine coupling. Another possibility is that the muonium would be located in the immediate surrounding of the octahedral iron complex. In that event a decrease in the crystal field would be expected which should stabilise the HS state. No such effect is observed, indicating that the muonium is likely to be localised in the vicinity of the aromatic ring rather than the iron complex.

The observed unusual diamagnetic behaviour of the muons in the paramagnetic surroundings of the HS state of the SC compound presumes a strong spin-spin interaction between the electron of the radical and the unpaired electrons of the SC molecule in the HS state, which is greater than the strength of the muon hyperfine interaction [7] and causes the electron's spin to fluctuate rapidly enough to decouple it from the muon. The fluctuation still causes a slow relaxation of order $\lambda = A^2/\nu$ meaning that the electron fluctuation rate must be at least ν=800 GHz at 300 K. The disappearance of unpaired spins during the SC transition allows us to observe the paramagnetic radical state which interacts with the static internal magnetic fields created by the nuclear dipole moments of the lattice atoms. No shift of the transition temperature (as measured from the asymmetry curve) to high temperatures was observed which mitigates against dynamic SC processes or molecular motion as being the

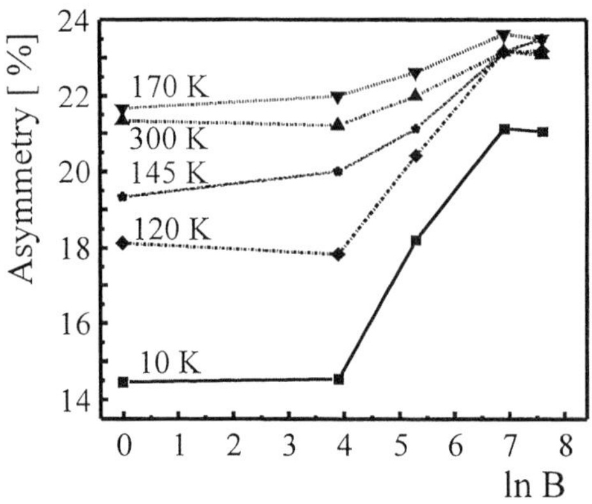

Figure 4. Dependence of the asymmetry parameter on magnetic field at the temperatures indicated. The lines act as a guide to the eye.

main reason for the unusual broadening observed in the Mössbauer spectra of $[Fe(ptz)_6](ClO_4)_2$. This broadening may occur as a result of a structural transformation in the vicinity of the SC. More detailed Mössbauer measurements around the transition region are required to clarify this behaviour.

4. Conclusion

Unusual line broadening has been observed in the Mössbauer spectra of the SC compound $[Fe(ptz)_6](ClO_4)_2$ around the transition temperature centred at ~ 125 K. The present μSR results have enabled us to delineate the temperature dependence of the asymmetry parameters in the spin transition of a SC compound $[Fe(ptz)_6](ClO_4)_2$ for the first time. Muonium has been shown to form and is likely to be localised at the aromatic ring rather than in the vicinity of the octahedral iron complex. Both the Mössbauer and muon measurements are consistent with a mixture of high and low spin Fe ions in the transition region. Further investigations of $[Fe(ptz)_6](ClO_4)_2$ and related SC compounds by Mössbauer spectroscopy and μSR measurements are in progress.

References

[1] P. Gütlich, A. Hauser, H. Spiering, Angew. Chem. Int. Ed. Engl. **33** (1994) 2024;
P. Gütlich, Y. Garcia, H. A. Goodwin, Chem. Soc. Rev. **29** (2000) 419.
[2] P. Poganiuch, Dissertation, Universität Mainz 1989;
J. Jung, F. Bruchhäuser, R. Feile, H. Spiering, P. Gütlich, Z. Phys. B. **100** (1996) 517.
[3] L. Wiehl, Acta Cryst. **B49** (1993) 289.
[4] G. Schmitt, Dissertation, Universität Mainz 1996.
[5] S. J. Blundell, Phil. Trans. R. Soc. Lond. A. **357** (1999) 2923.
[6] E. Roduner, Appl. Mag. Reson. **13** (1997) 1.
[7] N. Shioyasu, K. Kagetsu, K. Mishima, M. K. Kubo, T. Tominaga, K. Nishiyama and K. Nagamine, Hyper. Interact. **84** (1994) 477.

Acknowledgements

SJC acknowledges renewal of an Alexander von Humboldt Research Fellowship while at the Johannes Gutenberg-Universität, Mainz, and support from the Access to Major Research Facilities Program, Australian Nuclear Science and Technology Organisation. We acknowledge support from the EU under Framework 5 for access to the MUSR instrument at the ISIS Facility, Rutherford Appleton Laboratory, U.K., and support from the European Commission for granting the TMR_Network "Thermal and Optical Switching of Molecular Spin States (TOSS)", Contract No. ERB-FMRX-CT98-0199EEC/TMR. We also gratefully acknowledge the financial help from the *Fonds der Chemischen Industrie* and the *Materialwissenschaftliches Forschungszentrum der Universität Mainz*.

Hyperfine Interactions
International Conference on the Applications of the Mossbauer Effect
Oxford University, September 2-7, 2001

Investigation Into the Microstructure of Iron-Chromium Oxyhydroxide Fine Particles.

M.K. Fisher, A.M. Khasanov, J.G. Stevens
Department of Chemistry
The University of North Carolina at Asheville
One University Heights
Asheville, NC 28804, USA

Abstract

The study of nanoparticles is rapidly becoming an area of major interest. Iron oxide based materials are important in this field due to their many industrial uses, including magnetic storage capability, composite multi-layer film structures, and a new generation of catalysts. The goal of these studies is a greater understanding of the microscopic arrangement of individual atoms, and of the property differences of bulk materials and fine particles.

A series of iron and iron-chromium oxyhydroxides has been synthesized and studied using Mössbauer spectroscopy and x-ray diffraction. Particles of different size and structure were produced by variation of the synthesis conditions. The temperature kinetics of the structure evolution were studied by analyzing changes in the Mössbauer parameters (isomer shift and quadrupole splitting). Rapid diffusion was observed between the iron and chromium containing components. The electronic configuration of iron was shown to be affected by presence of chromium atoms in the nearest neighbor environment.

Experiment and |Results

Each iron sample was prepared from an aqueous solution of a $FeCl_3{\bullet}6H_2O$ and $Na_2CO_3{\bullet}9H_2O$. Chromium containing samples substituted $CrCl_3{\bullet}6H_2O$ for iron. For the purposes of this experiment, the midpoint of this range was targeted, and all samples were precipitated out of solution by adding to a pH of ≈8.6.

X-ray diffraction results

X-ray diffraction analysis of both iron and chromium-iron samples show a broad peak corresponding to an interplanar distance of 2.5 Å (35° of 2θ) and also a peak corresponding to 1.5 Å (62° of 2θ). The width of the peak can be correlated to the particle size, given by the Scherrer formula, $d=K\lambda/\beta\cos\theta$, where λ is the wavelength of the incident x-rays and θ is the Bragg angle. The full-line half maximum width β was obtained from direct measurement from the diffraction pattern, and the particle size was determined by using a K value of 1.107, as determined by Bødker and Mørup [3]. This yielded a particle size on the order of 5 nm. The pattern also indicates the nanocrystalline nature of the sample in question. This pattern is consistent with the two-line ferrihydrite model obtained by the forced hydrolysis of Fe (III) salts by a strong base [4]. The presence of chromium in these samples does not result in the growth, or apparent alteration, of the nanocrystalline structure. It is noted from the x-ray data that the peak corresponding to 2.5 Å is slightly suppressed, but nevertheless still present.

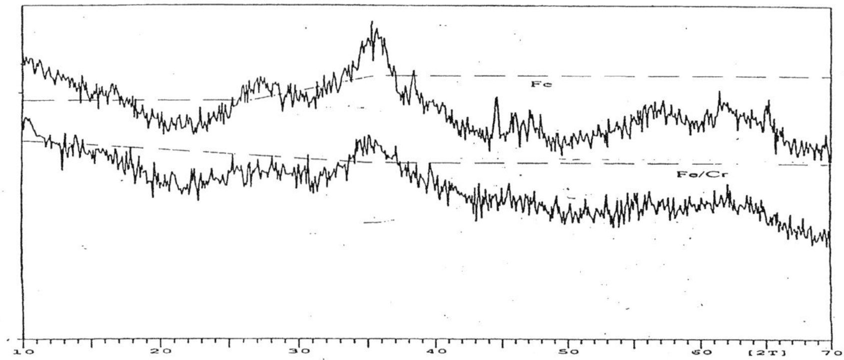

Figure 1 – X-Ray Diffraction Pattern for Fe and Cr/Fe Sample

3.2 Mössbauer spectroscopy results

The results of the Mössbauer spectra are listed in Table 1. All isomer shift values are relative to α-Fe and taken at room temperature. All spectra are contained in Figure 2 at the end of this paper.

Table 1 – Table of Mössbauer Parameters

Sample	Temperature of Heating	δ (mm/s)	Δ (mm/s)	Γ (mm/s)	Area (%)
Fe only		0.36	0.56	0.32	57
		0.36	0.94	0.37	43
	150	0.36	0.56	0.34	50
		0.35	0.96	0.41	50
	200	0.35	0.56	0.33	45
		0.34	0.96	0.47	55
	225	0.35	0.58	0.34	50
		0.34	0.98	0.43	50
	300	0.35	0.58	0.38	51
		0.33	1.00	0.51	49
	370	0.35	0.60	0.43	58
		0.34	1.02	0.50	52
50%/50% Fe/Cr		0.36	0.62	0.32	61
		0.35	0.98	0.35	39
	150	0.35	0.66	0.39	56
		0.34	1.12	0.40	44
	200	0.34	0.70	0.38	53
		0.33	1.16	0.43	47
	225	0.35	0.70	0.38	55
		0.34	1.16	0.43	45
	300	0.34	0.70	0.38	55
		0.34	1.17	0.42	45
	370	0.34	0.70	0.38	52
		0.33	1.18	0.43	48

Initially, each Mössbauer spectrum was attempted to be fit using one quadrupole doublet. This doublet exhibited approximate values for $\delta = 0.35$ mm/s, $\Delta = 0.75$ mm/s, $\Gamma = 0.40$ mm/s. The fitting program was able achieve convergence for this model, but the χ^2 values are on the order of 3-4. This model was dismissed based on the statistical unacceptability of the fit.

A system of two doublets was adopted for each spectrum. It is the current theory that the structure synthesized is comprised of chains of octahedra of ferrihydrite. The Mössbauer results obtained are in good agreement with those listed [5], with values for δ_1= 0.34 mm/s, Δ_1= 0.60 mm/s, and δ_2 = 0.33 mm/s, and Δ_2 = 0.99 mm/s. These chains consist of Fe $(O, OH)_6$ octahedra, connected at an edge sharing OH groups and at one corner. Each site along the chain differs slightly from its neighbors due to its relative location on the crystal. These slight differences are exposed experimentally by changes in the electric field gradient (EFG). The differences in the EFG produce slight differences in the value of Δ. A combination of all the distributions in EFG produces a "wider" doublet. A single doublet has been chosen to represent this distribution, since the separation distance is smaller than the natural line width of 0.26 mm/s.

Finally, each spectrum was attempted to be fit using three doublets. This model was capable of producing χ^2 values on the order of 1.0, however, the control parameters, which show the relative error associated with the parameter, for each value calculated did not suggest that the model was useable. The individual sites along the chains are too close to be resolved, due to the fact that the difference in Δ produced by slight changes in EFG differ by less than the natural line width.

There is a large difference that develops in Δ as the samples are heated. The experimental results are consistent with the initiation of diffusion of Cr^{+3} ions into the existing ferrihydrite structure. Cr^{+3} has a smaller atomic radius that Fe^{+3} (61.5 pm to 64.5 pm, respectively). This difference will produce distortions in the crystal structure as the Cr^{+3} ions diffuse into the lattice. This asymmetry is directly reflected in the EFG of the Fe octahedra. The diffusion is rapid, occurring at most over the 2 hour heating interval, and is apparently completed at a temperature of at most 200°C. Figure 2 shows the relationship between heating and Δ.

Additionally, contributions to the overall spectra from these sites increase with heating, and the width of the line of the higher quadrupole doublet is increased There is an increasing trend in all subspectra of both the Fe and the Cr/Fe samples.

This broadening can be attributed to the growth of the ferrihydrite chain, with the possible addition of chains to the sides of the structure. Addition to the middle of the chains, or to the length, decreases the influence of the ends, and this increasing contribution is supported experimentally by the decreasing ratios of areas associated with the two doublet sites. Increasing the length reduces the relative difference in neighboring sites since the neighbors are increasingly similar along the chain. Heating above 200°C does not produce any additional shift in the relative quantities of different iron sites.

The data also reveals that the addition of chromium affects the starting product very slightly. It is noted that the δ of the Cr/Fe sample is slightly higher, above error, than that of the Fe only sample (see Table 1). Based on the results of the samples prior to any heat treatment, initially iron and chromium precipitate out mostly separate. Noting the change in the value of δ, this indicates that there is some Cr^{+3} that has entered the ferrihydrite crystal structure, either during the coprecipitation, or during the heating of the precipitates prior to drying. This smaller radii forces the atoms shared jointly between neighbors to lie closer to the Cr center, and further from the Fe center.

In summary, it was found that the addition of chromium did not affect the precipitation of the ferrihydrite nanocrystals, as is indicated from the initial values of δ and Δ. Heating the chromium containing samples indicates the diffusion of chromium into the ferrihydrite structure, as indicated by a distinct increase in the value of Δ over that of the Fe containing samples. This diffusion occurs rapidly, and apparently does not influence the size of the particle, as is indicated from x-ray diffraction results.

4. References

[1] Dormann, J.L. and Fiorani, D., *Magnetic Properties of Fine Particles* (North-Holland, Amsterdam, 1992)

[2] Coe, E.M., Bowen, L.H., Bereman, R.D., Speer, J.A., Monte, W.T., and Scaggs, L. *J. Inorg. Biochem.*, **57**, 63-71, (1995)

[3] Bødker, F. and Mørup, S., *Europhys Lett,*. **52** (2), pp. 217-223 (2000)

[4] Schwertmann, U., Friedl, J., and Stanjek, H., *J. Coll. Interface Sci.*, **209**, pp. 215-223 (1999)

[5] Stevens, J.G., Khasanov, A. M., Miller, J. W., Pollak, H., and Li, Z., *Mössbauer Mineral Handbook*, 1998

Figure 2 – Mössbauer Spectra

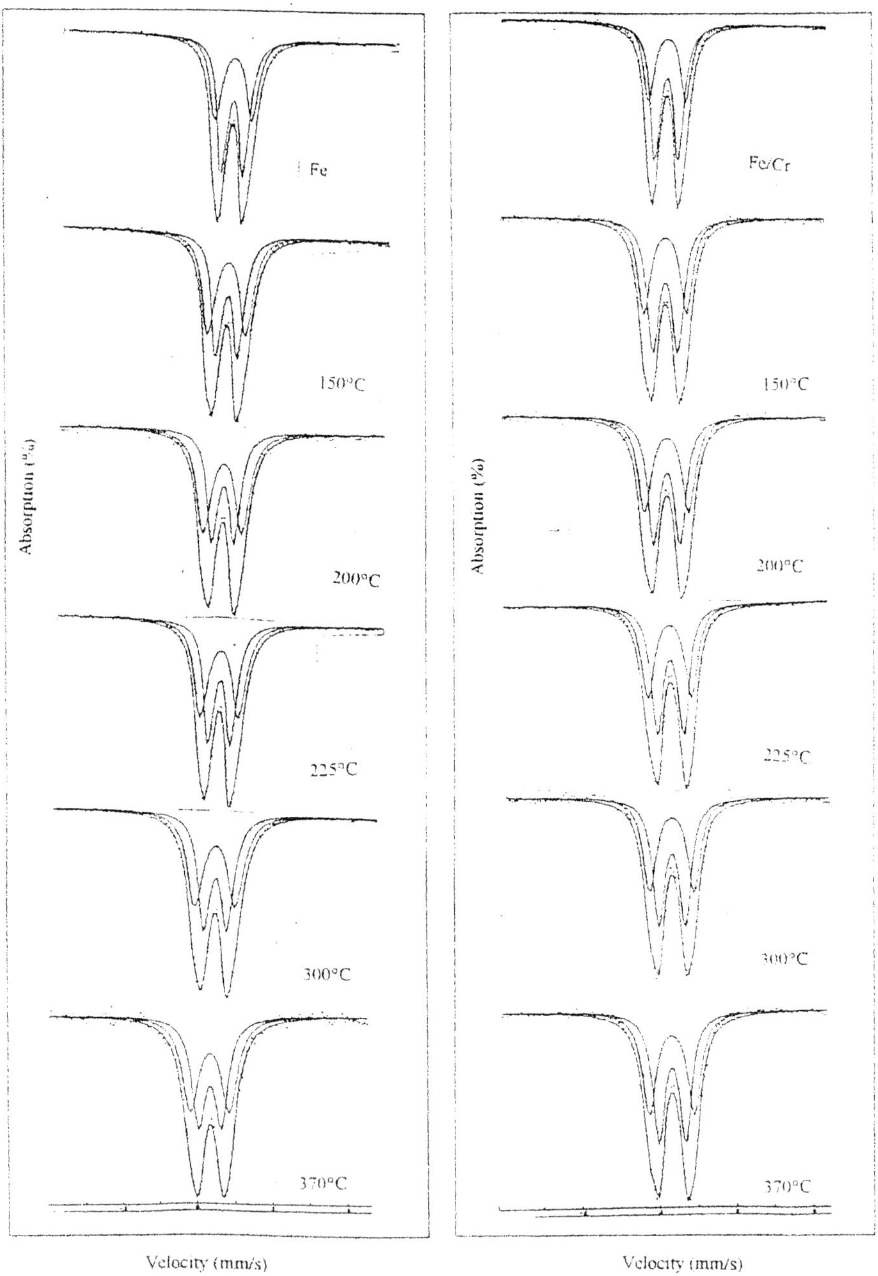

Mössbauer Spectral Study of Two Layered Honeycomb Molecular Magnets: $PPh_4Fe^{II}Fe^{III}(ox)_3$ and $NBu_4Fe^{II}Fe^{III}(ox)_3$

D. HAUTOT,[1] P. DAY,[1] G. J. LONG,[2] F. GRANDJEAN,[3] J. ENSLING,[4] and P. GÜTLICH[4]

[1]Royal Institution of Great Britain, London W1S 4BS, England;

[2]Department of Chemistry, University of Missouri-Rolla, Rolla, Missouri 65409-0010;

[3]Institute of Physics, B5, University of Liège, B-4000 Sart-Tilman, Belgium;

[4]Institut für Anorganische Chemie und Analytische Chemie, J. Gutenberg-Universität Mainz, Staudinger Weg 9, D-55099 Mainz, Germany

Abstract. The Mössbauer spectra of the layer ferrimagnets $PPh_4Fe^{II}Fe^{III}(ox)_3$ and $NBu_4Fe^{II}Fe^{III}(ox)_3$ have been measured between 1.9 and 315 K. The paramagnetic spectra exhibit both high-spin iron(II) and iron(III) doublets. The iron(II) quadrupole splittings are negative and decrease substantially upon cooling as a result of a low-symmetry crystal field splitting. In contrast, the iron(III) splittings are small, positive, and virtually independent of temperature. The respective magnetic ordering temperatures of 30 and 42.5 K agree well with the phase observed by bulk magnetometry and the magnetic sextets yield hyperfine fields of ca. 60 kOe for iron(II) and 540 kOe for iron(III). The unusually low iron(II) field results from the exceptionally large orbital and dipolar contributions, while the iron nuclei experience a static hyperfine field at 1.9 K and a relaxing hyperfine field between ca. 10 and 30 K.

1. Introduction.

The oxalate compounds, $AFe^{II}Fe^{III}(ox)_3$, are layered ferrimagnets, which can show unusual magnetic behavior such as a negative magnetization at low temperature. Thus the A = NBu_4 salt belongs to the Néel type N class of compounds whereas the A = PPh_4 salt is type Q [1]. To understand this difference we have analyzed Mössbauer spectra of both salts as a function of temperature from 1.9 to 293 K.

2. Spectra obtained above the critical temperature.

The Mössbauer spectra of both $PPh_4Fe_2(ox)_3$ and $NBu_4Fe_2(ox)_3$ at 293 K are shown in Fig. 1. These consist of two quadrupole doublets with hyperfine parameters which are reasonable for high-spin iron(II) and iron(III) in the distorted octahedral coordination environment found [2-5] earlier in $NBu_4Fe_2(ox)_3$ and related compounds.

The spectra of $PPh_4Fe_2(ox)_3$ demonstrate that it is paramagnetic between 293 and 40 K. Although the spectrum observed at 30 K is similar to the 40 K spectrum, the lines are broader, an indication of the onset of magnetic exchange correlations. In contrast, the spectra of $NBu_4Fe_2(ox)_3$, indicate that it is paramagnetic between 315 and 50 K. Although the spectrum observed at 42.5 K is similar to the 50 K spectrum, the lines are broader, an indication of the onset of magnetic exchange correlations at 42.5 K. Hence, the nature of the cation does influence the ordering temperature of these compounds.

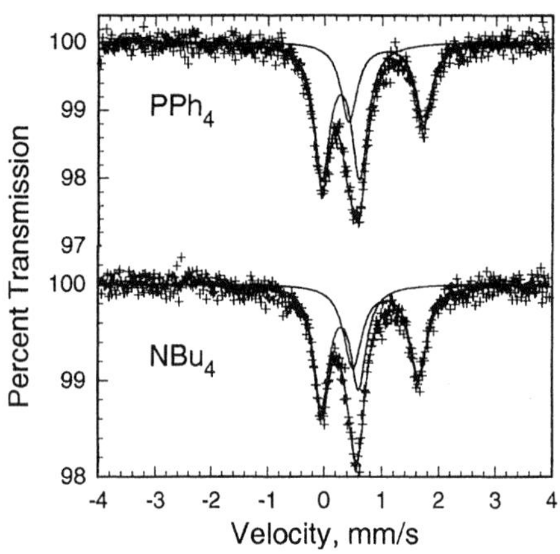

Fig. 1. The Mössbauer spectra of $AFeFe(ox)_3$ obtained at 293 K.

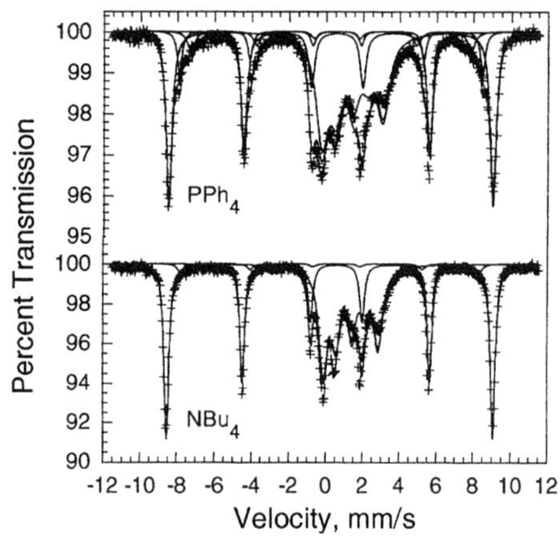

Fig. 2. The Mössbauer spectra of $AFeFe(ox)_3$ obtained at 4.2 K.

3. Spectra obtained below the critical temperature.

The Mössbauer spectra of both $PPh_4Fe_2(ox)_3$ and $NBu_4Fe_2(ox)_3$ obtained at 4.2 K are shown in Fig. 2. The iron(III) ions exhibit a sextet with a large hyperfine field, whereas the iron(II) ions exhibit a sextet with a relatively small hyperfine field. Similar results have been reported earlier [2-5] for $NBu_4Fe_2(ox)_3$. The presence of these sextets would seem to indicate long-range magnetic order in these complexes, in apparent contrast with the polarized neutron diffraction results of an earlier [6] study of $PPh_4Fe_2(ox)_3$, results which indicate the presence of *only* short range two-dimensional magnetic exchange correlations over a distance of ca. 65 Å at 5 K. Thus it would appear that the short range exchange correlations observed within the layers of $PPh_4Fe_2(ox)_3$ are sufficient, below 10 K, to yield a typical magnetic sextet. Between 10 and 22.5 K both paramagnetic doublets and broadened sextets coexist, a sign of slow paramagnetic relaxation. Thus the Mössbauer spectra reveal the dynamic nature of the short-range magnetic exchange correlations.

4. Hyperfine parameters.

The observed values of the isomer shifts are typical of high-spin iron(II) and iron(III) in a distorted octahedral environment. Further, the iron(II) quadrupole interactions are found to be negative for both $PPh_4Fe_2(ox)_3$ and $NBu_4Fe_2(ox)_3$. In contrast to the iron(III) sites, and as expected for a high-spin iron(II) ion with the approximate $t_{2g}^4e_g^2$ electronic configuration in a distorted octahedral coordination environment, the quadrupole splittings for the iron(II) sites in both $PPh_4Fe_2(ox)_3$ and $NBu_4Fe_2(ox)_3$ are rather different and increase dramatically with increasing temperature above ca. 50 K. This increase can be reproduced with the Ingall's model [7] for the quadrupole interaction at an iron(II) site.

The iron(III) average hyperfine field of ca. 540 kOe is typical of high-spin iron(III) and the ca. 60 kOe field for iron(II) is small because of the large orbital and dipolar contributions [8].

5. Conclusions.

The Mössbauer spectral isomer shifts and quadrupole interactions for both $PPh_4Fe_2(ox)_3$ and $NBu_4Fe_2(ox)_3$ indicate the presence of high-spin iron(II) and iron(III) in each compound. The iron(II) quadrupole interaction was found to be negative and indicated a greater distortion of the octahedral coordination environment of the iron(II) site in

$PPh_4Fe_2(ox)_3$. The hyperfine parameters obtained for the paramagnetic spectra have been essential in understanding the spectra obtained below the critical ordering temperature.

Below 10 K the iron(III) hyperfine field is ca. 540 kOe as expected while the iron(II) has only greatly reduced field, a reduction which is a direct consequence of the unusually large orbital and dipolar contributions. Between 10 K and the critical temperature, the Mössbauer spectra reveal the dynamic nature of the short-range magnetic exchange correlation which complements polarized neutron diffraction work which reveals the spatial extent of these correlations [6].

This work is supported by the EC TMR network 'molecular magnets'.

References

1. C. J. Nuttall and P. Day, Chem. Mater., 10 (1998) 3050.
2. S. Iijima, T. Katsura, H. Tamaki, M. Mitsumi, N. Matsumoto, and H. Okawa, *Mol. Cryst. Liq. Cryst.*, 233 (1993) 263.
3. S. Iijima, F. Mizutani, M. Mitsumi, N. Matsumoto, and H. Okawa, *Inorg. Chim. Acta*, 253 (1996) 47.
4. S. Iijima and F. Mizutani, *Mol. Cryst. Liq. Cryst.*, 335 (1999) 143.
5. S. Iijima and F. Mizutani, *Mol. Cryst. Liq. Cryst.*, 306 (1997) 227.
6. D. Visser, S. G. Carling, I. D. Watts, P. Day, and K. H. Andersen, *Physica B,* 267-268 (1999) 266.
7. R. Ingalls, *Phys. Rev.,* 133 (1964) A787.
8. T. E. Cranshaw and G. Longworth, in *Mössbauer Spectroscopy Applied to Inorganic Chemistry*, G. J. Long, (Plenum Press, New York, 1984) Vol. 1, p. 171.

Mössbauer studies on the metallic phases of Al Kidirate and New Halfa meteorites

Y. A. Abdu, T. Ericsson, H. Annersten, N. A. Dubrovinskaia, L. S. Dubrovinsky and A. M. Gismelseed[1]

Department of Earth Sciences, Uppsala University, Uppsala, SE-752 36, Sweden
[1]*Department of Physics, Sultan Qaboos University, Alkhoud, 123, Sultanate of Oman*

We have studied the metallic phases in two equilibrated ordinary chondrites Al Kidirate (H6) and New Halfa (L4) using Mössbauer spectroscopy and synchrotron X-ray diffraction. Mössbauer spectra on taenite-enriched samples from the two meteorites identify the following Fe-Ni phases: the ferromagnetic ordered taenite (*tetrataenite*) with ~ 50 % Ni, the ferromagnetic disordered taenite with ~ 50 % Ni, the low-Ni (~ 25 %) paramagnetic taenite, and small amounts of the α phase kamacite/martensite. The tetragonal unit cell parameters, in the space group *P4/mmm*, of tetrataenite from Alkidirate meteorite are: a = 3.5680 ± 0.0018 Å; c = 3.5901 ± 0.0015 Å; c/a = 1.0062.

1. Introduction

In spite of the considerable amount of work done on metals from iron, stony-iron, and stony meteorites, they still capture the attention of researchers from many fields. This is mainly because of their textural and compositional variability, and of the presence of the unusual Fe-Ni phases in their very complex microstructure. Mössbauer spectroscopy has played a major role in the study of Fe-Ni phases in meteorites, particularly in the discovery of the ordered Fe-Ni (50-50) phase in 1977 [1], which was later named tetrataenite [2].

In this paper we investigate the Fe-Ni phases in the metal particles from two equilibrated (*i.e.* petrologic type 4-6) ordinary chondrites; the Alkidirate (H6), recovered in 1983-Sudan, and the New Halfa (L4), recovered in 1994-Sudan, by Mössbauer spectroscopy and X-ray diffraction techniques.

2. Experimental

The metal particles were extracted from the bulk powdered samples of the meteorites by a hand magnet under acetone. Taenite-enriched samples were obtained by chemical treatment of the metal particles, to dissolve kamacite, using conc. HCl (6M) in ~ 50° C water bath for a few hours.

The Mössbauer spectra were recorded in transmission mode with a conventional constant acceleration spectrometer using ^{57}Co/Rh source. The spectrometer was calibrated with the room temperature spectrum of α-Fe foil. Low temperature measurements were performed using a liquid helium flow cryostat. The Mössbauer absorbers, which were in a form of self supporting foils, contained ~ 18 mg/cm^2 of the sample.

X-ray diffraction experiments were conducted at the European Synchrotron Radiation Facility (ESRF, Grenoble, France) on the beamline ID 30.

3. Results and discussion

From the weights of the magnetic and the non-magnetic fractions of the two meteorites, the metal fractions account for ~ 15 and 8 wt % of the bulk Alkidirate and New Halfa, respectively. The Alkidirate meteorite contains ~ 19 % (in mole fractions) of fayalite and ~ 16 % of ferrosilite and the average composition of plagioclase is ($Ab_{83}Or_6An_{11}$), calculated from [3], showing that it is an H6 chondrite [4].

The room temperature Mössbauer spectra of the taenite-enriched samples from Alkidirate (Alk-T) and New Halfa (NH-T) meteorites are shown in Fig. 1(a) and 1(b), respectively. The relevant hyperfine interaction parameters are presented in Table 1.

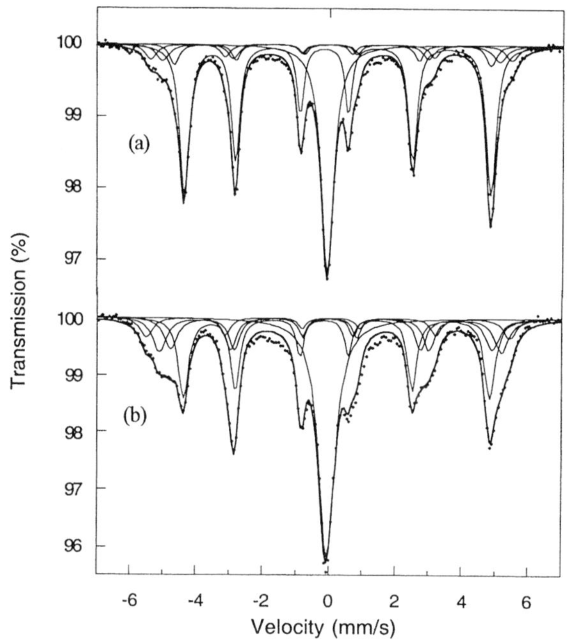

Fig. 1. Room temperature Mössbauer spectra of taenite-enriched samples from: (a) Alkidirate; (b) New Halfa meteorites.

On the basis of the weight change before and after the chemical treatment, Alk-T represents ~ 4 % of the metal fraction of Alkidirate meteorite, while NH-T accounts for ~ 8 % of the bulk metal of the New Halfa meteorite. This indicates that the New Halfa meteorite has higher taenite contents in its metal particles compared to Alkidirate meteorite.

It is evident from Fig. 1 and Table 1 that the low-Ni paramagnetic taenite is abundant in both samples. This phase seems to be common in equilibrated ordinary chondrites [5] and was recently proposed by Rancourt and Scorzelli as a new mineral (antitaenite) due to its distinct electronic structure [6]. The low temperature Mössbauer measurements on a taenite-enriched sample from New Halfa meteorite show that low-Ni paramagnetic taenite (width = 0.6 mm/s at room temperature) exhibits a continuous line broadening which reaches ~ 2.1 mm/s at 4 K. This could be due to relaxation effects, which may be the main reason for the misfit at the wings of the singlets, which is more pronounced in Fig. 1(b).

Table 1. Room temperature Mössbauer hyperfine parameters of taenite-enriched samples from Alkidirate and New Halfa meteorites.

sample	Fe-Ni phase	CS (mm/s)	ΔQ (mm/s)	B (T)	A (%)
Alk-T	ordered taenite	0.04	0.41	28.5	51.6
	partially ordered taenite	0.02	0.16	29.5	7.3
	disordered taenite	0.04	0.02	31.4	8.2
	paramagnetic taenite	-0.06	-	-	26.0
	kamacite/martensite	0.03	0.05	33.6	6.9
NH-T	ordered taenite	0.05	0.39	28.6	28.9
	partially ordered taenite	0.02	0.12	30.0	14.8
	disordered taenite	0.04	0.02	31.9	16.0
	paramagnetic taenite	-0.06	-	-	31.9
	kamacite/martensite	0.01	-0.08	33.9	8.5

CS = centroid shift relative to α-Fe at 295 K (± 0.01 mm/s); ΔQ = quadrupole interaction defined as $1/2\ (V_1 - V_2 - V_5 + V_6)$, where $V_1, \ldots V_6$ are the peak positions in the sextet with increasing velocities (± 0.01 mm/s); B = internal magnetic field (± 0.5 T); and A = relative area (± 5 %).

The Mössbauer hyperfine parameters (B = 28.5 T, ΔQ = 0.41 mm/s for Alkidirate, and B = 28.6 T, ΔQ = 0.39 mm/s for New Halfa (see table 1) are characteristic of a highly ordered Fe-Ni (50-50) phase [7]. This phase is believed to be formed when meteorites cool very slowly (~ 0.1 - 100 K/Ma) below 593 K which is the order-disorder transition temperature [8]. The disordered taenite, which is present in both Alk-T and NH-T, is fitted with two components, one shows partial ordering (*i.e.* $\Delta Q \neq 0$) and the other is completely disordered (*i. e.* $\Delta Q \approx 0$). This gives a better fitting and allows to resolve the small amounts of the α phase (kamacite/martensite) present in the samples. In the fitting procedure we have also used the parameter dB % which is a measure of the relative hyperfine field distribution with respect to the mean field B.

The Alk-T sample is selected for X-ray diffraction studies, as it contains an appreciable amount of tetrataenite. Fig. 2 shows a synchrotron powder X-ray diffraction pattern of the Alk-T sample. The following superstructure lines were detected: (001), (110), (201), (112), (221), (003), (310) and (312). To our knowledge, this is the first synchrotron powder X-ray diffraction pattern showing the superstructure lines of tetrataenite. The tetragonal unit cell parameters of tetrataenite in the space group *P4/mmm* are a = 3.5680 ± 0.0018 Å; c = 3.5901 ± 0.0015 Å; and c/a = 1.0062.

Based on the assumption that it is the ordering which causes the distortion of the cubic structure to the tetragonal structure, the tetragonality can be used to estimate the degree of order. By comparing the tetragonality we have obtained (c/a = 1.0062) in our sample with that of tetrataenite from the iron meteorites studied by Albertsen (c/a = 1.0036) [8], we can conclude that the degree of order in tetrataenite from Alkidirate meteorite is higher than that in the iron meteorites studied in [8]. Probably the cooling rate of Alkidirate parent body, at least locally, was extremely slow below the ordering temperature compared to the above mentioned iron meteorites.

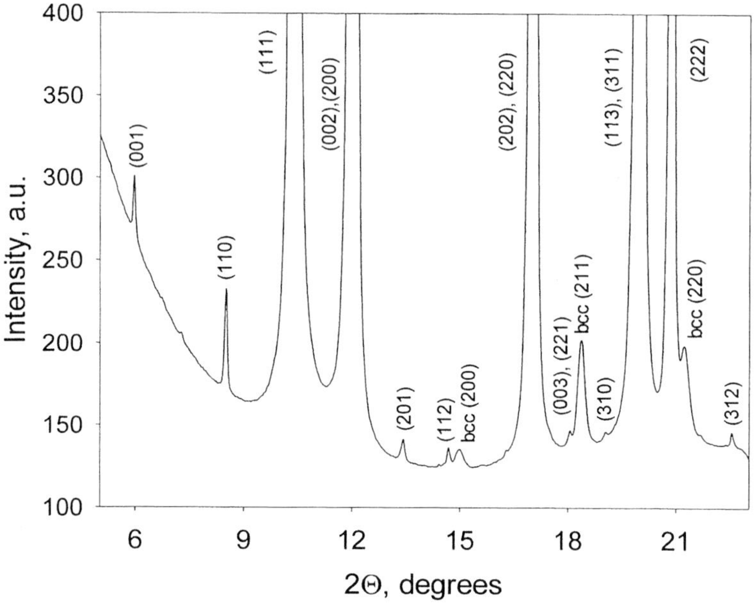

Fig. 2. Powder X-ray diffraction pattern of taenite-enriched sample from Alkidirate meteorite showing the weak superstructure lines (001), (110), (201), (112), (221), (003), (310), and (312) of tetrataenite, which also contributes to the strong fcc lines together with disordered taenite and low-Ni paramagnetic taenite. The lines marked bcc represent the small amounts of kamacite present in the sample.

References

[1] J.F. Petersen, M. Aydin and J.M. Knudsen, Phys. Lett. 62A (1977) 192.
[2] R.S. Clarke Jr. and E.R.D. Scott, Am. mineral. 65 (1980) 624.
[3] A.M. Gismelseed, F. Khangi, A. Ibrahim, A.A. Yousif, M.A. Worthing, A. Rais, M.E. Elzain, C.K. Brooks and H.H. Sutherland, Hyp. Int. 91 (1994) 551.
[4] R.T. Dodd, in: Meteorites, Cambridge University Press, 1981.
[5] P. Kong and M. Ebihara, Geochim. Cosmochim. Acta 60 (1996) 2667.
[6] D.G. Rancourt and R.B. Scorzelli, Am. Mineral. 81 (1996) 766.
[7] L. Larsen, H. Roy-Poulsen, N.O. Roy-Poulsen, L. Vistisen and J.M. Knudsen, Phys. Rev. Lett. 48 (1982) 1054.
[8] J.F. Albertsen, phys. Script. 23 (1981) 301.

^{57}Fe Mössbauer spectroscopic investigation of some coals of geographically different provenance and their fly ashes

J. Adetunji, J. M. Williams[†] and A. Mehra

Centre for Environmental and Applied Science Research, School of Environmental and Applied Sciences, University of Derby, Derby DE22 1GB UK;
[†]Department of Physics and Astronomy, University of Sheffield, Sheffield, S3 7RH, UK

Abstract

Coal samples of geographically different provenance and their undifferentiated and fractionated fly ashes were studied with a Mössbauer spectrometer at room temperature. The Mössbauer results were complemented with elemental concentration measurements, using Inductively Coupled Plasma Atomic Emission Spectroscopy (ICP-AES). A high concentration of arsenic, a toxic environmental pollutant, was observed in a coal sample from one of the sources. Mössbauer analysis showed that it also has high pyrite, FeS_2. Calculations of the ratios of Fe^{3+} to ΣFe in the coal fly ash aluminosilicate-glasses allowed us to assign structural positions of the ferric and ferrous ions in them.

1 Introduction

The disposal of large quantities of coal fly ash for various uses mobilises enriched elemental species into the atmosphere. Previous studies have highlighted the potential health and environmental hazards of toxic elements in coal fly ash and the problem of pyrite, which causes fouling of boiler pipes at power stations. The environmental consequences of these pollutants have been extensively reported [1-3]. It is known that the predominant transition metal found in atmospheric pollution is iron, most of which originates from coal fly ash [2]. The iron in the fly ash is mostly associated with aluminosilicate glasses. The amount of iron in the fly ash is dependent on the source of the parent coal and increases with decreasing particle size. Furthermore, fine-grained iron-bearing minerals in coals and their ashes are involved in lung damage after inhalation. An acknowledgement of this has encouraged continuing interest among environmental scientists in the study of coal fly ashes. For example, a recent report [3] has shown that the amount of iron found in treated lung cells is not only inversely related to the particle size of fly ash, but also varies with the coal. In addition, the removal of iron associated with glassy aluminosilicates in some coal fly ashes inhibits some effects of coal fly ash on human lung epithelial cells. As part of our extended study of the role of bioavailable iron and its compounds in coal fly ash, we carried out a Mössbauer spectroscopic investigation of three coal and fly ash samples of different geographic provenance. The aim is to determine other factors that could be used to characterise coals and, also, their influence on the structural distribution of ferric and ferrous ions in the glassy-silicate component of the coal fly ashes.

2 Sample collection

Two Indian coals R-Coal and I-Coal of different geographic provenance were collected from Rajghat and Indraprastha Power Houses, India, respectively. The corresponding undifferentiated fly ashes R-FA and I-FA were collected from the power stations' electrostatic precipitators. The fly ash samples were fractionated into < 32 μm particle size and respectively labelled R-FA32* and I-FA32*. The third coal sample, O-Coal, collected from Ostrava, Czech Republic, was ashed in the laboratory at low temperature (~ 400 ^{0}C) and the undifferentiated fly ash sample labelled as O-FA.

3 Experimentation

The determination of nine elemental concentrations in the samples was routinely carried out with an ICP-AES system. The Mössbauer measurements were done with a 25mCi ^{57}Co source in an Rh matrix, which was driven at a constant acceleration in a triangular mode at room temperature. Each absorber was made from 100 mg of the sample investigated. The spectra were recorded in 512 channels and the Lorentzian lines of the folded data were fitted using the least-square RECOIL computer programme. All isomer shift (δ) values are relative to α-iron.

4 Results and discussion

Coal samples

Al_2O_3 and SiO_2 are the major ingredient oxides in the formation of aluminosilicates [4]. The concentrations of these oxides and arsenic in the undifferentiated ash samples are shown in table 1. The concentration of Al_2O_3 in the Ostrava coal is low while its arsenic content is high compared with the other two samples.

Table 1: Concentrations of aluminium oxide, silica and arsenic in the coal samples.

Sample	Al_2O_3 %	SiO_2 %	As (mg kg^{-1}) Coal	Fly ash
R-coal	10.6	39.9	2.0	< 0.01
I-coal	16.6	30.1	< 0.01	< 0.01
O-coal	2.2	48.3	2906.0	187.0

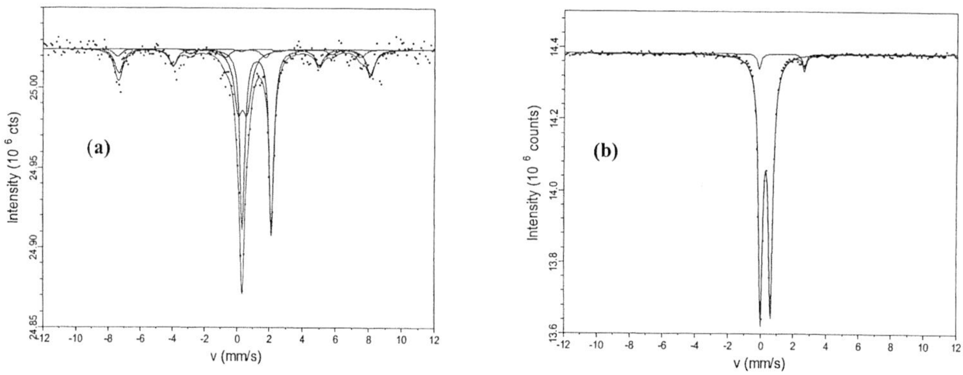

Fig 1: Mössbauer spectra of (a) R-coal sample (b) O-coal sample

The iron-bearing minerals were characterised using data from the Mössbauer Mineral Handbook [5] and parameters obtained in the literature [6, 7]. The room temperature Mössbauer spectra are shown in figure 1(a and b) and the parameters of all the coal samples, possible assignments and oxidation states of Fe are shown in table 2. On comparing the concentration of arsenic (table 1) with the relative percentage of pyrite (table 2) in the coal samples we notice a correlation between high arsenic content and high pyrite content. Is the high content of FeS_2 concomitant with high arsenic content in coals and also does it bear any relation to volatile organic compounds? Table 2 also shows that the R-Coal sample has 45.6% siderite ($FeCO_3$) and 25.5% pyrite; the rest is made up of magnetite. The iron compound in the I-Coal sample is 100% siderite and a trace amount of hematite.

Table 2: Mössbauer parameters obtained from spectra of coal the three samples.

Samples	δ mm s^{-1}	Δ mm s^{-1}	Hyp. Field kOe	Rel. Int. %	Possible assignments
R-Coal	1.24	1.79	-	45.6	Fe^{2+}, siderite, $FeCO_3$
	0.35	0.55	-	25.5	Fe^{3+}, pyrite, FeS_2
	0.45 0.74	-0.070 -0.70	487.7 463.0	21.9 7.0	Magnetite, Fe_3O_4
I-Coal	1.24	1.81	-	100	Fe^{2+}, siderite, $FeCO_3$
O-Coal	0.32	0.60	-	95.5	Fe^{3+}, pyrite-marcasite (10:1) mixture
	1.25	2.74	-	4.5	Fe^{2+}, Szomolnokite, $FeSO_4.H_2O$

The parameters of the O-Coal sample are in good agreement with those for the pyrite-marcasite mixture i.e. two forms of pyrite, FeS_2, namely pyrite 85.5% and ~10% of the more chemically stable marcasite. The minor component is 4.5% Szomolnokite, $FeSO_4.H_2O$. This type of sulphate is known to form on surfaces of pyrite crystals as a result of oxidation [6].

Fly ash samples

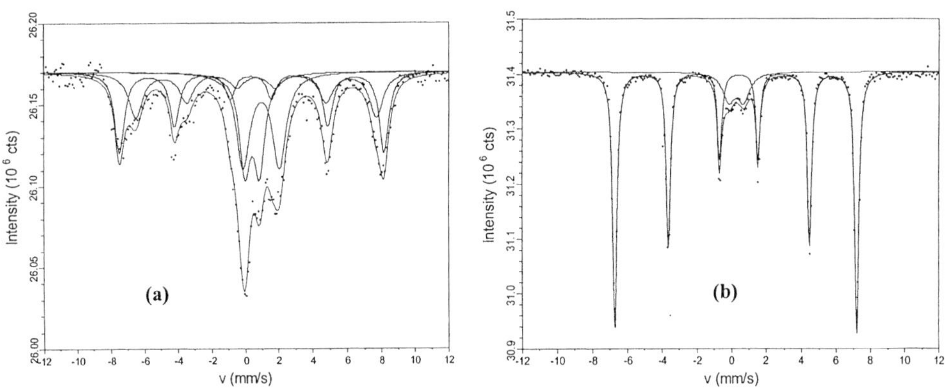

Fig 2: Mössbauer spectra of undifferentiated fly ash of (a) high temperature combusted R-Coal (b) low temperature combusted O-Coal.

Figure 2 shows the Mössbauer spectra of two of the investigated fly ash samples. Both the undifferentiated and fractionated fly ash samples contain significant amounts of magnetite relative to other iron compounds. However, the magnetic component in the Ostrava fly ash is mostly hematite (87%), being the only oxide that can be formed at a low temperature. The doublet in figure 2b for the undifferentiated Ostrava fly ash (O-FA) spectrum has $\delta = 0.25$ mm s^{-1} and $\Delta = 0.85$ mm s^{-1}. These values are in agreement with the values for super paramagnetic hematite particles [7]. An analysis of the total relative intensity values of the magnetic component in the undifferentiated and fractionated ash samples shows an increase of the magnetic component in the R-FA32* sample but a decrease in the I-FA32* sample [8]. The parameters of the doublets (table 3) in the undifferentiated R-FA and I-FA fly-ash spectra fall within the range of those reported for over fifty synthetic aluminosilicates. Our calculations of the ratios of Fe^{3+} to ΣFe in the doublets and the assigned structural positions of the Fe^{3+} and Fe^{2+} in them are summarised in table 3. The basis for these assignments follow an established reported procedure [4].

Table 3: A summary of Mössbauer parameters, ferric/total Fe ratio and co-ordination of Fe^{3+} in the aluminosilicate-glass component of fly-ash samples from two power stations.

Sample	Fe^{2+}			Fe^{3+}			$\frac{Fe^{3+}}{\Sigma Fe}$ and comment	
	δ mm s^{-1}	Δ mm s^{-1}	Rel. Int. %	δ mm s^{-1}	Δ mm s^{-1}	Rel. Int. %		
R-FA	0.97	2.17	24.7	0.43	0.87	20.1	0.45	Tetrahedral and octahedral co-ordinated Fe^{3+} coexist with Fe^{2+} acting as modifier of the network
I-FA	0.87	2.15	17.7	0.36	0.86	25.6	0.59	
R-FA32*	0.77	2.26	14.3	0.39	0.86	24.5	0.63	Fe^{3+} tetrahedrally co-ordinated.
I-FA32*	0.96	2.15	13.8	0.37	1.01	32.5	0.70	

The ratios obtained for the undifferentiated fly ash lie between 0.3 and 0.6, representing an average value for aggregate particle size in the fly ash and consistent with the type of co-ordination of Fe^{3+} we have assigned to them. However, the results 0.63 and 0.70 obtained for samples of particle less than 32 μm in diameter show that Fe^{3+} is tetrahedrally co-ordinated. These results also show that the type of co-ordination is particle size dependent. The absence of aluminosilicate minerals in the Ostrava fly ash, (O-FA), sample must have been due to the very low Al_2O_3 content in the parent coal and low combustion temperature.

5 Conclusion

The results from this work show that iron-bearing aluminosilcates can only be found in fly ashes from the high temperature combusted coals. The results also show that the structural positions of Fe^{2+} and Fe^{3+} ions in glass aluminosilicate structures is particle size dependent. Furthermore, the results obtained for the coal samples suggest a correlation between the high contents of arsenic and pyrite in coals. It would be interesting to ascertain whether the correlation could also be extended to their content of volatile organic compounds, known to be inversely related to the concentrations of bioavailable arsenic and iron in the coal fly ash. This could form a basis for further investigation requiring analyses of a large variety of coal samples.

Acknowledgement

We would like to thank Mrs Susan Elliott for the ICP-AES analysis of the samples.

References

[1] I. O. B. Ewa, J. Adetunji and S. B. Elegba, J. Environ. Sci. Health **A31** (5), (1996) 1089.
[2] W. P. Linak, J. O. L. Wendt, Fuel Processing Technology **39** (1) (1994) 173.
[3] B. R. Ball, K. R. Smith, J. M. Veranth and A. E. Aust, Inhal. Toxicol. **12** (2000) 209.
[4] B. O. Mysen and D. Virgo, American Mineralogist, **74** (1989) 58.
[5] J. G Stevens, A. M. Khasanov, J. W. Miller, H. Pollak and Z. Li, in: *Mössbauer Mineral Handbook* (Publ. Mössbauer effect Data Centre, Asheville, USA.1998).
[6] B. Morosin, R. A. Graham, P. Richards, Appl. Phys. Com. **5** (1-2) (1985) 1.
[7] K. S. Harchand and D. Raj. Nucl. Instrum. Meth. Phys. Res. **B76** (1993) 249.
[8] J. Adetunji, J. M. Williams and A. Mehra [Paper to be published].

The aging process in ceramics

A. Amulevičius [a], A. Daugvila [a], R. Davidonis [a] and G. Vaitkevičius [b]

[a] *Institute of Physics, Goštauto g. 12, LT-2001 Vilnius, Lithuania*
[b] *Institute of History, Kražių g. 5, LT-2001 Vilnius, Lithuania*

In this study, the influence of both natural ground dampness and ionising radiation (IR) on the process of aging of archaeological ceramics was investigated. The investigations were carried out using Mössbauer effect (ME), X-ray diffraction (XRD) and γ-ray spectroscopic methods. Accurately dated archaeological ceramic samples were used in the study as well as fired model samples from the same clay. Specially created equipment allowed the model samples to be aged the equivalent in real time of several hundred years. It was established that the effect of IR on ceramics depends on the firing temperature and decreases when the temperature and/or the firing time increases. In investigating the archaeological ceramic samples, it was established that the effect of IR obviously depends on whether the archaeologisational environment was dry or wet.

1. Introduction

One area for the use of ME is the study of aging processes in archaeological material. In the course of time of archaeologisation, the structure of the ceramic articles, which are under the influence of environmental factors, change: the shard hydrates, carbonises, and develops defects created by IR. The possibilities of ME are especially promising in this area. In order to use ME to study the aging processes, it is necessary to discover: 1) how the IR effect is reflected in the ME spectra parameters; 2) how much the radiation effect depends on the firing conditions of the ceramic shard; and 3) how much the radiation effect depends on the physical-chemical characteristics of the aging environment. This work is intended to solve these problems.

2. Experimental

In the course of the work, two series of samples were studied: old ceramics and model samples. The influence of different archaeologisation conditions was observed in the construction ceramics of the Missionary Church in Vilnius. This building was selected because the dates of all the construction stages are known; and this building has not experienced a fire, destruction, or abandonment. The samples from the nave had a constant indoor temperature and the wall was dry; the bricks at the bottom part of the tower were at a constant temperature and the wall was damp. The firing temperatures of the bricks studied, according to dilatometrical data, ranged from 880 to 950° C. Model samples were manufactured from the clay from the deposit used for the construction of the church. The samples were fired at 500 to 1000° C at intervals of 100° C; some samples were heated and gradually cooled, others were kept at the maximum temperature for 5 hours (figure 1). A comparison of the ME spectra of raw and annealed archaeological samples enables the size and character of the defects created during archaeologisation to be defined (figure 1). Under laboratory conditions the creation of the defects is also possible. For this purpose, aging equipment capable to irradiate a shard with the required IR dose in a short period was created. To determine the radiation dose absorbed by the samples of old ceramics, the spectra of the energy and intensity of IR of the location ceramic shards, were measured. This established the nature of the IR effect on the material. In order to assure the proximity of the experimental and natural conditions, isotope ^{226}Ra, was used as the source in the equipment.

In order to avoid systematic errors connected with the model for the ME spectrum description as well as to observe insignificant changes caused by IR, the differences of the spectra of the raw and irradiated samples were found and investigated [1].

3. Results and Discussion

The form of the ME spectra of the model (figure 1) and old ceramic samples (figure 3) slightly differed. It is possible to divide ME spectra into two parts: ferromagnetic and paramagnetic. The quantity of the magnetic phase grew from 5 to 60%, increasing the firing temperature and time. The central part of the spectrum is well defined by two quadrupole doublets. The isomeric shifts (IS) of both doublets were similar (0.3-0.4) mm/s while the quadrupole splitting was ΔQ1 (0.7-0.8) mm/s and ΔQ2 (1.8-1.9) mm/s. These parameters depend very little on the firing temperatures. The first doublet is attributed to Fe ions in tetrahedron; the second doublet in octahedron surrounding [2].

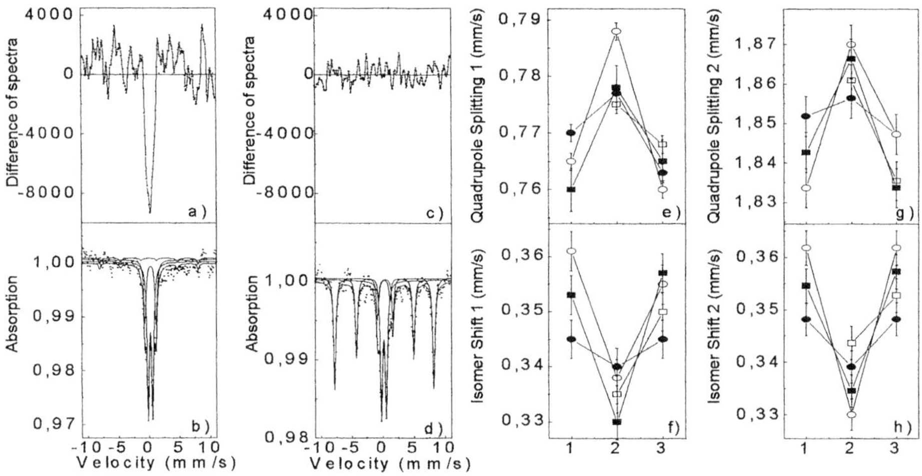

Figure 1. ME spectra of the samples: a) the difference in the spectra fired at 500° C for 0 h and aged 300 years; b) sample fired at 500° C for 0 h; c) the difference in samples fired at 1000° C for 5 hours and aged 300 years; d) sample fired at 1000° C for 5 hours. The IS (f, h) and ΔQ (e, g) changes of the ME parameters in the archaeological ceramic and model samples: 1 fired samples; 2 samples naturally and artificially aged 300 years; and 3 annealed samples. Samples ● fired at 800° C for 0 hours; ■ at 500° C for 5 hours; ○ at 500 for 0 hours; and □ wet brick.

An analysis of the γ-ray spectra shows that ^{40}K radiation, causes the largest part (70%) of IR (figure 2). The origin of ^{40}K in the environment is connected with human domestic activity: at that time, the principle fuel was firewood and a large quantity of ^{40}K was accumulated and settled around inhabited areas in the form of ashes and aerosol smoke particles [3]. In Vilnius, the concentration of ^{40}K is 6-10 times greater than in the periphery. The γ-rays themselves weakly act on the structure of the ceramics and the defects in it are caused by the intrinsic radiation of Compton electrons. The ME spectra react to the changes in the position of the Fe atom in respect to the surrounding atoms.

The IS and ΔQ changes of the Fe ions only insignificantly exceeded the experimental error (figure 1), however, the dynamics of the changes were themselves clearly observable. We see that the changes in the ME parameters and electron structures are very small.

The differences of the ME parameters of the fired and then irradiated model samples are shown in figure 1. These differential spectra (figure 1) reflect not small IS or ΔQ changes in the irradiated structure. This spectrum reflects any Fe ion area having a slightly different surrounding, the existence and parameters of which were not foreseen in the first model of the spectra description. The differential spectra are also well approximated by two quadrupole doublets. These changes in spectra parameters can be caused by several phenomena: change of 0.03-0.1 the elementary charge in the 4s electron charge density; change of 0.04-0.12 the elementary charge in the 3d electron charge density; or changes of 1.5-4 % in the distance between the Fe and other ions.

In studying the ME spectra, we established that the area of the differential spectrum decreases with the firing temperature and time as lnT. We think that the greater the firing temperature and time, the less metastable structure is created and this is reflected in the ceramic body resistance to IR effect.

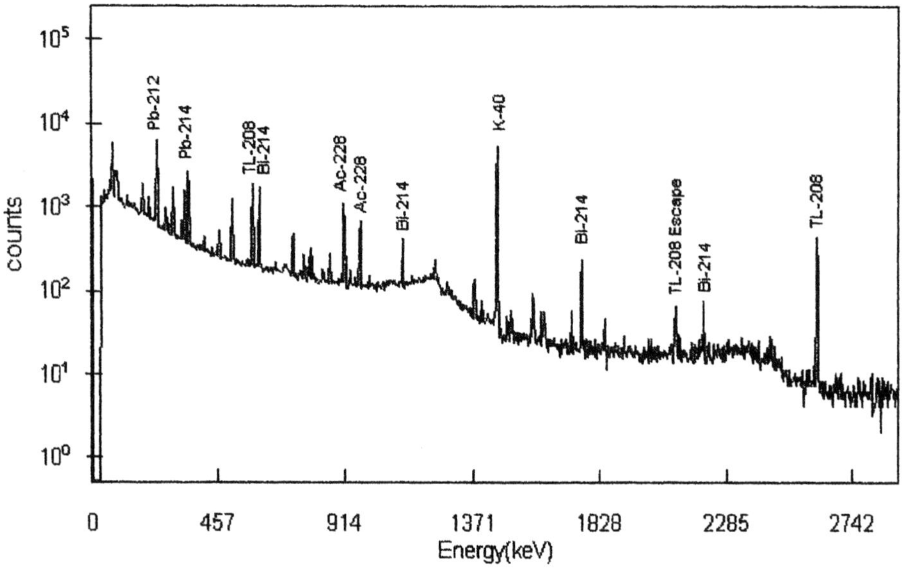

Figure 2. The γ-radiation spectra of the raw clay.

The long range atomic order does not practically change and this is reflected in the XRD spectra peak positions. The fact that the differential spectra of the model samples fired at 700° C and 0 h the IR effect is evident, but at over 700° C and 5 h the effect of IR disappears confirms the diffusion model of structure formations (figure 3).

An analysis of the ME and XRD spectra showed that the changes in the ME spectra are caused by small SRO changes in the fired and irradiated samples.

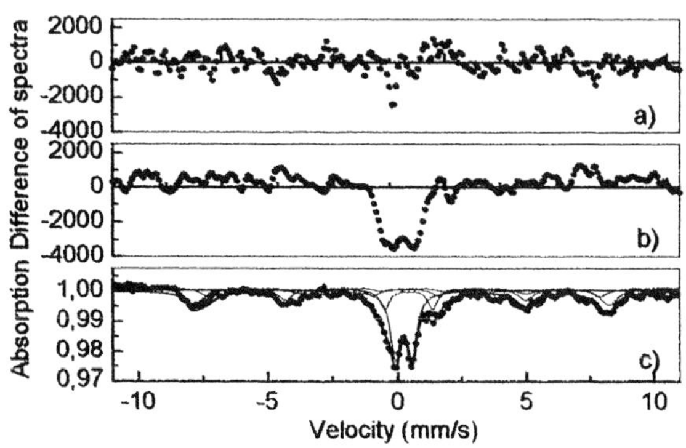

Figure 3. The difference of the ME spectra of the naturally aged bricks of the Missionary Church and annealed bricks: a) dry brick b) wet brick c) ME spectrum of naturally aged bricks.

In the bricks that were archaeologised in dry places the IR effect does not exceed the limits of measurement errors. At the same time for the bricks archaeologised in wet locations the IR effect is evident (figure 3). This is connected with the elimination of the OH groups from the ceramic shard at temperatures higher than 800° C. [4]. Water or the products of its hydrolysis, by interposing into the clay structure, increases its plasticity, increases the structure's tension, and stimulates the creation of defects as a consequence of IR. The samples were artificially aged and relaxed several times. The ME studies showed that the results of the multiple annealing-radiation effect were repeatable. In clay, Fe is not the main element; therefore the information obtained by ME reflects only the physics chemical-processes tendencies taking place in the clay.

4. Conclusions

The changes in the ME spectrum, which are due to IR, reflects the changes in the values of the HFI interactions and differential spectrum area. The IR effect decreases as the firing temperature and time increase, the dampness of the archaeologisation environment influences the IR effect.

Acknowledgements

We would like to thank the Lithuanian State Science and Studies Foundation for financial support.

References

[1] N. E. Cranshaw, Physics BC86-BC888 (1977) 391.
[2] D. C. Rancourt, ed, Hyp. Interact. (1998) 1-436.
[3] V. J. Aitken, Rep. Prog. Phys. 62 (1999) 1333.
[4] A. M. Pollard and C. Heron, ed., Archaeological Chemistry, the Royal Society of Chemistry.

CHEMICAL DYNAMICS OF Fe COMPOUNDS IN BRAZILIAN MANGROVE SEDIMENTS STUDIED BY MÖSSBAUER SPECTROSCOPY

G.T. Aragon[1], F.L. Silva[1]
[1]Lab. Ciências Ambientais, UENF, 28015-620 Campos dos Goytacazes, RJ, Brazil
A. Bustamante D[2,*], R.B. Scorzelli[2], A.A. Gomes[2]
[2]Centro Brasileiro de Pesquisas Físicas, Rua Xavier Sigaud 150, 22290-180 Rio de Janeiro, RJ, Brazil
and
J.D. Fabris[3]
[3]Dep. de Química, UFMG, Campus – Pampulha, 31270-901 Belo Horizonte, MG, Brazil

Urban contamination and diagenetic reactions motivate the study of the chemical dynamics of Fe compounds in mangrove sediments. Mössbauer data of samples collected down to about 50 cm below the surface of a mangrove sediment in the state of Rio de Janeiro, Brazil, led to the identification of the main iron-bearing minerals, and to a proposed mechanism involving pyrite (FeS_2) oxidation and increasingly formation of szmolnokite ($FeSO_4$) and iron (hydr)oxydes (mainly goethite, $\alpha FeOOH$) on going from the bottom towards the surface of this mangrove layer.

INTRODUCTION

Iron is one of the dominantly occurring elements of diagenetic reactions in tidal flat sediments. The great influence of reactive iron in marine sediments, in which the geochemical reactions involving iron dominate the pyrite formation, the release of phosphate and the storage of trace metals has been subject of considerable attention. In salt marsh sediments, iron undergoes a dynamical seasonal cycle, which is related to the sedimentary exposure to the atmosphere, plant growth and climate

---

[*]On leave of absence from Facultad de Ciencias Físicas, Universidad Nacional Mayor de San Marcos, Peru.

[1]. In tropical regions, on the other hand, the characteristic vegetation in tidal flats are mangrove perennial forests, which do not present drastic physiological seasonal changes, and iron geochemistry is less understood. Mangrove roots are able to oxidize the surrounding sediments [2], acting together with bioturbation to produce a remarkable spatial heterogeneity [3].

This Mössbauer study is an attempt to add more detailed information to the classical geochemical approach being currently used to identify the main iron compounds involved in the pyrite formation and in its alteration in the mangrove environment.

EXPERIMENTAL

The sediment was sampled during dry (samples A, B and C) and wet (samples D, E and F) seasons, each one representing a series (dry or wet season) of samples, in different depths, from the top down to 55 cm, in a sampling pit dug out in the Gargaú mangrove, located at the estuary of the Paraíba do Sul river (21° 36' 09'' S and 41° 02' 59'' W), state of Rio de Janeiro, Brazil. A set of freeze-dried samples was used as Mössbauer absorbers. Spectra were recorded at room temperature (RT, 298 K) and at liquid helium temperature (4.2 K) with a ^{57}Co/Rh source in a conventional transmission set up.

RESULTS AND DISCUSSION

RT and 4.2 K Mössbauer spectra of pairs of samples collected at 0 (top) - 5 cm (samples A and D), 20 - 25 cm (B and E) and 50 – 55 cm (C and F) exhibit spectral features of three iron-bearing minerals: pyrite (ideal chemical formula, FeS_2) in all samples, and szmolnokite ($FeSO_4$), and iron (hydr)oxydes (mainly goethite, αFeOOH), in the intermediate and uppermost samples. As no significant spectral difference was found between corresponding Mössbauer spectra from series of samples collected during dry (samples A, B and C) and wet (samples D, E and F) seasons, the present discussion will be mainly focused on results of samples from the dry series (Figure 1).

A common feature of all Mössbauer spectra of the two series of samples is that the characteristic pyrite spectrum (typical fitted RT parameters: isomer shift relative to the αFe, $\delta = 0.31$ mm s^{-1} and quadrupole splitting, $\Delta = 0.58$ mm s^{-1}; at

4.2 K, $\delta = 0.28$ mm s^{-1} and $\Delta = 0.61$ mm s^{-1}) is always present irrespective of the sampling depth, although its relative proportion to the whole spectral area decreases from about 80 % for the two deepest samples to about 30 % for the uppermost sample; ferrous sulfate (RT, $\delta = 1.05$ mm s^{-1} and $\Delta = 2.55$ mm s^{-1}; 4.2 K, $\delta = 1.142$ mm s^{-1} and $\Delta = 2.96$ mm s^{-1}) and goethite appear in all except in the deepest sample. Goethite is assumed to occur in very fine superparamagnetic particles so that no magnetic structure is observed in RT spectra. Spectrum obtained at 4.2 K can be fitted with and independent model of hyperfine field distribution, with fitted $\delta = 0.347$ mm s^{-1} and quadrupole shift, $2\varepsilon_Q = -0.19$ mm s^{-1}. Characteristic maximum probability (P) of the curve $P(B_{hf})$ is observed at hyperfine magnetic field, $B_{hf} = 49.2$ tesla.

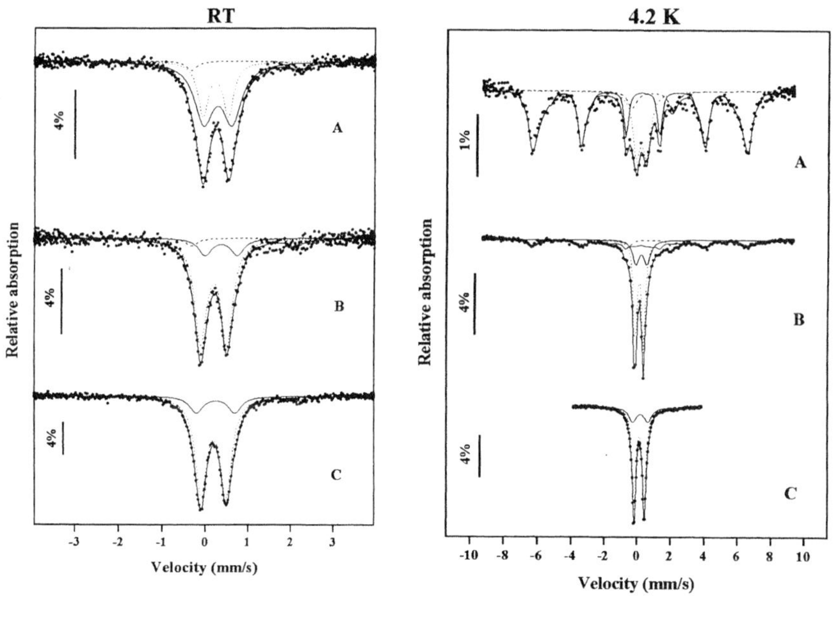

Figure 1. Mössbauer spectra at RT and 4.2K of samples collected during the dry season. Corresponding spectral patterns (not shown) for the wet season samples (D, E and F) series are essentially similar.

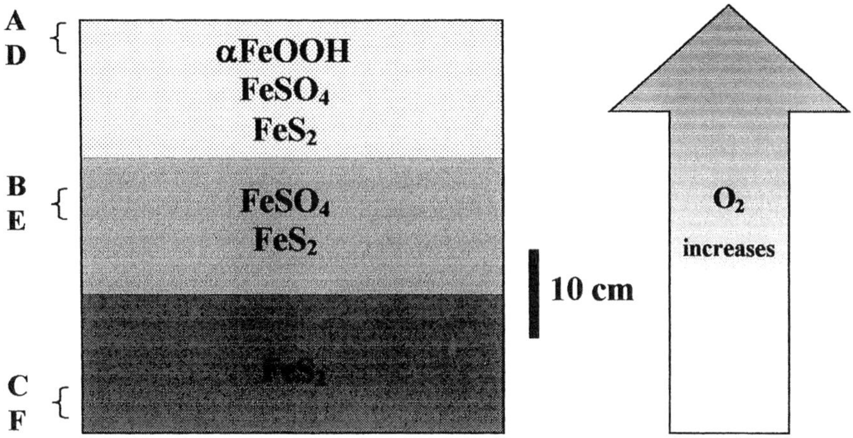

Figure 2. A proposed mechanism for the mineralogical alteration in the mangrove sediments profile. Mainly occurring minerals in each of the three layers are represented (the bar scale indicates approximate depth in the sampling pit) by their ideal chemical formulae: pyrite (FeS_2), szmolnokite ($FeSO_4$) and iron (hydr)oxydes (mainly goethite, αFeOOH). The arrow indicates increasing oxygen concentration.

A model for the biogeochemical reactions in a related system was recently proposed by Yeh et al. [4]. In the present case, the relative subspectral areas (RA) of the 4.2 K Mössbauer spectra strongly suggest that the pyrite found at the bottom sediments is accumulated after continuous formation, initiated at superficial sediments. Iron input to this sedimentary environment is probably dominated by iron (hydro)oxides, considering the large oxygen contents of the fluvial and marine waters. At the sedimentary environment the oxygen consumption during organic mater decomposition results in anoxic conditions and in sulfate reduction. These conditions favor the iron sulfides formation. The oxygen input associated to tidal cycles, root action and bioturbation at top layers can result in pyrite oxidation, releasing SO_3 to form firstly $FeSO_4$ [5] and then iron (hydr)oxides (mainly goethite, αFeOOH) at upper levels under higher oxidation conditions. This pathway is illustrated in Figure 2. The Mössbauer spectrum also reveals that a small amount of hematite (RA < 5 %) does exist in the uppermost sample.

CONCLUSIONS

The present Mössbauer analysis evidences a systematic relative decreasing of the contribution of the low spin ferrous component due to pyrite from deeper levels, the appearance of high spin iron due to ferrous sulfate and ferric (hydr)oxydes at intermediate and upper positions of the mangrove layer. The recently reported biogeochemical model [4], which considers several reaction time scales, need to be somehow extended to describe sulfur oxidation processcs, so to accommodate these information.

ACKNOWLEDGMENTS

We would like to thank financial support from CBPF, CNPq and FAPERJ. Fellowship from CLAF to ABD is gratefully acknowledged.

REFERENCES

[1] J.E. Kostka, G.W. Luther III, Geochim Cosmochim. Acta, 58 (1994) 1701.

[2] L.D. Lacerda, C.E.V. Carvalho, K. Tanizaki-Fonseca, A.R.C. Ovalle, C.E. Rezende, Biotropica 25 (1993) 252.

[3] G.T. Aragon, R.C. Ovalle, J.P. Carmouze, Mangrove and Salt Marshes 59 (1999) 1.

[4] Gour-Tsyh Yeh, William D. Burgos and John M. Zachara, Advances in Environmental Research 5 (2001) 219.

[5] P.G. Manning, J.D.H. Williams and M.N. Charlton, Nature 280 (1979) 134.

MÖSSBAUER SPECTROMETRY TO STUDY DIAGENETICAL PROCESSES OF RED BEDS AND SAND DIKES IN THE VALLÈS-PENEDÈS HALF GRABEN

Ll. Casas[1], *D. Parcerisa*[2], *D. Gómez-Gras*[2], *F. Calvet*[3], *A. Roig*[1] *and E. Molins*[1]

[1]Institut de Ciència de Materials de Barcelona, Bellaterra, Catalonia (Spain),
[2]Dpt. de Geologia,Universitat Autònoma de Barcelona, Bellaterra, Catalonia (Spain),
[3]Dpt. de Geoquímica, Facultat de Geologia, Univ. de Barcelona, Barcelona, Catalonia (Spain).

Abstract The Vallès-Penedès neogene half-graben (NE of Spain) is filled with a thick succession of terrigenous rocks, the lowermost unit of the basin consists of red-coloured clays, sandstones and conglomerates (red beds) formed in an alluvial fan environment. In the upper part of this unit some vertical joints are filled with blue sand (sand dikes). Sand dikes are interpreted as having been formed in an early stage after sedimentation. Despite the colour difference, the common sediment source of both, red-coloured sediments and blue-coloured sand dikes is evidenced by almost identical results obtained by using different techniques: Petrographic microscope, Cameca electron microprobe, XRD and SEM. Mössbauer spectrometry has proven to be a useful technique providing significant data to explain the colour difference and helping to understand the diagenetical processes involved herein.

1.Introduction

Red beds are very profuse in continental sediments around the world. The causes of their red coloration have been discussed for many years [1] but even now are not really understood. Taking this into account, the study of color changes within red beds becomes a true challenge. Many authors assign the red coloration only to iron oxides [2], some others to iron oxides associated with clay minerals [1]. Clay general structure is based on tetrahedral silicate sheets (Si_2O_5) strongly bonded to octahedral aluminium oxide/hydroxide layers forming sets of 2 or 3 layers, interlayers are occupied by water, hydroxyl groups and in general also by metallic cations. Iron may occur in the structure of clay minerals in many different forms [3].

The colour of iron containing minerals is often indicative of the Fe(II)/Fe(III) ratio [4], many of these minerals are pigments, having big influence on the rock colour. In the Vallès-Penedès neogene half graben (NE of Spain) the lowermost unit of the basin consists of red-coloured rocks affected by vertical joints filled with blue sandstones. Despite the colour difference, several analysis techniques show that both materials have the same constituents.

2.Geological considerations

In the Catalan Coastal Ranges (NE Spain) there is an extensional sedimentary basin (Vallès-Penedès graben) neogene in age (Figure 1). It is up to 100 km long and 12 km wide and is filled with a thick succession of terrigenous rocks deposited in continental and transitional paleoenvironments [5]. The lowermost unit of the basin is Aquitanian to Burdigalian in age and consists of lacustrine carbonates and red-coloured clays, sandstones and conglomerates (red beds) formed in an alluvial fan environment. The upper part of the unit is affected by numerous fractures, usually filled by blue sand (blue sand dikes). Sand dikes are vertical walls that cut all the stratigraphic layers. They have a large continuity in outcrop scale and show an irregular outline.

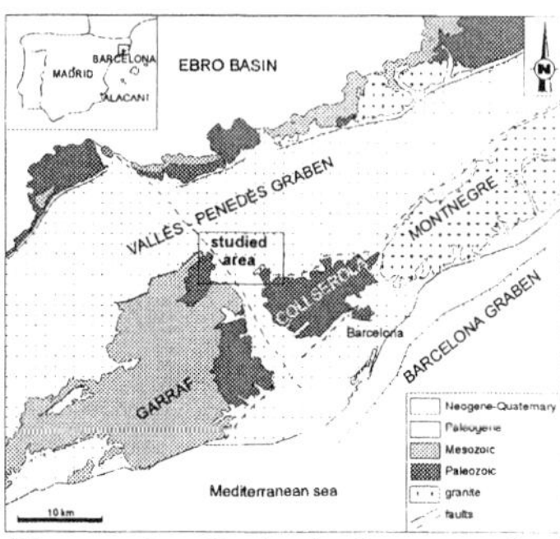

Figure 1. Geological map of the Catalan Coastal Ranges.

Sand dikes fossilise a network of N-S oriented fractures (joints and faults indistinctly). The contact between dikes and confining rocks is sharp without any structure, whereas detrital biotites appear vertically oriented within it. The emplacement of sand dikes is developed in an early stage of the basin's infill because they are affected by compaction phenomena. The orientation of biotites indicates that sand dike injection took place under the water table. Blue coloration of sand dikes and some stratigraphic layers (only at the bottom and/or the top) probably happened in a later stage due to fluid circulation.

Similar structures in other places have been explained as the result of two main mechanisms: i) sand liquefaction by earthquake mechanisms [6] and ii) sudden emplacement of turbiditic sediments in deep-sea fans [7]. In our geological setting (continental deposits in a tectonic active basin) sand dike formation is better explained by the first option.

3.Experimental

One sample of blue sandstone strata and five samples of sand dikes with five adjacent red sandstone strata (red beds) were collected in different sites of the sedimentary basin.

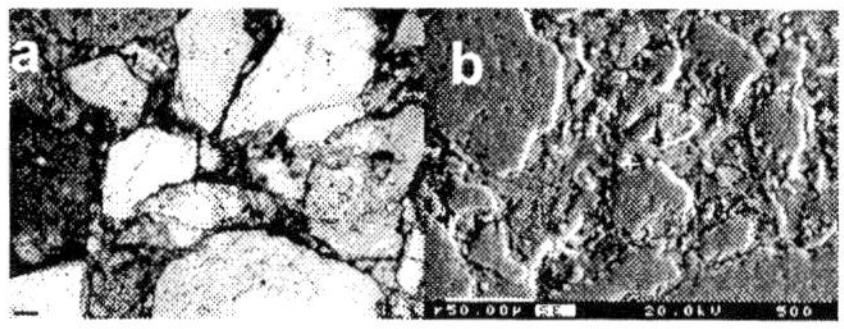

Figure 2. Microphotographs showing: (a) A detail of red beds with the red matrix around grains. Uncrossed polars (right). Scale 100 µm. (b) A backscattered photo of the plate-like particles in the matrix of red beds.

Thin sections of all the samples were analysed with a standard petrographic microscope. Three samples of both sand dikes and red beds were carbon-coated and studied using a Cameca model SX50 microprobe equipped with a backscattered electron detector. Two samples of red beds and one sample of sand dike were carbon-coated and examined with a JEOL JXA-840 scanning electron microscope (SEM).

Two representative samples of red beds and sand dikes were crushed and analysed with a Phillips 1710 X-ray diffractometer, XRD (Kα: λ=1.54060 Å) and with an X-ray fluorescence spectrometer (XRF). In the same samples the < 2 µm fraction was separated and transferred onto a glass slide to analyse the clay mineralogy using the Oriented Aggregates Diffraction analysis (OAD) [8]. The ^{57}Fe Mössbauer spectra were measured in a conventional transmission Mössbauer spectrometer, operating in a constant acceleration mode, with a ^{57}Co source in a Rh matrix. Spectra were recorded on all samples at 300K; some of the samples were also measured at low temperature (80 K and 70K).

4. Results

Petrographic microscope analysis of thin sections reveals that red beds and sand dikes have the same detritic and diagenetic components. They are fine to coarse-grained, moderately sorted litharenites made up of quartz, feldspar, phyllite and limestone fragments. The matrix (2 to 12 %) is formed by clay minerals and the cement is always Mn-rich calcite. The sole difference found is that the matrix appears red coloured for red bed samples (Figure 2a) whereas is not coloured for the blue samples. SEM images show plate-like morphologies (clays) for the crystals that form the matrix (Figure 2b).

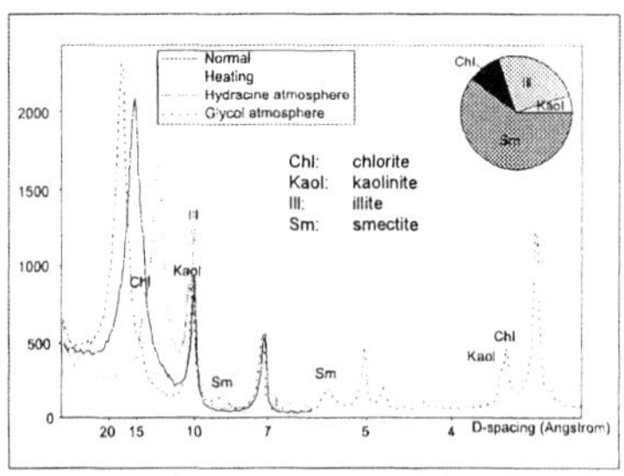

Figure 3 OAD pattern of the clay fraction of sand dikes and red beds

XRD patters show the same mineral bulk composition for red bed and sand dikes (not shown). The main minerals present are quartz, calcite, clays and traces of feldspar. X-ray analysis on the clay fraction (OAD) shows invariably the same clay mineralogy (Figure 3), the dominant clay mineral is smectite but there is also illite, chlorite and kaolinite. Some differences are revealed looking at XRF data, the main compounds of the whole rock sum up the same total percentage for both kinds of rocks but sand dikes exhibits lower Fe and Mg content and higher Si content (Table 1).

Table 1.-X-Ray Fluorescence (XRF) data for red bed and sand dike samples.

	Fe_2O_3	MnO	TiO_2	CaO	K_2O	P_2O_5	SiO_2	Al_2O_3	MgO	Na_2O	Total
Red beds	4.5	0.1	0.7	9.6	2.5	0.2	57.7	11.1	2.0	0.7	89.0
Sand dikes	2.5	0.1	0.5	10.1	2.7	0.1	60.8	10.1	1.4	0.7	89.0

Table 2.-Hyperfine parameters for fitted Mössbauer spectra.

	Γ (mm/s)	Area (%)	δ_{Fe} (mm/s)	$\Delta / 2\varepsilon$ (mm/s)	B_{hf} (T)
Red beds*					
Fe^{3+} doublet	0.70(10)	depends on the sample	0.44(2)	0.71(4)	--
Fe^{2+} doublet	0.43(3)		1.18(1)	2.51(3)	--
Fe_2O_3 sextet	0.70(15)		0.35(5)	-0.10(10)	50.1(5)
Sand dikes					
Fe^{3+} doublet	0.50(10)	55(3)	0.44(4)	0.66(3)	--
Fe^{2+} doublet	0.40(3)	45(5)	1.17(3)	2.53(4)	--
*No sextet is observed in blue samples at the top and/or bottom of some beds					

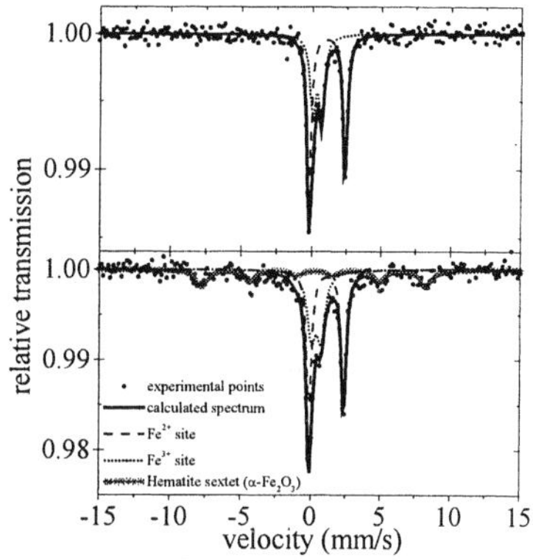

Figure 4.-^{57}Fe Mössbauer spectra taken at 300 K for red bed samples (bottom) and sand dike samples (top).

Mössbauer spectra also show differences between red beds and sand dikes (Figure 4). In both there are two partially overlapped doublets with very different quadrupolar splitting. Hyperfine values of doublets belongs to octahedral Fe(III) and Fe(II) and surely correspond to iron in a clay structure (smectite) [9]. Red samples shows an extra subspectrum consisting of a sextet with typical hyperfine values of hematite, α-Fe_2O_3, (Table 2). However, hematite is not observed in XRD patters. Two facts (Table 2) reveals the presence of superparamagnetic hematite: i) the increase of sextet area at low temperature, ii) the enlargement of the linewith and quadrupolar spliting of Fe(III) site with respect to the same site in sand dike samples, this enlargement could be due to the overlapping with a superparamagnetic doublet.

5. Discussion

Although there are similarities between blue sand dikes and red beds, some important differences exist, as a simple visual or petrographical test shows (Figure 2). However, the main part of analytical techniques cannot give significant information to know the exact causes of this difference, XRD and OAD analyses (Figure 3) reflect the same composition in the whole rock and in the lutitic fraction, respectively. XRF analyses (Table 1) reveal a higher percentage of Fe and Mg in red beds; but this technique does not provide any information about oxidation state and mineralogical location of these cations. Without looking at the Mössbauer data we can state that the responsible for the colour difference is basically Fe and that this Fe belongs to a clay matrix inside the sandstone. There are four possible sites for Fe related to the clay [3]: i)Tetrahedral sites in silica sheets, ii)Octahedral sites between silica sheets, iii)Dodecahedral interlayer sites and iv)Being absorbed at the edges of clay particles as Fe oxide/hydroxide. Mössbauer data indicate that Fe is located in position ii) for all samples and also in position iv) for red beds. The set of results points to a thin layer of Fe oxide absorbed around the clay particles as the responsible for the red colour, magnetic behaviour of this oxide (with a fraction being superparamagnetic at room temperature) suggests a nanometric size for this layer, although we cannot exclude the existance of separate ultrafine particles. The blue colour of sand dikes might be due to the grey-blue coloration of Fe (II)-bearing smectites.

In blue sand dikes X-Ray Fluorescence data show a decrease in some mobile cations, such as Fe and Mg (Table 1). This fact suggests that leaching phenomena are involved in this colour difference. To dissolve Fe from hematite reducing conditions are required, after dissolution, fluid circulation can leach the Fe. In our geological context sand dikes are vertical structures that connect different permeable layers (sandstone and conglomerate strata), hence they probably act as conduits for fluid flow. All these evidences points to a strong fluid circulation through sand dikes in reducing conditions as the main agent for the disappearance of hematite (=colour change). On the other hand, fluid flow through sandstone and conglomerate red beds is restricted by the presence of numerous interstratified impermeable lutitic layers being not enough to reduce and leach the Fe. Blue colorations on the top and/or the bottom of some beds are perhaps related with the existence of organic matter, which promotes the development of reducing conditions. Mössbauer data supports this idea of leaching in more restricted conditions because blue sandstone strata exhibit remnants of Fe oxides only in a superparamagnetic form (Table 2).

6.Conclusions

Common origin of both kinds of rocks is evidenced by the same mineralogical composition; even the clay composition is identical. The colour difference of sandstones and sand dikes is due to the presence of hematite (α-Fe_2O_3) in the sandstones. This iron oxide is associated to the clay particles of the matrix forming a nanometric layer around them.

Blue sandstone strata and sand dikes should have lost the iron oxide by reduction and leaching processes. In blue sandstones some traces of iron oxide are found by Mössbauer spectrometry in form of a superparamagnetic doublet. Leaching is also evidenced by XRF since the content of mobile cations (Fe and Mg) is lower for sand dikes.

Mössbauer spectrometry is a good technique to solve some geological questions, mainly those related with Fe-bearing minerals, when a question arises from all the analytical tools sometimes Mössbauer is successful where other techniques fail.

References

[1] F.B. Van Houten, Annual Review of Earth and Planetary Sciences 1 (1973) 39-61.
[2] J. Torrent and U. Schwertmann, J. Sed. Pet. 57, 4 (1987) 682-686.
[3] B.A. Goodman in *Advanced Techniques for Clay Minerals Analysis*, eds. J.J. Fripiat (Elsevier, Amsterdam-Oxford-New York, 1982) 113-137.
[4] P. Komadel, P.R. Lear and J.W. Stucki, Clay Clays Miner. 38, 2 (1990) 203-208.
[5] L. Cabrera and F. Calvet in *Tertiary Basins of Spain,* eds. P.T. Friend and C. Dabrio (Cambridge University Press, 1996) 97-105.
[6] M. A. Rodríguez-Pascua, J.P. Calvo, G. De Vicente and D. Gómez-Gras, Sedimentary Geology 135 (2000) 117-135.
[7] O. Parize, *Sills et dykes gréseux sédimentaires:paléomorphologie, fracturation précoce, injection et compaction.* Mem. Sc. de la Terre 7 (E. N. S. M. P., Paris, 1988) 335 p.
[8] G. Brown, *The X-ray identification and crystal structure of clay minerals* (Mineralogical Society, London, 1972) 544 p.
[9] L. Heller-Kallai and I. Rozenson, Phys. Chem. Minerals 7 (1981) 223-238.

Aknowledgements This work has been founded by the Dpt. per Universitats i Recerca de la Generalitat de Catalunya (1999-SGR-00205, 2000-SGR-00030, project 08831996-SGR-00086). Mr. Thiry of the École del Mines de Paris (CIG) is also gratefully acknowledge for diffraction analyses and interpretations.

MÖSSBAUER INVESTIGATION OF BIOLOGICALLY-INDUCED MINERALIZATION PROCESSES

N.I. Chistyakova, V.S. Rusakov, D.G. Zavarzina

Lomonosov Moscow State University, Vorobyevy gory, 119899 Moscow, Russia

Mössbauer investigations of the kinetics of iron-containing mineral formation in controlled conditions by thermophilic Fe(III)-reducing bacteria were carried out. Anaerobic bacterium (strain Z-0001) was isolated from the ferric sediments of hot springs Stolbovskie, Kunashir Island. This microorganism used the amorphous Fe(III)-oxide as an electron acceptor and acetate CH_3COO^- as an electron donor. Probes of solid phase processed by bacteria were sampled at different time intervals for the cultures incubated with 20 and 100% of CO_2 in gas phase and studied by Mössbauer spectroscopy. The influence of physical and chemical conditions on the mineral phase formation was also investigated.

Keywords: Mössbauer spectroscopy, Fe(III)-reducing bacteria, extracellular precipitation, magnetite Fe_3O_4, siderite $FeCO_3$.

A microbiological pathway of mineral formation has been intensively studied last decade. One of the possible ways of iron-containing mineral formation is their extracellular precipitation induced by dissimilatory Fe(III)-reducing bacteria first described by D. Lovely in 1988 [1]. In our study Mössbauer investigations of the kinetics of iron-containing mineral formation by thermophilic Fe(III)-reducing bacteria were carried out. Thermophilic anaerobic strain Z-0001 was isolated from the ferric sediments of hot springs Stolbovskie ($T=60\div65^0C$ and $pH=6.8\div7.0$), Kunashir Island. For energy donating process, needed for the anaerobic growth, this bacterium uses Fe(III)-oxide reduction. In our study synthesized amorphous Fe(III)-oxide (90 mmol/l of Fe(III)) was used by strain Z-0001 as an electron acceptor and acetate (CH_3COO^-) (20 mM) as an electron donor.

The Mössbauer study of solid phase samples that were incubated by bacteria in different time intervals for two series with 20% (series I) and 100% (series II) CO_2 in gas phase were carried out at room temperature. Sterile medium (SM) as well as medium, inoculated by strain Z -0001 and afterwards sterilized at 120^0C (IM) was used as controls. The Mössbauer spectra were processed by use of the program DISTRI [2] from MSTools Program Complex for simultaneous restoration of several distribution functions of hyperfine parameters for different partial spectra.

Mössbauer investigations (for series I and II) demonstrated that magnetite (Fe_3O_4) was the only solid magneto-ordered phase that was formed during the growth of strain Z-0001 [3]. The complete reduction of amorphous Fe(III)-hydroxide was revealed on the 5th day of the experiment and corresponded to the beginning of stationary phase of microbial growth. For series II (3.5-day probe) we observed the formation of siderite ($FeCO_3$), which was not formed in series I. These results could be explained by different conditions for the formation of magnetite because of different content of CO_2 in gas phase. Mössbauer spectra of both controls - SM and IM were similar to the initial amorphous Fe(III)-oxide for all variants thus demonstrating that under used conditions (T 60 ^{0}C, pH 7.0) no abiogenic changes occur up 120-days of incubation time.

Influence of physical and chemical condition on biogenic formation of iron-containing minerals was demonstrated by comparison of the Mössbauer spectra for 14-day incubation samples, which corresponds to the stationary phase of microbial growth. Gas phase composition and liquid-solid phase ratio was variable (figures 1 and 2). In series I with 20% of CO_2 and series II with 100% CO_2 in gas phase we used standard quantity of 90 mmol/l of Fe(III) in solid phase with 20 ml liquid and 40 ml gas phase per flask. In series III quantity of Fe(III)-oxide was two time more than normal; normal; 0.5 and 0.25 of normal. The influence of CO_2 content was demonstrated in series IV, where the quantity of gas phase was increased from 40 ml to 180 ml with content of CO_2 equal to 20% and 100% (see figure 2).

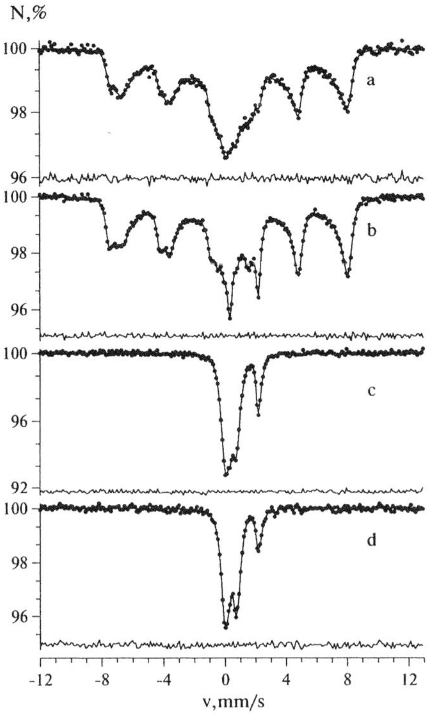

Figure 1. Mössbauer spectra of samples (series III) produced with 20% of CO_2 in gas phase with different quantity of amorphous Fe(III)-oxide: 2 time more than normal (a), normal (b), 0.5 of normal (c), 0.25 of normal (d).

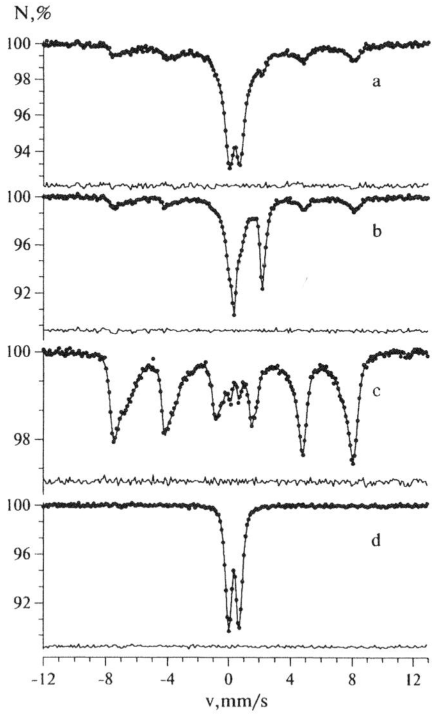

Figure 2. Mössbauer spectra of 14-days probes, obtained for normal solid-liquid ratio in case of 20% (a) and 100% (b) of CO_2 in gas phase (series IV), the spectra of sample from series I with 20% of CO_2 (c) and SM control (d).

The increase of more reduced phase (siderite) content with decreasing the initial Fe(III)-oxide concentration is observed (figure 1). When quantity of Fe(III)-oxide was normal or 2 time more than normal (figures 1(a) and 1(b)) the production of magnetite formed increased and remaining content of amorphous Fe(III)-oxide was about 10%. Practically we didn't observed the formation of magnetite when quantity of Fe(III)-oxide was less then normal (figures 1(c) and 1(d)). In these cases siderite was formed as the only reduced mineral phase and content of unreduced Fe(III)-oxide was large enough (about 60%).

The increase of siderite formation was observed in series IV with elevated content of CO_2 in gas phase from 10% (for 20% CO_2) to 41% (for 100% CO_2) (figure 2). In pure CO_2 atmosphere the content of unreduced amorphous oxide was appreciably less. The comparison of Mössbauer spectra of 14-day probes (figures 2(a) and 2(c)) both produced with 20% CO_2 in gas phase demonstrated that the increase of the gas-liquid phase ratio results in incomplete Fe(III)-oxide reduction. The Mössbauer spectrum of control probe (SM) for 20% CO_2 in gas mixture presented in figure 2(d). This spectrum is similar to the initial amorphous oxide.

Thus, reduction of amorphous Fe(III)- oxide is the result of vital functions of thermophilic Fe(III)-reducing bacteria and this process does not occurs without living cells activity. Magnetite (Fe_3O_4) is the only solid magneto-ordered phase that is formed during the growth of strain Z-0001. The physical and chemical conditions influence on the composition of solid phases formed. The increase of CO_2 pressure results in the predominant formation of Fe^{2+} carbonate – siderite ($FeCO_3$). When the amount of initial amorphous Fe(III)-oxide is decreased, the complete reduction of oxide and the increase of more reduced phase siderite content occur. Anaerobic bacteria (strain Z-0001) reduce Fe^{3+} to Fe^{2+} but the formation of predominant mineral phase depends on the physical and chemical conditions and, thus, is determined by abiogenetic parameters.

The present work was supported by RFBR, Grant 99-05-65267.

REFERENCES

[1] D.R. Lovley and E.J.P. Phillips, Appl. Environ. Microbiol. 54 (1988) 1472.

[2] V.S. Rusakov, Izv. RAN. Ser. Phys. 63 (1999) 1380 (in russian).

[3] N.I. Chistyakova, V.S. Rusakov, D.G.Zavarzina, and S.V. Kozerenko, in: Abstr. Euro-Asian Symp. "Trends in Magnetism", Ekaterinburg, 2001, p.375.

Two magnetic transitions at lower temperature in $(Fe_{1-x}Me_x)_2SiS_4$-thioolivines, Me = Mg or Mn

Tore Ericsson[a], Örjan Amcoff[a] and Per Nordblad[b]

[a]) Department of Earth Sciences, Villavägen 16, SE-75236 Uppsala, Sweden

[b]) Department of Materials Science, Box 534, SE-75121 Uppsala, Sweden

abstract. $(Fe_{1-x}Me_x)_2SiS_4$, Me = Mg and x = 0.1, 0.2, 0.3 or Me = Mn and x = 0.1, 0.2, having the orthorhombic *Pnma* structure, has been synthetisized using quartz-ampoule techniques. The samples have been studied using transmission Mössbauer spectroscopy down to ≈ 10 K and magnetization measurements (SQUID) from 5 K to room temperature. The samples show two magnetic transitions at lower temperature. Below 126 K there is a PM - AFM transition with decreasing T_N for increasing x, more depressed for Mg than for Mn. At ≈ 30 K there is an AFM - FiM transition (T_C), also sensitive to the composition. However, while it is depressed and smeared out by Mg-substitution, it is somewhat increased by Mn-substitution, to ≈ 40 K for x = 0.2.

Introduction. The olivine structure, orthorhombic *Pnma*, is quite common in nature. Ordinary olivine, $(Fe, Mg)_2SiO_4$, with the Fe-endmember fayalite and the Mg- endmember forsterite, is an important mineral in the crust and upper mantle of the Earth. Fe_2SiO_4 has a PM - AFM transition at T_N = 66 K and a spin-canting at ≈ 23 K [1]. Sarcopside, $Fe_3(PO_4)_2$, has a related structure to olivine, containing ordered vacancies, with T_N = 44 K. In the AFM-state 1/3 of the iron atoms are non-magnetic for T > 2.4 K, but below there are indications of a second magnetic transition involving all Fe atoms [2]. Both Fe_2SiS_4 and Fe_2GeS_4 have the olivine structure. In these two compounds there is also at lower temperatures T_C a second transition (AFM to FiM). In all compounds other 3d-elements can substitute iron to substantial degrees. We have now studied isostructural $(Fe, Me)_2SiS_4$, with Me = Mg or Mn with the aim to compare especially its magnetic properties with related compounds.

Crystal structure. Iron occupies two sites of equal occurence in the orthorhombic *Pnma* structure, M1 having an inversion center and M2 possessing a mirror-plane. In Fe_2SiS_4, a = 12.405 Å, b = 7.200 Å, c = 5.813 Å and the M1 octahedron is somewhat smaller compared to the M2 octahedron [3]. The unit cell increases with substitution of iron by Mg [4] or Mn [5], but decreases with substitution by Ni or Co [6]. Mg and Mn prefer the M2 site while Ni and Co both prefer the M1 site.

Results and discussion. At room temperature Mössbauer PM-spectra show two sharp doublets, of which the one with lower centroid shift CS (= isomer shift + second order Dopplershift) and higher magnitude of quadrupole splitting |QS| is assigned to iron at M1. Also in the AFM region two main patterns, representing iron at M1 and M2, are needed for reasonable fits. However, the spectra are broadened due to the metal substitutions, especially in the Fe/Mg-case, indicating distributions of hyperfine fields, see Figure 1.

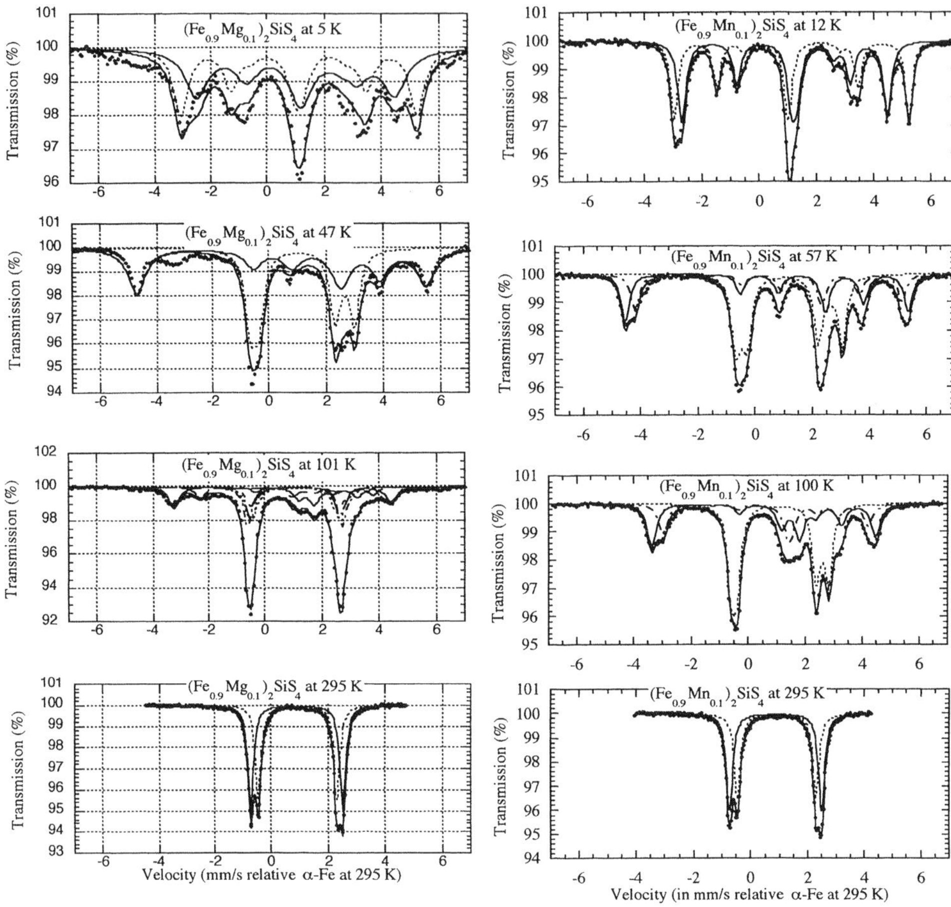

Figure 1. Mössbauer spectra of $(Fe, Mg)_2SiS_4$, left column and $(Fe, Mn)_2SiS_4$, right column recorded at room and lower temperatures.

The Mössbauer spectra have to be fitted using the full hamiltonian. At low temperature (≈ 30 K) there is an AFM - FiM transition, clearly seen in the spectra. QS has been defined in the same way in all PM, AFM and FiM spectra:

$$QS = \frac{eQV_{zz}}{2}\sqrt{1+\frac{\eta^2}{3}}$$

Due to the distributions in hyperfine fields in the AFM region, two Lorentzian subpatterns have been used to fit iron at M1 in the Mn-case with a difference of ≈ 2 T in magnetic fields. In the FiM region the spread in hyperfine fields are less in the Fe/Mn-case, but still substantial in the Fe/Mg-case. MS-parameters for $(Fe_{0.9}Mn_{0.1})_2SiS_4$ are shown in Figure 2.

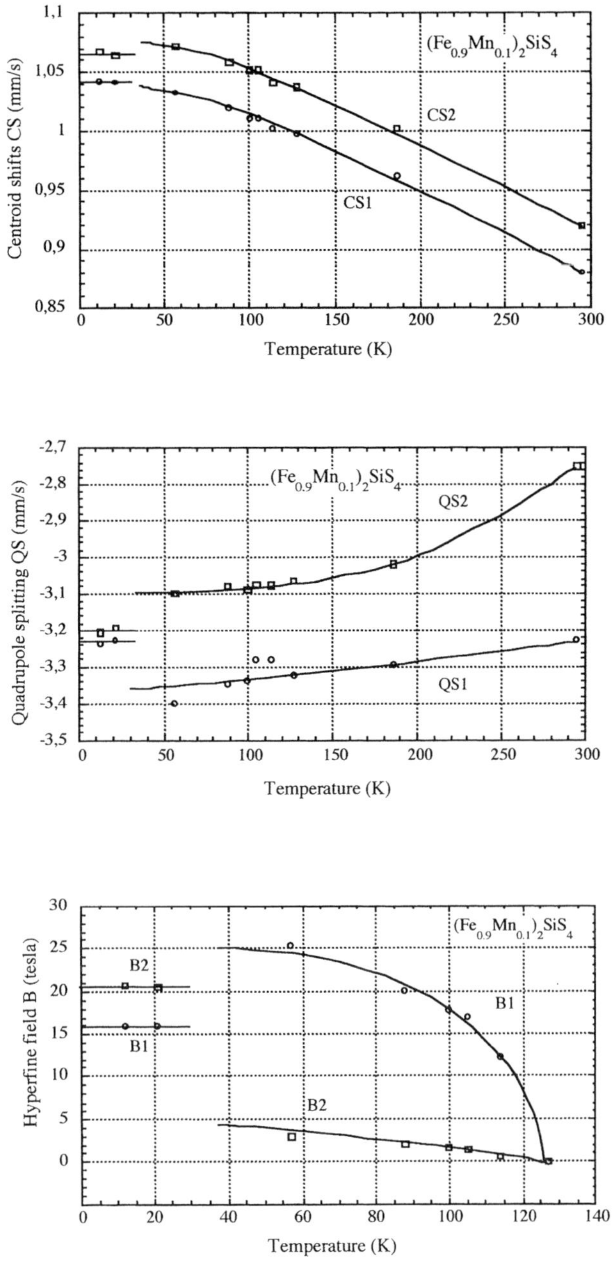

Figure 2: CS, relative α -Fe at room temperature, QS, as defined above, and hyperfine fields B1 (highest component for M1) and B2 for M2 vs temperature for $(Fe_{0.9}Mn_{0.1})_2SiS_4$. The lines are just guides for the eye.

The highest M1- and M2- hyperfine field values are, roughly, not affected by the Mn/Fe- or Mg/Fe-substitutions. Incorporation of 20 % and higher of Mg result in such broad spectra that it is of no use to fit them using the simple model above. Measurements were performed on powder samples using also a SQUID magnetometer at increasing temperatures, from 5 K to room temperature on zero field cooled (ZFC) and field cooled (FC) samples. T_N is given by marked kinks and increased irreversibility between these curves, T_C by the dramatic increase in magnetisation at $T < T_C$. Estimated values from MS and SQUID measurements (not shown) for $(Fe_{0.75}Me_{0.25})_2SiS_4$ are T_N = 155 K/ 128 K/ 110 K/ 55 K/ and T_C = < 5 K/ < 5 K/ ≈ 45 K/ coexisting AFM-FiM phases at 12 K/ for Me = Ni [6]/ Co [6]/Mn /Mg respectively (Fe_2SiS_4 has T_N = 126 K and T_C = 32 K [6]). Thus, there does not seem to be any clear relation between T_N and T_C, nevertheless two magnetic transitions of related types show up in all compared compounds. It is also interesting to note that $FeMgGeS_4$ has T_N well above 88 K, however in this compound Mg prefers the M1 site [8], and not the M2 site as Mg prefers in $(Fe,Mg)_2SiS_4$.

References

1. Lottermoser, W., et al., Powder- and single crystal Mössbauer spectroscopy on synthetic fayalite. Chem. Minerals, 1995. 22: p. 259-267.
2. Ericsson, T. and F. Khangi, An investigation of $Fe_3(PO_4)_2$-sarcopside between 1.6 K - 721 K: comparison with fayalite. Hyp. Int., 1988. 40: p. 783.
3. Baron, V., Ö. Amcoff, and T. Ericsson, Neutron powder diffraction study of the crystal and magnetic structure of Fe_2SiS_4. J. Magn. Magn. Mater., 1999. 195: p. 81-92.
4. Ericsson, T., Ö. Amcoff, and M. Kalinowski, Cation preferences in thio-olivines $(Fe_{1-x}Mg_x)_2SiS_4$, $x \leq 0.30$, studied by Mössbauer spectroscopy at room temperature. N. Jb. Miner. Mh., 1999. H. 11: p. 518-528.
5. Ericsson, T. and Ö. Amcoff, Cation preferences in thio-olivines $(Fe_{1-x}Mn_x)_2SiS_4$, $x \leq 0.50$, studied by Mössbauer spectroscopy at room temperature. N. Jb. Miner. Mh., 2001. H. 9: p 403-413.
6. Nordblad, P., Ö. Amcoff, and T. Ericsson, Magnetisation and Mössbauer studies of $(Fe_{1-x}Me_x)_2SiS_4$, $x \leq 0.25$ and Me = Co or Ni. Phys. Scr., 1999. 60: p. 373-380.
7. Ericsson, T., K. Holényi, and Ö. Amcoff, A Mössbauer study of Fe_2SiS_4, having the olivine structure: comparison with fayalite and related minerals. J. Phys. Cond. Matter, 1997. 9: p. 3943-3953.
8. Ericsson, T. and A.G. Nord, Cation partitioning in olivine-related $(Fe, M)_2GeS_4$ thiogermanates, M = Mg, Mn, Zn. N. Jb. Miner. Mh., 1995. H. 5: p. 202-210.

A Mössbauer Study of Meteorites
– A possible Criterion to Identify Meteorites from the same Parent Body?

S.D. Forder[1], P.A.Bland[2], J.Gałązka-Friedman[3], M.Urbański[3], Z. Gontarz[4], M. Milczarek[4], N. Bakun-Czubarow[5]

[1] *Materials Research Institute, Sheffield Hallam University, Sheffield, S1 1WB, U.K.*
[2] *Planetary Space and Sciences Research Institute, Open University, Milton Keynes, MK7 6AA, U.K.*
[3] *Faculty of Physics, Warsaw University of Technology, 75 Koszykowa Str., 00 662 Warsaw, Poland*
[4] *Faculty of Chemistry, Warsaw University of Technology, 3 Noakowskiego Str., 00 664 Warsaw*
[5] *Institute of Geological Sciences, Polish Academy of Sciences, 51/55 Twarda Str., 00 818 Warsaw*

Nine ordinary chondrites and synthetic troilite were investigated to test the hypothesis that the Mössbauer parameters of troilite could serve as a suitable criterion to establish whether various meteorites come from the same parent body. It was noticed that in five meteorites (four falls and one find) the coefficient of asymmetry is rather small ($\eta \leq 0.2$) and that angles ϕ and θ have similar values ($\phi < 7°$, $\theta \cong 49°$). In the other group of four meteorites (finds) η has a larger value, between 0.33 and 0.74. This spread of Mössbauer parameters for troilite raises the question of whether this is caused by different conditions of crystallisation in the parent body or by processes of alteration on the Earth's surface.

1. Introduction

It has been suggested that perhaps the Mössbauer parameters of troilite could serve as a criterion for whether meteorites originate from the same parent body [1]. This hypothesis was created because the parameters describing the quadrupole and magnetic interactions depend on the number of vacancies and the level of impurities in troilite and the rate of cooling of the material [2,3] and are characteristic of the conditions of crystallisation of the parent body of the meteorites (asteroids). To investigate this hypothesis nine meteorites have been studied, five found in Antarctica in the vicinity of Allan Hills (ALH88002 (L4), ALH88003 (L4), ALH88017 (H4), ALH88020 (H3.5) and ALH88021 (H5)). The other four are falls (Barwell (L6), Colby (L6), Farmington (L5) and Ohuma (L5)).

In fitting the subspectrum of troilite it has been assumed that the asymmetry parameter, η, is non-zero. This requires the introduction of two angles, ϕ and θ (the azimuthal and polar angles of H_{hf} in the EFG tensor principal axes coordinate system [4]), in order to determine the relationship between the direction of the magnetic field and the distribution of the electric field. Consequently the problem of 'non-uniqueness' of the Mössbauer fit has emerged [5].

Synthetic troilite was also analysed because the Mössbauer parameters are sensitive to the exact composition and structure of the FeS.

2. Experimental procedure

The preparation of the meteoritic samples, the synthesis of the iron sulphides and details of the Mössbauer experiments have been published previously [6]. All the Mössbauer spectra were fitted using Recoil software, with Lorentzian lines, using the full static Hamiltonian site analysis [7]. The non-troilite parameters were fixed, and three different methods were used to fit the troilite, allowing for asymmetry within the troilite.

3. Results and Discussion

In all the meteorites studied, four mineralogical phases (olivine CS = 1.13 mm/s, $e^2qQ/2$ = 2.13 mm/s, $\Gamma/2$ = 0.19 mm/s, pyroxene CS = 1.15 mm/s, $e^2qQ/2$ = 2.97 mm/s, $\Gamma/2$ = 0.18 mm/s, kamacite CS = -0.004 mm/s, H = 33.4 T, $e^2qQ/2$ = -0.004 mm/s, $\Gamma/2$ = 0.21 mm/s and troilite CS = 0.75 mm/s, H = 30.8 T, $e^2qQ/2$ = -0.75 mm/s, $\Gamma/2$ = 0.21 mm/s, η= 0.45, ϕ = 43° and θ = 48°) and ferric iron (CS = 0.38 mm/s, $e^2qQ/2$ = 0.71 mm/s, $\Gamma/2$ = 0.33 mm/s) were identified. These Mössbauer parameters are for meteorite ALH88017, figure 1.

The subspectra of troilite were fitted with η not equal to zero and the two angles, ϕ and θ. For each meteorite three sets of Mössbauer parameters were found with equally low values of χ^2 (table 1). Attempts were made to try to identify the correct solution, taking into consideration crystallographic structure. In ideal FeS axial symmetry exists [2] and η = 0. In natural troilite a small value of η can be expected. Five of the meteorites studied (Barwell, Colby, Farmington, Ohuma and ALH88003) possess a solution with $\eta \leq 0.2$, connected with a small value of ϕ, and a value of θ of approximately 49 ° (values in bold).

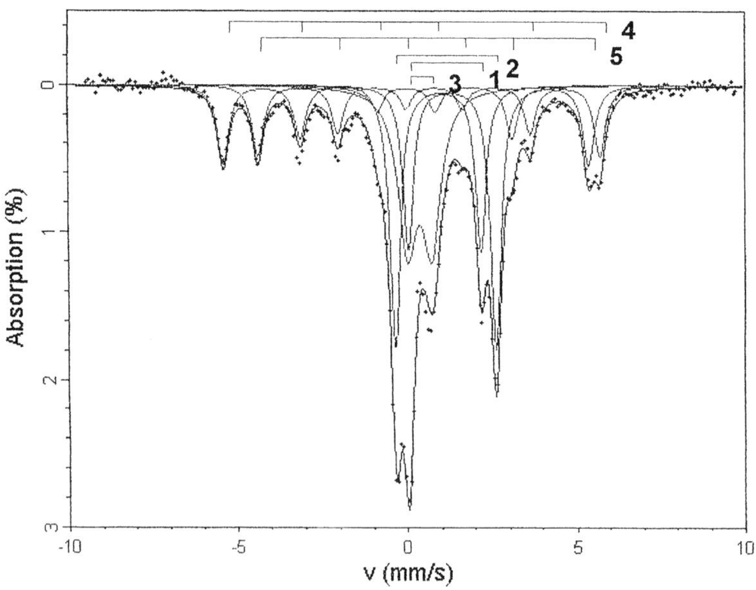

Figure 1 Mössbauer spectrum of ALH88017;
Subspectra 1 = olivine , 2 = pyroxene , 3 = Fe 3+, 4 = kamacite , 5 = troilite

Table 1. A comparison of the parameters of troilite within meteorites, fitted with different starting parameters for η, ϕ and θ

a δ and H fixed, then free. η = 0.01, ϕ = 1°, θ = 1°, ΔE_Q free with all other parameters
b η = 0.5, ϕ = 45°, θ = 45°, δ and H fixed, then free. ΔE_Q free with all other parameters
c η = 0.5, ϕ = 45°, θ = 45°, δ and H free with all other parameters from start of fit.

Name		χ^2	δ (mm s^{-1})	H (T)	ΔE_Q (mm s^{-1})	η	Γ/2 (mm s^{-1})	Area (%)	ϕ (°)	θ (°)
Barwell	**a**	0.799826	0.745(9)	31.0(2)	-0.74(5)	**0.19**(7)	0.161(8)	15.22	**5.7**(42)	**47.5**(7)
	b	0.799893	0.745(8)	31.0(1)	-0.66(5)	0.85(31)	0.160(6)	15.19	41.8(50)	45.38(7)
	c	0.804762	0.744(7)	31.0(2)	-0.66(4)	0.71(31)	0.162(7)	15.24	138(3)	45.3(5)
Colby	**a**	1.06879	0.750(7)	30.9(2)	-0.79(3)	**0.17**(3)	0.158(3)	22.4	**0.7**(0.9)	**48.8**(5)
	b	1.06897	0.750(7)	30.9(1)	-0.73(3)	0.68(24)	0.158(4)	22.4	39.9(18)	47.6(6)
	c	1.16212	0.751(7)	31.1(1)	-0.57(10)	0.046(86)	0.165(9)	22.8	180(88)	44.5(11)
Farmington	**a**	1.02209	0.748(10)	30.77(17)	-0.88(14)	**0.20**(13)	0.188(10)	15.36	**2.4**(5.3)	**50.0**(24)
	b	1.04779	0.75(1)	30.76(16)	-0.777 (86)	0.49(24)	0.19 (1)	15.35	40.0(35)	48.7(18)
	c	1.02893	0.749(11)	30.85(15)	-0.766(85)	0.07(25)	0.19(1)	15.35	120(37)	46.5(19)
Ohuma	**a**	1.10721	0.751(7)	30.96(19)	-0.765(37)	**0.16**(6)	0.149(4)	18.89	**3.0**(48)	**48.9**(7)
	b	1.10876	0.751(9)	31.00(25)	-0.684(24)	0.22(17)	0.149(5)	18.9	51.4(46)	45.8(7)
	c	1.10879	0.753(8)	30.98(18)	-0.687(41)	0.68(17)	0.149(4)	18.9	137.3(26)	46.8(4)
ALH88002	a	0.615421	0.736(11)	30.2(3)	-1.42(14)	**0.67**(11)	0.206(13)	16.2	**11.6**(19)	**58.4**(23)
	b	0.663006	0.734(15)	30.63(15)	-0.93(11)	**0.45**(35)	0.213(15)	16.2	**37.0**(45)	**50.1**(25)
	c	0.663393	0.736(18)	30.7(3)	-0.87(20)	**0.49**(31)	0.21(2)	16.0	**145.6**(66)	**50.5**(41)
ALH88003	**a**	0.684501	0.751(9)	30.95(15)	-0.69(6)	**0.20**(13)	0.176(9)	14.1	**7.5**(33)	**48.6**(15)
	b	0.692111	0.751(9)	30.95(14)	-0.607(1)	0.49(25)	0.176(9)	14.1	40.7(45)	46.9(12)
	c	0.685236	0.751(9)	30.97(22)	-0.58(5)	0.97(29)	0.177(7)	14.1	135.8(47)	45.3(9)
ALH88017	a	1.05769	0.752(15)	30.87(13)	-0.77(11)	0.06(22)	0.207(17)	15.5	31(38)	47.9 (14)
	b	1.05768	0.753(13)	30.87(13)	-0.76(11)	**0.45**(34)	0.207(19)	15.5	**42.6**(54)	**47.9**(13)
	c	1.05765	0.753(13)	30.87 (14)	-0.77(10)	**0.33**(40)	0.207(19)	15.5	**139**(13)	**48.1**(20)
ALH88020	a	1.1461	0.7610(97)	30.68(17)	-1.05(11)	**0.40**(11)	0.169(9)	12.9	**10.3**(26)	**52.5**(16)
	b	1.2002	0.7609(99)	30.76(13)	-0.855(89)	**0.49**(32)	0.170(10)	12.9	**37.1**(25)	**49.5**(13)
	c	1.15902	0.7603(97)	30.96(23)	0.628(41)	0.9211(26)	0.1681(14)	12.9	93.75(13)	54.01(84)
ALH88021	a	1.18763	0.7534(92)	30.63(23)	-1.12(11)	**0.74**(13)	0.184(12)	15.2	**23.4**(21)	**55.9**(20)
	b	1.23261	0.754(10)	30.91(15)	-0.764(65)	**0.49**(28)	0.184(15)	15.1	**42.8**(42)	**47.5**(16)
	c	1.2177	0.7546(95)	30.96(13)	-0.807(64)	0.0015(56)	0.1844(98)	15.1	116(68)	47.1(12)

In the other four meteorites, one solution gave a very small or large value of η and was linked with large values of the angles. It is thought that these situations are not crystallographically realistic. In the other solutions for these meteorites the value of η varies between 0.33 and 0.74 (values in bold). Without additional measurements it is not possible at this time to find a unique solution for fitting troilite in some meteorites.

The Mössbauer spectra of two samples of synthetic troilite (one doped with cobalt) show broadened lines, caused by the presence of four non-equivalent magnetic sites with the value of θ being different for each of the four sites (115 °, 60 °, 117 ° and 138 °) It was impossible to achieve one magnetic site for the iron because of the short cooling time (24 hours). Doping with small amounts of cobalt changes slightly the quadrupole interaction. These results show the influence of the cooling rate and impurities on the value of the Mössbauer parameters of troilite.

4. Conclusion

The nine meteorites studied can be divided into two groups in which the Mössbauer parameters of troilite appear to be different. In one group, consisting of four falls and one find, the coefficient of asymmetry $\eta \leq 0.2$, and the angles ϕ and θ are similar in each meteorite ($\phi < 7$ °, $\theta \cong 45$ °). In the second group, which consists of four finds, the value of η is higher between 0.33 and 0.74. This raises the question of whether the results are affected by processes of alteration on the Earth's surface or whether they are due to different conditions during the crystallisation and cooling in the parent bodies.

Acknowledgements

The authors acknowledge the financial support of the British Council (WAR/922/192) and by grant number 504a/1020/0156 from the Rector of Warsaw University of Technology.

References

[1] J. Gałązka-Friedman, E.R. Bauminger, I. Nowik, N. Bakun-Czubarow, M. Stępniewski and J. Siemiątkowski, Geological Quarterly, 45(3), (2001) 319-326
[2] O. Kruse, American Mineralogist, 75, (1990) 755-763
[3] O.Kruse, American Mineralogist, 77, (1992) 391-398
[4] G.M. Bancroft, Mössbauer Spectroscopy, An Introduction for Inorganic Chemists and Geochemists, McGrawHill, 1973, 23
[5] J.van Dongen Torman, R. Jagannathan and J.M. Trooster, Hyp. Int. 1 (1975) 135- 145
[6] J. Gałązka-Friedman, Z. Gontarz, J. Kalinowska, M. Milczarek, W. Tlaczała, M. Urbański, N. Bakun-Czubarow, J. Siemiątkowski, M.Stępniewski, Molecular Physics Reports (Mol. Phys. Rep.) 30 (2000) 52-58
[7] D.G.Rancourt and K.Lagarec, Recoil, Mössbauer Spectral Analysis Software for Windows, www.physics.uottawa.ca/~recoil

Mössbauer Study of Brazilian Emeralds

E. Kuzmann[ab], V. K. Garg[a], A. C. de Oliveira[a], G. C. B. Braga[a], J.A. Freitas[c], Jr. and R. Garg[a]

[a]*Institute of Physics, University of Brasília, 70910-900 Brasília, Brazil*
[b]*Department of Nuclear Chemistry, Eötvös University, Budapest, Hungary.*
[c]*Naval Research Laboratory, Washington, U. S. A.*

Emerald samples from two different locations in Brazil have been analyzed using transmission ^{57}Fe Mössbauer spectroscopy, X-ray diffraction, electron microprobe analysis and optical methods. Mössbauer spectra were decomposed into four subspectra associated with Fe ions at different sites. We found that Fe^{2+} substitutes both Al and Be sites in different proportions (varied with the origin of emerald samples). Moreover, about 1/3 of iron ions in Fe^{3+} valence state are located primarily in two different sites.
Keywords: emeralds, Mössbauer spectroscopy

1. Introduction

Emerald belongs to the beryl family. In natural beryl the occurrence of main constituent oxides are BeO (14%), Al_2O_3 (19%) and SiO_2 (67%). The iron content of emerald is usually between 0.3 - 3.14% [1, 2]. The basic structure, and general physical and chemical properties of beryls have already been studied [1], however, much less is known about the parameters exclusive to emerald, such as its cation distribution. Mössbauer spectroscopy is an excellent tool to determine the location and distribution of Fe ions with different valence states in iron-bearing minerals [3]. Iron-containing beryls have already been investigated through Mössbauer spectroscopy [4, 5], but none of the samples was ever identified as emerald. As far as we know this is the first report of ^{57}Fe Mössbauer spectroscopy of emerald.

Our aim was to study the sites and distribution of Fe ions in Brazilian emeralds. For this we performed ^{57}Fe Mössbauer spectroscopy, XRD, and optical measurements on gem-quality natural emeralds.

2. Experimental

We collected gem-quality emeralds of varied sizes from Campos Verdes, Goiás (GO) and Salinhas, Bahia (BA). The samples were identified as emerald by optical methods, X-ray diffractometry (X-ray diffractograms were recorded by means of a computer-controlled RIGAKU diffractometer using $Cu_{K\alpha}$ radiation and a monocromator), and electron microprobe analysis, based upon their characteristic color, photo luminescence absorption of Cr, crystal structure and composition. ^{57}Fe Mössbauer spectra of emerald samples (both powdered and single crystal) were recorded by a conventional Mössbauer spectrometer (WISSEL) using a proportional detector in transmission geometry, both at room temperature and at liquid nitrogen temperature (78K) using a 10^8Bq activity ^{57}Co(Rh) source. The samples were prepared by applying small force to press the powdered emeralds in the sample holder. To decrease the disturbing effect of texture, powdered samples with a surface density of 50 mg/cm^2 were used for the measurements. Some measurements were also made by putting the powdered sample of emerald in preferred orientation to the incident gamma rays. Due to the low iron content and to collect more than 10^7 counts in each channel, each spectrum was recorded over the course of 10 days. Isomer shifts in this paper are reported relative to the α-Fe. The Mössbauer spectra were evaluated by a least-square fitting MOSSWINN [6] program, and the quadrupole splitting distributions were obtained by modified Hesse-Rübartsch methods.

3. Results and discussion

The physical properties of emerald are strongly influenced by the chemicals present in the mining area. The intensity of the pale green color characteristic of the emerald depends on the concentration of Cr^{3+} in the crystal lattice. The addition of other impurities leads to the presence of additional bands of light absorption, which alter the emerald's typical color. Thus, identifying extra impurities allows one not only to confirm the stone's origin but also the feasibility of neutralizing the effect of the impurities through adequate treatment.

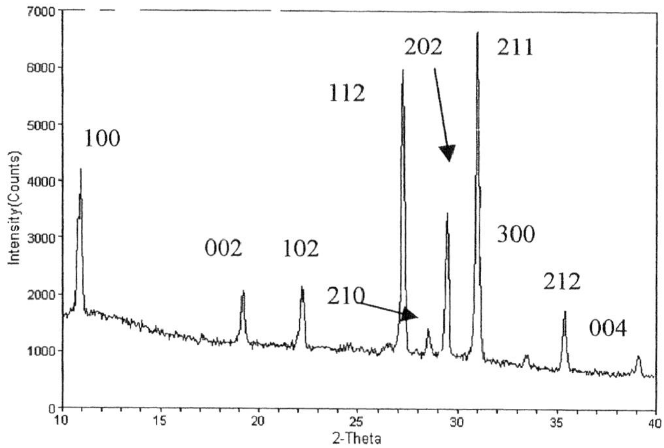

Figure 1. XRD of the emerald sample G003

The XRD results, Figure 1, shows that the structure of our samples agrees with that of beryl. Emerald is of green color, has a hexagonal structure of the *P6/mcc* group, with lattice parameters: c=9.17 Å e a=9.21 Å [7,8]. Fig. 2 shows the basic structure of emerald, a projection on 0001 plane, where dashed lines outline the unit cell. The hexagonal rings of six SiO_4 tetrahedra stacked above each other forming channels are parallel to the crystallographic c axis [9]. The channels are linked by Al octahedrally coordinated to six O ions and by Be coordinated to four O ions in a distorted tetrahedron. Al and Be cation sites can be substituted by Fe ions. Typical color is the result of absorption bands associated to the electronic transitions $^4A_{2g} \rightarrow {}^4T_{2g}$ and $^4A_{2g} \rightarrow {}^4T_{1g}$ of the impurities of Cr^{3+} ions, which substitute the Al^{3+} ions in the crystal sub-lattice, causing intense absorption bands in the yellow-orange and the blue respectively [10]. To detect and identify Cr^{3+} and other impurities, we measured photo luminescence (PL) in the temperature range of 6K to 300K in several samples from Goiás (GO) and Bahia (BA). The optical measurements confirmed the samples to be emerald.

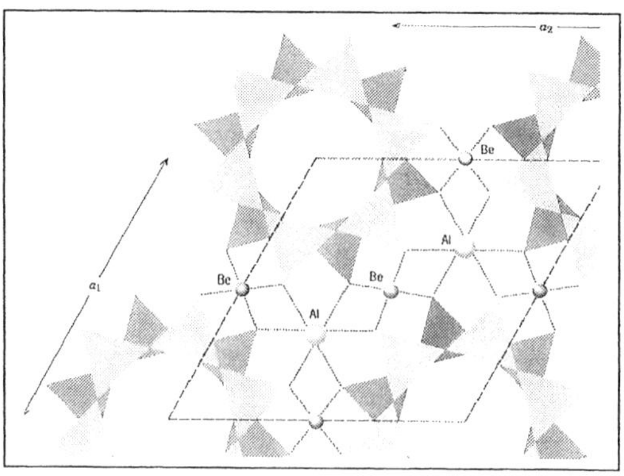

Figure 2. Structure of emerald

Fig. 3 depicts typical Mössbauer spectra of powdered Brazilian emerald. The spectra exhibit peaks in a velocity interval between -0.5 and 3 mm/s. Mössbauer spectra were decomposed by fitting a few quadrupole split subspectra of Lorentzian lines in the frame of an appropriately chosen model containing the constraints between the Mössbauer parameters of subspectra and the different iron sites. In the model the site assignment of the Mössbauer subspectra corresponds well to the distinct

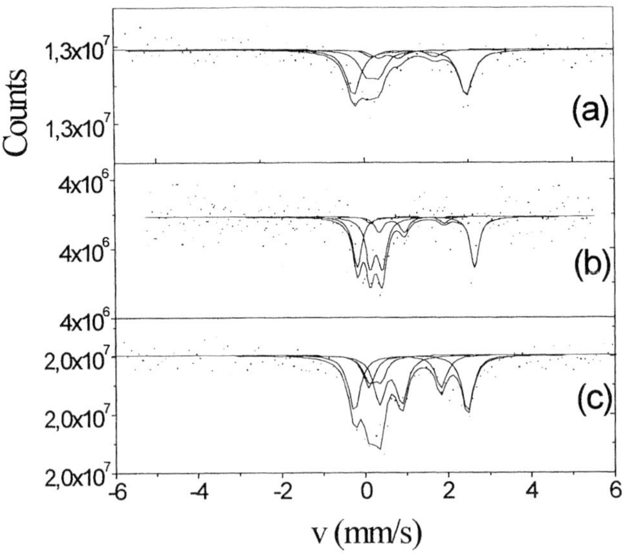

Figure 3. MS of emerald GO (295K) and BA at 295K(b) and 77K(c)

cation positions at different crystallographic sites. The result of this procedure depends first and foremost on correct knowledge of the number of iron positions distinguishable by Mössbauer spectroscopy. This number can be different from the exact number of crystallographic iron sites because different microenvironments of the same site (e.g., having different cation neighborhoods) can also be distinguished by the Mössbauer method. The presence of numerous microenvironments can make conventional evaluation unreliable. For such cases we can calculate the quadrupole splitting distribution derived from the Mössbauer spectrum. Correspondence can be obtained between the peaks of the distribution curve and the different microenvironments. With this method we can distinguish different microenvironments better than conventional spectrum evaluation.We used the quadrupole splitting distribution method to estimate the Mössbauer spectra of the emeralds.

Fig. 4. shows the distribution curves of Fe^{2+} and Fe^{3+} obtained for the GO and BA emeralds, determined by Mosswinn program [6]. Two distributions were calculated simultaneously and ternary polynomial functions were used to give the relation between the isomer shift and the quadrupole splitting. By decomposition of the quadrupole splitting distribution, we found four peaks (Fig. 4) associated to the four iron microenvironments.

Table 1. Mössbauer parameters of Brazilian emerald

	Mössbauer Parameters of Emerald		
	Goiás (GO) T =R	Bahia (BA) T=R	Bahia (BA) T=N
δ(D1) (mm/s)	1.10±0.01	1.11±0.01	1.22±0.01
δ (D2) (mm/s)	0.95±0.01	0.96±0.01	1.03±0.01
δ (D3) (mm/s)	0.58±0.01	0.59±0.02	0.65±0.01
δ (D4) (mm/s)	0.23±0.01	0.22±0.02	0.27±0.01
ΔE_Q (D1) (mm/s)	2.71±0.02	2.69±0.02	2.83±0.02
ΔE_Q(D2) (mm/s)	1.74±0.02	1.62±0.03	1.77±0.02
ΔE_Q. (D3) (mm/s)	0.55±0.02	0.51±0.02	0.60±0.02
ΔE_Q (D4) (mm/s)	0.28±0.02	0.30±0.02	0.30±0.02
A (D1) (%)	36±3	56±4	42±7
A (D2) (%)	21±2	8±1	6±1
A (D3) (%)	29±2	9±2	13±2
A (D4) (%)	14±1	27±3	39±4

R = room temperature, N = liquid nitrogen temperature

Table 2. Iron distribution in Brazilian emeralds

Fe	Goiás (GO) - Samples	Bahia (BA) - Samples
Fe^{3+}/Fe	0.43	0.36
Fe^{2+}/Fe	0.57	0.64
Fe^{2+}/ Fe^{3+}	1.33	1.77
Fe^{3+}/ Fe^{2+}	0.75	0.56
Fe^{2+} in Al site	0.36	0.56
Fe^{2+} in Be site	0.21	0.08
Fe^{3+} in Al site	0.29	0.09
Fe^{3+} in Be site	0.14	0.27
Fe in Al site	0.65	0.65
Fe in Be site	0.35	0.35
Fe_{Al}/Fe_{Be}	1.85	1.85

Based on the results of the quadrupole splitting distribution method, four doublets of Lorentzian lines were used for conventional decomposition of the obtained Mössbauer spectra.This way all Mössbauer spectrum could be decomposed into four doublets (D1, D2, D3 and D4) with corresponding parameters. Table 1 illustrates the Mössbauer parameters of these subspectra for a few cases. It should be noted that the spectra can also be fitted into three doublets, when the local parameters of doublets (isomer shift and quadrupole splitting) coincide to those obtained in the four doublets evaluations – which is another argument in favor of the four-doublet model. All results are fitted data based on the evaluation of measured spectra.The Mössbauer spectra of the samples, polished and ground down to 0.3 mm thickness, indicated similar texture effect as observed earlier with iron-bearing beryls [4]. Powdered samples also exhibited some texture effect. Nonetheless, much less temperature-dependent relaxation effect was observed compared to the effect reported with aquamarine [3].

The *site assignment* of the doublets observed in the Mössbauer spectra of emerald is based on the following:1) Knowledge of Mössbauer parameters in minerals, 2) Expectation of crystallographic sites for iron in the emerald structure and 3) Earlier Mössbauer and other studies of iron-containing beryl. Iron can be expected to be situated at the following positions in this structure:
-at substitutional cation sites: octahedrally co-ordinated Al site as Fe^{2+} or Fe^{3+}, tetrahedrally co-ordinated Be site as Fe^{2+} or Fe^{3+} and tetrahedrally co-ordinated Si site as Fe^{3+}.
Doublet D1 is attributed to Fe^{2+} located at octahedral Al site in the emerald lattice. The isomer shift and the quadrupole splitting of doublet D1 are typical for octahedrally coordinated Fe^{2+} ion in minerals. The Mössbauer parameters of doublet D1 agree well with those reported for another Fe-containing beryl [4, 5]. Doublet D2 is associated with Fe^{2+} ions occupying the tetrahedral Be site. The isomer shift is characteristic for an Fe^{2+} ion and the quadrupole splitting indicates a four-coordinated microenvironment. Be and Si are both in tetrahedral coordination, however, the substitution of Si by Fe^{2+} ion is unlikely; in that case the quadrupole splitting is expected to be much less than the measured. Price et al [4] have studied Mössbauer spectra of another beryl, and found a doublet with parameters similar to D2. Doublets D3 and D4 certainly represent two different Fe^{3+} microenvironments. The occurrence of Fe^{3+} at both Al and Be sites has already been predicted but not shown directly. The observed Mössbauer parameters of doublets D3 and D4 are not characteristic for any known mineral or compounds, which are usually part of an emerald ore. ESR measurements confirmed the presence of Fe^{3+} inside the lattice. Previous Mössbauer studies of iron-bearing beryls indicated a smaller contribution of Fe^{3+} [3-4]. The larger percentage of Fe^{3+} in Brazilian emeralds can be explained by the relative abundance of Fe in Brazilian minerals than those from elsewhere. This agrees with the Mössbauer analysis of bazzite – the natural scandium analog of beryl found in an iron-rich area of Kazakhstan – which displayed about 50% Fe^{3+} microenvironments [11]. We found no evidence of Fe ions in structural channels, where iron sites have been postulated earlier [12], probably because of their small recoilless fraction that might not be observable by the Mössbauer spectra. Considering the above arguments we can describe doublets D3 and D4 as following:

Doublet D3 can be associated with Fe^{3+} ion substituted into the octahedral Al site. The Mössbauer parameters of this doublet correspond to the octahedrally coordinated Fe^{3+} site. Similar parameters were derived from the magnetically split spectrum obtained with a beryl in an external magnetic field (6T) [4]. Doublet D4 can be associated with Fe^{3+} ion substituted into the tetrahedral Be site. The isomer shift and the quadrupole splitting confirm a tetrahedral coordination for Fe^{3+} ion. Lower values of isomer shift than these for Fe^{3+} substitution into Si site have been reported earlier [3, 13]. This site assignment agrees with the expectation that Fe^{3+} substituted Be (as mentioned above), even in beryls with lower iron content [4]. We found no evidence of Fe ions in structural channels probably because their small recoilless fraction might not be observable by Mössbauer spectra.

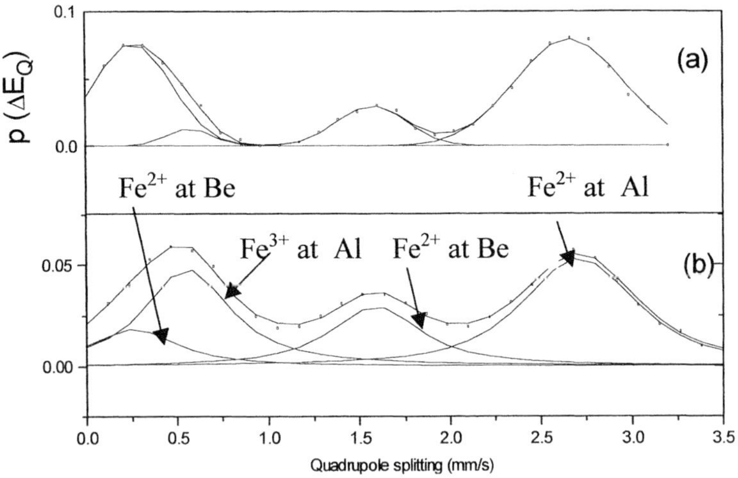

Figure 4
ΔE_Q distribution of GO (a) and BA (b)

Based upon this site assignment, Table 2 shows the typical iron distribution data of emeralds originating from two different regions in Brazil The Brazilian emeralds contain a large amount of Fe^{3+}, not found in any other iron-containing beryls in earlier works. Interestingly enough, the relative population of iron in Al sites compared to iron at Be sites was the same in every sample. The most striking difference between the two types of emeralds was that in GO samples Fe^{3+} preferred the Al site, while in the BA samples Fe^{3+} preferred the Be site. Fe^{2+} prefers the Al site in both cases, but this choice is more than three times stronger in BA than in GO emeralds. One can interpret these differences as consequence of how these emeralds were formed. Our measurements, show that the emeralds from Bahia (BA samples) contained vanadium and chromium as principal color centers, but those from Goiás (GO samples) had only chromium [14].

We can conclude that Mössbauer spectroscopy of natural Brazilian emeralds revealed two Fe^{2+} and two Fe^{3+} sites in the crystal. The Fe^{2+} and Fe^{3+} substitute both the octahedral Al and the tetrahedral Be sites. The distribution of Fe^{2+} and Fe^{3+} ions among the crystallographic sites is characteristic of the emerald location. The occurrence of the Fe^{3+} sites was found around 30% in the samples.

References

[1] Deer, W A, Howie R A, Zussman J, 1962 Rock Forming Minerals, (London: Longman), Vol. 1.
[2] G.M. Pulz, L.J.H.D. Silva, S.L. Barros Neto, C.A. Santos, T.M.M. Brum, V.P. Pereira, J.J. Silva, P. Barreto, P.L. Juchem, Journal of Gemmology, 26 (1998) 252.
[3] Kuzmann E, Nagy S, Vértes A, Weiszburg T G, Garg V K 1999 Geological and Mineralogical Application of Mössbauer effect, in Nuclear Methods in Geology, eds. A. Vértes, S. Nagy, K. Süvegh (New York: Plenum Press)
[4] Price D C, Vance E R, Smith G, Edgar A, Dickson B L 1976 J. Phys. (Paris) 37, C6-811
[5] Platonov A. N., Polshin E V, Taran M N 1979 Dokl. AN UkSSR 108, 726
[6] Klencsár Z, Kuzmann E, Vértes A 1996 J. Radional. Nucl. Chem. Lett. 210, 105
[7] Platonov A N, Taran M N, Minko O E, Polshin E V 1978 Phys. Chem. Miner. 3, 87
[8] Blak A R, Isotani S, and Watanabe S 1963 Phys. Chem. Minerals, 8 161
[9] Hurlbut, Jr. C S and Klein C 1071 Manual of Mineralogy (New York: John Wiley & Sons)
[10] Ohkura H, Hashimoto H, Mori Y, Chiba Y, and Isotani S 1987 Japan J. Appl. Phys. 26 1422
[11] Platanov A N, Polshin E V, Tsistyakova M B and Taran M N 1981 Geokhimiya 3, 393
[12] Wood, D L. Nassau K 1968 Amer. Mineral. 53, 777
[13] G. M. Bancroft-1973 Mössbauer Spectroscopy (London: McGraw Hill)
[14] Freitas, Jr. J A, Strom U, and Tran D C 1985 Mat. Sc. Forum, 6 627.

Key role of the kinetics of γ-FeOOH bioreduction on the formation of Fe(II-III) minerals

G. Ona-Nguema,[1] F. Jorand,[1] O. Benali[2], M. Abdelmoula,[1] J.-M. R. Génin,[1]* and J.-C. Block[1]

[1]*Laboratoire de Chimie Physique et Microbiologie pour l'Environnement, CNRS-Université Henri Poincaré-Nancy 1, 405, rue de Vandœuvre, 54600 Villers-lès-Nancy, France.*
[2] *Département de Chimie, Faculté des Sciences, Université Ibn Tofail, BP 133 Kénitra, Morocco.*

During bacterial reduction of γ-FeOOH lepidocrocite in the presence of formate as electron donor at initial pH 7.5 in bicarbonate buffer, the obtained Fe(II-III) minerals were characterised by transmission Mössbauer spectroscopy (TMS) and X-ray diffraction (XRD). Two types of kinetics, high and slow, of γ-FeOOH bioreduction were obtained. The high rate of ferric oxyhydroxide reduction produced magnetite whereas the slow rate yielded hydroxycarbonate green rust.

1. Introduction

Bacterial reduction of Fe(III) coupled with the oxidation of the organic matter is a significant process in anoxic environments [1]. It leads to the formation of Fe(II)-bearing minerals such as magnetite, siderite, vivianite, or green rusts. If the chemical composition of the culture medium, during iron bioreduction, explains the formation of siderite in presence of HCO_3^-, and vivianite in presence of PO_4^{3-} [2], the factors which lead towards the formation of green rust or magnetite are not yet known. In this study, the kinetics of ferric oxyhydroxide bioreduction is shown to be a factor that supports the formation of either green rust or magnetite. Transmission Mössbauer spectroscopy (TMS) and X-ray diffraction (XRD) were used to characterise the minerals.

2. Experimental method

γ-FeOOH synthesis. Lepidocrocite was prepared as follows: a 0.228 M $FeCl_2 \cdot 4H_2O$ solution was mixed with 0.4 M NaOH to precipitate $Fe(OH)_2$. The suspension was aerated by magnetic stirring ensuring a homogeneous oxidation and the reaction was monitored at 25 ± 0.5°C by recording the pH and redox potential of the solution. At the end of the oxidation process, the solution was poured into a container, completed with a Milli-Q® water and sealed; then the precipitate settled down. The ferric oxyhydroxide that formed in these conditions was lepidocrocite. It was washed twice with sterile Milli-Q® water to remove electrolytes, and was characterised by TMS (Fig.1a).

Bacterial culture. Shewanella putrefaciens CIP 8040 was aerobically cultivated in trypcase-soy broth (BioMérieux®, 51019) at 30°C. The cells were harvested at late log phase by centrifugation (16,000 g at 20°C for 10 minutes) and washed twice with sterile Milli-Q® water. Cells were anaerobically added to a culture medium of 2.6×10^{-2} M $NaHCO_3$, 10^{-4} M anthraquinone-2,6-disulfonate AQDS and 50 or 75 mM formate, mixed with 160 mM γ-FeOOH lepidocrocite and traces (< μM) of elements necessary for bacterial culture. All components were sterilised by autoclaving except $NaHCO_3$, AQDS, and formate, sterilised by filtration through 0.22 μm membrane. The pH was 7.5 after addition of the last component.

Analyses. HCl-extractable Fe(II) was obtained by pouring 1 mL of suspension directly into 1 mL of 1 M HCl. After at least 1 hour of contact, Fe(II) was determined by using the modified 1,10-phenanthroline method [3]. The formate and bicarbonate anions in solution were quantified by using capillary electrophoresis (Waters Capillary Ion Analyser).

* corresponding author; E-mail: jmg@lcpe.cnrs-nancy.fr

Transmission Mössbauer spectroscopy (TMS) and X-ray diffraction (XRD). Solid samples for analysis by TMS and XRD were extracted by filtration (0.45 μm) of 10 mL of slurry under anaerobic atmosphere. TMS was performed from 77 K to room temperature with a closed cycle-helium Mössbauer cryogenic workstation, equipped with vibration isolation stand manufactured by Cryo Industries of America®. The Mössbauer spectrometer consists of a conventional apparatus with a 50 mCi source of ^{57}Co in Rh, calibrated with a 25 μm foil of α-Fe at room temperature. The amount of materials being optimised (10 mg cm^{-2} Fe), the spectra can be approximated by a sum of Lorentzian-shape lines (thin absorption approximation). This proved to be appropriate in the case of GRs since each peak has a perfect Lorentzian shape. The parameters resulting from computer fitting were both mathematically (χ^2 minimisation) and physically significant. The samples were prepared under N_2 atmosphere in the glove box and quickly transferred to the cryostat under an inert He atmosphere before performing TMS measurements. Additional analysis were performed by X-ray diffraction (XRD). Patterns were recorded in the 1 - 40° range of θ using Co-Kα_1 radiation (λ = 0.17889 nm) in Bragg-Brentano geometry. Once filtered, the wet paste was spread out on a glass plate and coated with glycerol in order to avoid any oxidation [4].

3. Results and discussion

Formation of magnetite from high rate of lepidocrocite bioreduction. Iron reduction experiments with *S. putrefaciens* (3 × 10^9 cells mL^{-1}) were performed with 160 mM γ-FeOOH, and 50 mM formate in bicarbonate buffer (Fig.1b).

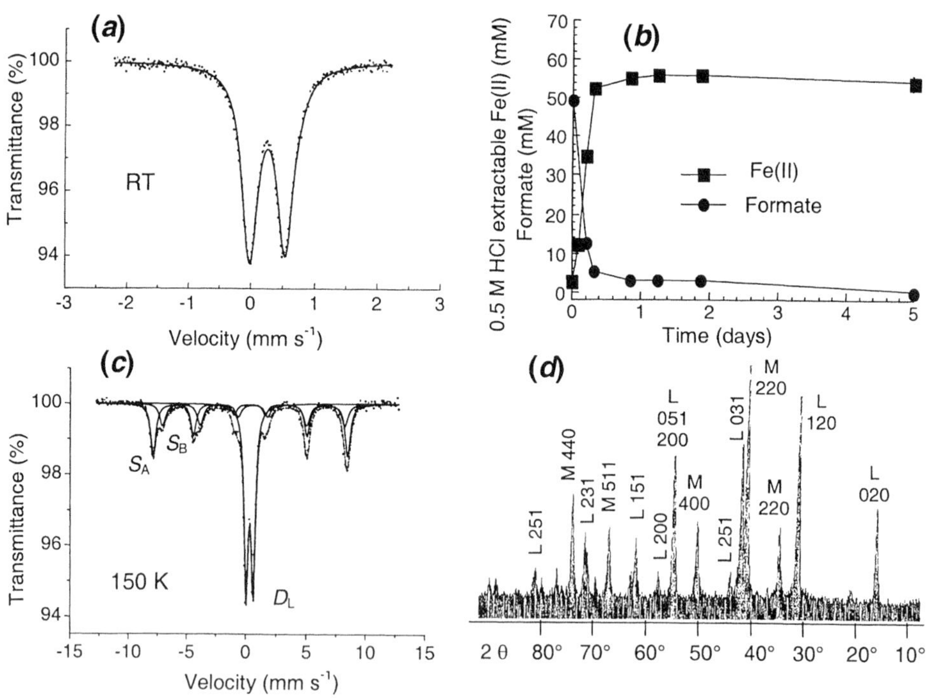

Fig. 1. (a) TMS spectrum at room temperature (RT) of initial reactant, γ-FeOOH (t_0). (b) Fe(II) production and formate consumption during high rate of γ-FeOOH bioreduction. (c) TMS spectrum at 150 K of a mixture of magnetite and unreduced γ-FeOOH after 1 day of culture. (d) XRD pattern of a mixture of magnetite and unreduced γ-FeOOH after 1 day of culture.

Lepidocrocite was characterised by TMS at room temperature (δ = 0.35 mm s^{-1} and Δ = 0.55 mm s^{-1}) (Fig.1a). The initial rate of iron reduction was 6.9 mM h^{-1} and after 1 day of culture 56 mM of Fe(III) were reduced and 50 mM of formate were consumed. The pH increased from 7.5 to 9.3 during the reduction illustrated by Fig. 1b. The characterisation of product by TMS and XRD exhibited a mixture of magnetite and lepidocrocite (Fig. 1c & d). The hyperfine parameters of TMS spectrum at 150 K are identical to those found in the literature for magnetite, *i.e.* S_A (δ = 0.41 mm s^{-1}; H = 508 kOe) and S_B (δ = 0.68 mm s^{-1}; H = 476 kOe) [5], and lepidocrocite, *i.e.* D_L (δ = 0.5 mm s^{-1} and Δ = 0.57 mm s^{-1}) [6].

Formation of hydroxycarbonate green rust from slow rate of lepidocrocite bioreduction. Microbial reduction of lepidocrocite was accompanied by the consumption of formate (Fig. 2a) when pH increased from 7.5 to 9.5. With initial concentration of 8×10^9 cells mL^{-1}, the formate (75 mM initially) was completely removed from the solution when 160 mM of γ-FeOOH were added to the culture whereas 116 mM of iron were reduced (Fig. 2a). Bacterial iron reduction started with an initial rate of 2.6 mM h^{-1}. According to the stoichiometry of the redox equation, 1 mole of formate oxidised by 2 moles of Fe(III), the bacteria oxidised 58 mM of formate with 116 mM of Fe(III). Thus, only 17 mM of formate cannot be attributed to its oxidation by Fe(III), probably due to the adsorption onto lepidocrocite or incorporation into biomass.

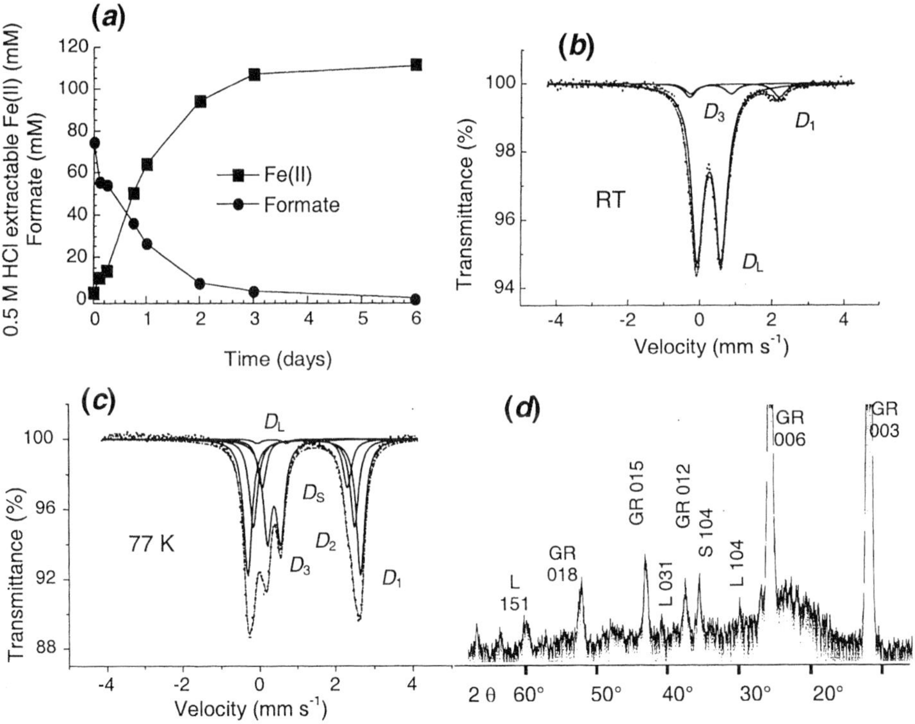

Fig. 2. (a) Fe(II) production and formate consumption during slow rate of γ-FeOOH bioreduction; (b) TMS spectrum at room temperature (RT) of GR and γ-FeOOH obtained after t = 18 hours of culture; (c) TMS spectrum at 77 K of the mixture of GR, siderite and traces of unreduced γ-FeOOH obtained after 6 days of culture; (d) XRD pattern of the mixture of GR, siderite and traces of unreduced γ-FeOOH obtained after 6 days of culture.

TMS spectra, measured at room temperature and 77 K, of precipitates sampled during bioreduction after 18 hours and 6 days (Fig. 2b & c; Table 1), respectively, exhibit quadrupole doublets typical of GR (D_1, D_2 & D_3) [7], lepidocrocite (D_L) [6], and siderite (D_S) [8]. The difference concerns not only the intensities of GR and lepidocrocite, which vary during bacterial reduction, but also the formation of siderite due to bicarbonate excess at the end of the culture.

Table 1. Mössbauer hyperfine parameters of products obtained during slow rate of bioreduction of γ-FeOOH (spectra displayed in Fig. 2).

	Sampled at t = 18 hours and analysed at room temperature			Sampled at t = 6 days and analysed at 77 K		
	δ	Δ	*RA*	δ	Δ	*RA*
D_1	1.07	2.46	8	1.28	2.94	36
D_2	-	-	-	1.27	2.64	24
D_3	0.41	1.12	5	0.49	0.35	26
D_L	0.38	0.66	87	0.44	0.76	2
D_S	-	-	-	1.29	2.22	12

δ (mm s^{-1}), isomer shift with respect to α-Fe; Δ (mm s^{-1}), quadrupole splitting; *RA* (%), relative abundance.

4. Conclusion

According to the used inoculum, two types of kinetics of γ-FeOOH bioreduction (high and slow) were obtained, leading to the formation of either magnetite or GR. When the kinetics of lepidocrocite bioreduction was high, a large Fe(II) amount is produced rapidly in a short time. This excess of Fe^{2+} ions adsorbs on the surface of the γ-FeOOH, which is not yet reduced. This adsorption involves the transformation of the solid phase *in situ* into magnetite (Fe_3O_4) through electron exchange inside the solid state [9]. When the kinetics of lepidocrocite bioreduction was slow, the formation of $[Fe^{II}_{(1-x)}\ Fe^{III}_{x}\ (OH)_2]^{x+} \bullet [(x/2)\ CO_3^{2-} \bullet y\ H_2O]^{x-}$ hydroxycarbonate green rust, was observed. These results show that the kinetics of iron reduction is an important factor governing the nature of the Fe(II)-bearing minerals that form.

References

[1] D. R. Lovley, in: Environmental microbe-metal interactions D. R. Lovley (ed.). ASM Press, Washington, D.C., 2000, p. 3.
[2] J. M. Fredrickson, J. M. Zachara, D. W. Kennedy, H. Dong, T. C. Onstott, N. W. Hinman and S.-M. Li. Geochim. Cosmochim. Acta, 62 (1998) 3239.
[3] H. Fadrus and J. Maly, Analyst, 100 (1975) 549.
[4] H. C. B. Hansen, Clay Miner., 24 (1989) 663.
[5] R. E. Vandenberghe and E. De Grave, in: Mössbauer Spectroscopy Applied to inorganic chemistry, Vol. 3, ed. G. J. Long and F. Grandjean, Plenum Press, New York, 1989, p. 59.
[6] E. Murad and U. Schwertmann, Mineral. Mag., 48 (1984) 507.
[7] S. H. Drissi, Ph. Refait, M. Abdelmoula and J.-M. R. Génin, Corros. Sci, 37 (1995) 2025.
[8] M. A. Ahmed, R. E. Vandenberghe, E. De Grave, N. A. Eissa and J. V. Ibarra, Fuel, 8 (1999) 453.
[9] Y. Tamaura, K. Ito and T. Katsura, J. Chem. Soc. Dalton Trans. (1983) 189.

Mössbauer spectra of circumstellar silicate dust analogs used as a Fischer-Tropsch catalyst

E. J. Hoffman[1], H. G. H. Hill[2], J. A. Nuth III[2], and D. Seifu[1]

[1]*Physics Department, Morgan State University, Baltimore, MD 21251, USA*

[2]*Astrochemistry Branch, Mail Code 691, NASA Goddard Space Flight Center, Greenbelt, MD 20771, USA*

Synthetic smoke produced by nonequilibrium gas-to-solid condensation in a Fe-SiO-H_2-O_2 vapor served as a catalyst for Fischer-Tropsch-type catalytic production of hydrocarbons, reactions likely occurring in protostellar nebulae. The freshly-synthesized smoke gave a Mössbauer spectrum showing just a hint of sharp absorption maxima; use as a catalyst brought about deepening and narrowing of the absorption peaks along with a 100-fold increase in mass density, apparently by hydration. Predominant in both raw and post-catalytic spectra is a doublet typical of ferric silicates and oxides. We conclude that the Fischer-Tropsch-type reactions enhanced development of ferric phases similar to those in interplanetary dust particles and chondritic meteorites.

1. Introduction

Astronomical UV to near-IR spectroscopy of the last century proved the existence in interstellar space of submicrometer-sized dust grains, "interstellar dust" (reviewed in [1]). By the late 1960's it was clear that broad IR bands centered at 10 and 18 μm correspond to fundamental vibrations of SiO_4 tetrahedra in amorphous constituents of the dust [1, 2]. Such amorphous silicate dust, or perhaps more accurately, smoke, has since also been found in the circumstellar envelopes of evolved, O-rich stars and in many other astrophysical settings [2]. Laboratory-based experiments with synthetic silicate grain analogs should thus be expected to provide chemical and physical constraints for natural grains in these environments.

Mg-Fe-Si smokes can indeed be synthesized in the laboratory [3], and they furthermore serve well as catalysts for Fischer-Tropsch-type catalytic production of hydrocarbons [4], reactions likely occurring in protostellar nebulae, including the precursor of our own solar system. These laboratory operations thus mimic two steps in a cosmic process critical for formation of a solar system with carbon-based life. We report results of ^{57}Fe-Mössbauer spectroscopy on these smokes in an attempt to follow the evolution of the iron oxidation states during changes the smoke/dust itself undergoes during the second of these steps, service as a catalyst.

2. Experimental

The raw material was a synthetic smoke produced by nonequilibrium gas-to-solid condensation in a Fe-SiO-H_2-O_2 vapor [5].

2.1 *Catalysis*

This smoke served as catalyst for ten several-day Fischer-Tropsch-type syntheses at 200-500° C. H_2 (ca. 750 torr) and CO (ca. 75 torr) were circulated at the start of each run. FTIR spectroscopy detected significant quantities of new species including CO_2 and CH_4, with H_2O as a by-product [4].

2.2 *Mössbauer Spectroscopy*

Transmission spectra were collected at room temperature for 1-mm-thick samples in an ASA spectrometer calibrated with iron foil.

3. Results and Discussion

The freshly-synthesized smoke gave a Mössbauer spectrum showing just a hint of sharp absorption maxima (Fig. 1, lower curve), consistent with its chemically random character [5]. The solid line is the envelope of an empirical fit of the data to a symmetrical doublet (dashed line) plus a broad singlet (dotted line). As discussed below, spectral parameters for the doublet are roughly in the range typical of many ferric minerals, and the additional absorption, here modeled by the singlet, suggests that several phases must be present.

Use of the smoke as a catalyst brought about a 100-fold increase in density compared to that of the raw dust, to about 75 mg cm^{-3}, apparently due to hydration by the H_2O by-product, as well as a change in Mössbauer spectrum [6]. The absorption peaks narrowed and deepened, although a broad singlet was again necessary for a reasonable fit, still indicating the presence of unresolved ferric or ferrous phases (Fig. 1, upper curve). As Table 1 shows, for the doublet in both spectra (dashed line) the isomer shift, $\delta = 0.3$ mm s^{-1}, and quadrupole splitting, $\Delta = 1.0$ mm s^{-1}, are in the range [7] for paramagnetic iron in ferric oxides, hydroxides, and silicates (mainly the sheet silicates). The fit required a bit less of the extra single peak (dotted line) than in the Raw Dust, so the catalysis seems to be encouraging formation of the product giving the main doublet, likely due to the aqueous environment. To investigate further we plan to take spectra at low temperature, to hydrate samples under non-catalytic conditions, and to examine smoke prepared under different conditions.

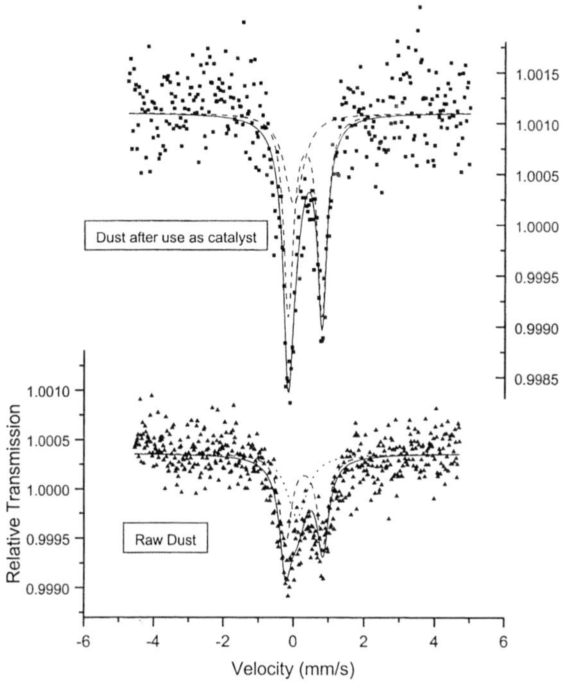

Fig. 1. Fe-Mössbauer spectrum of silicate dust (smoke) analogs, modeled by a paramagnetic doublet (dashed line) typical of Fe^{3+} ion in a variety of silicates and oxides [7]; need for an additional broad singlet (dotted line) suggests the presence of other unresolved phases. *Lower curve*: the sample before use as a catalyst shows shallow absorption peaks. consistent with its chemically random condition [5]. *Upper curve*: use as a catalyst enhances the peaks.

We thus conclude that the Fischer-Tropsch-type reactions resulted in further development of ferric phases, presumably some already found in the original material. The aqueous conditions and the increase in density suggest that post-catalysis phases may include ferric ion in sheet silicates [7], a possibility to be explored by Mössbauer spectroscopy at low temperature. Aqueous alteration is often apparent under the microscope in the minerals of primitive solar system materials such as interplanetary dust particles [8] and chondritic meteorites [9]. These results may help explain the origin of at least some of these altered minerals.

Table 1. Mössbauer parameters of empirical fit shown in Fig. 1for synthetic silicate dust (all but % Area in mm s^{-1})

	Raw Dust	Dust after use as catalyst	Fe^{3+} ion in many minerals [7]
Doublet:			
δ	0.3	0.3	0.1-0.5
Δ	1.0	1.0	0.2-2.0
$\Gamma/2$	0.4	0.3	
% Area	60	70	
Singlet:			
δ	0.1	0.0	
$\Gamma/2$	0.7	0.7	
% Area	40	30	

Acknowledgements

This work was supported by grant NAG 5-9710 from NASA to Morgan State University. We thank Dr. Fred Oliver and Dr. Leopold May for valuable discussion..

References

[1] J. Dorschner, Revs. Mod. Astronom. 6 (1993) 117.
[2] J. P. Simpson, Astrophys. J. 368 (1991) 570.
[3] J. A. Nuth III, S. L. Hallenbeck, and F. J. M. Rietmeijer, J. Geophys. Res. 105 (2000) 10387.
[4] H. G. M. Hill and J. A. Nuth III, Meteorit. Planet. Sci. 35 (2000) A73.
[5] F. J. M. Rietmeijer, J. A. Nuth III, and J. M. Karner, Astrophys. J. 527 (1999) 395.
[6] E. J. Hoffman, H. G. M. Hill, and J. A. Nuth III, Meteorit. Planet. Sci. 36(2001) A81.
[7] R. G. Burns, in: *Remote Geochemical Analysis: Elemental and Mineralogical Composition* (Cambridge Univ. Press, NY, 1993) p. 539.
[8] F. J. M., Rietmeijer, in *Planetary Materials*, (Mineralogical Society of America, Washington, 1998) p. 2-1.
[9] A. J. Brearley and R. H. Jones, in *Planetary Materials*, (Mineralogical Society of America, Washington, 1998) p. 3-1.

Rock paintings from Minas Gerais, Brasil, investigated by in-situ Mössbauer spectroscopy

G. Klingelhöfer[1], G.M. da Costa[2], A. Prous[3], B. Bernhardt[1]
[1]Inst.Anorganische und Analytische Chemie, Joh. Gutenberg-Universität Mainz, Germany;
[2]Departamento de Química, Universidade Federal de Ouro Preto, Ouro Preto, MG;
[3]Museu de História Natural, Universidade Federal de Minas Gerais, Belo Horizonte, MG.
E-mail: klingel@mail.uni-mainz.de

The archaeological site of Santana do Riacho is located at about 90 km north-eastern of Belo Horizonte, in the "Serra do Cipó" region. The recent development of a miniaturised Mössbauer spectrometer opened the possibility of making in-situ measurements of the rock paitings, and the results of such pioneer work is presented in this contribution. The spectrometer MIMOS II operates in backscattering geometry measuring the scattered 14.41 keV Mössbauer radiation and the 6.4 keV Fe X-rays. The dimensions of the whole instrument are approximately 110 mm (long) x 80 mm (diameter). The paintings were selected based on their colours and apparent thickness, and six spots were measured. The analysis of the data suggests that hematite and goethite seems to be the pigments responsible for the red and yellow colours, but no conclusion was yet found for an ochre painting.

Keywords: iron oxides, rock painting, scattering, instrumentation, methodology

Introduction

The manufacturing and composition of the paints from the prehistoric period are of great archaeological interest because it is possible to deduce aspects of the ancient cultures based on their abilities to produce works of arts. In addition, restoration and/or conservation can be better applied if a knowledge exists about the composition of the materials used. In most cases, for obvious reasons, it is impossible to remove samples, and therefore in-situ methods have to be applied. Many pigments used in ancient paintings are iron-based compounds, and hence Mössbauer spectroscopy could be an invaluable tool to characterize these materials. The recent development of a miniaturised Mössbauer spectrometer opened the possibility of making in-situ measurements, and the results of such pioneering work will be presented in this contribution. The spectrometer MIMOS II operates in backscattering geometry measuring the scattered 14.41 keV Mössbauer radiation and the 6.4 keV Fe X-rays [1,2]. The dimensions of the whole instrument are approximately 110 mm (long) x 80 mm (diameter).

The archaeological site of Santana do Riacho is located at about 90 km north-eastern of Belo Horizonte, in the "Serra do Cipó" region. Excavations of all sort of materials (bones, stones, pigmented materials, etc) have been performed over the years and a vast body of information was collected [3,4]. Several prehistorical paintings are found in this locality, but so far no conclusive proof exists about the nature of the materials (e.g. paints) that were used in these drawings. The vast majority of the paintings are red (70%) and yellow (20%), but there are others in orange, black, brown, and ochre. Iron oxides are most probably the pigments that were used [5].

The impressive paintings in Santana do Riacho, associated to the recent development of the spectrometer MIMOS II, prompted the authors to perform, for the first time, in-situ Mössbauer measurements on rock paintings.

Experimental

The natural shelter of the archaeological site of Santana do Riacho where the rock paintings were made is approximately 80 meters long and 5 meters high. Hundreds of figures are clearly visible all over the wall, and six spots were selected based on their colours and apparent thickness. The spectrum of one red sample was collected in the lab from a small piece originating from the wall.

The spectrometer, consisting of the sensor head and source, was rigidly mounted on a tripod. The acquisition of the data as well as the power supply were located in a separate box. The spectra were collected in 1024 channels during about 24-48 hours. As this experiments were meant to test the ability of MIMOS to operate in the field, and also due to the lack of time, the statistics of some spectra are relatively poor. The average daily temperature was 27-32° C , and 18-22° C during the night. This is much higher than the temperature on Mars the instrument is designed to. Because of this the detector noise level was higher than in the normal case.

Results and Discussion

The spectra of some of the samples investigated are displayed in figures 1.-3., showing the data points itself as well as the fitted individual components. As can be seen for instance in fig.3, in some cases the baseline is not exactly horizontal. This effect is due to a problem with the electronics system, in particular the single channel analyser design, causing a velocity (e.g. channel number) dependent count rate in the case of increased noise level, which was the case because of the elevated temperatures. This problem, which appeared for the first time during these experiments, could be fixed in the meantime and is no issue anymore. Six individual paintings with distinguished different colours have been investigated during about 5 days in the field. The locations of the paintings on the wall are well separated (up to a few meters), so we do not expect any interference, but it may well be that the paintings have different ages, and that paints may originate from different origin.

The derived hyperfine parameters for four samples are listed in Table 1. Unfortunately the statistics of the spectra of two out of the six samples have been rather low and it was not possible to obtain meaningful fits. In fig.1 the data and a preliminary fit are shown for a red painting in painel 1, page 232 [3,4]. The preliminary analysis indicates that the dominating Fe-oxide pigment is hematite, with an Fe^{3+} doublet component which may be attributed to either superparamagnetic hematite particles or other Fe-oxide pigments. Hematite has been identified also in one other red painting (see table 1). In fig.2 the data obtained from an ochre painting are shown. The preliminary fit with two sextets indicates the presence of Goethite. But the doublet component can not be attributed to an oxide because of the large splitting. The parameters are indeed compatible with Jarosite, an Fe sulphate [6]. This mineral is found in mine spoils and other acidic environments, like certain areas of Mauna Kea volcano. To confirm the presence of Jarosite, an elemental analysis of the paint would be needed. This could be done in-situ at the Santana do Riacho site by using a portable X-ray fluorescence spectrometer.

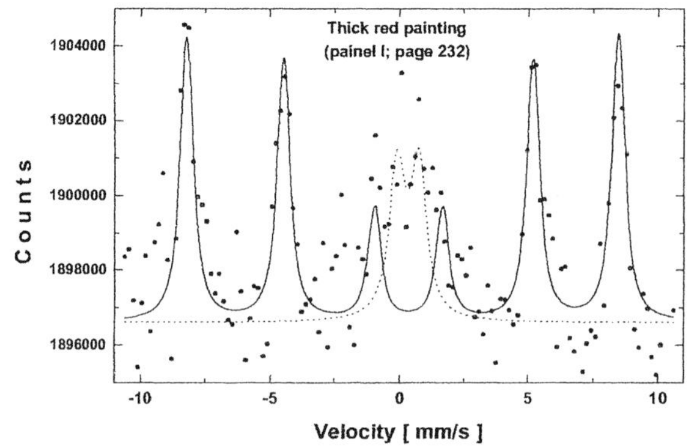

fig. 1.: Thick red painting located in painel I, page 232 of [3]

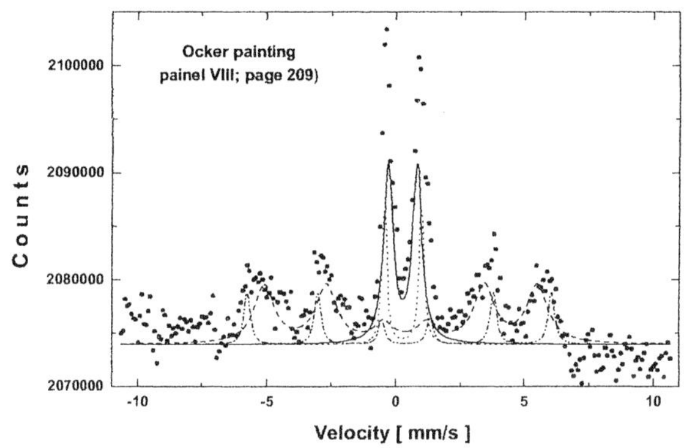

fig. 2.: Ochre painting located in painel VIII, page 209 of [3]

This first in-situ field Mössbauer experiment demonstrated that by the use of our miniaturized instrument MIMOS II it is possible now to obtain data with sufficient accuracy within an acceptable measurement time.

Acknowledgments: This work was funded by the Germany Space Agency DLR under contract 50QM99022, and by CNPq (Brazil).

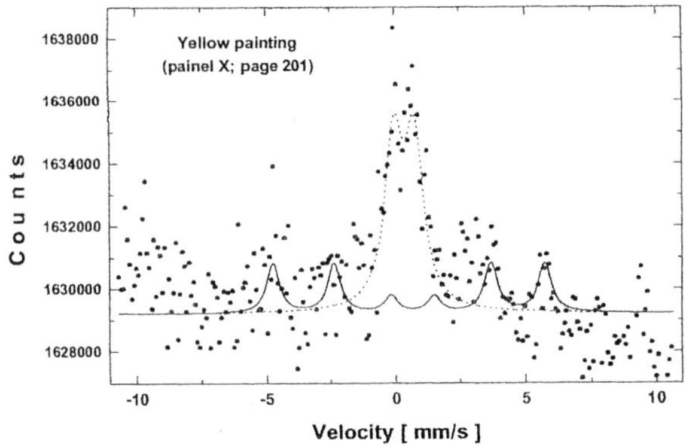

Fig. 3.: Yellow painting located in painel X, page 201 of [3]

Table 1. Hyperfine field H_{hf} (kOe), isomer shift δ (mm/s), quadrupole shifts $(2\varepsilon_Q)$/splittings(ΔE_Q)(mm/s) and relative areas RA (%) derived from the Mössbauer spectra.

Sample	H_{hf}	δ	$2\varepsilon_Q$	RA	ΔE_Q	δ	RA
Yellow	325	0.69	-0.16	34	0.80	0.45	66
Dark red	517	0.40	-0.22	17	0.85	0.40	36
	323	0.40	-0.17	8	1.36	0.38	39
Thick red	517	0.35	-0.22	80	0.63	0.51	13
					1.48	0.56	7
Ocher	365	0.38	-0.24	14	1.11	0.38	28
	328	0.40	-0.16	46	1.43	0.39	12

References

[1] G.Klingelhöfer, in: Mössbauer Spectroscopy in Materials Science, eds. M.Miglierini and D.Petridis, Kluwer Academic Publishers, 1999, p.

[2] G. Klingelhöfer, B. Fegley Jr., R.V. Morris, E. Kankeleit, P. Held, E. Evlanov, and O. Priloutskii, *Planet. Space Sci.*, Vol. 44, No. 11 (1996) 1277

[3] A. Prous, and I.M. Malta. Santana do Riacho, Tomo I. Arquivos do Museu de História Natural, Vol. 12, (1991) 384 p.

[4] A. Prous. Santana do Riacho, Tomo I. Arquivos do Museu de História Natural, Vols. 13/14 (1992) 417 p.

[5] G.M. da Costa, L.A. Cruz Souza, and M.F. de Jesus Filho, (1991) *Hyperfine Interactions* **67** (1991) 459.

[6] R.M. Cornell and U. Schwertmann, *The Iron Oxides*, VCH Verlagsgesellschaft, Weinheim; VCH Publishers, New York; 1996.

Trends in atmospheric iron measured by Mössbauer spectroscopy

B. Kopcewicz[1] and M. Kopcewicz[2]

[1]*Institute of Geophysics, Polish Academy of Sciences, Ks. Janusza 64, 01-452 Warsaw, Poland,*
[2]*Institute of Electronic Materials Technology, Wolczynska 133, 01-919 Warsaw, Poland.*

The Mössbauer measurements were performed on the atmospheric aerosol collected in Poznan (industrial city in central Poland) in period of 1972 – 1997. At this period the average iron concentration dropped from 1.9 μg m^{-3} to 0.4 μg m^{-3}, proportionally to the changes in general air pollution (from 0.075 mg m^{-3} to 0.015 mg m^{-3}). The changes of iron concentration were accompanied by changes in the chemical composition and structure of iron-containing aerosol (e.g. decrease of bulk components of iron oxides).

1. Introduction

Observation and monitoring of the aerosol background in the troposphere is very important for atmospheric physics. It is the first step in studying antropogenic components and their impact on the climate. Iron (both Fe(II) and Fe(III)) plays an important role in the multiphase atmospheric chemistry of S(IV) as a catalyst as well as an oxidant, and a photolytic source of OH radical [1]. In order to assess the extent in which the iron content in the troposphere may change and to which extent that change may be attributed to human activity, it is necessary to have a complete picture of the distribution of iron concentration and its variation. For these purposes the Mössbauer measurements of iron-containing atmospheric aerosols have been made for over 25 years at several sites in Poland [2]. In this paper we present results of measurements performed on the atmospheric aerosol collected in Poznan (industrial city in central Poland, LAT. 52.41, LON. 16.85) in the period of 1972 – 1997.

2. Experiment

Atmospheric air was pumped through each membrane filter for 24 h and the volume of air passed was recorded. The samples for the Mössbauer measurements were prepared using 15 filters obtained during 15 successive days. The volume of air pumped through the 15 filters varied from approximately 1000 to 1200 m^3. The samples were used in the Mössbauer measurements as absorbers. Measurements were performed at 300 and 78 K. The isomer shifts were related to the α-Fe standard. The general air pollution was determined (in μg m^{-3}) by weighing the amount of dust on each filter.

3. Results

The results of the Mössbauer measurements performed for samples collected in 1972, 1992 and 1997 in Poznan are presented. As an example Figure 1 shows Mössbauer spectra obtained for the samples collected at the first half of the December 1972 run at room temperature (Fig. 1a) and at 80 K (Fig. 1b). Similar Mössbauer spectra measured at room temperature (Fig. 1c) and at 80 K (Fig. 1d) were recorded for samples collected at the same site 25 years later, at the second-half of June 1997. The Mössbauer parameters determined for selected samples are shown in Table 1. Low temperature measurements (Fig. 1b and 1d) improve significantly the credibility of the phase identification because they allow

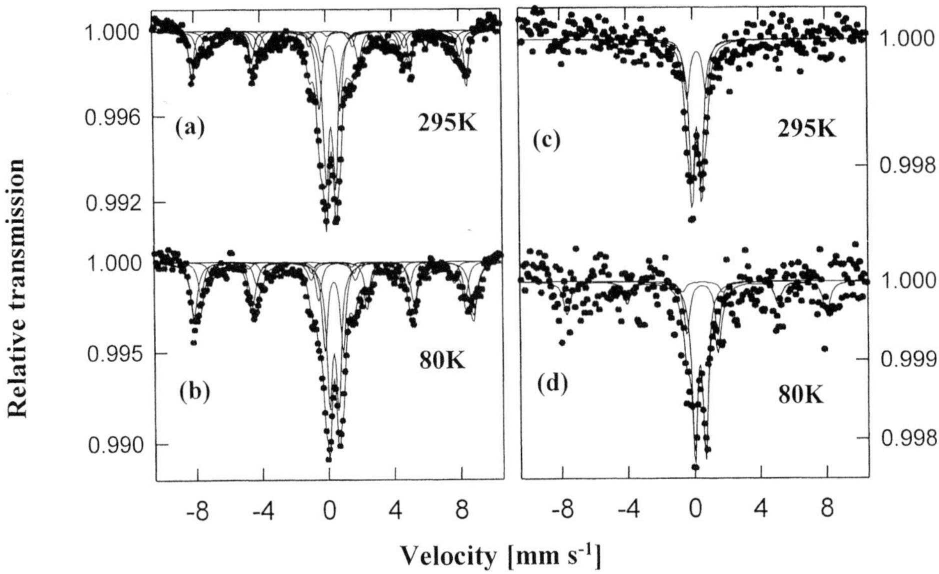

Fig. 1. Mössbauer spectra for aerosol samples collected in Poznan (December I/1972) measured at 295K (a), at 80K (b), and in June II/1997 measured at 295K (c) and 80K (d).

Table 1. Mössbauer parameters for selected samples collected in Poznan[a]

Sample	Component	δ [mm s^{-1}]		Δ [mm s^{-1}]		H_{hf} [T]		A^b [%]	
		RT	LN	RT	LN	RT	LN	RT	LN
	ϒ-FeOOH	0.39	0.48	0.55	0.61			31.9	25.9
	jurasite	0.33	0.49	1.23	1.11			17.1	17.0
	pyroxen	1.10	1.10	2.24	2.90			7.3	7.6
P1	Fe(2+) magnetite	0.57	0.92			45.5	46.6	9.6	6.3
	Fe(3+) magnetite	0.25	0.44			47.8	49.7	8.2	13.3
	hematite	0.37	0.46			51.3	52.4	21.4	27.2
	FeS	0.81	0.8			31.3	33.8	4.5	2.6
	ϒ-FeOOH	0.34	0.47	0.57	0.68			68.8	54.0
P2	jurasite	0.37	0.63	1.18	1.21			31.8	18.1
	hematite		0.48				52.2	0	27.9

[a] P1 and P2 are the samples for first half of December 1972 and second half of June 1997, respectively.
[b] A is the relative spectral area of the Mössbauer spectrum related to the concentration of the given compound.
RT and LN are the room temperature and liquid nitrogen temperature, respectively.

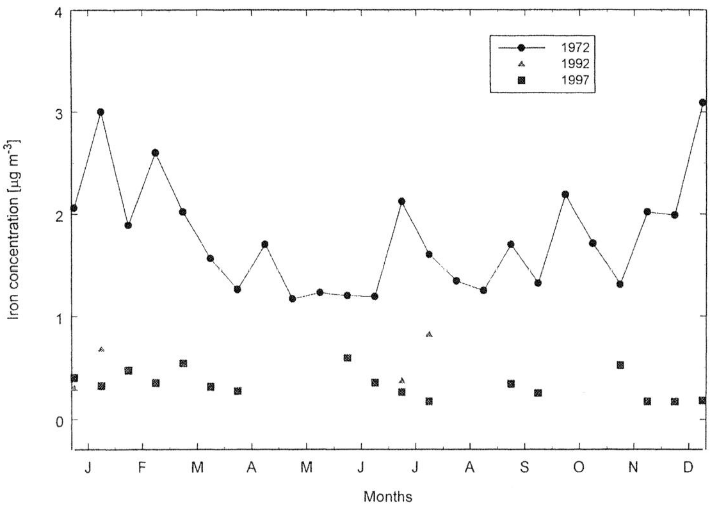

Fig. 2. Seasonal variations of iron concentration

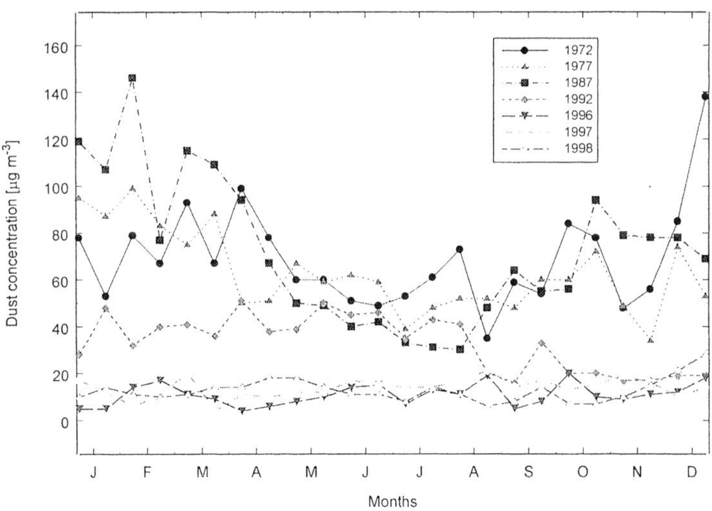

Fig. 3. Seasonal variations of air pollution

observation of the change of the magnetic behaviour of ultrafine α-Fe_2O_3 particles (transition from superparamagnetic, at room temperature, to ferromagnetic state at 80 K). The increase of the spectral area of the magnetic component at 80 K (Fig. 1b) at the expense of the quadrupole doublet reveals that about 6% of the spectral area assigned originally to iron oxyhydroxide corresponds to ultrafine α-Fe_2O_3 particles. Similar tendency was observed for sample at Fig. 1c and 1d.

The concentration of ^{57}Fe in all samples was calculated from experimental spectra using the "area method" [3] described in our earlier papers [2,4]. Assuming the natural abundance of iron in the sample (^{57}Fe amounts to 2.19% in natural iron) Fig. 2 shows the concentration of iron determined for samples collected in Poznan in 1972, 1992 and 1997.

Concentrations of airborne particles have been measured in Poznan over last 40 years. Recorded series of air pollution data (Fig. 3) showed high levels of total aerosol concentration until the early 1990s. Also, seasonal variations of aerosol concentration observed in 1970s and 1980s have disappeared, and now the level of dust concentration is fairly stable all over the year. Similar year to year variations of the seasonal pattern were observed for iron concentration (Fig. 2). The changes in the iron concentration were accompanied by changes in the chemical properties of the collected aerosol. In samples collected in 1990s we observed much less (if any) iron in the form of iron sulfides, which were the products of coal combustion.

4. Conclusions

1. Total atmospheric air pollution measured in Poznan in 1972-1998 shows a distinct downward trend (Fig. 3).
2. The same downward trend was observed for annual iron concentration measured by Mössbauer technique (Fig. 2).
3. The changes (both, in dust and iron concentration) are more distinct in Autumn and in Winter, meaning that the high air pollution was related to coal combustion (heating season), particularly that it was accompanied by changes of iron sulfides concentration.
4. In the period of 1972 – 1997 the average iron concentration dropped from 1.9 $\mu g\ m^{-3}$ to 0.4 $\mu g\ m^{-3}$, proportionally to the changes in total aerosol concentration (dropped from 0.075 $mg\ m^{-3}$ to 0.015 $mg\ m^{-3}$).

Acknowledgements

The filters and data concerning air pollution were kindly provided by the Institute of Meteorology and Water Management (Poland).

The financial support from the grant No. 6 PO4D 057 17 from the Polish Committee for Scientific Research is gratefully acknowledged.

References

[1] J. H.Seinfeld, in: *Air Pollution,* ed. A Wiley Interscience Publication (John Wiley & Sons, New York/Chichester/Brisbane/Toronto/Singapore, 1986), p. 220.
[2] B. Kopcewicz and M. Kopcewicz, in: *Analytical Chemistry of Aerosols*, ed. K. R. Spurny (Lewis Publishers, Boca Raton/London/New York/Washington. D.C., 1999), p.185.
[3] D.W. Hafemeister and E. Brooks Shera, Nucl. Instrum. Methods 41(1966)133.
[4] M. Kopcewicz and B. Dzienis, Tellus 23(1971)176.

Energetic Heavy Ion Radiation Effect in Brazilian Emeralds

E. Kuzmann**, V. K. Garg*, A. C. de Oliveira* , G. C. B. Braga*, Z. Klencsár** and

K. Havancsák***

*Institute of Physics, University of Brasília, 70910-900 Brasília, Brazil
**Department of Nuclear Chemistry and Research Group of Nuclear Methods in Structural Chemistry HAS, Eötvös University, Budapest, Hungary.
*** Department of Solid State Physics, Eötvös University, Budapest, Hungary

Emerald samples, originating from Goiás, Brazil, containing a significant portion of Fe^{3+} beside Fe^{2+} were irradiated with 209 MeV ^{89}Kr ions in order to study the effect of irradiation on the valence state of iron. Using ^{57}Fe transmission Mössbauer spectroscopy a valence state transition from Fe^{3+} to Fe^{2+} was clearly observed as a result of irradiation.

Keywords: Emerald, heavy ion irradiation, Mössbauer spectroscopy

1. Introduction

The dark green coloured emeralds are a Cr containing variety of beryl ($Be_3Al_2Si_6O_{18}$). Their colour is a result of Cr cations acting as colour centres in the structure. Emerald has a hexagonal crystal structure [1] in which Al^{3+} and Be^{2+} cations are coordinated with oxygen ions in an octahedral and tetrahedral configuration, respectively. Beryls can contain a small amount of iron as well. Price et al. applied ^{57}Fe Mössbauer spectroscopy in order to study the location of iron in the structure of beryl [2]. Their sample contained Fe^{2+} at two different sites in the structure of beryl. One of the sites was identified as the regular site of Be^{2+}, and the other as the regular site of Al^{3+}. Apart from Fe^{2+} they found a relatively small amount of Fe^{3+} in their samples. At the same time by applying a heat treatment of 600 °C for 1 h they observed an evidence for a heat induced reduction of Fe^{3+} to Fe^{2+}. Their results indicated that Fe^{3+} occupies Al^{3+} as well as Be^{2+} sites in the structure, just like Fe^{2+} cations do. Platonov et al. [3] have shown that in beryls containing iron the heat induced reduction of Fe^{3+} to Fe^{2+} is connected to the colour change of beryl observed simultaneously.

Emeralds originating from Brazil contain unusually large amounts of iron. In a recent paper [4] we have shown that in emeralds from Brazil about 33 % of iron is in the 3+ valence state. An attempt was made to transform Fe^{3+} to Fe^{2+} using a similar heat treatment (T = 600 C for 3 hours) as applied for other beryls [2,3]. Surprisingly, unlike in the above mentioned cases, we did not observe any change in the Fe^{3+}/Fe^{2+} ratio as a result of heat treatment. In order to study the Fe^{3+} to Fe^{2+} transformation in emeralds we applied energetic heavy ion irradiation on the Brazilian emerald samples. Radiation effects of light particles on beryls have already been investigated, and valence state changes due to irradiation were observed [5-7]. Mössbauer spectroscopy is a very sensitive tool to detect changes in the valence state even in minerals [8]. Consequently, we examined the effect of irradiation by using mainly ^{57}Fe transmission Mössbauer spectroscopy.

2. Experimental

For the present study gem-quality emeralds were collected from Campos Verdes (Goiás-GO-Brazil). The samples were identified as emerald by optical methods, X-ray diffractometry (X-ray diffractograms were recorded by means of a computer controlled RIGAKU diffractometer using

CuKα radiation and a monochromator) and electron microprobe analysis, based upon its characteristic colour, photoluminescence absorption of Cr, crystal structure and composition [9].

^{57}Fe Mössbauer spectra of emerald samples were recorded by a conventional Mössbauer spectrometer (WISSEL) using a proportional detector in transmission geometry both at room temperature and liquid nitrogen temperature (78 K) using a 10^8 Bq activity ^{57}Co(Rh) source. The procedure of sample preparation was similar to that described in our previous work [4]. The samples were measured at the same condition before and after the irradiation.

The samples were irradiated with 209 MeV Kr ions, available at the U-400 cyclotron of JINR, in Dubna. Irradiations were carried out at room temperature in a vacuum better than 10^{-6} Pa. As the irradiation flux was maintained at a relative low level (10^9 ions/cm^2.s), the temperature did not increase more than 15°C during the irradiation. The fluX of irradiation was 10^{12} ion/cm^2 . The thickness of the sample was in the range of the penetration depth of the applied Kr ions.

3. Results and Discussion

Fig. 1a and 1b show the room temperature ^{57}Fe Mössbauer spectrA of the emerald sample before and after the irradiation, respectively. The spectra can be decomposed into four doublets (D1, D2, D3 and D4) representing four distinct iron sites. The corresponding Mössbauer parameters are tabulated in Table 1. The site assignment of these spectral components in emerald are the following [4]:

Doublet D1 is attributed to Fe^{2+} located at octahedral Al site
Doublet D2 is associated with Fe^{2+} ions occupying the tetrahedral Be site
Doublet D3 is associated with Fe^{3+} ion substituted into the octahedral Al site
Doublet D4 is attributed to Fe^{3+} ion substituted into the tetrahedral Be site

Table1
Mössbauer parameters of Brazilian emerald

Mössbauer parameters	Emerald	
	non-irradiated	irradiated
δ (D1) (mm/s)	1.10±0.01	1.11±0.01
δ (D2) (mm/s)	0.95±0.01	0.96±0.01
δ (D3) (mm/s)	0.58±0.01	0.55±0.02
δ (D4) (mm/s)	0.23±0.01	0.22±0.02
ΔE_q (D1) (mm/s)	2.71±0.02	2.65±0.02
ΔE_q (D2) (mm/s)	1.74±0.02	1.73±0.03
ΔE_q (D3) (mm/s)	0.55±0.02	0.41±0.02
ΔE_q (D4) (mm/s)	0.28±0.02	0.20±0.02
A (D1) (%)	36±3	54 ±4
A (D2) (%)	21±2	37 ±3
A (D3) (%)	29±2	5 ± 2
A (D4) (%)	14±1	4 ±2

δ is the isomer shift
ΔE_q is the quadrupole splitting
A is the relative line area

Table 2
Distribution of iron in Brazilian emeralds derived from the Mössbauer data

Fe	non-irradiated sample	Irradiated sample
Fe^{3+}/Fe	0.43	0.10
Fe^{2+}/Fe	0.57	0.90
Fe^{2+}/ Fe^{3+}	1.33	9.00
Fe^{3+}/ Fe^{2+}	0.75	0.11
Fe^{2+}/Fe in Al site	0.36	0.53
Fe^{2+}/Fe in Be site	0.21	0.37
Fe^{3+}/Fe in Al site	0.29	0.05
Fe^{3+}/Fe in Be site	0.14	0.05
Fe^{2+}+ Fe^{3+}/Fe in Al site	0.65	0.58
Fe^{2+}+Fe^{3+}/Fe in Be site	0.35	0.42
$Fe^{2+}_{Al}/Fe^{2+}_{Be}$	1.71	1.43
$Fe^{3+}_{Al}/Fe^{3+}_{Be}$	2.07	0.72
$Fe^{2+}_{Al}/Fe^{3+}_{Al}$	1.24	10.39
$Fe^{2+}_{Be}/Fe^{3+}_{Be}$	1.5	7.71
Fe_{Al}/Fe_{Be}	1.85	1.39

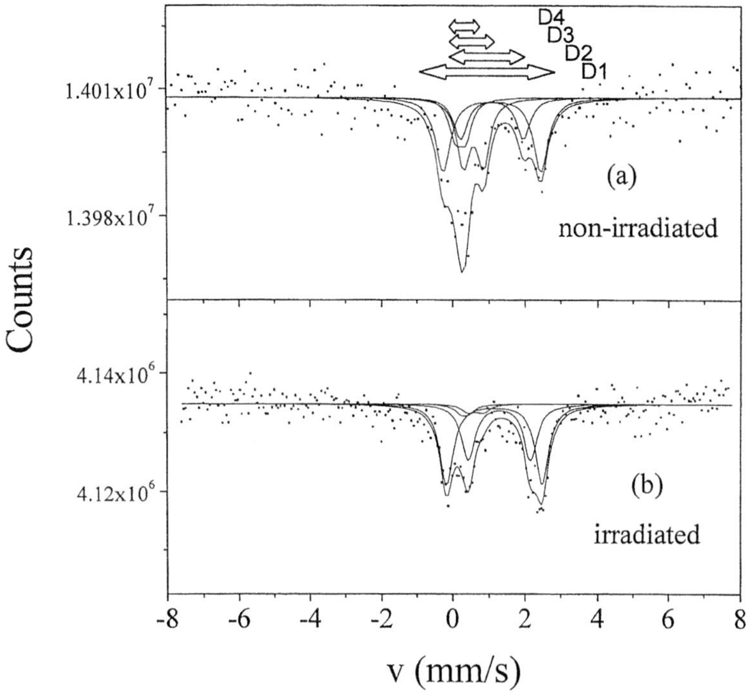

Figure 1. Room temperature Mossbauer spectra of emerald sample (a) non-irradiated, (b) irradiated with 209 MeV Kr ions with a dose of 10^{12} ions/cm^2

Comparing the relative intensities of subspectra belonging to the emerald before irradiation (Table 1), two main observations can be noted:

- Before the irradiation Fe^{2+} and Fe^{3+} distribute almost equally among Al and Be sites (Table 2).
- The relative occurrence of Fe^{3+} is very large in the unirradiated Brazilian emerald (43 %).

The ^{57}Fe Mössbauer spectrum of the irradiated emerald (Fig. 1b) differs significantly from that of the non-irradiated one (Fig 1a). The most striking changes are the decrease of the relative area of the subspectra belonging to Fe^{3+} . The corresponding area fractions are depicted in Table 1. It is clear that as a result of irradiation the relative area of the doublets D3 and D4, belonging to Fe^{3+}, diminished strongly at the expense of the relative area of doublets D1 and D2 representing Fe^{2+}. This finding indicates the reduction of Fe^{3+} to Fe^{2+} as the main result of the energetic heavy ion irradiation. This result is consistent with the earlier results obtained in connection with *light* ion irradiations [5-7].

Considering that the irradiation promotes the transformation of metastable states to the equilibrium one [10], the observed reduction of Fe^{3+} indicates that iron in the 3+ valence state can be considered to be metastable in emerald. The metastability of Fe^{3+} in emerald is in accordance with the heat treatment induced reduction of Fe^{3+} to Fe^{2+} as observed in some beryls [2,3]. However, given that the Fe^{3+} to Fe^{2+} transformation was not possible to realise via heat treatment at 600 °C for 3 hours, in emerald the Fe^{3+} state seems to be more stable than in other kind of beryls.

This increased stability of Fe^{3+} may be connected to the higher iron content found in the Brasilian emerald.

In the case of heavy ion irradiation the energy of the Kr ion is transferred to the target material in two different ways. By energy deposition in electronic excitations and by elastic collisions with the atoms of the target material. In the case of emerald being an insulator, the energy deposition by electronic excitations is expected to dominate. Accordingly, the reduction of Fe^{3+} to Fe^{2+} via energetic heavy ion irradiation in emerald can be explained as follows.

The energetic Kr ions transfer their kinetic energy to the emerald mainly by creating so called thermal spikes, where the local temperature can extend well over 600 °C. Although Kr ions will undoubtedly create new defects in the system, the resulted high local temperature can change the original defect structure around Fe^{3+} cations at both Al and Be sites, and consequently it can promote the transformation of Fe^{3+} to the more stable Fe^{2+}.

As the site occupancy of Fe (2+ and 3+) at the Al and Be sites did not change considerably as a result of irradiation (Table 2), the reduction process is likely to occur without considerable diffusion process. Thus, the irradiation effects mainly the electronic state of Fe^{3+} without altering its site occupancy considerably.

4. Conclusions

Using energetic heavy ion irradiation we have shown that the $Fe^{3+} \rightarrow Fe^{2+}$ transformation, observed earlier in some beryls due to thermal treatment, takes place in emerald at Al as well as at Be sites. The reduction process due to energetic heavy ion irradiation occurs locally, without considerable diffusion process.

Acknowledgments

The authors express their thankfulness for the financial support to FAPDF (project 0916-VKG, project 052/98 ACO), to CNPQ (project VKG 400134/97-7) and to OTKA (projects T014970 and T029537). The irradiation at JINR was financially supported by the Hungarian Academy of Sciences.

References

[1] W.A. Deer, R.A. Howie, J. Zussman: ROCK FORMING MINERALS, Longman, London (1962), Vol.1
[2] D.C. Price, E.R. Vance, G. Smith, A. Edgar, B.L. Dickson: *J. Phys. (Paris)* **37** (1976) C6 - 811.
[3] A.N. Platonov, M.N. Taran, O.E. Minko, E.V. Polshin: *Phys. Chem. Miner.* **3** (1978) 87.
[4] E. Kuzmann, V. K. Garg, A. C. de Oliveira and G. C. B. Braga: (Sent for publication in this volume)
[5] Charles T. Liddicoat: *Gems and Gemology* **24** (1988) 196.
[6] K. Nassau: GEMSTONE ENHANCEMENT - HISTORY, Science and State of Art, Butterworth-Heinemann, (1994) Oxford, U.K. p 47.
[7] K. Schmetzer:, *J. Gemm.* **23** (1993) 288.
[8] E. Kuzmann, S. Nagy, A. Vértes, T.G. Weiszburg, K.V. Garg: GEOLOGICAL AND MINERALOGICAL APPLICATIONS OF MÖSSBAUER SPECTROSCOPY, in *Nuclear Methods in Mineralogy and Geology* (eds. A. Vértes, S. Nagy, K. Süvegh) Plenum Press, NY (1998)
[9] J.A. Freitas, A.C. de Oliveira, K.V. Garg, G.C. Braga: *XXII Encontro Nacional de Física da Materia condensada, São Lourenço, Brazil,* (1999)
[10] Frank V. Nolfi, jr (ed.): PHASE TRANSFORMATION DURING IRRADIATION, Applied Science Publisher, London, NY (1983)

MÖSSBAUER SPECTROSCOPIC STUDY ON CHARACTERIZATION OF IRON IN THE PERMIAN TO TRIASSIC DEEP-SEA CHERT FROM JAPAN

M. Matsuo, K. Kubo and Y. Isozaki
Graduate School of Arts and Sciences, The University of Tokyo,
3-8-1 Komaba, Meguro-ku, Tokyo 153-8902, Japan.

Chemical states of iron in the Permian to Triassic deep-sea chert from Japan have been investigated by ^{57}Fe Mössbauer spectroscopy. In the ^{57}Fe Mössbauer spectra five iron species were identified, that is, hematite (α-Fe_2O_3), pyrite (FeS_2), goethite (α-FeOOH), paramagnetic Fe^{2+}, paramagnetic Fe^{3+}. In the Early to Middle Permian and Middle to Late Triassic cherts, main iron-containing mineral was hematite and less than 35 % of iron was in the ferrous state. This fact suggests that oxic conditions persisted in the deep-sea, such that ferric iron was stable in the sediments. On the other hand, in the Late Permian and Early to early Middle Triassic cherts, main iron-containing mineral was pyrite, and hematite disappeared completely, then more than 65 % of iron was in the ferrous state. This fact suggests that the redox condition of deep-sea open-ocean globally changed from oxic to anoxic for approximately 20 million years across the Permo-Triassic boundary.

1. INTRODUCTION

It has been known that the greatest mass extinction in the Phanerozoic occurred at the timing of Permo-Triassic (P-T) boundary (251 million years ago) [1,2]. Various hypotheses for the cause have been proposed and focused on changes in the ocean, including overturn of an anoxic ocean [3]. Recently, P-T boundary sections were discovered in deep-sea cherts that crop out extensively in the Jurassic accretionary complex in southwest Japan [4]. The cherts represent ancient pelagic sediments, which were primarily deposited in the deep-sea of the superocean Panthalassa and accreted to Japan in the Middle Jurassic [5]. Deep-sea chert, one of sedimentary rocks, has a record about the environmental changes in ancient open-ocean. Therefore, the continuous chert samples in southwest Japan are thought to give an important clue to clarifying a secular change of deep-sea open-ocean across the P-T boundary.

The ^{57}Fe Mössbauer spectroscopy is one of the most useful means for characterizing the chemical states of iron in a wide variety of natural samples [6]. This technique has been employed successfully in our previous investigations on urban estuarine sediments [7] and natural chromite samples from ophiolite complexes [8] as well as chert samples from Japan, in which we clarified that colors of chert derived from iron-containing minerals reflected the depositional condition [9,10]. In the present study, thus, we have applied ^{57}Fe Mössbauer spectroscopy to the Permian to Triassic deep-sea chert from Japan. Our attention is paid to investigating the chemical states of iron in the cherts and to examining a redox change of deep-sea open-ocean across the P-T boundary.

2. EXPERIMENTAL

The chert sections including the Permian-Triassic transition occur in the Mino-Tanba belt in southwest Japan. They have not suffered from high-grade metamorphism or deformation. The sedimentation rate of bedded chert was estimated to be very low, approximately 0.5 g/cm^2/1000 yr [5]. For this study, we collected 14 chert samples from the Permian and 34 chert samples from the Triassic. In addition, 5 samples of siliceous/carbonaceous claystone near P-T boundary were collected. The collected sample was cut with a diamond cutter and only fresh parts were used for the following experiments. These samples were firstly ground to coarse powder in a tungsten mortar, and ground into finer powder (100 mesh) in an agate mortar. Approximately 400 mg of each sample was mounted in a sample holder (16mm in diameter, 2 mm thick) for Mössbauer measurement.

Mössbauer spectra were measured with an Austin Science S-600 Mössbauer spectrometer using a 1.11 GBq $^{57}Co/Rh$ source at room temperature. The curve fitting of the obtained spectra was performed with a personal computer, assuming that the spectra were composed of peaks in Lorentzian line shape. Isomer shifts were expressed with respect to the centroid of the spectrum of metallic iron foil. Peak positions of pyrite (FeS_2) were constrained (using values of δ =0.307 mms^{-1}, ΔEq=0.610 mms^{-1} at room temperature) as in Manning and Ash [11], since the peak positions of paramagnetic high-spin Fe^{3+} and pyrite, containing low-spin Fe^{2+}, are in very close proximity. We judged that pyrite did not exist when the calculated peak area of pyrite was lower than that detection limit calculated as three times the standard deviation of the baseline count.

3. RESULTS AND DISCUSSION

Bedded chert consists of rhythmically alternating vitric siliceous layers (ca. 5 cm thick) and argillaceous layers (ca. 2 mm thick). The vitric layers have abundant radiolarian shells and their fragments, while the argillaceous layers consist mostly of fine-grained clays. The Early to Middle Permian and Middle to Late Triassic cherts either retain the original brick red color. On the other hand, Late Permian and Early to early Middle Triassic cherts show no reddish color but have gray to black primary colors.

We found a layer of yellow chert in the Middle Triassic. Above this layer all cherts showed reddish color, and below this layer the cherts mainly showed dark color such as gray and black. Figure 1 shows a composite of three typical Mössbauer spectra out of 34 spectra of the chert samples in the Triassic measured at room temperature. The fit curves shown as solid lines in Figure 1 well threaded the observed spectra, and the chi-squared values were sufficiently small, which justifies the validity of the applied model. Mössbauer spectra of brick red chert (Fig. 1a) consist of two overlapping doublets and one sextet. The doublet with a larger quadrupole splitting (δ =1.11 mms^{-1}, ΔEq=2.65 mms^{-1}) was ascribed to paramagnetic high-spin Fe^{2+}, whereas the doublet with a smaller quadrupole splitting (δ =0.369 mms^{-1}, ΔEq=0.614 mms^{-1}) was ascribed to paramagnetic high-spin Fe^{3+} according to Mössbauer parameters. These iron components may be contained in silicate

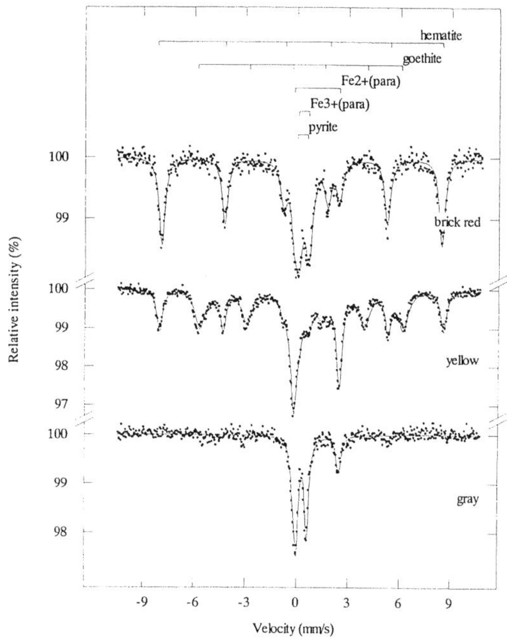

Fig. 1. Mössbauer spectra of the chert samples collected from the Triassic.

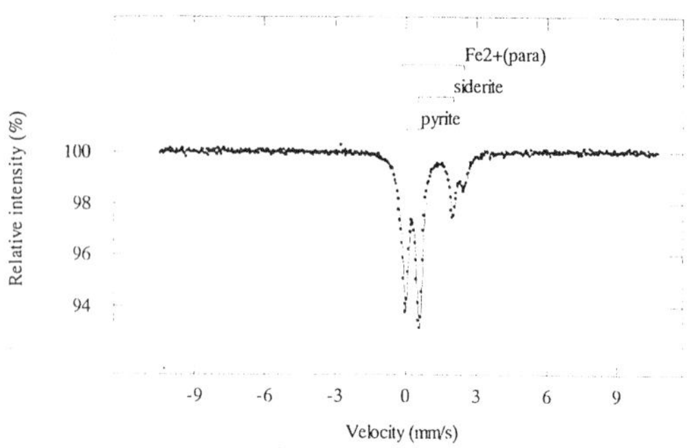

Fig. 2. Mössbauer spectra of siliceous/carbonaceous claystone collected from just below the Triassic gray chert.

Table 1. Relative content (%) of individual iron species in chert and siliceous/carbonaceous claystone samples calculated from Mössbauer peak area.

Sample	Color	Fe^{3+}				Fe^{2+}			
		hematite	goethite	para Fe^{3+}	total	pyrite	siderite	para Fe^{2+}	total
Triassic chert									
Upper	brick red	54.0	–	32.8	86.8	–	–	13.2	13.2
Middle	yellow	24.6	37.3	12.4	74.3	–	–	25.7	25.7
Lower	gray	–	–	–	0.0	69.4	–	30.6	100.0
Claystone (near P-T boundary)	gray	–	–	–	0.0	55.0	25.1	19.8	100.0
Permian chert									
Upper	gray	–	–	32.2	32.2	14.5	–	53.2	67.8
Lower	brick red	63.7	–	20.2	83.9	–	–	16.1	16.1

or clay minerals. The sextet whose hyperfine field was about 50.7 T was assigned to hematite (α-Fe_2O_3). The existence of hematite as a main component suggests that oxic conditions persisted in the deep-sea.

In addition to these three components, another sextet was found in the yellow chert (Fig. 1b). From the Mössbauer parameters (δ =0.377 mms^{-1}, Δ Eq=-0.249 mms^{-1}, Hi=36.8 T), this component could be assigned to goethite (α-FeOOH). In contrast, Mössbauer spectra of the gray chert showed quite different appearance, namely, no significant magnetic sextet was detected (Fig. 1c). The spectra consist of two doublets; one of them was assigned to paramagnetic Fe^{2+}, which was the same component as found in brick red and yellow cherts. Although the peak position of another doublet was very close to that of paramagnetic Fe^{3+}, we could clearly distinguish the doublet as pyrite, which contained low-spin Fe^{2+}, by the method described in the experimental part. The existence of pyrite and compete lack of hematite is suggestive of an anoxic depositional environment. Thus, we found five iron components in the chert samples collected from the Triassic. In the same way, we analyzed iron components in 14 chert samples collected from the Permian. Relative content was calculated as the peak area of individual iron species in the Mössbauer spectra (Table 1). The composition of iron species in the gray chert is totally different from the reddish chert. Ferrous iron and pyrite are dominated in the gray chert whereas ferric iron dominated in the reddish chert.

Very near to P-T boundary, there are no chert samples occurring, instead, siliceous/ carbonaceous claystone was found, of which thickness was approximately 20-35 m. The color of claystone was as dark as gray to black. Figure 2 shows one of Mössbauer spectra of siliceous/ carbonaceous claystone, which was collected from just below the Triassic gray chert. The spectra consist of three doublets; two of them were assigned to paramagnetic Fe^{2+} and pyrite. The third doublet (δ =1.24 mms^{-1}, ΔEq=1.50 mms^{-1}) was attributed to siderite ($FeCO_3$) or iron carbonate compound in which iron was partly substituted by other metals. Relative content of iron species in the claystone is also shown in Table 1. In this way, it was found that chemical state of iron in the siliceous/carbonaceous claystone was almost ferrous state.

It is considered that the chemical states of iron in chert and siliceous/carbonaceous claystone reflect the ancient redox condition of the depositional environment. Therefore, we calculated the relative content of total ferrous iron from Mössbauer data. Here, total ferrous iron means the sum of the percentages of pyrite, siderite and paramagnetic Fe^{2+}. Figure 3 shows stratigraphic redox change of the Permian to Triassic deep-sea pelagic sediments, evaluated by the relative content of total ferrous iron. In this figure, color of columnar section of chert and siliceous/carbonaceous claystone is also shown. In the Early to Middle Permian and Middle to Late Triassic cherts, main iron-containing mineral was hematite and less than 35 % of iron was in the ferrous state. This fact suggests that oxic conditions persisted in the deep-sea, such that ferric iron was stable in the sediments. On the other hand, in the Late Permian and Early to early Middle Triassic cherts, main iron-containing mineral was pyrite, and hematite disappeared completely, then more than 65 % of iron was in the ferrous state. In siliceous/carbonaceous claystones near P-T boundary, more than 72 % of iron was in the ferrous state. Although there are temporary oscillations between oxic and anoxic states both in the Late Permian and in the early Middle Triassic, the above facts suggest that the redox condition of deep-sea open-ocean became anoxic for approximately 20 million years across the P-T boundary.

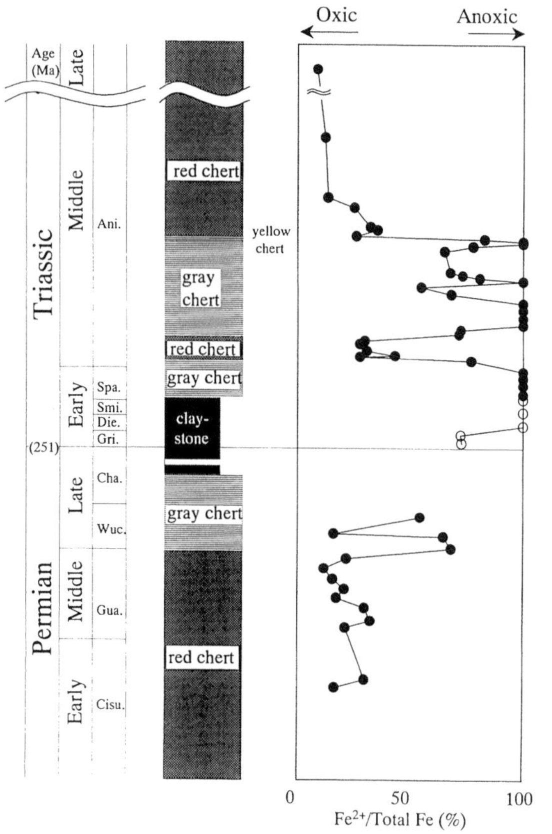

Fig. 3. Stratigraphic redox change of the Permian to Triassic deep-sea pelagic sediments, evaluated by the relative content of total ferrous iron.

4. CONCLUSIONS

The Mössbauer spectroscopic study on the ancient deep-sea pelagic chert is effective. The development of anoxic condition in the superocean Panthalassa across the P-T boundary was confirmed to have lasted for approximately 20 million years. It is suggestive of the coeval extinction of Paleozoic radiolarians in the Panthalassa.

ACKNOWLEDGEMENTS

We gratefully acknowledge the financial support from the Ministry of Education, Science, Sports and Culture, Japan through Grant-in-Aid for Scientific Research on Priority Areas (grant no. 09214204).

REFERENCES

[1] J. J. Sepkoski Jr., J. Geol. Soc. London 146 (1984) 7.
[2] D. H. Erwin, Nature 367 (1994) 231.
[3] A. H. Knoll, R. K. Bambach, D. E. Canfield and J. P. Grotzinger, Science 273 (1996) 452.
[4] Y. Isozaki, Science 276 (1997) 235.
[5] T. Matsuda and Y. Isozaki, Tectonics 10 (1991) 475.
[6] T. Tominaga and Y. Minai, Nucl. Sci. Appl. 1 (1984) 749.
[7] M. Matsuo, B. Takano and K. Sugimori, Il Nuovo Cimento D 50 (1996) 757.
[8] A. Kuno, R. A. Santos, M. Matsuo and B. Takano, J. Radioanal. Nucl. Chem. 246 (2000) 79.
[9] K. Nakao and Y Isozaki, J. Geol. Soc. Jpn. 100 (1994) 505.
[10] K. Kubo, Y Isozaki and M. Matsuo, J. Geol. Soc. Jpn. 102 (1996) 40.
[11] P. G. Manning and L. A. Ash, Can. Miner. 17 (1979) 111.

IRON MINERAL MODIFICATIONS IN THE SAND FRACTION OF A SUBTROPICAL SOIL FROM ARGENTINA

A. Mijovilovich[1], *H. Causevic*[2], *H. Morrás*[3] *and C.Saragovi*[2]
[1]EMBL c/o Desy. Notkestrasse 85. Geb 25A. 22603. Hamburg. Germany
[2]Dept of Physics, CNEA, Av. Gral. Paz 1499 (1650), Buenos Aires, Argentina
[3]INTA-CIRN, Institute of Soils, Villa Udaondo, (1712) Castelar, Argentina
Correspondent author: saragovi@cnea.gov.ar

Abstract The sand fraction of a B_{t22} horizon (105-155 cm depth) of a clayey red Ultisol in a subtropical forest environment (Misiones, Argentina) was studied by X-ray diffraction (XRD), saturation magnetization (σ_S), optical microscopy (OM) and Mössbauer spectroscopy (MS). Saturation magnetization for the whole sand fraction (wsf), the non magnetic (nmsf) and the magnetic fractions (msf) gave σ_S values of 10.79 $JT^{-1}Kg^{-1}$, 16.92 $JT^{-1}Kg^{-1}$ and 1.50 $JT^{-1}Kg^{-1}$ respectively. Mainly quartz, ilmenite, Al-substituted hematite, magnetite together with maghemite contribution and goethite were found. Magnetite/maghemite contents are high, being magnetite predominant in the msf. Results are compared with those from the upper B_1 horizon (10-35cm depth) of the same soil, in which a lower wsf σ_S value and higher values of msf σ_S and of (σ_S (msf)-σ_S(wsf)) values were measured. These results confirm the stability of magnetite in this soil contrasting with other results on soils from neighbouring áreas.

1.Introduction

Subtropical red-soils with a relatively high content of magnetic components are found in the province of Misiones (NE of Argentina). Some information concerning the Fe mineralogy and the development of a Typic Kandihumult from this area have been provided in previous studies [1,2,3,4]. An important amount of magnetite was found [1], at difference from soils developed in mafic pedodomains in neighbouring regions in Brazil where only maghemite was found in their coarse fraction [8, 9].

According to those former results, the B_1 horizon of these Ultisol contains magnetic compounds of rather high magnetization both in the sand and silt fractions. The sand fraction showed the highest magnetization value (24.3 $JT^{-1}Kg^{-1}$). In turn, the magnetic sand fraction showed lower magnetization values (14.3 $JT^{-1}Kg^{-1}$) in cultivated pedons than in the natural soils, even though the total magnetization for both pedons were similar (around 2.9 $JT^{-1}Kg^{-1}$), then suggesting some modifications on the iron mineralogy due to anthropic influence [1].

The objective of the present study is to further contribute to understand rock weathering and soil evolution processes in this area. With this aim, the magnetic mineralogy of a deeper horizon of a deeper horizon from the aforementioned Ultisol is studied by X-ray diffraction (XRD), saturation magnetization (σ_S), optical microscopy (OM) and Mössbauer spectroscopy (MS).

2.Experimental

A B_{t22} horizon (105-155 cm depth) of a clayey red Ultisol in a subtropical forest environment was selected for study and the sand (50 μm – 2000 μm), silt (2 μm – 50 μm) and clay fractions (< 2 μm) were obtained. A hand magnet was used to extract the magnetic sand fraction (msf) from the whole sand fraction (wsf). The msf accounts for 74 wt. % of the wsf and the remained 26 wt. % sand fraction was named the non-magnetic sand fraction (nmsf). Observations by optical microscopy were performed and, in particular observation and grain counting on the 250 – 500 μm sand sub-fraction, which is the most abundant and representative in this soil.

Magnetization measurements were performed using a commercial SQUID magnetometer at RT and in fields up to 5T.

Powder X-ray diffraction patterns were taken at RT using a Philips X-Pert PW3710 diffractometer with Cu graphite monochromatized radiation. The standard $5^{o} < 2\theta < 70^{o}$ range and a 0.02 -2θ step with a counting time of 1 s per step were used.

Mössbauer measurements were performed with the source at RT and the absorbers at RT and at 15K. Samples were carefully ground in an agate mortar, weighted and mixed with non-refined sugar to avoid texture effects [5] before assembling the "thin absorbers" [6]. All the spectra were fitted using the NORMOS program [7]. Crystalline and distribution sites were taken into account accordingly to the experimental spectra. The quality of each computer fit was checked by a χ^2 test. The relative populations of the Fe-bearing compounds were computed from their corresponding fitted area at 15 K in each sample.

3.Results and Discussion

The following main reflections were used for identification of the compounds present in the samples studied: for quartz (0.334, 0.426 nm), ilmenite (0.2754 nm), anatase (0.352 nm), hematite (0.27, 0.2519 nm) and magnetite (0.2532, 0.2967 nm) and maghemite (0.2514, 0.2953 nm). The X-ray diffraction patterns are shown in Figure 1, where Mt+Mh indicates that it was no possible to differentiate magnetite (Mt) from maghemite (Mh). Anatase is only suggested in the whole sand fraction pattern and Mt + Mh is hardly identified in the non-magnetic sand fraction pattern.

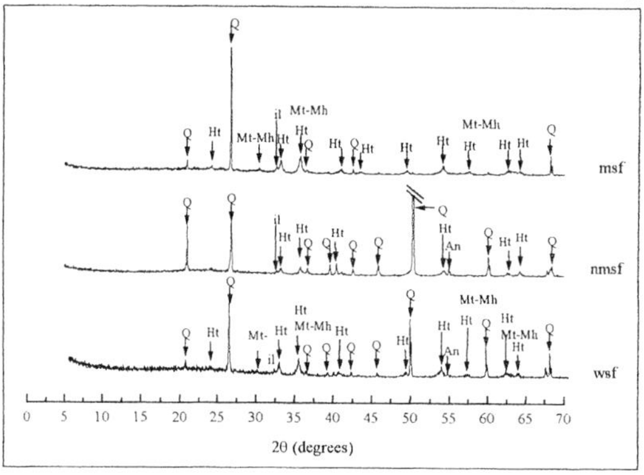

Figure 1: XRD diagram of the B22 Ultisol sand fractions

The values of the saturation magnetization obtained are: σ_S =10.79 $JT^{-1}Kg^{-1}$ for the wsf, σ_S =1.50 $JT^{-1}Kg^{-1}$ for the nmsf and σ_S =16.92 $JT^{-1}Kg^{-1}$ for the msf. The observation of the sand fractions under a polarizing microscope showed the presence of quartz grains, the typical opaque grains of magnetite and some hematite and probable goethite in different amounts for the different fractions.

Spectra at RT and 15K were fitted with two distributions of hyperfine fields (a narrow S1 distribution and a broad S2 one), one sextet (S3) and two doublets (D1 and D2). The behaviour of S1 and S2 when lowering the temperature helped to clarify the assignments. The S1, S2 and S3 parameters obtained at 15K and the corresponding normalized subspectral areas are shown in Table 1, in which the values of the highest maxima of the distribution profiles are indicated as H_M. The IS values are referred to α-Fe.

A characteristic fitted spectrum is displayed in Figure 2 a).

The paramagnetic doublets D1 (QS ≈ 0.78 mm/s, IS ≈ 1.4 mm/s at 15K) and D2 (QS ≈ 0.66 mm/s, IS ≈ 0.39mm/s at 15K) are due to the presence of Fe^{2+} and Fe^{3+} respectively; Fe^{2+} is located in the ilmenite detected by XRD. The high proportion of D2 could be due to contributions of Fe^{3+} from non-stoichiometric ilmenite and from superparamagnetic oxides even at 15K. The paramagnetic area is roughly 1/3 of the magnetic one. The paramagnetic region will be discussed elsewhere.

Table 1: Mössbauer parameters of the fitted sextets at 15K, H_M values are in Tesla; QS and IS are in mm/s. Areas are in percentages (%) of the total magnetic zone area.

Sample	S1				S2				S3			
	H_M	QS	IS	Area	H_M	QS	IS	Area	H_M	QS	IS	Area
Wsf	50.4	-0.03	0.28	25.8	49.5	-0.05	0.66	16.7	52.6	-0.20	0.43	57.5
Msf	50.0	-0.06	0.32	37.0	48.3	-0.10	0.79	19.2	52.4	-0.16	0.42	43.8
Nmsf	52.0	-0.05	0.35	25.8	43.0	-0.12	0.43	10.1	52.8	-0.20	0.44	64.1

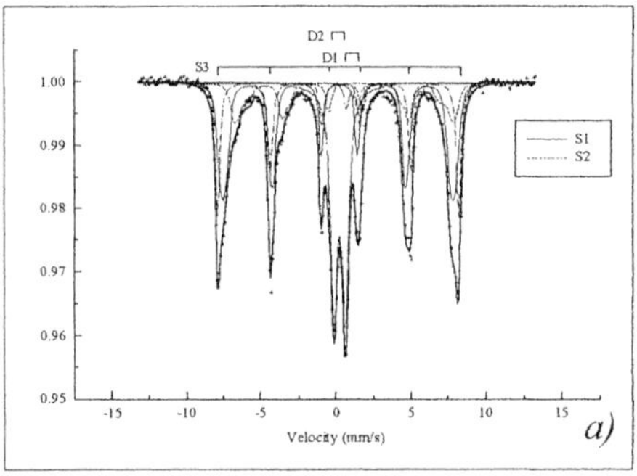

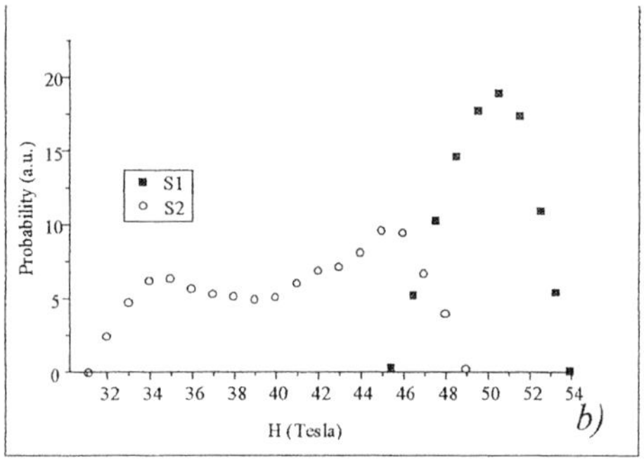

Figure 2: a) RT Mössbauer spectrum of the msf of a B_{22} horizon. b) Distributions profiles corresponding to nmsf at RT.

Sextet S3 is assigned to weakly ferromagnetic (WF) hematite for all the used temperature range since no Morin transition occurred. The absence of this transition is due to very small sized hematite and/or Al substitution. In the present highly weathered soil fractions, Al substitution is very likely to occur (total Al_2O_3 = 28.3%, total Fe_2O_3 = 16.1%). Hematite is present in all fractions, even in the msf pointing out to a partial separation achieved with the hand magnet. Contributions of highly substituted and or small sized hematite, which would be present in S1 and S2, would enlarge the values displayed in Table 1.

S1 distribution is assigned to Fe^{3+} from site A of magnetite and to a possible contribution of maghemite. Broad S2 distribution shows the contribution of $Fe^{2.5+}$ from sites B of magnetite and of Fe^{3+} from goethite. S1 and S2 areas proportions increase their values in the msf. The presence of Mt and Mh is also supported by the other results found; both are mainly concentrated in the msf, remaining even in the nmsf after the present magnetic separation. The IS value of the msf S2 distribution indicates $Fe^{2.5+}$ suggesting that the magnetite predominates over maghemite. The variation of the (S1+S2+S3)/(D1+D2) signal ratio when lowering T is significant in the nmsf (~24% vs ~5% in the wsf, ~8% in the msf) suggesting that poorly crystallized oxides, in particular goethite are mainly present in the nmsf. Ferrihydrite and highly substituted or very small sized hematite cannot be discarded. The shape of the S2 distributions at RT (Figure 2 b) show a maximum at about 35 T shifting to higher values when the temperature is lowered, which is consistent with the existence of goethite in the nmsf and furthermore, the optical micrographs showed "stained" coarser grains. Then, it is possible to conclude that poorly crystallized oxides as goethite are mainly coating the quartz and other silicate and iron oxides grains in the wsf. It should be mentioned that the S2 hyperfine field distributions corresponding to the msf show a smooth shape around 35 T. The σ_S values measured on this B_{t22} horizon were σ_S = 10.79 $JT^{-1}Kg^{-1}$ for the wsf and σ_S = 16.92 $JT^{-1}Kg^{-1}$ for the msf. For the B_1 (10-35 cm) horizon of the same profile [1], the values found were σ_S = 2.9 $JT^{-1}Kg^{-1}$ for the wsf and σ_S = 24.3 $JT^{-1}Kg^{-1}$ for the msf. In the B_{t22} horizon, the (σ_S (msf)-σ_S (wsf) value is 8 $JT^{-1}Kg^{-1}$ while in the B_1 [1], the corresponding difference is 21 $JT^{-1}Kg^{-1}$. In both horizons, the magnetic fractions account for approximately 70-74 wt. % of the corresponding whole sand fractions.

The magnetization values of the total fraction and the magnetic fraction for the B_{t22} horizon are rather similar. Thereby, the non-magnetic material is scarce and/or finely distributed among the magnetic material. On the contrary, the larger difference observed between the magnetization values of the wsf and the msf from the B_1 horizon, suggests a higher content of non-magnetic minerals (essentially quartz) and a bigger size and/or better crystallinity of the magnetic components in this more surficial horizon. In fact, microscopic analyses performed on the medium size sand fraction, showed a higher content of quartz in the B_1 horizon (magnetite: quartz relationship value around 5), than in the B_{t22} horizon (magnetite: quartz ratio around 8).

An intense process of weathering and some illuviation of colloidal particles characterise the studied soil. The results obtained show a higher proportion of quartz in the total sand and bigger and better crystallized magnetite in the magnetic sand fraction in the B_1 horizon, suggesting that weathering is more intense in the B_1, close to the surface, than in the B_{22t}. horizon. Besides, the presence of highly magnetic compounds at different depths in this soil is an interesting finding which differs from soils also developed on mafic parent rocks in neighbouring areas.

References

[1] A. Mijovilovich, H. Morrás, C. Saragovi, G. Santana and J. Fabris, Hyp. Int. C. 3 (1998) 332.

[2] A. Mijovilovich, H. Morrás, H. Causevic and C. Saragovi, Hyp. Int.122 (1999) 83.

[3] H. Causevic, Estudio Mössbauer de un suelo magnético de la Provincia de Misiones. Licenciatura Thesis, University of Buenos Aires (2001).

[4] H. Morrás, A. Mijovilovich, E. Leyva, C. Saragovi. XVI World Congress of Soil Science, Montpellier, France, (1998), (edited in CD).

[5] T. Ericsson and R. Wäppling, Journal de Physique 6, (1976) 719.

[6] G.J. Long, T.E. Cranshaw and G. Longworth Mössbauer Effect Data Ref. J., 6 (2), (1983) 42.

[7] R.A. Brand, NORMOS program, Internat. Rep. Angewandte Physik, Univ. Duisburg (1987).

[8] J.D. Fabris, M.D. de J Filho, J.M.D. Coey, W. da N. Mussel and A.T. Goulart, Hyp. Int. 110, (1997) 23.

[9] J.E.M. Allan, J.M.D. Coey, U. Schwertmann, G. Frederic, I.S. Sanders and A. Wickonski, Mineralog. Mag. 53 (1989) 299.

Acknowledgement Authors acknowledge Dr. R. Zysler (CAB, CNEA) for the magnetization measurements and A. Petragalli and A. Carnero (CNEA) for technical support.

Mössbauer and X-ray diffraction study of Mackinawite FeS_{1-x} air oxidation

M. Mullet*, S. Boursiquot, M. Abdelmoula, J.–J. Ehrhardt, J.-M. Génin

Laboratoire de Chimie Physique et Microbiologie pour l'Environnement,
UMR 7564 CNRS – Université Henri Poincaré-Nancy I
405 rue de Vandoeuvre, 54600 Villers-lès-Nancy, France

The dry oxidation of mackinawite FeS_{1-x} at room temperature and atmospheric pressure has been studied by X-ray diffraction (XRD), transmission electron microscopy (TEM) and transmission Mössbauer spectroscopy (TMS). After 7 days of air-oxidation, slow disappearance of mackinawite occurs with concomittant appearance of greigite Fe_3S_4, elemental sulphur S(0), as well as poorly crystallized iron (oxyhydr)oxides (magnetite Fe_3O_4 and probably goethite α-FeOOH). Greigite, which is an intermediate reaction product, and mackinawite have disappeared after 6 months of oxidation, and the final oxidation products are elemental sulphur and iron (oxyhydr)oxides.

1. Introduction

Mackinawite is a metastable sulphur-deficient iron(II) monosulphide, principally encountered in anoxic sediments and marine waters, where it is formed by biologic activity or hydrothermal processes. The mackinawite structure (P4/nmm space group) consists of a distorted cubic-close-packed array of sulphur atoms with iron occupying some of the tetrahedral interstices. This mineral is of particular interest with regard to its place among iron sulphides : mackinawite is the first formed in aqueous systems and is considered as a precursor in the sedimentary pyrite (FeS_2) formation.

In this study, we propose to examine the mechanism of dry mackinawite oxidation at room temperature and atmospheric pressure and to identify the oxidation products.

2. Experimental methods

The procedure for mackinawite synthesis is described in details elsewhere [1,2]. It relies on the reaction of metallic iron with a dissolved sulphide solution in an acetate buffer. The formed mackinawite is filtered and washed with de-aerated de-ionized water before being freeze-dried to prevent any uncontrolled oxidation of the starting material. Mackinawite samples are then naturally air-oxidized at room temperature and atmospheric pressure and the oxidation process is monitored by examining samples at regular times of air exposure (from 7 days to 6 months) using XRD, TEM and TMS.

* Corresponding author e-mail : mullet@lcpe.cnrs-nancy.fr

3. Results and discussion

Figure 1 presents the X-ray diffraction patterns collected after different times of air-exposure. The diffraction pattern of the starting material shows only reflections from mackinawite, with line intensities and positions in good agreement with previously reported data [3]. After 7 days of air exposure, new reflection lines corresponding to greigite (inverse spinel structure : $Fe_2^{3+}Fe^{2+}S_4^{2-}$, the sulphur analogue of magnetite Fe_3O_4) and elemental sulphur appear. As the new reflection line intensities increase, the mackinawite line intensities decrease. However, after 31 days of air-exposure, both line intensities of mackinawite and greigite decrease before disappearing : after 6 months, the only well crystalline solid is elemental sulphur.

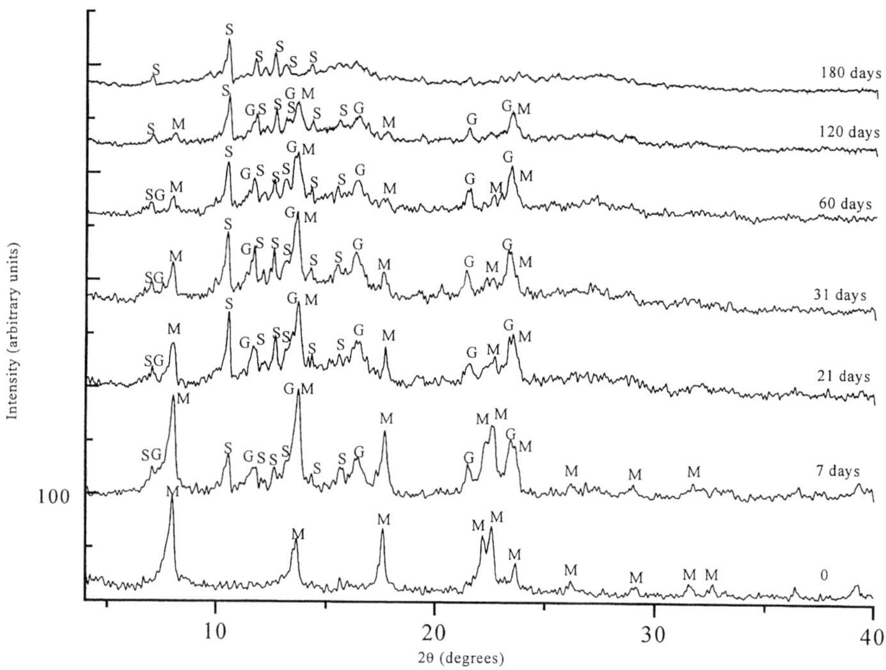

Fig. 1 : X-ray diffraction patterns for the initial material and samples of air-oxidation experiments. S : elemental sulphur, M : mackinawite, G : greigite. ($MoK\alpha_1$, $\lambda = 0.070930$ nm)

The TEM analysis of air-exposed samples supported the XRD analysis, showing the partial transformation of mackinawite to greigite. In addition to XRD results, TEM experiments also revealed the formation of poorly crystallized iron (oxyhydr)oxides i.e. magnetite Fe_3O_4 and probably goethite α-FeOOH. These new phases appeared after 7 days of air-exposure and were those remaining in TEM experiments after 4 months [2].

The TMS spectra of the initial mackinawite, for 14 days and 6 months air-exposed samples are presented in Figure 2. The room temperature spectrum of the starting mineral is composed of two doublets D_1 and D_2 and one singlet S_1. S_1, with an isomer shift of $\delta = 0.42$ mm s^{-1} corresponds to low spin Fe(II) in common tetrahedral sites in agreement with the mackinawite structure [4]. Doublet D_1, with $\delta = 0.13$ mm s^{-1} and a quadrupole splitting $\Delta = 0.38$ mm s^{-1} may likely characterize low spin Fe(III) in tetrahedral sites. Doublet D_2 with the same isomer shift of S_1 ($\delta = 0.41$ mm s^{-1}) characterizes low spin Fe(II) but its value of $\Delta = 0.64$ mm s^{-1}, indicates a new environment which may be caused by the presence of an Fe(III) ion as next nearest neighbour [2]. The presence of Fe(III) reveals the formation of a weathering layer, although drastic precautions were taken during synthesis.

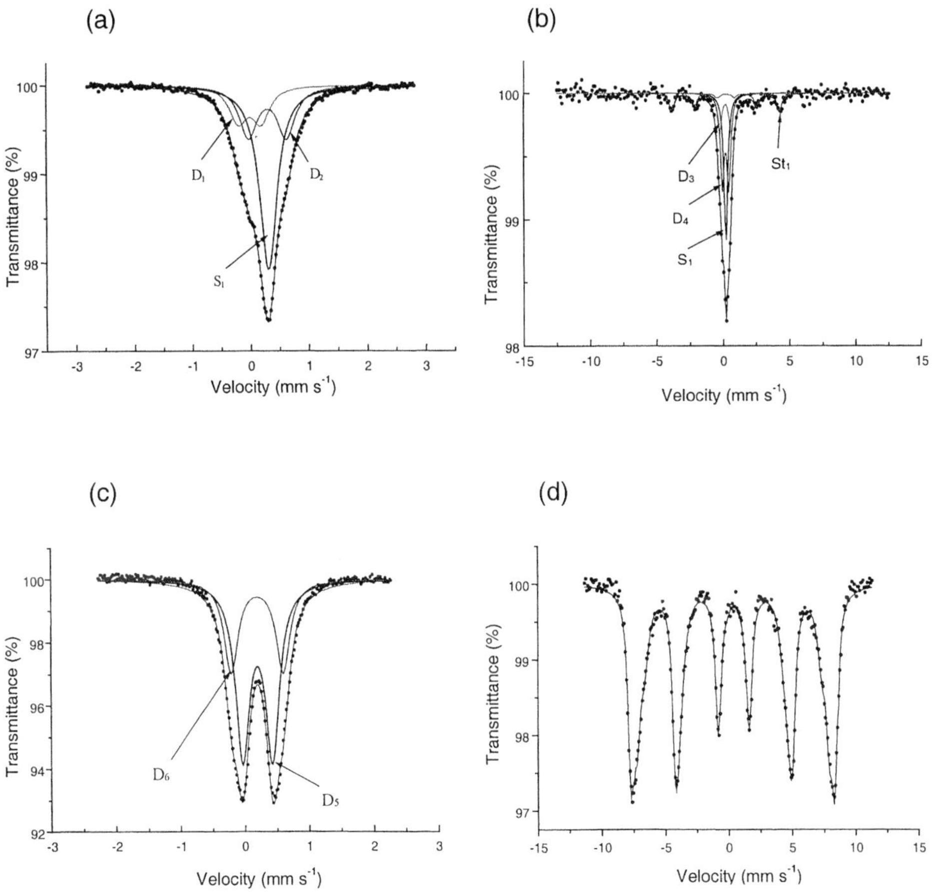

Fig. 2 : Mössbauer spectra of : (a) initial mackinawite measured at room temperature ; (b) 14 days' air-exposed sample measured at room temperature ; (c) 180 days' air-exposed sample measured at room temperature ; (d) 180 days' air-exposed sample measured at 12 K. Isomer shifts are given relative to that of α-iron at room temperature.

The room temperature spectrum of the 14 day air-exposed sample is fitted with one singlet S_1 ($\delta = 0.37$ mm s^{-1}), one sextet St_1 ($\delta = 0.32$ mm s^{-1}, $H = 252$ kOe) and two doublets

D_3 ($\delta = 0.34$ mm s^{-1}, $\Delta = 0.85$ mm s^{-1}) and D_4 ($\delta = 0.35$ mm s^{-1}, $\Delta = 0.85$ mm s^{-1}). The simultaneous presence of magnetic and non-magnetic components illustrate the heterogeneity of the sample ; according to XRD and TEM, at least three phases are associated : mackinawite, greigite and magnetite with eventually goethite. It should be noted that the room temperature spectrum is not able to distinguish all of the phases that probably induce an overlapping of the different components, but St_1, D_3 and D_4 are in rather good agreement with parameters of synthetic greigite [5]. Furthermore, S_1 may attest the presence of remaining mackinawite.

The room temperature spectrum of the 6 month air-exposed sample is fitted with two doublets D_5 and D_6. They have similar isomer shifts (0.32 mm s^{-1}) but different quadrupole splitting (0.46 and 0.8 mm s^{-1} respectively). This spectrum is characteristic of fine particles in superparamagnetic phases. A spectrum of this sample was taken at 12 K, in order to hinder relaxation effects ensuing from small particle size. It is rather complex, showing at least five sextets. However, it may be carefully assigned to non-ideal magnetite, but a mixture of magnetite and goethite cannot be excluded, according to the XRD and TEM results.

Based on these results, we propose the dominant following reaction sequences to explain the dry mackinawite oxidation by air in our experiments :

$$3FeS + 2O_2 \rightarrow Fe_3O_4 + 3S(0)$$

$$3FeS + S(0) \rightarrow Fe_3S_4$$

then $$Fe_3S_4 + 2O_2 \rightarrow Fe_3O_4 + 4S(0)$$

Similar secondary reactions may occur with adsorbed H_2O, resulting in the formation of iron oxyhydroxides, the most likely being goethite α-FeOOH [2].

4. Conclusion

The dry oxidation of mackinawite leads to the formation of greigite and elemental sulphur (TEM and XRD results) and poorly crystallized iron (oxyhydr)oxides (magnetite and goethite). It was shown that greigite was an intermediate reaction product. In fact, both mackinawite and greigite disappear with air-exposure and elemental sulphur and iron (oxyhydr)oxides are the final oxidation products.

References

[1] A.R. Lennie, S.A.T. Redfern, P.F. Schofield, D.J. Vaughan. Mineral. Mag. 59 (1995) 577-683.

[2] S. Boursiquot, M. Mullet, M. Abdelmoula, J.-M. Génin, J.-J. Ehrhardt. Phys. Chem. Miner. 28 (2001) 600-611.

[3] D.T. Rickard. Stockholm Contrib. Geol. 26 (1969) 67-95.

[4] D.J. Vaughan, M.S. Ridout. J. Inorg. Nucl. Chem. 33 (1971) 741-746.

[5] J.A. Morice, L.V.C. Rees, D.T. Rickard. J. Inorg. Nucl. Chem. 31 (1969) 3797-3802.

Characterization of iron sulphides from ore mining rejects

Wagner N. Mussel, José D. Fabris and Viviane G.Costa
Departamento de Química, Universidade Federal de Minas Gerais, Campus, Pampulha, 31270-901 Belo Horizonte, MG, Brazil

Walter A. P. Abrahão and Jaime W. V. Mello
Departamento de Solos, UFV, 36571-000 Viçosa, MG, Brazil

August 15, 2001

Abstract. Three (Fe,S)-rich samples were isolated from mining rejects of copper and coal ores of Camaquã and Butiá, RS, Brazil. Sulphides minerals were characterized by X-ray diffraction, scanning electron microscopy, Mössbauer spectroscopy and some aspects of their crystal structure are discussed. Mössbauer spectra at room temperature and 110 K showed characteristic patterns of chalcopyrite or pyrite. For the chalcopyrite-rich sample, a magnetically ordered Mössbauer pattern suggests the existence of two non-equivalent sites for Fe which is confirmed by the Rietveld analysis of the corresponding XRD pattern.

Keywords: Mössbauer, Chalcopyrite, Pyrite

1. Introduction

Sulphides form an important class of minerals present in many economically exploitable ores. Their general formula is X_mS_n in which X refers to the metallic ion. Pyrite, FeS_2, has a cubic structure, with covalently bonded sulphur pairs. Its polymorphic form is marcasite, which differs from pyrite by its orthorhombic crystal structure. Sulphides are also important in environmental chemistry. The so known acid mine drainage phenomenon is generated mainly via oxidation of sulphides. For chalcopyrite a characteristic six-line Mössbauer pattern of magnetically ordered structure, has been reportedly fitted with only a single Fe site (Frank, 1968); (Eissa et al., 1976). However reported Mössbauer spectra most commonly show clear asymmetric intensities, by comparing lines 1 (leftmost, at negative velocities) and 6 (rightmost). In this paper, some results of the mineralogical characterization of chalcopyrite and pyrite from mine rejects of copper and coal ores are presented.

2. Material and methods

2.1. Sample collection and separation

Two sulfur-rich samples were separated under a binocular lens from a set of materials collected from the copper ore deposit of Camaquã, near Caçapava do Sul (30° 30' 44" S 53° 29' 29" W); a third sample was separed from a coal mining area near Butiá (30° 07' 11" S 51° 57' 44" W), both in the state of Rio Grande do Sul, Brazil.

2.2. Laboratory methodology

The sulphide-rich samples but the chalcopyrite were submited to a set of selective chemical separation. They were firstly disaggregated with ultra sound, then were treated with NaOH 5 mol L^{-1} to remove silicates. The X-ray experiments were made with a Rigaku Geigerflex 3064 diffractometer, equipped with a graphite diffracted beam monochromator, using Cu(Kα) radiation. The average elemental composition for all samples (Table I) was obtained with a Jeol 733 scanning electron microscope, equipped with an energy-dispersive spectrometer.

Table I. Elemental chemical composition (wt%) for the chalcopyrite and pyrite samples.

Sample	Fe	Cu	S
Chalcopyrite	30.4(2)	34.3(4)	34.6(6)
Pyrite 1	50.1(2)		48.8(5)
Pyrite 2	50.5(4)		49.5(3)

The Mössbauer analysis was performed with a constant acceleration transmission spectrometer and a $^{57}Co/Rh$ source. The saturation magnetization of chalcopyrite was obtained with a portable soil magnetometer (Coey et al., 1992).

3. Results and Discussion

The XRD patterns presented in Figures (1) (a) and (b) show typical reflections of pyrite, marcasite, silicon and quartz. The two sequencial treatments, ultra sound and NaOH 5 mol L^{-1}, were not able to completely remove all silicates. The chalcopyrite XRD pattern (Figure 1(c)) was fitted by the Rietveld method with two crystallographic phases, one with a tetragonal cell, the α phase (according to Eissa et al. (1976)), representing 74% of the whole X-ray pattern, with lattice parameters a = 0.52855(1) and c = 1.04117(4) nm, and other with a cubic structure (β phase, 26%), with a = 0.5273(2) nm (Mussel et al., 2001). In

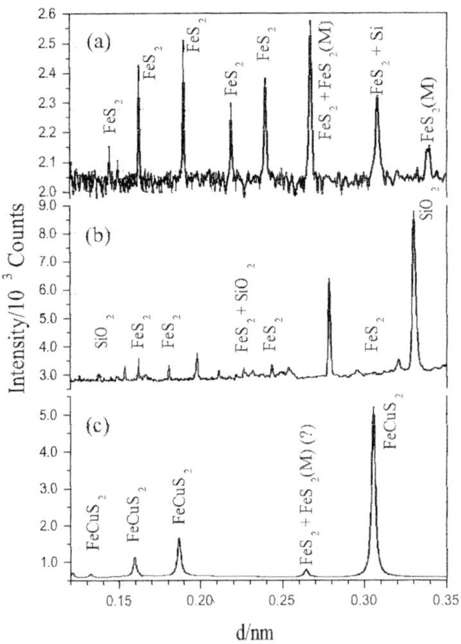

Figure 1. XRD patterns of pyrite-rich samples from (a) coal and (b) copper ores, showing reflections of marcasite, FeS_2(M), silicon (internal standard) and quartz (SiO_2) and (c) of chalcopyrite ($FeCuS_2$) sample from the copper ore.

this case, we assumed that the asymmetric intensities observed in the Mössbauer spectrum are due to two different sites for Fe (Table II). The relative spectral area ratio (Mussel et al., 2001) of the Mössbauer spectrum of the magnetically ordered sites is 82:18 (RT).

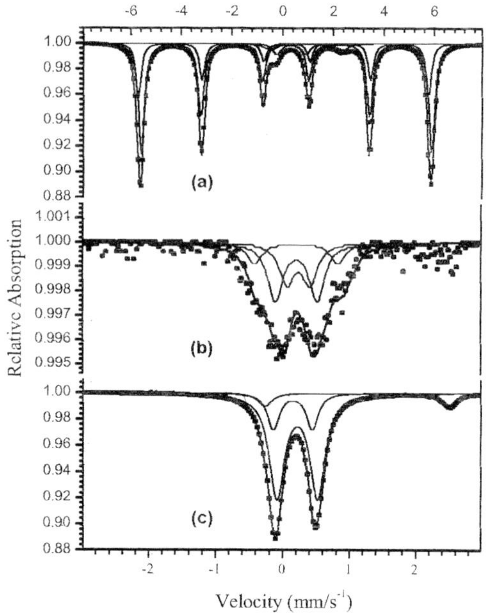

Figure 2. Mössbauer spectra of 110 K copper ore (a) chalcopyrite, (b) pyrite associated with quartz and (c) RT coal-pyrite with clear spectral contribution of Fe^{2+} from silicate.

Table II. Fitted Mössbauer parameters for chalcopyrite and pyrite samples. δ = isomer shift relative to αFe; $2\epsilon_Q$ = quadrupole shift; Δ = quadrupole splitting and B_{Hf} = hyperfine magnetic field.

Sample	δ/mm s^{-1}	$2\epsilon_Q$/mm s^{-1}	Δ/mm s^{-1}	B_{Hf}/tesla
Chalcopyrite (110 K)	0.275(4)	-0.066(5)		36.5(2)
	0.380(1)	0.016(1)		36.8(4)
Chalcopyrite (RT)	0.276(2)	-0.048(5)		36.7(2)
	0.387(1)	0.013(3)		36.9(4)
Pyrite 1 (110 K)	0.310(1)		0.624(2)	
Pyrite 2 (RT)	0.302(1)		0.602(1)	

4. Conclusions

Chemical, mineralogical and physical analyses of sulphide samples from rejects of Cu ore and coal spoils of Camaquã and Butiá, RS, Brazil, revealed that: (i) the pyrite samples are intimately associated with quartz and clay minerals; the ultra sound disaggregation was not able to completely remove quartz and silicates from the sample. (ii) samples treated with NaOH were enriched in Fe-bearing sulphide phases, namely pyrite and some marcasite, with a significant amount of silicate being removed from the raw sample. (iii) Mössbauer spectroscopy indicates two iron sites in the chalcopyrite structure, with relative subspectral areas ratio of 82:18. (iv) The saturation magnetization of the chalcopyrite sample was found to be 32.7 J T^{-1} kg^{-1}.

5. Acknowledgements

Work supported by FAPEMIG, CNPq and FINEP (Brazil). The authors wish to thank Dr. Enver Murad (Bayerisches Geologisches Landesamt, Marktredwitz, Germany) for his helpful discussion.

References

Eissa, N. A., Sallam, H. A., El-Ockr, M. M., Mahmoud, E. A., and Saleh, S. A. Mössbauer effect study of natural egyptian chalcopyrite de Physique Supl. **Vol. 12, Tome 37,(1976) C6** pp. 793-796.

Frank E. Comments on the interpretation of Mössbauer effect data in chalcopyrite Il Nuovo Cimento **LVIII B, N.2, (1968)** pp. 407-412.

W. N. Mussel, J. D. Fabris, W. S. Moreira, J. B. S. Barbosa, C. C. Murta, W. P. Abrahão, J. W. V. de Mello, V. K. Garg Characterization of chalcopyrite from a copper ore deposit (unpublished data)

J.M.D. Coey, O. Cugat, J. McCauley, J.D. Fabris Portable soil magnetometer Revista de Física Aplicada e Instrumentação **7 (1992)** pp. 25-29.

MAGNETIC CHARACTERIZATION OF SEDIMENTS OF URBAN SUPPLY LAKES IN SÃO PAULO, BRAZIL

C.S.M. Partiti, R.A.L. Imbernon[1], D.R. Franco, T.S. Berquó and J.E. Bevilacqua[2]

Dept. de Física dos Materiais e Mecânica, Instituto de Física da Universidade de São Paulo, C.P. 66318-CEP 05315-970, São Paulo, Brazil
[1]Núcleo de Ciências Ambientais, Universidade de Mogi das Cruzes, Mogi das Cruzes, Brazil
[2]Dept. de Educação Ambiental, Faculdade SENAC, São Paulo, Brazil

We report on Mössbauer spectroscopy, X-ray diffraction and chemical analysis studies of samples from urban supply lakes. Our results are in accordance with literature data that have recently shown that the presence of foreign metals and P inhibits the transformation from ferrihydrite (paramagnetic phase) to goethite and hematite (more stable magnetic phase).

1. Introduction

The identification and quantification of Fe-rich phases present in soils and sediments are important applications of geochemical environment studies [1]. Another important factor for geochemical environment studies is the presence of foreign metals in Fe-bearing phase structures, since these metals are indicative of pollution level in soils and lake sediments. Foreign metals (like Cu, Ni, Co, Zn and Pb) incorporated into the structure of Fe-minerals cause distortions in unit-cell dimensions, depending on ionic radius differences [2].

Ferrihydrite is commonly the iron hydroxide initially formed in soils and sediments [3] and is transformed to goethite or hematite, depending on solution conditions [4]. Metals sorbed to or co-precipitated with ferrihydrite can significantly inhibit transformation into more crystalline end products like goethite and hematite [4]. Laboratory experiments have shown that the presence of phosphate decreases the transformation rate of ferrihydrite to crystalline Fe oxides and influences both the type of oxide produced and its characteristics [5,6].

Identification of Fe-bearing phases in soils and sediments is usually done by X-ray diffraction. Mössbauer spectroscopy is another technique that has successfully been applied to study these minerals. It is especially suited to materials in which Fe phases are present in minor quantities or when they are too fine-grained and has been used extensively in the study of Fe compounds in soils [7,8,9].

Goethite is common in nature as a weathering product of iron minerals, particularly in soils and sediments. Mössbauer investigations concerning the influence of crystallinity and Al for Fe substitution in goethites have been carried out on synthetic samples in order to establish a reference against which natural samples might be compared [10]. Results show that both Al substitution and poor crystallinity reduce the magnetic hyperfine field [11]. The influence of foreign metals on Al is not well understood, but there are some indications that they could also reduce the magnetic hyperfine field [2].

In this work, X-ray diffraction, Mössbauer spectroscopy and chemical analysis are applied to samples from urban supply lakes. The main goal of this study is the application of different techniques in order to investigate the Fe-bearing phases and their impurities and to establish a methodology to further studies.

2. Experimental

We have performed our studies in four samples of sediments from both the Billings and Guarapiranga urban supply lakes (São Paulo, Brazil). These samples were collected from different locations on these lakes. They are: 5A: sample from the reversion point of Pinheiros river, Billings; 15A: sample from the Taquacetuba river border, Billings; 30A: sample from the mouth of Parelheiros river, Guarapiranga and 35A: sample from the river bed, Guarapiranga.

X-ray diffraction patterns were acquired using a Philips MPD diffractometer with Cu K_α radiation. Mössbauer spectra were measured with a constant-acceleration spectrometer and a $Rh(^{57}Co)$ source. Absorbers were prepared by pressing finely ground samples between two lucite plates to a sample thickness of 10 mg/cm^2 Fe. All Mössbauer measurements were performed at 77K in a liquid-nitrogen cryostat in order to minimize relaxation effects due to small particle size of the iron oxides.

3. Results and Discussion

3.1 Chemical analysis

TABLE 1 - Chemical analysis results. Numbers in ppm.

	Fe	Cu	Cr	Mn	Ni	Zn	P
5A	113000	467	395	479	302	1170	2300
15A	48000	44	121	526	91	134	1200
30A	56700	102	46	206	23	55	800
35A	68900	1860	90	915	35	83	2000

Comparing sample 5A with 15A, 5A has the higher Me/(Fe+Me) molar fraction, excepting Mn. The 30A sample has the lowest proportion of P and also the lowest Me/(Fe+Me) molar fraction, excepting Mn. Sample 35A presents the highest Mn and Cu Me/(Fe+Me) molar fraction, but the same parameter, relative to the other metals, lower than in the other samples. The differences relative to Mn and Cu can be explained with the hypothesis that Cu prefers associations with Mn rather than with Fe [12].

3.2 X-Ray diffraction

X-ray diffraction results show that goethite is present in all samples , but only in sample 30A, is it possible to identify peaks that can be related to hematite.

3.3 Mössbauer spectroscopy

TABLE 2 - Mössbauer spectral parameters: hyperfine field (B_{hf}), quadrupole splitting (Δ) and fractional resonant absorption area (A). Numbers in parentheses indicate standard deviation on the last significant figure.

	Δ (mm/s)	A_D (%)	B_{hf1} (T)	A_1 (%)	B_{hf2} (T)	A_2 (%)	B_{hf3} (T)	A_3 (%)
5A	0.74(1)	66	48.5(2)	26			40.0(4)	8
15A	0.77(1)	64	48.4(2)	24			43.5(2)	12
30A	0.74(2)	36	47.9(3)	39	52.9(1)	25		
35A	0.75(2)	65	48.6(4)	22	52.3(1)	13		

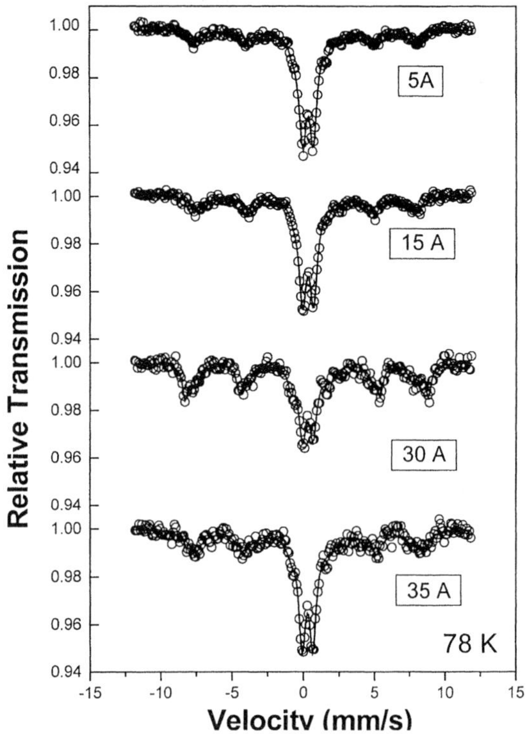

Figure 1 Mössbauer spectra at 78 K of sediments of urban supply lakes.

Each Mössbauer spectrum (Figure 1) shows one doublet that can be assigned to ferrihydrite (A_D in Table 2) [13] and one sextet that can be assigned to goethite (A_1 in Table 2) [14]. The 5A and 15A spectra show one sextet with a lower hyperfine field related to goethite representing the incorporation of foreign metals (8%, 12%, A_3 in Table 2). In fact, the samples with higher concentration of metals are expected to show lower hyperfine fields [2,10,11]. These results are in accordance with chemical analysis results (Table 1). The 30A and 35A spectra each show two sextets. The first one can be assigned to goethite and the other (A_2) to hematite [15]. The presence of hematite in sample 30A was confirmed by X-ray diffraction. On the other hand, only Mössbauer spectroscopy could detect hematite in sample 35A. In fact, X-ray diffraction did not indicate the presence of hematite, probably due to a concentration of this mineral too low for that method's sensivity. Sample 30A has the lowest concentration of metals and ferrihydrite, and the highest proportion of hematite. This is in accordance with the data of Ford et al. [4] that show that the presence of metals inhibits the transformation of the ferrihydrite to goethite and hematite [4]. It appears that the same conclusion could be drawn for the presence of P in sample 30A. As shown in Figure 2, the ferrihydrite fractional absorption area as a function of P/Fe is in accordance with Galvez et al. [5,6], and saturates above a certain value. The same behavior is observed in Figure 3, which shows the proportion of ferrihydrite as a function of the total Me/(Fe+Me) molar fraction.

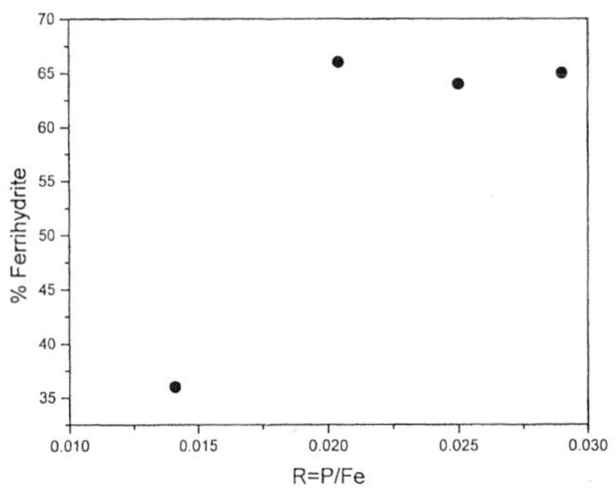

Figure 2 Ferrihydrite fractional absorption area as a function of P/Fe.

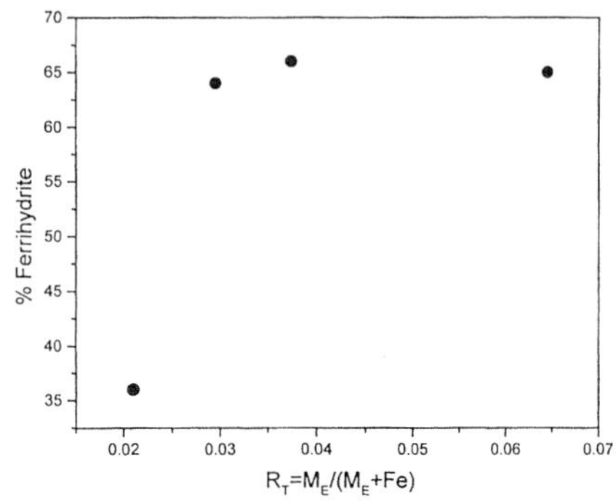

Figure 3 Ferrihydrite fractional absorption area as a function of total Me/(Fe+Me) molar fraction.

4. Conclusions

The mineral Fe-bearing phases of the analyzed samples were well identified through different techniques. The conclusions obtained from the different methods are consistent. Our results are in accordance with literature data that have recently shown that the presence of foreign metals and P inhibits the transformation from ferrihydrite (paramagnetic phase) to goethite and hematite (more stable magnetic phase) [4,5,6]. It can be considered that the ratio ferrihydrite/(goethite.hematite) would be an environment as parameter indicative of pollution. These results are encouraging and a detailed study in other urban supply lakes is in progress. Two laboratory experiments have recently reported that the presence of water and Al or even differences of temperature and pH play a role in hematite formation [16,17]. In our study of the Taiaçupeba Supply Lake that is in progress these parameters are taken into account.

Acknowledgements
FAPESP and CNPq. The authors are grateful to E. Galego for X-ray diffraction measurements.

References

[1] M. Drodt, A.X. Trautwein, I. König, E. Suess and C. Bender Koch, Phys. Chem. Minerals 24(1997)281.

[2] S.M.B. Oliveira, R.A.L. Imbernon, C.S.M. Partiti and H.R. Rechenberg, Geoderma 73(1996)245.

[3] L. Carlson and U. Schwertmann, Wat. Res. 21(1987)165.

[4] R.G. Ford, K.M. Kemner and P.M. Bertsch, Geochim. Cosmochim. Acta 63(1999)39.

[5] N. Gálvez, V. Barrón and J. Torrent, Clays Clay Mineral. 47(1999)375.

[6] N. Gálvez, V. Barrón and J. Torrent, Clays Clay Mineral. 47(1999)304.

[7] D.C. Golden, L.H. Bowen, S.D. Weed and J.M. Bigham, Soil Sci. Soc. Am. J. 43(1979)802.

[8] R.E. Vandenberge, E. De Grave, C. Landuydt and L.H. Bowen, Hyp. Inter. 53(1990)175.

[9] E. Murad, Min. Mag. 43(1979)335.

[10] E. Murad and U. Schwertmann, Clay Miner. 18(1983)301.

[11] R.K. Vempati, R.V. Morris, H.V. Lauer Jr. and J. Helmke, J. of Geophys. Res. 100(1995)3295.

[12] R.A.L. Imbernon, A. Blot, S.M.B. de Oliveira and P. Magat, Geochim. Brasil. 13(2)(1999)145.

[13] Murad and U. Schwertmann, Amer. Mineralogist. 65(1980)1044.

[14] J.B. Forsyth, I.G. Hedley and C.E. Johnson, J. Phys.C1 (1968)179.

[15] W. Kundig, H. Bommel, G. Costabaris and L.H. Lindquist, Phys. Rev. 142(1966)327.

[16] U. Schwertmann, J. Friedl and H. Stanjek, J. Colloid Interface Sci. 209(1999)215.

[17] U. Schwertmann, J. Friedl, H. Stanjek and D.G. Schulze, Clay Clays Miner. 48(2000)159.

MÖSSBAUER INVESTIGATION ON Ti-GARNETS FROM DIFFERENT GEOLOGICAL ENVIRONMENTS

G. Pedrazzi[1], *E. Schingaro*[2], *F. Scordari*[2]
[1]Istituto di Scienze Fisiche ed Unità INFM, University of Parma, Italy
[2]Dipartimento Geomineralogico, University of Bari, Italy

Abstract. The aim of the present work is to check whether Mössbauer-determined iron site distribution or iron-bearing substitution mechanisms occurring in garnet structures may be used as "fingerprint" of the geological environment. The results of a Mössbauer and crystallographic study on three natural Ti-garnets from different geological environment (intrusive, metamorphic) and provenance are reported and compared with those from previous investigations on volcanic samples.

1. Introduction

Ti-garnets constitute a subgroup of the garnet family (general formula $X_3Y_2Z_3O_{12}$) which can contain up to ≈ 20 wt% of TiO_2. At the X(8-fold) site divalent cations (predominantly Ca) can be located. The Y(6-fold) site can be occupied mainly by $Fe^{2+,3+}$, $Ti^{3+,4+}$, Al, Zr. Finally at the Z (4-fold) site Si may be substituted for Fe^{3+}, Al, Ti^{4+}, $4H^+$ and Fe^{2+}. In order to determine the cation distribution, chemical and structural investigations must be complemented by suitable spectroscopic techniques [1, 2].

Substitution mechanisms through which Fe and Ti enter the garnet structure are expected to be related to the geological environment in which they occur. It has been recently proposed that in samples from volcanic rocks schorlomite substitution $Ti^{4+}(Y) + Fe^{3+}(Z) \leftrightarrow Si^{4+}(Z) + Fe^{3+}(Y)$ is predominant, associated with a low degree of hydrogarnet component, $(SiO_4)^{4-} \leftrightarrow (O_4H_4)^{4-}$; on the contrary Ti-garnets from metamorphic rocks should be affected by morimotoite substitution $Ti^{4+}(Y) + Fe^{2+}(Y) \leftrightarrow 2Fe^{3+}(Y)$ and a high degree of hydrogarnet component [3]. However this hypothesis was not based on systematic investigations of Fe by Mössbauer spectroscopy.

In the present work the results of Mössbauer and crystallographic study on three natural Ti-garnets from different geological environment (intrusive, metamorphic) and provenance are reported, and compared with those from previous investigations on volcanic samples [1, 2, 5, 9] in order to check whether Mössbauer-determined iron distribution over the independent crystallographic sites of the garnet structure may be used as "fingerprint" of the geological environment.

2. Materials and methods

The following natural samples have been here considered: 1) sample W12, from the carbonatitic complex of Magnet Cove, Arkansas; it is representative of samples from intrusive rocks and was firstly studied by Howie and Woolley [4]; 2) sample NZALA, from Atlas Mountains, Morocco, and 3) sample ZER2 from Zermatt, Valais, Switzerland, both representative of metamorphic environment. These samples have been selected for publication from a larger suite of investigated Ti-garnets: 9 from volcanic rocks, out of which 2 unpublished, 5 from carbonatitic rocks, out of which 3 unpublished, 3 samples from metamorphic rocks, out of which 2 unpublished. Mössbauer spectra were recorded on powdered samples, at room temperature, in transmission geometry, using a source of $^{57}Co/Rh$ and a constant acceleration spectrometer. On sample W12 chemical analyses and crystallographic investigations were also performed. Chemical compositions were measured with a Cameca SX50 electron microprobe whereas X-ray diffraction data were collected with a Huber automated single crystal diffractometer equipped with graphite monochromated MoKα source. Details on experimental conditions have been published elsewhere [2, 5]. Chemical and crystallographic data for samples NZALA and ZER2 are taken from Armbruster et al. [3].

3. Results and discussion

The results of Elecron Probe Micro-Analyses (EPMA) are reported in Table 1. Each composition is the average of about 10 repeated analyses on different points. In Table 2 the main results of crystallographic investigation are presented, in terms of cell parameters, bond distance and mean atomic numbers (m.a.n.) as determined by the refinement of occupancies at X, Y and Z sites. Mössbauer investigation yielded the results illustrated in Table 3 and Figure 1, where the spectra of samples W12 and NZALA are shown.

Sample	CaO	Na_2O	MgO	MnO	FeO	Al_2O_3	TiO_2	SiO_2
W12	32.62	-	1.02	0.31	20.96	2.01	9.09	30.61
NZALA	32.51	0.16	0.59	0.44	23.69	0.98	6.89	32.66
ZER2	33.14	-	0.31	0.27	19.42	4.39	4.30	34.10

Table 1. EPMA analyses (oxide wt%)

Sample	a	<X-O>	Y-O	Z-O	m.a.n. X	m.a.n. Y	m.a.n. Z
W12	12.123(2)	2.443	2.017(2)	1.667(2)	58.9(5)	47.2(5)	44.3(5)
NZALA	12.105(1)	2.436	2.013(1)	1.666(1)	59.8(5)	48.2(5)	45.7(5)
ZER2	n.d	n.d	n.d	n.d.	n.d.	n.d.	n.d.

Table 2. Cell edges, main bond distances (Å) and mean atomic numbers (electrons, e-) for the analysed samples. N.d. stands for not determined

By combining and/or cross-checking the results of different techniques (see [2, 6] for a detailed description of the procedure) the following experimental cation distribution (in atoms per formula units, a.p.f.u.) can be obtained for the analysed samples:

W12: $(Ca_{2.98}\,Mn_{0.02})(Mg_{0.13}\,Al_{0.08}\,Ti^{3+}_{0.05}\,Ti^{4+}_{0.53}\,Fe^{3+}_{1.21})(Si_{2.60}\,Al_{0.12}\,Fe^{3+}_{0.28})O_{12}$
NZALA:$(Ca_{2.93}Mn_{0.03}Na_{0.03}\,Mg_{0.02})(Mg_{0.05}\,Al_{0.10}\,Ti^{4+}_{0.44}Fe^{2+}_{0.10}Fe^{3+}_{1.31})(Si_{2.74}Fe^{2+}_{0.05}Fe^{3+}_{0.21})O_{12}$
ZER2: $(Ca_{2.92}\,Mn_{0.02}\,Mg_{0.04})(\,Al_{0.43}\,\,Ti_{0.27}\,Fe^{3+}_{1.30})(Si_{2.80}\,)(OH)^{-}_{0.80}\,O_{11.20}$

For sample NZALA Armbruster et al. [3] proposed a cation distribution with 91% of Fe^{3+}(Y) and 9% of Fe^{3+}(Z), whereas for sample ZER2 those authors evaluated Fe^{2+}(Y) = 14% and Fe^{3+}(Y) = 86% on the basis of the postulated occurrence of the morimotoite substitution (see below). Furthermore Armbruster et al's distributions were based on the combination of EPMA, crystallographic, qualitative IR investigations and charge balance considerations. No Mössbauer investigation was performed by those authors. Finally the lack of closure to 3 a.p.f.u. of X-site for sample ZER2 suggests minor problems in the chemical analysis.

The analysed samples have quite different Ti-content (Table 1), which is reflected in the variation of their cell-edges, a correlation already pointed out by Howie and Woolley [4]. More recent investigations [6, 7] allowed to ascertain that Y and Z site contribute most the cell–edge variation, due to both compositional (Al, Fe, Ti substitutions) and to non-compositional effects (bond length distortions), whereas the geometrical variation of the X-site is induced by the extensive polyhedral edge-sharing occurring in the garnet structure.

Sample		site	δ (mm/s)	Δ (mm/s)	Γ (mm/s)	A %
W12	Fe^{3+}	Y	0.399(5)	0.617(6)	0.326(5)	81(3)
	Fe^{3+}	Z	0.221(2)	1.208(4)	0.401(4)	19(2)
			$\chi^2_r = 1.29$			
NZALA	Fe^{3+}	Y	0.402(5)	0.597(5)	0.331(6)	79(7)
	Fe^{3+}	Z	0.208(1)	1.253(3)	0.318(4)	12(7)
	Fe^{2+}	Z	0.70(1)	1.66(1)	0.25(9)	4(2)
	Fe^{2+}	Y	1.28(8)	2.91(2)	0.48(2)	5(3)
			$\chi^2_r = 1.11$			
ZER2	Fe^{3+}	Y	0.399(1)	0.582(3)	0.312(4)	100
			$\chi^2_r = 1.16$			

Table 3. Results of Mössbauer investigation at RT.

Taking into account the whole suite of analysed samples (see the materials and method section) and thus also the result of previous investigations [2, 5, 6, 8, 9] the following behaviour resulted from the fitting of the Mössbauer spectra: a) from 2 quadrupole doublets (Fe^{2+}(X), Fe^{3+}(Y)) to 5 quadrupole doublets (Fe^{2+}(X), Fe^{3+}(Y), Fe^{3+}(Z), Fe^{2+}(Y), Fe^{2+}(Z)) are necessary to fit Mössbauer spectra of samples from volcanic rocks; b) from 2 quadrupole doublets (Fe^{3+}(Y), Fe^{3+}(Z)) to 5 quadrupole doublets (Fe^{2+}(X), Fe^{3+}(Y), Fe^{3+}(Z), Fe^{2+}(Y), Fe^{2+}(Z)) were used to fit Mössbauer spectra of samples from carbonatite rocks; c) from 1 quadrupole doublet (Fe^{3+}(Y)) to 4 quadrupole doublets (Fe^{3+}(Y), Fe^{3+}(Z), Fe^{2+}(Y), Fe^{2+}(Z)) were used to fit Mössbauer spectra from metamorphic rocks samples.

On the other hand it must be taken into account that in Ti-garnets structure more than one of the following substitution mechanisms may occur: 1) Ti^{4+}(Y) + Fe^{3+}(Z) ↔ Si^{4+}(Z) + Fe^{3+}(Y), schorlomite substitution; 2) Zr^{4+}(Y) + M^{3+}(Z) ↔Si^{4+}(Z) + M^{3+}(Y) , kimzeyite substitution; 3) Ti^{4+}(Y) + Fe^{2+}(Y) ↔ 2Fe^{3+}(Y), morimotoite substitution; 4) Ti^{3+}(Y) ↔ M^{3+}(Y); 5) Ti^{4+}(Y) + Na^{+}(X) ↔ Fe^{3+}(Y) + Ca^{2+}(X); 6) Fe^{2+}(Z) + 2OH^{-} ↔ Si^{4+}(Z) + 2 O^{2-}; 7) Ti^{4+}(Z) + Fe^{2+}(Y) + OH^{-} ↔ Si^{4+}(Z) + Fe^{3+}(Y) + O^{2-} ; 8) $(SiO_4)^{4-}$↔ $(O_4H_4)^{4-}$, hydrogarnet substitution. Accordingly, combining Mössbauer derived Fe-site populations and the final cation distribution it can be observed that different distributions of Fe, Al, and Ti over Y and Z sites may characterise samples with the same origin. For instance in some volcanic samples the Z site is mainly composed of Si, Ti^{4+} and Al, while in others Fe^{3+}, Ti^{4+} and a little component of hydrogarnet make up for the Si deficiency. Similarly, in garnets from carbonatitic rocks tetrahedral site composition spans from quite simple (only Si + Fe^{3+}) to quite complex (Si + Fe^{3+} + Fe^{2+} + Ti^{4+} or Al). Finally out of the three Ti-garnets from metamorphic environments, one is anydrous with Z site containing Si, Fe^{3+} and minor Fe^{2+}, the other two have only Si at Z and substantial hydrogarnet substitution.

Since a large amount of data (chemical, crystallographic, spectroscopic, cation sites distributions) is now becoming available for Ti-bearing garnets, attempts are being performed to correlate Fe- and Ti-involving substitutional mechanisms to peculiar petrogenetic features. This is because Ti-bearing garnets have always been considered suitable mineral phases to be used as petrogenetic indicators for the host rocks in which they occur. However, their complicated crystal chemistry [1,2,6] has long prevented hitting this target. Recently Armbruster et al. [3] suggested the schorlomite substitution to be typical of most anhydrous Ti-garnets with volcanic origin, and the morimotoite substitution to be present in hydrated Ti-garnets from metamorphic rocks. The results of the present investigation are not in agreement, neither qualitatively nor quantitatively, with the conclusions of Armbruster et al. [3]. It has rather been found that different Fe, Ti and Al site partitions characterise samples with similar environment. Generally speaking, schorlomite substitution may be present in Ti-garnets from all the three geologic environments here considered. The same is true for morimotoite substitution, which always occurs as a minor one. Finally, hydrogarnet substitution seems generally low or absent in Ti garnets. The only exception is represented by two out of the three garnets from metamorphic environment, for which it is remarkable. In conclusion, even if Mössbauer derived Fe-site populations cannot identify the geological environment nevertheless they are critical to restrain the cation distribution and to work out the active substitution mechanisms.

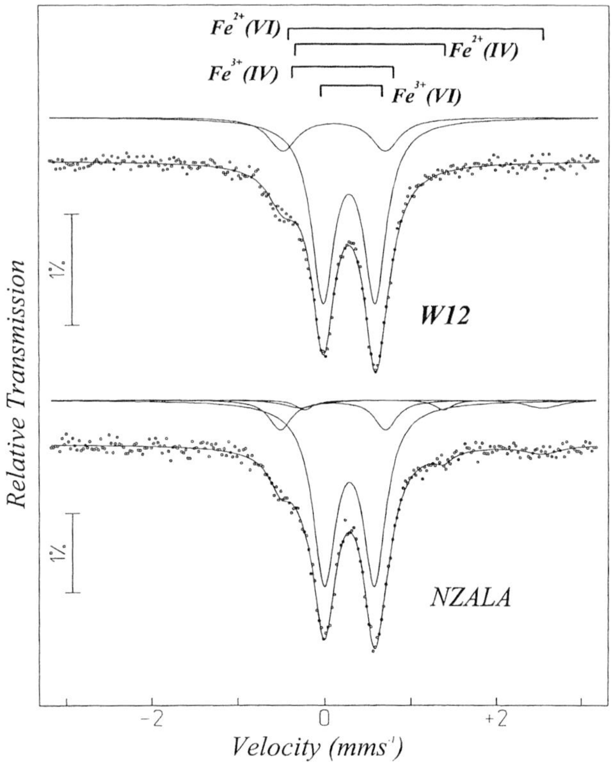

Figure 1. Mössbauer spectra of the samples W12 and NZALA at 297K.

References

[1] C. Malitesta, I. Losito, F. Scordari, E. Schingaro, Eur. J. Mineral. 7 (1995) 847

[2] F. Scordari, E. Schingaro, G. Pedrazzi, Eur. J. Mineral. 11 (1999) 855

[3] T. Armbruster, J. Birrer, Libowitzky, A. Beran, Eur.J.Mineral. 10 (1998) 907

[4] R.A. Howie and A.R. Woolley, Mineral. Mag. 59 (1968) 115

[5] G. Agrosi', E. Schingaro, G. Pedrazzi, E. Scandale, F. Scordari, Eur.J. Mineral. (2001) submitted

[6] E. Schingaro, F. Scordari, C. Malitesta, G. Pedrazzi, XI EUG Meeting, Strasbourg (2001) Book of Abstracts, 270

[7] M. Merli, A. Callegari, E. Cannillo, F. Caucia, M. Leona, R. Oberti, L. Ungaretti, Eur. J. Mineral. 7 (1995) 1239

[8] G. Pedrazzi, E. Schingaro, F. Scordari, Hyperfine Interactions. C 3 (1998) 321

[9] E. Schingaro, F. Scordari, G. Pedrazzi, A. Mottana, Plinius 12 (1998) 123

MÖSSBAUER STUDY OF IRON SULFIDE CRYSTALLIZATION PROCESSES

V.S. Rusakov[a], N.I. Chistyakova[a], S.V. Kozerenko[b]

[a]*Lomonosov Moscow State University, Vorobyevy gory, 119899 Moscow, Russia*
[b]*Vernadsky Institute of Geochemistry and Analytical Chemistry Russian Academy of Sciences, Kosygin Str. 19, 117975 Moscow, Russia*

The kinetics of iron sulfide crystallization from hydrotroilite was investigated by ^{57}Fe Mössbauer spectroscopy. Synthesis was carried out in the temperature range $20^0C<t<200^0C$. Investigated samples were exposed to aging from several days to several years. It was shown that the reaction proceeded in different ways depending on redox conditions. The hydrotroilite → pyrite transformation was realized when elemental sulfur was present in the solution. In more reducing conditions without elemental sulfur a sequence of metastable sulfides formed: hydrotroilite → mackinawite → greigite → pyrite and marcasite.

Keywords: Mössbauer spectroscopy, iron sulfide crystallization, pyrite, hydrotroilite, mackinawite, greigite, marcasite.

1. Introduction

Interest in sulfide minerals is connected with their wide abundance in nature. Sulfide forming reactions are considered an important factor in the global carbon, sulfur and iron cycles in Earth crust conditions [1]. Iron sulfide sedimentation is considered a factor in H_2S binding, which has governed sulfate and oxygen content in world ocean waters throughout terrestrial history [2]. In this paper processes of iron sulfide crystallization (pyrite FeS_2, mackinawite FeS_{1-x}, greigite Fe_3S_4) from X-ray amorphous waterbearing precipitates (hydrotroilite) were investigated by use of ^{57}Fe Mössbauer spectroscopy. This method permits to tracking of sulfide formation reactions and provides data on structure transformation.

2. Experimental

Synthesis was carried out within the temperature interval $20^0C<t<200^0C$ at pH=6.3-7.0 and concentration of sulfide sulfur near saturated under two sets of conditions - in excess and in the absence of excess of elemental sulfur in the mineral forming media. In this temperature range as our investigations have shown reaction velocities are increasing with the increase of synthesis temperature and the sequence of forming sulfides is the same. Therefore the data obtained for room temperature synthesis (when processes of phase formation are proceeding slowly) are presented in this paper for detail study of iron sulfide crystallization processes. Samples were aged from several days to several years. Sulfide sediments studied were initially amorphous as judged by X-ray diffraction. Mössbauer investigations were carried out at room temperature on the spectrometer operated in constant acceleration mode and equipped with ^{57}Co-source in Rh-matrix. Both superposition of parameter distribution functions and curve fitting were used for analysis of the Mössbauer spectra.

3. Results and Discussion

We found out that the reaction proceeded in depending on redox conditions. When there was an excess of elemental sulfur in the mineral forming media the following scheme was realized: hydrotroilite→pyrite. Mössbauer spectra of both initial substance and end product showed asymmetric quadrupole doublets; the initial spectrum had the distinctive broadening characteristics of amorphous materials. No intermediate phases were found and pyrite (FeS_2) was the final product of aging (7200 hours). We have found Mössbauer line and quadrupole shift distribution functions p(δ) and p(ε) for the initial and aged products (figure 1). Mössbauer investigations have shown that Fe^{2+} ions in hydrotroilite are in the low spin state and presumably exist in pyrite-like nearest surroundings [3, 4]. Our investigations demonstrated that the main factors in pyrite crystallization are sediment aging conditions, temperature and time interval.

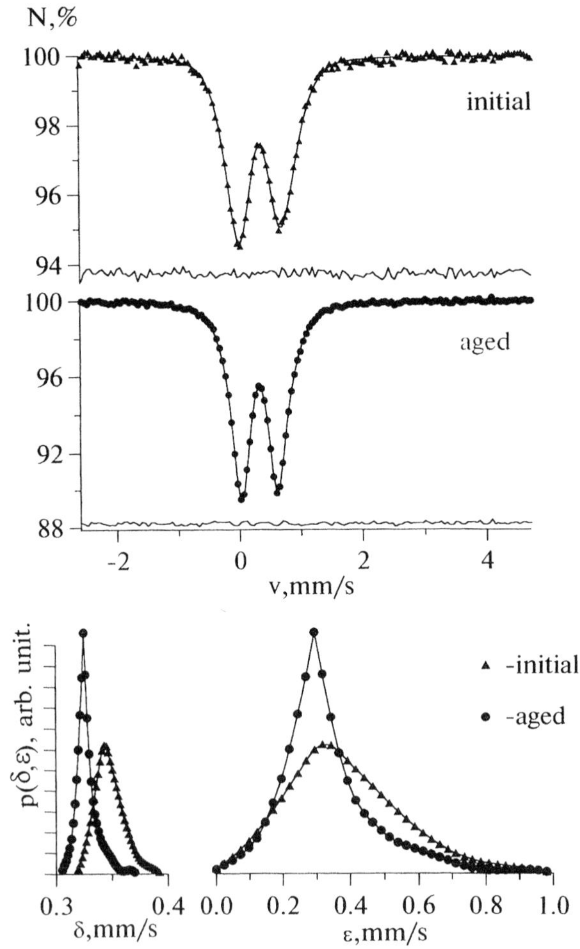

Figure 1. Mössbauer spectra and Mössbauer line and quadrupole shift distribution functions p(δ) and p(ε) for the initial and aged (7200 hours) in an excess of elemental sulfur products at room temperature.

In the absence of an excess of elemental sulfur in the mineral forming media several metastable sulfides formed in succession: hydrotroilite and mackinawite → greigite → pyrite and marcasite (figure 2). Comparison of quadrupole shift distribution functions showed that the first sample from this series (10 days of aging) was a mixture of hydrotroilite (94.3±1.6%) and mackinawite (5.7±0.4%). The Mössbauer spectrum of the 35-day sample corresponded to 45.9±1.3% greigite, 3.4±0.3% mackinawite approximately as before and the balance hydrotroilite. With the increase of synthesis time to 120 days we observed the increase of greigite content. At 260 days mackinawite decreased to 4.4±0.2% and greigite to 2.6±0.6%. The quadrupole doublct had a relative intensity of 93±5% with Mössbauer line and quadrupole shifts typical, as they were for hydrotroilite, of Fe^{2+} ions in low spin state. X-ray phase analysis showed that the sample was a mixture of pyrite and marcasite, minerals with similar parameters. Average values of δ and ε, full width at half maximum of the distribution and standard deviation showed this pyrite and marcasite mixture to be the end product of aging.

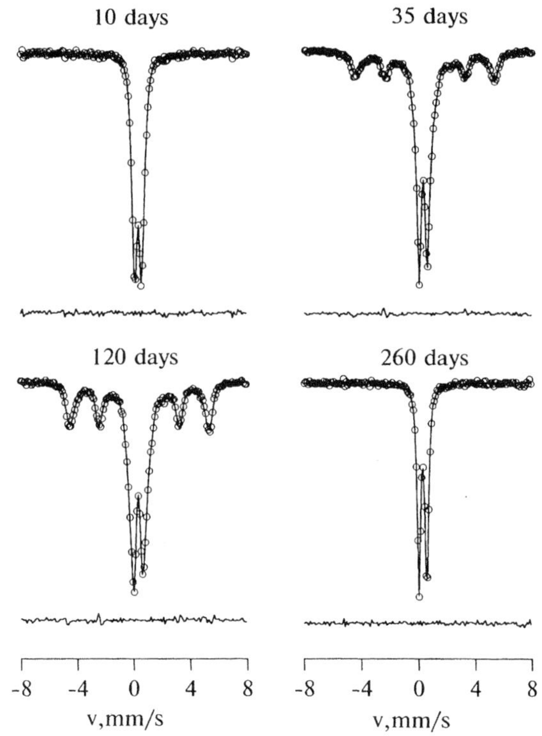

Figure 2. Mössbauer spectra of samples aged in the absence of excess elemental sulfur at room temperature.

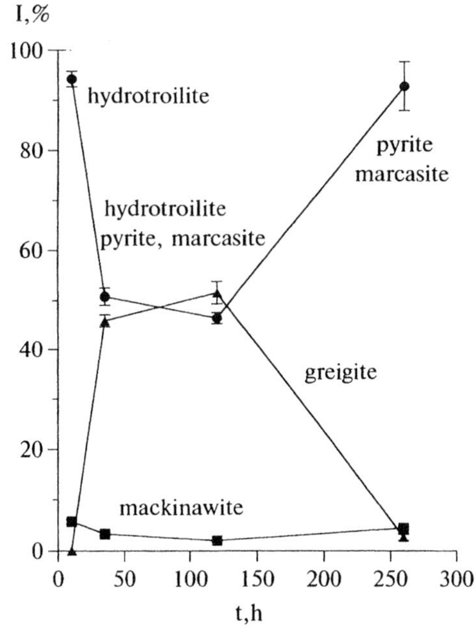

Figure 3. The relative content of synthesized phases as the function of aging time in the absence of excess elemental sulfur.

The processes of phase formation in this system could be divided into three stages - the decrease of hydrotroilite content with greigite formation, further formation of greigite from hydrotroilite and formation of pyrite and marcasite from greigite, which gradually disappears (figure 3). Mackinawite content was practically unaffected by time interval of synthesis. The nucleation barrier of mackinawite is not high, so details of its crystallization were primarily determined by the sedimentation conditions (pH, pS). However the mackinawite content was found to increase with increasing synthesis temperature.

This work was supported by RFBR, Grant 00-05-65372.

REFERENCES

[1] R.A.Berner, Geochim. et Cosmochim. Acta, 48 (1984) 605.

[2] H.D.Holland, in: The Chemistry of the Atmosphere and Ocean, Wiley and Sons, New York, 1978, p.351.

[3] S.V.Kozerenko, D.A.Khramov, V.V.Fadeev et al, Geochemistry International 33 (1996) 16.

[4] V.S.Rusakov, D.A.Khramov, N.I.Chistyakova et al, Phys. Stat. Sol.(a) 144 (1994) K45.

Fe-Ni METAL STRUCTURES OF MESOSIDERITES STUDIED BY MÖSSBAUER SPECTROSCOPY

I. Souza Azevedo, A. Bustamante D.[1] and R.B. Scorzelli
Centro Brasileiro de Pesquisas Físicas, Rua Xavier Sigaud 150, 22290-180
Rio de Janeiro, Brazil

Mesosiderites, which are breccias composed of roughly equal proportions of silicates and metallic Fe-Ni plus troilite, comprise a strange group of about 45 differentiated meteorites known. The cooling rates determined for mesosiderites are the slowest yet determined for meteorites. A systematic study has been performed by Mössbauer spectroscopy of this perplexing differentiated meteorite group.

INTRODUCTION

Mesosiderites, which are breccias composed of roughly equal proportions of silicates and metallic Fe-Ni plus troilite, comprise a strange group of about 45 differentiated meteorites known. The cooling rates determined for mesosiderites in the range 500-350 °C are the slowest yet determined for meteorites. This unique aspect suggests that mesosiderites originated in a different parent body, or bodies, than did most other meteorites. In a very recent paper [1] it has been proposed that mesosiderites formed when a 200-400 km diameter asteroid with a molten core was disrupted by a 50-150 km diameter projectile.

The silicate fraction of mesosiderites consists of mineral and lithic clasts in a fine-grained fragmental or igneous matrix. Metallic Fe-Ni in mesosiderites is mostly in the form of millimetre or submillimetre grains which are intimately mixed with similar-sized silicate grains. Metal-troilite-silicate textures indicate that except for rare metallic nodules, the metal crystallized after being mixed with silicate.

---

[1]On leave of absence from Facultad de Ciencias Físicas, Universidad Nacional Mayor de San Marcos, Peru.

Investigation of meteorites by Mössbauer spectroscopy has definitely enriched the understanding of complex meteoritic Fe-Ni phases. In slowly cooled meteorites, meteoritic metal contains a characteristic microstructure that is common in stony, stony-iron and iron meteorites and their differences are most likely a function of Ni content and cooling history at low temperatures. Mössbauer spectroscopy has played an important role in the study of these phases giving the first conclusive evidence for the existence of a ferromagnetic atomically ordered FeNi phase later called *tetrataenite (50 at% Ni)* [2]. Whereas *tetrataenite* and the other common phases in meteorites, *kamacite*, *martensite* and ordinary *taenite*, can occur in single phase regions, *antitaenite*, γ_{LS} (~ 25-30 at% Ni) whose presence is established by Mössbauer spectroscopy, always occurs in association with *tetrataenite* [3].

In this paper a systematic study has been undertaken on mesosiderites by Mössbauer spectroscopy expecting that, when their metal fractions have mean composition in the range ~ 25-30 at% Ni, the γ_{LS} would be present and the intergrowth *tetrataenite/antitaenite* (TT/AT) would be coarse [3]. Since the degree of ordering of *tetrataenite* depends on the cooling rate, the hyperfine parameters could reflect the unusual thermal history of this group of meteorites.

EXPERIMENTAL

Thin slices of the mesosiderites, Crab Orchard, Esterville, Vaca Muerta and Lowicz, were cut with a precision saw. A sample of Vaca Muerta was analysed using electron probe microanalyser (EPMA) and backscattered electron (BSE) images, and the other samples are still under study. Absorbers for Mössbauer measurements were prepared by polishing the slices to a thickness of about 70 μm. The spectra were recorded at RT with ^{57}Co/Rh source in a conventional spectrometer.

RESULTS AND DISCUSSION

The mesosiderites under study in this paper are Crab Orchard (8.83 wt% Ni), Estherville (8.57 wt% Ni), Vaca Muerta (8.50 wt% Ni) and Lowicz (7.52 wt% Ni).

A sample of the metallic part of the Vaca Muerta meteorite was analysed using an electron microprobe analyser (EPMA) and backscattered electron (BSE) images. The BSE images showed mainly two Fe-Ni regions, corresponding to the γ high-Ni phase (*tetrataenite*) and the α-phase, *kamacite*. The results of the chemical analysis of 10 positions using EPMA are in the range of 41.8-40.8 wt% Ni, for *tetrataenite* and 4.7-4.4 wt% Ni for *kamacite* [4].

Mössbauer spectroscopy has been used to analyse metallic samples of Crab Orchard, Estherville, Vaca Muerta and Lowicz meteorites without any chemical treatment. Mössbauer spectra of these untreated samples are shown in Fig. 1.

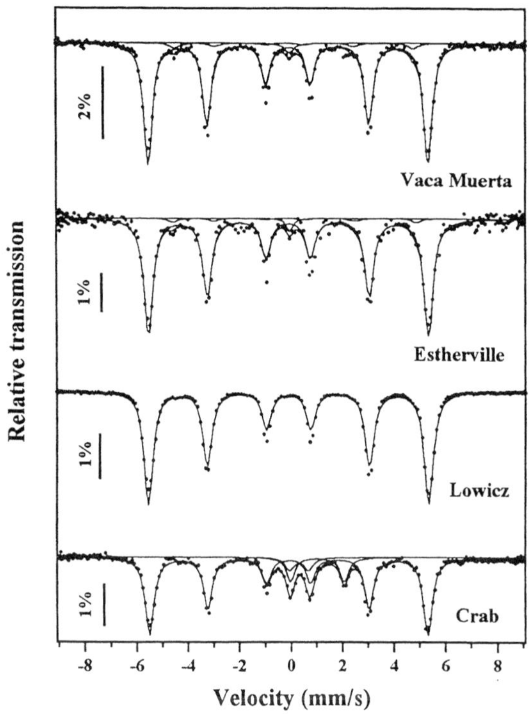

Fig. 1 - Mössbauer spectra at RT

In each of these spectra the main component is *kamacite* (bcc Fe-Ni 4-8% Ni). The *taenite* containing the intergrowth *tetrataenite/antitaenite* (TT/AT) seen in ataxites, octahedrites and metal particles of chondrites is present in Estherville and Vaca Muerta in very low proportion, and is apparently absent or not enough for Mössbauer detection in the Lowicz sample, in which the spectrum only shows the presence of *kamacite*. In the case of Crab Orchard, the samples was very oxidized and containing parts of silicate that could not be removed without a chemical treatment. The spectrum shows together with *kamacite*, doublets of *pyroxene* and Fe-oxides. Estherville and Vaca Muerta exhibit the same three spectral components: *kamacite* (H=34T), *tetrataenite* (H=29T) and the singlet corresponding to *antitaenite*.

In order to better determine the relative proportion of the intergrowth TT/AT, and evaluate the degree of ordering of *tetrataenite*, taenite fractions are being extracted in which the TT/AT intergrowth can be enriched by chemical etching (HCl 5N for one mounth), allowing therefore a more accurate determination of the hyperfine parameters that should reflect the reported very slow cooling rates of this unique group of meteorites [5].

CONCLUSION

From the results obtained up to now it can be seen that in untreated samples, in which *kamacite* is present in a high proportion, the intergrowth TT/AT can be evidenced only in some cases. Further studies are still in progress in chemically treated samples and in the silicates extracted from these meteorites.

ACKNOWLEDGMENTS

Financial support from CBPF, FAPERJ and CNPq is gratefully acknowledged by R.B.D. One of us (A.B.D.) wishes to thank the fellowship from CLAF.

REFERENCES

[1] E.R.D. Scott, H. Haack and S.G. Love, Meteoritics & Planetary Science 36 (2001) 869.

[2] J.F. Petersen, M. Aydin and J.M. Knudsen, Phys. Lett. 62A (1977) 192; J. Danon, R.B. Scorzelli and I. Souza Azevedo, C.R. Acad. Sci. Paris 287 (1978) B-199.

[3] D.G. Rancourt and R.B. Scorzelli, J. Magn. Mater., 150(1995) 30.

[4] B. Lavielle, E. Gilabert, M.R. Soares, M.A.Z. Vasconcellos, G. Poupeau, C. Canut de Bom, M. Cisternas and R.B. Scorzelli, Meteoritics & Planetary Science 33 (4) (1998) A92.

[5] W.D. Hopfe and J.I. Goldstein, Meteoritics & Planetary Science 36 (2001) 135.

WEATHERING OF VOLCANIC MATERIAL FROM PERHUMID COSTA RICA STUDIED BY MÖSSBAUER SPECTROSCOPY, INAA AND XRD

J. J. van Loef[1], Th. G. van Meerten[1], E. L. Meijer[2] and P. Buurman[2]

1 Interfacultair Reactor Instituut, Technische Universiteit Delft, Mekelweg 15, 2629 JB Delft, Nederland

2 Laboratorium voor Bodemkunde en Geologie, Universiteit Wageningen, Postbus 37, 6700AA Wageningen, Nerderland

Iron compounds are common constituents of volcanic rocks and soils. In rocks Fe is both divalent and trivalent, while in soils it is mostly trivalent. Samples were collected from the Turrialba volcano in Costa Rica at different altitudes on the slope below 1.000 m in a region with extreme precipitation and a large gradient in rainfall, and from the inside crater at 3020 m. The samples were dried at 40°C; the Fe content varied between 5 and 11 wt.%.

Mössbauer spectra (MS) were measured at RT and 85K and one sample at 77 and 4.2K in addition. Hematite and magnetite are identified by their hyperfine splitting. A Fe^{3+} -doublet dominates the MS. Its total linewidth increases at 85K and it broadens even more at 77K, indicating onset of magnetic ordering. In MS [4.2K], the doublet has practically disappeared and changed into a sextuplet with asymmetrically broadened spectral lines. Murad [1] found a similar behaviour in natural ferrihydrite. Computer analysis of MS [4.2K] shows a wide distribution of hyperfine fields characterized by a maximum of 49T and a halfwidth of 8T. A small Fe^{3+} -doublet, left in the center, is attributed to either aluminous ferrihydrite or iron substituting in allophane, the presence of which is identified by electron microscopy. In general ferrihydrite dominates MS of samples that were most weathered as indicated by their very low contents of alkalis and earth alkalines found by INAA. Ferrihydrite may primarily be located on the surface of the allophane particles [2].

Fe^{2+} -doublets, assigned to pyroxene, are measured in particular in those samples that were least weathered and its content is negatively correlated with ferrihydrite. XRD identified clay, gibbsite, quartz and pyroxene in the different samples; hematite was found once. Neither magnetite nor ferrihydrite were identified. In volcanic soils, recrystallization/transformation of the latter to goethite or hematite is hindered by high contents of allophane, dissolved silica and organic matter [3].

Among the weathering products of primary minerals, like pyroxene, iron(hydr)oxides and smectite are by far the most important secondary minerals. The very specific physico-chemical properties of the amorphous secondary products (allophane and ferrihydrite) dominate the soil properties of andisols. MS is one of the very few methods by means of which ferrihydrite can be identified. This work is part of an extensive study of soil formation of volcanic deposits under extreme weathering conditions [4,5].

[1] E. Murad, L. H. Bowen, G. J. Long and T. G. Quin, Clay Miner. 23 (1988) 161
[2] K. J. D. Mackenzie and C. M. Cardile, Thermochim.Acta 130 (1988) 259
[3] U. Schwertmann, J. Friedl, H. Stanjek and D. G. Schulze, Clay Miner. 35(2000) 613
[4] P. Buurman, K. de Boer and Th. Pape, Geoderma 78 (1997) 71
[5] E. L. Meijer and P. Buurman, Geoderma 80 (1997) 129

THE LEACHING OF A SULPHIDE ORE - A MÖSSBAUER AND SEM PERSPECTIVE

F.B.Waanders and D.S. Smit
School of Chemical and Minerals Engineering, PU for CHE, Potchefstroom, 2531, South Africa. email: chifbw@puknet.puk.ac.za

Abstract: Pyrometallurgical extraction is used in conventional sulphide concentrate processing, but due to high costs and pollution problems, hydrometallurgical processing is an attractive method for the selective extraction of valuable base metals, in particular of low grade ores.
In the present investigation the leaching mechanism of a Ni-Cu-Co-sulphide concentrate has been studied in an oxidative pressure-acid medium. The ore consisted of pyrrhotite, pentlandite and chalcopyrite with gangue minerals mostly pyroxenes. The dissolution of the valuable minerals was achieved effectively after mechanical activation, followed by leaching. From the Mössbauer spectra it was clear that the Ni-bearing pentlandite was leached within the first 15 minutes, whilst the Cu-bearing mineral, chalcopyrite, was only leached after a period of more than 30 minutes. The final residue was found to consist only of goethite. The values of the apparent activation energies of the Ni, Cu and Co, extracted from the sulphide concentrate, were calculated and found to be 20.6 ± 4.4 kJ/mol K, 33.6 ± 4.2 kJ/mol K and 17.4 ± 3.5 kJ/mol K respectively.

Keywords: *Hydrometallurgy, leaching, base metals, activation energy*

1. Introduction

Whether a metal is extracted in a water environment or at high temperatures is, in principle, immaterial. Each extractive situation must be assessed on its own merits by the consideration of such factors as capital and operating costs and the environmental impact of the two different options that are available [1]. Hydrometallurgy will have an increasing role to play in the future regarding the extraction of valuable metals from sulphide minerals. It is unique in its application to low-grade ores, which cannot be beneficiated economically and should be viewed as an alternative technology having high chemical specificity [2].
According to ref. [3] the hydrometallurgical route for the recovery of a metal, where dissolution (leaching), separation, concentration and reduction to the metal is carried out at near ambient temperature, is becoming more competitive with the conventional high temperature processes used in the smelting of metals.

2. Experimental

A freshly milled ore sample (80% -62μm) and a dry flotation concentrate from the Nkomati mine, South Africa, was obtained The concentrate consisted mainly of the minerals pentlandite $(FeNi)_9S_8$, pyrrhotite ($Fe_{1-x}S$), chalcopyrite ($CuFeS_2$) and traces of pyrite (FeS) and magnetite (Fe_3O_4), whilst the gangue minerals consisted mainly of pyroxenes.

A Svedala Stirred Media Detritor, constructed of stainless steel, lined with thick rubber and fitted with a rotor, composed of crossed rubber bars at the bottom of the shaft, was filled with a grinding medium of ceramic balls to produce the target grind sizes of 80% -10μm particles.

A laboratory autoclave, to perform the Activox® test work, was constructed of 904 L Austenitic stainless steel and fitted with an agitator, composed of 316 stainless steel and two impellers made from Ertelon. The agitator shaft, as well as the sampling dip tube, was coated with a Teflon (PTFE) layer to protect the metal from the acid conditions inside the vessel. Heating and cooling mechanisms were provided to control the exothermic reaction. The complete autoclave with a working volume of 2 litre was situated inside a steel frame to facilitate the complete experimental set-up.

The freshly ground sample, obtained from the ultra fine milling, was mixed with water to make up a slurry of 1.7 litres. The slurry was poured into the autoclave and all the parameters were set at target values and the slurry leached for 2.5 hours with intermediate sampling performed at 15, 30, 50, 90 and 150 minutes via a sampling dip tube. The slurry was filtered and the residue solids were dried, weighed and analysed.

All Mössbauer spectra were obtained with the aid of a Halder Mössbauer spectrometer, capable of operating in conventional constant acceleration mode using a proportional counter as detector. An initial 50 mCi ^{57}Co source, plated into a Rh-foil, was used to produce the γ-rays. The sample analyses were done at room temperature and data collected in a multi-channel analyser to obtain a spectrum of count rate against source velocity. The amount of each constituent present in a sample was determined from the areas under the relevant peaks.

3. Results and discussion

From the Mössbauer spectrum of the freshly milled ore sample (see Figure 1 (left)) and the Mössbauer parameters in Table 1, it was clear that the ore sample consisted of gangue material (35%) and the massive sulphide ore (65%). The main sulphide minerals identified were pyrrhotite (53%), pentlandite (7%) and pyrite (5%), whilst the gangue minerals consisted mainly of pyroxenes. The Mössbauer parameters, obtained for the gangue minerals (see Table 1), are in agreement with those found in literature [4].

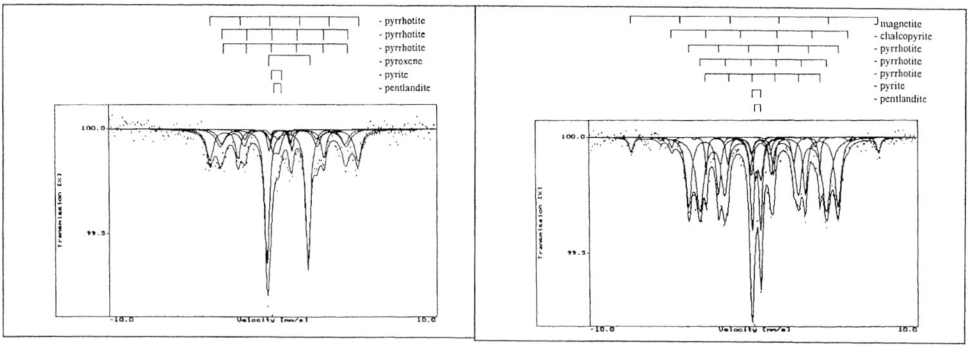

Figure 1: Mössbauer spectra of the ore sample (left) and floatation concentrate (right).

Floatation ensured that the gangue minerals were removed, as can be seen in the Mössbauer spectrum of this sample (Figure 1 (right)), where the large doublet (IS = 1.06 ± 0.02 mms^{-1}, QS = 2.50 ± 0.02 mms^{-1}) does not exist in the spectrum anymore. In addition to the minerals mentioned, magnetite and chalcopyrite were also identified in the float sample but due to their low relative abundance they were not detected in the ore sample where the doublet due to the gangue minerals shielded the contribution of these minerals.

Pyrrhotite, the most abundant Fe-sulphide mineral in the ore sample, gave rise to three sextets in the Mössbauer spectra, with parameters (see Table 1) in agreement with those reported in literature [4]. After 30 minutes leaching, only the strongest of the three sextets remained in the spectrum, due to complete dissolution, with the bulk of the residue consisting of goethite (IS = 0.44 ± 0.02 mms^{-1}, QS = 0.89 ± 0.02 mms^{-1}).

The main Ni- and Co-bearing mineral pentlandite, gave rise to the formation of a doublet (IS = 0.36 ± 0.02 mms^{-1}, QS = 0.35 ± 0.02 mms^{-1}), with parameters consistent with literature [4]. The pentlandite was almost completely dissolved after 30 minutes of leaching, as can be seen from the disappearance of this doublet in the spectra, in agreement with the leach solution reporting that most (75%) of the Ni appeared in the leach solution under optimum conditions.

Chalcopyrite, the Cu-bearing mineral, was characterised in a roasting process by Banduopadhyay *et al.*, [5]. They reported an isomer shift of between IS = 0.24 ± 0.02 mms^{-1} and IS = 0.30 ± 0.02 mms^{-1} for chalcopyrite, but in the present experiment a higher value of IS = 0.36 ± 0.02 mms^{-1} was found, still within the limits given in ref. [4]. The quadrupole splitting and hyperfine magnetic field strength values are in agreement with those reported by ref. [5].

It is clear from the spectra that primarily goethite remained after 50 minutes of leaching. The values of the isomer shift for goethite varied between (IS = 0.32 ± 0.02 mms^{-1} and IS = 0.45 ± 0.02 mms^{-1}) and this can be ascribed to the fact of the presence of impurities, incomplete leaching, or the fact that very finely formed goethite was found in the experiments. Ideally, goethite should be investigated at low temperatures (4K) as well, but this falls beyond the scope of this experiment.

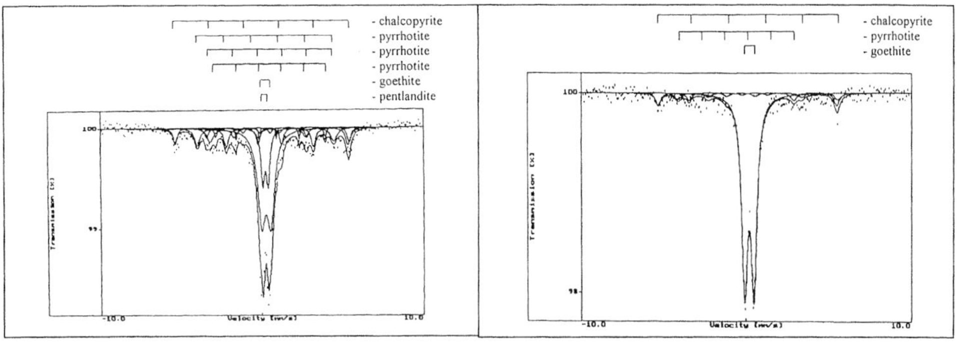

Figure 2: Mössbauer spectra of the leach product after 15 minutes of leaching (left) and the spectrum of the leach product after 30 minutes of leaching (right).

It is interesting to note that the chalcopyrite was the most difficult mineral to leach as it only disappeared from the leach residue after 30 minutes of leaching, whilst the pentlandite was already dissolved within the first 15 minutes of leaching (see Figure 2).

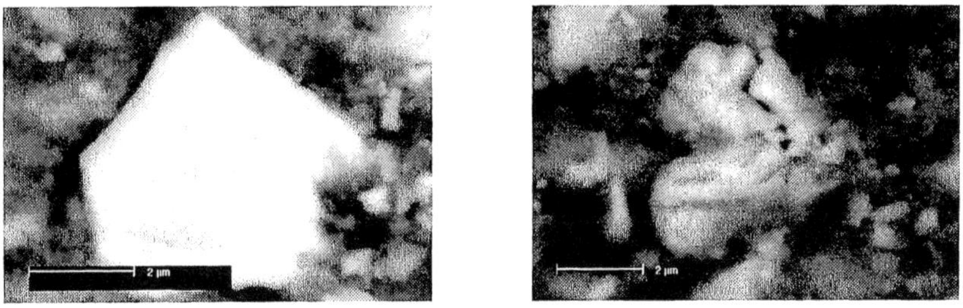

Figure 3: SEM micrograph of an euhedral chalcopyrite particle (left) and a pentlandite particle (right) after 15 minutes leaching, showing almost complete dissolution of the latter.

Theory suggests [6] that the reaction is initially controlled by the shrinking particle model and then by diffusion through the product layer model. Results from the present research however, suggest that these two models are competing from the start of the leaching process through to the end, and it is not recommended to fit these well established models separately to different stages of the leaching reaction. This is in agreement with an investigation by ref. [7] and it was thus appropriate to fit the data to a universal first order model [8] of the form:

$$x = 1 - e^{-kt}$$

where, k = reaction rate (1/min), x = leached fraction and t = time (min)

The activation energies for the Ni, Cu and Co, extracted from the sulphide concentrate were determined by fitting the apparent reaction rate constants to the Arrhenius equation:

$$\ln k = c - \frac{Q}{RT}$$

where, k = reaction rate constant (1/min), Q = activation energy (J/mol K).

The values of the apparent activation energies of Ni, Cu and Co were found to be 20.6 (± 4.4) kJ/mol K, 33.6 (± 4.2) kJ/mol K and 17.4 (± 3.5) kJ/mol K, respectively, in agreement with those found by ref. [8].

Table 1: The fitted Mössbauer parameters derived from the Mössbauer fits to the spectra, studied at 300K. (The values reported are relative to α- iron)

Sample	IS mms^{-1} (error±0.02)	QS mms^{-1} (error±0.02)	B Tesla (error±2)	Relative intensity (%)	Assignment
Ore sample	0.63	0.18	28.1	25	Pyrrhotite
	0.61	0.15	23.8	18	
	0.61	0.16	23.5	10	
	0.36	0.34	-	7	Pentlandite
	0.35	0.52	-	5	Pyrite
	1.06	2.50	-	35	Pyroxene
Float sample	0.65	0.10	28.4	32	Pyrrhotite
	0.63	0.14	24.0	25	
	0.61	0.16	21.5	15	
	0.37	0.37	-	10	Pentlandite
	0.37	0.53	-	11	Pyrite
	-0.15	0.70	46.9	4	Magnetite
	0.36	0.00	35.5	3	Chalcopyrite
Leach product 15 min	0.64	0.18	29.2	18	Pyrrhotite
	0.62	0.15	23.9	11	
	0.61	0.17	21.2	4	
	0.36	0.36	-	14	Pentlandite
	0.36	-0.01	33.5	8	Chalcopyrite
	0.39	0.88	-	45	Goethite
Leach product 30 min	0.62	0.18	30.3	6	Pyrrhotite
	0.35	-0.02	33.5	11	Chalcopyrite
	0.32	0.86	-	83	Goethite
Leach product 50 min	0.44	0.89	-	100	Goethite
Leach product 90 min	0.44	0.89	-	100	Goethite
Leach product 150 min	0.45	0.89	-	100	Goethite
Leach residue	0.32	0.88	-	100	Goethite

References

[1] Woollacott L.C. & Eric R.H., 1994, Mineral and Metal Extraction An Overview. Johannesburg : The South African Institute of Mining and Metallurgy. 331p.

[2] Osseo-Asare K. & Miller J.D., 1982. Hydrometallurgy research, development and plant practice. Warrendale, Pennsylvania : The metallurgical society of AIME. 412p.

[3] Bautista R.G., 1984, Hydrometallurgical process fundamentals. New York : Plenum Press. 221 p.

[4] Stevens *et al.*, 1998, Mössbauer Mineral Handbook. University of North Carolina, Asheville, USA : Mössbauer Effect Data Centre. 527p.

[5] Banduopadhyay *et al.*, May 1999, Technical note study of the roasting of chalcopyrite minerals by ^{57}Fe Mössbauer spectroscopy. Minerals Engineering, 13(8-9):973.

[6] Bredenhann R. & Van Vuuren C.P.J., February 1999, The leaching behaviour of a nickel concentrate in an oxidative sulphuric acid solution. Minerals Engineering, 12(6):687.

[7] Maurice D. & Hawk J.H., March 1999, Ferric chloride leaching of a mechanically activated pentlandite-chalcopyrite concentrate. Hydrometallurgy, 52:289.

[8] Amer A.M., September 1995, Investigation of the direct hydrometallurgical processing of mechanically activated complex sulphide ore, Akarem area, Egypt. Hydrometallurgy, 38:225.

Acknowledgements

The authors wish to thank AVMIN for financial support and Dr LR Tiedt for SEM analyses and photography

Sicán kilns and furnaces: Field firings and archaeometric studies

U. Wagner [a], I. Shimada [b], D. Goldstein [b], and W. Häusler [a]

[a] *Physik-Department E15, Technische Universität München, 85747 Garching, Germany,*
[b] *Southern Illinois University, Department of Anthropology, Carbondale 69207, USA.*

A number of pottery kilns and a cluster of metal working furnaces were found in a 1000 years old Sicán workshop, that was excavated in 1999 at Huaca Sialupe at the north coast of Peru. In 2000, field firings were conducted in a replica kiln built to scale from local raw materials. Mössbauer spectroscopy and X-ray diffraction were used to characterize both replicas and ancient material. In an attempt to reconstruct ancient firing procedures, local clay fired in the laboratory under controlled conditions was compared with ancient sherds and replica ceramics from the field firings. Mössbaur spectra observed in sherds of ancient ceramics could be emulated in the laboratory and in the field firings.

Keywords: Mössbauer spectroscopy, X-ray diffraction, northern Peru, middle Sián culture, pottery making kilns, furnaces, replica field firings

1. Introduction

During the excavation of a 1000-year old Middle Sicán workshop (ca. 1000 A.C.) at Huaca Sialupe on the north coast of Peru, at least two types of kilns, numerous molds, and other evidence of intensive pottery production were found. Within the same workshop several inverted ceramic urns were located. Middle Sicán period metal working tools, slags, and other remains of metal working were found together with inverted urns. In 2000 the Sicán Archaeological Project conducted a series of field firing experiments using a replica kiln in order to reconstruct Sicán pottery-making techniques under controlled conditions. Mössbauer spectroscopy supported by X-ray diffraction [1] was applied to correlate the firing conditions with the physical properties of the products.

2. Field firing

The replica kiln (Fig. 1) was built from local clay by José Sosa, a potter from Chulcanas Piura. He also made a large number of replica vessels using molds and paddle-and-anvil forming techniques. Additionally, the potential use of the inverted urns as furnaces for metal working was tested. The field firing experiments

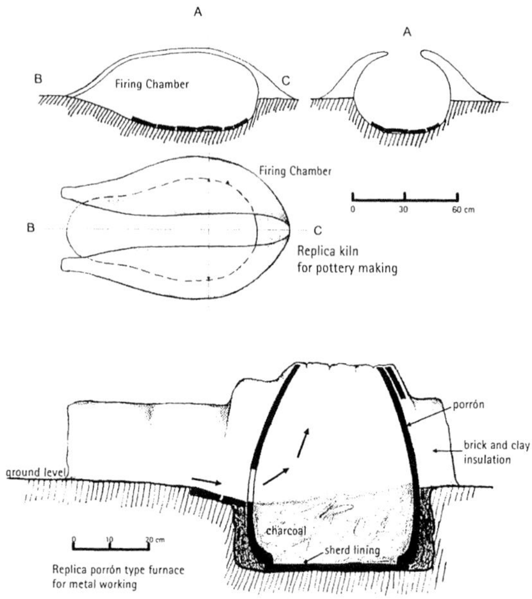

Figure 1. Pottery making kiln (top) and metal-working porrón-type furnace (bottom).

were monitored with NiCr/Ni-thermocouples, whose tips were embedded in balls of clay to measure time-averaged temperatures. Temperatures of 700 to 800 °C necessary to fire pottery could be reached and sustained in oxidizing as well as reducing environment for periods long enough to emulate the ancient ceramics. Locally available zapote (*Capparis angulata*) and algorrobo (*Prosopis pallida*) wood was used as fuel. Sufficient draft was created by the wind from the onshore breeze in the afternoon. Llama and dried cow dung was used to obtain a reducing atmosphere.

In a second experiment a replica furnace built from a porrón (Fig. 1), a large ceramic vessel, was heated with charcoal. Temperatures of 1000 °C and above were reached in less than 30 minutes on the bottom, while the temperature at the top reached 800 °C, which is sufficient for annealing sheet metals.

3. Study of local clay

Two different batches of recent clay were studied in laboratory experiments. One batch is coarse, while the other is much finer and contains a comparatively large amount of 14 Å minerals. Three firing series were performed in the laboratory. Samples of clay were fired (i) in air for 48 h at increasing temperatures (O), (ii) fired in air for 48 h after a preceding 3h reduction with charcoal at

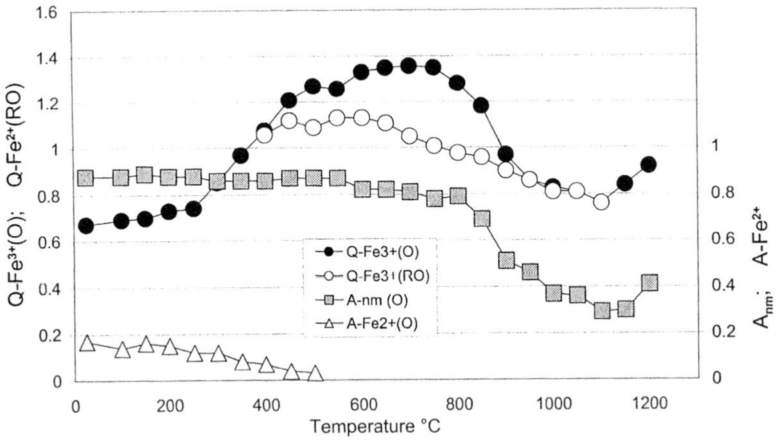

Figure 2. Change of quadrupole splitting of Fe^{3+}-species (Q-Fe^{3+}) on firing clay Sialupe in the laboratory without (full circles, O), and with (open circles, RO), a preceding reduction at 800 °C with charcoal. The quadrupole splittings are given in mm/s. The fractional areas of the nonmagnetic components (A_{nm} = (A-Fe^{3+} + A-Fe^{2+})) and the area of A-Fe^{2+} are also plotted.

800 °C in a closed vessel (RO) and (iii) fired for 3 h under reducing conditions produced with charcoal (R). The most characteristic data obtained after firing in air, the quadrupole splitting of the Fe^{3+} fraction and the fractional area of the nonmagnetic species are plotted in Fig. 2 together with the quadrupole splitting observed after firing in air following a preceding reduction. The characteristic increase of Q-Fe^{3+} during dehydroxilation from 0.72 to 1.37 mm/s at temperatures between 300 and 500 °C can be seen clearly. This increase is absent after a preceding reduction (RO) around 800 °C. The material thus remembers the reducing pretreatment, which helps to distinguish between oxidizing firing only and firing after a preceding reduction. The detailed data will be published elsewhere.

4. Sicán ceramics

RT Mössbauer spectra of the inner part of two sherds of Sián blackware are shown in Fig. 3 (left and middle), together with their respective XRD scans. The Mössbauer spectra of the surface layers of these sherds are similar. Sicán ceramics are famous for their intriguing black sheen, which is due to graphite deposition [2] on the burnished surfaces and not to magnetite formation as observed by Maniatis et al. [3] in the black gloss of Attic ceramics. The spectrum on the right of Fig. 3 is from a sherd of late middle Sicán blackware from the previously excavated site of Huaca Loro. It still contains 14 Å minerals and was only heated to about 300 °C. The black color of its surface is caused by soot that penetrated

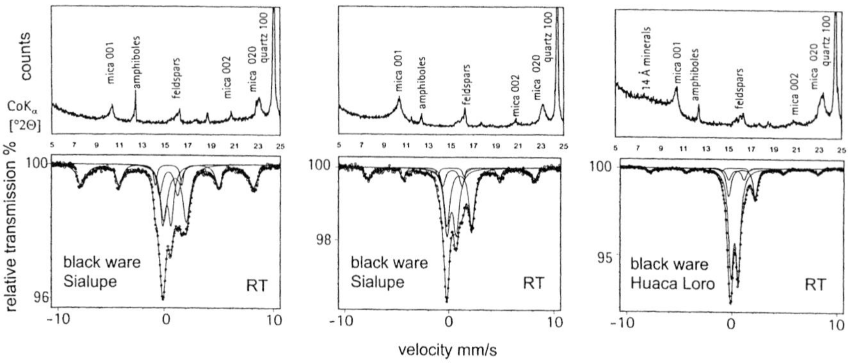

Figure 3. X-ray diffraction diagrams (top row) measured with Co K_α radiation and room temperature Mössbauer spectra of the cores of three sherds of Sicán blackware, two from Sialupe (left and middle) and one from Huaca Loro (right).

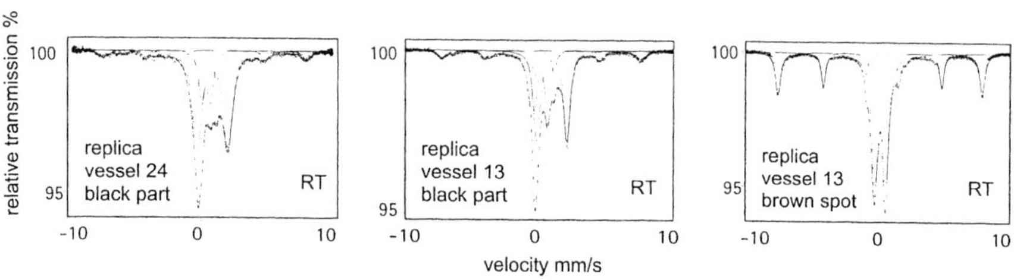

Figure 4. RT Mössbauer spectra of the black material from two replica vessels from the field firing experiments. The spectrum of a misfired, brownish spot is shown on the right.

into the clay. Replica vessels from the field firing experiments give rise to similar Mössbauer patterns as Sicán blackware (Fig. 4).

The study of their artifacts shows that Sicán craftsmen were expert artisans. They maintained the desired kiln temperature by using charcoal and controlling the draft. Their technical abilities were highly developed and their tradition lasted until the time of the Spanish conquest of Peru in the 16th century.

Support by the German Research Council and the National Geographic Society is thankfully acknowledged.

References

[1] U. Wagner, F.E. Wagner, W. Häusler and I. Shimada, in "Radiation in Art and Archaeometry", D.E. Creagh, and D.A. Bradley, eds, Elsevier, Amsterdam, 2000, p.417-443.
[2] I. Shimada and U. Wagner, MRS Bulletin, 26 (2001) 25-30.
[3] Y. Maniatis, E. Aloupi, A.D. Stalios, Archaeometry, 35 (1993) 23-34.

^{57}Fe Mössbauer Spectroscopy in Coal with Different Caloric Values

C.I. Wynter[1], L.May[2], F.W. Oliver[3], E. Hoffman[3], D. Seifu[3], and J.A. Hall[4]

[1]*Department of Chemistry, Nassau Community College, Garden City, N.Y. 11530, U.S.A.*
[2]*Department of Chemistry, The Catholic University of America, Washington, DC 20064, U.S.A.*
[3]*Department of Physics, Morgan State University, Baltimore, MD 21251, U.S.A.*
[4]*Environmental Department, Ontario Power Generation, Toronto, Ontario, Canada M5G 1X6*

The environmental concern associated with mining and burning of coal has long been discussed in the literature. The present study deals with the four different ranks of coal with varying calorific values (CV) aiming to determine if these values are influenced by the sulfur content and/or the forms of iron-sulfur minerals associated with these types of coal. It has been established that Mössbauer spectroscopy is undoubtedly the best analytical tool for discerning iron minerals. The four different types of coal had CVs ranging from 32,403 to 16,000 kJ/kg and sulfur concentrations ranging from 0.25 to 2.5%. There was no direct correlation between the CV and the percentage of pyrites and ferrous sulphate. However, in general, as the CV increases, the percentage of sulfur increases.

Keywords: Coal, Caloric value, Pyrites, Ferrous Sulphate, Mössbauer spectroscopy, Sulphur.

1. Introduction

Coal can be classified or ranked into four groups [1] based upon their caloric values (CV) measured in kilojoules/kilogram (kJ/kg); this includes anthracite or semi-anthracite with the highest CV, followed by bituminous coal, sub-bituminous coal, and lignite. This ranking also reflects the hardness of coal with anthracite being the hardest, and then lignite is the softest.

The CV and the percentage of inorganic sulfur are important, as the price of coal is dependent on the CV and the percentage of sulfur; namely the higher the CV, the higher the price. This is offset by a corresponding sulfur value, that is, lower inorganic sulfur increases the price of coal.

Several workers [2-7] have used Mössbauer spectroscopy to look at the environmental effects of mining and burning coal; included are M. Fajardo, et al [2] who used Mössbauer spectroscopy and X-Ray diffraction [XRD] to look at iron minerals in particular and other compounds in Colombian coal. Kopcewiez and Kopcewiez [3] used Mössbauer spectroscopy to compute iron sulfides [FeS, FeS_2] concentrations in coal combustion. Harchand and Raj [4] used Mössbauer spectroscopy and XRD to characterize the iron-bearing phases in coal and ash at 500°C. Wynter [5] and his colleagues used Mössbauer spectroscopy to investigate low-iron coals and their flyash while Clark [6] and his co-workers observed mild acid treatment of virgin coals using Mössbauer spectroscopy.

In 1993, Traneja, Aggarwall, and Brar [7] used Mössbauer spectroscopy and XRD to study coal and coal ash whereas, as early as 1977, Montano [8] quantitatively determined pyrites in coal. The present study deals with the first attempt to link the ranking of coal (price of coal) with the percentage of (inorganic) sulfur by Mössbauer spectroscopy as this technology is as reliable as the conventional (wet chemistry) method. Presently, the price of coal is determined by an on-line analysis [9] including the determination of CV and the percentage of inorganic sulfur.

2. Experimental

Eight coal samples were provided by Ontario Power Generation. The CVs of the coal samples were determined by using a bomb calorimeter, and the inorganic sulfur contents were measured by precipitating barium sulfate ($BaSO_4$). The Mössbauer spectra of ground samples were measured using an Austin Scientific Model ASA with moving source geometry calibrated with iron foil at room temperature. The resultant spectra were analyzed by constrained least-square fit to Lorenzian shaped lines. The percentage of the pyrites and ferrous sulfate were determined by the method of Montano [8].

3. Results and Discussion

Table 1 gives the CV, % sulfur, %pyrites and %ferrous sulfate. Typical spectra are shown in Fig. 1, in which the features of the two types of spectra used in the analysis of the sulfur containing minerals in the coal samples. In general, the CV decreases as its sulfur content decreases. In contrast, the percentage of the pyrites increases as the CV decreases. This suggests that the more ferrous or the reduced species is present the lower is the CV. Further work is in progress to confirm these preliminary observations.

Table 1. Caloric Values (CV) and Sulfur Components of Coal

SAMPLE	RANK	CV [kJ/kg]	%S	%Pyrites	%$FeSO_4$
1	Anthracite	30,600	2.5	95.61	4.39
2	Lignite	16,100	0.4	100	0.00
3	SubBituminous	25,500	0.57	69.04	30.96
4	Lignite	20,350	0.25	100	0.00
5	Anthracite	32,403	2.47	72.8	27.2
6	SubBituminous	24,800	0.68	100	0.00
7	Bituminous	28,100	0.35	100	0.00
8	Bituminous	29,800	0.65	66.55	33.45

References

[1] Communication with the Coal Association of Canada, Suite 502, 205-9 Ave. S.E., Calgary, Alberta T2GOR3 Canada.

[2] M. Fajardo, J. Mohica, H. Barraza, G.A.Perez Alcazar and J.A. Tabares, Hyp. Int. 122(1999)129.

[3] B. Kopcewiz and M. Kopcewiz, Hyp.Int.126 (2000) 131.

[4] K.S. Harchand and D. Raj, Nucl. Instr. Meth.Phys. Res. B76 1993)249.

[5] C.I. Wynter, TM. Clark, B.J. Evans, M.A. Edwards, D. Radcliffe, and J.A. Hall, Hyp. Int. 112 (1998) 231.

[6] T.M. Clark, B.J. Evans, C.I. Wynter, and H. Pollack, Hyp. Int. 112 (1998)227.

[7] S.P. Taneja, M.P. Aggarwall, and A.S. Brar, Nucl. Instr. Meth. Phys. Res. B 76 (1993) 236.

[8] P.A. Montano, Fuel, 56 (1977), 397.

[9] R.C. Woodward, and B.Lee, Coal Age 102 (March 1997) 22.

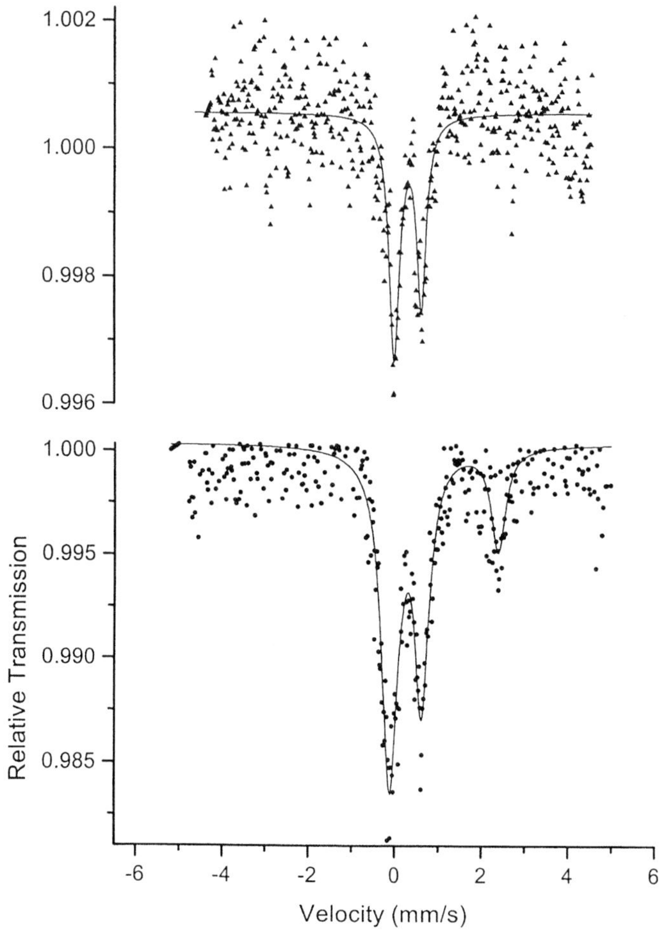

Fig. 1. Typical Mössbauer spectra of coal. Top. Pyrites only. Bottom. Pyrites and ferrous sulphate.

STUDY OF OXIDE-SUPPORT INTERACTIONS IN SILICA-SUPPORTED IRON OXIDE PRECURSORS

J.F. Bengoa, S.G. Marchetti, M.V. Cagnoli, A.M. Alvarez, N.G. Gallegos, and R.C. Mercader

CINDECA, CONICET, CICPBA, IFLP, Facultad de Ciencias Exactas, Facultad de Ingeniería,

Universidad Nacional de La Plata, 47 N°257, 1900 La Plata, Argentina

ABSTRACT

Mössbauer studies of α-Fe_2O_3 particles in an Fe/SiO_2 system calcinated in air for 8, 144, and 344 h at 698 K, reveal average diameters of less than 28 Å for the sample calcinated 8 h, and of about 50Å for longer calcination times. In the frame of Debye's model applied to the thermal dependence of the relative f-factors, we found that the smaller particles have the strongest oxide-support interaction with an infinite force-constant between the oxide particles and silica.

INTRODUCTION

The Fe/SiO_2 system has been extensively studied to be used in hydrocarbon synthesis from CO and H_2. In the reduction step towards the preparation of the catalyst, the iron oxide-silica interaction modifies the Fe reducibility and consequently the activity in the Fischer-Tropsch reaction [1]. Earlier [2], we have studied the influence of the calcination temperature on the degree of the iron-oxide/SiO_2 interaction. In this paper, we use Mössbauer spectroscopy to investigate the modification of the degree of the iron-oxide/SiO_2 interaction caused by different calcination times.

EXPERIMENTAL

Samples were prepared by the impregnation method with a solution of $Fe(NO_3)_3.9H_2O$ on silicagel so as to obtain an oxide with an Fe content of 5% wt/wt. The product was divided in three batches, which were calcined in air at 698 K for 8, 144, and 344 hours, called S8, S144, and S344, respectively. Mössbauer spectra were recorded as described elsewhere [2]. All spectra were fitted with a programme capable to analyse hyperfine parameters distributions. The samples were also characterised by X-ray diffraction and N_2 adsorption.

RESULTS AND DISCUSSION

a)Species and particle sizes. Fig. 1 displays the Mössbauer spectra taken at 15 and 298 K. Table I shows the hyperfine parameters of the fitted spectra at 15 K.

Table I: Most likely values of hyperfine parameters of the samples at 15 K.

Species	Parameter[a]	S8	S144	S344
α-Fe_2O_3 (sp)	δ (mm/s)	0.42 ± 0.01	0.25 ± 0.06	0.2 ± 0.2
	Δ (mm/s)	0.73 ± 0.01	———	———
α-Fe_2O_3 (relaxing)	H (T)	37 ± 1	———	———
	δ (mm/s)	0.4 ± 0.2	———	———
α-Fe_2O_3(core)	H (T)	———	49.9 ± 0.1	50.0 ± 0.1
	δ (mm/s)	———	0.51 ± 0.01	0.49 ± 0.01
α-Fe_2O_3(shell)	H (T)	———	46.2 ± 0.2	46.3 ± 0.2
	δ (mm/s)	———	0.45 ± 0.01	0.45 ± 0.01

[a] All isomer shift are referred to α-Fe at 298 K. (sp): superparamagnetic.

The hyperfine parameters are characteristic of Fe^{3+} in α-Fe_2O_3. Only one doublet is observed at 298 K for all samples. S8 spectrum at 15 K exhibits one quadrupole signal and a curved background which belong to superparamagnetic α-Fe_2O_3 [3] and magnetically split relaxing α-Fe_2O_3 –simulated by a sextet in the fitting– respectively. Instead, S144 and S344 show a non-resolved doublet assigned to superparamagnetic α-Fe_2O_3 and two magnetic signals corresponding to the "core" and "shell" of α-Fe_2O_3 particles at 15 K [4].

Table II shows the oxide diameters and the blocking temperatures (T_B). The diameters were estimated through Kündig's model [5], using the blocking temperatures and a magnetic anisotropy constant value of 0.5×10^5 Jm^{-3} [6]. It can be concluded that 144 h is a calcination time long enough to achieve the maximum particle size.

X-ray diffraction patterns only display the signal corresponding to amorphous silica, which is consistent with the Mössbauer results. The sizes are small enough to be located inside the silica pores. This location is confirmed by N_2 adsorption, since the iron-oxide/silica pore diameters are smaller than the pure silica ones.

b) Recoilless fraction. We followed the methodology used previously [2], i.e., after verifying the criterion of the thin absorber limit, we compare the thermal variation of the normalised spectral areas, A(T)/A(0),

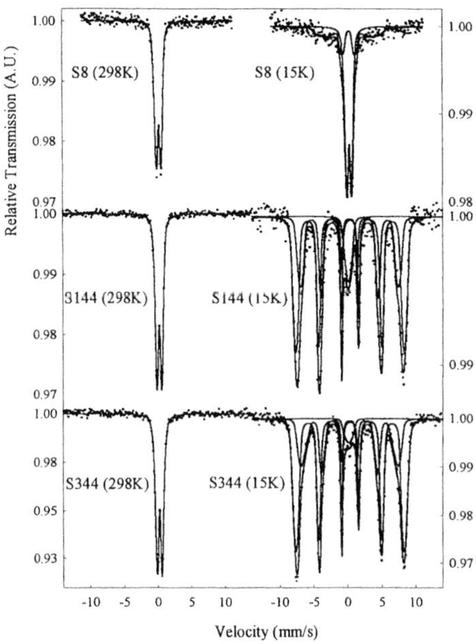

Figure 1: Mössbauer spectra of the samples.

with the asymptotic behaviour predicted by Debye's model.

The temperature dependence of ln A(T)/A(0) (Fig. 2) is well fitted with straight lines at all temperatures for S8 and up to 50 K for S144 and S344. This behaviour is different from the one normally observed for larger crystals for which a linear temperature dependence is found only for $T \geq \theta_D/2$. The Debye temperatures, θ_D, obtained are shown in Table II. θ_D of the crystallites must be similar to those of bulk specimens as reported in [7]. However, only S8 shows a θ_D value equal to that of bulk α-Fe_2O_3. The differences in θ_D for S144 and S344 can be assigned to the motion of the particle as a whole when the particle-support interaction is weak [2].

Assuming that the lattice and particle vibrations are uncoupled, the measured f-factor can be written as: $\mathbf{f_{TOT} = f_{LATT} \cdot f_{PART}}$. For these systems, if only f_{LATT} were used instead of f_{TOT}, the θ_D obtained would be lower than the bulk value. Assuming that the vibrational coupling between the particles can be neglected – this is a good assumption for these samples with low iron content and a high surface area support– the vibration of the particles as a whole is described by a localised mode and the Einstein model is applicable [8]. In this case, the particle vibrations as a whole can be described by a harmonic oscillator, whose angular frequency, ω, particle-support force-constant, q, and Einstein temperature, θ_E, can be estimated (Table II) for data obtained at temperatures above θ_E. This condition is fulfilled at T≥15K.

Table II: Parameters describing properties and vibrations characteristics of the sample particles.

Sample	θ_D(K)	T_B (K)	<D> (Å)	q(J/m^2)	ω(s^{-1})x10^{-11}	θ_E (K)
S8	508 ± 9	<15	<28	∞	---	---
S144	223 ± 2	89	50	34.2	3.16	2.4
S344	251 ± 1	96	51	44.5	3.47	2.6

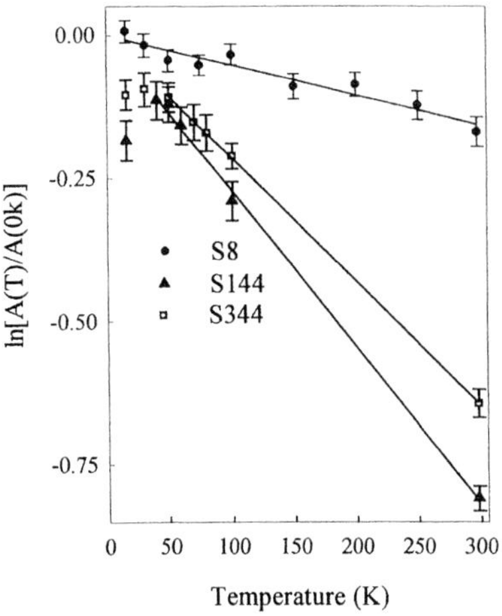

Figure 2: Thermal dependence of the normalised spectral areas

The infinite q value for S8 indicates –coherent with its very small size– a strong oxide-support interaction that impairs the particle vibrations. The force constant increased about 20% in S344 with respect to S144. Since both samples have the same particle size, we believe that a longer calcination time changes the crystallites morphology leading to a stronger linking with the support.

CONCLUSIONS

The calcination time does not alter the iron species present on the silica but influences the degree of interaction modifying the size and/or the morphology of the oxide particles. The temperature dependence of the f-factors is another piece of evidence that the higher area/volume ratio of the crystals increases the strength the particle-support interaction.

ACKNOWLEDGEMENTS

This work was supported by CONICET (PIP 4326), and ANPCyT (PICT 14-04315).

REFERENCES

[1] M. V. Cagnoli *et al.* J. Catal., 123 (1990) 21.

[2] S. G. Marchetti *et al.* J. Phys. Chem. Solids, 58 (1997) 2119.

[3] G. B. Raupp and W. N. Delgass, J. Catal., 58 (1979) 337.

[4] M. Vazquez-Mansilla *et al.* J. Magn. Magn. Mat., 204 (1999) 29.

[5] W. Kündig *et al.* Phys. Rev. 142 (1966) 27.

[6] S. Mørup and H. Topsøe, Appl. Phys. 11 (1976) 63.

[7] P. J. Picone *et al.* J. Phys. C: Solid State Phys. 15 (1982) 317.

[8] M. P. A. Viegers and J. M. Trooster, Phys. Rev. B, 15 (1977) 72.

On-Line and In-Situ Identification of Air Pollution *

Paulo A. de Souza Jr., G. Klingelhöfer
Institut für Anorganische und Analytische Chemie
Johannes Gutenberg - Universität Mainz, 55099 Mainz, Germany

T. Morimoto
Departamento de Meio Ambiente e Comunicação
Companhia Siderúrgica de Tubarão, 29164-280 Serra, Brazil

Abstract. The elemental, iron-phase, and oxidation state analysis of urban aerosol samples, collected from different sites in Vitória metropolitan region, ES, Brazil, was carried out using Mössbauer (MB) Spectroscopy (MBS) and X-ray fluorescence (XRF) techniques. A portable XRF spectrometer for elemental analysis from Na to Fe, and the miniaturized MB spectrometer MIMOS-II have been installed together using a 100 mCi $^{57}Co(Rh)$ MB source for XRF excitation. Both measurements can be performed simultaneously. For continuous and fast characterization purposes a software package was developed for Mössbauer and XRF analysis. The software is based on genetic algorithm, fuzzy logic and artificial neural networks.

1. Introduction

The identification of air pollution sources is one of the main goals of any monitoring and controlling of the air quality in an industrialized urban region. Investment priorities to reduce the impact of these sources can only be identified with a good assessment. The faster a particular pollution is monitored and identified, the wider the utility will be. Most of the monitored urban areas in the world are recording only dust concentrations as on-line results. The characterization itself is performed in the laboratory. The later the characterization is obtained, the harder the source identification and assessment will be. As a consequence the control of potentially problematic sources of pollution will not be efficient, or even might seem to be unnecessary. Nowadays, an improved and more efficient control is possible by on-line and in situ monitoring. A portable MB spectrometer, developed for extraterrestrial studies [1], opens up new interesting applications, but its on-line use has also to be accompanied by a fast and automated data analysis. A portable X-ray detection system, similar to the one used by the Alpha-Proton-X-Ray Spectrometer (APXS) in the 1997 Mars-Pathfinder mission [2], has been integrated into an air pollution monitoring setup together with the MIMOS II instrument using the 6.4 keV, 14.4 keV, and 122.2 keV

* Supported by CAPES/Brasília, Brazil

X- and gamma rays of the $^{57}Co(Rh)$ Mössbauer source for fluorescence excitation. Both resonant and non-resonant fluorescence measurements were performed simultaneously. One drawback for the wider application of Mössbauer spectroscopy in industry is the complexity of the analysis of Mössbauer data and the interpretation of the derived parameters. The analysis is time-consuming and requires most often a dedicated specialist. Some attempts to build routines for automatic analysis of Mössbauer data using genetic algorithms, fuzzy logic and artificial neural networks have been published [4, 5]. These systems are becoming feasible at least for specific applications, as for instance that one discussed in this paper.

Table I. Typical Mössbauer parameters obtained from airborne particles collected at VMR from 1995 to 2000 (Fe^{2+}/Fe^{3+} ratio from 0 to 0.032).

Identified Fe-substance	δ (mm/s)	Δ (mm/s)	B_{eff} (Tesla)	Relative Area (%)	*Remarks*
hematite	*0.37*	*-0.21*	*51.8*	*85 to 96*	always present
fine hematite	*0.37*	*0.70*		*up to 8*	grain size$\leq 50nm$
gothite	*0.36*	*0.40*	*38.1*	*3 to 15*	soil
pyrite	*0.31*	*0.61*		*0.7 to 4.7*	mineral coal
silicates		varies		*1 to 7.3*	soil/quarry
magnetite	*0.36* *0.67*	*0.01* *-0.01*	*49.2* *46.0*	*up to 6*	steelwork plant

2. Experimental Section

2.1. Samples

Samples for test runs and calibrations were collected at Vitória metropolitan region (VMR), southeastern Brazil. VMR embraces five cities, holds an area of $1461km^2$, has a population over 1.2 million inhabitants, is undergoing an expansion and has to bear an intense traffic. Previous studies [6] show that there is an equivalent distribution of industrial and human source emissions at VMR at EAMES (Vila Velha City) and Ilha do Boi (Vitória city). Some specific regions (Dório Silva Hospital and UFES), quite away from the industrial area, have stronger contributions from human activities (vehicles, quarry, civil constructions and vegetation fire), but equivalent concentrations of airborne particles.

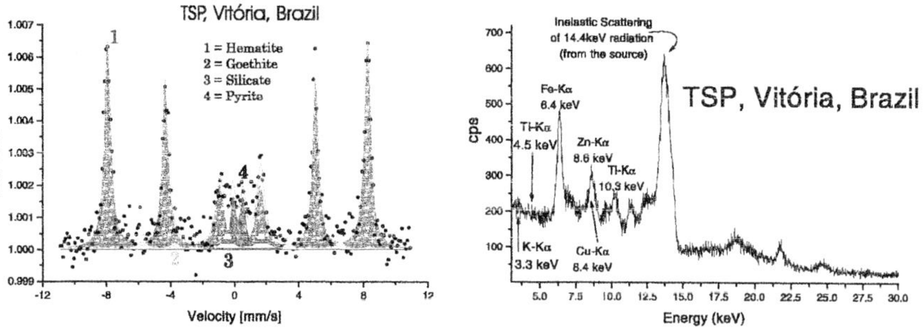

Figure 1. Mössbauer and X-ray fluorescence spectra obtained from total suspended particles from VMR.

The new system will be installed in these selected places during the 2002 summer (critical season for those experiments).

2.2. MÖSSBAUER SPECTROMETER MIMOS II AND X-RAY FLUORESCENCE SPECTROSCOPY

Reduced in size, weight, and power consumption in comparison to ordinary laboratory setups, the MIMOS-II sensor head just weights approximately 400g, with a volume of (50x50x90)mm^3[1]. Operating in backscattering geometry no sample preparation is required enabling dust sampling and simultaneous in situ MB measurements. Typical results of Mössbauer parameters and corresponding Fe-bearing phases identified in samples of suspended particles collected at VMR are given in Table I and Figure 1.

The XRF detector is a 500 μm thick Si-PIN-Diode mounted on a peltier cooler, with a 0.25 mm thick Be window to permit soft X-ray detection. The power supply, amplifiers and the 16k multichannel analyzer (MCA) are installed in a separate box. The measured FWHM for the Cu-Kα line is 250 eV and for the Fe high energy lines smaller than 700 eV. This XRF detection system is portable (size of a computer mouse for the sensor head).

The fundamental parameter method [7] was used to calculate the fluorescence intensities of a given composition. This iterative method lacks some uncertainties originating from uncertainties in the tabulated mass absorption coefficients, stopping power and ionization cross-sections [8]. The exact values are not known and the implemented algorithm has to bear these uncertainties. Genetic algorithm (GA)is able to find minima in a given input data space. An uncertainty of $\tilde{1}5\%$ of the tabulated reference values [8] was allowed by the GA to "select" those values that give the best results for a given well-known sample composition. A preliminary analysis (Figure 1) shows that airborne particles are rich

in Fe, Zn, Ti, K and Cu. Quantitative results will be reported later after an extensive calibration procedure.
Both MIMOS-II and the portable XRF instruments were installed inside a total suspended particles sampler (TSP, $\varnothing \leq 100\mu m$)called *hi-vol.* The vibrations ($\sim$ 300 Hz) of the air pump system have been mechanically isolated and did not affect substantially the Mössbauer drive system performance. The system can also be installed inside a PM_{10} ($\varnothing \leq 10\mu m$) sampler.

3. Conclusions and Outlook

Miniaturized spectrometers were installed inside a particle sampler. Genetic algorithms, fuzzy logic and artificial neural networks have been used for automated analysis of its Mössbauer and X-ray spectroscopy data, overcoming at least partially the limitations of the fundamental parameters method related to uncertainties of the tabulated physical parameters. This experimental setup is portable, easy to use and of low cost. The preliminary results obtained are very promising and encourage further development of the system. Optimal working and geometry conditions will be determined by further calibrations. The most interesting and promising results from the on-line and in situ airborne particles characterization are the fast and automated answers obtained. These recent developments have already produced many relevant information for both Brazilian industries and environmental agencies.

4. References

[1] G. Klingelhöfer, B. Fegley Jr., R. V. Morris, E. Kankeleit, P. Held, E. Evlanov, O. Priloutskii, Planet. Space Sci. 44 (1996)1277.
[2] R. Rieder, T. Economou, H. Wänke, A. Turkevich, J. Crisp, J. Brückner, G. Dreibus, H. Y. McSween Jr., Science 278, 5 Dec. (1997)1771.
[4] P. A. de Souza Jr., Hyp. Int. 113(1998)47; and Laboratory Robotics and Automation, 11(1999)3.
[5] E. O. T. Salles, P. A. de Souza Jr., and V. K. Garg, Nucl. Instrum. Meth. Phys. Res. B 94(1994)499 and J. Radioanal. Nucl. Chem. 190(1995)439.
[6] P. A. de Souza Jr., R. S. de Queiroz, T. Morimoto, A. F. Guimarães, V. K. Garg, J. Radioanal. Nucl. Chem., 246(2000)85.
[7] J. W. Criss, L. S. Birks, Anal. Chem. 40(1968)1080.
[8] J. H. Hubbell, X-ray Spectrom. 28(1999)215.

CEMS study of corrosion product : γ-FeOOH

A. Nakanishi and T. Kobayashi
Department of Physics, Shiga University of Medical Science,
Otsu, Shiga 520 - 2192, JAPAN

Corrosion products formed on an iron foil by sulfuric acid water have been investigated with Conversion electron Mössbauer spectroscopy (CEMS). The resultant species of corrosion products are strongly influenced by oxygen concentration during the corrosion process. In order to investigate the corrosion products when oxygen is sufficiently supplied during the corrosion process, a natural iron foil was immersed in the acid water and air was blown into the water during the immersion of the foil. Conversion electron Mössbauer spectra were taken at temperatures between 15 K and room temperature. The hyperfine field of the observed sextet at 15 K suggests that the corrosion products consist of γ-FeOOH. The corrosion products formed on the weathering steel were also investigated.

1. Introduction

Conversion electron Mössbauer spectroscopy (CEMS) is a powerful tool to characterize the corrosion products because the Mössbauer spectrum can be observed without separating the corrosion products from the substrate materials. Conversion electron Mössbauer (CEM) measurements are usually carried out at room temperature (RT). Since the corrosion products consist of a lot of small particles, ferric doublets are mostly observed in a Mössbauer spectrum at RT. In order to investigate the properties of corrosion products in detail, therefore, it is necessary to carry out CEM measurements at lower temperature. Corrosion products formed on an iron foil by sulfuric acid water have been studied by CEMS at temperatures between 15 K and RT [1,2]. From previous studies, CEM spectra at 15 K suggest that the corrosion products consist of α-FeOOH, γ-FeOOH and γ-Fe_2O_3. The produced amount of these species is influenced strongly by the oxygen concentration during the corrosion process. The amount of γ-FeOOH becomes larger as the oxygen concentration increases. In this study the produced amount and the characteristic of γ-FeOOH were investigated when oxygen was sufficiently supplied during the corrosion process.

2. Experimental

The acid water was prepared by diluting the sulfuric acid to obtain 3mg SO_4^{2-} per liter. A natural iron foil washed by acetone was immersed in the acid water. Air was blown into the water during the immersion of the foil in order to saturate oxygen concentration in the water. Three samples were prepared by changing the immersion time, 3 hours (sample A), 6 hours (sample B) and 12 hours (sample C). Two samples of weathering steel were also prepared in a similar way. The immersion times are 3 hours (sample D) and 6 hours (sample E). CEM spectra were taken at temperatures between 15K and RT, with the gas filled proportional counter described elsewhere [3, 4]. Hydrogen gas and the 98 % He - 2 % CH_4 gas mixture were used as the counter gas at low temperatures and RT, respectively. X-ray diffraction (XRD) analyses were also performed using a Philips diffractometer (PW1710) with Co Kα radiation.

3. Results and Discussion

The CEM spectra of the samples A, B and C at 15 K and RT are shown in Figs. 1 and 2. In the spectra at RT, a ferric doublet in addition to the sextet of the substrate is observed. As the immersion time increases, the intensity of the substrate sextet decreases, which indicates the corrosion layer becomes thicker with increasing of the immersion time.

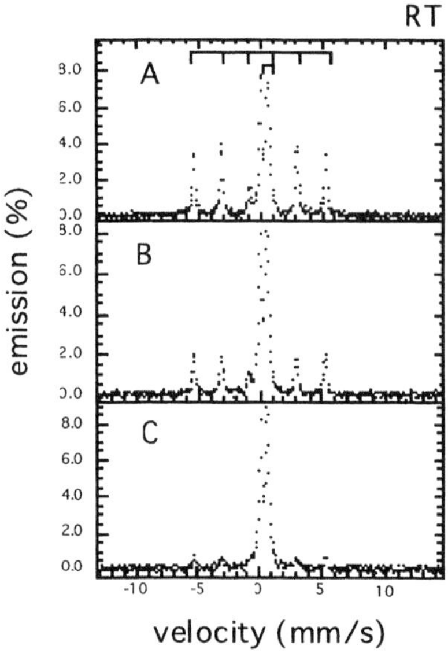

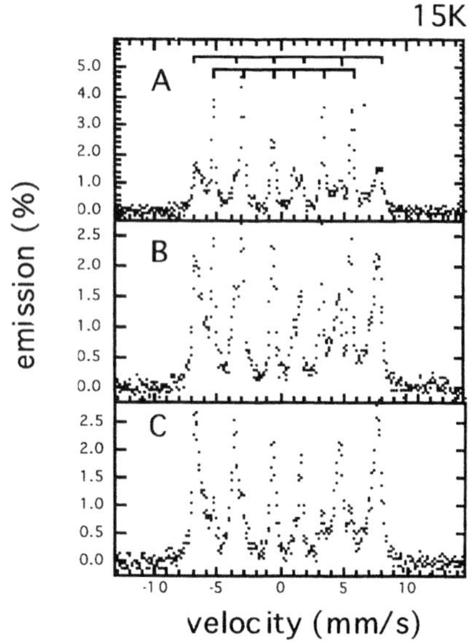

Fig.1.CEM spectra of samples at RT. Fig.2.CEM spectra of samples at 15 K.

In the spectra at 15 K, the substrate sextet and a sextet with broad line are observed. No other sextet and no doublet are noticed. From the hyperfine field and the isomer shift of the observed sextet, it can be assigned to γ-FeOOH. Sextets of α-FeOOH and γ-Fe_2O_3 are not observed, therefore, the corrosion products consist of γ-FeOOH. The particle sizes of γ-FeOOH were estimated from the width of the intense diffraction line in the XRD patterns using the Scherrer's equation. Assuming the spherical shape of γ-FeOOH, the average particle sizes were found to be 14 – 18 nm and it is also found a slight increase of the particle size with increasing the corrosion layer thickness. Since the sextet of γ-FeOOH with a broad line is noticed, its hyperfine field is distributed. The hyperfine field distribution is calculated with the method of Wivel and Mørup [5] at temperatures between

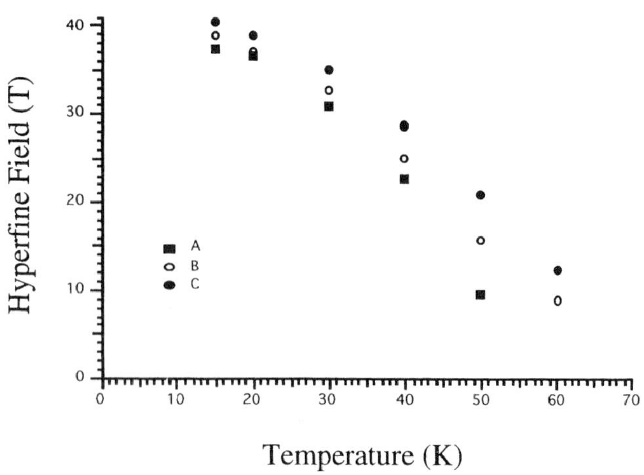

Fig.3. Temperature dependence of the mean hyperfine field of samples.

References
[1] A.Nakanishi, K.Fukumura and T.Kobayashi,
Conf. Proc. (I.Ortalli Ed., SIF, Bologna, 1996) vol.50, 669.
[2] A.Nakanishi, T.Kobayashi and K.Fukumura,
Hyperfine Interactions **112** (1998) 43.
[3] K.Fukumura, R.Katano, T.Kobayashi, A.Nakanishi and Y.Isozumi,
Nucl. Instr. Meth. **A301** (1991) 871.
[4] K.Fukumura, A.Nakanishi and T.Kobayashi,
Nucl. Instr. Meth. **B86** (1994) 387.
[5] C.Wivel and S.Mørup, J.Phys. **E14** (1981) 605.
[6] L.H.Bowen, E.DeGrave and R.E.Vandenberghe, in *Mössbauer Spectroscopy Applied to Magnetism and Materials Science*, G.Long and F.Grandjean eds., Vol.1, p.115, Plenum Press, New York (1993).
[7] E.DeGrave, R.M.Persoons, D.G.Chambaere, R.E.Vandenberghe and L.H.Bowen,
Phys.Chem.Mineral. **13** (1986) 61.
[8] J.Dávalos, J.F.Marco, M.Gracia and J.R.Gancedo,
Hyperfine Interactions **66** (1991) 63.

Mössbauer spectroscopic study of iron oxyhydroxides and rust formed on the steel surfaces

Saburo Nasu, Takayuki Kamimura, Takashi Tazaki, Kaori Kuzushita and Syotaro Morimoto
Division of Materials Physics, Department of Physical Science Graduate School of Engineering Science, Osaka University, Toyonaka, Osaka 560-8531, Japan

Abstract: The rust formed on weathering steel and mild steel exposed to an industrial environment for a long time was characterized by means of ^{57}Fe Mössbauer spectroscopy. The rust consists of the crystalline α-FeOOH, γ-FeOOH, $Fe_{3-\delta}O_4$ (γ-Fe_2O_3) and an X-ray amorphous phase. γ-Fe_2O_3 is found only in the rust formed on mild steel. The amount of the X-ray amorphous phase exceeds 50% of the total amount of the rust. The ^{57}Fe Mössbauer spectra observed at 10 K indicate that the rust contains only α-FeOOH, γ-FeOOH and γ-Fe_2O_3. The x-ray amorphous phase in the rust layer, which gives no peak in the X-ray diffraction pattern, is mainly α-FeOOH which shows a superparamagnetic doublet due to small grain size at room temperature. The grain size of α-FeOOH is widely distributed.

Key words: ^{57}Fe Mössbauer spectroscopy, Rust, Weathering steel, Mild steel, α-FeOOH, γ-FeOOH, X-ray amorphous phase, Superparamagnetism

1. Introduction

Steels exposed to atmospheric conditions, where they experience we/dry cycles, form rust. It consists of oxyhydroxides, α-FeOOH, γ-FeOOH and β-FeOOH, magnetite and X-ray amorphous substance [1,2]. β-FeOOH is formed in the case of existence of chloride. The rust substance, which does not show any sharp X-ray diffraction peaks, is called the X-ray amorphous substance. It is widely believed that the rust layer formed, which may act as coatings, greatly affect the resistance of the corrosion because rust layers can separate base-steels from corrosives. Under various environmental conditions it is therefore of great importance to investigate rust formed on steels.

In this investigation, the constitution of corrosion products formed on weathering steel and mild steel exposed to an industrial environment for a long time are characterized by means of X-ray diffraction and Mössbauer spectroscopy. Phase analysis of the rust formed on steel surfaces exposed to atmospheric conditions by Mössbauer spectroscopy has been already reported [3-7], since the Mössbauer spectroscopy has a great advantage of identification of poorly crystalline or amorphous corrosion products.

2. Experimental

Plates (60 x 100 x 4 mm^2) of weathering steel and mild steel were exposed in an industrial atmospheric environment (Amagasaki, Japan) at an angle of 30 degree facing south. Expose periods were 32 years for weathering steel and 15 years for mild steel. The chemical compositions of the steels used for the exposure test are listed in Table 1.

Table 1. Chemical composition of the steels used (mass %) for W. Steel (weathering steel) and M. Steel (mild steel).

	C	Si	Mn	P	S	Cu	Ni	Cr	Sol Al	N
W. Steel	0.11	0.24	0.75	0.014	0.005	0.33	0.12	0.49	0.038	0.006
M. Steel	0.12	0.32	1.47	0.023	0.023	0.01	0.02	0.04	0.045	0.007

Specimens were cut perpendicular to the plate surface with a dimension of 15 x 10 mm^2. A scrap of steel was first affixed to the surface of the specimens with rust by hard adhesives and then fixed by an epoxy resin. The cross-section of the fixed specimen was mechanically polished by emery papers and a diamond paste. The corrosion products formed on the steel surface were scraped off by a razor and then ground into powder. The outer rust was scraped off up to a depth of 10 μm from the rust surface, which was carefully removed parallel to the steel surface and then remainder rust (inner rust) was scraped off from the steel. The powdered samples were characterized by means of X-ray diffraction technique. Powder patterns with a Co radiation have been obtained and the quantitative determination of the rust constituents comprising α-FeOOH, γ-FeOOH and the X-ray amorphous substance was carried out from the diffraction intensities. The diffraction intensities of (011) reflection of α-FeOOH and (020) reflection of γ-FeOOH were measured and referred to (100) reflection of ZnO powder that has been well mixed with rust samples as an internal standard. The cross-section of the surface corrosion products was investigated by the observation of the reflection behavior of polarized light in an optical microscope, laser Raman spectroscopy by Ar^+ laser with 514.5 nm and 0.1 mW and an electron probe microanalyzer. For the ^{57}Fe Mössbauer spectroscopy, the measurements have been performed in transmission geometry with a specimen absorber and a ^{57}Co in Rh source in the temperature range between 300 K and 10 K. Velocity scale is relative to α-Fe at 300 K.

2. Results and Discussion

The surface of the steels was covered with dark-brown rust. By microscopic observation using reflected polarized light and crossed nicols, it was found that the rust layer formed on the weathering steel can be divided into two layers; one corresponds to the outer layer which is optically active (illuminated) and the inner layer which is optically isotropic (darkened). The large cracks reaching the steel substrate are clearly observed for the mild steel, whereas no crack is observed for the weathering steel. Figure 1 shows the X-ray diffraction patterns obtained from rust formed on mild steel exposed for 15 years (lower spectrum in the figure) and rust formed on weathering steel for 32 years (upper spectrum in the figure). It is quite obvious from the x-ray diffraction pattern that rust layers formed on the weathering steel and the mild steel were mainly composed of α-FeOOH and γ-FeOOH as crystalline constituents, as shown in Fig. 1, whereas Fe_3O_4 can be hardly detected in the steels used. Quantitative phase-analysis obtained from the diffraction intensities indicated that the rust on weathering steel exposed for 32 years contained 25% of α-FeOOH, 5% of γ-FeOOH and 70% of X-ray amorphous substance and the rust formed on mild steel exposed for 15 years contained 20% of α-FeOOH, 15% of γ-FeOOH and 65% of X-ray amorphous substance. In order to perform the qualitative phase analysis of the rust layer, the outer and the inner layers of Rust formed was

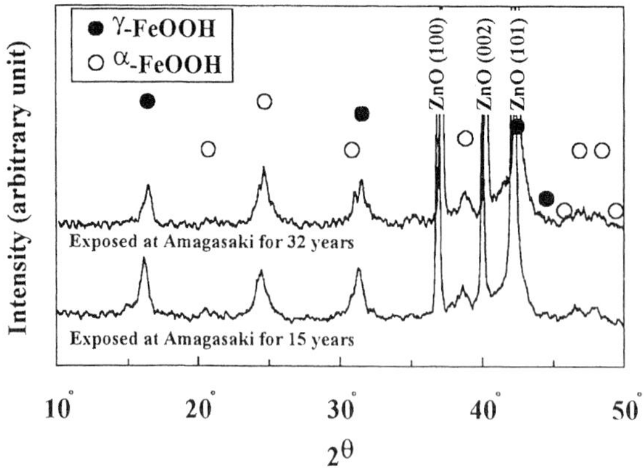

Figure 1. X-ray diffraction patterns obtained from weathering steel exposed for 32 years (upper) and from mild steel exposed for 15 years (lower). Co radiation was used.

characterized by the Raman spectra. The observed spectra were characterized by comparing the spectra of reference iron oxides. Raman spectra of several different positions of the outer layer were almost identical. The major bands observed at the vicinities of 255, 380, 530 and 655 cm^{-1} are clearly due to γ-FeOOH. For the inner layer observed as an optically isotropic (darkened) by a polarizing microscope, the bands at the vicinities of 245, 300, 397, 485, 550 and 685 cm^{-1} were mostly due to α-FeOOH. The broad bands, which may correspond to the X-ray amorphous substance, were observed in some cases. Distribution of the small additional elements, Cu, Cr, Ni and P has been investigated by using an electron probe microanalyzer. The striking feature is the enrichment of Cr observed only in the inner layer, whereas the enrichment of neither Cu nor P was observed. Ni existed throughout the rust.

Figure 2 and 3 show the typical ^{57}Fe Mössbauer spectra measured at room temperature, 78 K and 10 K obtained from outer rust of weathering steel exposed for 32 years and the spectra obtained from outer rust of mild steel exposed for 15 years, respectively.

The outer rust of weathering steel whose thickness is about 10 μm may correspond to the illuminated region (optically active) observed by the reflection behavior of polarized light.

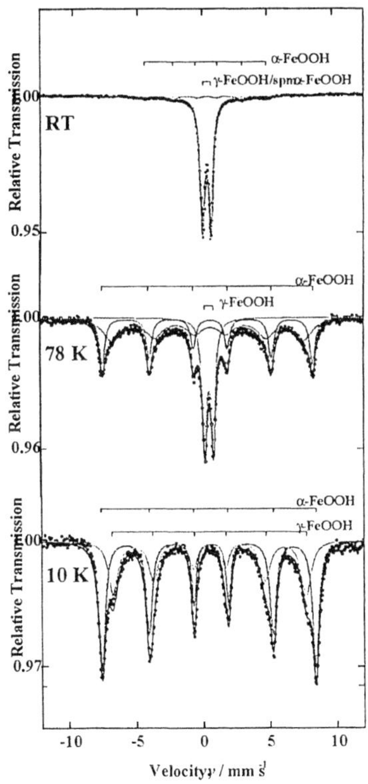

Figure 2. Typical ^{57}Fe Mössbauer spectra measured at room temperature, 78 K and 10 K obtained from the outer rust formed on weathering steel exposed for 32 years. Velocity scale is relative to a-Fe at 300 K.

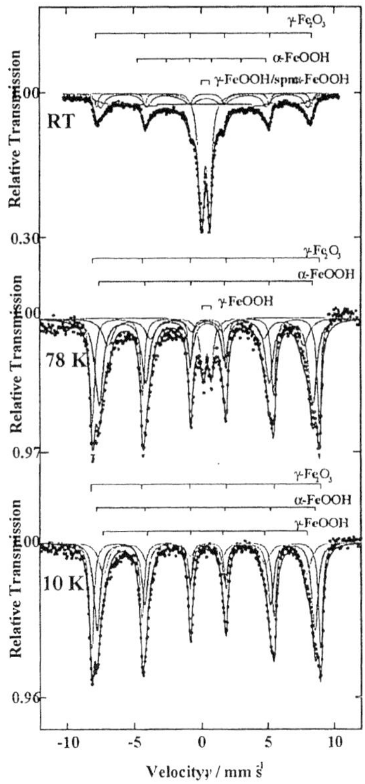

Figure 3 Typical ^{57}Fe Mössbauer spectra measured at room temperature, 78 K and 10 K obtained from the outer rust formed on mild steel exposed for 15 years. Velocity scale is relative to α-Fe at 300 K.

The room temperature spectrum for the outer rust obtained from weathering steel shows an intense quadrupole-split doublet and an extremely weak sextet component, as shown in Fig. 2. The room temperature spectrum for the outer rust obtained from mild steel shows somewhat different feature to the case of weathering steel. Spectrum consists of sharp quadrupole-split doublet superposed with rather strong magnetically split sextets whose magnitude of the hyperfine fields is large. Furthermore, broad components also exist and make broad background absorption as shown in Fig. 3. For the spectra at 78 K, the intensity of the central doublet decreased and the intensities of the sextet lines increased. These results suggest that the central paramagnetic component at 78 K may be due to the paramagnetic γ-FeOOH whose Néel temperature is reported to be 73 K [38] and also due to the superparamagnetic α-FeOOH whose Néel temperature is much higher than 78 K. This result is significant and leads to evidence of the existence of superparamagnetic small particles in the rust layer. For the spectra at 10 K the central doublet completely vanished, which implies that the superparamagnetic fluctuation of magnetic moments in a small particle become to be static at 10 K. The spectra observed at 10 K are rather sharp and superposition of the well distinguishable sextets. This result is also significant and leads to evidence there is no amorphous phase in the rust layer, because Mössbauer spectrum of amorphous magnetic substances should show a broad distribution of the hyperfine magnetic fields. Spectra obtained from the rust formed on weathering steel consist of two sextets; one is due to α-FeOOH having B_{int}=50 T, and the other is due to γ-FeOOH having B_{int}=45 T. Spectra obtained from the rust formed on mild steel consist of three sextets: one is α-FeOOH, second component is γ-FeOOH and third one is γ-Fe_2O_3 having B_{int}=52 T. Rust formed on weathering steel does not contain γ-Fe_2O_3 but the rust formed on mild steel contains γ-Fe_2O_3. The reason why the rust of weathering steel does not contain γ-Fe_2O_3 is important to understand the mechanism to form the protective layer and further investigation is necessary.

3. Conclusions

The rust formed on weathering steel and mild steel exposed to an industrial environment for a long time was characterized by means of ^{57}Fe Mössbauer spectroscopy. The amount of the X-ray amorphous phase exceeds 50% of the total amount of the rust. The ^{57}Fe Mössbauer spectra observed at 10 K indicate that the rust contains only α-FeOOH, γ-FeOOH and γ-Fe_2O_3. γ-Fe_2O_3 is found only in the rust formed on mild steel. The x-ray amorphous phase in the rust layer, which gives no peak in the X-ray diffraction pattern, is mainly α-FeOOH which shows a superparamagnetic doublet due to small grain size.

References

[1] M. Yamashita., H. Miyuki, T. Matsuda, H. Nagano and T. Misawa, Corros. Sci., 36 (1994) 283.

[2] H. Kihara, S. Ito and T. Murata, Corros. Sci., 31 (1990) 383.

[3] D. C. Cook, Hyperfine Int., 28 (1986) 891.

[4] K. Nomura and Y. Ujihira, Hyperfine Int., 29 (1986) 1467.

[5] J. Davalos, M. Gracia, J. F. Marco and J. R. Gancedo, Hyperfine Int., 69 (1991) 871.

[6] D. C. Cook, R. Balasubramanian, S. J. Oh and M. Yamashita, Hyperfine Int., 122 (1999) 59.

[7] J. F. Marco, M. Gracia, J. R. Gancedo, M. A. Martín-Luengo and G. Joseph, Corros. Sci., 42 (2000) 753.

IRON PHASES IN $FeO-Fe_2O_3-CaO-SiO_2$ METALLURGICAL SLAGS

D. Orlokh[1], J. D. Cashion[1] and S. Jahanshahi[2],

[1] School of Physics and Materials Engineering, Monash University, VIC 3800, Australia
[2] CSIRO Minerals, Clayton South, VIC 3169, Australia

Abstract:
The iron phase distribution in a $FeO-Fe_2O_3-CaO-SiO_2$ metallurgical slag have been analysed by Mössbauer spectroscopy. The slag was equilibrated at 1350°C before exposure to oxygen from one end. The dominant species were magnetite at the top and wustite at the bottom, but up to six iron phases were identified in the samples, including some glassy phases. The fraction of ferrous iron was determined as a function of position and found to be systematically slightly higher than that determined by chemical analysis.

Keywords: slag, magnetite, wustite, glass, ferrous/ferric ratio, Mössbauer effect

1. Introduction

Physico-chemical properties of slags play a central role in controlling the metallurgical performance of smelting and refining processes that are used for extraction of valuable metals from ores. New information on the structure and properties of molten phases is required for development and validation of new models of such processes. Over the past decade the Melt Chemistry Group at CSIRO Minerals has undertaken interactive experimental and modelling studies on thermodynamic, structural and transport properties of high temperature phases such as slags. A structurally related model of the thermodynamics and viscosity of silicate and aluminate melts has been developed and its application has been extended to complex multi-component slags [1]. Ongoing work is aimed at extending the application of these multi-phase reactive models to cover kinetics of reactions and other key transport properties such as diffusivity and conductivities of the melts.

Iron oxide enters most smelting and refining processes through the ore and in most cases a portion of it is oxidised and reports to the slag phase. Studies on the physico-chemical properties of molten slags have demonstrated the strong and in some cases, dominating, effect of iron oxide content on various properties of silicate based slags [2]. For example due to its multiple valency, iron oxide plays a key role in determining the mechanism (ionic or electronic) and hence ease of electrical conduction in slags [3,4]. Similarly, it has a considerable effect on the diffusivity of oxygen within the molten slag phase [5]. Apart from the valency state of iron cations in the network of silicate anions, the coordination of the cations are thought to play a significant role in determining some of transport properties of such melts. For example recent molecular dynamics simulation of alumina-silicate melts has revealed that, depending on the basicity of such melts, alumina can form different "species" or bonds with the oxygen anions [6]. The formation of tri-cluster species causes some hindering of movements of ionic species in the melts and hence slag viscosity will be affected. Given that iron oxide in silicate melts can also exhibit different coordination, then one would expect a similar effect on transport properties.

In order to provide quantitative data on the problem, we have been determining the distribution of iron valencies and phase concentrations in quenched FeO-Fe_2O_3-CaO-SiO_2 slags.

2. Sample preparation

The experimental technique used in this work has been described in some detail elsewhere [7]. Master slags were prepared by mixing appropriate amounts of high purity Fe_2O_3, $CaCO_3$ and SiO_2. The mixtures were then pre-melted in a Pt-Rh crucible, quenched, pulverized and remelted to ensure good homogeneity. The water quenched product from the pre-melted slag was then pre-reduced in an atmosphere of CO-CO_2 to ensure that the starting composition of the samples for the diffusion runs was close to the required equilibrium composition.

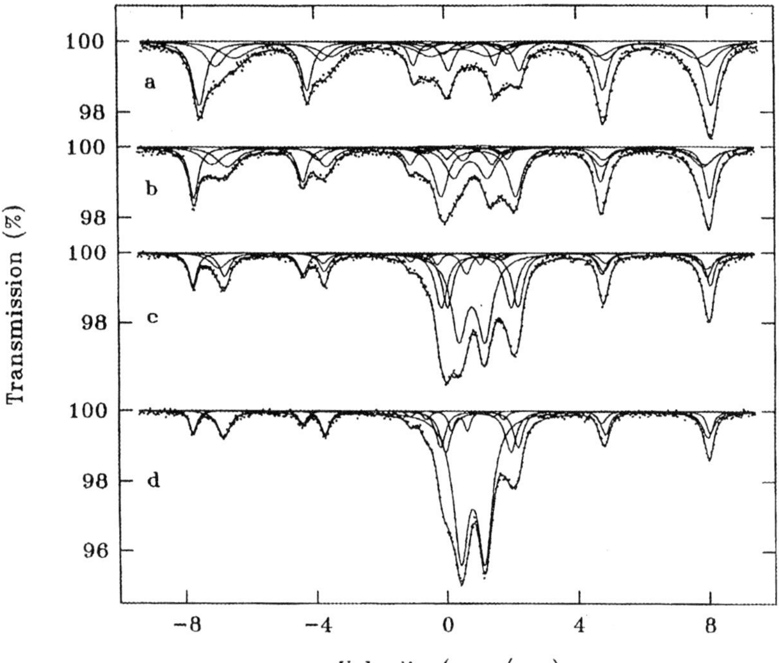

Fig. 1. ^{57}Fe Mössbauer spectra of the four samples at room temperature. The sections were located at (a) 0.5cm, (b) 1.0cm, (c) 2.0cm and (d) 4.0 cm from the top of the sample.

The apparatus used for the diffusion runs consisted of a vertical tube furnace with $MoSi_2$ resistance heating and a gas purification and mixing unit. The furnace tube was fitted with appropriately placed thermal shields to obtain a hot zone of about 10 cm long for the experiments. The temperature was monitored with a Pt-6%Rh/Pt-30%Rh thermocouple, which was in contact with the bottom of the crucible. Slag samples were held in a platinum lined magnesium capillary crucible (7 mm ID and 100 mm high) which was suspended from the bottom of an alumina gas delivery tube. The gas delivery tube had an alumina nozzle of 3 mm ID and the tip of the nozzle was kept about 20 mm above the surface of the melt.

Each run involved equilibrating a 5 to 6 cm deep melt in the crucible under a controlled atmosphere at 1350 °C, prior to oxidation by pure oxygen for a pre-determined period. The crucible containing the melt was then quickly withdrawn from the furnace and water quenched. The solidified rod of the slag was sectioned into discs which were analysed for total and ferrous iron contents by titration technique and for other elements by the ICP technique. Samples for Mössbauer spectroscopy were prepared by cutting slices perpendicular to the cylinder axis and then crushing them in a mortar and pestle. Concentration profiles of ferrous and ferric iron were determined from each analysis technique and formed the base for further analysis to extract oxygen diffusion data. Quenched samples do not always represent the high temperature phases due to the possibility of crystallisation, phase separation and electron hopping during quenching. However, we expect good representation of the oxygen diffusion profile.

3. Results

The room temperature spectra of the four disc sections are shown in Fig. 1. The spectra been fitted with three sextets which represent the magnetite phase, and five doublets comprising two doublets due to ferrous glassy phases, one doublet for wustite, FeO, and two ferric doublets. Isomer shifts are referenced to α-iron at room temperature.

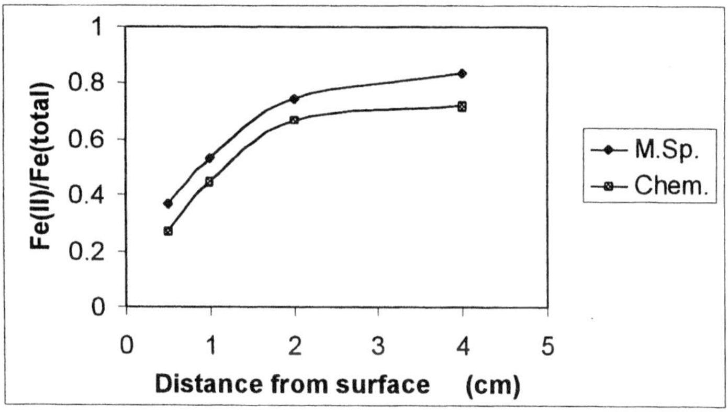

Fig. 2. Proportion of ferrous iron by Mössbauer spectroscopy and chemical analysis

It can be seen in Fig. 1 that the sextet structure has a well defined outer sextet in all spectra, but the definition of the inner sextet improves considerably as one progresses through the series. The ratio of the A:B site areas increased from 1:1.8 to 1:1.79. The hyperfine field of the outer sextet increases slightly from 48.3T to 48.9T through the series and the linewidth of the outer lines drops from 0.50 mm/s to 0.28 mm/s. The mean hyperfine field for the two sextets representing the B sites also increases from 45.0T to 45.8T through the series. The isomer shift (IS) of the outer sextet decreases slightly from 0.30(1) mm/s to 0.27(1) mm/s while the mean IS of the B sites increased from 0.57(2) mm/s to 0.68(2) mm/s. The two ferrous glassy phases have reasonably consistent parameters of IS = 1.17(6) mm/s, QS = 2.25(10) mm/s and IS = 1.02(6) mm/s and QS = 2.0(2) mm/s corresponding to octahedral coordination. The wustite contribution increases rapidly with depth and accounts for just over half of the iron in the fourth sample. Spectra taken at 82K and 5K were very difficult to fit because of the complicated pattern of magnetically ordered wustite.

The ratio of Fe(II)/Fe(total) calculated from the results of the Mössbauer analyses are compared with chemical analyses in Fig. 2. It can be seen that there is a systematic difference between the two results for which one can consider two possible causes. The Mössbauer analyses assume a 50% ferrous occupation of B sites and are based on the spectral areas, which assumes that all the species have the same recoilless fraction. Experiments are being conducted to measure these parameters. Secondly, chemical analysis of ferrous species is prone to the possibility of oxidation during analysis and the observed difference is consistent with this.

4. Conclusions

The results showed that the proportion of magnetically ordered phases, principally magnetite, decreases with depth in the samples, whereas the quantity of wustite, FeO, shows the opposite variation. The glassy phase distribution did not have a systematic variation with depth. There appears to be a systematic difference between the proportion of ferrous iron as determined by Mössbauer spectroscopy and conventional chemical analysis.

It clear that Mössbauer spectroscopy can give unique information about the phase distribution in metallurgical slags. The spectra show that there are at least six iron species present, including up to four glassy sites. This information is not available by other means. Thus the use of Mössbauer spectroscopy can not only provide ferrous/ferric ratios but also considerably more information on the nature of the iron phases which can be fed into the modelling of the processes.

Acknowledgments

We are pleased to acknowledge the financial support of a grant from the Australian Research Council for this work.

References

1. L. Zhang, S. Jahanshahi, S. Sun, M. Lim, B. Bourke, S. Wright and M. Somerville; Materials Forum, **25**(2001)201.
2. E.T. Turkdogan; Physicochemical Properties of Molten Slags and Glasses, The Metal Society, London; 1983.
3. E.A. Pastukhov, O.E. Esin and S.K. Chuchmarev; Soviet. Electrochem., **2**(1966)193.
4. H.J. Engell and P. Vygen; Ber. der Bunsengesell. fur Physik. Chem., **72**(1968)5.
5. S. Sun and S. Jahanshahi; "Redox Equilibria and Kinetics of Gas-Slag Reactions"; Belton Memorial Symposium; Met. Mat Trans B, 2000, (31), pp. 9237-43.
6. L. Zhang, S. Sun and S. Jahanshahi; J. Non-Cryst. Solids, 2001, (182), pp.24-29.
7. Y. Sayadyaghoubi, S. Sun and S. Jahanshahi, Met. & Materials Trans.B, **26**(1995)795.

The role of silicon atoms inside the Brownmillerite phase of cement clinker

M.Y. Hassaan, M.M. El-Desoky[1], M. Katada[2] and T. Nishida[3]

Department of Physics, University of Al- Azhar, Cairo, 11884, Egypt
[1]Department of Physics, University of Suez Canal, El-Arish, Egypt
[2]Graduate School of Science, University of Tokyo Metropolitan, Minami-osawa, Hachioji, Tokyo, Japan
[3]Department of Biological and Environmental Chemistry, University of Kinki, Iizuka, 820-8555, Japan

Seven samples of the ferrite phase of cement clinker (Brownmillerite) were prepared at 1350 °C. The first sample is a pure Brownmillerite, while the other six samples are doped with 3, 6, 9, 12, 15 and 30 mol% SiO_2. Structural role and solubility limit of SiO_2 in the Brownmillerite phase were investigated by Mössbauer effect, X-ray diffraction (XRD) and scanning electron microscope (SEM). Above the solubility limit of SiO_2, dicalcium silicate phase of cement clinker was observed in the XRD pattern, as a result of the interaction of SiO_2 with CaO. Mössbauer spectrum measured at room temperature shows that the pure Brownmillerite has three iron sites; two of them are magnetic and the third is superparamagnetic because of its small particle size. The superparamagnetic doublet disappears and replaced by a sextet when measured at liquid N_2 temperature.

1. Introduction

It is well known that cement clinker is composed of four main phases: tetracalcium aluminum ferrite (brownmillerite) C_4AF^*, tricalcium aluminate C_3A, dicalcium silicate C_2S^* (belite) and tricalcium silicate C_3S. These phases prepared together in the form of a solid solution are called cement clinker. The mixture of lime, silica, alumina and iron oxide are the raw materials, which are used in producing cement clinker. In laboratory, pure Brownmillerite phase was prepared from 4 C + 1 A + 1 F, while pure dicalcium silicate phase was prepared from 2 C + 1 S. Brownmillerite is formed at about 1400°C. Because the C_4AF phase is the only phase that contains iron in cement clinker, extensive investigations have been done on this phase and on cement clinker by using Mössbauer spectroscopy [1-6]. Neubauer *et al.* [7] investigated the effect of Si and Mg on the brownmillerite. They found that no introduction of Si into C_4AF is possible under their preparation conditions and the whole SiO_2 content in the sample can be calculated as C_2S (belite). Kurdowski *et al.* [8] prepared C_2S by means of low-temperature synthesis. Maki *et al.* [9] studied the formation of belite clusters from quartz grains in Portland cement clinker. Nishida *et al.* [10] heat treated 60 $CaO.27Al_2O_3.13Fe_2O_3$ glass, and observed a precipitation of antiferromagnetic dicalcium ferrite, C_2F, and mayenite, $C_{12}A_7$.

This work deals with the structural role and the solubility limit of SiO_2 inside the brownmillerite phase of cement clinker by using Mössbauer, X-ray diffraction (XRD) and scanning electron microscope (SEM).

2. Experimental

Seven samples of brownmillerite were prepared at 1350 °C with and without adding SiO_2, as indicated in Table 1. They were prepared according to the composition: 4 mole C + one mole A + one mole F. Then SiO_2 was added to each mixture, as indicated in table 1. Each mixture in a platinum crucible was heated in air in an electric furnace at 1350°C for half an hour, and then quenched in air. After having been ground, each sample was heated again at the same temperature for additional half an hour in order to ensure that all the oxides reacted with each other to form brownmillerite. The prepared samples were ground again and measured by XRD, Mössbauer and SEM. Seven samples were chemically analyzed using the insoluble residue method to determine the concentration of free SiO_2.

The XRD measurements were performed using CuK_α radiations. SEM measurements were carried out for powder samples using SEM Model Philips XL 30 attached with EDX unit, with accelerating voltage 30 kV, magnification 10x up to 400,000x and resolution for width 3.5 nm. The samples were coated with carbon.

---

* The cement abbreviation symbols; C = CaO, S = SiO_2, A = Al_2O_3, F = Fe_2O_3.

Mössbauer measurements were obtained using $^{57}Co(Rh)$ source of 925 MBq. The spectrometer is a Wissel constant acceleration type. Temperature control was made by using Helitran LT-300. A sheet of metallic iron foil was used as the reference for ME parameters.

Table 1. The added SiO_2 and the free SiO_2 in each prepared samples

Sample no.	Added SiO_2 mol %	Free SiO_2 mol %
1	0	--
2	3	0.27
3	6	0.63
4	9	1.01
5	12	1.59
6	15	1.65
7	30	5.18

3. Results and discussion

3.1. Chemical Analysis Results

Table 1 shows the concentration of free SiO_2 (in mol%) presented in each sample. We can understand from Table 1 that the amount of free SiO_2 increases as the added SiO_2 mol% increases. This may be due to the decrease in the burnability due to the increase in the SiO_2 content in the C_4AF sample. The authors consider that the combined SiO_2 may be found as C_2S phase of cement clinker, which may be composed on the account of C_4AF phase percentage. The presence of free SiO_2 means that it did not react with other oxides in the mixture, but it is still free as interstitials in the C_4AF lattice.

3.2. XRD results

The XRD patterns are exhibited in figures 1(A) and 1(B) for the pure C_4AF, sample no. 1, and for the sample no. 7 (30 mol% SiO_2), respectively. Pattern (A) shows that the C_4AF phase is formed. The XRD patterns of the samples nos. 2-6 show the same feature, while the pattern (B) of sample no. 7 shows the appearance of C_2S phase along with an increase in the intensity of some peaks of C_4AF, which overlapped with the C_2S peaks at 2θ=32 and 47.2 °. At the same time, new peaks represent the C_2S phase at 2θ=31.5 and 33.1 °. This result means that the SiO_2 addition below 30 mol% does not show the C_2S peaks in the XRD patterns due to its low percentage, but the C_2S phase begins to appear in the XRD pattern in the sample containing 30 mol% SiO_2. This may be due to the addition of SiO_2 at low percentage, playing an important role in forming the glass phase, which appears as a small hump in the XRD pattern in addition to the C_2S phase which has small content so that it is not observed in the XRD patterns. In the case of the sample containing 30 mol% SiO_2, the C_2S phase is formed besides the formation of glass hump, as observed in figure 1(B).

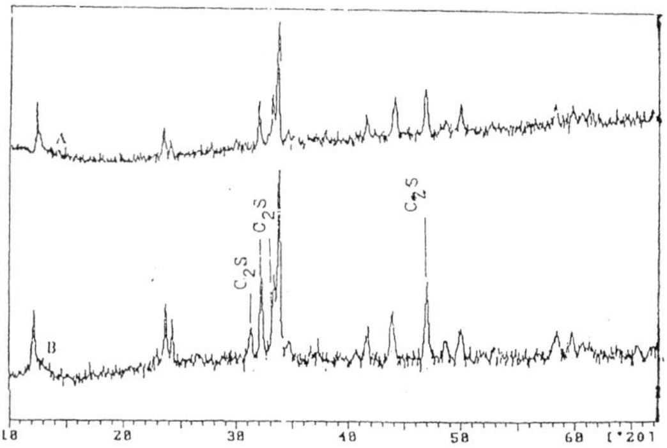

Figure 1. XRD patterns of samples nos. 1 (A) and 7 (B).

3.3. SEM results

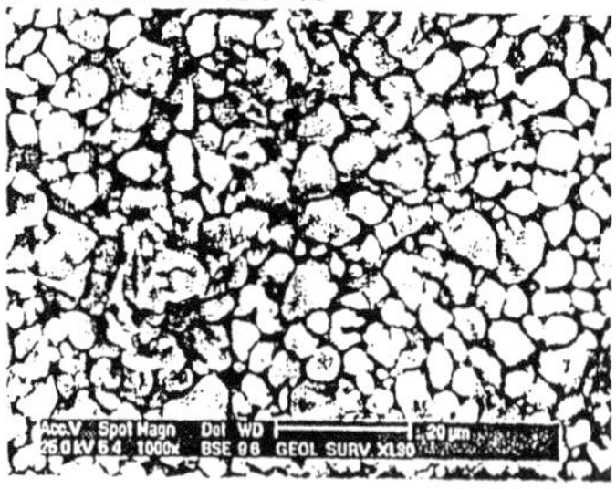

Figure 2. SEM photo of sample no. 2

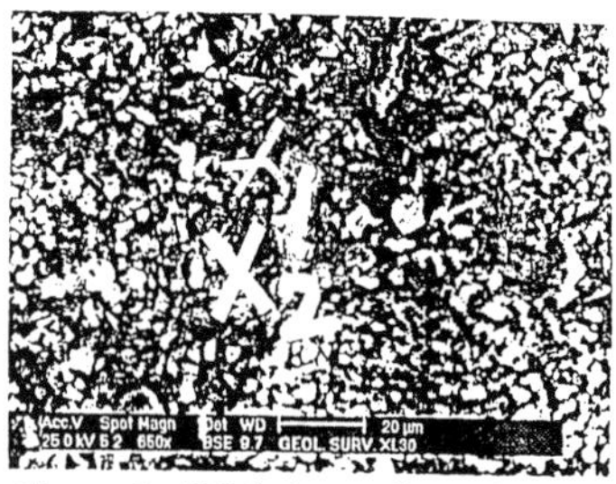

Figure 3. SEM photo of sample no. 6

SEM results are shown in figures 2-4. Figure 2 exhibits the C_4AF structure of sample no. 2 with 24.29 wt% A, 45.42 wt% C, 28.58 wt% F and 1.70 wt% S. When the added SiO_2 amounts to 15 mol% in sample no. 6, C_4AF phase appears and at the same time, SiO_2 forms an amorphous phase, which may be aluminosilicate glass containing iron, by combining with some nonreacted oxides present in the sample mixture such as A and F, as observed in figure 3. Position x_1 shown in Fig. 3 represents a high content of silica, *i.e.* C_2S. Position x_2 represents a low amount of silica, *i.e.* free silica. Figure 4 represents C_4AF phase, glassy phase and C_2S phase which locates at interstitial sites. These two results are confirmed by the XRD study.

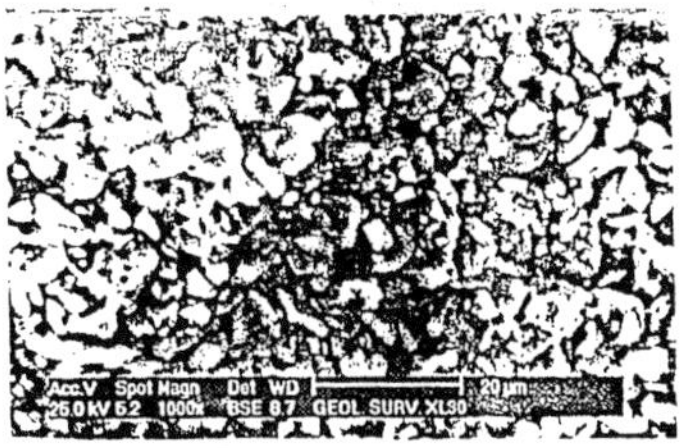

Figure 4. SEM photo of sample no. 7

3.4. Mössbauer results

Representative room temperature Mössbauer spectra are shown in Figure 5. The Mössbauer parameters of the analysis are exhibited in table 2. From the figure and the table, one can see that each spectrum contains three sites, two of them are magnetic and the third one is superparamagnetic. The first site has nearly constant values of isomer shift (δ), 0.28-0.33 mms^{-1}, and quadrupole splitting (Δ) ranging from -0.66 up to -0.58 mm s^{-1}. The magnetic field (B) also has nearly constant values of about 43 Tesla. The relative absorption area shows a gradual decrease from the sample no. 1 to sample no. 5, then an increase in the sample nos. 6 and 7.

The second magnetic site in the room temperature Mössbauer spectra exhibits a nearly constant magnetic field and increasing values of the relative absorption area up to sample no. 5, then it decreases again. The third site is superparamagnetic due to nanocrystals of C_4AF.

Table 2. ***Mössbauer parameters of seven samples at room temperature.***

Sample no.	Site 1				Site 2				Site 3		
	δ mm s^{-1}	Δ mm s^{-1}	B Tesla	Area %	δ mm s^{-1}	Δ mm s^{-1}	B Tesla	Area %	δ mm s^{-1}	Δ mm s^{-1}	Area %
1	0.33	-0.60	44.50	56.3	0.13	0.73	37.2	36.00	0.18	0.74	7.70
2	0.32	-0.58	43.77	44.9	0.21	0.27	36.1	51.70	0.21	0.80	3.30
3	0.28	-0.66	43.19	41.1	0.16	0.17	35.65	56.69	0.24	0.74	2.20
4	0.29	-0.58	42.77	37.1	0.16	0.16	35.47	57.64	0.26	0.87	5.25
5	0.30	-0.62	42.53	1.99	0.16	0.14	36.41	92.83	0.17	0.68	5.18
6	0.30	-0.59	43.13	63.77	0.09	0.52	36.41	30.90	0.20	0.87	5.34
7	0.29	-0.58	43.80	40.02	0.29	0.30	35.84	55.74	0.19	0.91	4.24

Table 3. ***Mössbauer parameters of seven samples at liquid N_2 temperature.***

Sample no.	Site 1				Site 2				Site 3		
	δ mm s^{-1}	Δ mm s^{-1}	B Tesla	Area %	δ mm s^{-1}	Δ mm s^{-1}	B Tesla	Area %	δ mm s^{-1}	Δ mm s^{-1}	Area %
1	0.45	-0.56	51.51	57.62	0.27	0.74	45.11	34.36	0.26	0.82	8.02
2	0.45	-0.59	52.04	68.56	0.25	0.77	45.64	31.44	--	--	--
3	0.44	-0.62	51.56	66.76	0.25	0.79	45.27	33.24	--	--	--
4	0.42	-0.61	51.76	66.92	0.22	0.77	45.40	33.08	--	--	--
5	0.44	-0.62	51.40	66.94	0.24	0.78	45.26	33.06	--	--	--
6	0.45	-0.58	50.98	66.75	0.26	0.77	44.62	33.25	--	--	--
7	0.46	-0.60	52.62	65.35	0.27	0.79	46.35	34.65	--	--	--

When Mössbauer spectra were measured for the seven samples at liquid N_2 temperature (see Fig. 6 and Table 3), the superparamagnetic site appears only in sample no.1, while in other samples it disappears because it behaves at low temperatures as a magnetic site and its B value coincides with the value of the first site. This means that the site three has the same *B* as the site one, but the site three has much smaller particle size than that of site one.

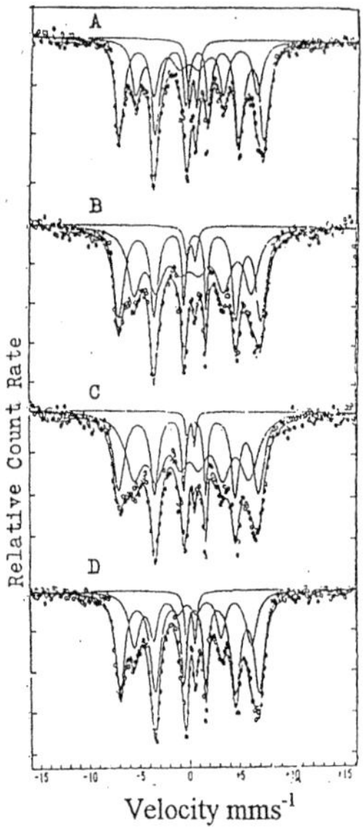

Figure 5. Room-temperature Mössbauer spectra of: sample nos. 1 (A), 2 (B), 3 (C) and 6 (D).

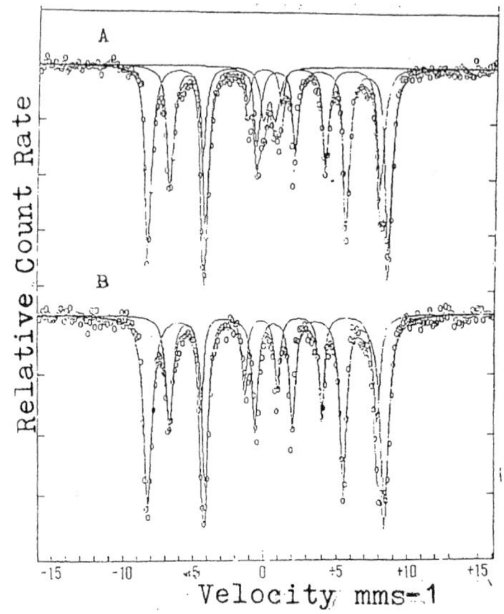

Figure 6. Liquid N_2-Mössbauer spectra of: sample nos. 1 (A) and 4 (B).

4. Conclusions

It was found that SiO_2 additions to Brownmillerite with increasing percentages appear in the structure of this phase free as interstitials, and also they form aluminosilicate glass containing iron, and finally at high percentage of SiO_2 the C_2S phase also appear.

References

[1] M.Y. Hassaan and N.A. Eissa, Am. Ceram. Soc. Bull., 66 [12] (1987) 1947-54.

[2] M. B. Marinho and F. P. Glasser, Cement and Concrete Research, Vol. 14 (1984) 360-368.

[3] G. Kakali and G.Parissakis, Cement and Concrete Research, Vol. 25, No. 1 (1995) 79-85.

[4] M.Y. Hassaan, N.A. Eissa and S.H. Salah, TIZ International Powder Magazine, 114, No. 6 (1990) 401-403.

[5] M.Y. Hassaan, Mat. Sci. Eng., A123 (1990) 153-157.

[6] M.Y. Hassaan, D. Panteleev and S.H. Salah, Hyp. Int., 71 (1992) 1389-1394.

[7] J. Neubauer, R. Sieber, H.J. Kuzel and M. Ecker, Cement and Conc. Res., 26, No. 1 (1996) 77-82.

[8] W. Kudrowski, S. Duszak and B. Trybalska, Cement and Conc. Res., 27, No. 1, (1997) 51-62.

[9] I. Maki, K. Fukuda, T. Imura, H. Yoshida and S. Ito, Cement and Conc. Res., 25, No. 4 (1995) 835- 840.

[10] T. Nishida, S. Kubuki, M. Shibata, Y. Maeda and T. Tamaki, J. Mater. Chem., 7, No. 9 (1997) 1801-1806.

ON MAGNETITE FORMATION AS A CORROSION PRODUCT OF STEEL

C.A. Barrero[1,2], A.L. Morales[1,2], M.I. Mejia[1,2], and C.E. Arroyave[2]
[1]Solid State Group, Institute of Physics, University of Antioquia.
[2]Corrosion and Protection Group, University of Antioquia.
P.O. Box 1226, Medellín, Colombia

A Mössbauer study on magnetites formed under the influence of differing concentrations of sulfate and chloride ions and also under the presence Cu, Mn and Cr ions is presented. It is found that different concentrations of sulfate and chloride ions in the synthesis procedure does not influence the physical and chemical properties of the magnetite. On the other hand, the concentrations of these ions in accelerated corrosion tests influence mainly the amount of magnetite formed. Mössbauer results suggest that the physical and crystallographic properties and amounts of the final products can be differently affected by the combined effect of sulfate and chloride ions plus the presence of alloying elements.

Introduction

In corrosion science it is very important to understand the mechanisms of formation of the different iron oxides as atmospheric rust components. This study includes the understanding of the role played by pollutants and alloying elements on rust formation. One of the main rust components is magnetite, which is very common in steel exposed to chloride rich atmospheres.

The aim of the present study is to look into the magnetite formation under the influence of varying concentrations of chloride and sulfate ions and also under the combined effect of these ions plus cations such as Cu^{2+}, Mn^{2+} and Cr^{3+}. The effect of the chloride and sulfate ions was investigated by using two approaches. The first one consisted in preparing magnetites with different types of precursors and the second method, by performing alternating immersion/emersion tests on low carbon steels.

Experimental procedure

The synthetic pure samples were prepared using a hydrolysis method similar to the one proposed by Schwertmann and Cornell [1], which uses ferrous sulfate and alkaline solutions as precursors. In some instances, the ferrous sulfates were replaced partly or totally with ferrous chlorides, while keeping constant the contribution of ferrous ions (see Table 1). In other cases, the amount of ferrous ions was varied, but without changing the amount of sulfates or chlorides. The first case was achieved by replacing ferrous sulfate by sodium sulfate, and the second case was accomplished by replacing ferrous chloride by sodium chloride.

The effect of pollutants was also studied through accelerated corrosion tests [2,3]. For that purpose, AISI 1008 low carbon steel coupons were subjected periodically to the influence of several combinations of chloride and sulfate ions during 65 days (see Table 2). The coupons rotated continuously over a period of 41 min, 13 min of which they were immersed in the solutions and the remaining time they were outside at a temperature of about 45 °C. The rust obtained was classified into inner (adherent) and outer (loosely bound) layers. The inner rust was investigated by room temperature Mössbauer spectroscopy.

Table 1. Some selected examples about the mixture of solutions employed in the preparation of magnetites in pure form and in the presence of the Cu^{2+} solutions. The percentages are given in mole %.

In pure Form		With cationic solutions	
Sample	Precursors	Sample	Precursors
1	100% $FeSO_4$	6	95% $FeCl_2$ + 5% $CuCl_2$
2	50% $FeSO_4$ + 50% $FeCl_2$	7	95% $FeSO_4$ + 5% $CuCl_2$
3	100% $FeCl_2$	8	95% $FeCl_2$ + 5% $CuSO_4$
4	50% $FeCl_2$ + 50% Na_2SO_4	9	95% $FeSO_4$ + 5% $CuSO_4$
5	50% Fe SO_4 + 50% NaCl	10	99% $FeCl_2$ + 1% $CuCl_2$

Magnetites were also prepared in the presence of copper, chromium or manganese sulfate (or chloride) solutions (see Table 1). For this purpose, 1, 5 and 10 mole % of Fe, in the nominal composition, was replaced by Cu, Cr, or Mn [4,5].

Room temperature Mössbauer spectra were obtained in the transmission mode using a constant acceleration drive and a triangular reference signal. A $^{57}Co/Rh$ source with initial activity of 25 mCi was used. All absorbers were prepared by mixing the material with pure sugar to achieve a homogeneous thickness of about 10 mg Fe/cm^2.

Results and discussion

The synthesis methods produce magnetites with relatively good crystallinities and stoichiometries. It has been found that the magnetic hyperfine fields for the A and B sites, $B_{hf,A}$ and $B_{hf,B}$ respectively, in samples obtained from sulfate and chloride precursors are the same between the error bars of about ± 0.1 T. Additionally, $B_{hf,A}$ and $B_{hf,B}$ and the other hyperfine parameters were not noticeably affected in the magnetites obtained using sodium chloride or sodium sulfate as secondary precursors. These results suggest that the type and concentration of precursors have little or no effect on the crystallographic, chemical or physical properties of the pure magnetites.

As mentioned in the introduction, the effect of sulfate and chloride ions was also studied through accelerated corrosion tests. The results of these investigations are summarized in Table 2.

Table 2. Pollutant concentration for the different corrosion tests, relative amount of magnetite present at the inner most layer in the corrosion products, and the mean corrosion rate after 65 days coupons exposure to the solutions [2].

Corrosion tests	1	2	3	4	5	6
Cl^- (10^{-3} M)	1.00	1.00	1.25	1.50	1.75	3.00
SO_4^{2-} (10^{-4} M)	1.00	1.25	1.00	1.00	1.00	2.00
Magnetite (%)	9.8	5.6	58	10.2	26.2	42.9
Corrosion rate (μm/y)	9.8	5.6	58	10.2	26.2	42.9

Morales et al. [2] have found a correlation between the corrosion rate and the relative magnetite content for coupons exposed during 65 days, i.e. higher corrosion velocities produce a higher quantity of magnetite. From Table 2, it is possible to see that magnetite is more abundant in corrosion tests 3, 5 and 6. The results mean that the pollutant concentration affects the amount of magnetite being formed. It is found that the pollutant concentration in corrosion test 3 favors the formation of large amounts of magnetite, whereas pollutant concentration as in case 4 inhibits it. A proper interpretation of these results requires further investigation and it is beyond the goal of the present paper. It is important to keep in mind that the content of Cu, Cr and Mn in AISI 1008 is very small, and its presence does not affect the magnetite formation.

By combining the results of the magnetite prepared using different types of precursors with the magnetite obtained through accelerated corrosion tests, it is suggested that the pollutants strongly influence the amount of magnetite being formed, but that they have little effect on the crystallographic, chemical or physical properties of the samples.

In contrast to pure magnetites, the magnetites formed under the presence of both, sulfate and chloride ions, and cations such as Cu^{2+}, Mn^{2+} and Cr^{3+} are strongly affected in their physical properties, especially for the copper cations. We have found that $B_{hf,B}$ (see Fig. 1) is slightly smaller in the magnetites obtained from the solutions $FeCl_2$ + $CuCl_2$ than the ones prepared from $FeSO_4$ + $CuSO_4$ (see Table 1 for concentrations). On the other hand, the B site center shift, the width of the hyperfine field distribution, and the area of an impurity doublet, which is partly assigned to non-magnetic goethite, are greater in the magnetites prepared from $FeCl_2$ + $CuCl_2$ (MC) than the ones prepared from $FeSO_4$ + $CuSO_4$ (MS). All these findings may suggest that copper can be incorporated more effectively on the octahedral sites in the samples prepared from copper and iron chloride solutions. Furthermore the magnetite formed from both MS and MC present also a magnetic goethite component, with larger hyperfine fine, $B_{hf,I}$, in the first case and therefore more degree of crystallinity.

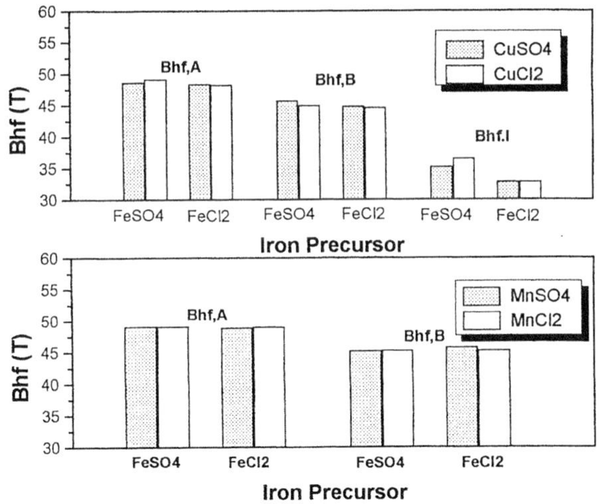

Figure 1. Comparison of hyperfine fields for the A- and B-magnetite sites and a magnetic goethite impurity phase (I) formed in the samples prepared in the presence of alloying elements.

The inclusion of Cr ions in the solution appears to introduce less goethite in the final product than the inclusion of Cu and Mn, even for sulfate rich and chloride rich precursors. The other contribution to the impurity doublet observed in the Mössbauer spectrum of the magnetite formed in the presence of Cu ions can be assigned to another phase whose chemical nature is not yet known; XRD rules out the presence of delafossite [4]. Two possible explanations for this phase have been raised. Copper can destroy the morphology and produces magnetites with a broad distribution of particle sizes [6], including perhaps the formation of superparamagnetic magnetite particles. However, this phase can also originate from magnetite particles for which the Cu distribution is not uniform and is more concentrated towards the surface layers of the particles [7]. More measurements to shed light into this problem are required. It is important to mention that Cu degrades more strongly the physical and crystallographic properties of magnetite than Mn and Cr, since it exhibits the Jahn-Teller effect and because is more electronegative than Fe and Mn [7].

From the discussion above it is concluded that the magnetite physical and chemical properties are more affected by the Cu^{2+} presence than for Mn^{2+} and Cr^{3+}, and that these cations favors the formation of other phases. The sulfates and chlorides play a secondary role.

Conclusions

For pure magnetites, the different concentrations of sulfates and chlorides do not affect their properties in an important way. The effect of these ions upon the magnetite formation is more clearly noticed in the accelerated corrosion tests. In fact, it has been found that the sulfate and chloride ionic concentration affect the amount of magnetite formed at the carbon steel surfaces, in such a way that for 65 days exposure time, magnetite is most abundant in the corrosion test 3. The study of the mechanisms for this behaviour is still under way. On the other hand, the alloying elements affect differently the physical and crystallographic properties of the magnetite. The addition of Cu^{2+} in the solutions promotes the formation of large amounts of impurities in comparison to Mn^{2+} and Cr^{3+}.

Acknowledgements. This work was supported by CODI (University of Antioquia), and COLCIENCIAS (Colombia) under project number 1115-05-10113. The authors thank to the referee for valuable comments on the manuscript.

References

[1] U. Schwertmann and R.M. Cornell: Iron Oxides in the laboratory-preparation and characterization, VCH, Weinheim, 1991, p. 73.
[2] A.L. Morales, D. Cartagena, J.L. Rendon, and A. Valencia, Phys. Stat. Sol. (b) 220 (2000) 351.
[3] A.L. Morales, D. Cartagena, J.L. Rendon, A. Valencia, C.A. Barrero, and C. Dauwe, Hyp. Int. (C) 2 (1997) 143.
[4] C.A. Barrero, A.L. Morales, J. Restrepo, G. Perez, J. Tobón, J. Mazo-Zuluaga, F. Jaramillo, D.M. Escobar, C.E. Arroyave, R.E. Vandenberghe, and J.M. Greneche, Hyp. Int., in press.
[5] A.L. Morales, H. Mosquera, and C. Arroyave, Hyp. Int. (C) 4 (1999) 34.
[6] G.B. McGarvey and D.G. Owen, J. Mater. Sci., 31 (1996) 49.
[7] P.S. Sidhu, R.J. Gilkes, and A.M. Posner, (1978) J. Inorg. Nucl. Chem., 40 (1978) 429.

A Mössbauer study of steel and its ore originating from Northern Iran.

H. Aghamohammadzadeh, J. M. Williams and J. Adetunji†

Department of Physics and Astronomy, University of Sheffield, Sheffield, S3 7RH;
†Centre for Environmental and Applied Science Research, School of Environmental and Applied Sciences, University of Derby, Derby DE22 1GB UK.

Abstract

Room temperature Mössbauer spectroscopic analyses of iron ore from Northern Iran shows that it contains four components of iron bearing oxides with magnetite as the major component. The other three components, hematite, goethite and lepidocrocite are either oxidised or weathering products of the magnetite. The goethite was fitted with a sextet indicating it is well crystalline. The results of the elemental composition analysis show that Al and Ca are present in appreciable quantity to constitute the main elements partially substituting for Fe in the non-stoichiometric magnetite in the ore sample.

1 Introduction

Developing countries have long acknowledged that a basic requirement for technological take off is self-sufficiency in locally produced iron and steel, for domestic and industrial manufacturing processes. In addition, more foreign exchange can be earned by processing the raw materials before export. Large deposits of the raw iron ore, usually in the form of iron oxides, are widespread in many developing countries. Iran is no exception, being endowed with large deposits of the ore in the Chadermalu region, along with large reservoirs of natural gas, which could be used for reducing the oxide ores directly to produce iron. Currently, the country is producing enough iron and steel for local industrial consumption and is exporting large quantities of the surplus to earn foreign exchange.

Hitherto, all studies carried out on Iranian iron oxides and steel have involved the use of various chemical and physical analytical techniques, but not Mössbauer Spectroscopy (MS). Among the analytical methods employed so far, only X-ray analysis is non-destructive and capable of determining iron compounds in the ore, provided they are crystalline and constitute at least 1.0% by weight of the analysed sample. However, the Mössbauer spectroscopic method has the advantage of high sensitivity to both crystalline and amorphous iron-bearing compounds, even for very low quantities, ~ 0.01%, in the samples and, it is also non-destructive. In addition, because the various iron compounds possess a small particle morphology, MS is often much more direct and complimentary to other experimental characterisation techniques. It can also be used to deduce the intrinsic properties of the various components, such as particle size, crystallinity, isomorphic substitution, etc. [1].

In this study, we have analysed an iron oxide sample from Chadermalu region, Northern Iran, by the Mössbauer spectroscopic method, with the aim of providing complete information on the constituent iron-bearing compounds. An original steel ingot produced from the same source was similarly analysed. Knowledge of all the constituent iron compounds in the iron ore could help to determine the best, most efficient method of reducing the ore [2]. A qualitative discussion is given, of the mineralization and oxidation processes involved in the genesis of the ore. An analysis of the sample, for elemental composition, was also carried out.

2 Sample preparation and analysis

The iron oxide sample and steel ingot were supplied by the Iranian Steel Company. The base ore is clearly made up of two components of differing colours: A brownish-red component and a dark component, which we consider to belong to the hematite and magnetite-chromate Class IV oxide group. They are labelled BR-ore and DK-ore respectively. The two components were separated and each component ground to powder with an agate mortar and pestle. A portion of each sample was routinely analysed for a range of elements by Inductively Coupled Plasma Atomic Emission Spectrometer (ICP-AES), using the reference standard GBW 08401 obtained from the National Research Centre for Certified Reference Materials (NRCM).

Mössbauer Transmission absorbers were prepared from 80mg of each component oxide. The transmission absorber of the original steel ingot of dimensions 4cm×4cm×1cm was prepared by grinding to 0.1mm in thickness. It was further reduced to 70 –100μm on a lapping machine. The Mössbauer measurements were done with a 25mCi ^{57}Co source in an Rh matrix, which was driven at a constant acceleration in a triangular mode at room temperature. The spectra were recorded in 512 channels and the Lorentzian lines of the folded data were fitted using the least-square Recoil computer programme [3]. All centre shift (CS) values are relative to α-iron.

3 Results

Chemical Analysis

The elemental compositions of the two component samples, BR-ore and DK-ore, are shown in table 1a. For comparison, the percentages of some oxide components of the Chadermalu iron ore concentrate supplied by the Iranian Steel Company are shown in table 1b.

Table 1a: Elemental concentrations in the two components of the iron oxide ore in mg/kg.

Elements	Al	As	Ca	Cd	Co	Fe	K	Mg	Na	Ni	P	Pb	Ti	Zn
BR-Ore	1252	8	663	<0.01	16.1	547750	172	232	363	35.5	<0.01	204	173	154.0
DK-ore	2147	5	1510	<0.01	77	561850	240	424	621	68	<0.01	227	144	206

Table 1b: Concentrations of some metallic oxides and moisture in the Chadermalu iron ore concentrate.

Ore sample	Al_2O_3	CaO	FeO	MgO	SiO_2	H_2O
Base ore concentrate (%)	0.258	0.410	13.22	0.14	0.714	6.35

Mössbauer Spectroscopic analysis

Mössbauer spectra of the samples, Brown ore (BR-ore) and Dark ore (DK-ore), are shown in figure 1(a, b). Each spectrum was fitted as a superposition of four sextets and a doublet. The spectrum for the original steel ingot of the base ore is shown in figure 1c and has been fitted with a single magnetic sextet.

Figure 1: Room temperature Mössbauer spectra of (a) BR-ore (b) DK-ore and (c) steel ingot.

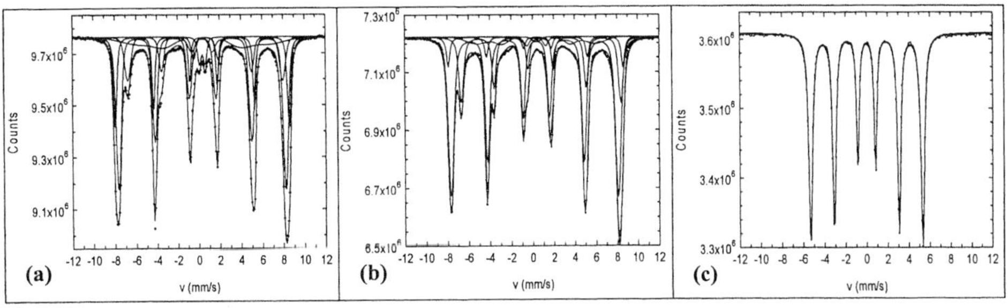

The constituent iron oxide and oxy-hydroxide minerals in the ore were identified using data from the literature [1, 4, 5]. Mössbauer parameters for the spectra in figure 1(a-c) and possible assignments are shown in table 2.

Table 2: Mössbauer parameters of room temperature spectra in figure 1(a-c).

Sample	CS mms^{-1}	QS mms^{-1}	Hyp. Field (kOe)	Rel. Int. %	Possible assignments
BR-ore	0.36	0.53	-	3.0	Lepidocrocite, γ-FeOOH
	0.37	-0.09	513	21.0	Hematite, α-Fe_2O_3
	0.64 0.33	-0.07 -0.05	456 492	15.0 47.0	Magnetite, Fe_3O_4
	0.37	-0.04	381*	14.0	Goethite, , α-FeOOH
DK-ore	0.36	0.53	-	0.5	Lepidocrocite, γ-FeOOH
	0.38	-0.06	512	8.0	Hematite, α-Fe_2O_3
	0.78 0.28	-0.05 -0.06	467 489	22.5 61.0	Magnetite, Fe_3O_4
	0.37	0.03	347*	8.0	Goethite, α-FeOOH
Steel ingot	0.00	0.00	330	100	bcc Steel

** Indicates a component fitted using a broad magnetic hyperfine field distribution.*

4 Discussion

The major components of the two samples were identified as magnetite and hematite. The BR-ore contains 62% and 21% of magnetite and hematite respectively while the values are 83% and 8% in the DK-ore. The Mössbauer parameters of the magnetite in our samples are in good agreement with non-stoichiometric magnetite in which Fe has been partially substituted by other elements such as Al, Mg and Ti [2]. Hematite, α-Fe_2O_3, is often found in pure form in nature. Its presence in the samples can be easily explained because it is usually formed by spontaneous transformation from magnetite. It is also a common weathering product of pyrite although no evidence of this is apparent in our samples. Therefore magnetite, being the dominant constituent

mineral of the base ore, was probably the start material for the production of the hematite which is known to be an oxidising product of lithogenic magnetite [6]. Intermediary minerals with constituents of some of the elements in table 1a could also have acted as oxidising agents in the phase transformation. Titanium does not normally occur in native state but as oxides such as TiO_2 or distributed in other minerals such as magnetite. The parameters of the fitted spectra did not indicate the more common ilmenite $FeTiO_3$ that could have acted as a oxidising agent for the phase transformation.
The small amounts of goethite, α-FeOOH, were identified in both samples. The room temperature Mössbauer spectrum of goethite is, in most cases, fitted with a doublet. However, the spectra of the goethite in our two samples were fitted with broad hyperfine field distributions indicating well crystalline phases. Vandenberghe et al. [1] have reported a similar observation for goethite found in soils. It is reported that when sextets are observed for goethite, they usually exhibit considerably broadened lines. Various broadening mechanisms such as superparamagnetic relaxation, surface effects, cluster ordering and inter-particle interactions are suggested for this phenomenon. Furthermore, because of partial substitution of Al and possible particle size distributions, it is difficult to consider a priori the various effects involved when fitting spectra of natural samples containing goethite. It is less frequently a weathering product of hematite and since there is no evidence of iron bearing sulphides in the base ore from which it could be formed, we therefore attribute its origin to the weathering of the magnetite in the base ore. Lepidocrocite, γ-FeOOH, found in the samples is much rarer than goethite. It is formed preferentially from pyrite, which as indicated above, is lacking in our samples. However, this oxide is generally associated with goethite on which it frequently occurs as coarse grain crusts [6]. A great number of occurrences of lepidocrocite have been studied but almost every where its quantity is smaller than goethite [6]. In addition to weathering product of pyrite, large amounts of it are found in the ores of the Pretoria Iron Mine, South Africa, in the ores from the lower Khan-Valley in South West Africa, in 'brown iron' from Waldrennbach, Germany and also in the weathering products of fahlore from Djebel Quénza, Algeria [6, 7].

The Mössbauer parameters for the steel ingot show no measurable departure from those obtained for pure α-iron, indicating little perturbation to the iron atom sites due to the added dopants. Two types of steel can usually be distinguished on the basis of their structure and Mössbauer parameters. One is the austenite steel having a *fcc* structure of Fe with dissolved carbon. This is paramagnetic, exhibiting a single line in its Mössbauer spectrum. The second type, martensite, has the basic *bcc* structure of Fe with addition of small amounts of carbon and its Mössbauer spectrum is similar to that of pure iron. We have therefore assigned martensite steel to the ingot sample, as it is magnetically ordered with a similar hyperfine field to that of pure α-Fe.

References

[1] R. E. Vandenberghe, C. A. Barrero, G. M. da Costa E. Van San and E. De Grave, Hyperfine Interactions **26** (2000) 247-259.
[2] M. S. Ellid and F. Fallagh, Hyperfine Interactions **92** (1994) 1007-1011
[3] Recoil: Mossbauer Spectral Analysis Software for Windows. ©1998 K. Lagarec and D. G. Rancourt.
[4] J. G Stevens A. M. Khasanov J.W. Miller H. Pollak and Z. Li, *Mössbauer Mineral Handbook* (Publ. Mössbauer effect Data Centre, Asheville, USA.1998)
[5] E. Murad and J. H. Johnson in *Mössbauer Spectroscopy Applied to Inorganic Chemistry Vol. 2*, G. J. Long, ed., Plenum Press, New York, 1984, p. 509.
[6] P. Ramdohr in: *The Ore Minerals And Their Intergrowths,* Pergamon Press, Oxford, 1969.
[7] P. Picot and Z. Johan in: *Atlas of Minerals*, B. R. G. M. Elsevier, 1982

Mössbauer and Structural Studies of $GdFe_{2-x}Ti_x$ Alloys

I.A. Al-Omari[(a)] and H.H. Hamdeh[(b)]
[a]*Department of Physics, College of Science, Sultan Qaboos University, P.O. Box 36, Al-Khodh, Postal Code 123, Muscat, Sultanate of Oman*
[b]*Department of Physics, Wichita State University, Wichita, Kansas, 67260, USA*

Bulk Samples of $GdFe_{2-x}Ti_x$, with x=0, 0.05, 0.10, 0.15, 0.20, and 0.30, were prepared by arc-melting. The x-ray diffraction shows that these samples have the cubic Cu_2Mg type structure, and a small amounts of Fe-Ti and Fe-Gd appear as minor phases for large values of x. The lattice parameter (a) and the unit cell volume (V) are found to increase with increasing Ti concentration. Room temperature Mössbauer spectra show that all the samples are magnetically ordered. The spectra consist of two magnetic components consistent with the two magnetic sites in RFe_2 compounds, and one or two paramagnetic components are added in the fitting for the samples with large values of x corresponding to the minor phases. The average hyperfine field is found to decrease with increasing the Ti concentration.

1. Introduction

Titanium alloys are of great technological importance, their physical metallurgy is quite complex, and modern methods are currently used for their analysis. In addition to this, compounds of rare-earth elements (R) and 3d transition elements (T) have attracted considerable attention in recent years because of the possibility of using these materials for permanent magnets, magnetic recording and optical recording.

Duc *et al.* [1,2] studied the R-T compounds and found that the strength of the interaction between R and T depends on the rare-earth and transition-metal nature, and on the rare-earth concentration. Buschow [3] studied the magnetic and structural properties of $GdFe_2$ and found that it has the same structure as YFe_2, a C15-type laves phase with lattice prameter a=0.7396 nm, and it has a magnetic ordering temperature of 796 K and a ferrimagnetic alignment between Gd and Fe moments with a saturation moment of 3.55 μ_B/f.u.

The effects of Ag substitution for Fe in Gd-Ag-Fe on the magnetic and structure properties of these alloys have been studied by Saleh *et al.* [4]. This study revealed that these alloys form three phases, GdAg phase, α-Gd phase, and $GdFe_2$ phase, with Néel temperature of 140 K for GdAg phase, and Curie temperatures of 300 K for α-Gd phase and 820 K for $GdFe_2$ phase. Yamada *et al.* [5] studied the magnetic properties of $Ti(Fe_{1-x}Al_x)_2$ and found that there is an antiferromagnetism in this system for $x<0.1$, while ferromagnetism appears for $0.1<x<0.5$.

Duc et al. [2] studied the effects of Ti substitution for Fe in $R(Fe_{1-x}Ti_x)_2$ alloys (x=0 and 0.1), R=Gd, Y, and Lu. They analyzed the magnetic interactions in terms of Fe-Fe and R-Fe interactions and found that the Gd-Fe interactions increase with increasing the Ti content. They found that a substitution of Fe by 10% Ti enhances the exchange parameter (A_{GdFe}) from $16.2\text{x}10^{-23}$ to $22.4\text{x}10^{-23}$ J and reduces the iron magnetic moment (M_{Fe}) from 1.6 μ_B/at to 1.25 μ_B/at. This study suggests that the large value of A_{GdFe} must be from the increase in the number of electrons participating in the 5d-3d(Fe) hybridization, which leads to the observed increase in R-Fe coupling and the decrease in the Fe moment. Recently, Al-Omari *et al.* [6] studied the magnetization of $GdFe_{2-x}Ti_x$ ($0\leq x\leq 0.3$) and found that the saturation magnetization (M_s) increases with increasing x at room temperature and at 5 K.

In this work we use x-ray diffraction and Mössbauer spectroscopy to study the effects of Ti substitution for Fe in $GdFe_{2-x}Ti_x$ alloys on the magnetic and structure properties of these alloys. This study is intended to complement the previous investigation of this system and to shed more light on the magnetic properties of this system.

2. Experimental Procedure

Bulk samples of $GdFe_{2-x}Ti_x$, with x=0, 0.05, 0.10, 0.15, 0.20, and 0.30, were prepared by arc-melting the elements (Gd, Fe, and Ti) of at least 99.9% purity in a water-cooled copper boat in a flowing argon gas atmosphere. The ingots were melted four to five times to ensure homogeneity. After melting the samples, each sample was wrapped separately in a tantalum foil and annealed in vacuum below $3\text{x}10^{-6}$ Torr at a temperature of 875 °C for 36 hours and subsequently quenched in water.

The phase purity for all the samples was checked by x-ray diffraction measurements on the powder samples using Cu-K_α radiation.

The samples for Mössbauer spectroscopy were prepared by sprinkling a thin layer of the powder of $GdFe_{2-x}Ti_x$, with a grain size of about 30 μm in diameter, on a piece of tape. The samples were studied with a Ranger Mössbauer spectrometer, model MS1200. The velocity drive of this spectrometer operates in the constant acceleration mode over 1024 channels. A 50 mCi Co^{57} in Rh matrix was used as the γ-ray source in this experiment. In this study we used α-iron spectrum for calibration and all the isomer shifts were measured relative to the centroid of α-iron spectrum at room temperature.

3. Results and Discussion

Figure 1 shows the x-ray diffraction patterns for $GdFe_{2-x}Ti_x$ alloys. From this figure we see that the samples crystallize in the cubic laves $MgCu_2$-type structure and small amounts of Fe-Ti and Fe-Gd appear as minor phases at large values of x. The shift in the peaks with increasing the Ti concentration is due to the volume expansion because of the larger atomic radius of Ti (0.200 nm) compared with the atomic radius of Fe (0.172 nm). This behavior is in agreement with previous observations by Azez *et al.* [7] for similar compounds. The lattice parameter for $GdFe_2$ is equal to 0.7430 nm. This value is in good agreement with the value of 0.7396 nm obtained by Buschow [3], and of 0.7404 nm by Saleh *et al.* [4]. We find that the lattice parameter (a) increases linearly from 0.7430 for x=0 to 0.7447 nm for x=0.3. The linear increase in the lattice parameter as the concentration of Ti increases suggests a simple dilution process.

Figure 2 shows the room temperature Mössbauer spectra (dots) and the fitting (solid curves) for $GdFe_{2-x}Ti_x$ alloys with x=0, 0.05, 0.10, 0.15, 0.20, and 0.3. The spectra show magnetic ordering for all values of x, with different subspectra having different magnetic hyperfine fields. The spectra were fitted

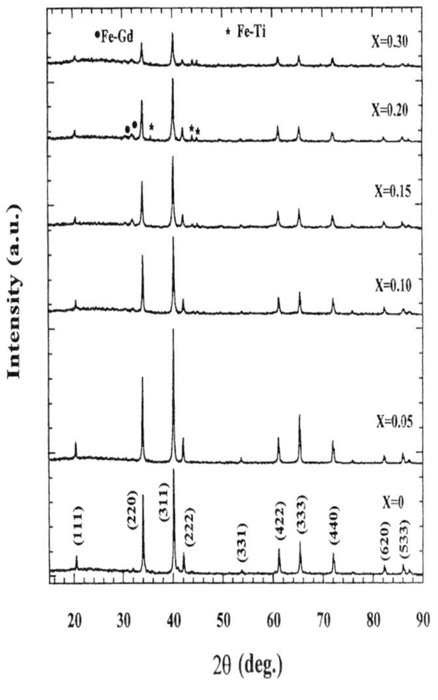

Figure 1. X-ray diffraction patterns for $GdFe_{2-x}Ti_x$ alloys.

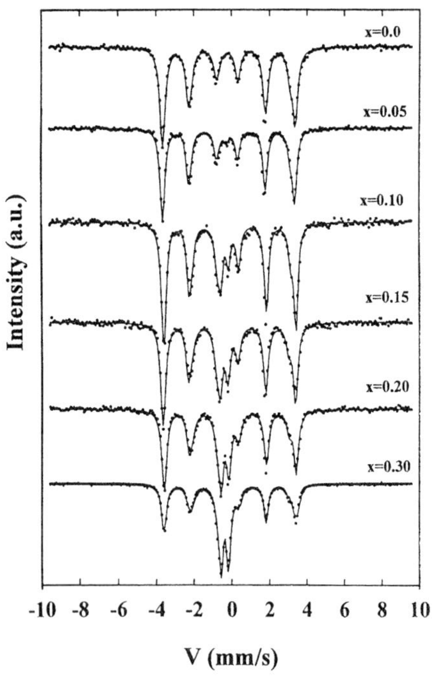

Figure 2. Room temperature Mössbauer spectra of $GdFe_{2-x}Ti_x$ alloys. The solid curves represent the fitting.

with two magnetic components and one or two paramagnetic components. Since the iron atoms occupy the corner-sharing of the tetrahedral networks with a three fold axis along the [111] direction which is the same as the easy direction for the magnetic field, then the local magnetic field and electric field gradient tensor axes will not be oriented identically on all the Fe sites and two six line patterns in the ratio 1:3 are expected. In our fitting we used this model, where the relative intensity for the two magnetic components are 3:1 for $GdFe_2$ and this relative intensity increases with increasing the Ti concentration reaching the value of 3.3 for x=0.3. The ratio of 3:1 for the parent compound, $GdFe_2$, is in agreement with previous results by Bowden *et al.* [8], Wertheim and Wernick [9], and Wertheim *et al.* [10] for $GdFe_2$. The increase in the relative intensity with increasing x indicates that Ti occupies preferentially one of the two iron sites (the one with the lower intensity and lower magnetic hyperfine field).

The relative intensity of each subspectra in the Mössbauer spectrum give the percentages of the different Fe-phases. Figure 3 shows the percentages of the different phases in each sample as a function of the Ti concentration. It can be seen from this figure that the percentage of the Fe-Ti phase increases with increasing x. These results are in qualitative agreement with our x-ray diffraction results. The average magnetic hyperfine field for the different Ti concentrations is shown in figure 4. We see from this figure that the average magnetic hyperfine field decreases with increasing x. The average magnetic hyperfine field for the parent compound is 218 kOe. This value is in agreement with the value of 215 kOe for $GdFe_2$ found by Bowden *et al.* [8]. The decrease in the average magnetic hyperfine field with increasing the Ti concentration can be attributed to the substitution of magnetic element (Fe) by the non-magnetic element (Ti). This behavior is in agreement with the magnetization results by Al-Omari *et al.* [6], where the Fe magnetic moment decreases with increasing the Ti concentration.

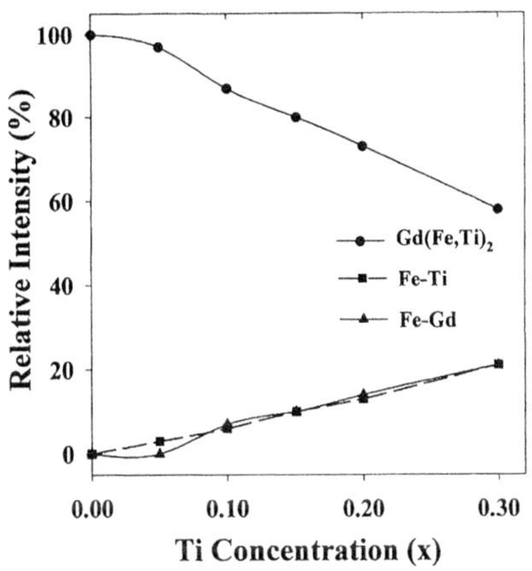

Figure 3. The percentages of the different phases in each sample of $GdFe_{2-x}Ti_x$ as a function of the Ti concentration.

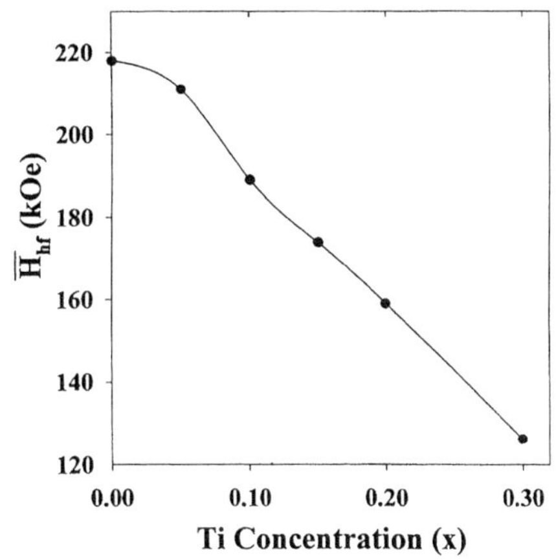

Figure 4. Dependence of the average hyperfine field for $GdFe_{2-x}Ti_x$ alloys on the Ti concentration (x), at T=295 K.

Figure 5 shows the average isomer shift relative to α-iron as a function of x. It is clear from this figure that the average isomer shift decreases with increasing Ti concentration. This figure shows also that the average isomer shift is negative and it decreases with increasing x. The negative value of the isomer shift in $GdFe_2$ is an indication of the increase in the s-electron density at the Fe nucleus. The average isomer shift for $GdFe_2$ is equal to –0.08 mm/s. The decrease in the isomer shift with increasing x can be ascribed to the transfer of electrons from the Ti atom to the Fe atom; this is in agreement with other observations for similar alloys [11,12]. These transferred electrons are partially s-like and partially d-like, but the net result is to produce an increase in the s-electron density at the Fe nucleus, and hence a decrease in the isomer shift.

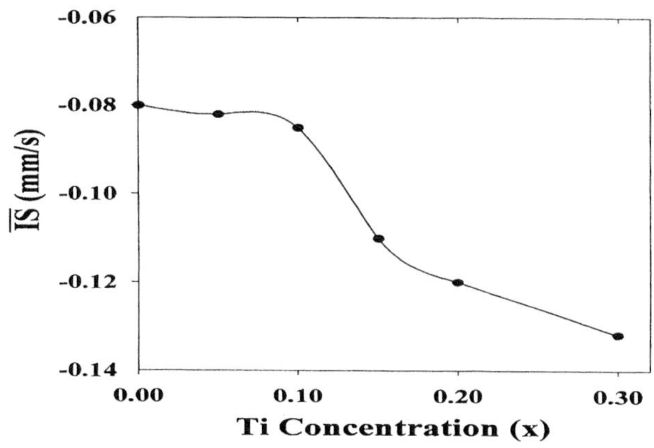

Figure 5. Dependence of the average isomer shift for $GdFe_{2-x}Ti_x$ alloys on the Ti concentration (x), at T=295 K. The solid line represent a guide to the eye.

4. Conclusions

The x-ray diffraction shows that the alloy samples under investigation have the cubic Cu_2Mg type structure and small amounts of Fe-Ti and Fe-Gd appear as minor phases for large values of x. The lattice parameter (a) is found to increase linearly with increasing Ti concentration. Room temperature Mössbauer spectra show that all the samples are magnetically ordered. The average hyperfine field is found to decrease with increasing the Ti concentration. The isomer shift is found to decrease with increasing the titanium concentration.

Acknowledgments

The authors would like to acknowledge Sultan Qaboos University, Jordan University of Science and Technology, and Wichita State University for the support. I.A. Al-Oamri would like to thank Professor Sami Hahmood, Physics Department/Yarmouk University, for the helpful assistant and discussion.

References

[1] N.H. Duc, J.D. Hien, D. Givord, J.J. M. Franse, F.R. de Boer, J. Magn. Magn. Mater. **124** (1993) 305.
[2] N.H. Duc, M.M. Tan, N.D. Tan, D. Givord, J. Teillet, J. Magn. Magn. Mater. **177-181** (1998) 1107.
[3] K.H.J. Buschow, Rep. Prog. Phys. **40** (1977) 1179.
[4] A.S. Saleh, L.A. Saleh, N.m. Nayfeh, S.H. Mahmood, and I. Abu-Aljaraysh, J. Appl. Phys.**71** (1992) 1893.
[5] Y. Yamada, K. Kuroda, A. Sakata, T. Murakami, J. Magn. Magn. Mater. **177-181** (1998) 1397.
[6] I.A. Al-Omari, M.A. Al-Akhras, M.K. Hasan, J. Shobaki, K.A. Azez, J. Alloys and Compounds **319** (2001) 34.
[7] K.A. Azez, I.A. Al-Oamri, J. Shobaki, M.K. Hasan, G.M. Al-Zoubi, and H.H. Hamdeh, Submitted to J. Alloys and Compounds.
[8] G.J. Bowden, D.St.P. Bunbury, A.P.Guimaras and R.E. Snyder, J. Phys. C **1** (1968) 1376.
[9] G.K. Wertheim and J.H. Wernick, Phys. Rev. **125** (1962) 1937.
[10] G.K.Wertheim, V. Jaccarino, and J.H. Wernick, Phys. Rev. **135** (1964) A151.
[11] I. Omari, A. S. Saleh and S. H. Mahmood, J. Magn. Magn. Mater. **78** (1989) 183.
[12] S.M. Dubiel and W. Zinn, J. Magn. Magn. Mater. **45** (1984) 298.

THE EFFECT OF SHORT OXIDATION TIME ON Zr $(Fe,Cr)_2$ LAVES PHASES

[+]*P.B. Bozzano*, **P. Vázquez*, **F. Saporiti*, **C.Ramos*, [+]*R. Versaci, and* [++]*C. Saragovi*
[+] *Dep. of Materials, CNEA,* [++] *Dep. of Physics, CNEA, *Inst. de Tecnología "Prof. J. A.Sabato", UNSAM-CNEA, Av. Gral. Paz 1499 (1650), Buenos Aires, Argentina.*
Correspondent author: saragovi@cnea.gov.ar

Abstract

Two Laves phases Zr $(Cr_{0.40}\ Fe_{0.60})_2$ ((C14) hexagonal structure) and $Zr(Cr_{0.15}\ Fe_{0.85})_2$ ((C15) cubic structure) when oxidized 3 minutes in an open furnace were characterized by X-Ray diffraction (XRD) and Mössbauer Spectroscopy (MS) techniques. Oxidation modified them but part of the original Laves phases remained. $Zr_{2-x}O_x$, Cr oxides, α-Cr, (Fe, Cr) oxides, α-Fe and α-Fe_2O_3 were formed. The presence of α-Fe sustains a model previously suggested for the oxidation of the Laves phases. On the other hand, the existence of α-Fe_2O_3 together with α-Fe suggests that 3 minutes is longer than the necessary time to follow the oxidation kinetics.

Introduction

Oxidation of the Laves phase Zr $(Cr, Fe)_2$ is an important topic to study when considering Zry-4 oxidation. As it is well known, Zry-4 (Sn 1.45- 1.5 wt. %, Fe 0.18-0.24 wt. %, Cr 0.07- 0.13 wt. %; O 1400 ppm. with Zr the balance) is widely used as fuel cladding material in nuclear industry. The mentioned Laves phase is present in Zry-4 as precipitates of intermetallic compounds, which occur in hexagonal (C14) and/or cubic (C15) structures.

Many studies were carried out in order to evaluate the evolution of the precipitates when Zry-4 is oxidized. Previous work [1,2] showed that precipitates undergo chemical composition changes after open furnace oxidation. EDS (Energy Dispersive Spectroscopy) showed the presence of Zr, Cr and Fe within both the non-oxidized and partially oxidized precipitates, but no Fe was detected within the oxidized ones. An empirical model has also been proposed for the Laves phase oxidation [3].

In a prior work [4] three Laves phases (C14) Zr $(Cr_{0.40}Fe_{0.60})_2$, (C15) Zr $(Cr_{0.95}\ Fe_{0.05})_2$ and (C15) $Zr(Cr_{0.15}Fe_{0.85})_2$ were oxidized for 7 minutes in an open furnace at 923 K and characterized by Analytical Electron Microscopy (AEM), X-Ray diffraction (XRD) and Mössbauer Spectroscopy (MS) techniques. Oxidation modified them but the corresponding Laves phases remained. $Zr_{2-x}O_x$, Cr oxides, α-Cr, (Fe, Cr) oxides, α-Fe and α-Fe_2O_3 were also formed in the oxidized Fe-rich phases. The existence of α-Fe but also α-Fe_2O_3 suggests that 7 minutes was longer than the necessary time to follow the oxidation kinetics. On the contrary, in the Cr-rich one only $(Cr, Fe)_x\ O_y$ appeared, showing that the same interval was insufficient for α-Fe to appear.

In the present work, with the aim of further contribution to the mentioned experimental evidence, the Laves phase (C14) $Zr(Cr_{0.4}Fe_{0.6})$, which is the most frequently observed in Zry-4 precipitates, and (C15) Zr $(Cr_{0.15}Fe_{0.85})_2$ (Fe-rich) were oxidized for 3 minutes in an open furnace at 935K. Mössbauer Spectroscopy together with X- Ray diffraction techniques were used to analyze the samples.

Experimental

$Zr(Cr_{0.4}Fe_{0.6})_2$ and $Zr(Cr_{0.15}Fe_{0.85})_2$ Laves phases were prepared in an arc furnace with a non-consumable W electrode in an Ar atmosphere. Samples were melted from commercial available Fe, Cr and Zr of nuclear purity. The as-cast structures were characterized by XRD. All the samples were powdered and oxidized in an open furnace at 923 K for about three minutes. Powder XRD was performed using a Cu anticathode in a 10° to 120° 2θ range at a speed of 2°/min. Patterns were analyzed using the Rietveld method [5].

Room temperature (RT) Mössbauer spectra were collected using a conventional constant-acceleration spectrometer, in standard transmission geometry with a ^{57}Co source in Rh. All the spectra were fitted with sextets (S) and doublets (D) considering crystalline sites and/or distributions of hyperfine field according to the experimental spectra. The quality of each computer fit was checked by a χ^2-test. The relative population of the Fe-bearing compounds was estimated from their corresponding fitted areas in each sample.

Results

The as-cast powder samples were characterized by XRD and analyzed using the Rietveld method. A C14 structure was found in $Zr(Cr_{0.4}Fe_{0.6})_2$ with the following lattice parameters: a= 0.50160 +/- 0.0002 nm and c= 0.82183 +/- 0.00007 nm. In the as-cast Fe-rich sample, a C15 structure was found with a= 0.70810 +/- 0.00008nm.

Figure 1 shows the XRD patterns obtained for the oxidized powder samples. The first two columns of Table 1 and 2 display the identified phases and the lattice parameters obtained with the Rietveld method for the oxidized $Zr(Cr_{0.4}Fe_{0.6})_2$ and $Zr(Cr_{0.15}Fe_{0.85})_2$ samples respectively.

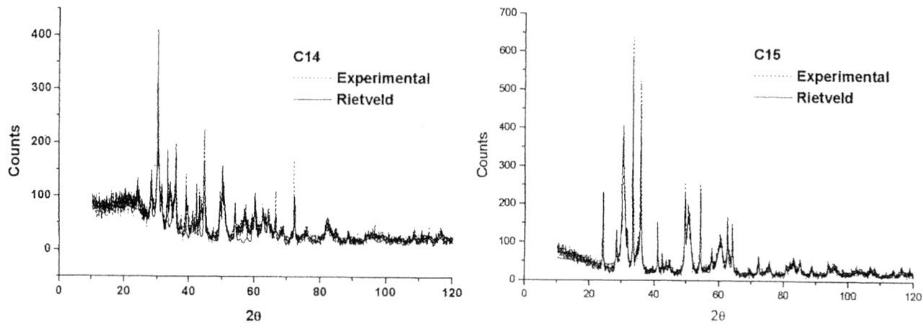

Figure 1: XRD patterns for the oxidized powder samples.

RT Mössbauer spectra of the non-oxidized and oxidized samples are shown in Fig. 2. The parameters for the non-oxidized samples were coincident with those found in the literature [6, 7]. The low velocity C14 sample spectrum was fitted as in [7] but in the higher velocity spectrum, an "average" doublet was used. The non-oxidized C15 sample spectrum was fitted with a hyperfine field distribution ($H_1 \sim 5$ T and $H_2 \sim 15$ T respectively for the decreasing maxima with QS = 0 mm/s and IS = -0.15 mm/s) and with a doublet (QS = 0.29 mm/s and IS = -0.14 mm/s). This fit shows the existence of a polytypic structure, which has been reported previously [8]. The parameters of the oxidized C14 and C15 samples are summarized in the 4th column of Tables 1 and 2 whilst the last column displays the assignments considering the Mössbauer parameters and the percentile of each subspectrum area [9,10].

Discussion

MS results, complemented by XRD analysis, sustain the described model in literature [3]. The $Zr(Cr_{0.4}Fe_{0.6})_2$ and $Zr(Cr_{0.15}Fe_{0.85})_2$ Laves phases underwent a selective oxidation process upon 3 minutes oxidation in an open furnace. A part of the

Table 1: Identified phases and lattice parameters obtained with XRD, Mössbauer parameters and assigned phases for the C14 $Zr(Cr_{0.4}Fe_{0.6})_2$ sample.

X – Ray Diffraction			Mössbauer Spectroscopy	
Identified Phases	Lattice Parameters (nm)	Phases	Parameters	Phases
Cubic I	a= 0.2871 +/- 0.0008	α-Fe α-Cr	IS = 0 mm/s QS = 0 mm/s H = 33 T	α-Fe 25%
C14	a= 0.5015 +/- 0.0005 c= 0.8206 +/- 0.0005	C14	IS = -0.19 mm/s QS = 0.24 mm/s	C14 34%
M_2O_3	a=0.5026 +/- 0.0006 c=0.1371 +/- 0.0002	Fe_2O_3 Cr_2O_3 $(Fe, Cr)_2O_3$	IS = 0.37 mm/s QS = -0.20 mm/s H = 51.2 T	$(Fe, Cr)_2O_3$ 31%
Monoclinic	a= 0.5145 +/- 0.0005 b= 0.5184 +/- 0.0005 c= 0.5314 +/- 0.0005	Monoclinic ZrO_2	----------------	------------
Tetragonal	a= 0.3601 +/- 0.0002 c= 0.5174 +/- 0.0005	Tetragonal ZrO_2	---------------	-------------
Cubic F	a=0.5101 +/- 0.0008	Cubic ZrO_2	---------------	-------------
---------------	-------------	-------------	IS = 0.69 mm/s	$(Fe, Cr)_xO_y$ 10%

Table 2: Identified phases, lattice parameters obtained with the Rietveld method, Mössbauer parameters and assigned phases for the rich -Fe sample.

X – Ray Diffraction			Mössbauer Spectroscopy	
Identified Phases	Lattice Parameters (nm)	Phases	Parameters	Phases
Cubic I	a=0.2870 +/ 0.0006	α-Fe α-Cr	IS=0 mm/s QS=0 mm/s H=33 T	α-Fe 12%
M_2O_3	a= 0.5027 +/- 0.0006 c =1.3707 +/- 0.0002	Fe_2O_3 Cr_2O_3 $(Fe, Cr)_2O_3$	IS= 0.34 mm/s QS= -0.18 mm/s H= 51.5 T	α-Fe_2O_3 61%
C15	a = 0.7066 +/- 0.0005	C15	IS=-0.21 mm/s QS=0 mm/s H=15 T	C15 22%
Monoclinic	a = 0.5131 +/- 0.0005 b = 0.5179 +/- 0.0005 c = 0.5305 +/- 0.0005	Monoclinic ZrO_2	----------	-------------
Tetragonal	a = 0.3572 +/- 0.0002 c = 0.5192 +/- 0.0005	Tetragonal ZrO_2	----------	-----------
----------	-----------	-----------	IS=0.65 mm/s	$(Fe, Cr)_xO_y$ 5%

original Laves phases remained and the rest of it presented important modifications becoming more important as the Fe- content increased.

Large amounts of Zr_xO_{2-x}, α-Cr, α-Cr_2O_3, α-Fe, α-Fe_2O_3 and $(Fe, Cr)_xO_y$ were formed in both samples upon oxygen introduction. The existence of α-Fe together with α-Fe_2O_3 indicates that an oxidation time of 3 minutes is still longer than necessary to follow the oxidation kinetics. The last two subspectra evidence a partial Fe–substitution by Cr in both lattices. Moreover, it can be seen that the amounts of the Fe-bearing phases vary with Fe-content. However, it is not possible to suggest any influence of the structure (C14 or C15) in the oxidation process.

The aforementioned results are similar to those reported in a previous work [4]

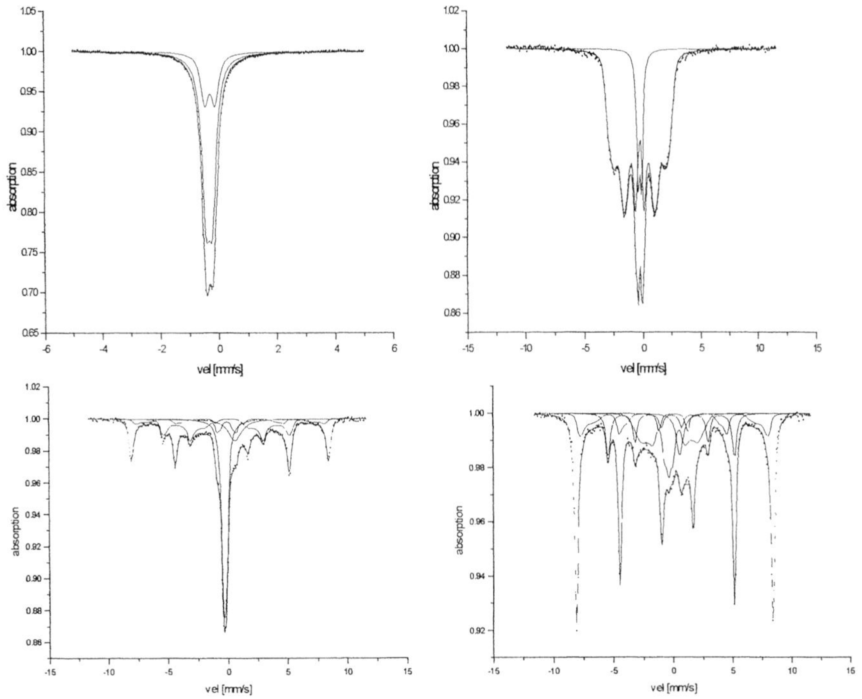

Figure 2: Room temperature Mössbauer spectra of non-oxidized (up left) and oxidized (down left) C14 and of non-oxidized (up right) and oxidized (down right) C15 samples.

in which similar samples were oxidized for 7 (seven) minutes. The 3 minutes interval chosen for the oxidation process is still longer than needed to follow the oxidation kinetics when the Fe-content of the Laves phases is higher (≥1.2%at). Further work is in progress using different oxidation times.

References

[1]P.B.Bozzano, M.Ipohorski and R.A.Versaci, Acta Microscopica, Vol.5(B)(1996) 238.

[2]R.AVersaci, P.B.Bozzano and M.Ipohorski, Proc. Conference on Materials and Nuclear Power. EUROMAT 96, Bournemouth, U.K. (1996) 411.

[3]P.B.Bozzano, M.Ipohorski and R.A. Versaci, Proc. Electron Microscopy Symp. ICEM 14 (1998)105.

[4] F. Saporiti, P. Bozzano, R. Versaci, C. Ramos, P. Vázquez, I. Raspini and C. Saragovi, ISIAME 2000, Virginia, USA.

[5] L.Lutterotti and P.Scardi, J.Appl.Cryst. 25 (1992) 459.

[6] F.Labenski de Kanter, C.Saragovi, M.Granovsky and D.Arias in Applications of the Mössbauer Effect,eds E. Baggio-Saitovich, E.Galvao da Silva and H.R. Rechenberg, World Scientific, Singapore, 1992, p246.

[7] J.A.H. Coaquira, H.R. Rechenberg and J. Mestnik Filho, Journal of Alloys and Compounds 288 (1999) 42-49.

[8] X.M.Buranu and D.O.Northwood, J.Less- Comm.Met. 170(1991) 27.

[9]MEDC Information services and activities report 1990, Univ. of North Carolina, Ashville, N. Carolina.

[10] R.E. Vandenberghe, C.A. Barrero, G.M. da Costa, E. Van San and E. De Grave, Hyperfine Interactions 126 (2000) 247-259

RARE EARTH ADDITIONS IN A FERRITIC STEEL: A MÖSSBAUER STUDY

P. Bruna[1], *I. Mejia*[2], *J. Cabrera*[3] *and T. Pradell*[4]

[1]*Departament de Física Aplicada, Universitat Politècnica de Catalunya. Campus Nord UPC, Mòdul B4, 08034-Barcelona, SPAIN*

[2] *Instituto de Investigaciones Metalúrgicas, Universidad Michoacana de San Nicolás de Hidalgo, Edificio "U", Ciudad Universitaria, C.P. 58060, Morelia, Mich. MÉXICO*

[3]*Departament de Ciència dels Materials i Enginyeria Metal·lúrgica, E.T.S.E.I.B. – Universitat Politècnica de Catalunya. Av. Diagonal 647, 08028-Barcelona, SPAIN*

[4]*E.U.E.T.A.B, Universitat Politècnica de Catalunya. Urgell 187, 08036-Barcelona, SPAIN*

Abstract. Rare earth metal additions in carbon steels promote better control of the inclusions morphology, resulting in better mechanical properties at room temperature. However, their hot flow behaviour has received scarce attention in the literature. In the present study, uniaxial hot compression tests are carried on two similar carbon containing steels with and without Ce and La (<0.1 wt%). As a general trend, the rare earth containing steel (RE) evidenced larger stress values than the rare earth–free steel (REF), particularly at testing temperatures below 1000°C. The rare earth metals may appear either as precipitates or dissolved into the austenitic phase. A Room Temperature Mössbauer study was performed to clarify the role played either as precipitate or solute by the rare earth elements. For this purpose samples of both steels were austenized at 900°C and 1000°C and quenched at room temperature. Hyperfine parameters of the ferritic and austenitic phases were obtained and compared. Mössbauer results indicate that rare earth metals are mainly affecting the flow stresses through a drag solute mechanism, and consequently are basically dissolved in the austenitic phase.

1. Introduction

Inclusion shape control is a well established industrial practice during steelmaking operations[1]. Calcium and rare earth metal (REM) additions were traditionally used to modify the shape of MnS inclusions and promote formation of oxysulfides and/or sulfides, resulting in better final mechanical properties of a given steel. Although extensive research was performed over the period 1960-1980, the expensive cost of using REM declined its application in western countries. However, countries like China, with large rare earth resources, still employ such steelmaking route [2].

On the other hand, less investigation has been addressed to derive the influence of the REM additions on the hot working behaviour of REM-treated steels. In principle RE metals, mainly Ce and La, can be either precipitated or dissolved in the austenitic phase. In the first case, REM can be found within MnS inclusions, or even in the form of Ce_2O_3 or CeS. Particularly, in the precipitated state provides grain refinement[3]. Nevertheless, whether precipitated or dissolved, some hardening must be expected under hot working conditions. The aim of the present study is to show the additional degree of hardening of carbon steel due to REM additions and to analyse by Mössbauer spectroscopy if REM are mainly dissolved or precipitated.

2. Mechanical testing

Two commercial low carbon steels were selected in this work, namely, a rare earth containing steel (RE) and a rare earth free steel (REF). Both steels are microalloyed with similar amounts of V and Ti (about 0.1 wt%) and also, similar quantities of Mn (1.0 wt%) and Si (0.2

wt%). The REF steel shows a slightly larger carbon amount than the RE steel, 0.032 and 0.026 wt% respectively. The total amount of "impurities", excluding the rare earth additions, is of 1.528 wt% and 1.517 wt% for the REF and RE steels respectively. The REM additions for the RE steel are 0.013 wt% of Ce and 0.010 wt% of La.

The hot working behaviour was analysed by high temperature uniaxial compression tests. Samples were extracted from the supplied material along the rolling direction. They were cylindrical (11 mm in height and 7 mm in diameter). After austenitization at 1100°C, the samples were deformed in an electromechanical testing machine at constant temperature (900°C, 1000°C and 1100°C) at strain rates ranging from 0.1 to 0.001 s^{-1}. The flow curves were available immediately after test.

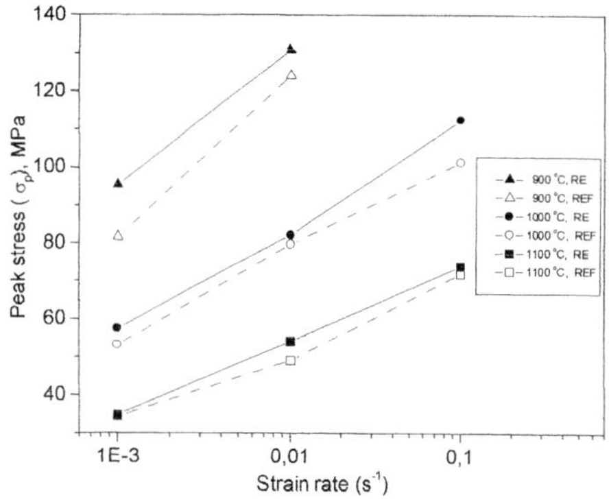

Figure 1 *Dependence of the peak stresses in the flow curves of the strain rate at a given temperature in the RE and REF steels.*

The nature of the hot flow curves indicates that dynamic softening mechanism is acting along the deformation process, i.e., dynamic recovery and dynamic recrystallisation. It has been shown [4] that dynamic recristallisation is the predominant softening mechanism, and is the responsible of the peak stresses observed. Larger stresses are readily apparent in the RE steel than in the REF, as it can be seen in the plot of the peak stresses against the strain rate (*figure 1*). The hardening effect is relatively independent on strain rate. At 900°C is of about +10MPa and at 1000°C is lower, of about +2.5MPa. Bearing in mind the low additions (23 ppm of rare earth metals to the RE steel), the hardening by mass unit of rare earth is extraordinarily large.

However, the latter results do not clearly state the role played by Ce and La, as hardening on the flow curve can be obtained either by precipitation and solute drag. Consequently, a Mössbauer study was carried out in order to clarify their role.

3. Mössbauer analysis

For the Mössbauer study, both steels were first austenized at 1100°C, then cooled down to 900°C and 1000°C maintained for 15 minutes and then quenched in water at room temperature. Quenched samples were cut, ground and polished to a thickness between 40μm and 50μm. The measures were made in a Mössbauer spectrometer with a full width at half maximum (FWHM) of 0.25 mm/s measured over standards. The velocity was calibrated by using as standard an α-Fe foil of 25 μm thickness. The spectra were fitted by using a Voigt profile shape (with a Lorentzian width equal to the source width, 0.25 mm/s). The use of Voigt profiles accounts for the presence of impurities in the second coordination shell, which are not able to brake the cubic symmetry of the austenite or to reduce in a significative amount the magnetic hyperfine field in the ferrite.

The Mössbauer spectra of the austenized RE and REF steels are shown in figures 2 and 3. As expected because the quench, ferrite is the main phase and only an extremely small quantity of austenite is found (below 3 at% of the total iron). The hyperfine fitted parameters corresponding to both phases are given in Table I. The spectra consist in a paramagnetic component (doublet) and a ferromagnetic component (two sextets). The paramagnetic component corresponds to the austenite and the doublet is caused by the presence of either interstitials in the vicinity of the iron atoms or iron substitutions (Si, Mn, Ti, V) resulting in a local deviation of the cubic symmetry which leads to the quadrupolar splitting. It has not been possible to fit a singlet related also with the austenite phase, this may be due either to the lack of resolution, due to the small quantity of austenite present

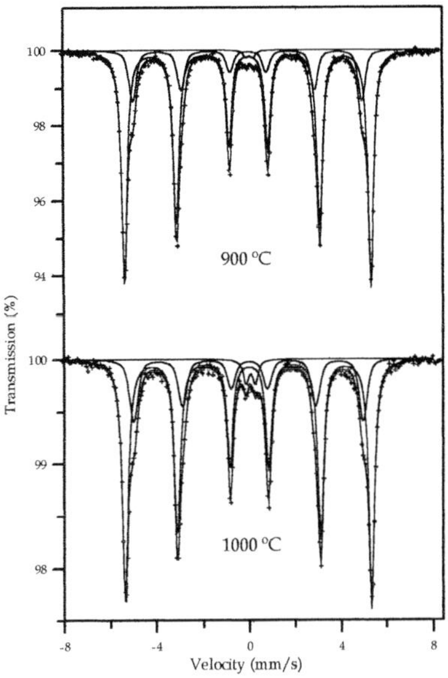

Figure 2 *Mössbauer spectra of the RE free steel.*

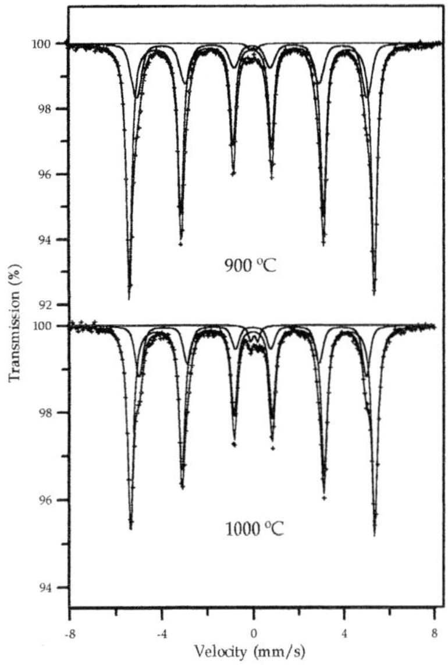

Figure 3 *Mössbauer spectra of RE steel.*

	T (°C)	phase		IS (mm/s)	QS (mm/s)	BHF (T)	area %	(f-II/ferrite) %
REF	900	ferrite	f-I	-0.009(1)	0.004(1)	33.15(1)	79.1(9)	20.1(6)
			f-II	-0.009(2)	0.008(4)	30.80(2)	19.9(5)	
		austenite		0.06(2)	0.30(3)	-	0.9(1)	
	1000	ferrite	f-I	-0.010(1)	0.007(1)	33.17(1)	73.7(9)	24.1(8)
			f-II	-0.009(2)	0.004(6)	30.87(4)	23.4(7)	
		austenite		0.05(1)	0.40(1)	-	2.9(1)	
RE	900	ferrite	f-I	-0.008(1)	0.006(1)	33.23(1)	74.4(9)	25.1(8)
			f-II	-0.005(2)	0.012(4)	31.20(4)	24.9(8)	
		austenite		0.04(3)	0.28(4)	-	0.7(1)	
	1000	ferrite	f-I	-0.009(1)	0.004(1)	33.17(1)	75.9(9)	22.6(8)
			f-II	-0.006(3)	0.004(6)	30.79(3)	22.2(7)	
		austenite		0.04(1)	0.31(2)	-	2.0(1)	

Table 1 *Mössbauer fitted parameters of both austenized steels, given with their corresponding standard deviation. The isomer shift (IS) is given relative to a-Fe; QS corresponds to the quadrupolar splitting; BHF to the magnetic hyperfine field.*

in the samples, but also to the absence of a perfect cubic site, that is absence of interstitial atoms/substitutions in the first coordination shell of the austenite. The two ferromagnetic sextets are related to two different sites around the Fe atoms in ferrite; the f-I corresponds to absence of either interstitial and substitutional atoms in the first coordination shell and f-II corresponds to either interstitial atoms in the first coordination shell and substitutional element in either the first or second coordination shells. Interstitials in the second coordination shell are only responsible of the Voigt shape of the lines.

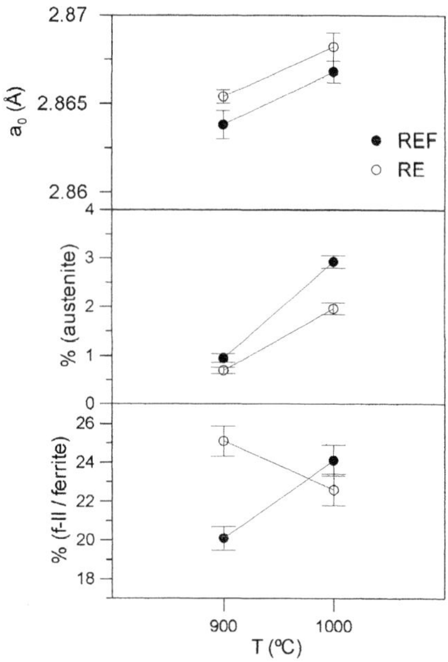

Figure 4 *Cell parameter of the ferrite phase (top), at% of Fe in the austenite phase (center) and fraction of f-II ferrite (bottom) as a function of austenizing temperature.*

As expected, both steels show an increase of the austenite content at increasing the annealing temperatures -see figure 4-, the REF steel showing slightly larger amounts of retained austenite, probably due to its slightly higher C content. The hyperfine parameters (IS and QS) related to austenite are similar for both steels at 900ºC, although in the case of 1000ºC the QS value of the REF steel shows a sensibly higher value than the other. In global terms, these results would indicate a relatively similar dissolution of atoms in the austenite phase in all the cases. Moreover, present results suggest that in case that Ce and La are kept into solution, they are not affecting significantly the untransformed austenite after quenching, and probably the phase transformation during cooling.

The ferrite shows also similar IS, QS and also BHF values for both subspectra, f-I and f-II. At 900ºC the ratio f-II/(ferrite) is higher for the RE steel -see figure 1-, while this behaviour switches at 1000ºC. Larger amounts of f-II indicate more atoms put into solution in the ferritic phase. According to the present chemical compositions, one should expect larger amount of f-II ferrite in the REF steel as the Fe amount is lower than in the RE. Indeed, this is observed at 1000ºC. The switching behaviour at 900ºC can be explained in terms of the microalloying nature of both steels. According to the literature[5], the precipitation of V-Ti carbonitrides occurs at temperatures below 950ºC, and it should be more extensive in the steel with larger carbon amounts. Consequently, a more significant amount of C, V and Ti are precipitated in the REF steel than in the RE steel at 900ºC, promoting, in the overall balance, that more elements are put into solution in the latter steel than in the REF one, and therefore, more f-II signal should be obtained

Consequently, the results indicate that Ce and La are mostly kept into solution at the temperatures involved in the present study. Moreover, X-ray diffraction analysis of the ferritic phase was also performed and the crystallographic cell parameter was determined by a least square fit of the peak positions. Figure 1 shows that the RE ferrite has also a bigger cell parameter than the REF ferrite, independent of the temperature considered. These results also indicate the dissolution of REM additions and the microalloying nature of both steels.

4. Conclusions

Mechanical testing demonstrates that rare earth additions on low carbon steels promote a strengthener behaviour even at high temperatures. Although the chemical complexity of the present steels may promote precipitation of rare particles, the Mössbauer study indicates that Ce and La are more probably kept into solution.

Acknowledgements

This work was financed by CICYT, grant MAT98-0672-C02-02.

References

[1] Gladman, T. in The Physical Metallurgy of Microalloyed Steels. The Institute of Materials, 305-310,1997.

[2] Zongsen, Y. in HSLA Steels: Processing, Properties and Aplications. Ed. by Geoffrey Tither and Zhang Shouhua. The Minerals, Metals & Materials Society, 267-271, 1992.

[3] Guo, M. and Suito, H., ISIJ International, 39(11), 1169-1175, 1999.

[4] Sakai,T. and Jonas, J.J., Acta Metallurgica, 32(2), 189-209, 1984.

[5] Weiss, I. and Jonas, J.J., Metallurgical Transactions, 10A, 831-840, 1979.

Mossbauer study of the Effect of Milling Period on the Dynamic Behavior of $Fe_{30}Ti_{70}$ Alloy

N.A.Eissa[a], N.H.Sheta[a], M.Sherif El-Eskandarany[b], A.H.Ashour[c], T.Z.Amer[a] and H.A.Sallam[a]

[a]*Mossbauer lab., Physics Department., Faculty of Science, Al Azhar University, Cairo, Egypt.*
[b]*Mining and Petrolem Engineering Department, Faculty of Engineering, Al Azhar University, Cairo, Egypt.*
[c]*National Center For Radiation Research and Technology, Cairo, Egypt.*

$Fe_{30}Ti_{70}$ alloy was synthesized by mechanical alloying (MA) of pure elemental Fe and Ti powders for different period (up to 200h) using a high energy ball mill. The MA powders were investigated by X-ray diffraction (XRD) and Mossbauer spectroscopy. Milling for 200h led to the formation of 48.4%Fe in FeTi alloy at room temperature (RT)measurement, but at 80K this ratio decreased to 40.7%, this may be due to the overlap between the superparamagnetic iron and FeTi alloy. Nanocrystalline structure phases were confirmed from the decrease in the values of Debye temperature (θ_D) after different milling periods.

1. Introduction

In recent years Fe-Ti alloys have been studied extensively because of their importance as hydrogen storage system[1-4]. The mechanically produced Fe-Ti alloy has very rough surface which is favorable for hydrogen storage [6]. In the present work, Mossbauer measurements at different temperatures and after different milling periods allowed tracing the different phases as well as their dynamic behavior.

2. Experimental details

Pure elemental powders of Fe(60μm, 99.7%) and Ti (50μm, 99.9%) were mixed to give the nominal composition of $Fe_{30}Ti_{70}$, in a glove box under purified argon atmosphere free from O_2 and H_2O, were less than 1 ppm and were sealed in a cylindrical vial together with ten stainless steel balls. The ball to powder weight ratio was about (10:1). The milling procedure has been carried out at room temperature (RT) by mounting the vial in a high energy ball mill with a milling speed of 500rpm.

XRD measurements were performed using Cu-K_α radiation. Mossbauer spectra (MS) were measured at different temperatures ranging from 50 to 500K. using 1.8GBq-^{57}Co source diffused in a rhodium matrix. Isomer shift (IS) was measured relative to metallic iron. A least square fitting program based on Voigt was used to determine the Mossbauer parameters. Absorption areas and the distribution of both magnetic hyperfine field (MHF) and quadrupole splitting (QS) were made by the distribution program[5]. The milling process was interrupted at regular intervals and small amount of the milled powder was exctrated.

3. Results and discussions

The XRD patterns (figure 1) of the milled $Fe_{30}Ti_{70}$ samples showed a drastic decrease in the major Ti peak after milling for 24h. This indicates that MA enhanced the solid solubility of Ti in Fe. Milling for 100h, the pattern indicated the formation of an intermetallic FeTi phase[4,8], residual α-Fe and Ti crystals. After 150h, these peaks became broader with a diffused halo suggesting the formation of

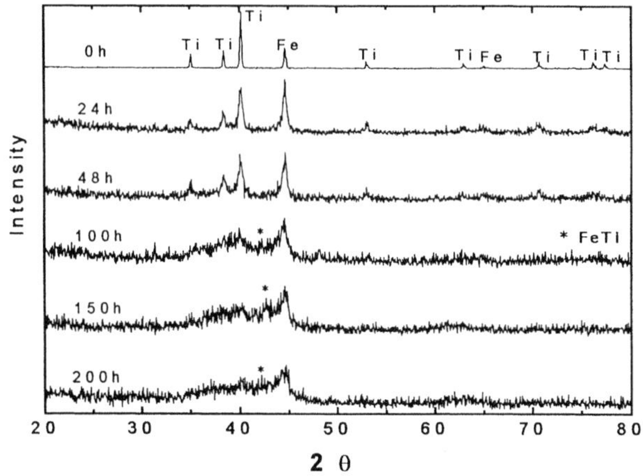

Figure 1 The XRD patterns of $Fe_{30}Ti_{70}$ samples at different milling time

an amorphous phase coexisting with intermetallic crystalline FeTi phase and residual α-Fe crystals. No significant changes in the diffraction pattern could be observed after 200h of MA. Using the Scherrer's equation, the average grain size of Fe was found to be about 5nm after 200h of milling.

The MS of the milled samples are shown in figure 2,where the spectrum of the starting sample shows the six-line pattern (S_1) corresponding to α-Fe.

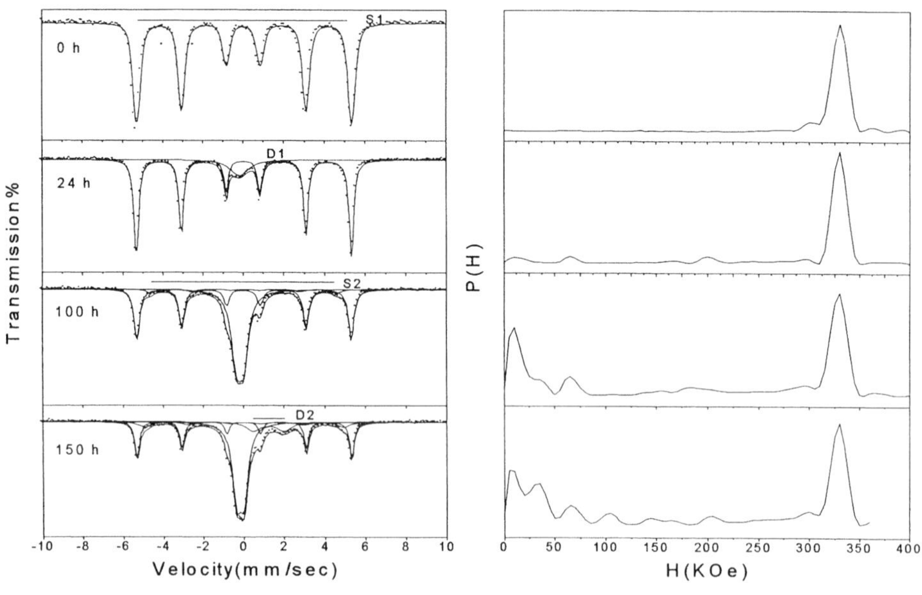

Figure 2. MS of $Fe_{30}Ti_{70}$ at RT after milling at the times indicated, with the corresponding P(H) for the magnetic, nanocrystalline and paramagnetic phases.

After 24h milling period, a broad paramagnetic doublet (D_1) due to Fe^{3+} ions appeared. This indicates the formation of an intermetallic FeTi phase (14% of the total iron which was increased with increasing the milling period). A broad six-line pattern (S_2), having smaller HMF and larger IS, appeared after milling for 100h. The large line width of this component indicates a broad distribution of the HMF which reflects a different interatomic spacing[6]. This result is confirmed by the high value of IS of this component compared with that of α-Fe, this reveals a reduction in the electron density of this component and reflects a volume expansion of this phase. Hence, this phase (S_2) was attributed to the grain boundary regions and to the strain field resulting from the prolonged time of milling[7]. A new broad Fe^{2+}doublet (D_2) appeared after milling for 150h which could be due to the formation of an amorphous Fe^{2+} phase of Fe and Ti. By increasing the milling period to 200h, this phase increased only by 5.5%, (from 11-16.5%). This indicates that the MA does not proceed under metastable equilibrium conditions. This result agrees with Eckert et al.,[8], who reported that the smaller negative mixing enthalpy in Fe-Ti system (-17 KJ/mol) is not sufficient for the formation of completely amorphous material. The variation of the relative area of Fe in each phase with the milling period is shown in figure 3.

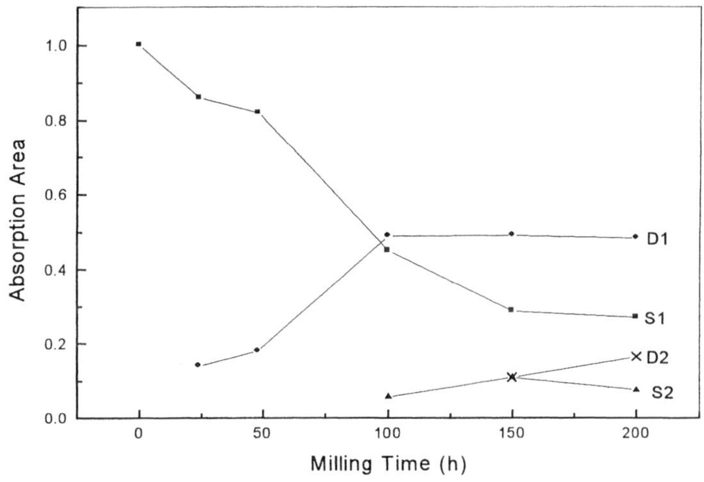

Figure 3. The variation of relative area of each phase (at RT) with the milling time.
S1 unprocessed iron phase
S2 Grain boundary phase.
D1 Amorphous paramagnetic Fe^{3+} phase.
D2 Amorphous paramagnetic Fe^{2+} phase.

The distribution P(H) of MHf (figure 2) after milling for 24h showed small probability at the low values of the field. This probability showed a pronounced increase in the sample milled for 100h. These low field probabilities were attributed to the formation of submicrometer grain structure as well as the grain boundary phases of nanocrystalline and paramagnetic structures. The formation of these nanocrystalline phases showed superparamagnetic behavior at 80K which was confirmed also from MS, where the ratio of S_2 was increased at the expense of D_1.

On the other hand, distribution of the QS of the D_1 Fe^{3+} phase (figure 4) for the sample milled for 24h to 100h showed a bimodal distribution that exhibits one maximum at around 0.41mm/s due to the Fe-Ti alloy and another maximum at low QS values (around 0.084mm/s) due to the nanocrystalline phases. After milling for 150h, only a broad peak appeared instead of the two peaks denoting that the milling process led to continuity and homogeneity of the resulted mechanically alloyed phases.

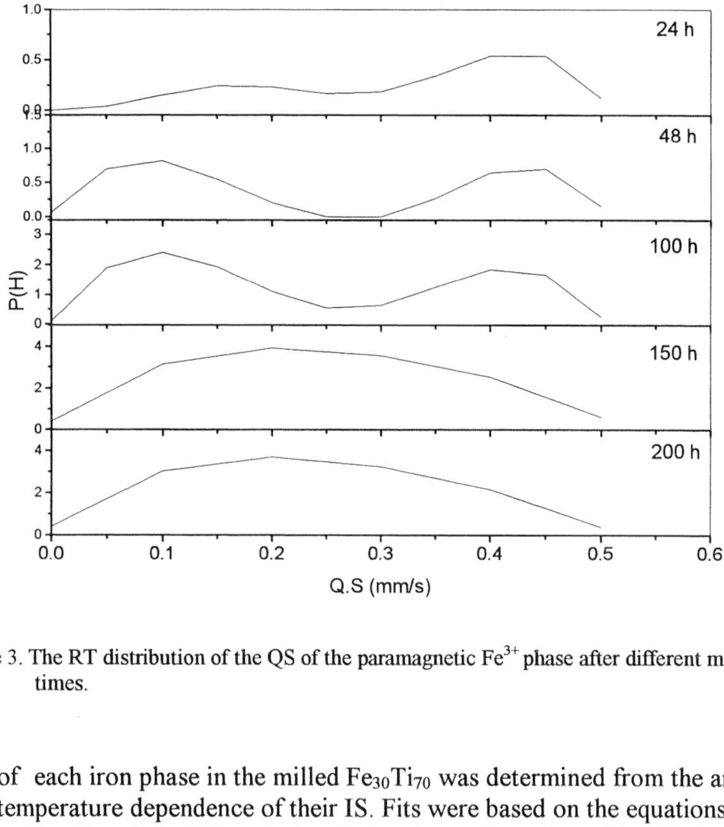

Figure 3. The RT distribution of the QS of the paramagnetic Fe^{3+} phase after different milling times.

θ_D of each iron phase in the milled $Fe_{30}Ti_{70}$ was determined from the analysis of the temperature dependence of their IS. Fits were based on the equations of the second order Doppler shift and the Debye integrals evaluated by Wickman[9]. Table 1 presents θ_D values of each phase after milling for 100h and 200h. The reduction of θ_D from 410K to 340K for the grain boundary phase (compared with 476K for the unprocessed α-iron) indicates the reduced number of Fe near-neighbor atoms at the grain boundaries[10] and confirms the formation of nanostructure. The increase in θ_D for intermetallic phase indicates the development of the FeTi alloy in crystalline structure.

Table 1. θ_D values of the different phases after 100h and 200h milling periods.

	α-Fe metal phase	Grain boundary phase	intermetallic Fe^{3+} phase	Fe^{2+} phase
100 h	430 K	410 K	377 K	—
200 h	415 K	340 K	490 K	310 K

4. Conclusion

MA of $Fe_{30}Ti_{70}$ for 200h measured at RT produced 48.4 % Fe of the total iron in the form of FeTi alloy, 16.5% in the form of Fe^{2+} alloy, 7.6% as nanocrystalline Fe and 27.5 % in the form of residual iron metal. MA for 200h at 80K indicated that the Fe in FeTi alloy was only 40.7 % of the total iron and this may be due to the overlap between the super paramagnetic Fe and the FeTi alloy.
The formation of nanophases is confirmed from the reduction of θ_D of the various phases after different milling periods.

References

[1] J.J.Reilly and R.H.Wiswall, J. Inorg. Chem., 13(1974)218.
[2] H.Gleiter, Prog. Mater. Sci., 33(1989)223.
[3] L.Zaluski, P.Tessier, D.H.Ryan, C.B.Doner, A.Zaluska, J.O.Strom-Olsem, M.L.Trudeau and R.Schultz, J. Mater. Res., 8(1993)3059.
[4] L.Sun, H.Liu, D.H.Bradhurst and S.Dou, J.Mater. Sci. Lett. 17(1998)1825.
[5] R.E.Vandenberghe, E.DeGrave and P.M.A.deBakkerm, Hyperfine Interactions, 83(1994)29.
[6] S.J.Campbell and H.Gleiter "Mossbauer Spectroscopy Applied to Magnetism and Materials Science", eds G.J.Long and F.Grandjean, Plenum Press, New York, Vol1, (1993) 241.
[7] T.Tanaka, S.Nasu, B.Huang, K.N.Ishihara and P.H.Shingu, Nucl. Instr. Meth., Phys. Res. B76(1993)195.
[8] J.Eckert, L.Schultz and K.Urban, J.Non Cryst. Sol., 127(1991)90.
[9] H.H.Wickman, In Mossbauer Effect Methodology Vol. 2, I.J.Gruverman, Plenum Press: New York (1966).
[10] H.Kuwano, H.Ouyang and B.Fultz, Mater. Sci. Forum, 88-90(1992)561.

MAGNETIC PROPERTIES OF IRON-COPPER ALLOYS

M.E. Elzain[1], A. Al Rawas[1], A. Gismelseed[1], A. Yousif[1], Xialion Ren[2], F J Berry[2] and H M Widatallah[3]
[1]Depertment of Physics, Box 36, Sultan Qaboos University, Al Khod 123, Sultanate of Oman
[2]Department of Chemistry, The Open University, Milton Keynes, MK7 6AA, UK
[3]Department of Physics, Khartoum University, Khartoum 11115, Sudan

Abstract. The discrete-variational method in the local density approximation is used to calculate the local properties at the iron central site in iron-copper alloys. The local magnetic moment, hyperfine field and the isomer shift at iron sites are calculated versus the number of copper atoms at the neighboring shells. It is found that the magnetic moment increases with increasing number of copper atoms in the first coordination shell, whereas the magnetic hyperfine field decreases. Assuming a binomial distribution of copper atoms the average properties are calculated and compared to experimental results.

1. Introduction

The solubility of iron in copper or copper in iron is minute. There are no intermediate phases of the Fe-Cu systems, however, metastable and homogeneous alloys can be formed over the whole concentration range. There a several methods to produce metastable solid solutions and recently mechanical alloying was used to prepare solid solutions of $Fe_{1-x}Cu_x$ [1-3]. These solutions are in the bcc phase for $x<0.4$ and in the fcc phase for $x>0.6$. In the intermediate concentrations $0.4<x<0.6$, the two phases coexist.

Gonser and Ron [4] identified a number of Fe-environments in the $Fe_{1-x}Cu_x$ alloys. These include configurations such as Fe associations, Fe in coherent fcc-precipitates, Fe in bcc environment and many more. It was noticed that Fe in the fcc-environment has a small magnetic hyperfine field and a positive isomer shift relative to α-Fe. In the as-milled mechanically alloyed samples of $Fe_{1-x}Cu_x$ alloys, the Mössbauer results were interpreted in terms of 3 phases, [5]. The largest hyperfine field of about 32 T was associated with the bcc Fe-rich phase, while the hyperfine field of about 22 T was associated with the fcc Fe-rich phase and fields below 10 T were related to fcc Cu-rich alloys.

The saturation magnetization versus temperature curve for annealed samples was found to be very flat at the low temperatures end, which is characteristic of bcc Fe [3]. The magnetization was also analyzed in terms of three phases composed of bcc Fe, fcc Fe and fcc Cu, such that the latter two phases contributed no net magnetic moment to the system. The magnetic moment per Fe-atom in the as-milled fcc sample was found to be around 2.35 μ_B. Local moment in the order of 2.38 μ_B was determined using polarized neutron scattering in the bcc $Fe_{1-x}Cu_x$ alloy [6].

A number of *ab initio* calculations were performed on $Fe_{1-x}Cu_x$ [7] with emphasis being on the magnetic moments in the ferromagnetic state. The local moment at α-Fe site was found to increase with Cu concentration in agreement with experiment. The hyperfine fields and isomer shifts were not considered. In addition the effect of the presence of AFM-coupling in the fcc structures was not studied.

In this contribution, we used the discrete variational method to solve the Kohn-Sham equation in the local density spin polarized approximation. The $Fe_{1-x}Cu_x$ alloys are represented by clusters of 51 atoms with bcc and fcc configurations. The local properties at the central site were calculated for different cluster configurations with a varying number of nearest and next-nearest neighbors of Cu atoms. The average properties were obtained using binomial distribution. The magnetic hyperfine field was assumed to result from the Fermi contact term and was split into two parts. The core term was assumed to be directly proportional to the local 3d moment with a constant of proportionality of -10 T/μ_B [8]. The valence term was directly calculated from the contact valence spin density. The calculated properties are presented and discussed in the following section, whereas a short conclusion is given in the summary.

2. Results and discussions

In the bcc structure the Fe-Cu systems are ferromagnetic. The properties of bcc clusters were calculated for various configurations with different numbers of nearest (n) and next nearest (m) neighboring Cu-atoms. The type of atoms outside the first and second shells of neighbors makes little change on the properties at the central site. The magnetic moment, the contact charge density and contact hyperfine field for a particular configuration are denoted by $\mu(n,m)$, $\rho(n,m)$ and $B_{hf}(n,m)$ respectively. On the other hand, the fcc structure is more complex. Various magnetic phases including antiferromagnetic (AFM), modulated spin-density waves and ferromagnetic were reported [9]. The AFM ground state of the γ-Fe is closely represented by double layers along the [001]-direction, denoted by AFMD, which have intra-layer ferromagnetic coupling and inter-double-layer AFM coupling. The magnetic moment at the central Fe-atom in the fcc clusters is parallel to the moment of eight of the nearest neighboring Fe-atoms, and opposite to moment of the remaining four. The local properties vary with the number of Cu-atoms in the first shell as well as on the relative orientations of the central Fe and nearest neighbor (NN) Fe local moments. Henceforth, the local properties are again denoted as above with n and m now being the number of Fe-atoms with moment parallel to and opposite to that of the central atom respectively.

The average moment and distribution of the hyperfine field were calculated assuming a binomial distribution of Cu in the bcc lattice and of Fe in the fcc lattice. In particular the probability distribution of the hyperfine field was calculated using the relation

$$P(H) = \sum \delta(H - B_{hf}(n,m))p(n,m) \qquad (1)$$

where $p(n,m)$ is the binomial probability distribution for the configuration with n and m atoms at neighboring shell as defined above for bcc and fcc structures.

2.1 bcc structures

Consider Fe-Cu-systems in bcc structure. In Table 1 the local moment, the contact charge and spin densities and the hyperfine field at the central Fe for m = 0 and n = 0 to 8 are presented. We note that the total moment increases with increasing number of Cu atoms at NN position, whereas the contact charge and spin densities and the contact field decrease in magnitude. In Figure 1 we depict the average moment per atom and per Fe-atom and compare the results with simple dilution and experimental results taken from ref 2. The experimental results follow the simple dilution, but lie below the calculated curve since, in clusters, moments are usually large. The average moment per Fe-atom is large and increases with Cu-concentration. Since the contact charge density decreases with increasing Cu-contents, it is expected that the isomer shift becomes more positive at higher Cu-concentration. On the other hand the average hyperfine field at Fe-site decreases with Cu-concentration in contrast to the local moment.

Table 1. The local magnetic moment μ (in μ_B,) the contact charge and spin densities ρ and $\delta\rho$ (in a.u) and the magnetic hyperfine field B_{hf} (in T) at the central Fe-site for number of Cu atoms at NN positions equal to 0, 2, 4, 6 and 8.

Fe	[0,0]	[2,0]	[4,0]	[6,0]	[8,0]
μ	2.30	2.46	2.62	2.69	2.75
ρ	5.75	5.59	5.44	5.27	5.08
$\delta\rho$	-0.23	-0.17	-0.10	-0.07	0.0
B_{hf}	-36	-34	-31	-29	-26

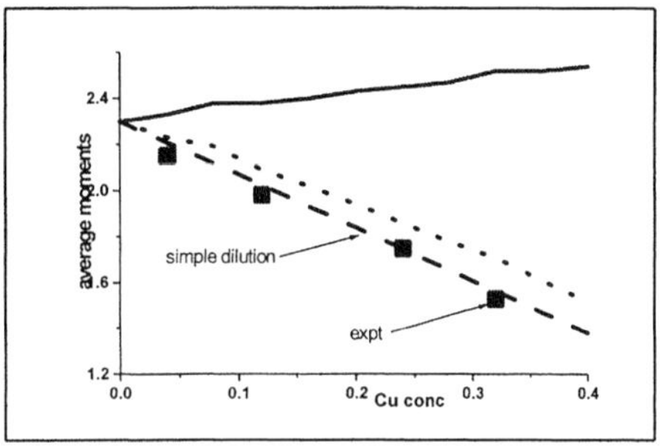

Figure 1. Average moment per Fe-atom (solid), per atom (dotted), simple dilution (dashed) and experiment (square)

2.2 fcc structures

For the Fe-Cu alloys in the fcc phase FM and AFMD order was considered. The results for the fcc FM phase follow the same trends observed for the bcc structure. On the other hand, the results for the AFM phase do not vary only with the number of NN Cu-atoms, but also with the nature of magnetic coupling between the central Fe atoms and Fe atoms at the NN positions. In general, the moment increased and the contact charge density decreased with increasing Cu-contents. The changes in the hyperfine field are drastic. In Table 2 representative results for 4 and 8 Cu atoms at NN positions with different orientation of Fe atoms are shown. The large changes in the hyperfine field resulted from large changes in the valence contact spin densities.

Figure 2 shows the hyperfine field distribution calculated using equation (1) for different phases at x = 0.4. It is clear that the distributions due to bcc and FM fcc structures are almost identical with peaks around high negative fields. There is no way to relate the spectra with low fields to such structures. On the other hand, the field distribution of the AFM fcc-alloy is very broad and cover all the range of low fields. Hence, we attribute the low fields observed in the mechanically alloyed samples to phases where AFM Fe-Fe coupling exit. For higher

Cu-concentration the peaks of all distributions shifted to smaller fields, and in addition the AFMD phase distribution developed a larger peak around zero fields.

Table 2. The local moment μ (in μ_B,) the contact charge and spin densities ρ and $\delta\rho$ (in a.u) and the hyperfine field B_{hf} (in T) at Fe-site for 4 and 8 NN Cu-atoms with parallel orientations for NN Fe atoms

n Cu	4		8	
n Fe	8	4	4	0
μ	2.70	2.53	2.71	2.80
ρ	5.29	5.35	5.06	5.15
$\delta\rho$	-0.07	0.29	0.02	0.47
B_{hf}	-31	-8	-25	0

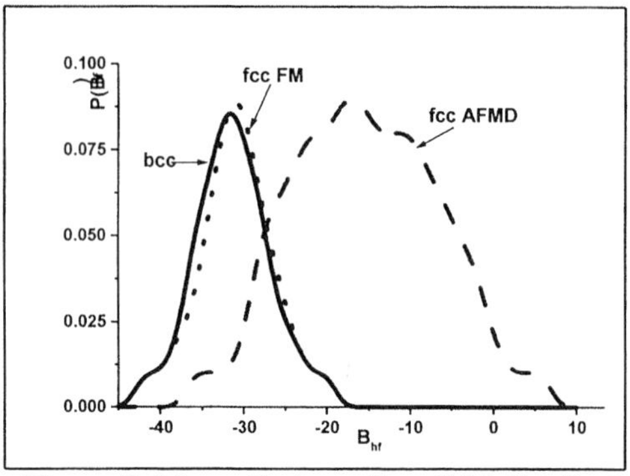

Figure 2. Magnetic hyperfine field distribution bcc (solid), FM fcc (dashed) and AFMD (dotted)

3. Summary

The local properties at the central Fe-site in Fe-Cu alloys were found to have the following trends. In both the bcc and fcc ferromagnetic structures, the magnetic moment increased with increasing number of Cu atoms at the NN site, whereas the contact charge and spin densities and the magnetic hyperfine field decreased. The hyperfine field distributions are peaked around large negative values. The average magnetic moment per atom decreased almost linearly with increasing Cu-concentration, in agreement with experimental results. The reduction in contact charge density resulted in positive isomer shifts. Similar trends were observed for the fcc double layer AFM structure except that, the changes in the magnetic hyperfine field are now more sensitive to the Fe-Fe coupling. In presence of AFM coupling the distribution of the magnetic hyperfine field is broad and extended to low field regions. The low field component of the Mössbauer spectrum, would be attributed to regions where AFM coupling exited.

References

1. J. Eckert, J. C. Holzer, C. E. Krill III and W. L. Johnson, J. Appl. Phys. 73 (1993) 2794
2. E. Ma, M. Atzmon and F. E. Pinkerton, J. Appl. Phys. 74 (1993) 955
3. J. Z. Jiang, Q. A. Pankhurst, C. E. Johnson, C. Gente and R. Bormann, J. Phys. (Condens. Matter) 6 (1994) L227;
4. U. Gonser and M Ron, Application of Mossbauer Spectroscopy, Vol. II, ed. R. L. Cohen, Academic Press, NY, 1980 p. 281
5. Y. Li, T. Li, Y.Zhang, L. Qu, C. Gao and W. Liu, J. Phys. 8 (1996) 7191
6. F. Kajzar and G,. Parette, J. Appl. Phys. 50 (1979) 1966
7. A. F. Tatarchenko, V. S. Stepanyuk, W. Hergert, P. Rennert, R. Zeller and P. H. Dederichs, Phys. Rev. B 57 (1998) 5213
8. M E Elzain, D E Ellis and D Guenzberger Phys. Rev. B 34 (1986) 1430
9. Y.Zhou, W. Zhang, L. Zhong and D. Wang, J. Magn. Magn. Mater. 145(1995) 273.

AUSTEMPERING TRANSFORMATION IN 0.11wt% Mn VERMICULAR CAST IRON.

K. Laneri, J. Desimoni, R.C. Mercader
Departamento de Física Facultad de Ciencias Exactas, Universidad Nacional de La Plata, IFLP, Calle 47 y 115 CC 67, 1900 La Plata, Argentina.
R. W. Gregorutti, J. L. Sarutti
Laboratorio de Entrenamiento Multidisciplinario para la Investigación Tecnológica Calle 52 e/121 y 122, 1900 La Plata, Argentina.

Abstract

The kinetics of the austempering transformation in vermicular cast iron with 0.11wt% Mn has been studied in the temperature range 573K-673K by Mössbauer spectroscopy. The Johnson-Mehl-Avrami approximation was used to analyse the transformation kinetics in stage I. The fitted value of the exponent n (≈ 1.6) indicates a localised nucleation and a phase transformation controlled by an interface reaction. The values of the reaction rate constant k don't display an Arrhenius-type dependence on austempering temperature, behaviour which proves independent of the Mn content.

1.Introduction

Vermicular (or compacted graphite) cast irons display high-thermal cycling resistance with good mechanical properties and are used to manufacture pieces of vehicles and engines. Over the last decade their mechanical properties have been improved by austempering heat-treatments, whose kinetics of transformation depends on the Mn content of the alloy.

In earlier works, we have used Mössbauer Spectroscopy to study the kinetics of the first stage of the austempering transformation, during which *austenite* transforms into *high carbon austenite* and *ferrite*. We investigated the reaction both for low and high Mn content (0.11 and 0.58 wt%) at a constant austempering temperature of 623K [1]. When the austempering temperatures are in the range 573K to 673K for irons with high Mn content we found, using Johnson-Mehl-Avrami (J-M-A) approach, that the isothermal phase transformation is controlled by a diffusion process at the interface and does not follow an Arrhenius-type dependence [2,3]. This behaviour was explanained qualitatively through the balance between nucleation and growth processes of ferrite in austenite.

In this work, we have completed the Mössbauer analysis of the kinetics of vermicular cast irons with 0.11wt% Mn content in the austempering temperature range 573K-673K. We discuss the results in terms of nucleation and growth processes within frame of the J-M-A approximation.

2.Experimental

Alloys (final composition 3.52 wt.% C, 2.1 wt.% Si, 0.11 wt.% Mn, 0.03 wt.% Cu, 0.01 wt.% P and 0.03 wt.% S) were prepared in a medium frequency induction furnace [1,2].

Heat-treatments consisted of austenitising at 1173K during 30 minutes and then quenching in a salt bath held at different temperatures in the range 573- 673K for different times up to 30 minutes. The samples were specially prepared for Mössbauer spectroscopy by conventional grinding techniques to reduce their thickness down to $\approx 70\mu m$, using diamond paste of 6, 1 and $0.1\mu m$ for final polishing. Mössbauer spectra were taken in transmission geometry in a standard

512 channels constant acceleration spectrometer. Velocity calibration was performed against a 12 μm thick α- Fe foil. All isomer shifts are referred to this standard at 298K. Spectra were fitted with Lorentzian line-shapes and a non-linear least-squares program with constraints. For the effective thickness of the samples analysed no Voigt line-shape correction was necessary [4].

3.Results and discussion

Mössbauer spectra belonging to samples austempered for 1, 10 and 20 minutes are plotted in Figure 1. The spectra exhibit six lines belonging to the ferrite/martensite magnetic phases [5] and a central sub-spectrum associated to austenite. As the result of the fitting procedure the following average values were obtained: $B_1 \approx 33\pm1T$, $\delta_1 \approx 0.01\pm0.01mm/s$; $B_2 \approx 31\pm1T$, $\delta_2 \approx 0.05\pm0.01mm/s$; $B_3 \approx 27\pm1T$, $\delta_3 \approx 0.08\pm0.01mm/s$.

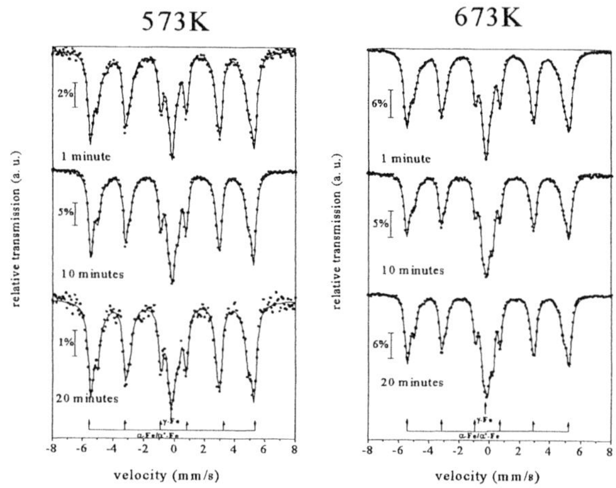

Figure 1: *Mossbauer spectra of vermicular cast iron with 0.11wt.% Mn austempered at 573K and 673K for austempering times of 1, 10 and 20 minutes.*

The austenite pattern was simulated with a quadrupole doublet and a single line, resulting in the average values of $\delta_4 \approx 0.02\pm0.01mm/s$, $\Delta_4 \approx 0.62\pm0.01mm/s$; $\delta_5 \approx 0.03\pm0.01mm/s$ [6, 7].

The *stage I* of austempering transformation was analysed following the evolution of the austenite relative fraction (f_γ) with time, assuming an equal recoilless fraction for all phases. The f_γ relative fraction displays a sigmoidal behaviour (see Fig.2) and can be analysed in the frame of nucleation and growth theories [3, 8]. In a first approximation the linearised J-M-A equation, $\log\log\,[1-X(t)]^{-1} = (n \log k + \log\log e) + n \log t$, was used where X(t) is the transformed fraction. This is defined as $X(t) = [f_\gamma(t) - f_\gamma(0)]/[\,f_\gamma(f) - f_\gamma(0)]$, where $f_\gamma(0)$ is the austenite fraction extrapolated to austempering time zero, while $f_\gamma(t)$ and $f_\gamma(f)$ are the relative fractions of austenite at time t and after the completion of the transformation, respectively. The time exponent *n* is related to the particular nucleation and growth conditions, while the constant rate of the reaction *k* is associated to the rate of nucleation and growth of the new phase.

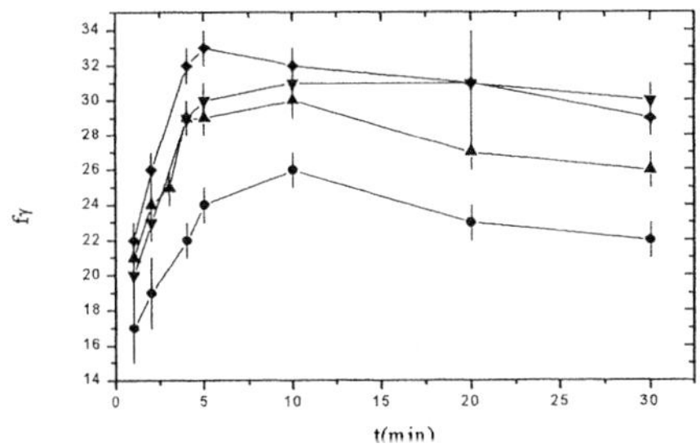

***Figure** 2: Austenite fraction vs. austempering time. Circles, up and down triangles and diamonds correspond to 598K, 623K, 648K, 673K austempering temperatures respectively.*

The resulting fitted n values, listed in Table I, agree with a localised nucleation, and a phase transformation controlled by an interface reaction [3]. After the nucleation sites get saturated, a diffusion process controls the development of the transformation [3]. Fig. 3 shows the obtained k values as a function of the austempering temperature. Two main features can be observed. The increase in the Mn content lessens the nucleation and growth rates, *i.e.*, the transformation slows down.

This effect is mainly due to the decrease in the austenite C diffusion coefficient caused by the Mn [9]. In addition, the parameter k displays a maximum at ≈ 623 K, like it does for low-Mn alloys.The temperature dependence of the nucleation process suggests that as the undercooling ΔT increases, the volume free energy change increases, promoting a higher nucleation rate [10]. Concurrently, according to ref.[11], the growth rate is proportional to the diffusion coefficient of C in austenite. Hence, at high transformation temperatures, the growth rate is high and the nucleation rate is slow; consequently, this last process is the one that controls mainly the transformation rate. The contrary situation occurs at low temperatures, and the reaction becomes mainly dependent on

Temperature	***n***	***k*** **(s^{-1})**
673K	1.7_1	$6.33\ 10^{-3}$
648K	1.5_1	$6.37\ 10^{-3}$
623K	1.5_2	$7.83\ 10^{-3}$
598K	1.3_1	$4.93\ 10^{-3}$

***Table I:** Values of k y n obtained by the Johnson-Mehl's equation for all the temperatures. Errors in n are quoted as subindex.*

***Figure** 3: Lnk vs. 1/T. Circles and squares represent samples with 0.11wt.%Mn and 0.58wt.% Mn, respectively*

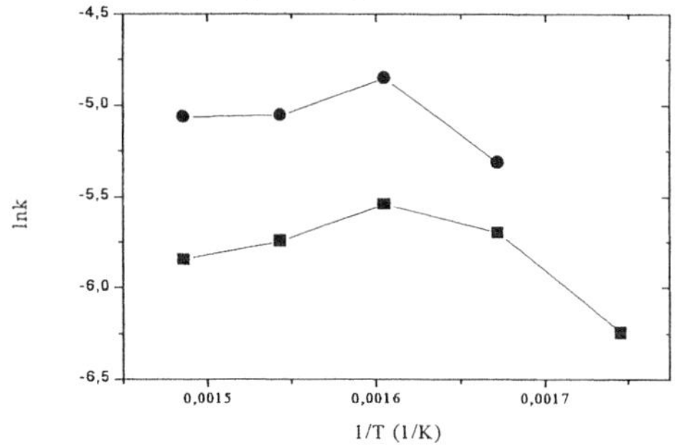

diffusion. The balance between both trends implies that the transformation rate parameter k, should display a maximum at some temperature. This behaviour resembles closely the characteristic "C" curve of the T-T-T diagram [11].

4. Conclusions

We have completed a systematic study of the kinetics of the *stage I* transformation of compacted graphite cast irons for two different Mn contents (0.11wt.% and 0.58 wt.%) for austempering temperatures from 573K to 673 K.

For both alloys and in the whole temperature range, the mechanisms that govern the transformation are a localised nucleation and a phase transformation controlled by an interface reaction.

The growth rate is slowed down for the higher Mn content alloy, probably caused by the decrease in the austenite C diffusion coefficient.

The current results suggest that at low temperatures the transformation is controlled by the growth process, while at higher temperatures it is controlled by nucleation. This evolution is in close relation with the characteristic "C" curve of the T-T-T diagrams.

Acknowledgments

This work was partially supported by PIP 4326 of Consejo Nacional de Investigaciones Científicas y Técnicas (CONICET), PICT 1277 of the Agencia Nacional de Promoción Científica y Tecnológica (ANPCyT), Fundación Antorchas, CICPBA, and Laboratorio de Entrenamiento Multidisciplinario para la Investigación Tecnológica (L.E.M.I.T.-C.I.C.) Argentina.

References

[1] J. Desimoni, R. Gregorutti, K. Laneri, J. L. Sarutti and R. C. Mercader, Metallurgical and Materials Transactions A, , vol 30A, pp. 2745-2752, 1999.

[2] K. Laneri, J. Desimoni, R. C. Mercader, R. Gregorutti and J. L. Sarutti, Metallurgical and Materials Transactions A, vol 32A, p 51, 2001.

[3] J. W. Christian, *The Theory of Transf. in Metals and Alloys*, Pergamon Press, Oxford, U. K., pp 525 – 548, 1965

[4] D. G. Rancourt, A. M. Mc Donald, A. E. Lalonde and J. Y. Ping, American Mineralogist 78, pp 1-7, 1993.

[5] M. Ron: *Applications of Mössbauer Spectroscopy*, vol. II, R. L. Cohen, Academic Press, New York, pp 329 – 388, 1976.

[6] O.N.C.Uwakweh, J.P.Bauer and J.M.Génin: *Metall. Trans. A*21, 589, 1990.

[7] K. Oda, H. Fujimura and H. Ino, J. Phys.: Condens Matter 6), 679,1994.

[8] J. W. Christian, *Physical Metallurgy*, R. W. Cahn editor, North Holland Publishing Company, pp 516-527,1970.

[9] S. S. Babu, H. K. D. H. Bhadeshia, J. of Mat. Scie. Lett. 14, 314, 1995.

[10] R. D. Doherty, *Physical Metallurgy*, vol. 2, R. W. Cahn and P. Haansen editors, North Holland, pp 1363 - 1505, 1996.

[11] J. W. Christian, *Physical Metallurgy*, R. W Cahn editor, North Holland Publishing Company, 1970, pp 516-527.

Ball milling of Fe_3Al-Si powder mixtures

M. Meyer, S. Gialanella[1], A. Maddalena, and G. Principi

Settore Materiali and INFM, Dipartimento di Ingegneria Meccanica, Università di Padova, Italy
[1]Dipartimento di Ingegneria dei Materiali, Università di Trento, Italy

High energy ball milling of powders of a Fe_3Al based alloy obtained by nitrogen gas atomisation mixed with 5 and 10 wt% Si has been studied by Mössbauer spectroscopy and X-ray diffraction. It is observed that milling produces several structural changes depending on milling time and composition of the powder mixture. The starting mixtures display a bcc diffraction pattern with B2 superlattice peaks that vanish after milling for short time. Prolonged milling leads to loss of magnetic order for the higher silicon content, with the formation of a disordered Fe-Al-Si solid solution, which remains almost unchanged by annealing at 1000 K. For the lower silicon content the magnetic order is not lost by milling, which produces a metastable structure partially ordered by subsequent annealing.

1. Introduction

High energy ball milling, initially developed by Benjamin [1] in 1970 for the production of superalloys strengthened by oxide dispersion, has been extensively exploited in recent years for the synthesis of intermetallic compounds [2–5]. This processing technique is inexpensive because performed at room temperature using a simple equipment, can yield comparatively large quantities of material, and produces powders with nanometer-sized grains.

Several papers have been published on ball milling of binary Fe–Al powders [6–11]. Liu et al. [6] reported that a disordered bcc Fe(Al) solid solution instead of the FeAl intermetallic compound tended to form by milling a $Fe_{50}Al_{50}$ powder, due to the low formation enthalpy of the FeAl compound and the relatively high diffusivity of Al in α-Fe. Wolski et al. [7] proposed a local Al melting mechanism to describe the formation of the disordered Fe(Al) solid solution. A bcc Fe(Al) solid solution was reported by Bonetti et al. [8] and Venkatasuwamy et al. [9] also for a milled $Fe_{75}Al_{25}$ powder. Interdiffusion of both elements into each other was claimed by Cardellini et al. [10] and Enzo et al. [11] in the ball milling of $Fe_{75}Al_{25}$ and $Fe_{50}Al_{50}$ systems.

The formation of two ordered phases with B2 and B20 structures has been claimed by milling elemental $Fe_{50}Al_{25}Si_{25}$ powders [12], while two-phase ordered+disordered structures are reported to result by milling low Si, Fe_3Al-based alloys [13]. The aim of this work is to study the influence of different contents of silicon on the final phase formed by ball milling the Fe_3Al-Si system.

2. Experimental

The samples were obtained by milling up to 96 hours in a Fritsch Pulverisette 5 planetary ball mill, under argon atmosphere, powder mixtures of a $Fe_{70}Al_{28}Cr_2$ alloy (kindly supplied by C. G. McKamey of Oak Ridge National Lab.) with 5 wt% and 10 wt% Si (<45μm grain size, 99.5% purity), giving the atomic compositions $Fe_{63}Al_{21}Si_{16}$ and $Fe_{69}Al_{23}Si_8$, respectively. The ball to powder weight ratio was about 13. Samples milled for 96 hours were annealed in argon for 15 minutes at 1000 K.

X-ray diffraction patterns were taken with Bragg-Brentano geometry and data analysed using a full pattern fitting program based on the Rietveld method [14]. Mössbauer spectroscopy measurements were performed in transmission geometry by means of a conventional spectrometer with constant acceleration movement using a ^{57}Co:Rh source having activity of about 5 mCi. The isomer shifts of spectral components were referred to that of an α-Fe standard foil.

3. Results and discussion

The crystalline structure of the Fe_3Al based starting powder resulted from XRD to be a B2 type partially disordered, as a consequence of composition and synthesis method. In fact,

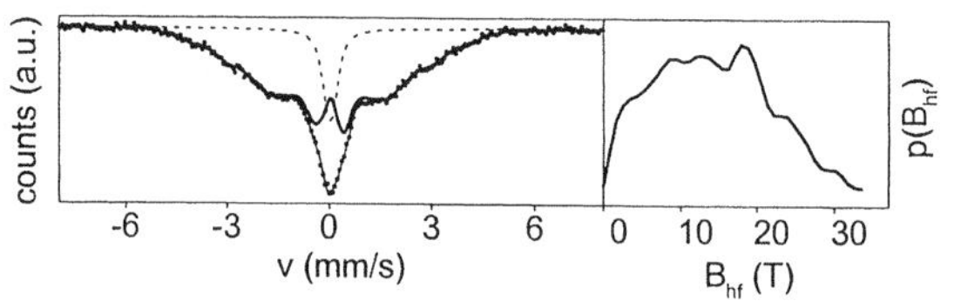

Figure 1. Mössbauer spectrum of Fe_3Al-based starting powder (left) and corresponding hyperfine field distribution (right). Symbols are: data points (circles), best fit (thin line), B2 (dash), hyperfine field distribution (thick line).

nitrogen gas atomisation route is able to quench the high temperature B2 structure of the Fe-Al compound. A high number of iron atoms in antisite positions (replacing aluminium atoms) and vacancies (mostly of iron) are present in the material. The corresponding Mössbauer spectrum, shown in figure 1, is a superposition of a broad non magnetic component and a broad sextet. The non magnetic component can be attributed to the B2 structure, while the broad magnetic component can be attributed to iron atoms in the same structure having iron-rich environments. The spectrum of the same sample, taken at 12.5 K (not reported), is by all means similar to the previous one, so that a superparamagnetic behaviour can be ruled out.

The evolution with milling time of the Mössbauer spectra is very different for the two compositions, as displayed in figures 2 and 3, where also the hyperfine field distributions are shown. As the milling proceeds, the $Fe_{63}Al_{21}Si_{16}$ sample gradually becomes almost fully paramagnetic, while the $Fe_{69}Al_{23}Si_8$ sample becomes fully magnetic. The XRD structural data give as final state in both cases a disordered bcc solid solution with a minor WC contamination (<4%), due to the milling tools. Vanishing of crystalline Si peaks after the first 12 h of milling was observed in both samples.

The Mössbauer spectra of $Fe_{63}Al_{21}Si_{16}$ sample milled up to 48 h can be fitted with a magnetic distribution and a doublet, with isomer shift $\delta\sim0.10$ mm/s and quadrupole splitting $\Delta E_Q\sim0.34$ mm/s, which gradually increases its relative area and becomes predominant after 96 h milling. The almost complete loss of magnetic character as milling proceeds, can be attributed to a progressive incorporation of Si atoms into the intermetallic structure, replacing Fe atoms, in an amount high enough to hinder Fe-Fe magnetic coupling. The Mössbauer spectrum of this sample annealed for 15 minutes at 1000 K can be fitted with an unresolved doublet having $\delta=0.20$ mm/s and $\Delta E_Q=0.24$ mm/s, close to the values of a partially ordered B2 structure [15]. After annealing the Mössbauer spectrum displays a sharpening of the paramagnetic component, as compared to that of the sample just milled. Moreover, the ferromagnetic contribution, still present after 96 h milling, completely disappears. This is an indication of ordering promoted by the thermal treatment. Milling defects are annealed out and Si atoms diffuse and occupy more efficiently the vacancies surrounding Fe atoms, which result therefore less magnetically coupled.

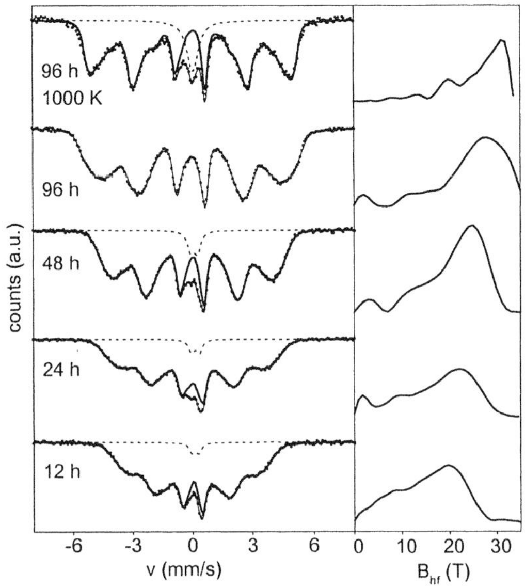

Figure 2. Mössbauer spectrum of $Fe_{69}Al_{23}Si_8$ samples milled for different times (left) and corresponding hyperfine field distributions (right). Symbols are: data points (circles), best fit (thick line), B2 (dash), hyperfine field distribution (thick line). Also the spectrum of sample milled 96 h and annealed is shown.

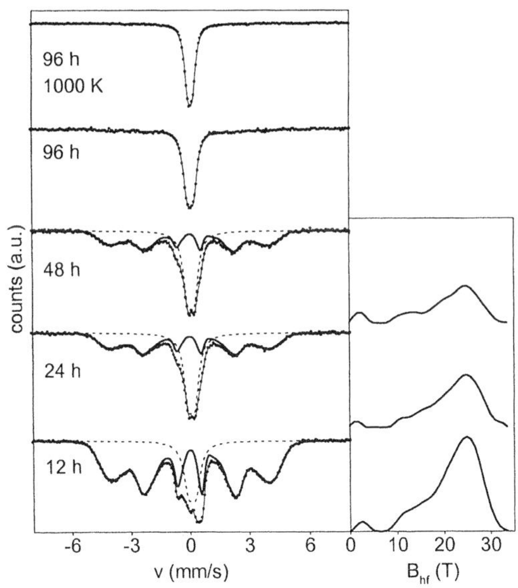

Figure 3. Mössbauer spectrum of $Fe_{63}Al_{21}Si_{16}$ samples milled for different times (left) and corresponding hyperfine field distributions (right). Symbols are: data points (circles), best fit (thin line), B2 (dash), hyperfine field distribution (thick line). Also the spectrum of sample milled 96 h and annealed is shown.

As noted earlier, XRD data indicate that also the $Fe_{69}Al_{23}Si_8$ sample turns into a bcc random solid solution with milling. To take into account the slight asymmetry in the Mössbauer spectra, a correlation between hyperfine magnetic field and isomer shift was used in the fitting procedure. In this way it is found that a decrease with milling time of average isomer shift corresponds to an increase of average hyperfine magnetic field. This is consistent with a higher disorder that leads to an increment in the Fe-Fe coupling.

Figure 3 also shows the Mössbauer spectrum of the $Fe_{69}Al_{23}Si_8$ sample milled up to 96 hours and annealed for 15 minutes at 1000 K, which can be fitted with two components, a hyperfine field distribution and an unresolved doublet, while the spectrum of the sample just milled displays only a hyperfine field distribution. The doublet parameters are close to those of the B2 phase. The hyperfine field distribution sharpens and shifts to the right, indicating an enrichment in Fe around Fe atoms. Two phases are then formed under annealing, an ordered B2 structure (the doublet) and a bcc solid solution (the distribution).

In the case of $Fe_{69}Al_{23}Si_8$ composition, therefore, milling is unable to take the system to an equilibrium state, yielding to the formation of a disordered Fe-Al-Si solid solution which is then turned into a partially ordered B2 phase by annealing. For the sample $Fe_{63}Al_{21}Si_{16}$ the condition achieved after 96 h milling is just refined by annealing without significant microstructural changes.

4. Conclusions

A Fe_3Al based alloy, having a partially ordered B2 structure, has been ball milled mixed with different amounts of Si. It is found that addition of 10 wt% Si gives rise to a non magnetic structure, while addition of 5 wt% Si results in a completely magnetically ordered structure. In both cases the crystalline structure results disordered after milling. Annealing of samples ball milled for long times produces significant microstructural changes only in the sample with less silicon content.

Acknowledgements

This work has been carried out with the financial support of Padova University, Progetti di Ricerca di Ateneo 2000, project entitled "Leghe polifasiche ottenute con macinazione ad alta energia". One of the authors (M.M.) acknowledges a fellowship by CONICET, Argentina.

References

[1] J.S. Benjamin, Metall. Trans. 1 (1970) 2943.
[2] S.J. Hwang, P. Nash, M. Dollar and S. Dymek, Mater. Sci. Forum 88–90 (1992) 611.
[3] M. Jain, T. Christan, Acta Metall. 42 (1994) 1901.
[4] G.H. Chen, C. Suryanarayana and F.H. Froes, Metall. Mater. Trans. 26A (1995) 1379.
[5] C.C. Koch and J.D. Whittenberger, Intermetallics 4 (1996) 339.
[6] Z.G. Liu, J.T. Guo, L.L. He and Z.Q. Hu, Nanostr. Mater. 4 (1994) 787.
[7] K. Wolski, G. Le Caer, P. Delcroix, R. Fillit, F. Thevenot and J. Le Coze, Mater. Sci. Eng. A207 (1996) 97.
[8] E. Bonetti, G. Scipione, G. Valdré, S. Enzo, R. Frattini and R.P. Macrì, J. Mater. Sci. 30 (1995) 2220.
[9] M.A. Venkatasuwamy, J.A. Schneider, J.R. Groza, A.K. Mukher-jee, K. Yamazaki and K. Shoda, Mater. Sci. Eng. A207 (1996) 153.
[10] S. Enzo, R. Frattini, R. Gupta, P.P. Macrì, G. Principi, L. Schiffini and G. Scipione, Acta Mater. 44 (1996) 3105.
[11] F. Cardellini, V. Contini, R. Gupta, G. Mazzone, A. Montone, A. Perin and G. Principi, J. Mater. Sci. 33 (1998) 2519.
[12] V.I. Fadeeva, I.A. Sviridov, G.A. Kochetov and Yu. V. Baldokhin, Mater. Sci. Forum 343-346 (2000) 296.
[13] D.G. Morris and S. Gunther, Mater. Res. Soc. Symp. Proc. 460 (1997) 301.
[14] L. Lutterotti, S. Matthies and H.R. Wenk, Int.Union of Crystallography (IUCr) CPD Newsletter 21 (1999) 14.
[15] M. Meyer, L. Mendoza-Zelis and F.H. Sanchez, Hyperfine Interactions C2 (1997) 213.

MÖSSBAUER SPECTROSCOPY STUDY OF IRON REMOVAL PROCESS DURING ZINC EXTRACTION

A.F. Mulaba-Bafubiandi[1,*], *J. Claassen*[2] *and FB Waanders*[3]

1: Department of Metallurgy, School of Process and Mechanical Engineering, Technikon Witwatersrand, P.O. Box 526, Wits 2050, Johannesburg South Africa
2: Zinc Corporation of South Africa Ltd., P.O. Box 218, Springs 1560, South Africa
3: School of Chemical and Minerals Engineering, PU for CHE, Potchefstroom, South Africa
*: To whom correspondences should be addressed.

Abstract: Zinc metal is commonly produced from the sulphide ore, which is subjected to a roast-leach-electrowinning process. During leaching, metals such as iron are also dissolved, and have to be removed before electrowinning. Most of these oxides are dissolved but mixed zinc-ferrite is produced during roasting, leading to a loss of the precious metal in the tailings. Atomic absorption spectroscopy, Induced Plasma spectroscopy and X-ray techniques identify the crystallised phases present, but Mössbauer spectroscopy assists in the further identification of amorphous iron phases present.
This paper enlightens the added values of the Mössbauer spectroscopy to the identification of crystallised as well as amorphous iron precipitates produced during zinc extraction. Results on zinc-ferrite formation during the roasting process are presented along with the characterisation of iron precipitates removed from the zinc-rich pregnant solution.
Keywords: *Zinc extracting, roasting, electrowinning*

1. Introduction

In South Africa zinc metal is commonly produced from its sulphide ore, Sphalerite (Zn,Fe)S. The concentrate is subjected to the conventional roast-leach-electrowinning route. It contains zinc sulphide with associated iron sulphide being one of the main impurities. The concentrate is roasted (converted from sulphide to oxide) and subsequently leached with dilute sulphuric acid to solubilise the zinc.

During leaching, other metals such as iron are also dissolved, which have to be removed from the zinc-rich solution prior to the electrowinning step. Dissolved iron may be precipitated with any alkaline hydroxide or with the calcine. In the latter case, the roasted concentrate is used as a neutralising agent. Most of the oxides in the calcine are dissolved during this process, but conventional and/or mixed zinc-ferrite ($(Zn_{1-x},M_x)\ OFe_2O_3$) produced during roasting are relatively stable under these conditions [1,2] and subsequently reports to the tailings dam along with the precipitated iron.

Because of the heterogeneous nature of most iron precipitates, various analytical techniques are required to study iron removal processes. X-ray diffraction assists in the identification of crystalline phases present in the material produced during various stages of zinc extraction and recovery but it allows only a semi-quantitative approach and is limited by the identification of amorphous or poorly crystallised phases. Atomic absorption spectroscopy (AAS), Induced Coupled Plasma spectroscopy (ICP) along with X-ray Fluorescence (XRF) techniques assist in the quantification of metals/phases in the integrated zinc metal extraction plant. This is the first report of a Mössbauer investigation on a plant scale and Chen and Dutrizac [1] performed previous laboratory work.

2. Materials and Method

Zinc concentrate, produced by bulk flotation of the sulphide ore, the calcine product obtained from the roasting of the concentrate at 950 ^{0}C for 2hrs and the precipitated product (pH = 2.5) were investigated with XRD (CuK_α) and with transmission Mössbauer Spectroscopy (MS) at room temperature and at 7K. MS experiments were carried out with about 15mg/cm^2 of sample and a Co (Rh) source. The hyperfine parameters are reported with respect to α-Fe.

3. Results and Discussion

The as received run off mine (ROM), the concentrate, the calcine and iron precipitates were measured with the spectroscopic techniques described. The quartz 100% peak ($2\vartheta \sim 16.03^0$) observed in the X-ray diffractogram of the ROM has been substantially reduced in the concentrate, and later extinguished in the calcined product roasted at 900 ^{0}C for 4hrs (see Figure 1). Additional zinc-iron mixed oxide peaks (2ϑ= 20^{0}-65^{0}) are observed in the diffractogram of the roasted product, indicating the presence of zinc-ferrite (Franklinite). An optimal sulphur removal (i.e. 98% sulphur removed) was observed when roasting the concentrate at 900 ^{0}C for 4hrs. This is indicative of a rather dead roasting of the sulphide concentrate

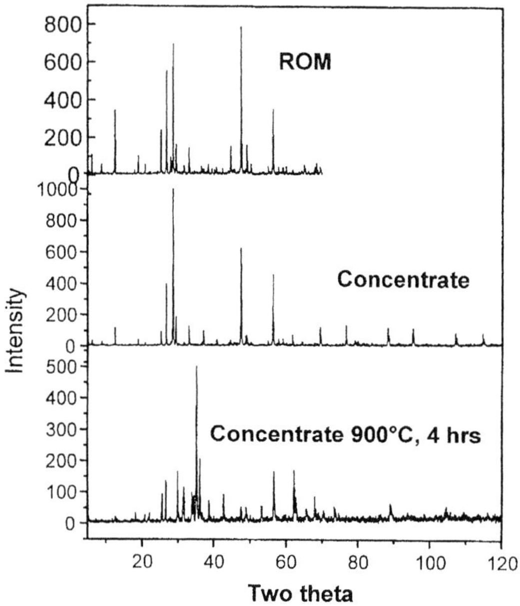

Figure 1: XRD diffractogram of ROM, concentrate and the calcine (900^{0}C, 4hrs). Franklinite (zinc-ferrite) peaks are observed (2ϑ=20^{0} – 65^{0}).

Each MS spectrum was recorded at room temperature (see Figure 2 top) and showed a typical broadened [Γ= 0.33-0.47 mm/s] non-magnetic or paramagnetic doublet indicative of a superposition of highly overlapping sub-component doublets. Identified doublets displayed hyperfine interaction parameter similar to Franklinite, Schwertmannite, Ferrihydrite and Jarosite. In the 7K spectrum (see Figure 2 bottom) one can clearly see the resolution into four sextets for these minerals. The Mössbauer parameters (see Table 1) are in accordance with those reported in literature [3].

It is worth noticing that the MS spectrum of the calcine as it is received from the roasting furnace depicts only Franklinite as the sole iron bearing mineral ($\delta_{/Fe}$=0.35 mm/s, Δ=0.40 mm/s and Γ=0.7 mm/s). This is in agreement with the XRD results and reflects the dead roasting stage achieved at a temperature of 950 ^{0}C for 2hrs.

Water addition to the calcine (slurry formation) and air drying results in a chemical reaction between water and the calcine yielding broader doublets which could be attributed to Franklinite, Schwertmannite and an unknown phase. These latter two phases are probably amorphous as they were hardly detectable from the XRD patterns where peak broadening was observed.

Maintaining acidic conditions (pH=2.5-3) was essential to the dissolution of zinc. Iron compounds were precipitated by addition of the calcine to the pregnant solution. MS spectra of precipitates show at least three components:

(a) Zinc-ferrite as the most abundant phase in most of the cases.
(b) A second component with a larger quadrupole splitting value at RT, ascribed to Jarosite or a combination of Jarosite and a unknown poorly crystalline/amorphous phases.

TABLE 1 : Phase identification from nuclear hyperfine interaction parameters obtained from fits to the data of various samples

	297 K (RT- room temperature)						7 K (LT – liquid-helium)					
Sample	$\delta^{*}_{/Fe}$ (mm/s)	Δ (mm/s)	Γ (mm/s)	B_{hf} (Tesla)	Abundance (%)	Attribution	$\delta_{/Fe}$ (mm/s)	Δ (mm/s)	Γ (mm/s)	B_{hf} (Tesla)	Abundance (%)	**Attribution**
Precip 1 pH=2.5	0.32 0.35 0.35	0.41 0.66 1.07	0.33 0.34 0.39	- - -	29 48 23	Franklinite Schwertmannite Jarosite+ unknown	0.43 0.46 0.46 0.46	-0.03 -0.15 -0.12 0	0.67 0.62 0.40* 0.40	49 45 42 39	29 47 15 9	Franklinite Schwertmannite Jarosite+ unknown
Precip 2 pH=3	0.34 0.38 0.32	0.42 0.69 1.05	0.33 0.37 0.39	- - -	18 59 23	Franklinite Schwertmannite Jarosite+ unknown	0.42 0.44 0.43 0.50	-0.13 -0.21 -0.25 -0.24	0.6 0.60 0.40* 0.40	49 45 43 39	24 53 15 8	Franklinite Schwertmannite Jarosite+ unknown
Precip 3 pH=2.5	0.35 0.37 0.37	0.41 0.69 1.1	0.35 0.36 0.47	- - -	25 45 30	Franklinite Schwertmannite Jarosite+ unknown						
Calcine (oven)	0.35	0.45	0.37	-	100	Franklinite $Zn_xFe_{3-x}O_4$						
Calcine slurried +air dried	0.34 0.37 0.37	0.42 0.69 1.10	0.35 0.37 0.40	- - -	40 42 18	Franklinite Schwertmannite Unknown						

Notes:

* Nuclear Hyperfine Interaction Parameters

δ/Fe = isomer shift wrt Fe, Δ = quadrupole splitting, Γ = linewidth, B_{hf} = internal magnetic field

Typical errors are 5-10 %

δ: Isomer shift values are quoted relative to the α-Fe.

Γ: In an ideal sample which is highly crystalline and homogeneous, linewidth values of Γ~0.28 mm/s are expected, as determined by properties of the Mössbauer nucleus and instrumental broadening effects.

•Abundance, obtained from the integrated area under the spectral component.

Typical errors may range from 5 – 15 % of the quoted value.

* Fixed parameter value, to obtain consistency between fits at RT and LT. The two sextets with fixed relatively narrow LW values of ~0.4 mm/s may be replaced by a single sextet with a much broader linewidth.

(c) At low temperature the Jarosite and an additional unknown phase associated with components having the smallest internal magnetic fields of ~42 T and ~39 T respectively. These values are smaller than the typical values for Jarosite cited in the literature, and therefore may be an indication of the effect of poor crystallinity or the effect of impurities within the local environment of the Fe atom, compared to the ideal cases

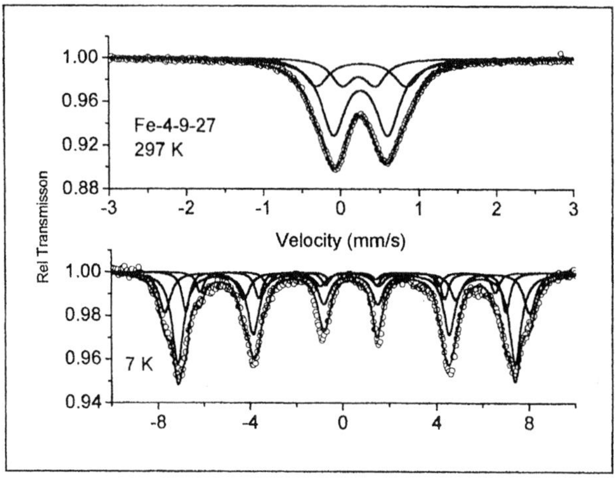

Figure 2. Room temperature and 7K Mössbauer spectroscopy spectra of the pH=2.5 precipitate. Jarosite has been observed in addition to Franklinite and Schwartzmannite and Ferrihydrite

4. Conclusions

The iron removal process in a sulphide ore could be studied with the combination of XRD and Mössbauer spectroscopy. The concentrate was produced by flotation and a dead roasting process was carried out. The produced calcine was acid leached and the same calcine was used as the neutralizing agent. Zinc and iron were dissolved and iron precipitated by calcine addition due to the pH change.. Franklinite present in the initial calcine was observed in all the precipitates in addition to the jarosites. Hyperfine interaction parameters similar to plumbo-jarosite were observed indicating that the ore contained lead as one of the primary impurities.

Acknowledgement

The logistic support of the University of the Witwatersrand is greatly appreciated. Thanks also goes to F. Sibiya for performing some of the experimental work performed.

References

[1] Chen, T.T. and Dutrizac, J.E., Practical; Mineralogical Techniques for the characterisation of hydrometallurgical products, Process Mineralogy IX, edited by William Petruk et al., The Minerals, Metals & Materials Society, (1990), 289.

[2] Nii, K. and Hisamatsu, Y., The promotion of acid dissolution of zinc ferrite, Transaction of National research institute for metals vol.. 8 no 5, (1966), 27.

[3] Stevens, J.G., Khasanov, A.M., Miller, J.W., Pollak, H. and Li, Z., Mössbauer Mineral Handbook, Mössbauer Effect Data Center, (1998), 527p.

EFFECT OF HEAT TREATMENT ON THE MECHANICAL ALLOYED Fe_2B

E. C. Passamani*, J. R. B. Tagarro and C. Larica
Universidade Federal do Espírito Santo – Depto. de Física –Vitória/ES/Brazil, 29060900

High energy milling process has been used to prepare the Fe_2B alloy. X-ray, DSC and Mössbauer data of the 310 h milled alloy (sample A) show the presence of three phases: small fractions of α-Fe (n-Fe) and Fe_2B (n-Fe_2B) nanocrystallines phases and a large amount of amorphous Fe_2B (a-Fe_2B) phase (73%). The heat treatment of the milled alloy was done either in a high vacuum furnace (sample B), with in situ Mössbauer measurements, or by annealing the sample sealed in a quartz tube (sample C). Mössbauer data of sample B display an increasing content of the n-Fe and n-Fe_2B phases for heat treatment temperatures (HTT) below 573K; an intriguing temperature stabilization of these phases seems to occur in the range from 573 K up to 673 K. For HTT≥ 673 K, a reduction of n-Fe phase fraction is observed while the n-Fe_2B phase remains almost unchanged. On the other hand, x-ray and Mössbauer data of sample C show a continuously increase of the nanocrystalline phase fractions with HTT. Thus, the presence of oxygen in the residual atmosphere during the heat treatment may have enhanced the segregation effects, possibly due the boron oxidation. In both annealed samples, B and C, it has been observed the presence of tet-Fe_3B phase.

Crystalline and amorphous Fe/B and related alloys exhibit attractive soft magnetic properties being an important system to be investigated. Fe/B amorphous alloys (a-Fe/B) have been produced by a variety of methods [1-2]. Thermal treatments applied to a-Fe/B alloys have yielded formation of α-Fe and Fe_3B phases [3]. The mechanism of segregation and crystallization observed is not well known. In this paper, it will be shown the results of mechanical alloyed Fe/B, in the 2:1 composition and its annealing, prepared under different conditions of atmosphere and temperature, focused on the obtained amorphous phase.

A nanostructured $Fe_{67}B_{33}$ (Fe_2B) alloy was obtained by milling 14 g of a mixture of high purity (99,999%) elemental powders of crystalline α-Fe and amorphous B, during 310 h (sample A). The milling process has been done inside a hard steel vial, clamped in a commercial vibrating frame machine, using a massive cylinder made of the same material as the vial for milling tool. More details of the milling tools and experimental procedures have been already reported by Larica et al.[4].

Part of sample A was sealed inside a high vacuum Mössbauer furnace and heat treated in the temperature range from 350 K up to 823 K with a turbo molecular pump continuously running to keep the vacuum better than $7x10^{-6}$ Torr during the annealing. The sample was submitted for 2 h heat treatments, at several temperature values apart 50 K and, after each temperature step the sample was cooled down to 300 K for Mössbauer measurement (sample B). Three other parts of sample A were enveloped in Ta foil and then encapsulated in separated quartz tubes, with initial vacuum better than $7x10^{-6}$ Torr, for heat treatment at 423 K, 623 K and 923 K, in a tubular resistive furnace, during approximately 2 h (sample C).

EDS analysis has been done in sample A and no Cr contamination has been detected. X-ray diffraction patterns (XRD) were obtained for all samples with the Kα-Cu radiation from a Rigaku diffractometer.

All ^{57}Fe Mössbauer spectra of the as milled and the annealed Fe_2B powders were measured at 300 K with a 25 mCi ^{57}Co:***Rh*** radioactive source in conventional transmission geometry. All spectra shown here are relative to the gamma source and mentioned isomer shift values are relative to α-Fe.

The XRD pattern of the sample A (not shown) displays broad reflection lines at angular positions of the α-Fe and Fe_2B phases. However, DSC measurement (also not shown) presents an exothermic peak at about 835 K, which is generally associated with an amorphous-crystalline transition.

Fig. 1 displays the Mössbauer spectra taken at 300 K of the sample A and sample B, the latter annealed at different temperatures. Sample A was fitted with three magnetic subspectra: two sextets (S1 and S2), having its hyperfine parameters close to those of α-Fe (IS $\approx$ 0.0 mm/s and $B_{hf} \approx$ 33 T) and Fe_2B (IS $\approx$ 0.17 mm/s, $B_{hf} \approx$ 23.5 T and QS $\approx$ 0.04 mm/s) phases, respectively, in agreement with the XRD results, plus a magnetic hyperfine field distribution subspectrum, having the most probable B_{hf} and IS values close to those hyperfine values of the S2 subspectrum (Fe_2B). Therefore, the latter one may be associated with an a-Fe_2B, consistent with the DSC exothermic peak observed at 835 K.

The 300 K Mössbauer spectra of sample B have been fitted with the same three magnetic subspectra used to fit the sample A, but varying their relative volume fraction for different annealing temperatures.

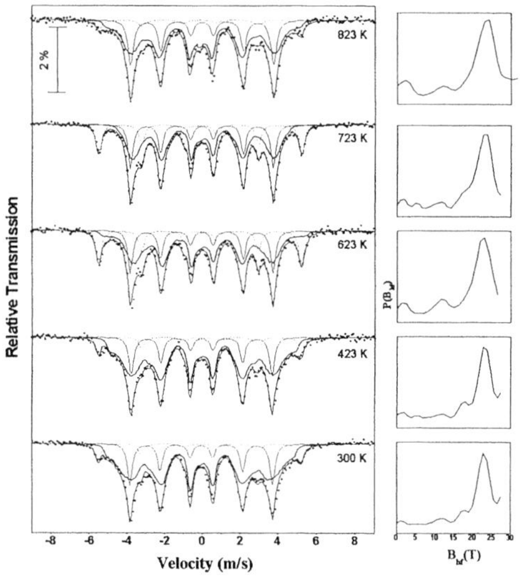

Figure 1 – 300 K Mössbauer spectra for as milled $Fe_{67}B_{33}$ alloy (sample A) and the same sample annealed at different temperature in high vacuum Mössbauer furnace (sample B).

Fig. 2 summarizes the thermal behavior of the alloy observed phases, during the annealing process, obtained from in situ Mössbauer measurements performed in a high vacuum furnace. The results indicate that the a-Fe_2B phase has its relative volume fraction reduced from 73 % to about of 50 %, while the fraction of bcc-Fe phase has increased from 10 % to 23 %, for annealing temperature values up to 573 K. For annealing temperatures above 675 K, the fraction of the bcc-Fe phase decreases to 12 %, while the relative area associated with the a-Fe_2B phase increases back to about 60 %. The high temperature increase of the a-Fe/B phase content is intriguing and may be due to: (a) the n-Fe phase may had reacted with free boron atoms in the annealing process, forming a rich Fe/B phase or (b) the Fe atoms from the n-Fe phase may had diffused into a-Fe_2B phase, increasing the Fe alloy content. These arguments may be

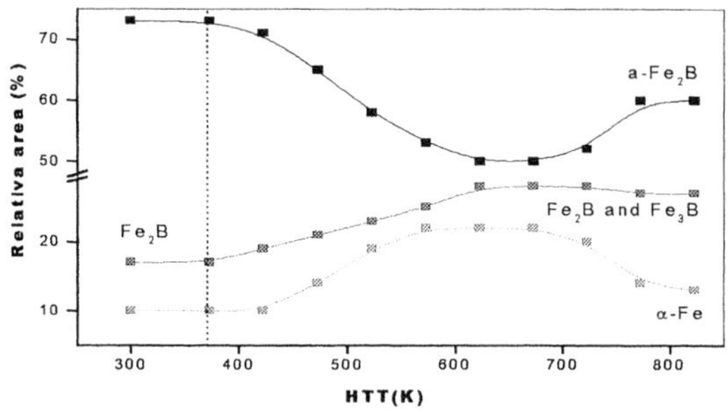

Figure 2 – Temperature dependence behavior of the Mössbauer relative absorption area for different phases observed in the as milled $Fe_{0.67}B_{0.33}$ alloy submitted to an annealing process in high vacuum Mössbauer furnace (sample A). The vertical dot line separates the region of n-Fe_2B from that one, where coexist the two phases: Fe_2B and Fe_3B. The lines connecting the points are guide to the eyes.

related to the observed slightly broader magnetic hyperfine field distribution curve of sample B annealed at 823 K compared, for example, with that annealed at 723 K (Fig. 1 – at right side). On the other hand, the initial n-Fe_2B phase shows a systematic increase of its relative volume fraction for annealing temperatures lower than 623 K, but it remains unchanged for higher temperatures. This temperature behavior observed for the S2 subspectrum is mainly associated with the appearance of the Fe_3B phase, detected by XRD pattern of the sample B after the annealing at 823 K.

Fig. 3 shows the Mössbauer spectra of the sample C obtained at 300 K, for three different annealing temperatures.

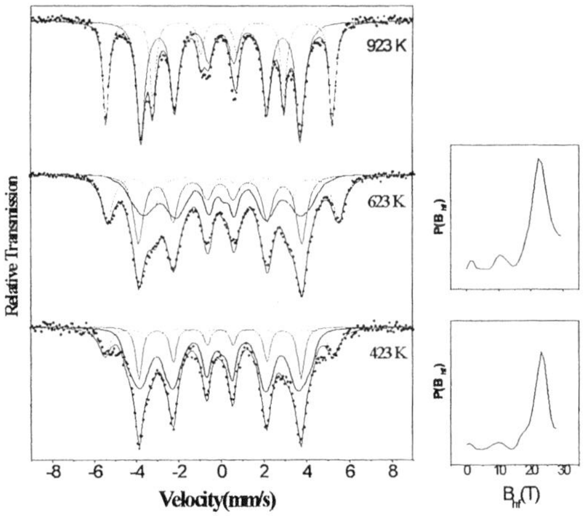

Figure 3 – 300 K Mössbauer spectra of the as milled $Fe_{67}B_{33}$ alloy sealed under vacuum in a quartz tube and annealed at different temperatures in a tubular resistive furnace (sample C).

The Mössbauer spectra of sample C, annealed at temperatures lower than 923 K, are analyzed with the same three subspectra used to fit the Mössbauer spectra of the samples A and B, but with different Mössbauer relative volume fractions. In this case, the content of α-Fe phase increases with the annealing temperature (Fig. 3).

Furthermore, comparing the relative volume fraction of α-Fe phase observed in the samples B and C, at similar annealing temperatures, for an example at 623 K, the alloy annealed in a quartz tube presents a Fe segregation effect higher than in the case of the alloy annealed in the Mössbauer furnace. This observation may indicates that the Fe segregation effect may had been enhanced by the presence of oxygen in the residual atmosphere, during the annealing process, possibly leading to an oxidation of boron atoms. Accordingly, Linderoth [3] has also observed the boron oxidation process in ultrafine a-Fe/B alloy prepared by chemical method and annealed at similar temperature range. Thus, better residual pressure expected for sample B suggests a lower Fe segregation and Fe_3B crystallization comparing with that observed also in sample C.

Yet, concerning the annealing process of the samples B and C, it deserves further comments: (a) the XRD diffractograms give evidences of the Fe_3B phase present in the samples annealed at temperatures about 623 K. However, from the correspondent Mössbauer spectra obtained at 300 K (Figs. 1 and 3) it cannot be distinguished between the Fe_2B and Fe_3B contributions due to the similarity of their hyperfine parameter values [5]. In this case, the Mössbauer fittings were done with one broad sextet, convoluting these two-phase subspectra. The formation of the Fe_3B phase may have occurred in the temperature range from 373 K to 623 K in sample B since, in these temperatures, the relative volume fraction of S2 subspectrum is enhanced for increasing temperature. By assuming the early surface crystallization effect indicated by the DSC analysis (lower exothermic peaks) and the presence of residual oxygen in the local annealing atmosphere, the boron oxidation process may have occurred at temperatures lower than 600 K, yielding rich Fe phases, i.e. α-Fe and Fe_3B ones.

In the present work, thermal studies of Fe_2B alloy composed by 73 % of a-Fe_2B, 10 % of n-Fe and 17 % of n-Fe_2B have been done in two different annealing environments. Higher Fe segregation effect is observed in the case of samples heat treated in sealed quartz tube (sample C) when compared with the case where the sample atmosphere has been continuously pumped during the heat treatment (sample B). Thus, the higher segregate Fe fraction reported in the literature for Fe_2B alloy obtained by milling technique [6], or yielded during its annealing process [3], may had been enhanced due to boron oxidation caused by contaminated residual atmosphere.

REFERENCES:

[1] – C.L. Chien, D. Musser, E.M. Gyorgy, R.C. Sherwood, H.S. Chen, F.E. Luborsky and J.L. Walter, Phys. Rev. B20 (1979) 283.
[2] – Y.Z. Yang, Q.S. Li, Y.L. Zhu, X.M. Ma and Y.D. Dong, J. of Mat. & Technology 15 (1999) 137.
[3] – S. Linderoth , J. Mag. and Magn. Materials **104-107** (1992) 128.
[4] – C. Larica, K.M.B. Alves, E. Baggio-Saitovitch, A.P. Guimarães, J. Mag. and Magn. Materials 145 (1995) 306.
[5] – F.H. Shánchez, Y.D. Zhang, J.I. Budwick and R. Haseguawa, J. Appl. Physics 66 (1989) 1671.
[6] – J. Balogh, L. Bujdoso, T. Kemeny and I. Vincze, J. of Phys.: Condensed Matter **9**, (1997) L503.

Mössbauer effect studies of $Dy(Fe_{0.7-x}Al_xCo_{0.3})_2$ and $Dy(Fe_{0.7}Co_{0.3-x}Al_x)_2$ compounds

J. Pszczoła[1], P. Stoch[1], L. Dąbrowski[2], J. Suwalski[2] and A. Pańta[3]

[1]Solid State Physics Department, AGH, Al. Mickiewicza 30, 30-059 Kraków, Poland

[2]Institute of Atomic Energy, 05-400 Świerk-Otwock, Poland

[3]Department of Metallurgy and Materials Engineering, AGH, Al. Mickiewicza 30, 30-059 Kraków, Poland

It was previously found that magnetic hyperfine fields H_{hf} (observed at ^{57}Fe nuclei, 4.2K) in the $Dy(Mn_{1-x}Fe_x)_2$ and $Dy(Fe_{1-x}Co_x)_2$ series form a Slater - Pauling curve with a maximum hyperfine field appearing for a $Dy(Fe_{0.7}Co_{0.3})_2$ compound. An Fe/Al (or Co/Al)-substitution is a method to modify hyperfine interactions and thus to probe 3d-bands of the compounds. In order to study a significance of iron or cobalt, the series $Dy(Fe_{0.7}Co_{0.3-x}Al_x)_2$ was prepared and ^{57}Fe Mössbauer effect measurements were performed. The obtained magnetic hyperfine fields create new branch of the Slater-Pauling curve. The data are compared to the branch observed previously for the $Dy(Fe_{0.7-x}Al_xCo_{0.3})_2$ series. Possible conclusions related to the 3d-band structure are considered.

1. Introduction

Ferrimagnetism of the heavy rare earth - transition metal compounds is a result of the coexistence between the 4f (5d) and 3d magnetism [1]. Magnetic hyperfine field $\mu_0 H_{hf}$ (μ_0 is the magnetic permeability), studied at ^{57}Fe nuclei (4.2K) in the $Dy(M\text{-}N)_2$ (M-N) = Mn-Fe, Fe-Co) intermetallics, treated as a function of the average number n of 3d electrons, forms a Slater - Pauling curve with a maximum field appearing for the $Dy(Fe_{0.7}Co_{0.3})_2$ compound [2,3]. A significance of iron atoms in the series $Dy(Fe_{0.7-x}Al_xCo_{0.3})_2$ with the Fe/Al substitution was previously studied using the ^{57}Fe Mössbauer effect [4]. In order to study a significance of cobalt atoms the ^{57}Fe Mössbauer effect measurements were performed for the $Dy(Fe_{0.7}Co_{0.3-x}Al_x)_2$ series. The $\mu_0 H_{hf}$(n)-dependencies for the Fe/Al and the Co/Al substituted series are compared and discussed.

2. Experimental

The series $Dy(Fe_{0.7}Co_{0.3-x}Al_x)_2$ (x = 0, 0.10, 0.15 and 0.20) was prepared by arc melting and X-ray diffraction patterns (300K) confirm the pure cubic, Fd3m (C15) Laves crystal phase for all compounds. The lattice parameter changes from 7.3008(8)Å for x=0 to 7.4510(16)Å for x=0.25. The average hyperfine interaction parameters were obtained from the good quality Mössbauer effect spectra (4.2K) following procedure described elsewhere [5]. The magnetic hyperfine field $\mu_0 H_{hf}$ decreases from 24.93(5)T for $Dy(Fe_{0.7}Co_{0.3})_2$ to 14.64(13)T for x=0.25. The isomer shift IS (with respect to pure Fe-metal, at 300K) increases from 0.061(6)mm/s for x=0 to 0.125(23)mm/s for x= 0.25. The quadrupole interaction parameter [5] eqQ/2 = 0.251(6)mm/s for x=0 and is constant with x.

3. The Slater-Pauling curve

The Slater-Pauling curve $\mu_0 H_{hf}(n)$ observed for the $Dy(M\text{-}M)_2$ compounds is presented in figure 1 (curve 1) for a comparison with the data of the studied series.

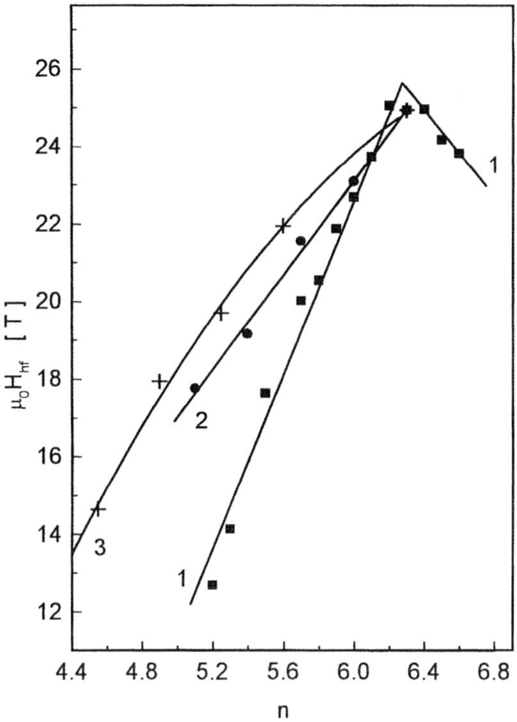

Figure 1. Magnetic hyperfine fields $\mu_0 H_{hf}(n)$ (4.2K) for series: $Dy(M\text{-}N)_2$ (M-N=Mn-Fe, Fe-Co) (1), $Dy(Fe_{0.7-x}Al_xCo_{0.3})_2$ (2) and $Dy(Fe_{0.7}Co_{0.3-x}Al_x)_2$ (3).

The fields $\mu_0 H_{hf}(n)$ for the $Dy(Fe_{0.7-x}Al_xCo_{0.3})_2$ series (curve 2) and for the studied $Dy(Fe_{0.7}Co_{0.3-x}Al_x)_2$ series (curve 3) form new branches of the Slater-Pauling curve. Curves are fits. It can be seen that the field $\mu_0 H_{hf}(n)$ [curve 2, $n = (0.7-x)*6+0.3*7$, where 6,7 is the number of 3d electrons for Fe, Co atom, respectively] observed for the $Dy(Fe_{0.7-x}Al_xCo_{0.3})_2$

series is more strongly reduced with n as compared to the field $\mu_0 H_{hf}(n)$ [curve 3, n=0.7*6+(0.3-x)7] observed for the $Dy(Fe_{0.7}Co_{0.3-x}Al_x)$ series.

4. Summary

The Al-substitution reduces the average number u(x) of the magnetic nearest neighbours surrounding the probed Fe-atom and thus reduces the energy shift $\Delta E \sim J_{M\text{-}M}\, u(x)\, m_M$ beetween the 3d-subbands, where $J_{M\text{-}M}$ is the exchange integral and m_M is an average magnetic moment per 3d atom. As a result, the 3d electrons should become gradually redistributed over the 3d-subbands and the difference between the ρ^+_{3d} (spin-up) and ρ^-_{3d} (spin-down) densities should become reduced step by step with x. Consequently the magnetic moment m_M and thus the magnetic hyperfine field $\mu_0 H_{hf}$ should decrease and finally the branch of the Slater-Pauling curve, is observed. The Fe/Al substitution reduces more strongly $\mu_0 H_{hf}(n)$ with the n-decreasing as compared to the Co/Al substitution.

Literature

[1] K.N.R. Taylor, Adv. Phys. 20(1971)551-675.

[2] B. Gicala, J. Pszczoła, Z. Kucharski and J. Suwalski, Phys. Lett. A 185(1994)491-494.

[3] B. Gicala, J. Pszczoła, Z. Kucharski and J. Suwalski, Solid State Commun. 96(1995)511-515.

[4] J. Pszczoła, P. Stoch, L. Dąbrowski, J. Suwalski and A. Pańta, J. Alloys and Comp. (in print).

[5] J. Pszczoła, B. Winiarska, J. Suwalski and Z. Kucharski, J. Alloys and Compounds, 265(1998)15-22.

Structural and magnetic characteristics of $Fe_{1-x}Co_x$ alloys containing 12.5 at % Nb

M.T. Raposo[1], A.I.C. Persiano[2*] and J.D. Ardisson[3]

1- DCNAT-FUNREI – CP 110 , 36.300-000 – São João Del Rei, MG. Brazil.
2- Departamento de Fisica,- UFMG, CP 702, 30123-970 Belo Horizonte-MG, Brazil.
3- CNEN-CDTN, CP 941, 30.123-970 – Belo Horizonte, MG. Brazil.

The effects on the structure and magnetic properties of the phases in $(Fe_{1-x}Co_x)_{0.875}Nb_{0.125}$ alloys due to the replacement of Fe by Co are studied by ^{57}Fe Mössbauer spectroscopy, X-ray diffraction, electron probe microanalysis and vibrating sample magnetometry. The results reveal the presence of two phases: α, based on the bcc α-Fe for $x \leq 0.75$ (or γ-Co for x=1; both ferromagnetic at RT) and ε, paramagnetic at RT, based on the $(Fe \text{ or } Co)_2Nb$ Laves phases with distinct characteristics that change with x. The EPMA technique shows that Nb has a strict solubility (< 0.5 at%) in the Fe+Co α phase up to x~0.75 while γ-Co dissolves about 4 at% Nb. Up to x=0.75 the Mössbauer data obtained at RT show that α is magnetic with maximum B_{hf} at x~0.25 while ε is paramagnetic exhibiting two non equivalent iron sites with population 3:1, for x=0, that drops with increasing x. The MS spectra for the samples in that range, measured at 20K, show that ε is magnetic, with increasing B_{hf} up to x=0.50, which shows that the presence of Co also plays an important role on T_c and the magnetic coupling of that phase. The VSM measurements show that the average magnetic moments per atom in α-Fe, α-(Fe+Co) or γ-Co phases at RT decrease linearly with increasing % Nb dissolved in those phases.

1. Introduction

The Fe-Co system is the base of interesting magnetic alloys that can be used in many applications such as the design of electric vehicles motors and generators operating at high temperatures. Its high Curie temperatures, T_c ~ 900°C and top saturation magnetization M_s, make the material adequate for that and other high M and T applications [1]. Its well known brittleness is due to ordering that transforms α (bcc) into α′ (B2 superstructure) which is overcome by ternary additions like V [2] and Nb [3-5] The former is the basis of the Hiperco family and the latter is object of our interest in this work. In this investigation we produced alloys $(Fe_{1-x}Co_x)_{0,875}Nb_{0.125}$ with x = 0, 0.25, 0.50, 0.75 and 1.0 in order to determine the phases present in this composition range and characterize them structurally and magnetically.

* Corresponding author: persiano@fisica.ufmg.br

2. Experimental

The primary constituents with 99.99% purity were arc melted under Ar atmosphere. The buttons were encapsulated into quartz tubes, annealed at 1,100°C for 240h followed by an additional annealing at 800°C for 48h and quenched in iced water. Measurements at room temperature (RT) by Electron Probe Microanalysis (EPMA), X-Ray Diffraction (XRD), Vibrating Sample Magnetometry (VSM) and ^{57}Fe Mössbauer Spectroscopy (MS) at RT and 20K in a conventional spectrometer with constant acceleration using a ^{57}Co in Rh source.

3. Results and discussion

The XRD and EPMA data revealed the existence of 2 phases with distinct characteristics dependent on x. For x=0: α-Fe (bcc) and εFe_2Nb (hcp); between x=0.25 and 0.75: α-(Fe+Co) and ε-$(Fe+Co)_2Nb$ and for x=1: γ-Co (fcc) and Co_2Nb. The micrographs in Fig.1 (a-e) are secondary electrons images showing the referred phases in conditions close to the eutetic composition (~ 12.5 at% Nb, independent of x). The dark phases correspond to α-(Fe/Fe+Co) or γ-Co while the light phases correspond to ε-$(Fe/Fe+Co/Co)_2Nb$.

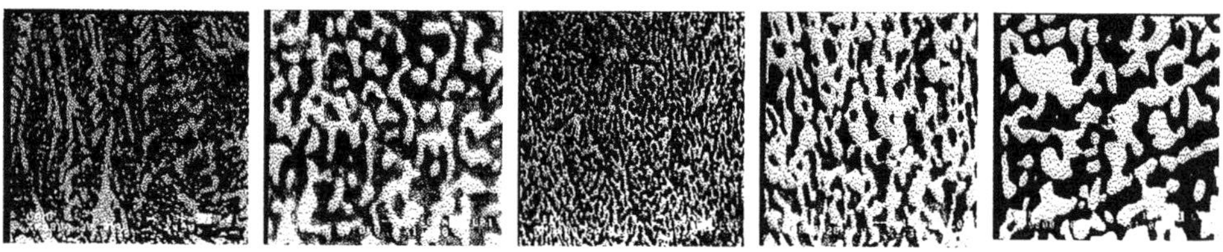

(a) x=0.00 (b) x=0.25 (c) x=0.50 (d) x=0.75 (e) x=1.00

Figure 1. Secondary electrons images (400 X) of $(Fe_{1-x}Co_x)_{0,875}Nb_{0.125}$ for x = 0.00, 0.25, 0.50, 0.75 and 1.00. The dark phases correspond to α-(Fe/Fe+Co) or γ-Co while the light phases correspond to ε-$(Fe/Fe+Co/Co)_2Nb$.

The EPMA results revealed that Nb has a very limited solubility in the α-phase - about 0.4 at.% - while γ-Co dissolves about 4 at.% Nb. Those solute contents in the matrix of α-(Fe/Fe+Co)+Nb and γ-Co+Nb can explain the drops observed in the saturation magnetization M_s of the ternary phases relative to the respective binary $Fe_{1-x}Co_x$ alloys. Indeed the VSM results for the average magnetic moment of $Fe_{1-x}Co_x$ (+ ~ 0.4 at%Nb) alloys in the range $0 \leq x \leq 0.75$ follow closely (~ 0.02 μ_{Bohr} below) the well known Slater-Pauling Plot for the binary $Fe_{1-x}Co_x$ series; on the other hand, for x = 1, the drop observed in that parameter follow the same trend, corresponding to a constant rate of - 0.09 μ_{Bohr} / at%Nb dissolved in both structures.

The presence of 2 phases was also observed in both families of ^{57}Fe Mössbauer spectra taken at RT and 20K - see Figures 2 and 3, respectively.

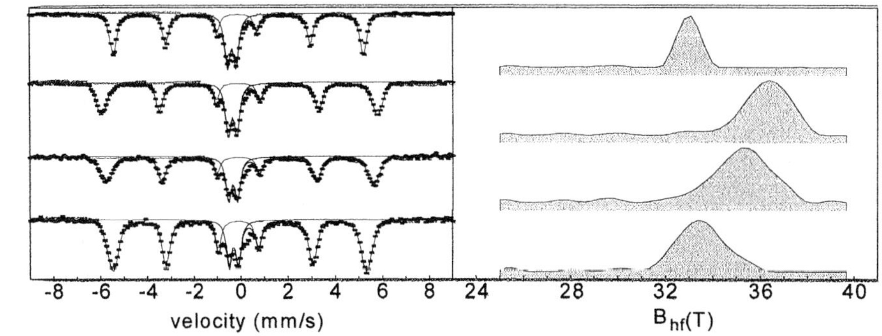

Figure 2. ^{57}Fe Mössbauer spectra and hyperfine field (B_{hf}) distribution histogram of the phases in $(Fe_{1-x}Co_x)_{0.875}Nb_{0.125}$ alloys at RT. The 2 sets of subspectra reveal the α (sextets) and ε (doublets) phases.

The evolution of the average $<B_{hf}>$ of α with x (see Fig. 2 and Tab. 1) exhibits a maximum at x ~ 0.25, typical of the Fe-Co system; the large breadth of the B_{hf} distribution histograms observed for x > 0 reflects the variety of environments Co/Fe in the bcc structure and demonstrates that the material is in the disordered condition. Despite the reliable EPMA results, the MS spectral analyses did not evidence the presence of Nb dissolved in the α.phase. The data displayed in Tab.1 suggest that the ε phase observed in the ternary system ($0.25 \leq x \leq 0.75$) is similar to the Fe_2Nb Laves phase. This was also confirmed by the XRD. The Mössbauer spectrum of this phase was analyzed in terms of two sub-spectral components corresponding to distinct iron sites with relative population 3:1, for x=0, that drops with increasing x - see Table1.

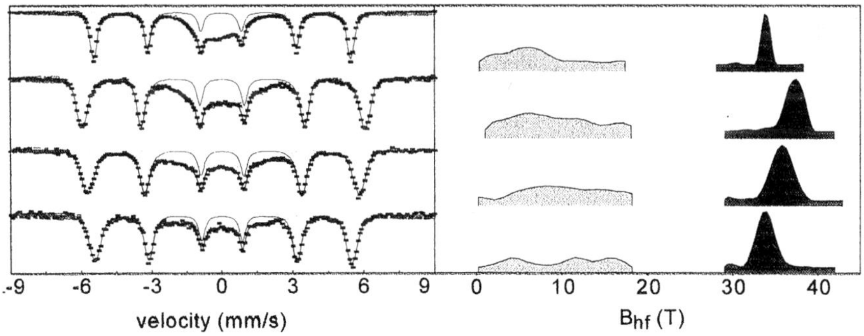

Figure 3. ^{57}Fe Mössbauer spectra and hyperfine field (B_{hf}) distribution histogram of the phases present in $(Fe_{1-x}Co_x)_{0.875}Nb_{0.125}$ alloys at 20K. Note that ε is magnetic at this temperature. Although not well defined, the B_{hf} distribution of ε indicates that the Co content plays an important role on the magnetization of that phase-see Table 1.

300K	Alpha phase				Epsilon phase - site 1			Epsilon phase - site 2		
x	IS (mm/s)	QS (mm/s)	$\langle B_{hf} \rangle$ (T)	Spectr area %	IS (mm/s)	QS (mm/s)	Spectr area %	IS (mm/s)	QS (mm/s)	Spectr Area %
0.00	-0,11	0,01	32,9	66	- 0,36	0,38	27	- 0,14	0,43	7
0.25	-0,08	0,01	37,0	68	- 0,35	0,35	23	- 0,14	0,42	9
0.50	-0,05	0,02	35,5	70	- 0,34	0,34	21	- 0,09	0,44	9
0.75	-0,06	0,02	33,4	78	- 0,32	0,32	14	- 0,01	0,48	8
20K	Alpha phase				Epsilon phase – sites 1 and 2					
x	IS (mm/s)	QS (mm/s)	$\langle B_{hf} \rangle$ (T)	Spectr area %	IS (mm/s)	QS (mm/s)	$\langle B_{hf} \rangle$ (T)	Spectr area %		
0.00	0,01	0,01	33,7	60	- 0,22	0,05	5,4	40		
0.25	0,04	0,02	36,9	64	- 0,28	0,04	8,9	36		
0.50	0,08	0,00	35,5	68	- 0,56	0,24	10,4	32		
0.75	0,07	0,02	33,9	77	- 0,10	0,06	10,1	23		

Table 1. The main hyperfine parameters of the α and ε phases at 300K and at 20K. The typical errors of the hyperfine parameters are: IS $\pm$ 0.02 mm/s; QS $\pm$ 0.02 mm/s; $\langle B_{hf} \rangle \pm 0.2$T; spectral areas $\pm$ 3%

Note that ε is magnetic at 20K and the average $\langle B_{hf} \rangle$ has a significant dependence on x. This shows that Co also plays an important role on the magnetization of that phase as it does in the alpha phase. Such a dependence of B_{hf} on the presence of Co atoms in the first or in the second coordination spheres relative to the ^{57}Fe probe atoms in alpha can be well established by using a semi-empirical model developed by ourselves and discussed elsewhere [6].

4. Conclusions

The $(Fe_{1-x}Co_x)_{0.875}Nb_{0.125}$ alloys system has two distinct phases: α, ferromagnetic - based on the bcc α-Fe (or γ-Co) and ε, paramagnetic at room temperature - based on the $(Fe/Co)_2Nb$ Laves phase. Nb has a strict solubility (~ 0.5 at %) in the Fe+Co α phase up to x ~ 0.75 while γ-Co (at x=1) dissolves about 4 at % Nb. At RT and up to x=0.75 α is magnetic with maximum B_{hf} at x~0.25 while ε is paramagnetic exhibiting two iron sites with population 3:1 for x = 0 that drops with increasing x. At 20K ε is likewise magnetic, with increasing B_{hf} up to x = 0.50, which shows that the presence of Co also plays an important role to T_c and the magnetic coupling of that phase

Acknowledgements

The authors would like do acknowledge Mr. W.T.Soares for the EPMA analyses and the Brazilian agencies Fapemig, CNPq and CAPES for the financial support.

References:

[1] R.H. Yu, et al. IEEE Transactions on Magnetics, vol. 36, No.5, September (2000) .3388-3393,
[2] C.W. Chen, Magnetism and Metal. of Soft Magn. Mater. Dover Publ. Inc., New York, (1986).p 92
[3] K.J. Kawahara, , J.Mat.Sciences, 18 (1983) 1709.
[4] C.M. Orrock, PhD Thesis, University of London, UK (1985).
[5] A.I.C Persiano .and R.D. Rawlings, J.Mat.Sci., 26 (1991) 4026
[6] M.T.Raposo and A.I.C.Persiano, to be published

MÖSSBAUER STUDY OF THE α' PHASE FORMATION IN Fe-Cr-BASED DUPLEX STAINLESS STEEL

H. R. Rechenberg[1], *S. D. Brandi*[2] *and A. F. Padilha*[2]

[1]Instituto de Física, Universidade de São Paulo, São Paulo, Brazil
[2]Escola Politécnica, Universidade de São Paulo, Brazil

Mössbauer spectroscopy was used to investigate the mechanism of the Cr-rich α' phase formation in a ferritic-austenitic stainless steel with duplex microstructure. Specimens were solubilized at 1050 °C for 30 min and heat treated at 475 °C for times ranging from 6 min to 500 hr. Mössbauer spectra were taken in the CEMS mode. The ferrite mean hyperfine field evolved from 24.4 to 28.6 T upon annealing, indicating a rise in the Fe:Cr ratio owing to Cr diffusion toward the nonmagnetic α' phase. Such evolution was significant only for times larger than 6 min, indicating the existence of an incubation time and hence a nucleation-and-growth mechanism for the α' phase formation in this steel.

INTRODUCTION

Ferritic-austenitic steels with a duplex microstructure exhibit a favorable combination of mechanical and corrosion properties, in addition to good workability and weldability. On the other hand, such steels are subject to three sorts of embrittlement mechanisms: i) α'-phase formation in ferrite ("475 °C embrittlement"); ii) σ-phase precipitation (mostly in ferrite), and iii) carbide formation (chiefly in the austenite), in high-carbon alloys. Out of these, α'-phase formation is most detrimental to the use of such materials at temperatures above 300 °C since it reduces their tenacity and corrosion resistance. α' particles are rich in Cr, have a bcc structure, are coherent with ferrite, and resist coalescence even for long treatments in the 350-550 °C range. Their small size hinders direct observation by TEM and x-ray or electron diffraction analysis. One controversial point on the α'-phase formation is its basic mechanism, whether by nucleation and growth or by spinodal decomposition [1-4]. In this work we present a contribution to this problem using Mössbauer spectroscopy.

EXPERIMENTAL DETAILS

The investigated steel (DIN 1.4467) had the following composition (in wt. %): 22.51 Cr, 5.60 Ni, 3.03 Mo, 1.60 Mn, 0.134 N, 0.60 Si, 0.06 Cu, 0.03 C, bal. Fe. Specimens cut from an ingot were polished, solubilized at 1050 °C for 30 min and heat treated at 475 °C for times ranging from 6 min to 500 hr. Mössbauer spectra were collected in the CEMS geometry, using a He+5 % methane mixture as ionization gas. An α-iron foil was used for velocity calibration and as an isomer shift standard.

RESULTS

Fig. 1 shows a representative spectrum. An intense unsplit line, corresponding to nonmagnetic austenite (γ phase) is superposed on a magnetically split sextet originating from

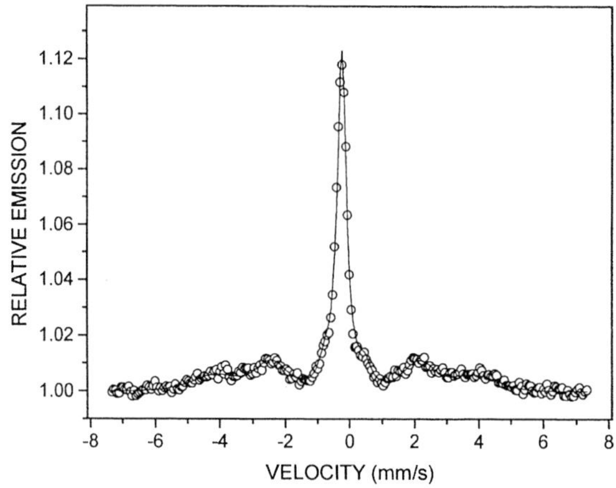

FIG. 1. CEMS spectrum of duplex steel solubilized at 1050 °C.

ferrite. The lines of the latter are strongly overlapping due to the presence of a hyperfine field distribution $P(B_{hf})$ caused by a multiplicity of Fe atomic environments. For binary Fe-Cr alloys, Schwartz and Chandra [5] have shown how a B_{hf} distribution can be associated to nearest-neighbor and next-nearest-neighbor configurations using binomial distributions. In the present case we obtained satisfactory fits by assuming a gaussian distribution function with two adjustable parameters: the mean $\langle B_{hf} \rangle$ and the gaussian width σ. A fixed 0.25 mm/s linewidth was assumed for each elementary magnetic sextet. A single lorentzian line was added to describe the austenite peak.

The isomer shift of the austenite and ferrite spectra did not change upon heat treatment: $\delta(\gamma) = -0.117(2)$ mm/s and $\delta(\alpha) = -0.008(5)$ mm/s. The austenite singlet linewidth was also constant, $\Gamma(\gamma) = 0.346(5)$ mm/s. Fitted parameters pertinent to the ferrite phase are given in Table 1.

TREATMENT TIME (hr)	$\langle B_{hf} \rangle$ (T)	FWHM (T)	FERRITE FRACTION
0	24.4(2)	10.4(7)	0.45(2)
0.1	24.6(2)	9.7(7)	0.43(2)
1	26.2(2)	10.8(7)	0.44(2)
10	27.1(2)	11.3(7)	0.43(2)
100	28.3(2)	9.9(7)	0.43(2)
500	28.5(2)	9.4(7)	0.42(2)

TABLE 1. Fitted hyperfine parameters for the ferrite phase: mean hyperfine field, full width of Gaussian distribution (FWHM = 2.355 σ), and ferrite fractional spectrum area.

The <B_{hf}> dependence on heat treatment time is shown in Fig. 2. It can be seen that the mean B_{hf} steadily increases with treatment time. This is a direct indication that the ferromagnetic α phase is being depleted in Cr as the aging proceeds. If our alloy had been binary Fe-Cr, it would be possible to follow its change in concentration by comparison with the known B_{hf} vs. x relationship in the $Fe_{1-x}Cr_x$ system [6], but the presence of other elements modifies the hyperfine field in a manner which is not known. It is noteworthy, however, that our B_{hf} values, both before and after aging, were very similar to those found in Ref. 7 for a steel with a composition close to ours.

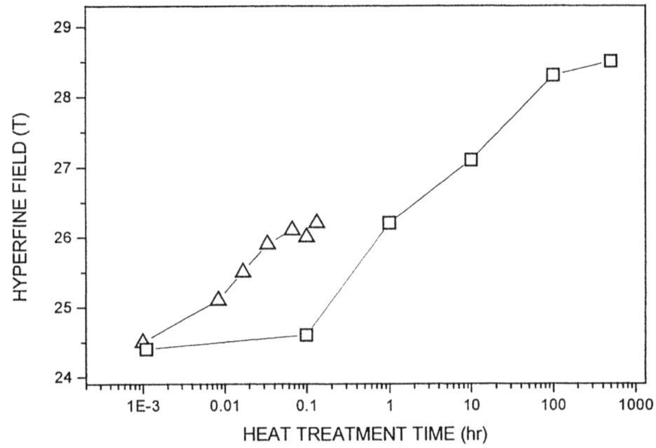

FIG. 2. Hyperfine field against time of 475 °C treatment. Squares: our data. Triangles: Fe-28Cr-5Ni data from Ref. 1. The leftmost data points stand for the samples' initial state.

The fractional area of the ferrite phase did not show any appreciable change within error limits. This result does not contradict the increase of the nonmagnetic α' phase (which should contribute to the central peak, although with a slightly different isomer shift) because the latter has a high Cr:Fe ratio, thus making only a small contribution to the spectral resonance area.

It is apparent in Fig. 2 that the α-phase hyperfine field does not change appreciably until t ≈ 6 min, suggesting the existence of an incubation time of order 2-3 × 10^2 s. This is a strong evidence in favor of a nucleation-and-growth mechanism for α' phase formation. Indeed, such a process is a thermally activated one which requires the formation of nuclei of a minimum critical size. In contrast, no incubation time is involved in spinodal decomposition. For the sake of comparison, we show in the same figure data by Kuwano *et al.* [1] for an Fe-28Cr-5Ni (wt. %) alloy subjected to 475 °C aging: for the latter, substantial changes are already induced by a treatment time as short as 30 s, which was taken as evidence for spinodal decomposition, corroborated by TEM data.

As noted by Kuwano *et al.* [1], an additional feature of spinodal decomposition in Fe-Cr-Ni alloys is a rapid increase in the P(B_{hf}) distribution width. The FWHM data in Table 1 do not show any consistent change within experimental accuracy. The absence of such a broadening in our experiment provides further evidence against spinodal decomposition in this alloy.

DISCUSSION

Both experimental work [1,4] and thermodynamic calculations [8] have suggested that the addition of Ni displaces the spinodal boundary towards lower Cr concentrations in Fe-Cr alloys. For example, it has been shown that α' phase formation at 475 °C takes place by nucleation and growth in Fe-28Cr and by spinodal decomposition in Fe-28Cr-5Ni [1]. According to [4], the expected spinodal boundary is at about 25 at. % Cr for a 5 at. % Ni content. The steel used in the present investigation has an Fe/Cr/Ni atomic ratio of 69.3/25.2/5.5 (normalized to 100 %), which is very close to the predicted spinodal line. It must be borne in mind, however, that the α and γ components in a duplex steel do not have identical compositions. Although the exact α and γ phase compositions in the DIN 1.4467 steel are not known, it can be expected that the Ni concentration in its ferrite fraction is lower than the nominal one since, in duplex steels, an excess of Ni in the γ phase is required to ensure its stability [7]. With less nickel in the α phase, the spinodal boundary is then likely to be shifted to a slightly higher Cr concentration. In this way, the present steel would be located beyond the spinodal boundary, such that α'-phase precipitates would form by the nucleation and growth mechanism, in agreement with the present experimental results.

ACKNOWLEDGEMENTS: This work was financially supported by FAPESP and CNPq.

REFERENCES

[1] H. Kuwano, Y. Ishikawa, T. Yoshimura and Y. Hamaguchi, Hyperfine Int. 69 (1991) 501
[2] K. Morimoto, A. Kirihigashi and S. Nasu, Nucl. Instr. Meth. B 76 (1993) 295
[3] H. Kuwano, Y. Ishikawa, T. Yoshimura and Y. Hamaguchi, Nucl. Instr. Meth. B76 (1993) 290
[4] H. Kuwano, Y. Nakamura, K. Ito and T. Yamada, Nuovo Cimento D 18 (1996) 259
[5] L. H. Schwartz and D. Chandra, Phys. Stat. Sol. (b) 45 (1971) 201
[6] F. B. Waanders, H. Pollak and S. W. Vorster, Hyperfine Int. (C) 3 (1998) 281
[7] H. D. Solomon and L. M. Levinson, Acta Met. 26 (1978) 429
[8] Y. Y. Chuang and Y. A. Chang, Metall. Trans. A 18 (1987) 733

Mössbauer spectral study of $PrTiFe_{11-x}Co_x$ ($x \leq 2$)

S. Sab, L. Bessais, C. Djega-Mariadassou
LCMTR, UPR209 CNRS, 2/8 rue Henri Dunant, B.P. 28, F-94320 Thiais, France

and J. M. Greneche
Laboratoire de Physique de l'état condensé, UMR CNRS 6087, Universit du Maine, F-72085 Le Mans, France

October 11, 2001

Abstract. $PrTi(Fe_{1-x}Co_x)_{11}$ alloys ($x \leq 2$) have been prepared by melting and subsequent annealing. Structure properties of these compounds have been investigated by x-ray analysis with Rietveld method. The tetragonal $ThMn_{12}$-type structure is observed for all samples. Substitution of Co for Fe leads to a contraction of the unit cell volume and enhancement in Curie temperature. The hyperfine parameter assignment, on the basis of the Wigner Seitz volume calculation, in the tetragonal structure assesses for the Co $8f$ preferential occupation. The mean hyperfine field increase, with Co content, might be correlated to the compensation of the core electron polarization field by the $4s$ polarization field produced by the $3d$ moment.

1. Introduction

Attention has been focused on the ternary intermetallics $R(\text{Fe,M})_{12}X_3$ (R = rare earth, M = Ti, V, Mo, $\cdots$, X = N) owing to their interesting fundamental characteristics and their potential applications as permanent magnets. These compounds show relatively high Curie temperature, appropriate saturation magnetization and favorable magnetocrystalline anisotropy [1]. Since the earlier investigations from [2, 3], extensive studies were performed with Nd and Pr as the rare earth element and Mo as the stabilizing metal. However scarce works are devoted to ternary $PrFe_{11}Ti$, so far, because it is difficult to stabilize the 1:12 phase with Pr and Ti [4]. In order to improve the magnetic properties and eventually the stability of $PrFe_{11}Ti$, we undertook a substitution of Co for Fe. X-ray diffraction does not help to discern the Co position due to the close diffraction factor of Fe and Co atoms, but Mössbauer spectroscopy appears as a powerful technique to define the preferential Co location. This paper presents a Mössbauer spectral study of $PrFe_{11-x}Co_xTi$ alloys ($x \leq 2$) with a view to investigate the effect of a partial substitution of Fe for Co.

2. Experimental

The alloys were prepared by induction melting high purity materials (at least 99.9%) in a water-cooled copper boat under a highly purified Ar atmosphere. The ingots were melted ten times to insure their homogeneity. As-cast alloys were annealed (6 days at 1373K) and quenched. The sample composition was checked by microprobe measurements with an accuracy of $\pm 5\%$. X-ray diffraction (XRD) patterns were registered on a Brucker diffractometer with Cu Kα radiation. The thermomagnetic analysis was performed with a differential sample Manics magnetometer, in a weak external magnetic field (100 mT) in the temperature range 300-850 K. The spectra refinement was performed with the FULLPROF computing code based on the Rietveld technique. ^{57}Fe Mössbauer spectra were collected at temperature between 77 and 300K using a conventional constant-acceleration spectrometer. The γ source was ^{57}Co in Rh matrix. Calibration was performed with α-Fe at room temperature with a line-width of 0.25 mm/s. The spectra have been fitted as explained below. The estimated error is $\pm$ 0.1 T for hyperfine fields H_{HF}, $\pm$ 0.01 mm/s for isomer shifts δ and $\pm$ 0.01 mm/s for quadrupole shifts 2ϵ.

3. Results and discussion

XRD patterns show that all samples crystallize in the $I4/mmm$ $ThMn_{12}$-type structure with a monotonous reduction of the unit cell parameters. Weak lines are assigned to α-Fe(Ti-Co), Fe_2Ti and Pr. The total amount of these additional phases defined by the Rietveld fit decreases from 7 vol % for $x = 0$ down to 2 vol % for $x = 2$ in agreement with the microprobe measurements which likewise corroborates their composition. However the microprobe analysis on the $I4/mmm$ phase reveals that the nominal composition describes correctly its composition. It results that the Co substitution stabilizes the 1:12 phase. The Curie temperature increases from 550 K for $x = 0$ to 740 K for $x = 2$. This behavior reflects an enhancement of the Fe-Fe interactions which does not result from a magneto-volume effects since all interatomic distances decrease, but from the filling of the 3d band by the Co additional electron [5]. Figure 1a shows the Mössbauer spectra at 77 K for all x content. They result from the convolution of numerous sub-spectra. It is well known that in the $ThMn_{12}$-type structure, Ti substitutes in 8i site. This implies that with the Co substitution, complex Mössbauer spectra are expected. In such case several sets of hyperfine parameters may lead to an excellent representation of the spectra. The choice of one solution must be supported by pertinent theoretical and experimental considerations. For the Mössbauer analysis, first, we have to take into account the small abundances of the extra phases observed on the XRD patterns. We must keep in mind that the Lamb-Mössbauer f factors of Fe_2Ti and α-Fe(Ti-Co), are larger than for the $PrFe_{11-x}Co_xTi$ phase. The contribution of these additional phases should not be neglected. Secondly, we have performed the Mössbauer analysis on the basis of two assumptions: i) we have excluded Co substitution in 8i site, already statistically occupied by Ti atoms. The variations of the relative abundance and the coherency of the hyperfine parameters of the 8i sextets as a function of x,

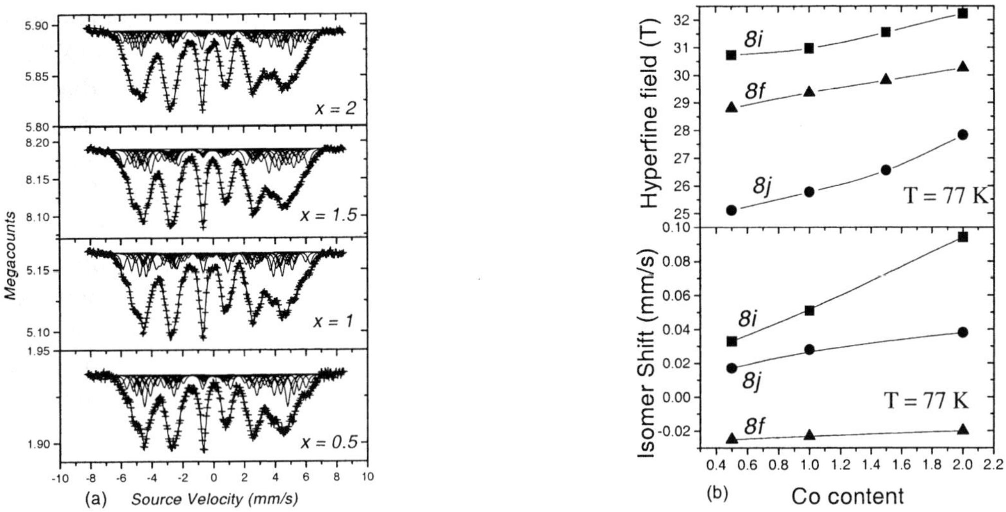

Figure 1. (a) The 77 K Mössbauer spectra of $PrFe_{10}CoTi$. (b) The compositional dependence of the isomer shifts and the hyperfine fields at 77 K.

will justify, a posteriori, this assertion; ii) the assignment of the sets of hyperfine parameters to a k site will be governed by the correlation between the Wigner-Seitz cell (WSC) volumes and the isomer shift δ: the larger the WSC volume, the larger the δ value. Consequently, in a first step we have calculated the relative abundances of the Fe families in the various crystallographic sites $8i$, $8j$, and $8f$ with the assumption of Ti in $8i$ site, Co in $8j$ or $8f$ sites with the multinomial law with the various environments given in Table I for the $ThMn_{12}$ structure. One must emphasize the symmetric role played by $8j$ and $8f$ sites and the impossibility to decide, at this step, upon the assignment of the various hyperfine parameter sets. During all the fitting procedure, these abundances were considered as fixed parameters, they were normalized after neglecting those lower than 2%, and the Lamb-Mössbauer factors were considered the same for the three sites. All other parameters viz. δ, H_{HF}, and 2ϵ were free parameters, just like the line-width, considered equal for all lines of all sextets. Moreover as no texture was observed on the XRD patterns, the relative intensities of each sextet line were taken in the ratio 3:2:1. The calculated WSC volumes have been calculated with atomic radii, respectively equal to 0.126, 0.125, 0.182, and 0.147 nm respectively for Fe, Co, Pr, and Ti. It results the following sequence for the WSC volumes $8i > 8j > 8f$ and the recurring relationships $\delta\{8i\} > \delta\{8j\} > \delta\{8f\}$. In the second step of the refinement, the averaged isomer shifts values have been assigned to each site of the $8i$, $8j$, $8f$ families. Finally in the last step of the fit, all hyperfine parameters were free. The Co content dependencies of both δ and H_{HF} at 77 K as examples are reported on Figure 1.b. At all temperatures, for a given x value, the relation $\delta\{8i\} > \delta\{8j\} > \delta\{8f\}$ is always obeyed, according to our assumption derived from the WSC volumes. $\delta\{8i\}$ and $\delta\{8j\}$ show a noticeable increase with Co substitution, while $\delta\{8f\}$ remains quasi constant. This result can be understood in terms of the preferential Co atom occupation. The $8i$ and $8j$ site Fe atoms have four adjacent $8f$ neighbors while the $8f$ site atoms have only two. The δ behavior of both $8i$ and $8j$ iron sites are attributed to the additional $3d$ electrons brought by $8f$ Co neighbors. They enhance the shielding of the $4s$ electrons, reduce the s charge density at the nucleus and result in an increasing isomer shift. On the contrary, if $8f$-atom surrounding is poorly affected by the Co substitution, the s charge density is then not modified. No change is observed for the $8f$-isomer shift. The difference for the $8f$-site behavior then corroborates the preferential occupation of Co atoms on this site. This later result is in agreement with the results obtained for $YFe_{11-x}Co_xTi$ [6] and $SmFe_{11-x}Co_xTi$ [7]. The H_{HF} variation corroborates too the validity of the field assignment as reported as a function of Co content x (Figure 1b). The weighted average hyperfine field increases with x in agreement with Curie temperature variations. One must outline that $H_{HF}\{8f\}$ is larger than $H_{HF}\{8j\}$, this result is in agreement with analysis performed previously on $CeFe_{11}Ti$ [8].

4. Conclusion

The unit cell parameter c decreases more significantly than a. The increase of Curie temperature T_C and hyperfine field H_{HF} attest for the Co continuous substitution. The substitution of Co, which contributes to complete the $3d$ subband induces the Curie temperature enhancement. The hyperfine parameters were assigned on the basis of the calculation of WSC volumes. The isomer shift behavior of the $8i$ and $8j$ site Fe atom is attributed to the additional $3d$ electrons brought by $8f$ Co neighbors. On the contrary no change is observed for the $8f$-isomer shift. The specific $8f$-site behavior corroborates the preferential occupation of Co atoms on this site in the 1:12 structure. The mean hyperfine field enhancement, with cobalt content, might be correlated to the compensation of the core-electrons polarization field by the $4s$ polarization field produced by the $3d$ moment.

References

1. K. H. J. Buschow, J. Magn. Magn. Mater. 100 (1991) 79.
2. Y. Z. Wang, L. Song, K. Y. Wang, B. P. Hu, G. C. Liu, W. Y. Lai, J. Magn. Magn. Mater. 140 (1995) 1019.
3. R. Rani, H. Hegde, A. Navarathna, F. J. Cadieu, J. Appl. Phys. 75 (1994) 6006.
4. M. Akayama, H. Fujii, K. Yamamoto, K. Tatami, J. Magn. Magn. Mater. 130 (1994) 99.
5. D. Givord, P. Courtois, J. Magn. Magn. Mater. 196 (1999) 684.
6. Z. W. Li, X. Z. Zhou, A. H. Morrish, J. Appl. Phys. 69 (1991) 5602.
7. L. Bessais, C. Djga-Mariadassou, Phys. Rev. B 63 (2001) 54412
8. G. Long, D. Hautot, F. Grandjean, O. Isnard, S. Miraglia, J. Magn. Magn. Mater. 202 (1999) 100.

Atomic Ordering at Grain Boundaries in Fe–Si and Fe–Al Alloys

O. Schneeweiss, Y. Jirásková, J. Čermák
Institute of Physics of Materials, Academy of Sciences of the Czech Republic, Žižkova 22, CZ-61662 Brno, Czech Republic

P. Lejček
Institute of Physics, Academy of Sciences of the Czech Republic, Na Slovance 2, CZ-18221 Praha, Czech Republic

Abstract. Atomic ordering at grain boundaries (GBs) in iron rich Fe-Si and Fe-Al alloys was studied using emission Mössbauer spectroscopy. The results on Fe-Si show differences in the chemical composition in grain interiors (GIs) and at GBs. An increase in Si concentration at GBs and atomic ordering corresponding to the $D0_3$ superstructure were found. GBs in Fe-Al samples exhibit more complicated phase and chemical composition. The components of the emission spectra can be interpreted as an inhomogeneous increase in Al concentration at GBs and a lower symmetry than the bcc lattice of GIs.

Keywords: grain boundaries, atomic ordering, iron alloys, Mössbauer spectroscopy

1. Introduction

Structure and topology of the atomic arrangement along grain boundaries (GBs) distinguish from that in crystal interiors and mostly it is accomplished by different chemical compositions, e.g., by segregation of impurities (Lejček and Hofmann, 1995). Impurities at GBs may cause structural transformation in comparison with that of the same pure material (Watanabe, 1993). Fe-Si and Fe-Al alloys with less than 25 at.% Si/Al crystallize in α phase (bcc) and above $\approx$10 at.%Si/20 at.%Al the superstructures B2 and $D0_3$ are formed (Massalski et al., 1986). Information about the structure and chemical composition of GBs can be obtained using the emission Mössbauer spectroscopy. This method was already applied for investigations of GBs in some metals, e.g., in (Schneeweiss et al., 1998; Schneeweiss et al., 1999). In this paper, the emission Mössbauer spectroscopy is applied to the study of GBs in polycrystalline samples of iron rich Fe-Si and Fe-Al alloys. The aim of the investigation is to describe atomic ordering and chemical composition at GBs.

2. Experimental details

Polycrystalline samples of the Fe-10at.%Al, Fe-22at.%Al , Fe-12at.% Si, and Fe-24at.%Si were used as matrices for preparation of ^{57}Co sources. The temperature for diffusion annealing was chosen so that the diffusion ^{57}Co atoms along GBs was dominating and their diffusion into GIs was negligible. The more detailed description of source preparations was published, e.g., in (Schneeweiss et al., 1998; Schneeweiss et al., 1999). The emission ^{57}Co Mössbauer spectra were taken at room temperature using scintillation detector doped by $Na_4Fe(CN)_6 \times 10H_2O$. The standard transmission ^{57}Fe spectra of polycrystalline foils and CEMS spectra were taken using a ^{57}Co in Rh source. Isomer shift δ, quadrupole splitting σ and hyperfine induction (B_{hf}) are calibrated against B_{hf} of α–iron (3.9156 mm/s corresponding to 33 T). Examples of the spectra are shown in Fig. 1.

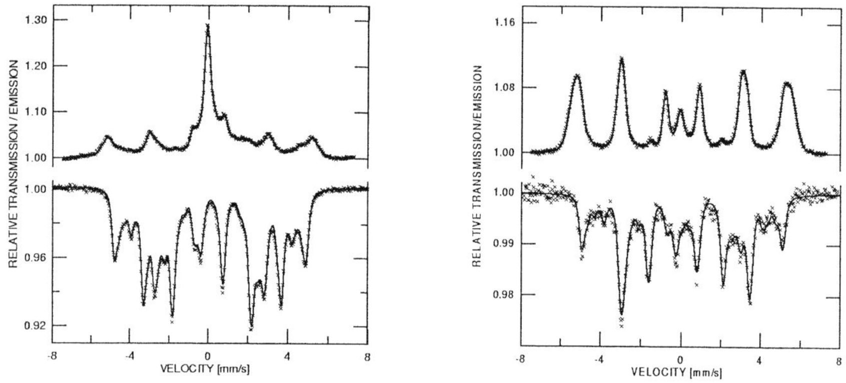

Figure 1. The emission spectra of the ^{57}Co in Fe–22at.% Al (upper left) and Fe–24at.%Si (upper right) alloy sources, and transmission spectra of the Fe–22at.% Al (lower left) and Fe–24at.%Si (lower right) samples using ^{57}Co in Rh source.

3. Results and Discussion

Experimental spectra were analyzed using standard procedure by fitting of set of independent sextets (components). The components were ascribed to the iron atoms with different numbers of Al/Si atoms in the nearest neighborhood. Theoretical models of iron atom surroundings were calculated for different atomic arrangements in the bcc structure and the model data were compared with the experimental intensities of the fitted components. The procedure was described previously (Schneeweiss et al., 1999). Besides the sextets a single line component is present in the emission spectra which was ascribed to deposited ^{57}Co atoms remaining on surface (Schneeweiss et al., 2000). The sextets in emission spectra exhibit important line broadening. It reflects the chemical and topological disordering along GBs.

The comparison of the experimental values obtained from the emission and transmission spectra of the Fe–Si alloy with the model results show that atomic ordering in GIs is in both alloys in good agreement with probabilities calculated for the model of the $D0_3$ superstructure. The emission data indicate that higher content of Si appears at GBs. For the Fe–12at.%Si, the concentration of silicon at GBs was determined to be 18±2 at.%Si. The emission spectrum of the Fe–24at.%Si alloy differs substantially from the transmission spectrum of this alloy. In accordance with the previous observation (Schneeweiss et al., 1999) it can be interpreted as strong site preference of the diffusing ^{57}Co for the lattice positions in the $D0_3$ superstructure with eight Fe atoms in the nearest neighborhood. In the ordered stoichiometric Fe_3Si alloy they are occupied by Si atoms only. It indicates occurrence of structural vacancies which substitute Si atoms missing to stoichiometry concentration. Their positions are used for diffusion of ^{57}Co along GBs. It is in agreement with the accepted diffusion mechanism along GBs (Girifalco, 1964). The values of isomer shift and quadrupole splitting of the component of the emission spectrum indicate that the positions of ^{57}Co atoms have a surrounding of cubic symmetry and the s-like electron density is close to that in bcc iron.

Information derived from the spectra of Fe–Al alloys were interpreted in the same way. Comparison of the intensities of the components of the transmission and emission spectra of Fe-Al alloys showed that there is a substantial difference between atomic ordering in GIs and at GBs. The intensities of the components of the transmission spectrum, which represent GIs, showed that the atomic arrangement in the GIs agrees with $D0_3$ superstructure partially disordered towards B2 superstructure. Different components were identified in the emission spectra of the Fe-Al samples. They are similar to those observed in the spectra of Fe–Al alloys with Al content higher than 27at.% after plastic deformation (Jirásková et al., 1993). Therefore a higher Al content at GBs and the atomic arrangements close to the structure with high defect density after severe deformation can be expected along GBs in our Fe-Al samples. The components (sextets) characterized by high values of isomer shifts ($0.4 \div 0.6$ mm/s) and low values of B_{hf} ($6 \div 8$ T), found only in emission spectra, indicate probably presence of the iron with more than four Al atoms in the nearest neighborhood. Two doublets and one singlet were also identified there and they can be ascribed to Al clusters, because the values of δ=0.35±0.03 mm/s are similar to that observed in the Fe–Al alloys with high Al concentration. The large quadrupole splitting of the doublets σ=0.4±0.1 mm/s can be explained by substantial deviation from the cubic symmetry of the basic lattice.

In summary, the comparison of the results derived from the transmission and emission spectra with the model of atomic arrangements in the bcc lattice, B2, and $D0_3$ superstructures of the studied alloys shows that GBs in Fe–Si and Fe–Al alloys differ in atomic ordering. In the Fe–Si alloys the atoms at GBs occupy regular $D0_3$ superstructure sites and positions of Si atoms missing to stoichiometry concentration are completed by structural vacancies. Higher Si concentrations at GBs than nominal alloy compositions were found. The atoms at GBs in Fe–Al alloys have surrounding of a lower symmetry in comparison with those in the GIs lattice. The components which were identified in emission spectra only, correspond well to Al rich clusters having structure similar to bcc lattice with high defect density. Effects of oxide particles cannot be excluded.

This work was supported by the Grant Agency of the Academy of Sciences of the Czech Republic (project No. K1010104) and the Grant Agency of the Czech Republic (project No. 106/99/0183).

References

Lejček, P. and S. Hofmann. Thermodynamics and structural aspects of grain boundary segregation. *Critical Review in Solid State and Materials Sciences*, 20:1–85, 1995.

Watanabe, T. Grain boundary design and control for high temperature materials. *Materials Science and Engineering*, A166:11–28, 1993.

Massalski, T. B., J. L. Murray, L. H. Bennet, and H. Baker (eds.). *Binary alloy phase diagrams*, page 1108, American Society for Metals, Metals Park, Ohio, 1986.

Schneeweiss O., I. Turek, J. Čermák, and P. Lejček. Properties of iron atoms at grain boundaries in Fe and $Fe_{72}Al_{28}$. *MRS Symp. Proc. Ser.*, 527:273–278, 1998.

Schneeweiss, O., S. Havlíček, and T. Žák. Site preference of ^{57}Co in Fe-Si alloys after grain boundary diffusion. Proc. NATO ASI Kluwer, Dordrecht 1999, pp. 337-348.

Schneeweiss O., J. Čermák, I. Turek, and P. Lejček. Investigations of grain boundaries in copper using emission Mössbauer spectroscopy. Proc. ICAME '99, *Hyperfine Int.* 126:215–218, 2000.

Girifalco, L. A. Atomic Migration in Crystals. pages 117–123, Blaisdell Publishing Company, New York, 1964

Jirásková, Y., M. Kočová, N. Pizúrová, T. Žák, and O. Schneeweiss. Recovery of highly strained Fe–28at.%Al alloy. *Materials Science and Technology*, 9:397–402, 1993.

Effects of Cold Rolling and Heat Treatments on Fe Precipitation in High Purity Aluminum Studied by Mössbauer Spectroscopy

M. Shiga, M. Kawai* and S. Kubota+
Department of Materials Science and Engineering, Kyoto University, Sakyo-ku, Kyoto 606-8501, JAPAN
* *Technical Development and Administration Division, Mitsubishi Aluminum Co., Ltd., Minato-ku, Tokyo 105-8546, JAPAN*
\+ *Technical Development Center, Mitsubishi Aluminum Co., Ltd., Susono-shi, Shizuoka 410-1127, JAPAN*

Abstract. Effects of cold reduction and heat-treatments on the Fe precipitation behavior in high purity aluminum with 50 ppm Fe were investigated by Mössbauer spectroscopy using 90% enriched ^{57}Fe. It has been revealed that the volume fraction of Fe precipitates becomes maximum by a heat treatment at 310 °C in supersaturated solid solutions with strains introduced by 96 % cold reduction. The precipitates in the 310 °C aging sample are probably metastable Al_6Fe compound. Surprisingly, it was found that the precipitation starts even at 120 °C.

Keywords: aluminium, iron, precipitation, Mössbauer spectroscopy

1. Introduction

The quality of high purity aluminum foils for electrolytic capacitors sensitively depends on manufacturing factors such as cold rolling and heat treatments. It is believed that the degree of precipitation of impurities, in particular, that of iron impurity is one of the most important factors. Therefore, establishing the methodology for estimating the degree of iron precipitation is highly desired not only for metallurgical interest but also for managing fabrication processes. In this paper, we report on effects of cold reduction and heat-treatments on the Fe precipitation behavior in high purity aluminum with 50 ppm Fe investigated by Mössbauer spectroscopy using 90% enriched ^{57}Fe.

2. Experiments

We prepared 1 kg high purity aluminum ingot using 99.995% pure aluminum adding small amount of Si, Cu and 90 % enriched ^{57}Fe, so that the final impurity compositions become to 50ppm Si, 10ppm

Cu and 50 ppm Fe, which are near to the material for electrolytic capacitors. The melt was cast into a book-shaped mold. A part of ingot was subjected to heat treatments: heated at 620 °C for 17hr, cooled down to 530 °C and then hot rolled into a plate of 7mm thickness. The Mössbauer samples were prepared by cold-rolling this plate into 0.1 to 1 mm thickness. In these treatments, 90 % of Fe exist as solutes as shown later.

Mössbauer measurements were carried out at room temperature using 370 MBq ^{57}Co doped in a metallic Rh foil. Rectangular sheets of 20x20 mm foils were set into the sample holder so as the total thickness to be about 0.8 mm. Data are accumulated for 7 to 10 days, giving a fairy clear spectrum even for such a dilute Fe sample.

The spectra thus obtained were fitted by Lorentzians corresponding to a singlet of Fe solute and a doublet of Al_6Fe precipitate. All of the Mössbauer parameters of the singlet were determined by least square fitting. On the other hand, the fixed parameters were used for the doublet using the values obtained for Al_6Fe compound, which is believed to be the precipitation products in dilute Al(Fe) alloys[1,2], slightly modified to obtain better fitting for the present results, namely, the isomer shift, δ = 0.20 mm/s, the $e^2qQ/4$ = 0.15 mm/s and the half width of each Lorentzian Γ = 0.28 mm/s. Neglecting the difference in recoil-free fractions between solute Fe and Al_6Fe compound, we regarded the ratio of integrated intensities of the both components as the fractions of Fe atoms of solute and in precipitates. This procedure might give rise to somewhat over-estimation of precipitates because the Debye temperature of Al_6Fe should be higher than that of solute Fe.

3. Results and discussion

Figure 1 shows the Mössbauer spectrum of the as-rolled sample (4 sheets of 0.2 mm thickness foils). As seen in the figure, the spectrum consists mostly of the singlet. Analyzing the spectrum, it becomes evident that approximately 90 % of Fe exist as solutes in the plate prepared by hot-rolling.

Figure 2 shows the Mössbauer spectrum of an aged sample at 310 °C for 6hr after cold rolling the plate sample up to 97% reduction in thickness (7mm to 0.2mm). It is clearly seen that the doublet component growth, indicating the development of precipitation of the Al_6Fe compound. The Fe fraction in precipitates thus estimated for the samples aged for 6h at various temperatures after cold rolling is shown in figure 3. It was found that the precipitation develops at surprisingly low temperatures even at 120 °C. The maximum precipitation occurs

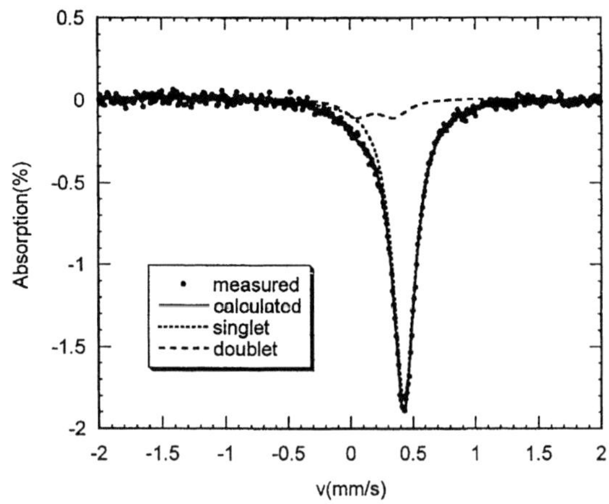

Figure 1. Mössbauer spectrum of the sample hot rolled from 530 °C. • : absorption rate (%), solid line : calculated value, chain line(singlet): solute Fe, dotted line(doublet): precipitated Fe.

at 310 °C. Above 310 °C, the Fe fraction in precipitates decreases with increasing temperature and becomes almost 0 at 500 °C. In this temperature range, the alloy should be in the thermally equilibrium state and the solubility limit of Fe in Al is 50 ppm at 500 °C.

Open circles in figure 3 indicate Fe fraction in precipitates for the samples aged for 6 hrs at various temperatures after solidification treatment, namely, rapidly cooled the Mössbauer sample from 550 °C. In this case, no precipitation takes place even at 300 °C, indicating that high density dislocations are necessary for developing precipitation.

So far, we have considered that the precipitates developed in the present conditions are Al_6Fe. In fact, the fitting of the Mössbauer spectra by the singlet of solute Fe and the doublet for Al_6Fe is fairly good. However, looking carefully, a tiny misfit is observed around the central dip of spectrum as observed in the spectrum of figure 2. We noted that this misfit becomes more distinct for the samples aged at lower temperature than 300°C. Furthermore, we found that the precipitates growth at the lower temperatures are rather unstable and easily destroyed by additional cold rolling, although such a phenomenon is not observed for the samples aged above 250 °C. Probably, the precipitates developed at the low temperatures are not complete Al_6Fe but something like precarious clusters leading to the formation of Al_6Fe compound with a very small size. Since we can not know the size of precipitates from the Mössbauer spectrum, some other experiment like TEM is desirable.

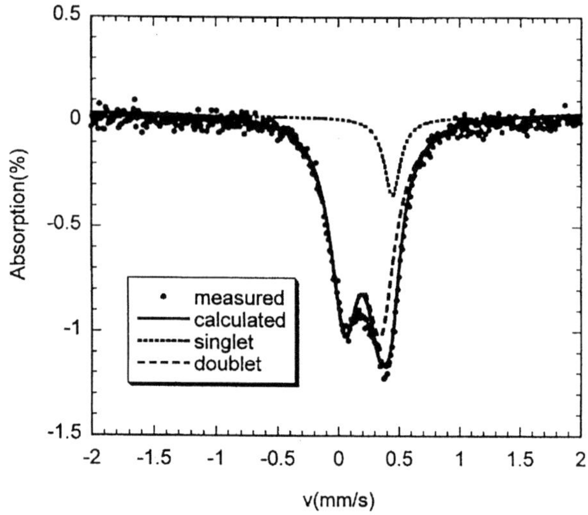

Figure 2. Mössbauer spectrum of the sample aged 6hrs at 310 °C after cold rolling by 97 %reduction. • : absorption rate(%), solid line : calculated value, chain line(singlet): solute Fe, dotted line (doublet): precipitated Fe.

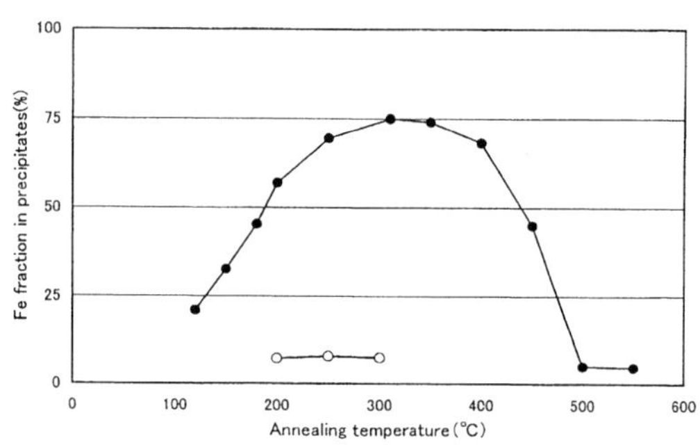

Figure 3. Fe fraction in precipitates for the samples aged 6 hrs at various temperatures from 120 °C to 550 °C. •: aged after 90 % cold rolling, ○ : aged without rolling

References

[1] C. A. Stickels and R. H. Bush, Metall. Trans., 2 (1971) 2031.

[2] T. Suzuki, K. Arai, M. Shiga and Y. Nakamura, Metall. Trans., 16A (1985) 27.

A Mössbauer study of the RENiSn (RE=La, Nd, Tb and Ho) intermetallic compounds and their deuterides

T. Spataru, V.A. Yartys[1], P. Blaha[2], G. Principi

Settore Materiali and INFM, DIM, Università di Padova, Padova, Italy
[1]Institute for Energy Technology, P.O.Box 40, N-2027, Kjeller, Norway
[2]Institute of Physical and Theoretical Chemistry, Vienna University of Technology, Vienna, Austria

The intermetallic RENiSn (RE = La, Nd, Tb and Ho) compounds and their corresponding deuterides have been studied by ^{119}Sn Mössbauer spectroscopy. The Mössbauer spectra measured at 300 K and 12.5 K exhibit paramagnetic doublets for tin atoms. An increase of the quadrupole splitting values of Sn atoms is observed in deuterides compared to initial phases. This effect is particularly pronounced for tin atoms partially substituting Ni and is associated with deuterium insertion into the RE_3Ni sites leading to changes of the local structure. The increase of quadrupole splitting for both tin sites in $LaNiSnD_2$ is particularly high, corresponding to an anisotropic deformation of the crystal structure due to the change from the orthorhombic (LaNiSn) to the hexagonal ($LaNiSnD_2$) unit cell.

1. Introduction

A significant improvement in the cycle stability of the nickel metal hydride electrodes is characteristic for the Sn-substituted $LaNi_5Sn_x$ intermetallics compared to the non-doped alloys [1-3]. In a crystallographic study of $LaNi_{5-x}Sn_x$ ($0.2 \leq x \leq 0.5$) alloys and their hydrides it has been found that tin atoms partially substitute nickel in the intermetallic $CaCu_5$-type compounds and that LaNiSn intermetallic compound is formed as the minor constituent of these alloys [4]. Hydrogenation and crystallographic properties of LaNiSn and corresponding deuteride $LaNiSnD_2$ have been studied recently [5]. In the equiatomic intermetallic compound and its deuteride a partial mutual substitution of Ni and Sn was also observed.

More data are needed to explain how tin contributes to influence favourably hydrogen storage properties of the intermetallics of rare earth metals and nickel.

The aim of this work is to obtain local structural information on RENiSn (RE = La, Nd, Tb and Ho) intermetallic compounds and their deuterides by means of ^{119}Sn Mössbauer spectroscopy.

2. Experimental

The alloys were prepared by arc melting the mixtures of high purity La, Nd, Ho and Tb (99.9 %), nickel (99.9 %) and tin (99.9 %). The alloy specimens were homogenised by annealing at 770 K for 4 weeks. Deuteration was performed at 670 K at a deuterium loading pressure of 1 bar.

The Mössbauer spectra were acquired by a constant acceleration spectrometer with symmetrical velocity waveform using a ^{119}Sn source in Ca_2SnO_3 matrix. A close-cycle RIAL cryostat was used for low temperature measurements. A current minimisation routine was used to fit the spectra. The isomer shifts are referred to SnO_2.

3. Results and discussion

According to powder diffraction data [5-7], the prepared alloys contained RENiSn (RE = La, Nd, Tb and Ho) intermetallic compounds as the major constituents. These crystallise with the orthorhombic TiNiSi-type of structure, space group Pnma. Both Sn and Ni sites are located in an almost similar environment, inside the trigonal prisms RE_6. A mutual substitution Sn ↔ Ni of about 10 at. % was observed for La and Nd compounds. Nd sample contained small amounts of $NdNi_2Sn_2$.

From figure 1 it can be seen that all Mössbauer spectra display broad or unresolved doublets and, on the basis of the structural data, can be fitted as the superposition of two components (LaNiSn, NdNiSn and TbNiSn), or as one component only (HoNiSn). The calculated values of the Mössbauer parameters for the Sn-based site in all the compounds are close to each other (see Table 1). Nevertheless, the rather high value of the quadrupole splitting of this site for TbNiSn, compared to LaNiSn, points out some differences in the local structure of these two compounds. The higher values of the quadrupole splitting for the Sn atoms substituting Ni atoms, compared to the Sn-based site, prove that the structure arrangement is highly anisotropic.

All structural and hyperfine parameters differences have their origin in the nature of the RE metal, since all the considered compounds have the same symmetry. Preliminary electron structure calculations using the WIEN97 code [8] show that the main differences among the RENiSn compounds consist in a different hybridisation of electron shells of atoms and different electron transfer charges from RE atoms to Ni and Sn.

The crystal structure of the intermetallic phase is considered as an important factor influencing hydrogen storage properties of intermetallic hydrides [9,10]. The Mössbauer spectra measured at RT and 12.5 K show that there are marked differences between the hyperfine parameters of both Sn sites in deuterides and as prepared intermetallics. On the basis of the structural data [5,7], we suggest that the increase of quadrupole splitting of Sn partially

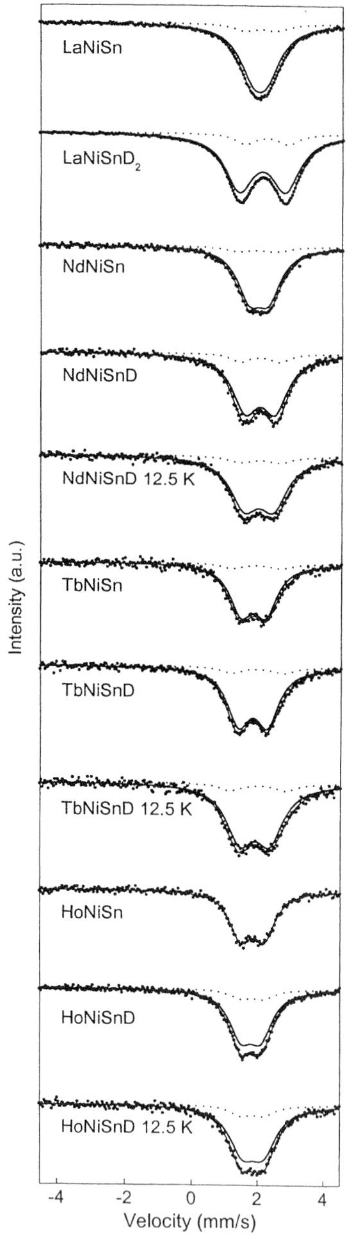

Figure 1. ^{119}Sn Mössbauer spectra (at RT, unless otherwise specified) of studied compounds and their deuterides. The components (except HoNiSn, see Table 1) are: Sn site (solid), Sn substituting Ni atoms (dot).

substituting Ni can be associated with the insertion of deuterium atoms into the RE_3Ni site. The increase upon deuteration of the quadrupole splitting for tin atoms in all samples, except for HoNiSnD, indicates a decrease of the environment symmetry at Sn and Ni sites. This increase is particularly high in $LaNiSnD_2$ for both Sn sites and corresponds to an anisotropic deformation of the crystal structure during a transformation from the orthorhombic (LaNiSn) to the hexagonal ($LaNiSnD_2$) unit cell [5]. Note that the more pronounced change in the local structure for $LaNiSnD_2$ can be associated with the highest deuterium storage capacity of this compound.

Table 1. ^{119}Sn Mössbauer parameters of tin sites of RENiSn intermetallics and their deuterides: δ, isomer shift; ΔE_Q, quadrupole splitting; A, area ratio; Γ, linewidth.

Compound	T (K)	Site	δ (mm/s)	ΔE_Q (mm/s)	Γ (mm/s)	A (%)
LaNiSn	RT	Sn/Ni	2.00	0.43	1.05	0.84
		Ni/Sn	2.02	0.98	0.91	0.16
$LaNiSnD_2$	RT	Sn/Ni	2.08	1.40	1.06	0.84
		Ni/Sn	2.21	1.31	0.76	0.16
NdNiSn	RT	Sn/Ni	1.98	0.59	0.91	0.86
		Ni/Sn	1.94	1.16	1.00	0.14
NdNiSnD	RT	Sn/Ni	2.01	0.91	0.97	0.86
		Ni/Sn	2.01	1.20	0.97	0.14
NdNiSnD	12.5	Sn/Ni	1.99	0.90	1.04	0.86
		Ni/Sn	2.04	1.20	0.94	0.14
TbNiSn	RT	Sn/Ni	1.84	0.75	0.85	0.89
		Ni/Sn	1.93	1.48	0.85	0.11
TbNiSnD	RT	Sn/Ni	1.84	0.93	0.86	0.89
		Ni/Sn	1.98	1.58	0.79	0.11
TbNiSnD	12.5	Sn/Ni	1.90	0.91	1.03	0.89
		Ni/Sn	2.00	1.68	0.81	0.11
HoNiSn	RT	Sn	1.84	0.72	0.81	1.00
HoNiSnD	RT	Sn	1.80	0.62	0.86	0.84
		Sn*	1.84	0.72	0.81	0.16
HoNiSnD	12.5	Sn	1.87	0.65	0.98	0.84
		Sn*	1.84	0.72	0.81	0.16

*The composition is 84% HoNiSnD and 16% HoNiSn [7]

According to the Mössbauer spectra, the tin states in the studied deuterided compounds are paramagnetic even at 12.5 K. Therefore, the RKKY type interactions between tin and magnetic atoms do not influence the Sn nuclear levels, in agreement with their low temperature ordering exhibited by magnetic measurements [11]. This is in agreement with the data of Łątka et al. [12,13], who have recently studied ^{119}Sn Mössbauer spectra of similar compounds, REAgSn, observing a magnetic splitting only below 9 K.

In the case of the $LaNiSnD_2$ compound, the quadrupole splitting values of both doublets are about equal (table 1). Therefore, the symmetry environment is about the same for both tin sites. This observation agrees with the crystal structure data showing that the two types of trigonal prisms, namely $NiLa_6$ and $SnLa_6$, have the same geometry [5].

4. Conclusions

^{119}Sn Mössbauer spectroscopy study of the RENiSn (RE = La, Nd, Tb and Ho) intermetallics and their corresponding deuterides $LaNiSnD_2$ and RENiSnD (RE = Nd, Tb, Ho) showed that the deuteration process is accompanied by an increase of the quadrupole splitting values of Sn atoms. This effect is significant for the tin atoms partially substituting nickel and can be associated with deuterium insertion into the RE_3Ni sites, in agreement with the powder neutron diffraction data. Similar increased values of quadrupole splitting for both tin sites in $LaNiSnD_2$ are due to a transformation from an original orthorhombic (intermetallic compound) to the hexagonal (deuteride) symmetry. The nature of the RE elements plays an important role in the modification of the structural, electronic and magnetic properties of the materials studied.

Acknowledgements

This work has been carried out in the frame of the project entitled "Idrogeno e scienza dei materiali: verso una nuova fonte di energia", supported by the Ministero della Ricerca Scientifica e Tecnologica (MURST). Financial support also from Norsk Hydro and the Research Council of Norway is acknowledged. We are grateful to T. Olavesen and Dr. H.W. Brinks (IFE) for their help in the experimental work.

References

[1] R.C. Bowman, C.H. Luo, C.C. Ahn, C.K. Witham and B. Fultz, J. Alloys Comp. 217 (1995) 185.
[2] B.V. Ratnakumar, C. Witham, B. Fultz and G. Halpert, J. Electrochem. Soc. 141 (1994) 189.
[3] S. Luo, W. Luo, J.D. Cleawley, T.B. Flanagan and R.C. Bowman, J. Alloys Comp. 231 (1995) 473.
[4] J.M. Joubert, M. Latroche, R. Černý, R.C. Bowman Jr., A. Percheron-Guégan and K. Yvon, J. Alloys Comp. 293-295 (1999) 124.
[5] V.A. Yartys, T. Olavesen, B.C. Hauback, H. Fjellvåg and H.W. Brinks, J. Alloys Comp. (in press).
[6] R.V. Skolozdra, in *Handbook on the Physics and Chemistry of Rare Earths*, Ch.164 (1997) 399.
[7] V.A. Yartys, T. Olavesen, B.C. Hauback and H.Fjellvåg, to be published.
[8] P. Blaha, K. Schwarz and J. Luitz, *WIEN97: A Full Potential Linearized Augmented Plane Wave Package for Calculating Crystal Properties* (Karlheinz Schwarz, Techn. Universität Wien, Austria), (1999) ISBN 3-9501031-0-4.
[9] K. Yvon, P. Fischer, Topics in Applied Physics, 63 (1988) 87.
[10] V.A. Yartys, V.V. Burnasheva and K.N. Semenenko, Sov. Adv. Chem., 52 (1983) 529.
[11] T. Olavesen, H.W. Brinks, V.A. Yartys and B.C. Hauback, to be published.
[12] K. Łątka, R. Kmieć and J.Gurgul, J. Alloys Comp. 319 (2001) 43.
[13] K. Łątka, E.A. Görlich, J. Gurgul and R. Kmieć, Hyperfine Interactions 126 (2000) 299.

MONTE CARLO SIMULATION OF THE MÖSSBAUER SPECTRA IN TERNARY FeCoMo ALLOYS

N. Ayari and S. Alleg

Laboratoire de Magnétisme et de Spectroscopie des solides (LM2S) - Département de Physique - Faculté des Sciences – Université de Annaba B. P.12 (23000) Algérie.

Abstract

Monte Carlo simulations of the local environment and the Mössbauer spectra for binary $Fe_{(1-X)}Mo_X$, $Fe_{50}Co_{50}$ and ternary $(FeCo)_{(1-X)}Mo_X$ (with x = 2, 3, 4, 5 and 6 at.%) alloys, were performed at a temperature of T = 650 °C. The results are in good agreement with the experimental ones and allow us to calculate the configuration distributions linked to the most probable sites and to predict the shape of the Mössbauer spectra.

Keywords : Monte Carlo simulation, FeCoMo system, Mössbauer spectra.

1. Introduction

In previous Mössbauer studies [1-3], it has been shown that in iron-based alloys containing a few percent of Si, Ni, Mo and other substitution alloying elements, the magnetic hyperfine field (HMF) at the iron atom can be calculated by the additive perturbations model given by the following expression :

$$\mathbf{H}_{\mathrm{Fe}} = \mathbf{H}_0 + \sum_{i} \sum_{m_i, n_i} \Delta\mathbf{H}_i^1 \mathbf{m}_i + \Delta\mathbf{H}_i^2 \mathbf{n}_i \qquad (1)$$

where H_0 = 33.0 T is the HMF for pure iron ; m_i and n_i are the number of atoms of the ith alloying element as near (NN) and next near (NNN) neighbours respectively ; ΔH_i^1 and ΔH_i^2 are the increments of the hyperfine magnetic field caused by the ith element.

For the non-dilute $Fe_{(1-x)}Co_x$ alloys (with x ≥ 25 at.%), where the additive perturbations model is inappropriate, the HMF is given by [3]:

$$\mathbf{H}_{Fe} = \mathbf{70}\mu_{Fe} + \mathbf{8.4}\sum_{m}\mu_m \qquad (2)$$

with μ_m the magnetic moment of the NN atom of the Fe atom and μ_{Fe} the magnetic moment of Fe atom which is a function of the order type in the alloy [3].

As the effect of impurities in the second sphere cannot be neglected in the ternary alloys containing a few amount of the third element [4], we have considered the FeCoMo alloys as pseudo-binary (FeCo)Mo and we have taken into account the effect of Mo atoms as NNN. The HMF is calculated using the proposed model :

$$\mathbf{H}_{Fe} = \mathbf{70}\mu_{Fe} + \mathbf{8.4}\sum_{m}\mu_m + \Delta\mathbf{H}^2_{Mo}\mathbf{n}_{Mo} \qquad (3)$$

The aim of the present work is to check the validity of equation 3 for FeCoMo alloys using Monte Carlo simulations. The atomic rearrangements were simulated at a temperature of T = 650°C within a rigid bcc lattice with 2×24^3 lattice sites and periodic boundary conditions were applied. The exchange between nearest neighbours atoms is accomplished according to Metropolis form [5].

2. Results and discussions

Figure 1 shows the hyperfine magnetic field distributions and corresponding Mössbauer spectra for the ternary $(FeCo)_{(1-x)}Mo_x$ alloys. One can observes that the Mo addition to $Fe_{50}Co_{50}$ alloy gives rise to satellites lines corresponding to reduced HMF values. In fact, it is known that the presence of one Mo atom as NN or NNN of iron atom decreases the HMF of about 3.52T [6]. Thus, the main component of the HMF distribution which is centered on H = 34.10 T is attributed to the FeCo-B_2 phase, and the other ones are due to

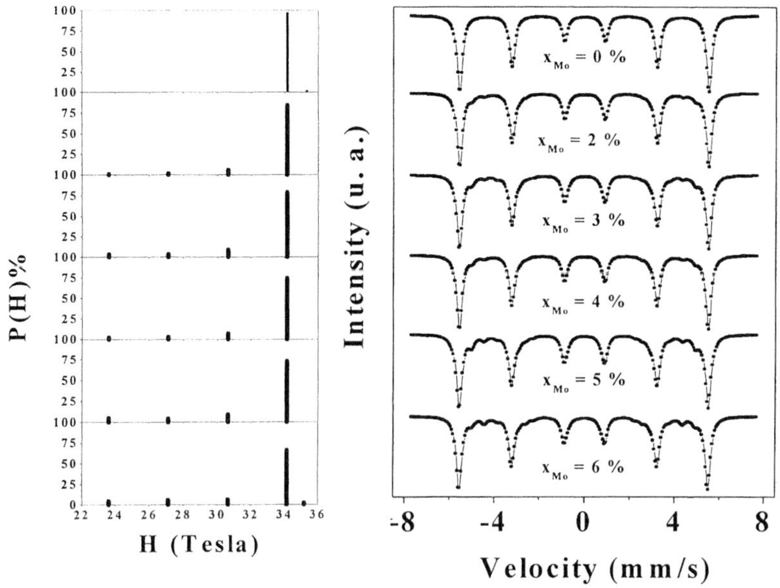

(a) (b)

Figure 1 : Simulated HMF distribution and corresponding Mössbauer spectra for $(FeCo)_{(1-X)}Mo_X$ alloys.

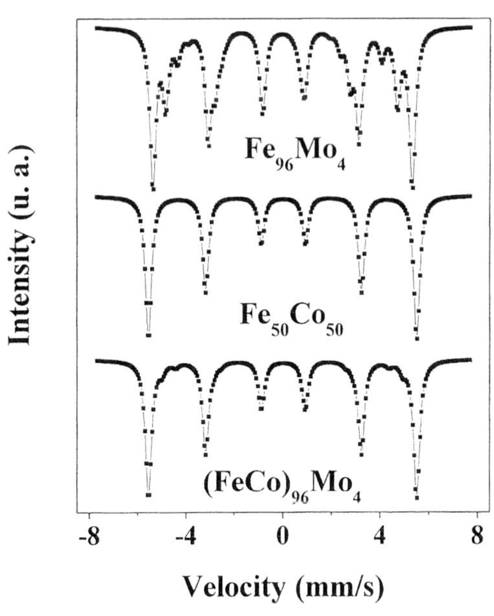

Figure 2 : Simulated Mössbauer spectra of the binary $Fe_{96}Mo_4$, $Fe_{50}Co_{50}$ and ternary $(FeCo)_{96}Mo_4$ alloys.

different atomic configurations where the Fe atom is surrounded by 1, 2 and 3 Mo atoms as NN and/or NNN.

Comparatively to binary FeMo alloys, the FeCoMo alloys, however, show not well-resolved satellites peaks in their Mössbauer spectra (figure 2). The decrease in the Mo satellite intensity is attributed to ordering which takes place and to the move away of Mo atoms from the NN sphere of the Fe atom to Co NN environment leading to the formation of the Co-Mo clusters. In fact, the probability to find one Mo atom as NN diminishes from 24.5% to 8.7 in the binary $Fe_{98}Mo_2$ and ternary $(FeCo)_{98}Mo_2$ alloys respectively, while the probability to find more than 2 Mo atoms as third nearest neighbours increases from 1.2 % to 3.4 % for the same alloys. These results are in good agreement with both the metallic theory of Miedema [7] according to which Co-Mo bonding is preferred to Fe-Mo bonding and previous experimental studies [6, 8].

3. References

[1] C. E. Johnson, M. S. Ridout and T. E. Cranshaw, Proc. Phys. Soc. 81 (1963) 1078.

[2] C. D. Mariadassou, L. Bessais and C. Servant, Phys. Rev. B51 (14) (1995) 8830.

[3] J. P. Eymery and P. Moine, J. Phys. Lett. 39 (1978) L-23.

[4] B. Fultz, in " Mössbauer Spectroscopy Applied to Magnetism and Materials Science", vol. 1, ed. J. L. Gary, F. Grandjean, Plenum, New York (1993).

[5] N. Metropolis, A. W. Rosenbluth, M. N. Rosenbluth and A. H. Teller, J. Chem, Phys. 21 (1953) 1087.

[6] B. Fultz, H. H. Hamdeh and D. H. Pearson, Acta Metall. 37(11) (1989) 2841.

[7] A. R. Miedema, P. F. de Chatel and F. R. deBoer, Physica B 100 (1980) 1.

[8] S. Sarkar and C. Bansal, Acta Mater. 49 (2001)1789.

Structural changes of an austenitic stainless steel induced by rf plasma nitriding

Yvonna Jirásková, Milan Svoboda, Oldřich Schneeweiss and Carsten Blawert*
Institute of Physics of Materials, Academy of Sciences of the Czech Republic, Žižkova 22, CZ-616 62 Brno, Czech Republic, E-mail:jirasko@ipm.cz
**Zentrum für Funktionswerkstoffe gGmbH Clausthal, Sachsenweg 8, D-38678 Clausthal-Zellerfeld, Germany*

Abstract. RF plasma nitriding at 400 °C for 24 h was used to modify structural properties of the X5CrNi189 austenitic stainless steel. The results of the surface and bulk phase composition studied by Mössbauer spectroscopy are correlated with the results of cross-sectional transmission electron microscopy.

Keywords: rf plasma nitriding, iron nitrides, stainless steel, Mössbauer spectroscopy, TEM

1. Introduction

The properties and structural stability of an austenitic stainless steel (ASS) depend on chemical composition. In general, ASS shows low hardness and wear resistance but good corrosion resistance given by presence of chromium in solid solution. A binding of Cr in compound, e.g. Cr nitrides and/or carbides, would deteriorate the resistance to corrosion, mainly intergranular one, in a substantial way. This is important to keep in mind by a choice of the treatment conditions. Besides the processes of precipitation the austenite structure is also susceptible to transformation into martensite either due to a temperature decrease below M_s or as a consequence of high plastic deformation. The nitriding of ASS is a well-known process allowing a sufficient nitrogen uptake in the surface layers (Wei et al., 1996) influencing the structural properties in a positive way. The martensitic transformation as well as the phase composition of nitrogen treated samples can be effectively studied by Mössbauer spectroscopy used here to follow the structural properties of the X5CrNi189 ASS prior and after rf plasma nitriding at 400 °C/24 h.

2. Experimental methods

A commercial 29 μm thick X5CrNi189 ASS foil of nominal composition [wt.%] 0.05C-1Si-2Mn-18Cr-9Ni-remainder Fe was used for sample

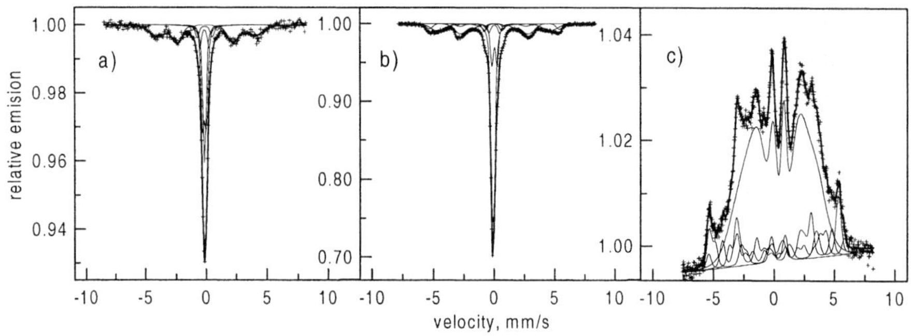

Figure 1. Mössbauer γ-ray transmission spectra of X5CrNi189 austenitic stainless steel foil; a) as-prepared b) rf plasma nitrided at 400 °C/24 h, c) conversion electron Mössbauer spectrum of nitrided surface.

preparation. The studies were performed in the as-prepared (AP) and plasma nitrided (PN) states.

Nitriding was done in a hot wall chambre in the N plasma generated by 300 W of rf power applied to an internal antenna. Details are described in (Jirásková et al., 2001). The microstructure was studied using the Philips CM12 TEM microscope working at 120 kV. The thin foils were prepared either by a standard ion thinning procedure for the AP state or by a cross-sectional method for the PN. The phase composition was studied at room temperature by γ-ray transmission and conversion electron Mössbauer spectroscopy (TMS, CEMS) using ^{57}Co in Rh matrix. An α-Fe was used for the velocity scale calibration. The computer processing of spectra was done using the CONFIT program package (Žák, 1999) giving the hyperfine parameters (hyperfine induction B [T], isomer shift δ [mm/s] and quadrupole splitting ϵ [mm/s]).

3. Results and discussion

The TMS spectrum of AP sample in Fig. 1a exhibits a superposition of the subspectra, the single- ($\delta = -0.11$ mm/s) and double-line ($\delta = -0.10$ mm/s, $\epsilon = 0.64$ mm/s) of total intensity I of 56 % for the austenite and the six-line (B = 24.8 T) for the martensite. Besides these components two more doublets of total intensity of 9 % ($\delta_1 = -0.09$ mm/s, $\epsilon_1 = 1.48$ mm/s and $\delta_2 = -0.08$ mm/s, $\epsilon_2 = 2.22$ mm/s) have to be used for appropriate spectrum fit. They could represent the ϵ-martensite which was detected by microscopic observations. TEM of AP sample, Fig. 2a, has shown that the microstructure is formed by austenite (fcc) grains of lattice parameter a = 0.36 nm separated by thick grain boundaries of highly deformed material with high dislocation density. The

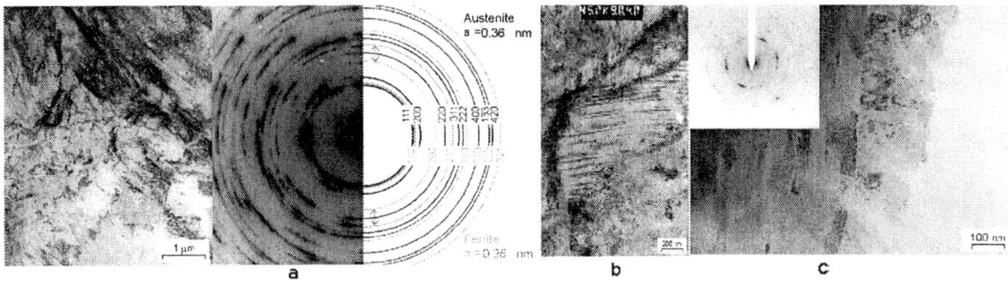

Figure 2. TEM micrographs and diffraction patterns of X5CrNi189 austenitic stainless steel: (a) as-prepared state; (b) ϵ-martensite needles; (c) rf plasma nitrided at 400 °C for 24 h.

grain size was determined to be $\leq 1\,\mu$m. The presence of α'-martensite (bct) is unambiguously demonstrated by 200 and 211 circular lines in diffraction pattern (Fig. 2a). Moreover the ϵ-martensite (hcp) needles are well visible in some austenite grains (Fig. 2b). It is difficult to get their diffraction with respect to high deformation. Nevertheless, a needle-like morphology is in good agreement with that one observed in (Gladman and Folaris, 1997). The high value of c/a ratio (1.64) could be a reason of arising the large quadrupole splitting of doublets detected by TMS.

A spectrum of the AP sample after additional annealing at 400 °C for 24 h, which should simulate an influence of the temperature and time used in process of the rf plasma nitriding, has shown an increase in B (29 T) accompanied by slight decrease in I due to a partial annealing out of defects. A change in charge density in the surrounding of resonanting iron atoms due to changes in configuration and redistribution of alloying elements causes an increase in δ. The cross section of PN sample (Fig. 2c) shows clearly different microstructure of the bulk and surface confirmed by Mössbauer measurements. While the ring patterns from the surface can be ascribed to expanded austenite with lattice parameter of 0.39 nm, the bulk is formed mainly by austenite and ferrite. Small dark spikes, concentrated mainly in the surface layers, correspond to CrN precipitates. In the TMS spectrum of PN, Fig. 1b, the paramagnetic austenite dominates (63 %). It is represented by three doublets. A slight increase in δ ($0.05 \div 0.1$ mm/s) and ϵ ($0.04 \div 1.4$ mm/s) in comparison to AP state is due to a higher content of N atoms and also C atoms that are pushed into the bulk by implanted N atoms (Blawert et al., 2000). This fact together with a high defect density are the reasons for the formation of ferrite which is represented by the broadened ferromagnetic subspectra. Visible different spectrum was measured by CEMS, Fig. 1c. The high content of nitrogen atoms in the surface layers is the reason that mainly ferromagnetic phases were observed in the surface layers. The best fit was obtained by distribution of hyperfine inductions with mean value of B = 16.8 T representing the major EA phase (I = 67 %) and by setup of discrete six-line components. According to literature data it was possible to ascribed the components with B = 34.8 – 22.9 T of total I = 9 % to γ'-Fe_4N and with B = 28.1 – 19.3 – 9.9 T of total I = 12 % to ϵ-Fe_3N phase. Approximately 10 % of the surface is formed by ferrite with CrN precipitates which cannot be directly proved by Mössbauer phase analysis. Their presence in ferrite influences the values of δ (0.30 mm/s) as well as ϵ (−0.19 mm/s).

4. Conclusions

The microstructure of as-prepared CrNi austenitic stainless steel foil is formed by 1 μm large austenite grains separated by thick grain boundaries of deformed material. High stress relief has evoked a partial transformation of austenite into ϵ- and subsequently into α'-martensite as detected by Mössbauer spectroscopy and TEM. Rf plasma nitriding leads predominantly to a modification of the surface layers containing the expanded austenite and minor γ'-Fe_4N, ϵ-Fe_3N, and ferrite/CrN.

Acknowledgements

This work was supported by the Czech Ministry of Education, Youth and Sports (ME373), by AS CR (S2041105) and by DFG (Mo319/30-1).

References

Wei, R., J.J. Vajo, J.N. Matossian, P.J. Wilbur, J.A. Davis, D.L. Williamson and G.A. Collins: A comparative study of beam ion implantation, plasma ion implantation and nitriding of AISI 304 stainless steel. *Surface and Coatings Technology*, 83:235–242, 1996.

Jirásková, Y., M. Svoboda, O. Schneeweiss, T. Žák, C. Blawert: Structural Properties of Austenitic Stainless Steel in the Initial State and After rf Plasma Nitriding. *Czech Journal of Physics*, 51: 2001 accepted for publication.

Žák, T. Confit for Windows 95. In *Mössbauer Spectroscopy in Material Science*, Kluwer Academic Publishers, Dordrecht, 385–390, 1999.

Gladman, G. and T. Foularis: Microscopical Characterisation of Martensite Formation in a Metastable Austenitc Stainless Steel. *Journal de Physique IV France*, C5:423–428, 1997.

Blawert, C., B.L. Mordike, G.A. Collins, K.T. Short, Y. Jirásková, O. Schneeweiss, V. Peřina: Characterisation of duplex structures produced by simultaneous implantation of nitrogen and carbon into austenitic stainless steel X5CrNi189. *Surface and Coatings Technology*, 128–129:219–225, 2000.

Mössbauer study of radiation effects on swift heavy ion irradiated Fe-Ni-Cr multilayers and electrodeposited alloys

E. Kuzmann[a], S. Stichleutner[a], G. Principi[b], C. Tosello[c], M. El-Sharif[d], C. U. Chisholm[d], K. Havancsák[e], I. Gerőcs[e], Z. Klencsár[a], Z. Homonnay[a] and A. Vértes[a]

[a]*Department of Nuclear Chemistry and Research Group of Nuclear Methods In Structural Chemistry HAS, Eötvös Loránd University, Budapest, Hungary*
[b]*Settore Materiali and INFM, DIM, Università di Padova, Italy*
[c]*Department of Physics and INFM, Trento University, Trento, Italy*
[d]*Glasgow Caledonian University, Glasgow, U.K.*
[e]*Department of Solid State Physics, Eötvös Loránd University, Budapest, Hungary*

^{57}Fe conversion electron Mössbauer spectroscopy (CEMS) and X-ray diffractometry (XRD) were applied to study radiation effects on Fe-Ni-Cr electrodeposited alloys and vacuum deposited multilayers. As an effect of irradiation new sextet and doublet components appear in the CEM spectra of irradiated multilayers. They correspond to the highly disordered microcrystalline ferromagnetic and paramagnetic metastable phases formed in Fe-Ni-Cr alloys. Although these phases never occur in thermally prepared alloys, they are similar to those observed in electrodeposited Fe-Ni-Cr coatings of the same composition [1, 2]. The metastability of the phases was also shown by aging of the irradiated multilayers. Small but significant dose dependent changes were observed in the CEM spectra of irradiated electrodeposits existing in the form of amorphous or microcrystalline paramagnetic solid solution phase. This finding can be discussed in terms of irradiation induced change in the short range order and phase composition due to irradiation of the alloys.

Keywords: electrodeposits, multilayers, ^{57}Fe conversion electron Mössbauer spectroscopy, swift heavy ion irradiation

1. Introduction

Fe-Ni-Cr alloy coatings have remarkable scientific and technological interest. In previous investigations we have found that electrochemical deposition [2], ion beam mixing of multilayers by medium energy heavy ions [3, 4] as well as ball milling preparation [1] are fairly good methods to prepare Fe-Ni-Cr alloys containing metastable ferromagnetic and paramagnetic phases with excellent physical, chemical and mechanical properties from the point of view of technology.

The aim of our present work was to study the radiation effects of energetic heavy ions in Fe-Ni-Cr multilayers and in electrodeposited Fe-Ni-Cr coatings. We intended to get information on whether the energetic heavy ion irradiation can or cannot be an alternative route for the preparation of Fe-Ni-Cr alloys in comparison with electrochemical deposition. For these studies ^{57}Fe CEMS and XRD were used.

2. Experimental

The Fe-Ni-Cr electrodeposits were prepared in a continuous flow plating system at conditions depicted in Table 1. The samples contained 89% Cr and 3% Ni.

Table 1
Parameters of electrodeposition

Bath constituent		Electrodeposition Conditions	
$[Cr(H_2O)_4Cl_2]Cl.2H_2O$	0.80 mol/l	Time of deposition	5 min
$NiCl_2.6H_2O$	0.20 mol/l	Electrolyte temperature	20 °C
$FeCl_2.4H_2O$	0.05 mol/l	Electrolyte velocity	0.67 m/s
NaCl	0.50 mol/l	Anode (40x60 mm)	graphite
NH_4Cl	0.50 mol/l	Cathode (30x10 mm)	copper
$B(OH)_3$	0.15 mol/l	Current density	40 A/dm^2
$AlCl_3.6H_2O$	0.45 mol/l		
Glycine	1.00 mol/l		

50%Fe-25%Ni-25%Cr multilayered samples were prepared by physical vacuum deposition using e-gun, at a pressure of about $6\cdot10^{-6}$ Pa, according to the following scheme: upilex/Ni(4.6)/^{57}Fe(10)/Cr(5)/Ni(4.6)/^{57}Fe(10)/Cr(5)/Al(2). The numbers in parentheses represent the layer thickness in nanometers.

The irradiation of samples was carried out with 246 MeV energy $^{86}Kr^{8+}$ ions at room temperature, at a current density of 0.5 μA/cm^2 and at a vacuum of about 10^{-3} Pa at the U-400 cyclotron of the Laboratory of Nuclear Reactions, JINR, Dubna, Russia. For the irradiation of multilayers fluences of $1\cdot10^{14}$ ions·cm^{-2} and $3\cdot10^{14}$ ions·cm^{-2} were applied. The electrodeposits were irradiated with fluences of $1\cdot10^{13}$ ions·cm^{-2}, $5\cdot10^{13}$ ions·cm^{-2} and $1\cdot10^{14}$ ion·cm^{-2}. The temperature of the target did not exceed 15 °C during irradiation.

The irradiated multilayers were heat treated isothermally at temperatures of 450 °C and 600 °C for 3 h in an evacuated quartz tube.

CEM spectra were recorded, at room temperature, by a conventional Mössbauer spectrometer (WISSEL) with a flowing gas (96% He, 4% CH_4) proportional counter and a ^{57}Co(Rh) source of 1.85 GBq activity. Isomer shifts are given relative to α-Fe. The evaluation of Mössbauer spectra was performed by least-square fitting of the lines using the MOSSWINN program [5].

XRD measurements were performed by Philips X'Pert powder X-ray diffractometer using $Cu_{K\alpha}$ (λ=0.154 nm) radiation and graphite monochromator as well as by a DRON-2 diffractometer using $Co_{K\alpha}$ (λ=0.179 nm) radiation with β-filter. The X-ray diffractograms were evaluated by the computer program ExRay developed at the Eötvös Loránd University, Budapest.

3. Results and discussion

Fig. 1 shows the CEM spectra of 50%Fe-25%Ni-25%Cr multilayered samples before and after the irradiation with different fluences, as well as after the subsequent heat treatments. While the spectrum of non-irradiated multilayer consists mainly of a sextet belonging to α-iron, new spectral components appear in the spectra of irradiated samples. Namely, a sextet (δ = –0.1 mm/s, B_{HF} = 26.1 T) and a doublet (δ = –0.1 mm/s, ΔE_Q = 0.51 mm/s) dominate in the spectra recorded after irradiation with fluences of 10^{14} ions·cm^{-2} and $3\cdot10^{14}$ ions·cm^{-2}. These subspectra are assigned to the highly disordered, metastable ferromagnetic and paramagnetic solid solution phases in the Fe-Ni-Cr alloys [1]. These phases never occur in the thermally prepared alloys of the same composition. On the other hand, all these phases correspond well to those obtained and identified in Fe-Ni-Cr samples prepared by other non-equilibrium techniques like electrochemical deposition, ball milling, ion beam mixing and laser beam irradiation of multilayers. The amount of metastable phases increases with the dose. The Mössbauer spectra of the sample irradiated with a fluence of $3\cdot10^{14}$ ions·cm^{-2} exhibit considerable changes after heat treatments at 450 °C and 600 °C. While the magnetically split components disappear, a singlet (δ = –0.1 mm/s) characteristic of the equilibrium paramagnetic phase develops gradually with the aging temperature. This process is revealed also by the corresponding XRD showing the gradual appearance of the equilibrium crystalline fcc phase [6]. The composition of the fcc phase was roughly estimated to be around 20%Cr and 20%Ni. Thus, the transformation of the radiation induced metastable phases to the stable alloy phase has been observed. The swift heavy ion irradiation of vacuum deposited multilayers proved to be an excellent tool to prepare Fe-Ni-Cr alloys by ion beam mixing. Comparing our results with those obtained with the ion beam mixed multilayers irradiated with 100 keV energy Xe ions [3, 4], the differences in the alloy formation can be attributed to the difference in energy transfer mechanisms from the ion beam to the target. Namely, the electronic excitation dominates in the case of high energy (246 MeV) irradiation, while the nuclear stopping is responsible mainly for energy transfer mechanism in the case of irradiation with 100 keV ions [6].

Fig. 2 shows the details of XRD patterns of Fe-3%Ni-89%Cr electrodeposited alloys before and after irradiation. The non-irradiated sample is in a highly disordered

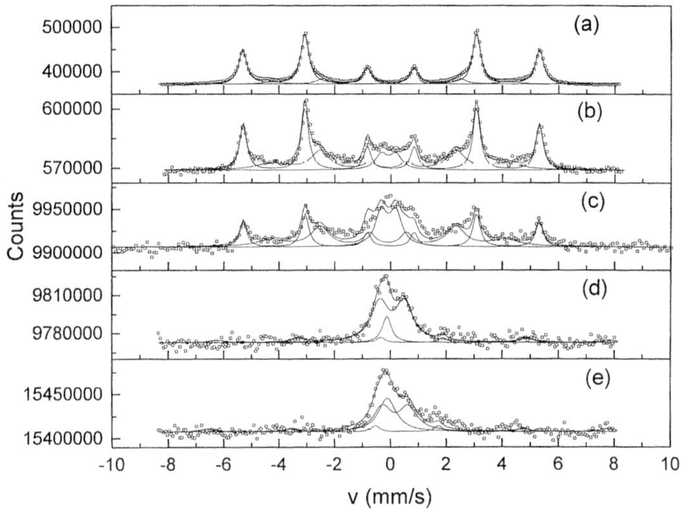

Figure 1. CEM spectra of Fe-25%Ni-25%Cr multilayers before (a) and after irradiation with 246 MeV Kr ions with fluences of $1 \cdot 10^{14}$ ions·cm^{-2} (b) and $3 \cdot 10^{14}$ ions·cm^{-2} (c) as well as after subsequent heat treatments at 450 °C (d) and 600 °C (e) of the latter.

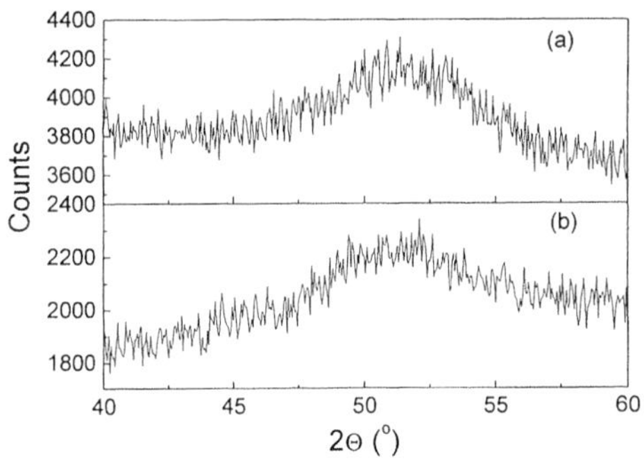

Figure 2. Details of XRD patterns of Fe-3%Ni-89%Cr electrodeposited alloys before (a) and after irradiation with 246 MeV Kr ions with fluences of $1 \cdot 10^{14}$ ion·cm^{-2} (b).

amorphous state. No transformation into the crystalline equilibrium phase occurs due to irradiation. The comparison of XRD patterns can indicate that the amorphous state becomes more disordered after irradiation.

CEM spectra and quadrupole splitting distributions of Fe-3%Ni-89%Cr electrodeposited alloys before and after irradiation are depicted in Fig. 3. The spectrum of the non-irradiated sample consists of a doublet ($\delta = -0.18$ mm/s, $\Delta E_Q = 0.34$ mm/s) representing electrodeposit existing in highly disordered metastable paramagnetic solid solution Fe-Ni-Cr phase in the high Cr concentration region. The energetic heavy ion irradiation causes small but significant dose dependent changes in the spectra. The changes can be analysed by the appearance of a new doublet in the spectra, which corresponds to the gradual development of a new peak in the quadrupole splitting distribution. The new component can be associated with a modifed Fe microenvironments in electrodeposits being in amorphous paramagnetic solid solution phase. The irradiation-induced transformation corresponds to a significant change in the chemical short range order. The transformation rate increases with the dose of

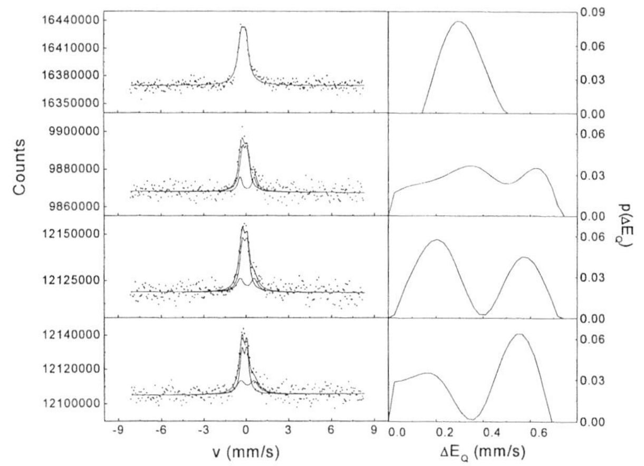

Figure 3. CEM spectra and quadrupole splitting distributions of Fe-3%Ni-89%Cr electrodeposited alloys before (a, e) and after irradiation with 246MeV Kr ions with fluences of $1\cdot10^{13}$ ions·cm^{-2} (b, f), $5\cdot10^{13}$ ions·cm^{-2} (c, g) and $1\cdot10^{14}$ ions·cm^{-2} (d, h).

irradiation. Thus, the irradiation of Fe-Cr-Ni electrodeposits by 246 MeV energy $^{86}Kr^{8+}$ ions results in partial transformation from the as-deposited state to a more disordered metastable paramagnetic phase due to ion beam mixing. Our results are consistent with those obtained with other amorphous alloys irradiated with swift heavy ions [7].

Acknowledgements

This work was supported by the Hungarian OTKA Fund (project No 029537), the Hungarian-Italian Science and Technology Program (project No I-21/98) and Fund of Hungarian Academy of Science (project No AKP 2000-143 2,4). The irradiation performed at the JINR in Dubna was financed by the Hungarian Academy of Sciences.

References

[1] E. Kuzmann, M. El-Sharif, C.U. Chisholm, G. Principi, C. Tosello, K. Havancsák, A.Vértes, K. Nomura,V.K. Garg and L. Takács, Electrodeposited, ion beam mixed and ball milled FeNiCr alloys, Mössbauer Spectroscopy in Materials Science (Eds. M. Miglierini and D. Petridis), Kluwer, 1999, 203-214.

[2] E. Kuzmann, I. Czakó-Nagy, A. Vértes, C.U. Chisholm, A. Watson, M. El-Sharif, J. Kerti and G. Konczos, Hyp. Int. 45 (1989) 397.

[3] G. Principi, R. Gupta, C. Tosello, L.M. Gratton, E. Jannitti, E. Kuzmann and I. Varga, Suppl. Nuovo Cim. 50 (1996) 707.

[4] A. Perin, R. Gupta, G. Principi, C. Tosello, L.M. Gratton, E. Jannitti, E. Kuzmann and Z. Klencsár, Surf. and Coat. Techn. 103-104 (1998) 93.

[5] Z. Klencsár, E. Kuzmann and A. Vértes, J. Radional. Nucl. Chem. Lett . 210 (1996) 105.

[6] E. Kuzmann, G. Principi, C. Tosello, K. Havancsák, S. Stichleutner, I. Gerőcs, Z. Homonnay and A. Vértes, Nucl. Instr. Meth. B (2001) in print.

[7] Kuzmann, E., Spirov, I.N, J. Nucl. Mater. 37, (1985) 22.

Ultra-Thin Fe_xMn_{1-x} Alloy Films on Ag(100)

R. Paniago, J. L. López and H.-D. Pfannes
Departamento de Física- Universidade Federal de Minas Gerais,
C. P. 702, 30123-970, Belo Horizonte – M.G., Brazil

Abstract

Fe_xMn_{1-x} ultra-thin alloy films were grown epitaxially on Ag(100) under ultra-high vacuum conditions by co-evaporation of ^{57}Fe and Mn. Structural and magnetic characteristics of the alloy films were investigated by *in situ* Conversion Electron Mössbauer Spectroscopy as a function of Fe content ($0.4<x<1$) at two thicknesses (3 and 6 monolayers). X-ray photoelectron spectra and Mössbauer spectra were characteristic of Fe-Mn alloying. As indicated by Mössbauer spectra of the 6 monolayers thick films the ferromagnetic α-Fe_xMn_{1-x} phase could be stabilized up to 28% of Mn content, higher than the maximum value of 20% reported previously for epitaxially grown films on GaAs(001) (C. Jing et al., J. Magn. Magn. Mater., 198-199 (1999) 270.). For higher Mn contents ($x<0.5$) the films exhibit typical γ-Fe_xMn_{1-x} (and eventually ε-phase) spectra. However in strained alloy films a complete distinction between the bcc and fcc phase is not possible. Most probably the films suffer an increasing tetragonal deformation with increasing Mn content, which could be responsible for the magnetic transition observed in spectra.

1. Introduction

The structural and magnetic properties of ultra-thin films of iron and iron alloys are currently object of many investigations, motivated by the need to understand their basic properties as well as by possible applications in electronic and magnetic devices. The correlation between lattice structure and magnetic behavior in strained thin alloy films is of great interest. Such systems can exhibit novel physical properties, e. g. when new phases, different from those of the bulk, are stabilized at room temperature by pseudomorphic growth.

Bulk Fe_xMn_{1-x} alloys are found to be in the γ-phase (fcc) over a large concentration range, $0.3<x<0,7$, and are antiferromagnetic (AFM) around room temperature. As the Fe content increases, the alloy undergoes a fcc to hcp (γ-ε) transformation at $x=0.7$. The hcp phase exists up to $x=0.9$, where a transformation from hcp to bcc (ε-α) occurs, i.e. according to the phase diagram of Fe-Mn, the ferromagnetic (FM) α-phase is stable only when the Mn content is <10% [1]. Epitaxy of γ- and α-Fe_xMn_{1-x} thin films on GaAs(001) has been reported for low ($x=0.35$) and high ($x=0.8$) Fe content, respectively [2]. In that case the pure single-crystal α-Fe_xMn_{1-x} phase was extended to 20% of Mn content.

In this work we report on an *in situ* investigation of Fe_xMn_{1-x} ultra-thin films with thicknesses of 3 and 6 monolayers (ML), which were grown on a Ag(100) substrate by molecular-beam epitaxy (MBE). Besides in the alloy formation itself, we were interested in the magnetic properties of those films. We chose to grow Fe_xMn_{1-x} alloy films on Ag(100), since Fe and Mn are known to grow epitaxially on this substrate. The lattice parameter of bulk Fe (2.87 Å) is just 0.8% smaller then the surface unit cell (2.89 Å square) of Ag(100) so that α-Fe(100) can be grown on it. On the other hand Mn films on Ag(100) exhibit a body centered tetragonal (rather than cubic) structure, similar to that of quenched γ-Mn [3].

2. Experimental

The Ag(100) crystal was previously cleaned in ultra high vacuum (UHV) by argon ion sputtering with an energy of 900eV at 300 K and subsequent annealing at 670 K for 10 minutes. The 3 and 6 monolayers thick Fe_xMn_{1-x} films were deposited on the clean Ag(100) substrate by co-evaporation of Fe (95% enriched in ^{57}Fe) and Mn from two separate sources in UHV, typically at a rate of 2 monolayers per minute. High purity (>99.9%) ^{57}Fe and Mn were used as source materials. After each growth and *in situ* measurements the film was removed with ion sputtering (2 keV), and the Ag(100) surface was again prepared, repeating the same procedure. Chemical and structural analysis of the films was performed by X-ray photoelectron spectroscopy (XPS) and reflection high-energy electron diffraction (RHEED). The composition of the alloy films, within an error of ±1%, was determined by peak intensities from XPS-spectra. Chemical shifts obtained from those spectra (not shown here) were typical of Fe-Mn alloying. During the growth of the Fe_xMn_{1-x} films ($0.4<x<1$) RHEED oscillations could only be seen for the first 2 monolayers. However, at a coverage of 6 monolayers still a streaked RHEED pattern with some degradation was observed (fig. 1). Microscopic structural and magnetic properties were then studied by *in situ* Conversion Electron Mössbauer Spectroscopy (CEMS) at 100-300 K.

(a) (b) (c)

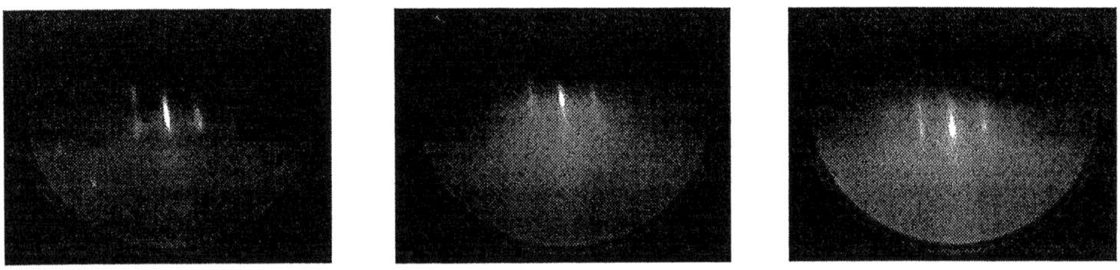

Fig.1: RHEED patterns at T=390 K along the [010] azimuth, of Ag(100) (a), and after growth of 3 ML–$Fe_{0.39}Mn_{0.61}$/Ag(100) (b), and 6 ML –$Fe_{0.56}Mn_{0.44}$/Ag(100) (c).

3. Results and Discussion

The Mössbauer spectra (fig. 2) were fitted with field distributions characteristic of disordered alloys. The magnetic orientation is mainly in-plane for all samples since the relative intensities of lines 1(6), 2(5) and 3(4) of the main sextets in Mössbauer spectra were approximately 3:4:1. For comparison a Mössbauer spectrum of Fe/ Ag(100) was recorded (fig. 2a) and could be fitted with two sextets (32.2 and 31.3 T, respectively), which can be attributed to bulk-like Fe, and Fe at the Fe/Ag interface. The B_{hf}=3 T peak in the field distribution curves for Fe_xMn_{1-x} in fig. 2b and 2c does not stem from a superparamagnetic state of small islands, since low temperature spectra do not change at all (fig.3). We attribute it to the amount of Fe_xMn_{1-x} alloy film in the antiferromagnetic γ-phase, which has approximately this field for bulk alloys [4]. The spectra show clearly that the amount of this phase increases with decreasing x and decreases with increasing film thickness from 3 to 6 ML. We concentrate our attention on the 6ML films, since they can be more easily compared to bulk phases. The singlet we included to fit the spectra of fig.

2c we attribute to a paramagnetic (PM) γ-phase or even to an antiferromagetic (AFM) ε-phase with a field of 1T (see ref. [4]), which are hardly distinguishable from each other by Mössbauer spectroscopy.

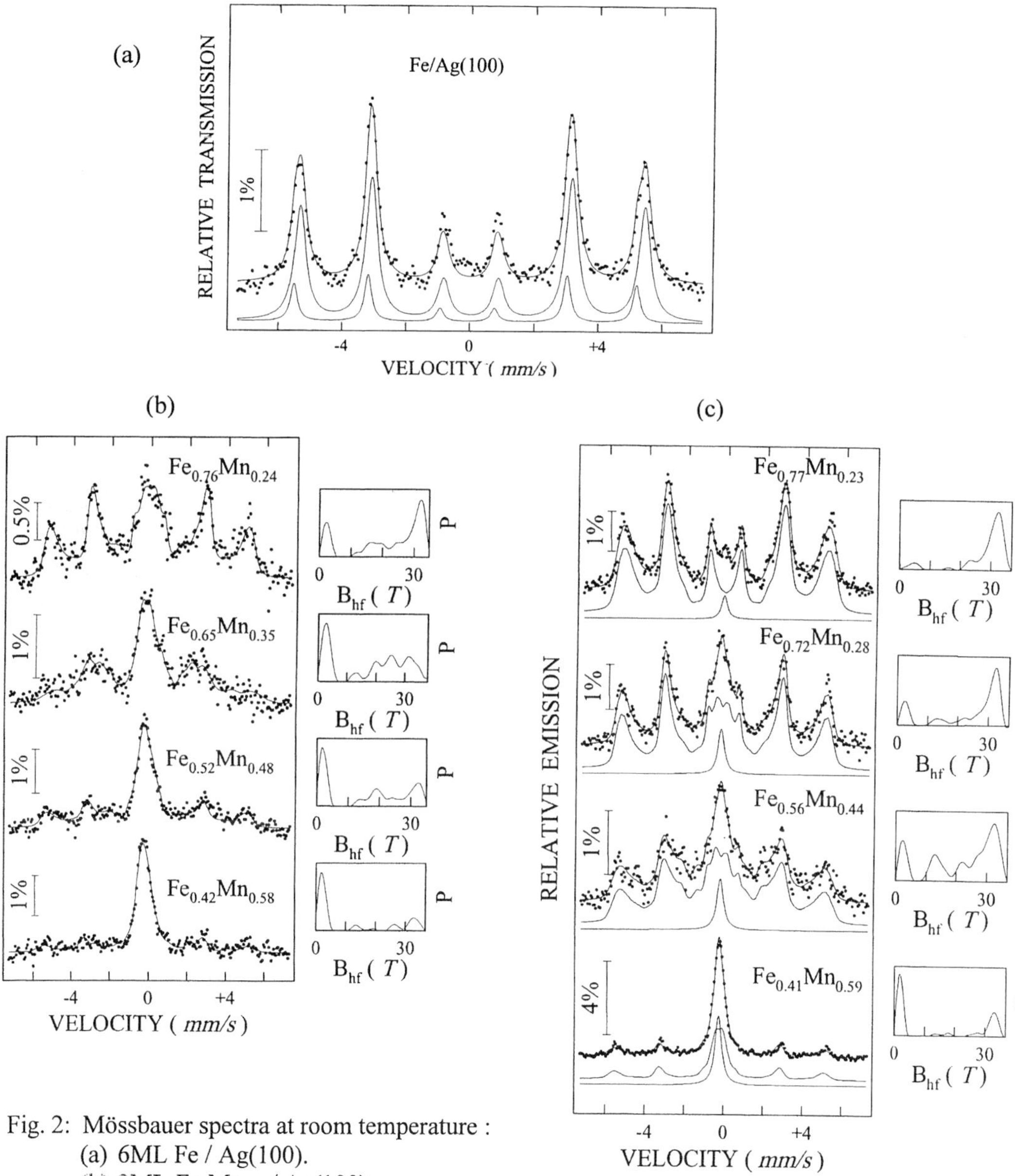

Fig. 2: Mössbauer spectra at room temperature :
(a) 6ML Fe / Ag(100).
(b) 3ML-Fe_xMn_{1-x} / Ag(100).
(c) 6ML-Fe_xMn_{1-x} / Ag(100).

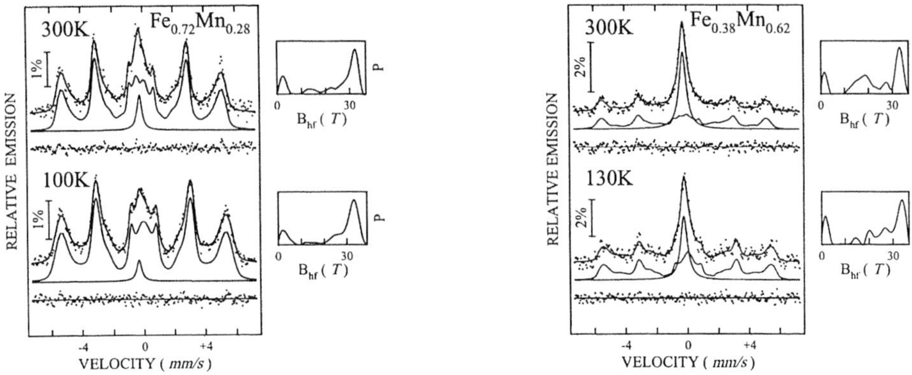

Fig 3: ^{57}Fe Mössbauer spectra of 6 ML- $Fe_{0.72}Mn_{28}$/Ag(100) (left) and 4.5 ML-$Fe_{0.38}Mn_{0.62}$/Ag(100) (right) at the indicated temperatures.

As indicated by the 6ML-Fe_xMn_{1-x}/Ag(100) Mössbauer spectra with $x>0.72$ (fig.2c), which exhibit field distributions concentrated in the B_{hf}=30T region, we conclude that the FM bcc-Fe_xMn_{1-x} phase could be stabilized up to 28% Mn (x=0.72). This is higher than the value of 20% reported previously for epitaxially grown films on GaAs(001) [2]. However we point out that in strained alloy phases a complete distinction between the bcc and fcc phase is not possible. Since the RHEED pattern does not change at all, specially the distance between the streaks is constant, we believe that we obtained a pseudomorphic growth of Fe_xMn_{1-x} on top of the Ag(100) crystal. However we can say nothing about the distance between layers, in other words the film could exhibit a tetragonal distortion in z-direction. Studies of $[Fe_xMn_{1-x}/Ir(100)]_{20}$ superlattices (x=0.7, 0.9) for example indicate that the alloy structure deviates significantly from the cubic symmetry by a tetragonal deformation [5]. In another paper [6] is claimed that for superlattices with x=0.75 in the AFM (or PM) region, the alloys have a distorted fcc structure, whereas in the FM region, they have a distorted bcc one. In fact the bcc and fcc structure are two special cases of the general tetragonal lattice - the fcc structure may be viewed as a continuous distortion of the bcc structure, and vice versa. We believe that in our case the Fe_xMn_{1-x} bcc film structure suffers an increasing tetragonal deformation with increasing Mn content, which could be responsible for the magnetic transition observed in the spectra.

Acknowledgements

This work was supported by FAPEMIG and CNPq (Brazil).

References

[1] M. Hansen, K. Ardenko, Constitution of Binary Alloys, McGraw-Hill, New York, 1958.
[2] C. Jing, Y.Z. Wu, Z. X. Yang, G. S. Dong and X. F. Jin, J. Mag. Mag. Mat. **198-199** (1999) 270.
[3] B. T. Jonker, J. J. Krebs, G. A. Prinz, Phys. Rev. B **39** (1989) 1399.
[4] T. Hinomura, S. Nasu, Y. Tomota, J. Jap. Inst. Metals, **62** (1998) 635.
[5] A. Déchelette, J.M. Tonnerre, M.C. Saint Larger, F. Bartolomé, L. Sève, D. Raoux, H. Fischer, M. Piecuch, V. Chakarian, C. C. Kao Phys. Rev B **60** (1999) 6636.
[6] A. Déchelette, M. C. Saint Larger, J. M. Tonnerre, G. Patrat, D. Raoux, H. Fischer, S. Andrieu, and M. Piecuch, Phys. Rev B **60** (1999) 6623.

Interface structure and magnetism of Fe/Cr multilayers

I. Sakamoto[a], S. Honda[b], M. Koike[a], H. L. Shen[a], and H. Tanoue[a]
[a] *National Institute of Advanced Industrial Science and Technology, 1-1-1 Umezono, Tsukuba, Ibaraki 305-8568, Japan*
[b] *Shimane University, 1060 Nishikawatsu, Matsue , Shimane 690-8504, Japan*

The interface structure and magnetism of Fe/Cr multilayers (MLs) have been investigated as a function of the thickness of Fe and Cr layers by conversion electron Mossbauer spectroscopy (CEMS), magnetization and magnetoresistance (MR) measurements. The CEM spectra show the change of the interface structure in $[Fe\ 0.3\ nm/Cr\ ]_{200}$ MLs with Cr layer thickness and the highest average value of distributed hyperfine fields (H_{hf}) for 1.2 nm Cr layer thickness. On the other hands, the average H_{hf} in $[Fe\ 0.9\ nm/Cr\ ]_{67}$ MLs are almost same values for the all Cr layer thickness measured here. The average H_{hf} and the saturation magnetization decrease with decreasing Fe layer thickness. Furthermore, the magnetization and the MR curves behave differently in 0.3 nm and 0.9 nm Fe layer thickness depending on Cr layer thickness, most probably due to the different interface structures.

Keywords: GMR, multilayers, interface structure, CEMS

1. Introduction

The giant magnetoresistance (GMR) properties are strongly affected by the interface structure between the magnetic and non-magnetic layers in metallic multilayers (MLs). The results of conversion electron Mossbauer spectroscopy (CEMS) measurements and the hyperfine field (H_{hf}) calculations in Fe/Cr MLs have been frequently reported in relation to the interface roughness, atomic interdiffusion and their dependence on the growth [1-5]. However the study of GMR properties related to the interface structure estimated from Mossbauer measurements is a few [6-8].

In this paper, we investigate systematically the GMR properties and the interface structures of Fe/Cr MLs using CEMS measurements as a function of the thickness of Fe and Cr layers.

2. Experimental

We have prepared [Fe 0.3-1.2 nm/Cr 0.6-1.8 nm] MLs on Si (100) substrates by helicon plasma sputtering method. The deposition rates of Fe and Cr layers were 0.04 and 0.08 nm/sec, respectively. The pressure of Ar gas was 5.5×10^{-4} Torr. The total Fe layer thickness is 60 nm for all samples. A vibrating sample magnetometer (VSM) was used to measure the magnetization curves at room temperature (RT) up to 15 kG. Resistivity was measured at RT by 4 points probe method. The CEMS measurement was performed at RT with a Mossbauer spectrometer using 370 MBq ^{57}Co γ-ray source (Rh matrix), and the conversion electrons were detected with a proportional counter flowed with He + 10% methane mixture gas. Conversion electron Mossbauer (CEM) spectra were analyzed by least squares fitting assuming the overlapped Lorentzian curves of sextet peaks.

3. Results and discussion

The magnetization and MR curves strongly depend on Fe and Cr layer thickness. The magnetization curves of [Fe 0.3 nm/Cr] and [Fe 0.9 nm/Cr] MLs indicate the antiferromagnetic coupling between Fe layers for Cr 0.6 nm and 1.2 nm and the ferromagnetic coupling for Cr 1.8 nm. The saturation magnetization decreases with decreasing Fe layer thickness as seen for Fe 0.3 nm. This decrease of the saturation magnetization seems to be due to a breaking of continuous Fe layers. The MR curves depend on the behavior of the magnetization curves and the MR ratios are 10% and 12% at 13 kG in $[Fe\ 0.3\ nm/Cr\ 0.6\ nm]_{200}$ and $[Fe\ 0.9\ nm/Cr\ 1.2\ nm]_{67}$ MLs, respectively. As shown in Fig.1, the CEM spectra change extremely with Fe layer thickness, and the total spectral area decreases with increasing Cr layer thickness owing to the depth dependence of the CEMS sensitivity. In Fig.1, the dependence of the CEM spectra on Cr layer thickness is clearly seen for 0.3 nm Fe layer thickness. Furthermore, we can observe the additional small peak near 0 mm/sec in the CEM spectrum of $[Fe\ 0.3\ nm/Cr\ 1.8\ nm]_{200}$ MLs, which implies a existence of non-magnetic Fe atoms in Cr layers. Assuming the H_{hf} distribution for the broad peak width and the intensity ratio (P1:P2:P3 = 3:4:1 corresponding to H_{hf} parallel to the film plane) of the sextet peaks, the hyperfine parameters were estimated and the average H_{hf} is tabulated in Table 1 with the MR ratio at 13 kG. The values of the average H_{hf} decrease with decreasing Fe layer thickness for 1.2 nm Cr layer thickness as described in the magnetization curves, and change with Cr layer thickness in 0.3 nm Fe layer thickness; the $[Fe\ 0.3\ nm/Cr\ 1.2\ nm]_{200}$ MLs shows the highest average H_{hf}. As described above, the MR depends on the layer thickness of Fe and Cr layers. The Cr layer thickness dependence of the MR ratio is due to the oscillation of the exchange interaction as a function of the Cr layer thickness which has a period of 1.8 nm [9]. In this case, however, the oscillation behavior of the MR ratio is different between 0.3 nm Fe and 0.9 nm Fe layer thickness as expected from the change of the interface structure like the discontinuities in 0.3 nm Fe layers.

Most probably, the decrease of average H_{hf} with decreasing Fe layer thickness suggests the increase in the contribution of the interface region. In the cases of 0.3 nm Fe layer thickness, the average H_{hf} changes with Cr layer thickness; that is, the highest value

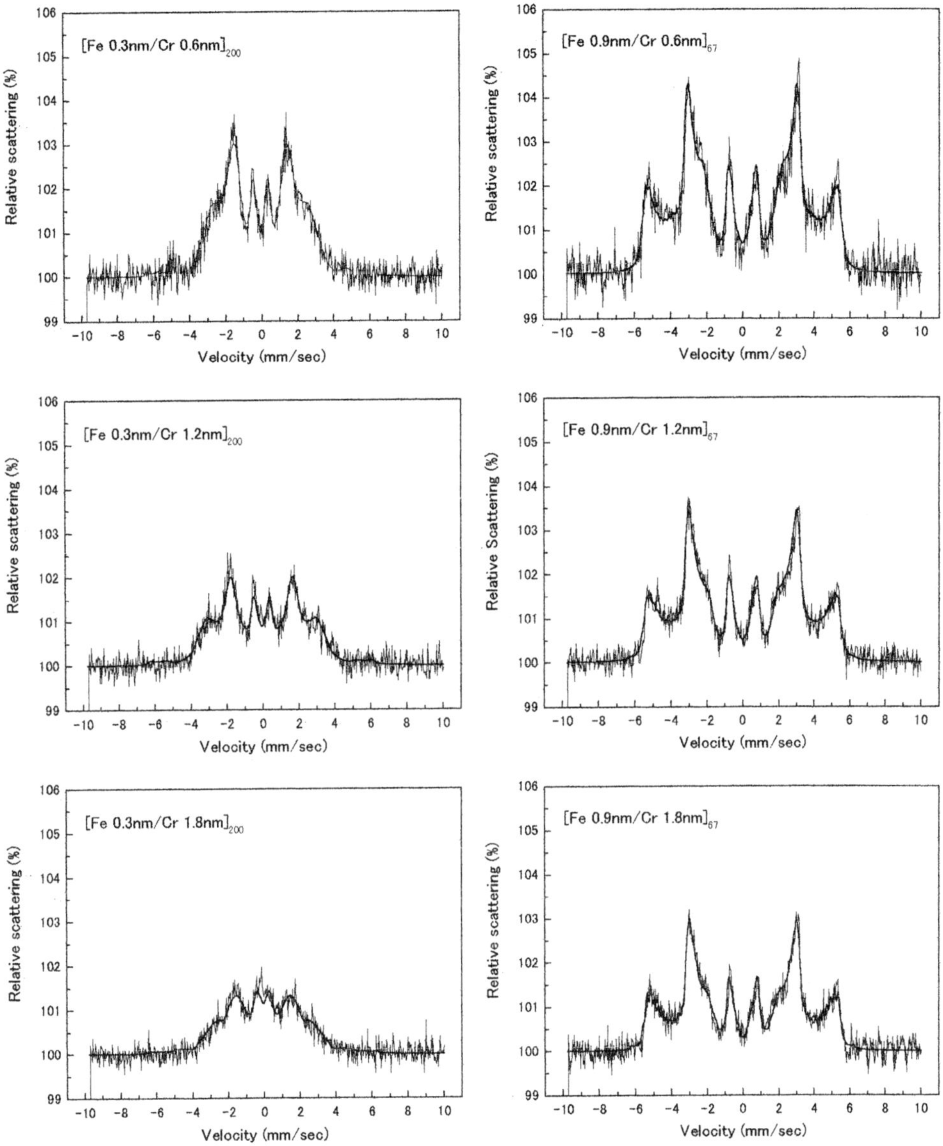

Fig.1 CEM spectra of [Fe 0.3 nm/Cr] and [Fe 0.9 nm/Cr] MLs.

for 1.2 nm Cr layer thickness. The rough interface structure tends to have more step sites than terrace sites in the interface region as modeled by Uzdin et al. [4]. The ^{57}Fe probe atoms at the step sites face to more Cr atoms than the terrace sites. This means smaller H_{hf} values at the step sites than the terrace sites in relation to Fe-Cr alloy [10]. Therefore, the interface structure of [Fe 0.3 nm/Cr $]_{200}$ MLs is considered to be smoother in 1.2 nm Cr layer thickness and rougher in 1.8 nm Cr layer thickness. Probably, these different interface structures in 0.3 nm Fe layer thickness affect the oscillation behavior of the MR ratio, which is different with the case of 0.9 nm Fe layer thickness.

Table 1 The average H_{hf} together with the MR ratio at 13 kG as a function of Fe and Cr layer thickness.

	0.6 nm Cr		1.2 nm Cr		1.8 nm Cr	
	H_{hf} (kG)	ΔR/R (%)	H_{hf} (kG)	ΔR/R (%)	H_{hf} (kG)	ΔR/R (%)
0.3 nm Fe	163	10.0	177	5.4	153	2.1
0.5 nm Fe			231	11.1		
0.9 nm Fe	268	7.0	265	12.0	266	0.4
1.2 nm Fe			281	10.1		

4. Conclusions

We conclude that the decrease of the average H_{hf} estimated from the CEM spectra of Fe/Cr MLs indicates the increase of the interface components and the change of the interface structure depending on Fe and Cr layer thickness. The MR ratio of [Fe 0.9 nm/Cr $]_{67}$ MLs as a function of Cr layer thickness shows clearly the oscillation behavior with a period of 1.8 nm but [Fe 0.3 nm/Cr $]_{200}$ MLs does not. This different oscillation behavior relates to the different interface structures of the ultra-thin Fe/Cr MLs.

References

[1] J. Landes, Ch. Sauer, R.A. Brand, W. Zinn, S. Mantl and Zs. Kajcsos, J. Magn. Magn. Mater. 86 (1990) 71.

[2] J. Zukrowski, G. Liu, H. Fritzsche, U. Gradman, J. Magn. Magn. Mater. 145 (1996) 57.

[3] F. Klinkhammer, Ch. Sauer, E.Yu. Tsymbal, S. Handschuh, Q. Leng, W. Zinn, J. Magn. Magn. Mater. 161 (1996) 49.

[4] V.M. Uzdin, W. Keune, H. Schror, and M. Walterfang, Phys. Rev. 63 (2001) 63.

[5] M. Kubik, T. Slezak, M. Przybylski, W. Karas, J. Korecki, Vacuum 63 (2001) 337.

[6] N.M. Rensing, B.M. Clemens, D.L. Williamson, J. Appl. Phys. 79(1) (1996) 7757.

[7] R. Schad, P. Belien, G. Verbanck, K. Temst, H. Fischer, S. Lefebvre, M. Bessiere, D. Bahr, J. Falta, J. Dekoster, G. Langouche, V.V. Moshchalkov, Y. Bruynseraede, J. Magn. Magn. Mater. 198-199 (1999) 104.

[8] M. Kopcewicz, T. Lucinski, F. Stobiecki, G. Reiss, J. Appl. Phys. 85(8) (1999) 5039.

[9] S.S.P. Parkin, N. More, and K.P. Roche, Phys. Rev. Lett. 64 (1990) 2304.

[10] S.M. Dubiel and J. Zukrowski, J. Magn. Magn. Mater. 23 (1981) 214.

Spin structures in exchange-biased thin films studied by Mössbauer Spectroscopy

V.Kuncser[1,2*], M.Doi[1,3], M.Acet[4], W.Keune[1]
1. Laboratorium für Angewandte Physik, Gerhard-Mercator-Universität , 47048 Duisburg, Germany
2. National Institute for Materials Physics, Bucharest-Magurele, Romania
3. Department of Materials Science and Engineering, Nagoya University, Japan
4. Laboratorium für Tieftemperaturphysik, Gerhard-Mercator-Universität , 47048 Duisburg, Germany
* Correspondence: Laboratorium für Angewandte Physik, Gerhard-Mercator-Universität , 47048 Duisburg, Germany

Abstract: The interfacial spin structure in an exchange biased Fe(6nm)/FeRh system was studied in correlation with the micro-structural behaviour by conversion electron Mössbauer spectroscopy and magnetic measurements. The in plane distribution of the interfacial Fe was analysed by taking the CEMS spectra with the γ radiation oriented at 30° vs. the interface plane. An experimental proof for a spin-flop coupling mechanism, with the interfacial spins of the ferromagnetic phase perpendicular to the spins of the antiferromagnetic phase, was found.

Keywords: CEMS, exchange-bias, spin-flop coupling, spin distribution

1. Introduction

Exchange bias is one of the phenomena associated with the unidirectional anisotropy induced at the interface between a ferromagnetic (FM) and an antiferromagnetic (AF) material, when the system is cooled down in an applied magnetic field through the Neel temperature of the AF phase. A macroscopic trace of the effect consists in a shift of the system hysteresis loop by the exchange bias field, H_E . Even though this effect was discovered more than four decades ago [1], its microscopic origin has not yet been fully understood. Recent applications of the exchange biased thin films as magnetic recording media[2], domain stabilisers in recording heads[3] and especially as spin-valve devices[4,5] have renewed the interest in the effect and in its underlying physics. On the other hand, the up to date theoretical models [6-11] are still controversial, especially due to a lack of understanding of the microscopic origin of the effect. In this respect, the information related to the spin structure in connection with the microstructural changes at the interface becomes of basic interest. Because the systems presenting exchange bias effects contain generally iron, we propose ^{57}Fe conversion electron Mössbauer spectroscopy (CEMS) within the tracer layer technique as a suitable tool for studying the interface spin configuration. Our present work is the first one from a series devoted to such studies and reports on the microscopic investigation of the interface spin structure in potential exchange-biased Fe/FeRh systems.

The equiatomic composition of the FeRh alloy presents an ordered body-centred-cubic (bcc) CsCl-type structure and is AF at room temperature (RT). By heating above RT it undergoes a first order magnetic phase transition to a FM order [12,13] accompanied by an abrupt lattice expansion. With a strong corrosion resistance and a high pinning force, FeRh alloy could be a potential candidate for the AF phase in an exchange bias coupled system. The preparation of suitable ordered equiatomic FeRh thin films is quite difficult because both the structure and the magnetic properties are quite sensitive to the stoichiometry [14,15]. Therefore we focused our attention for the moment on the Fe/FeRh system with the AF phase prepared as ribbon, tacking so the advantage of the well known properties for the bulk AF phase.

2. Experimental details

The equiatomic FeRh ribbons were prepared by the melt spinning method. A piece of such ribbon (0.0123 g) was selected and introduced in the UVH system with a main vacuum of $3x10^{-10}$ mbar. The flat face of the ribbon was Ar-ion sputtered at RT for 90 min at $5x10^{-5}$ mbar. Fe top layers were subsequently grown by MBE at RT with a deposition rate of 0.003 nm/s. A first layer of only 1 nm ^{57}Fe was covered by 5 nm natural Fe (resulting in 6 nm of FM phase) and finally by a 4 nm Pt protective layer. The top section of the coupled system presents 6x3 mm^2 area. The magnetic measurements were done by a SQUID magnetometer (with the in plane field applied longitudinal to the 6 mm long side of the ribbon surface) whereas the CEMS spectra were obtained at RT with the sample plane perpendicular ($\Phi = 90°$) and at an angle $\Phi = 30°$, respectively, vs. the direction of the γ ray . The 6mm long side of the sample plane was fixed at $\varphi = 90°$ (namely along the *Ox* axis in Fig.1).

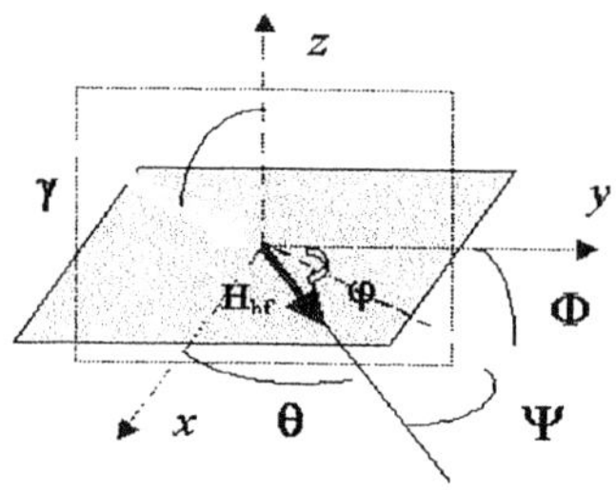

Fig.1: Geometrical arrangement of the CEMS exoeriment

3. Results and Discussion

The magnetisation vs. T curve in an applied field of 0.5 T is due especially to the thick FeRh ribbon and proves a sharp AF -> FM transition around 370 K. Moreover, the magnetisation in the FM state is only 10 times higher than in the AF state, suggesting the presence of some uncompensated Fe spins in the AF state where no magnetic moments are carried by the Rh atoms. The hysteresis loops (obtained at 5 K and 290 K, respectively) after cooling down the sample through the FM -> AF transition temperature in an applied field of 5 mT are shown in Fig.2. At a first view they look quite symmetrical, standing so for no offset field H_E. At a careful analysis, a magnetisation at saturation of around 0.035 emu as is shown by the 5 K loop can not be related to the Fe ferro-magnetic layer only. Indeed, if we consider a 6 nm thick Fe film covering 18 mm^2 and a lattice parameter of 0.285 nm for the bcc Fe lattice, it results $\approx 4.5 \times 10^{15}$ unit cells with 9×10^{15} Fe atoms carrying $18 \cdot 10^{15}$ μ_B at saturation, namely $\approx 1.7 \times 10^{-4}$ emu. This value is almost 200 times lower than the saturation value in the hysteresis loop proving that the signal does not come only from the FM layer but rather from the uncompensated FeRh thick substrate. In order to evidence the contribution of an eventual asymmetric signal originating from the 6 nm FM film, we estimated the left part area (negative fields) and the right part area (positive fields), respectively, of the available hysteresis loops. For symmetrical loops the left area (A_l) should be equal to the right area (A_r) within a computation error range of 1% for the relative area. By contrast, for both analysed hysteresis loops, A_l was found to be higher than A_r with a relative ratio, $A_{rel} = \Delta A/\langle A \rangle$, of 9% at 5 K and 5% at 290 K. This results stand clearly for a negative offset of the FM contribution in the experimental loops and additionally for a decreasing effect vs. temperature, in agreement with the usual behaviour of the exchange biased systems. The increases of the coercive field from only 24 Öe at 5K to 50 Öe at 290 K could be explained by

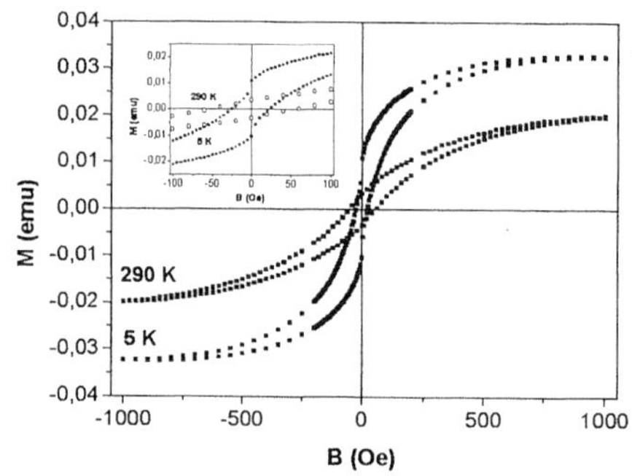

Fig.2: Hysteresis loops of the exchange biased sytem

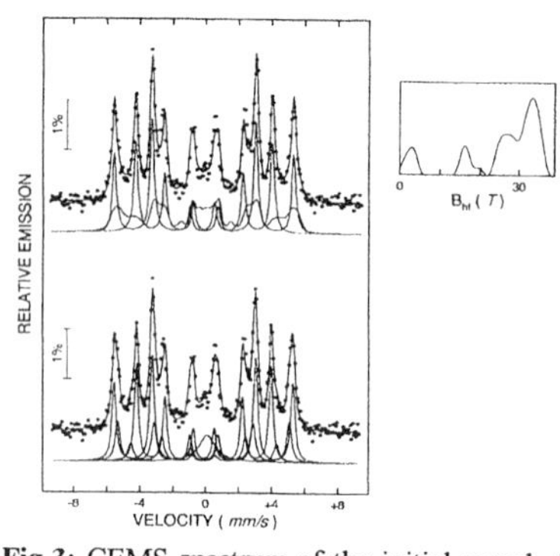

Fig.3: CEMS spectrum of the initial sample in perpendicular geometry (Φ=90°). Two fitting procedures are mentioned.

approaching the typical AFM - > FM transition temperature in the FeRh ribbons.

Previous transmission Mössbauer measurements performed on the considered FeRh ribbons have shown a simple magnetic sextet with an hyperfine field of 27.2(1) T at 4.2 K and 25.5(1) T at RT [13]. An intensity ratio $R_{23} = I_2/I_3$ close to 2 stands for randomly distributed Fe spins in the AF phase. A CEMS spectrum obtained at RT on the same ribbons evidences, besides of the typical sextet with almost the same hyperfine parameters as in the transmission measurements, an additional central pattern (less than 7% relative area) probably due to a face centred cubic (fcc) phase retained on the flat ribbon face in contact with the cold roller.

Fig.3. presents the CEMS spectrum of the analysed Fe/FeRh system with the sample surface perpendicular to the direction of the γ radiation (Φ=90°). In addition to the magnetic six line pattern due to the FeRh ribbons (the inner sextet), another sextet with hyperfine parameters specific to the bcc-Fe phase is easy evidenced. The fitting procedure of the experimental spectrum requires an additional complex pattern giving rise to the hyperfine field distribution shown in the same figure (right side). The most predominant peaks of this distribution are at 27 T and 32 T respectively. These values, close to those of the FeRh and main bcc-Fe suggest that the hyperfine field distribution may be related to interfacial Fe atoms. Supposing an expected diffusion of both Fe and Rh atoms at the Fe/FeRh interface, the 32 T hyperfine field can be assigned to interfacial Fe atoms in the bcc-Fe layer sensing 1 Rh atom in the nearest neighbourhood, whereas the 27 T hyperfine field is assigned to ^{57}Fe atoms in the FeRh phase sensing 1 Fe neighbour. Such a result gives us the opportunity to separate the bulk contribution from the interfacial one. Moreover, in order to analyse independently the spin structure in the two interface phases (one in the FM and the other in the AF surface layer) we proceed to a new fitting of the experimental spectrum by including 4 sextets (2 for the bulk and 2 for the interface phases) and an additional central pattern ($\approx$ 5% rel. area) related to the paramagnetic phase already evidenced at the FeRh ribbon surface. The hyperfine fields and intensity ratios of the magnetic phases obtained in this way are presented in TABLE I. The relative areas are: 35(1)% for the bulk bcc-Fe, 17(1)% for the interfacial bcc-Fe, 29(1)% for the bulk bcc-FeRh and 9(1)% for the ^{57}Fe in the interfacial bcc-FeRh. Consequently, equivalent ^{57}Fe thicknesses of 0.31 nm and 0.16 nm may be deduced for the interfacial Fe in the bcc-Fe and FeRh phases, respectively. The spin configurations can be analysed starting from the intensity ratio R_{23}. For Fe atoms presenting hyperfine fields directed at an angle Ψ vs. the γ ray direction, $R_{23} = 4\sin^2\Psi/(1+\cos^2\Psi)$. Special attention should be paid to the situation with in plane distributed spins. When the spins are located in the *Oxy* plane (sample plane), we may write (see Fig.1) : $\cos\Psi=\cos\Phi\cos\varphi$. If the spin directions follow a certain distribution, $P(\varphi)$, in the plane, the intensity ratio becomes:

TABLE 1: Hyperfine fields and intensity ratios for the analysed sample before (init.) and after (ex.b.) inducing exchange bias

Φ	Phase -> B_{hf}(T);R_{23}	Bulk bcc-Fe	interfacial bcc-Fe	bulk bcc-FeRh	Interfacial bcc- FeRh
90°	B_{hf}	33.5(1)	32.2(1)	25.3(1)	27.2(2)
init.	R_{23}	4.0(1)	2.9(1)	2.0(1)	3.8(1)
30°	B_{hf}	33.3(1)	31.9(2)	25.3(1)	26.7(2)
init.	R_{23}	2.0(1)	2.8(2)	2.0(1)	3.5(2)
30°	B_{hf}	32.8(2)	32.3(2)	25.2(2)	23.5(2)
ex.b.	R_{23}	2.1(1)	1.2(2)	2.0(1)	4.3(3)

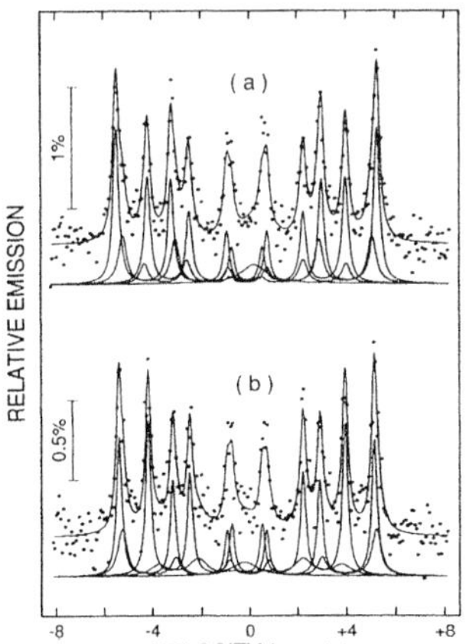

Fig. 4: CEMS spectra at Φ = 30° for the initial (a) and the exchange biased (b) system

$$R_{23} = 4\int_{\varphi} \frac{1-\cos^2\Phi\cos^2\varphi}{1+\cos^2\Phi\cos^2\varphi} P(\varphi)d\varphi \qquad (1)$$

It is easy to see that in perpendicular geometry (Φ=90°), R_{23} =4 and is insensitive to the in plane spin configuration, whereas for $\Phi\neq$90° R_{23} becomes sensitive to the form of $P(\varphi)$.

For spectra taken with the sample plane perpendicular to the γ radiation, we may conclude that the spins are in the sample plane for $R_{23}=4$, and that there is a tendency for an in plane alignment for $2<R_{23}<4$ and a tendency for a perpendicular alignment for $0<R_{23}<2$. For the analysed sample (TABLE 1) we may consider therefore a strong in plane alignment of the Fe spins in the bcc-Fe layer and in the interfacial FeRh phase as well as a moderate tendency for an in plane alignment of the interfacial bcc-Fe. By contrast, the Fe spins in the first 100 nm of the FeRh AF phase are randomly distributed. The next step is to obtain some information about the in plane distribution of the interfacial spins before and after inducing the exchange biasing field. To do this, we used the opportunity to take CEMS spectra with the sample plane oriented at $\Phi=30°$ vs. the γ direction. Details about the Mössbauer methodology for evaluating the in plane spin distribution via relation (1) will be given in a forthcoming paper. We like to emphasise that for $\Phi=30°$ and a random spin distribution in the sample plane, R_{23} approach 2.1 whereas for a δ-type distribution, R_{23} approach 0.7 for $\varphi_0= 0$ and 4 for $\varphi_0 = 90°$ (see Fig.1). The Mössbauer spectra taken at $\Phi=30°$, before and after inducing the exchange bias, are presented in Fig.4 and the corresponding fitted hyperfine fields and intensity ratios are given in TABLE 1 . It can be observed that the spins of the bcc-Fe layer are randomly distributed in the sample plane, both in the initial state as well as after inducing exchange bias ($R_{23} \approx 2$). For the initial state, the in plane projection of the bcc-Fe interfacial spins presents some tendency along the *Ox* direction, most probable due to coupling to the interfacial spins in the AF phase which show clearly a strong preference for the same direction ($R_{23} \approx 3.5$). To the contrary, by inducing the exchange bias effect, the bcc-Fe interfacial spins reorients along the *Oy* direction ($R_{23} \approx 1.2$), namely perpendicular on the interfacial AF spins which remains along the *Ox* direction ($R_{23} \approx 4$). Another interesting aspect concerns the decrease of the hyperfine field of the Fe atoms in the interfacial FeRh phase by about 3 T. Correlated also with a substantial increase of the sextet linewidth , such a result stands for spin relaxational effects in the interfacial AF phase of the exchanged biased system.

4. Conclusions

An FM/AF system was prepared by growing 6 nm Fe on FeRh ribbon. The system was cooled down through the FM->AF transition temperature in a 5 mT applied field. An exchange bias effect was evidenced in the final state by a careful analysis of the system hysteresis loops at 5 K and 290 K. The spin structure at the interface was studied in the initial state as well as after inducing exchange bias by taking CEMS spectra with the sample plane at 90° and 30° vs. the γ ray direction. A perpendicular coupling of the FM and AF interfacial Fe spins was found after cooling down the system in low applied field, in agreement with a "spin-flop" coupling mechanism[9,10].

Acknowledgements: Work supported by Deutsche Forschungsgemeinschaft (SFB491). The financial support through Alexander von Humboldt Stiftung is highly acknowledged by one of the authors (V.Kuncser). The authors wish to thank U. von Hörsten and E.Duman for valuable technical assistance.

References:

[1] W.H.Meiklejohn, C.P.Bean, Phys.Rev. 102 (1956) 1413
[2] S.Gangopadhyay, G.C.Hadjipanayis,C.M.Sorensen,K.J.Klabunde, IEEE Trans.Magn 28 (1992) 3174
[3] T.Lin,G.L.Gorman,C.Tsang, IEEE Trans. Magn. 32 (1996) 3443
[4] B.Dieny, J.Magn.Magn.Mater. 136 (1994) 335
[5] J.C.S.Kools, IEEE Trans.Magn. 32 (1996) 3165
[6] W.H.Meiklejohn, J.Appl.Phys. 33 (1962)1328
[7] D.Mauri,H.C.Siegmann,P.S.Bogus,E.Kay, J.Appl.Phys. 62 (1987) 3047
[8] A.P.Molozemoff, J.Appl.Phys. 63 (1988) 3874
[9] N.C.Koon, Phys.Rev.Lett. 78 25 (1997) 4865
[10] T.C.Schulthess and W.H.Butler, J.Appl.Phys. 85 8 (1999) 5510
[11] U.Nowak, A.Misra and K.D.Usadel, J.Appl.Phys. 89 11 (2001) 7269
[12] M.Fallot, R.Hocart, Rev.Sci 77 (1939) 498
[13] M.Rosenberg, V.Kuncser, O.Crisan, A.Hernando, E.Navarro, G.Filoti, J.Magn.Magn.Mater. 177-181 (1998) 135
[14] Y.Ohtany and I.Hatakeyama, J.Appl.Phys. 74 (1993) 3328
[15] S.Yuasa, M.Nyvlt, T.Katayama, S.Suzuki, J.Appl.Phys. 83 11 (1998) 3328